VOLUME FOUR HUNDRED AND FORTY-EIGHT

METHODS IN ENZYMOLOGY

RNA Turnover in Eukaryotes: Nucleases, Pathways and Anaylsis of mRNA Decay

METHODS IN ENZYMOLOGY

Editors-in-Chief

JOHN N. ABELSON AND MELVIN I. SIMON

Division of Biology
California Institute of Technology
Pasadena, California, USA

Founding Editors

SIDNEY P. COLOWICK AND NATHAN O. KAPLAN

VOLUME FOUR HUNDRED AND FORTY-EIGHT

Methods in
ENZYMOLOGY

RNA Turnover in Eukaryotes: Nucleases, Pathways and Anaylsis of mRNA Decay

EDITED BY

LYNNE E. MAQUAT
Department of Biochemistry and Biophysics
University of Rochester
School of Medicine and Dentistry
Rochester, NY, USA

MEGERDITCH KILEDJIAN
Department of Cell Biology and Neurosciences
Nelson Biology Labs
Piscataway, NJ, USA

WITHDRAWN
FAIRFIELD UNIVERSITY
LIBRARY

AMSTERDAM • BOSTON • HEIDELBERG • LONDON
NEW YORK • OXFORD • PARIS • SAN DIEGO
SAN FRANCISCO • SINGAPORE • SYDNEY • TOKYO
Academic Press is an imprint of Elsevier

ELSEVIER

Academic Press is an imprint of Elsevier
525 B Street, Suite 1900, San Diego, California 92101-4495, USA
84 Theobald's Road, London WC1X 8RR, UK

Copyright © 2008, Elsevier Inc. All Rights Reserved.

No part of this publication may be reproduced or transmitted in any form or by any means, electronic or mechanical, including photocopy, recording, or any information storage and retrieval system, without permission in writing from the Publisher.

The appearance of the code at the bottom of the first page of a chapter in this book indicates the Publisher's consent that copies of the chapter may be made for personal or internal use of specific clients. This consent is given on the condition, however, that the copier pay the stated per copy fee through the Copyright Clearance Center, Inc. (www.copyright.com), for copying beyond that permitted by Sections 107 or 108 of the U.S. Copyright Law. This consent does not extend to other kinds of copying, such as copying for general distribution, for advertising or promotional purposes, for creating new collective works, or for resale. Copy fees for pre-2008 chapters are as shown on the title pages. If no fee code appears on the title page, the copy fee is the same as for current chapters. 0076-6879/2008 $35.00

Permissions may be sought directly from Elsevier's Science & Technology Rights Department in Oxford, UK: phone: (+44) 1865 843830, fax: (+44) 1865 853333, E-mail: permissions@elsevier.com. You may also complete your request on-line via the Elsevier homepage (http://elsevier.com), by selecting "Support & Contact" then "Copyright and Permission" and then "Obtaining Permissions."

For information on all Elsevier Academic Press publications visit our Web site at elsevierdirect.com

ISBN-13: 978-0-12-374378-7

PRINTED IN THE UNITED STATES OF AMERICA
08 09 10 11   9 8 7 6 5 4 3 2 1

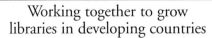

Working together to grow
libraries in developing countries

www.elsevier.com | www.bookaid.org | www.sabre.org

ELSEVIER   BOOK AID International   Sabre Foundation

Contents

Contributors	xv
Preface	xxi
Volumes in Series	xxiii

Section I. Decapping Analyses　　　　　　　　　　1

1. Analysis of mRNA Decapping　　　　　　　　　　3
Shin-Wu Liu, Xinfu Jiao, Sarah Welch, and Megerditch Kiledjian

1. Introduction	4
2. Measuring Decapping Activities of Recombinant and Endogenous Dcp2	6
3. Measuring DcpS Activity	15
4. Migration of Cap Analogs with Various Thin-layer Chromatography Running Buffers	17
Acknowledgment	19
References	19

2. A Kinetic Assay to Monitor RNA Decapping Under Single-Turnover Conditions　　　　　　　　　　23
Brittnee N. Jones, Duc-Uy Quang-Dang, Yuko Oku, and John D. Gross

1. Introduction	24
2. Kinetic Equations	25
3. Materials	28
4. Kinetic Assay	30
5. Summary	36
Acknowledgments	37
References	38

3. Purification and Analysis of the Decapping Activator Lsm1p-7p-Pat1p Complex from Yeast　　　　　　　　　　41
Sundaresan Tharun

1. Introduction	42
2. Purification of the Lsm1p-7p-Pat1p Complex	43

3. Analysis of RNA Binding by the Lsm1p-7p-Pat1p Complex	50
4. Concluding Remarks	53
Acknowledgments	53
References	53

4. Reconstitution of Recombinant Human LSm Complexes for Biochemical, Biophysical, and Cell Biological Studies — 57

Bozidarka L. Zaric and Christian Kambach

1. Introduction	58
2. Cloning	61
3. Protein Expression and Purification	65
4. LSm Complex Reconstitution	68
5. LSm Complex Functional Assays	70
6. Outlook	72
References	72

Section II. Polyadenylation/Deadenylation Analyses — 75

5. Regulated Deadenylation *In Vitro* — 77

Aaron C. Goldstrohm, Brad A. Hook, and Marvin Wickens

1. Introduction	78
2. *In Vitro* Deadenylation Systems	80
3. Troubleshooting	100
References	101

6. Cell-Free Deadenylation Assays with *Drosophila* Embryo Extracts — 107

Mandy Jeske and Elmar Wahle

1. Introduction	107
2. Preparation of *Drosophila* Embryo Extracts	110
3. Preparation of Substrate RNA	112
4. Deadenylation Assay with *Drosophila* Embryo Extracts	113
5. Characterization of Sequence-Dependent Deadenylation in *Drosophila* Embryo Extracts	115
Acknowledgments	116
References	116

7. Measuring CPEB-Mediated Cytoplasmic Polyadenylation-Deadenylation in *Xenopus laevis* Oocytes and Egg Extracts — 119

Jong Heon Kim and Joel D. Richter

1. Introduction	120
2. Monitoring Polyadenylation of Cyclin B1 RNA in *Xenopus laevis* Oocytes	121

3. Detection of Nuclear and Cytoplasmic Polyadenylation
   of *X. laevis* Cyclin B1 mRNA 126
4. Monitoring Deadenylation and Re-adenylation
   of *X. laevis* Cyclin B1 RNA 128
5. Detection of Endogenous Cytoplasmic Poly(A) Polymerase Activity 130
6. Overexpression of Exogenous mRNAs and Measuring
   Polyadenylation-Deadenylation 132
7. Preparation of *X. laevis* Egg Extracts and Measuring
   Polyadenylation-Deadenylation with Cyclin B1 RNA 133
Acknowledgments 137
References 137

8. The Preparation and Applications of Cytoplasmic Extracts from Mammalian Cells for Studying Aspects of mRNA Decay 139
Kevin J. Sokoloski, Jeffrey Wilusz, and Carol J. Wilusz

1. Introduction 140
2. Preparation of HeLa-Cell Cytoplasmic Extracts 142
3. Preparation of RNA Substrates 148
4. Evaluating mRNA Decay with Cytoplasmic Extracts 152
5. Concluding Remarks 161
Acknowledgments 162
References 162

Section III. Nucleases in mRNA Decay 165

9. *In Vitro* Assays of 5' to 3' Exoribonuclease Activity 167
Olivier Pellegrini, Nathalie Mathy, Ciarán Condon, and Lionel Bénard

1. Introduction 168
2. Purification of *Xrn*1 170
3. *In Vitro* RNA Substrate Synthesis 171
4. Degradation of RNA by XRN1 Depends on the Nature of the 5'-End 173
5. Determining the Directionality of Decay 176
6. Conclusions and Prospects 180
References 181

10. Reconstitution of RNA Exosomes from Human and *Saccharomyces cerevisiae*: Cloning, Expression, Purification, and Activity Assays 185
Jaclyn C. Greimann and Christopher D. Lima

1. Introduction 186
2. Cloning Strategies for Recombinant Protein Expression 190
3. PCR and Subcloning Protocols 191
4. PCR and Subcloning Protocols for Yeast cDNA 191
5. PCR and Subcloning Protocols for Human cDNA 194

6.	Expression and Purification of Yeast Exosome Proteins	195
7.	Expression and Purification of Human Exosome Proteins	198
8.	Reconstitution and Purification of Human and Yeast Exosomes	201
9.	Exoribonuclease Assays	206
10.	Comparative Exoribonuclease Assays with Different RNA Substrates	207
11.	Conclusions	208
	Acknowledgments	208
	References	209

11. Biochemical Studies of the Mammalian Exosome with Intact Cells — 211

Geurt Schilders and Ger J. M. Pruijn

1.	Introduction	212
2.	Identifying Protein–Protein Interactions by the Mammalian Two-Hybrid System	213
3.	Characterization of Different Exosome Subsets by Glycerol Sedimentation	218
4.	Studying Exosome Function with RNAi	222
	References	224

12. Determining *In Vivo* Activity of the Yeast Cytoplasmic Exosome — 227

Daneen Schaeffer, Stacie Meaux, Amanda Clark, and Ambro van Hoof

1.	Introduction	228
2.	Is My Favorite RNA Degraded and/or Processed by the Exosome?	229
3.	Is the Cytoplasmic Exosome Active in my Mutant or Under my Conditions?	232
	References	238

13. Approaches for Studying PMR1 Endonuclease–Mediated mRNA Decay — 241

Yuichi Otsuka and Daniel R. Schoenberg

1.	Introduction	242
2.	Identification of Endonuclease Cleavage Sites within mRNA	244
3.	Analysis of PMR1-Containing Complexes	251
4.	Affinity Recovery of PMR1-Containing Complexes	257
5.	Analysis of PMR1 Activity *In Vivo* and *In Vitro*	260
6.	Summary	262
	Acknowledgments	262
	References	262

Section IV. Measuring mRNA Half-life *In Vivo*   265

14. Methods to Determine mRNA Half-Life in *Saccharomyces cerevisiae*   267
Jeff Coller

1. Introduction   268
2. The Use of Inducible Promoters   269
3. Measuring mRNA Decay by Use of Thermally Labile Alleles of RNA Polymerase II   276
4. Measuring mRNA Decay with Thiolutin   277
5. RNA Extractions   277
6. Northern Blot Analysis   280
7. Loading Controls   280
8. Determination of mRNA Half-Lives   280
9. Concluding Remarks   282
Acknowledgments   282
References   282

15. mRNA Decay Analysis in *Drosophila melanogaster*: Drug-Induced Changes in Glutathione *S*-Transferase D21 mRNA Stability   285
Bünyamin Akgül and Chen-Pei D. Tu

1. Introduction   286
2. Materials and Methods   287
3. Concluding Remarks   295
Acknowledgment   296
References   296

16. Measuring mRNA Stability During Early *Drosophila* Embryogenesis   299
Jennifer L. Semotok, J. Timothy Westwood, Aaron L. Goldman, Ramona L. Cooperstock, and Howard D. Lipshitz

1. Maternal mRNAs and Early *Drosophila* Development   300
2. Gene-by-Gene Analysis of mRNA Decay   308
3. Genome-Wide Analysis of mRNA Decay   318
4. Concluding Remarks   331
Acknowledgments   331
References   332

17. Messenger RNA Half-Life Measurements in Mammalian Cells — 335
Chyi-Ying A. Chen, Nader Ezzeddine, and Ann-Bin Shyu

1. Introduction — 336
2. General Considerations of mRNA Half-Life Measurements — 337
3. Determining mRNA Decay Constant — 338
4. Methods for Measuring mRNA Half-Life — 339
5. Concluding Remarks — 354
Acknowledgments — 355
References — 355

18. Trypanosomes as a Model to Investigate mRNA Decay Pathways — 359
Stuart Archer, Rafael Queiroz, Mhairi Stewart, and Christine Clayton

1. Introduction — 359
2. Genetic Manipulation in Trypanosomes: Down-Regulating Expression of Proteins Involved in mRNA Decay — 361
Acknowledgments — 375
References — 375

19. Cell Type–Specific Analysis of mRNA Synthesis and Decay *In Vivo* with Uracil Phosphoribosyltransferase and 4-thiouracil — 379
Michael D. Cleary

1. Introduction — 380
2. Experimental Design Considerations — 382
3. Materials — 386
4. Methods — 388
Acknowledgments — 405
References — 405

Section V. Defining Degradative Activities — 407

20. Analysis of Cytoplasmic mRNA Decay in *Saccharomyces cerevisiae* — 409
Dario O. Passos and Roy Parker

1. Introduction — 410
2. Measuring mRNA Half-Life — 410
3. Determination of mRNA Decay Pathways — 415
4. Combination of the Preceding Approaches — 425
References — 425

21. Transcriptome Targets of the Exosome Complex in Plants — 429
Dmitry Belostotsky

1. Exosome: At the Nexus of the Cellular RNA Transactions — 429
2. Unique Features of the Plant Exosome — 432
3. Resources for the Mutational Analyses of the Plant Exosome — 434
4. Transcriptome-wide Mapping of Targets of the Plant Exosome Complex — 436
Acknowledgments — 440
References — 440

22. Sensitive Detection of mRNA Decay Products by Use of Reverse-Ligation–Mediated PCR (RL-PCR) — 445
Thierry Grange

1. Introduction — 446
2. Footprinting of RNA–Protein Interaction — 450
3. RL-PCR with Ligation of an RNA Linker — 452
4. Circularization RL-PCR to Analyze mRNA Decay Involving Modification of the 5′ and 3′-Ends — 458
5. Concluding Remarks — 465
Acknowledgments — 465
References — 465

23. Tethering Assays to Investigate Nonsense-Mediated mRNA Decay Activating Proteins — 467
Niels H. Gehring, Matthias W. Hentze, and Andreas E. Kulozik

1. Introduction — 468
2. Plasmid Cloning — 470
References — 480

24. Assays for Determining Poly(A) Tail Length and the Polarity of mRNA Decay in Mammalian Cells — 483
Elizabeth L. Murray and Daniel R. Schoenberg

1. Introduction: Poly(A) Tail Length Assays — 484
2. Introduction: Invader RNA Assay — 492
Acknowledgments — 504
References — 504

Section VI. Cell Biology of RNA Decay (i.e., Translational Repression/P Bodies)   505

25. Analyzing P-bodies in *Saccharomyces cerevisiae*   507
Tracy Nissan and Roy Parker

1. Introduction   508
2. Determining Whether a Specific Protein Can Accumulate in P-Bodies   508
3. Monitoring Messenger RNA in P-Bodies   514
4. Determining Whether a Mutation or Perturbation Affects P-Body Size and Number   515
5. Quantification of P-Body Size and Number   518
Acknowledgments   519
References   519

26. Real-Time and Quantitative Imaging of Mammalian Stress Granules and Processing Bodies   521
Nancy Kedersha, Sarah Tisdale, Tyler Hickman, and Paul Anderson

1. Introduction   522
2. Experimental Rationale   524
3. Experimental Considerations   524
4. Selection Criteria   527
5. Transfection   528
6. Properties of Representative Stable Lines   533
7. Environmental Control   541
8. Microscope Hardware: Widefield vs Confocal   543
9. Useful Microscopy Internet Resources   547
10. Sample Protocols   547
11. Conclusions   551
References   551

27. Cell Biology of mRNA Decay   553
David Grünwald, Robert H. Singer, and Kevin Czaplinski

1. Introduction   554
2. FISH Probe Design   556
3. Hybridization   557
4. Image Acquisition   557
5. FISH Protocol   558
6. Colabeling Protein with IF and RNA with FISH   560
7. IF-FISH Protocol   561
8. Following mRNA in Living Cells   562
9. Live Single-Molecule Detection   563

10. Single mRNA Data Analysis; What You Can Observe	564
11. How Do You Know That You See Single Molecules?	566
12. The Secret to Getting Good Data: More Photons, Less Noise	568
13. Setting Up a Microscope for Single Molecule Detection	569
14. Conclusion	571
Acknowledgments	571
References	575

Author Index	*579*
Subject Index	*597*

Contributors

Bünyamin Akgül
Department of Molecular Biology and Genetics, Izmir Institute of Technology, Izmir, Turkey, and Department of Biochemistry and Molecular Biology, The Pennsylvania State University, University Park, Pennsylvania, USA

Paul Anderson
Division of Rheumatology, Immunology and Allergy, Brigham and Women's Hospital, Boston, Massachusetts, USA

Stuart Archer
Zentrum für Molekulare Biologie der Universität Heidelberg (ZMBH), Heidelberg, Germany

Lionel Bénard
CNRS UPR 9073 (affiliated with Université de Paris 7—Denis Diderot), Institut de Biologie Physico-Chimique, Paris, France

Dmitry Belostotsky
Division of Molecular Biology and Biochemistry, School of Biological Sciences, University of Missouri, Kansas City, Missouri, USA

Chyi-Ying A. Chen
Department of Biochemistry and Molecular Biology, The University of Texas Medical School at Houston, Houston, Texas, USA

Amanda Clark
University of Texas Health Science Center-Houston, Department of Microbiology and Molecular Genetics, Houston, Texas, USA

Christine Clayton
Zentrum für Molekulare Biologie der Universität Heidelberg (ZMBH), Heidelberg, Germany

Michael D. Cleary
University of California, Merced School of Natural Sciences, Merced, California, USA

Jeff Coller
Center for RNA Molecular Biology, Case Western Reserve University, School of Medicine, Cleveland Ohio, USA

Ciarán Condon
CNRS UPR 9073 (affiliated with Université de Paris 7—Denis Diderot), Institut de Biologie Physico-Chimique, Paris, France

Ramona L. Cooperstock
Program in Developmental and Stem Cell Biology, Research Institute, Hospital for Sick Children, Toronto, Ontario, Canada, and Department of Molecular Genetics, University of Toronto, 1 King's College Circle, Toronto, Ontario, Canada

Kevin Czaplinski
Department of Biochemistry and Cell Biology, Stony Brook University, Center for Molecular Medicine 542, Stony Brook, New York, USA

Nader Ezzeddine
Department of Biochemistry and Molecular Biology, The University of Texas Medical School at Houston, Houston, Texas, USA

Niels H. Gehring
Department of Pediatric Oncology, Hematology and Immunology, University of Heidelberg, Heidelberg, Germany, and Molecular Medicine Partnership Unit, EMBL/University of Heidelberg, Heidelberg, Germany

Aaron L. Goldman
Program in Developmental and Stem Cell Biology, Research Institute, Hospital for Sick Children, Toronto, Ontario, Canada, and Department of Molecular Genetics, University of Toronto, 1 King's College Circle, Toronto, Ontario, Canada

Aaron C. Goldstrohm
Department of Biochemistry, University of Wisconsin—Madison, Madison, Wisconsin, USA

Thierry Grange
Institut Jacques Monod du CNRS, Université Paris 7, Paris, France

Jaclyn C. Greimann
Structural Biology Program, Sloan-Kettering Institute, New York, NY, USA

John D. Gross
Department of Pharmaceutical Chemistry, University of California, San Francisco, California, USA

David Grünwald
Albert Einstein College of Medicine, Anatomy and Structural Biology and Gruss-Lipper Singer Biophotonics Center, Bronx, New York, USA

Matthias W. Hentze
European Molecular Biology Laboratory, Heidelberg, Germany, and Molecular Medicine Partnership Unit, EMBL/University of Heidelberg, Heidelberg, Germany

Contributors

Tyler Hickman
Division of Rheumatology, Immunology and Allergy, Brigham and Women's Hospital, Boston, Massachusetts, USA

Brad A. Hook
Promega Corporation, Madison, Wisconsin, USA

Mandy Jeske
Institute of Biochemistry and Biotechnology, Martin-Luther-University Halle-Wittenberg, Halle, Germany

Xinfu Jiao
Rutgers University, Department of Cell Biology and Neuroscience, Piscataway, New Jersey, USA

Brittnee N. Jones
Program in Chemistry and Chemical Biology, University of California, San Francisco, California, USA

Christian Kambach
Biomolecular Research, Paul Scherrer Institut, Villigen PSI, Switzerland

Nancy Kedersha
Division of Rheumatology, Immunology and Allergy, Brigham and Women's Hospital, Boston, Massachusetts, USA

Megerditch Kiledjian
Rutgers University, Department of Cell Biology and Neuroscience, Piscataway, New Jersey, USA

Jong Heon Kim
Research Institute, National Cancer Center, Goyang, Gyeonggi, South Korea

Andreas E. Kulozik
Department of Pediatric Oncology, Hematology and Immunology, University of Heidelberg, Heidelberg, Germany, and Molecular Medicine Partnership Unit, EMBL/University of Heidelberg, Heidelberg, Germany

Christopher D. Lima
Structural Biology Program, Sloan-Kettering Institute, New York, NY, USA

Howard D. Lipshitz
Program in Developmental and Stem Cell Biology, Research Institute, Hospital for Sick Children, Toronto, Ontario, Canada, and Department of Molecular Genetics, University of Toronto, 1 King's College Circle, Toronto, Ontario, Canada

Shin-Wu Liu
Rutgers University, Department of Cell Biology and Neuroscience, Piscataway, New Jersey, USA

Nathalie Mathy
CNRS UPR 9073 (affiliated with Université de Paris 7—Denis Diderot), Institut de Biologie Physico-Chimique, Paris, France

Stacie Meaux
University of Texas Health Science Center-Houston, Department of Microbiology and Molecular Genetics, Houston, Texas, USA

Elizabeth L. Murray
Department of Molecular and Cellular Biochemistry and The RNA Group, The Ohio State University, Columbus, Ohio, USA

Tracy Nissan
The University of Arizona, Department of Molecular and Cellular Biology and Howard Hughes Medical Institute, Tucson, Arizona, USA

Yuko Oku
Department of Neuroscience, Johns Hopkins University School of Medicine, Baltimore, Maryland, USA

Yuichi Otsuka
Department of Molecular and Cellular Biochemistry and The RNA Group, The Ohio State University, Columbus, Ohio, USA

Roy Parker
The University of Arizona, Department of Molecular and Cellular Biology and Howard Hughes Medical Institute, Tucson, Arizona, USA

Dario O. Passos
The University of Arizona, Department of Molecular and Cellular Biology and Howard Hughes Medical Institute, Tucson, Arizona, USA

Olivier Pellegrini
CNRS UPR 9073 (affiliated with Université de Paris 7—Denis Diderot), Institut de Biologie Physico-Chimique, Paris, France

Ger J. M. Pruijn
Department of Biomolecular Chemistry, Nijmegen Center for Molecular Life Sciences, Institute for Molecules and Materials, Radboud University Nijmegen, Nijmegen, The Netherlands

Duc-Uy Quang-Dang
Department of Pharmaceutical Chemistry, University of California, San Francisco, California, USA

Rafael Queiroz
Zentrum für Molekulare Biologie der Universität Heidelberg (ZMBH), Heidelberg, Germany

Joel D. Richter
Program in Molecular Medicine, University of Massachusetts Medical School, Worcester, Massachusetts, USA and Research Institute, National Cancer Center, Goyang, Gyeonggi, South Korea

Daneen Schaeffer
University of Texas Health Science Center-Houston, Department of Microbiology and Molecular Genetics, Houston, Texas, USA

Geurt Schilders
Department of Biomolecular Chemistry, Nijmegen Center for Molecular Life Sciences, Institute for Molecules and Materials, Radboud University Nijmegen, Nijmegen, The Netherlands

Daniel R. Schoenberg
Department of Molecular and Cellular Biochemistry and The RNA Group, The Ohio State University, Columbus, Ohio, USA

Jennifer L. Semotok
Program in Developmental and Stem Cell Biology, Research Institute, Hospital for Sick Children, Toronto, Ontario, Canada, and Department of Molecular Genetics, University of Toronto, 1 King's College Circle, Toronto, Ontario, Canada

Robert H. Singer
Department of Biochemistry and Cell Biology, Stony Brook University, Center for Molecular Medicine 542, Stony Brook, New York, USA, and Albert Einstein College of Medicine, Anatomy and Structural Biology and Gruss-Lipper Biophotonics Center, Bronx, New York, USA

Kevin J. Sokoloski
Department of Microbiology, Immunology and Pathology, Colorado State University, Ft. Collins, Colorado, USA

Mhairi Stewart
Zentrum für Molekulare Biologie der Universität Heidelberg (ZMBH), Heidelberg, Germany

Ann-Bin Shyu
Department of Biochemistry and Molecular Biology, The University of Texas Medical School at Houston, Houston, Texas, USA

Sundaresan Tharun
Department of Biochemistry, Uniformed Services University of the Health Sciences (USUHS), Bethesda, Maryland, USA

Sarah Tisdale
Division of Rheumatology, Immunology and Allergy, Brigham and Women's Hospital, Boston, Massachusetts, USA

Chen-Pei D. Tu
Department of Biochemistry and Molecular Biology, The Pennsylvania State University, University Park, Pennsylvania, USA

Sarah Welch
Rutgers University, Department of Cell Biology and Neuroscience, Piscataway, New Jersey, USA

Ambro van Hoof
University of Texas Health Science Center-Houston, Department of Microbiology and Molecular Genetics, Houston, Texas, USA

Elmar Wahle
Institute of Biochemistry and Biotechnology, Martin-Luther-University Halle-Wittenberg, Halle, Germany

J. Timothy Westwood
Department of Cell and Systems Biology and Canadian *Drosophila* Microarray Centre, University of Toronto, Mississauga, Ontario, Canada

Carol J. Wilusz
Department of Microbiology, Immunology and Pathology, Colorado State University, Ft. Collins, Colorado, USA

Marvin Wickens
Department of Biochemistry, University of Wisconsin—Madison, Madison, Wisconsin, USA

Bozidarka L. Zaric
Institut Curie, UMR 7147, Equipe: Recombinaison et Instabilité Génétique, Paris, France

Preface

The expression of protein-encoding genes in eukaryotes is regulated at multiple levels, including the initial synthesis of pre-mRNA in the nucleus, the various nuclear processing events that convert pre-mRNA to mRNA, mRNA transport to the cytoplasm, and mRNA translation and, ultimately, degradation in the cytoplasm. All steps are exquisitely controlled to orchestrate the production of protein at the appropriate time and at a suitable level. It follows that a critical step in this orchestration is the proper maintenance of mRNA stability, which is often regulated to influence the amount of encoded protein. Thus, determinants of the rate at which an mRNA is degraded are important regulators of gene expression. As a consequence, the availability of methods and tools to analyze how and when individual mRNAs are either stabilized or turned over is essential to understand this exciting area of biology.

Determinants that dictate the stability of an mRNA include general elements, such as the 5' methylguanosine cap and the 3' poly(A) tail, and transcript-specific elements that are recognized by protein factors or noncoding RNAs. Regulation of mRNA stability is largely imparted by the ability of nucleases to access and remove the terminal elements that protect the mRNA. The bulk of mRNA decay appears to proceed through defined pathways that are initiated by deadenylation of the poly(A) tail. Deadenylated mRNA is subsequently subjected to one of two exonucleolytic decay pathways. One pathway continues to degrade the RNA from the 3' end to generate a capped oligonucleotide that is subsequently hydrolyzed. Alternately, deadenylation can trigger decapping of the mRNA to expose the 5' end to 5'-to-3' exonucleolytic decay. Interestingly, recent evidence indicates that these pathways are not necessarily mutually exclusive and can also occur simultaneously. Moreover, mRNAs can be shunted to exonucleolytic decay pathways after initial cleavage by an endonuclease. The rate at which different *trans*-acting factors and *cis*-acting elements coordinate the demise of an mRNA dictates the chemical half-life of the mRNA, which in eukaryotes can range from several minutes to several days. Thus, methods to analyze mRNA levels, mRNA half-lives, mRNA decay pathways, mRNA decay intermediates, and the associated nucleases and their activities are critical for a comprehensive understanding of mRNA turnover.

Significant advances have been made in recent years in ways to analyze both genome-wide mRNA decay and the decay of specific mRNAs in mammalian systems and in numerous model organisms. This volume is focused on six broad areas of mRNA turnover. The first section relates to methods to study mRNA decapping and consists of four chapters that range from the analysis of decapping activity and the kinetics of decapping to the reconstitution of complexes that promote decapping. The next section includes four chapters that present *in vitro* methods to analyze deadenylation. Methods to reconstitute and test the activities of nucleases that function in mRNA decay are presented in the third section, which addresses how to characterize and detect exoribonucleases and endoribonucleases.

A fundamentally important aspect of mRNA turnover studies involves methods that measure mRNA chemical half-life, and the fourth section consists of six chapters presenting state-of-the-art methods for such measurements in cells of a variety of organisms. Approaches aimed at detecting decay intermediates are detailed in the fifth set of chapters, which describe transcript-specific and genome-wide analyses of mRNA decay products and substrates of nucleases. Last, an exciting new development has been the realization that mRNA decay factors can concentrate in discrete cellular foci. These foci seem to consist of mRNAs undergoing decay and/or mRNAs that are translationally silenced. The final two chapters present approaches to detect and follow the formation of two such foci, processing bodies and stress granules.

All chapters begin with a brief overview of a specific area of mRNA turnover and progress to a detailed description of relevant methods. This volume is part of a three-volume series. The first companion volume (Vol. 448) focuses on RNA turnover in bacteria, archaea, and organelles, and the last volume (Vol. 450) covers the analysis of specialized and quality control RNA decay in eukaryotes. Cumulatively, this series presents all the latest approaches and methods to assess how RNA decay contributes to gene regulation in many commonly used organisms. The series aims to provide a valuable arsenal of tools with which one can study the exciting and rapidly expanding arena of mRNA turnover as it pertains not only to basic research but also to therapeutics.

Methods in Enzymology

VOLUME I. Preparation and Assay of Enzymes
Edited by SIDNEY P. COLOWICK AND NATHAN O. KAPLAN

VOLUME II. Preparation and Assay of Enzymes
Edited by SIDNEY P. COLOWICK AND NATHAN O. KAPLAN

VOLUME III. Preparation and Assay of Substrates
Edited by SIDNEY P. COLOWICK AND NATHAN O. KAPLAN

VOLUME IV. Special Techniques for the Enzymologist
Edited by SIDNEY P. COLOWICK AND NATHAN O. KAPLAN

VOLUME V. Preparation and Assay of Enzymes
Edited by SIDNEY P. COLOWICK AND NATHAN O. KAPLAN

VOLUME VI. Preparation and Assay of Enzymes *(Continued)*
Preparation and Assay of Substrates
Special Techniques
Edited by SIDNEY P. COLOWICK AND NATHAN O. KAPLAN

VOLUME VII. Cumulative Subject Index
Edited by SIDNEY P. COLOWICK AND NATHAN O. KAPLAN

VOLUME VIII. Complex Carbohydrates
Edited by ELIZABETH F. NEUFELD AND VICTOR GINSBURG

VOLUME IX. Carbohydrate Metabolism
Edited by WILLIS A. WOOD

VOLUME X. Oxidation and Phosphorylation
Edited by RONALD W. ESTABROOK AND MAYNARD E. PULLMAN

VOLUME XI. Enzyme Structure
Edited by C. H. W. HIRS

VOLUME XII. Nucleic Acids (Parts A and B)
Edited by LAWRENCE GROSSMAN AND KIVIE MOLDAVE

VOLUME XIII. Citric Acid Cycle
Edited by J. M. LOWENSTEIN

VOLUME XIV. Lipids
Edited by J. M. LOWENSTEIN

VOLUME XV. Steroids and Terpenoids
Edited by RAYMOND B. CLAYTON

VOLUME XVI. Fast Reactions
Edited by KENNETH KUSTIN

VOLUME XVII. Metabolism of Amino Acids and Amines (Parts A and B)
Edited by HERBERT TABOR AND CELIA WHITE TABOR

VOLUME XVIII. Vitamins and Coenzymes (Parts A, B, and C)
Edited by DONALD B. MCCORMICK AND LEMUEL D. WRIGHT

VOLUME XIX. Proteolytic Enzymes
Edited by GERTRUDE E. PERLMANN AND LASZLO LORAND

VOLUME XX. Nucleic Acids and Protein Synthesis (Part C)
Edited by KIVIE MOLDAVE AND LAWRENCE GROSSMAN

VOLUME XXI. Nucleic Acids (Part D)
Edited by LAWRENCE GROSSMAN AND KIVIE MOLDAVE

VOLUME XXII. Enzyme Purification and Related Techniques
Edited by WILLIAM B. JAKOBY

VOLUME XXIII. Photosynthesis (Part A)
Edited by ANTHONY SAN PIETRO

VOLUME XXIV. Photosynthesis and Nitrogen Fixation (Part B)
Edited by ANTHONY SAN PIETRO

VOLUME XXV. Enzyme Structure (Part B)
Edited by C. H. W. HIRS AND SERGE N. TIMASHEFF

VOLUME XXVI. Enzyme Structure (Part C)
Edited by C. H. W. HIRS AND SERGE N. TIMASHEFF

VOLUME XXVII. Enzyme Structure (Part D)
Edited by C. H. W. HIRS AND SERGE N. TIMASHEFF

VOLUME XXVIII. Complex Carbohydrates (Part B)
Edited by VICTOR GINSBURG

VOLUME XXIX. Nucleic Acids and Protein Synthesis (Part E)
Edited by LAWRENCE GROSSMAN AND KIVIE MOLDAVE

VOLUME XXX. Nucleic Acids and Protein Synthesis (Part F)
Edited by KIVIE MOLDAVE AND LAWRENCE GROSSMAN

VOLUME XXXI. Biomembranes (Part A)
Edited by SIDNEY FLEISCHER AND LESTER PACKER

VOLUME XXXII. Biomembranes (Part B)
Edited by SIDNEY FLEISCHER AND LESTER PACKER

VOLUME XXXIII. Cumulative Subject Index Volumes I–XXX
Edited by MARTHA G. DENNIS AND EDWARD A. DENNIS

VOLUME XXXIV. Affinity Techniques (Enzyme Purification: Part B)
Edited by WILLIAM B. JAKOBY AND MEIR WILCHEK

VOLUME XXXV. Lipids (Part B)
Edited by JOHN M. LOWENSTEIN

VOLUME XXXVI. Hormone Action (Part A: Steroid Hormones)
Edited by BERT W. O'MALLEY AND JOEL G. HARDMAN

VOLUME XXXVII. Hormone Action (Part B: Peptide Hormones)
Edited by BERT W. O'MALLEY AND JOEL G. HARDMAN

VOLUME XXXVIII. Hormone Action (Part C: Cyclic Nucleotides)
Edited by JOEL G. HARDMAN AND BERT W. O'MALLEY

VOLUME XXXIX. Hormone Action (Part D: Isolated Cells, Tissues, and Organ Systems)
Edited by JOEL G. HARDMAN AND BERT W. O'MALLEY

VOLUME XL. Hormone Action (Part E: Nuclear Structure and Function)
Edited by BERT W. O'MALLEY AND JOEL G. HARDMAN

VOLUME XLI. Carbohydrate Metabolism (Part B)
Edited by W. A. WOOD

VOLUME XLII. Carbohydrate Metabolism (Part C)
Edited by W. A. WOOD

VOLUME XLIII. Antibiotics
Edited by JOHN H. HASH

VOLUME XLIV. Immobilized Enzymes
Edited by KLAUS MOSBACH

VOLUME XLV. Proteolytic Enzymes (Part B)
Edited by LASZLO LORAND

VOLUME XLVI. Affinity Labeling
Edited by WILLIAM B. JAKOBY AND MEIR WILCHEK

VOLUME XLVII. Enzyme Structure (Part E)
Edited by C. H. W. HIRS AND SERGE N. TIMASHEFF

VOLUME XLVIII. Enzyme Structure (Part F)
Edited by C. H. W. HIRS AND SERGE N. TIMASHEFF

VOLUME XLIX. Enzyme Structure (Part G)
Edited by C. H. W. HIRS AND SERGE N. TIMASHEFF

VOLUME L. Complex Carbohydrates (Part C)
Edited by VICTOR GINSBURG

VOLUME LI. Purine and Pyrimidine Nucleotide Metabolism
Edited by PATRICIA A. HOFFEE AND MARY ELLEN JONES

VOLUME LII. Biomembranes (Part C: Biological Oxidations)
Edited by SIDNEY FLEISCHER AND LESTER PACKER

VOLUME LIII. Biomembranes (Part D: Biological Oxidations)
Edited by SIDNEY FLEISCHER AND LESTER PACKER

VOLUME LIV. Biomembranes (Part E: Biological Oxidations)
Edited by SIDNEY FLEISCHER AND LESTER PACKER

VOLUME LV. Biomembranes (Part F: Bioenergetics)
Edited by SIDNEY FLEISCHER AND LESTER PACKER

VOLUME LVI. Biomembranes (Part G: Bioenergetics)
Edited by SIDNEY FLEISCHER AND LESTER PACKER

VOLUME LVII. Bioluminescence and Chemiluminescence
Edited by MARLENE A. DELUCA

VOLUME LVIII. Cell Culture
Edited by WILLIAM B. JAKOBY AND IRA PASTAN

VOLUME LIX. Nucleic Acids and Protein Synthesis (Part G)
Edited by KIVIE MOLDAVE AND LAWRENCE GROSSMAN

VOLUME LX. Nucleic Acids and Protein Synthesis (Part H)
Edited by KIVIE MOLDAVE AND LAWRENCE GROSSMAN

VOLUME 61. Enzyme Structure (Part H)
Edited by C. H. W. HIRS AND SERGE N. TIMASHEFF

VOLUME 62. Vitamins and Coenzymes (Part D)
Edited by DONALD B. MCCORMICK AND LEMUEL D. WRIGHT

VOLUME 63. Enzyme Kinetics and Mechanism (Part A: Initial Rate and Inhibitor Methods)
Edited by DANIEL L. PURICH

VOLUME 64. Enzyme Kinetics and Mechanism
(Part B: Isotopic Probes and Complex Enzyme Systems)
Edited by DANIEL L. PURICH

VOLUME 65. Nucleic Acids (Part I)
Edited by LAWRENCE GROSSMAN AND KIVIE MOLDAVE

VOLUME 66. Vitamins and Coenzymes (Part E)
Edited by DONALD B. MCCORMICK AND LEMUEL D. WRIGHT

VOLUME 67. Vitamins and Coenzymes (Part F)
Edited by DONALD B. MCCORMICK AND LEMUEL D. WRIGHT

VOLUME 68. Recombinant DNA
Edited by RAY WU

VOLUME 69. Photosynthesis and Nitrogen Fixation (Part C)
Edited by ANTHONY SAN PIETRO

VOLUME 70. Immunochemical Techniques (Part A)
Edited by HELEN VAN VUNAKIS AND JOHN J. LANGONE

VOLUME 71. Lipids (Part C)
Edited by JOHN M. LOWENSTEIN

VOLUME 72. Lipids (Part D)
Edited by JOHN M. LOWENSTEIN

VOLUME 73. Immunochemical Techniques (Part B)
Edited by JOHN J. LANGONE AND HELEN VAN VUNAKIS

VOLUME 74. Immunochemical Techniques (Part C)
Edited by JOHN J. LANGONE AND HELEN VAN VUNAKIS

VOLUME 75. Cumulative Subject Index Volumes XXXI, XXXII, XXXIV–LX
Edited by EDWARD A. DENNIS AND MARTHA G. DENNIS

VOLUME 76. Hemoglobins
Edited by ERALDO ANTONINI, LUIGI ROSSI-BERNARDI, AND EMILIA CHIANCONE

VOLUME 77. Detoxication and Drug Metabolism
Edited by WILLIAM B. JAKOBY

VOLUME 78. Interferons (Part A)
Edited by SIDNEY PESTKA

VOLUME 79. Interferons (Part B)
Edited by SIDNEY PESTKA

VOLUME 80. Proteolytic Enzymes (Part C)
Edited by LASZLO LORAND

VOLUME 81. Biomembranes (Part H: Visual Pigments and Purple Membranes, I)
Edited by LESTER PACKER

VOLUME 82. Structural and Contractile Proteins (Part A: Extracellular Matrix)
Edited by LEON W. CUNNINGHAM AND DIXIE W. FREDERIKSEN

VOLUME 83. Complex Carbohydrates (Part D)
Edited by VICTOR GINSBURG

VOLUME 84. Immunochemical Techniques (Part D: Selected Immunoassays)
Edited by JOHN J. LANGONE AND HELEN VAN VUNAKIS

VOLUME 85. Structural and Contractile Proteins (Part B: The Contractile Apparatus and the Cytoskeleton)
Edited by DIXIE W. FREDERIKSEN AND LEON W. CUNNINGHAM

VOLUME 86. Prostaglandins and Arachidonate Metabolites
Edited by WILLIAM E. M. LANDS AND WILLIAM L. SMITH

VOLUME 87. Enzyme Kinetics and Mechanism (Part C: Intermediates, Stereo-chemistry, and Rate Studies)
Edited by DANIEL L. PURICH

VOLUME 88. Biomembranes (Part I: Visual Pigments and Purple Membranes, II)
Edited by LESTER PACKER

VOLUME 89. Carbohydrate Metabolism (Part D)
Edited by WILLIS A. WOOD

VOLUME 90. Carbohydrate Metabolism (Part E)
Edited by WILLIS A. WOOD

VOLUME 91. Enzyme Structure (Part I)
Edited by C. H. W. HIRS AND SERGE N. TIMASHEFF

VOLUME 92. Immunochemical Techniques (Part E: Monoclonal Antibodies and General Immunoassay Methods)
Edited by JOHN J. LANGONE AND HELEN VAN VUNAKIS

VOLUME 93. Immunochemical Techniques (Part F: Conventional Antibodies, Fc Receptors, and Cytotoxicity)
Edited by JOHN J. LANGONE AND HELEN VAN VUNAKIS

VOLUME 94. Polyamines
Edited by HERBERT TABOR AND CELIA WHITE TABOR

VOLUME 95. Cumulative Subject Index Volumes 61–74, 76–80
Edited by EDWARD A. DENNIS AND MARTHA G. DENNIS

VOLUME 96. Biomembranes [Part J: Membrane Biogenesis: Assembly and Targeting (General Methods; Eukaryotes)]
Edited by SIDNEY FLEISCHER AND BECCA FLEISCHER

VOLUME 97. Biomembranes [Part K: Membrane Biogenesis: Assembly and Targeting (Prokaryotes, Mitochondria, and Chloroplasts)]
Edited by SIDNEY FLEISCHER AND BECCA FLEISCHER

VOLUME 98. Biomembranes (Part L: Membrane Biogenesis: Processing and Recycling)
Edited by SIDNEY FLEISCHER AND BECCA FLEISCHER

VOLUME 99. Hormone Action (Part F: Protein Kinases)
Edited by JACKIE D. CORBIN AND JOEL G. HARDMAN

VOLUME 100. Recombinant DNA (Part B)
Edited by RAY WU, LAWRENCE GROSSMAN, AND KIVIE MOLDAVE

VOLUME 101. Recombinant DNA (Part C)
Edited by RAY WU, LAWRENCE GROSSMAN, AND KIVIE MOLDAVE

VOLUME 102. Hormone Action (Part G: Calmodulin and Calcium-Binding Proteins)
Edited by ANTHONY R. MEANS AND BERT W. O'MALLEY

VOLUME 103. Hormone Action (Part H: Neuroendocrine Peptides)
Edited by P. MICHAEL CONN

VOLUME 104. Enzyme Purification and Related Techniques (Part C)
Edited by WILLIAM B. JAKOBY

VOLUME 105. Oxygen Radicals in Biological Systems
Edited by LESTER PACKER

VOLUME 106. Posttranslational Modifications (Part A)
Edited by FINN WOLD AND KIVIE MOLDAVE

VOLUME 107. Posttranslational Modifications (Part B)
Edited by FINN WOLD AND KIVIE MOLDAVE

VOLUME 108. Immunochemical Techniques (Part G: Separation and Characterization of Lymphoid Cells)
Edited by GIOVANNI DI SABATO, JOHN J. LANGONE, AND HELEN VAN VUNAKIS

VOLUME 109. Hormone Action (Part I: Peptide Hormones)
Edited by LUTZ BIRNBAUMER AND BERT W. O'MALLEY

VOLUME 110. Steroids and Isoprenoids (Part A)
Edited by JOHN H. LAW AND HANS C. RILLING

VOLUME 111. Steroids and Isoprenoids (Part B)
Edited by JOHN H. LAW AND HANS C. RILLING

VOLUME 112. Drug and Enzyme Targeting (Part A)
Edited by KENNETH J. WIDDER AND RALPH GREEN

VOLUME 113. Glutamate, Glutamine, Glutathione, and Related Compounds
Edited by ALTON MEISTER

VOLUME 114. Diffraction Methods for Biological Macromolecules (Part A)
Edited by HAROLD W. WYCKOFF, C. H. W. HIRS, AND SERGE N. TIMASHEFF

VOLUME 115. Diffraction Methods for Biological Macromolecules (Part B)
Edited by HAROLD W. WYCKOFF, C. H. W. HIRS, AND SERGE N. TIMASHEFF

VOLUME 116. Immunochemical Techniques
(Part H: Effectors and Mediators of Lymphoid Cell Functions)
Edited by GIOVANNI DI SABATO, JOHN J. LANGONE, AND HELEN VAN VUNAKIS

VOLUME 117. Enzyme Structure (Part J)
Edited by C. H. W. HIRS AND SERGE N. TIMASHEFF

VOLUME 118. Plant Molecular Biology
Edited by ARTHUR WEISSBACH AND HERBERT WEISSBACH

VOLUME 119. Interferons (Part C)
Edited by SIDNEY PESTKA

VOLUME 120. Cumulative Subject Index Volumes 81–94, 96–101

VOLUME 121. Immunochemical Techniques (Part I: Hybridoma Technology and Monoclonal Antibodies)
Edited by JOHN J. LANGONE AND HELEN VAN VUNAKIS

VOLUME 122. Vitamins and Coenzymes (Part G)
Edited by FRANK CHYTIL AND DONALD B. MCCORMICK

VOLUME 123. Vitamins and Coenzymes (Part H)
Edited by FRANK CHYTIL AND DONALD B. MCCORMICK

VOLUME 124. Hormone Action (Part J: Neuroendocrine Peptides)
Edited by P. MICHAEL CONN

VOLUME 125. Biomembranes (Part M: Transport in Bacteria, Mitochondria, and Chloroplasts: General Approaches and Transport Systems)
Edited by SIDNEY FLEISCHER AND BECCA FLEISCHER

VOLUME 126. Biomembranes (Part N: Transport in Bacteria, Mitochondria, and Chloroplasts: Protonmotive Force)
Edited by SIDNEY FLEISCHER AND BECCA FLEISCHER

VOLUME 127. Biomembranes (Part O: Protons and Water: Structure and Translocation)
Edited by LESTER PACKER

VOLUME 128. Plasma Lipoproteins (Part A: Preparation, Structure, and Molecular Biology)
Edited by JERE P. SEGREST AND JOHN J. ALBERS

VOLUME 129. Plasma Lipoproteins (Part B: Characterization, Cell Biology, and Metabolism)
Edited by JOHN J. ALBERS AND JERE P. SEGREST

VOLUME 130. Enzyme Structure (Part K)
Edited by C. H. W. HIRS AND SERGE N. TIMASHEFF

VOLUME 131. Enzyme Structure (Part L)
Edited by C. H. W. HIRS AND SERGE N. TIMASHEFF

VOLUME 132. Immunochemical Techniques (Part J: Phagocytosis and Cell-Mediated Cytotoxicity)
Edited by GIOVANNI DI SABATO AND JOHANNES EVERSE

VOLUME 133. Bioluminescence and Chemiluminescence (Part B)
Edited by MARLENE DELUCA AND WILLIAM D. MCELROY

VOLUME 134. Structural and Contractile Proteins (Part C: The Contractile Apparatus and the Cytoskeleton)
Edited by RICHARD B. VALLEE

VOLUME 135. Immobilized Enzymes and Cells (Part B)
Edited by KLAUS MOSBACH

VOLUME 136. Immobilized Enzymes and Cells (Part C)
Edited by KLAUS MOSBACH

VOLUME 137. Immobilized Enzymes and Cells (Part D)
Edited by KLAUS MOSBACH

VOLUME 138. Complex Carbohydrates (Part E)
Edited by VICTOR GINSBURG

VOLUME 139. Cellular Regulators (Part A: Calcium- and Calmodulin-Binding Proteins)
Edited by ANTHONY R. MEANS AND P. MICHAEL CONN

VOLUME 140. Cumulative Subject Index Volumes 102–119, 121–134

VOLUME 141. Cellular Regulators (Part B: Calcium and Lipids)
Edited by P. MICHAEL CONN AND ANTHONY R. MEANS

VOLUME 142. Metabolism of Aromatic Amino Acids and Amines
Edited by SEYMOUR KAUFMAN

VOLUME 143. Sulfur and Sulfur Amino Acids
Edited by WILLIAM B. JAKOBY AND OWEN GRIFFITH

VOLUME 144. Structural and Contractile Proteins (Part D: Extracellular Matrix)
Edited by LEON W. CUNNINGHAM

VOLUME 145. Structural and Contractile Proteins (Part E: Extracellular Matrix)
Edited by LEON W. CUNNINGHAM

VOLUME 146. Peptide Growth Factors (Part A)
Edited by DAVID BARNES AND DAVID A. SIRBASKU

VOLUME 147. Peptide Growth Factors (Part B)
Edited by DAVID BARNES AND DAVID A. SIRBASKU

VOLUME 148. Plant Cell Membranes
Edited by LESTER PACKER AND ROLAND DOUCE

VOLUME 149. Drug and Enzyme Targeting (Part B)
Edited by RALPH GREEN AND KENNETH J. WIDDER

VOLUME 150. Immunochemical Techniques (Part K: *In Vitro* Models of B and T Cell Functions and Lymphoid Cell Receptors)
Edited by GIOVANNI DI SABATO

VOLUME 151. Molecular Genetics of Mammalian Cells
Edited by MICHAEL M. GOTTESMAN

VOLUME 152. Guide to Molecular Cloning Techniques
Edited by SHELBY L. BERGER AND ALAN R. KIMMEL

VOLUME 153. Recombinant DNA (Part D)
Edited by RAY WU AND LAWRENCE GROSSMAN

VOLUME 154. Recombinant DNA (Part E)
Edited by RAY WU AND LAWRENCE GROSSMAN

VOLUME 155. Recombinant DNA (Part F)
Edited by RAY WU

VOLUME 156. Biomembranes (Part P: ATP-Driven Pumps and Related Transport: The Na, K-Pump)
Edited by SIDNEY FLEISCHER AND BECCA FLEISCHER

VOLUME 157. Biomembranes (Part Q: ATP-Driven Pumps and Related Transport: Calcium, Proton, and Potassium Pumps)
Edited by SIDNEY FLEISCHER AND BECCA FLEISCHER

VOLUME 158. Metalloproteins (Part A)
Edited by JAMES F. RIORDAN AND BERT L. VALLEE

VOLUME 159. Initiation and Termination of Cyclic Nucleotide Action
Edited by JACKIE D. CORBIN AND ROGER A. JOHNSON

VOLUME 160. Biomass (Part A: Cellulose and Hemicellulose)
Edited by WILLIS A. WOOD AND SCOTT T. KELLOGG

VOLUME 161. Biomass (Part B: Lignin, Pectin, and Chitin)
Edited by WILLIS A. WOOD AND SCOTT T. KELLOGG

VOLUME 162. Immunochemical Techniques (Part L: Chemotaxis and Inflammation)
Edited by GIOVANNI DI SABATO

VOLUME 163. Immunochemical Techniques (Part M: Chemotaxis and Inflammation)
Edited by GIOVANNI DI SABATO

VOLUME 164. Ribosomes
Edited by HARRY F. NOLLER, JR., AND KIVIE MOLDAVE

VOLUME 165. Microbial Toxins: Tools for Enzymology
Edited by SIDNEY HARSHMAN

VOLUME 166. Branched-Chain Amino Acids
Edited by ROBERT HARRIS AND JOHN R. SOKATCH

VOLUME 167. Cyanobacteria
Edited by LESTER PACKER AND ALEXANDER N. GLAZER

VOLUME 168. Hormone Action (Part K: Neuroendocrine Peptides)
Edited by P. MICHAEL CONN

VOLUME 169. Platelets: Receptors, Adhesion, Secretion (Part A)
Edited by JACEK HAWIGER

VOLUME 170. Nucleosomes
Edited by PAUL M. WASSARMAN AND ROGER D. KORNBERG

VOLUME 171. Biomembranes (Part R: Transport Theory: Cells and Model Membranes)
Edited by SIDNEY FLEISCHER AND BECCA FLEISCHER

VOLUME 172. Biomembranes (Part S: Transport: Membrane Isolation and Characterization)
Edited by SIDNEY FLEISCHER AND BECCA FLEISCHER

VOLUME 173. Biomembranes [Part T: Cellular and Subcellular Transport: Eukaryotic (Nonepithelial) Cells]
Edited by SIDNEY FLEISCHER AND BECCA FLEISCHER

VOLUME 174. Biomembranes [Part U: Cellular and Subcellular Transport: Eukaryotic (Nonepithelial) Cells]
Edited by SIDNEY FLEISCHER AND BECCA FLEISCHER

VOLUME 175. Cumulative Subject Index Volumes 135–139, 141–167

VOLUME 176. Nuclear Magnetic Resonance (Part A: Spectral Techniques and Dynamics)
Edited by NORMAN J. OPPENHEIMER AND THOMAS L. JAMES

VOLUME 177. Nuclear Magnetic Resonance (Part B: Structure and Mechanism)
Edited by NORMAN J. OPPENHEIMER AND THOMAS L. JAMES

VOLUME 178. Antibodies, Antigens, and Molecular Mimicry
Edited by JOHN J. LANGONE

VOLUME 179. Complex Carbohydrates (Part F)
Edited by VICTOR GINSBURG

VOLUME 180. RNA Processing (Part A: General Methods)
Edited by JAMES E. DAHLBERG AND JOHN N. ABELSON

VOLUME 181. RNA Processing (Part B: Specific Methods)
Edited by JAMES E. DAHLBERG AND JOHN N. ABELSON

VOLUME 182. Guide to Protein Purification
Edited by MURRAY P. DEUTSCHER

VOLUME 183. Molecular Evolution: Computer Analysis of Protein and Nucleic Acid Sequences
Edited by RUSSELL F. DOOLITTLE

VOLUME 184. Avidin-Biotin Technology
Edited by MEIR WILCHEK AND EDWARD A. BAYER

VOLUME 185. Gene Expression Technology
Edited by DAVID V. GOEDDEL

VOLUME 186. Oxygen Radicals in Biological Systems (Part B: Oxygen Radicals and Antioxidants)
Edited by LESTER PACKER AND ALEXANDER N. GLAZER

VOLUME 187. Arachidonate Related Lipid Mediators
Edited by ROBERT C. MURPHY AND FRANK A. FITZPATRICK

VOLUME 188. Hydrocarbons and Methylotrophy
Edited by MARY E. LIDSTROM

VOLUME 189. Retinoids (Part A: Molecular and Metabolic Aspects)
Edited by LESTER PACKER

VOLUME 190. Retinoids (Part B: Cell Differentiation and Clinical Applications)
Edited by LESTER PACKER

VOLUME 191. Biomembranes (Part V: Cellular and Subcellular Transport: Epithelial Cells)
Edited by SIDNEY FLEISCHER AND BECCA FLEISCHER

VOLUME 192. Biomembranes (Part W: Cellular and Subcellular Transport: Epithelial Cells)
Edited by SIDNEY FLEISCHER AND BECCA FLEISCHER

VOLUME 193. Mass Spectrometry
Edited by JAMES A. MCCLOSKEY

VOLUME 194. Guide to Yeast Genetics and Molecular Biology
Edited by CHRISTINE GUTHRIE AND GERALD R. FINK

VOLUME 195. Adenylyl Cyclase, G Proteins, and Guanylyl Cyclase
Edited by ROGER A. JOHNSON AND JACKIE D. CORBIN

VOLUME 196. Molecular Motors and the Cytoskeleton
Edited by RICHARD B. VALLEE

VOLUME 197. Phospholipases
Edited by EDWARD A. DENNIS

VOLUME 198. Peptide Growth Factors (Part C)
Edited by DAVID BARNES, J. P. MATHER, AND GORDON H. SATO

VOLUME 199. Cumulative Subject Index Volumes 168–174, 176–194

VOLUME 200. Protein Phosphorylation (Part A: Protein Kinases: Assays, Purification, Antibodies, Functional Analysis, Cloning, and Expression)
Edited by TONY HUNTER AND BARTHOLOMEW M. SEFTON

VOLUME 201. Protein Phosphorylation (Part B: Analysis of Protein Phosphorylation, Protein Kinase Inhibitors, and Protein Phosphatases)
Edited by TONY HUNTER AND BARTHOLOMEW M. SEFTON

VOLUME 202. Molecular Design and Modeling: Concepts and Applications (Part A: Proteins, Peptides, and Enzymes)
Edited by JOHN J. LANGONE

VOLUME 203. Molecular Design and Modeling: Concepts and Applications (Part B: Antibodies and Antigens, Nucleic Acids, Polysaccharides, and Drugs)
Edited by JOHN J. LANGONE

VOLUME 204. Bacterial Genetic Systems
Edited by JEFFREY H. MILLER

VOLUME 205. Metallobiochemistry (Part B: Metallothionein and Related Molecules)
Edited by JAMES F. RIORDAN AND BERT L. VALLEE

VOLUME 206. Cytochrome P450
Edited by MICHAEL R. WATERMAN AND ERIC F. JOHNSON

VOLUME 207. Ion Channels
Edited by BERNARDO RUDY AND LINDA E. IVERSON

VOLUME 208. Protein–DNA Interactions
Edited by ROBERT T. SAUER

VOLUME 209. Phospholipid Biosynthesis
Edited by EDWARD A. DENNIS AND DENNIS E. VANCE

VOLUME 210. Numerical Computer Methods
Edited by LUDWIG BRAND AND MICHAEL L. JOHNSON

VOLUME 211. DNA Structures (Part A: Synthesis and Physical Analysis of DNA)
Edited by DAVID M. J. LILLEY AND JAMES E. DAHLBERG

VOLUME 212. DNA Structures (Part B: Chemical and Electrophoretic Analysis of DNA)
Edited by DAVID M. J. LILLEY AND JAMES E. DAHLBERG

VOLUME 213. Carotenoids (Part A: Chemistry, Separation, Quantitation, and Antioxidation)
Edited by LESTER PACKER

VOLUME 214. Carotenoids (Part B: Metabolism, Genetics, and Biosynthesis)
Edited by LESTER PACKER

VOLUME 215. Platelets: Receptors, Adhesion, Secretion (Part B)
Edited by JACEK J. HAWIGER

VOLUME 216. Recombinant DNA (Part G)
Edited by RAY WU

VOLUME 217. Recombinant DNA (Part H)
Edited by RAY WU

VOLUME 218. Recombinant DNA (Part I)
Edited by RAY WU

VOLUME 219. Reconstitution of Intracellular Transport
Edited by JAMES E. ROTHMAN

VOLUME 220. Membrane Fusion Techniques (Part A)
Edited by NEJAT DÜZGÜNEŞ

VOLUME 221. Membrane Fusion Techniques (Part B)
Edited by NEJAT DÜZGÜNEŞ

VOLUME 222. Proteolytic Enzymes in Coagulation, Fibrinolysis, and Complement Activation (Part A: Mammalian Blood Coagulation Factors and Inhibitors)
Edited by LASZLO LORAND AND KENNETH G. MANN

VOLUME 223. Proteolytic Enzymes in Coagulation, Fibrinolysis, and Complement Activation (Part B: Complement Activation, Fibrinolysis, and Nonmammalian Blood Coagulation Factors)
Edited by LASZLO LORAND AND KENNETH G. MANN

VOLUME 224. Molecular Evolution: Producing the Biochemical Data
Edited by ELIZABETH ANNE ZIMMER, THOMAS J. WHITE, REBECCA L. CANN, AND ALLAN C. WILSON

VOLUME 225. Guide to Techniques in Mouse Development
Edited by PAUL M. WASSARMAN AND MELVIN L. DEPAMPHILIS

VOLUME 226. Metallobiochemistry (Part C: Spectroscopic and Physical Methods for Probing Metal Ion Environments in Metalloenzymes and Metalloproteins)
Edited by JAMES F. RIORDAN AND BERT L. VALLEE

VOLUME 227. Metallobiochemistry (Part D: Physical and Spectroscopic Methods for Probing Metal Ion Environments in Metalloproteins)
Edited by JAMES F. RIORDAN AND BERT L. VALLEE

VOLUME 228. Aqueous Two Phase Systems
Edited by HARRY WALTER AND GÖTE JOHANSSON

VOLUME 229. Cumulative Subject Index Volumes 195–198, 200–227

VOLUME 230. Guide to Techniques in Glycobiology
Edited by WILLIAM J. LENNARZ AND GERALD W. HART

VOLUME 231. Hemoglobins (Part B: Biochemical and Analytical Methods)
Edited by JOHANNES EVERSE, KIM D. VANDEGRIFF, AND ROBERT M. WINSLOW

VOLUME 232. Hemoglobins (Part C: Biophysical Methods)
Edited by JOHANNES EVERSE, KIM D. VANDEGRIFF, AND ROBERT M. WINSLOW

VOLUME 233. Oxygen Radicals in Biological Systems (Part C)
Edited by LESTER PACKER

VOLUME 234. Oxygen Radicals in Biological Systems (Part D)
Edited by LESTER PACKER

VOLUME 235. Bacterial Pathogenesis (Part A: Identification and Regulation of Virulence Factors)
Edited by VIRGINIA L. CLARK AND PATRIK M. BAVOIL

VOLUME 236. Bacterial Pathogenesis (Part B: Integration of Pathogenic Bacteria with Host Cells)
Edited by VIRGINIA L. CLARK AND PATRIK M. BAVOIL

VOLUME 237. Heterotrimeric G Proteins
Edited by RAVI IYENGAR

VOLUME 238. Heterotrimeric G-Protein Effectors
Edited by RAVI IYENGAR

VOLUME 239. Nuclear Magnetic Resonance (Part C)
Edited by THOMAS L. JAMES AND NORMAN J. OPPENHEIMER

VOLUME 240. Numerical Computer Methods (Part B)
Edited by MICHAEL L. JOHNSON AND LUDWIG BRAND

VOLUME 241. Retroviral Proteases
Edited by LAWRENCE C. KUO AND JULES A. SHAFER

VOLUME 242. Neoglycoconjugates (Part A)
Edited by Y. C. LEE AND REIKO T. LEE

VOLUME 243. Inorganic Microbial Sulfur Metabolism
Edited by HARRY D. PECK, JR., AND JEAN LEGALL

VOLUME 244. Proteolytic Enzymes: Serine and Cysteine Peptidases
Edited by ALAN J. BARRETT

VOLUME 245. Extracellular Matrix Components
Edited by E. RUOSLAHTI AND E. ENGVALL

VOLUME 246. Biochemical Spectroscopy
Edited by KENNETH SAUER

VOLUME 247. Neoglycoconjugates (Part B: Biomedical Applications)
Edited by Y. C. LEE AND REIKO T. LEE

VOLUME 248. Proteolytic Enzymes: Aspartic and Metallo Peptidases
Edited by ALAN J. BARRETT

VOLUME 249. Enzyme Kinetics and Mechanism (Part D: Developments in Enzyme Dynamics)
Edited by DANIEL L. PURICH

VOLUME 250. Lipid Modifications of Proteins
Edited by PATRICK J. CASEY AND JANICE E. BUSS

VOLUME 251. Biothiols (Part A: Monothiols and Dithiols, Protein Thiols, and Thiyl Radicals)
Edited by LESTER PACKER

VOLUME 252. Biothiols (Part B: Glutathione and Thioredoxin; Thiols in Signal Transduction and Gene Regulation)
Edited by LESTER PACKER

VOLUME 253. Adhesion of Microbial Pathogens
Edited by RON J. DOYLE AND ITZHAK OFEK

VOLUME 254. Oncogene Techniques
Edited by PETER K. VOGT AND INDER M. VERMA

VOLUME 255. Small GTPases and Their Regulators (Part A: Ras Family)
Edited by W. E. BALCH, CHANNING J. DER, AND ALAN HALL

VOLUME 256. Small GTPases and Their Regulators (Part B: Rho Family)
Edited by W. E. BALCH, CHANNING J. DER, AND ALAN HALL

VOLUME 257. Small GTPases and Their Regulators (Part C: Proteins Involved in Transport)
Edited by W. E. BALCH, CHANNING J. DER, AND ALAN HALL

VOLUME 258. Redox-Active Amino Acids in Biology
Edited by JUDITH P. KLINMAN

VOLUME 259. Energetics of Biological Macromolecules
Edited by MICHAEL L. JOHNSON AND GARY K. ACKERS

VOLUME 260. Mitochondrial Biogenesis and Genetics (Part A)
Edited by GIUSEPPE M. ATTARDI AND ANNE CHOMYN

VOLUME 261. Nuclear Magnetic Resonance and Nucleic Acids
Edited by THOMAS L. JAMES

VOLUME 262. DNA Replication
Edited by JUDITH L. CAMPBELL

VOLUME 263. Plasma Lipoproteins (Part C: Quantitation)
Edited by WILLIAM A. BRADLEY, SANDRA H. GIANTURCO, AND JERE P. SEGREST

VOLUME 264. Mitochondrial Biogenesis and Genetics (Part B)
Edited by GIUSEPPE M. ATTARDI AND ANNE CHOMYN

VOLUME 265. Cumulative Subject Index Volumes 228, 230–262

VOLUME 266. Computer Methods for Macromolecular Sequence Analysis
Edited by RUSSELL F. DOOLITTLE

VOLUME 267. Combinatorial Chemistry
Edited by JOHN N. ABELSON

VOLUME 268. Nitric Oxide (Part A: Sources and Detection of NO; NO Synthase)
Edited by LESTER PACKER

VOLUME 269. Nitric Oxide (Part B: Physiological and Pathological Processes)
Edited by LESTER PACKER

VOLUME 270. High Resolution Separation and Analysis of Biological Macromolecules (Part A: Fundamentals)
Edited by BARRY L. KARGER AND WILLIAM S. HANCOCK

VOLUME 271. High Resolution Separation and Analysis of Biological Macromolecules (Part B: Applications)
Edited by BARRY L. KARGER AND WILLIAM S. HANCOCK

VOLUME 272. Cytochrome P450 (Part B)
Edited by ERIC F. JOHNSON AND MICHAEL R. WATERMAN

VOLUME 273. RNA Polymerase and Associated Factors (Part A)
Edited by SANKAR ADHYA

VOLUME 274. RNA Polymerase and Associated Factors (Part B)
Edited by SANKAR ADHYA

VOLUME 275. Viral Polymerases and Related Proteins
Edited by LAWRENCE C. KUO, DAVID B. OLSEN, AND STEVEN S. CARROLL

VOLUME 276. Macromolecular Crystallography (Part A)
Edited by CHARLES W. CARTER, JR., AND ROBERT M. SWEET

VOLUME 277. Macromolecular Crystallography (Part B)
Edited by CHARLES W. CARTER, JR., AND ROBERT M. SWEET

VOLUME 278. Fluorescence Spectroscopy
Edited by LUDWIG BRAND AND MICHAEL L. JOHNSON

VOLUME 279. Vitamins and Coenzymes (Part I)
Edited by DONALD B. MCCORMICK, JOHN W. SUTTIE, AND CONRAD WAGNER

VOLUME 280. Vitamins and Coenzymes (Part J)
Edited by DONALD B. MCCORMICK, JOHN W. SUTTIE, AND CONRAD WAGNER

VOLUME 281. Vitamins and Coenzymes (Part K)
Edited by DONALD B. MCCORMICK, JOHN W. SUTTIE, AND CONRAD WAGNER

VOLUME 282. Vitamins and Coenzymes (Part L)
Edited by DONALD B. MCCORMICK, JOHN W. SUTTIE, AND CONRAD WAGNER

VOLUME 283. Cell Cycle Control
Edited by WILLIAM G. DUNPHY

VOLUME 284. Lipases (Part A: Biotechnology)
Edited by BYRON RUBIN AND EDWARD A. DENNIS

VOLUME 285. Cumulative Subject Index Volumes 263, 264, 266–284, 286–289

VOLUME 286. Lipases (Part B: Enzyme Characterization and Utilization)
Edited by BYRON RUBIN AND EDWARD A. DENNIS

VOLUME 287. Chemokines
Edited by RICHARD HORUK

VOLUME 288. Chemokine Receptors
Edited by RICHARD HORUK

VOLUME 289. Solid Phase Peptide Synthesis
Edited by GREGG B. FIELDS

VOLUME 290. Molecular Chaperones
Edited by GEORGE H. LORIMER AND THOMAS BALDWIN

VOLUME 291. Caged Compounds
Edited by GERARD MARRIOTT

VOLUME 292. ABC Transporters: Biochemical, Cellular, and Molecular Aspects
Edited by SURESH V. AMBUDKAR AND MICHAEL M. GOTTESMAN

VOLUME 293. Ion Channels (Part B)
Edited by P. MICHAEL CONN

VOLUME 294. Ion Channels (Part C)
Edited by P. MICHAEL CONN

VOLUME 295. Energetics of Biological Macromolecules (Part B)
Edited by GARY K. ACKERS AND MICHAEL L. JOHNSON

VOLUME 296. Neurotransmitter Transporters
Edited by SUSAN G. AMARA

VOLUME 297. Photosynthesis: Molecular Biology of Energy Capture
Edited by LEE MCINTOSH

VOLUME 298. Molecular Motors and the Cytoskeleton (Part B)
Edited by RICHARD B. VALLEE

VOLUME 299. Oxidants and Antioxidants (Part A)
Edited by LESTER PACKER

VOLUME 300. Oxidants and Antioxidants (Part B)
Edited by LESTER PACKER

VOLUME 301. Nitric Oxide: Biological and Antioxidant Activities (Part C)
Edited by LESTER PACKER

VOLUME 302. Green Fluorescent Protein
Edited by P. MICHAEL CONN

VOLUME 303. cDNA Preparation and Display
Edited by SHERMAN M. WEISSMAN

VOLUME 304. Chromatin
Edited by PAUL M. WASSARMAN AND ALAN P. WOLFFE

VOLUME 305. Bioluminescence and Chemiluminescence (Part C)
Edited by THOMAS O. BALDWIN AND MIRIAM M. ZIEGLER

VOLUME 306. Expression of Recombinant Genes in Eukaryotic Systems
Edited by JOSEPH C. GLORIOSO AND MARTIN C. SCHMIDT

VOLUME 307. Confocal Microscopy
Edited by P. MICHAEL CONN

VOLUME 308. Enzyme Kinetics and Mechanism (Part E: Energetics of Enzyme Catalysis)
Edited by DANIEL L. PURICH AND VERN L. SCHRAMM

VOLUME 309. Amyloid, Prions, and Other Protein Aggregates
Edited by RONALD WETZEL

VOLUME 310. Biofilms
Edited by RON J. DOYLE

VOLUME 311. Sphingolipid Metabolism and Cell Signaling (Part A)
Edited by ALFRED H. MERRILL, JR., AND YUSUF A. HANNUN

VOLUME 312. Sphingolipid Metabolism and Cell Signaling (Part B)
Edited by ALFRED H. MERRILL, JR., AND YUSUF A. HANNUN

VOLUME 313. Antisense Technology (Part A: General Methods, Methods of Delivery, and RNA Studies)
Edited by M. IAN PHILLIPS

VOLUME 314. Antisense Technology (Part B: Applications)
Edited by M. IAN PHILLIPS

VOLUME 315. Vertebrate Phototransduction and the Visual Cycle (Part A)
Edited by KRZYSZTOF PALCZEWSKI

VOLUME 316. Vertebrate Phototransduction and the Visual Cycle (Part B)
Edited by KRZYSZTOF PALCZEWSKI

VOLUME 317. RNA–Ligand Interactions (Part A: Structural Biology Methods)
Edited by DANIEL W. CELANDER AND JOHN N. ABELSON

VOLUME 318. RNA–Ligand Interactions (Part B: Molecular Biology Methods)
Edited by DANIEL W. CELANDER AND JOHN N. ABELSON

VOLUME 319. Singlet Oxygen, UV-A, and Ozone
Edited by LESTER PACKER AND HELMUT SIES

VOLUME 320. Cumulative Subject Index Volumes 290–319

VOLUME 321. Numerical Computer Methods (Part C)
Edited by MICHAEL L. JOHNSON AND LUDWIG BRAND

VOLUME 322. Apoptosis
Edited by JOHN C. REED

VOLUME 323. Energetics of Biological Macromolecules (Part C)
Edited by MICHAEL L. JOHNSON AND GARY K. ACKERS

VOLUME 324. Branched-Chain Amino Acids (Part B)
Edited by ROBERT A. HARRIS AND JOHN R. SOKATCH

VOLUME 325. Regulators and Effectors of Small GTPases (Part D: Rho Family)
Edited by W. E. BALCH, CHANNING J. DER, AND ALAN HALL

VOLUME 326. Applications of Chimeric Genes and Hybrid Proteins (Part A: Gene Expression and Protein Purification)
Edited by JEREMY THORNER, SCOTT D. EMR, AND JOHN N. ABELSON

VOLUME 327. Applications of Chimeric Genes and Hybrid Proteins (Part B: Cell Biology and Physiology)
Edited by JEREMY THORNER, SCOTT D. EMR, AND JOHN N. ABELSON

VOLUME 328. Applications of Chimeric Genes and Hybrid Proteins (Part C: Protein–Protein Interactions and Genomics)
Edited by JEREMY THORNER, SCOTT D. EMR, AND JOHN N. ABELSON

VOLUME 329. Regulators and Effectors of Small GTPases (Part E: GTPases Involved in Vesicular Traffic)
Edited by W. E. BALCH, CHANNING J. DER, AND ALAN HALL

VOLUME 330. Hyperthermophilic Enzymes (Part A)
Edited by MICHAEL W. W. ADAMS AND ROBERT M. KELLY

VOLUME 331. Hyperthermophilic Enzymes (Part B)
Edited by MICHAEL W. W. ADAMS AND ROBERT M. KELLY

VOLUME 332. Regulators and Effectors of Small GTPases (Part F: Ras Family I)
Edited by W. E. BALCH, CHANNING J. DER, AND ALAN HALL

VOLUME 333. Regulators and Effectors of Small GTPases (Part G: Ras Family II)
Edited by W. E. BALCH, CHANNING J. DER, AND ALAN HALL

VOLUME 334. Hyperthermophilic Enzymes (Part C)
Edited by MICHAEL W. W. ADAMS AND ROBERT M. KELLY

VOLUME 335. Flavonoids and Other Polyphenols
Edited by LESTER PACKER

VOLUME 336. Microbial Growth in Biofilms (Part A: Developmental and Molecular Biological Aspects)
Edited by RON J. DOYLE

VOLUME 337. Microbial Growth in Biofilms (Part B: Special Environments and Physicochemical Aspects)
Edited by RON J. DOYLE

VOLUME 338. Nuclear Magnetic Resonance of Biological Macromolecules (Part A)
Edited by THOMAS L. JAMES, VOLKER DÖTSCH, AND ULI SCHMITZ

VOLUME 339. Nuclear Magnetic Resonance of Biological Macromolecules (Part B)
Edited by THOMAS L. JAMES, VOLKER DÖTSCH, AND ULI SCHMITZ

VOLUME 340. Drug–Nucleic Acid Interactions
Edited by JONATHAN B. CHAIRES AND MICHAEL J. WARING

VOLUME 341. Ribonucleases (Part A)
Edited by ALLEN W. NICHOLSON

VOLUME 342. Ribonucleases (Part B)
Edited by ALLEN W. NICHOLSON

VOLUME 343. G Protein Pathways (Part A: Receptors)
Edited by RAVI IYENGAR AND JOHN D. HILDEBRANDT

VOLUME 344. G Protein Pathways (Part B: G Proteins and Their Regulators)
Edited by RAVI IYENGAR AND JOHN D. HILDEBRANDT

VOLUME 345. G Protein Pathways (Part C: Effector Mechanisms)
Edited by RAVI IYENGAR AND JOHN D. HILDEBRANDT

VOLUME 346. Gene Therapy Methods
Edited by M. IAN PHILLIPS

VOLUME 347. Protein Sensors and Reactive Oxygen Species (Part A: Selenoproteins and Thioredoxin)
Edited by HELMUT SIES AND LESTER PACKER

VOLUME 348. Protein Sensors and Reactive Oxygen Species (Part B: Thiol Enzymes and Proteins)
Edited by HELMUT SIES AND LESTER PACKER

VOLUME 349. Superoxide Dismutase
Edited by LESTER PACKER

VOLUME 350. Guide to Yeast Genetics and Molecular and Cell Biology (Part B)
Edited by CHRISTINE GUTHRIE AND GERALD R. FINK

VOLUME 351. Guide to Yeast Genetics and Molecular and Cell Biology (Part C)
Edited by CHRISTINE GUTHRIE AND GERALD R. FINK

VOLUME 352. Redox Cell Biology and Genetics (Part A)
Edited by CHANDAN K. SEN AND LESTER PACKER

VOLUME 353. Redox Cell Biology and Genetics (Part B)
Edited by CHANDAN K. SEN AND LESTER PACKER

VOLUME 354. Enzyme Kinetics and Mechanisms (Part F: Detection and Characterization of Enzyme Reaction Intermediates)
Edited by DANIEL L. PURICH

VOLUME 355. Cumulative Subject Index Volumes 321–354

VOLUME 356. Laser Capture Microscopy and Microdissection
Edited by P. MICHAEL CONN

VOLUME 357. Cytochrome P450, Part C
Edited by ERIC F. JOHNSON AND MICHAEL R. WATERMAN

VOLUME 358. Bacterial Pathogenesis (Part C: Identification, Regulation, and Function of Virulence Factors)
Edited by VIRGINIA L. CLARK AND PATRIK M. BAVOIL

VOLUME 359. Nitric Oxide (Part D)
Edited by ENRIQUE CADENAS AND LESTER PACKER

VOLUME 360. Biophotonics (Part A)
Edited by GERARD MARRIOTT AND IAN PARKER

VOLUME 361. Biophotonics (Part B)
Edited by GERARD MARRIOTT AND IAN PARKER

VOLUME 362. Recognition of Carbohydrates in Biological Systems (Part A)
Edited by YUAN C. LEE AND REIKO T. LEE

VOLUME 363. Recognition of Carbohydrates in Biological Systems (Part B)
Edited by YUAN C. LEE AND REIKO T. LEE

VOLUME 364. Nuclear Receptors
Edited by DAVID W. RUSSELL AND DAVID J. MANGELSDORF

VOLUME 365. Differentiation of Embryonic Stem Cells
Edited by PAUL M. WASSAUMAN AND GORDON M. KELLER

VOLUME 366. Protein Phosphatases
Edited by SUSANNE KLUMPP AND JOSEF KRIEGLSTEIN

VOLUME 367. Liposomes (Part A)
Edited by NEJAT DÜZGÜNEŞ

VOLUME 368. Macromolecular Crystallography (Part C)
Edited by CHARLES W. CARTER, JR., AND ROBERT M. SWEET

VOLUME 369. Combinational Chemistry (Part B)
Edited by GUILLERMO A. MORALES AND BARRY A. BUNIN

VOLUME 370. RNA Polymerases and Associated Factors (Part C)
Edited by SANKAR L. ADHYA AND SUSAN GARGES

VOLUME 371. RNA Polymerases and Associated Factors (Part D)
Edited by SANKAR L. ADHYA AND SUSAN GARGES

VOLUME 372. Liposomes (Part B)
Edited by NEJAT DÜZGÜNEŞ

VOLUME 373. Liposomes (Part C)
Edited by NEJAT DÜZGÜNEŞ

VOLUME 374. Macromolecular Crystallography (Part D)
Edited by CHARLES W. CARTER, JR., AND ROBERT W. SWEET

VOLUME 375. Chromatin and Chromatin Remodeling Enzymes (Part A)
Edited by C. DAVID ALLIS AND CARL WU

VOLUME 376. Chromatin and Chromatin Remodeling Enzymes (Part B)
Edited by C. DAVID ALLIS AND CARL WU

VOLUME 377. Chromatin and Chromatin Remodeling Enzymes (Part C)
Edited by C. DAVID ALLIS AND CARL WU

VOLUME 378. Quinones and Quinone Enzymes (Part A)
Edited by HELMUT SIES AND LESTER PACKER

VOLUME 379. Energetics of Biological Macromolecules (Part D)
Edited by JO M. HOLT, MICHAEL L. JOHNSON, AND GARY K. ACKERS

VOLUME 380. Energetics of Biological Macromolecules (Part E)
Edited by JO M. HOLT, MICHAEL L. JOHNSON, AND GARY K. ACKERS

VOLUME 381. Oxygen Sensing
Edited by CHANDAN K. SEN AND GREGG L. SEMENZA

VOLUME 382. Quinones and Quinone Enzymes (Part B)
Edited by HELMUT SIES AND LESTER PACKER

VOLUME 383. Numerical Computer Methods (Part D)
Edited by LUDWIG BRAND AND MICHAEL L. JOHNSON

VOLUME 384. Numerical Computer Methods (Part E)
Edited by LUDWIG BRAND AND MICHAEL L. JOHNSON

VOLUME 385. Imaging in Biological Research (Part A)
Edited by P. MICHAEL CONN

VOLUME 386. Imaging in Biological Research (Part B)
Edited by P. MICHAEL CONN

VOLUME 387. Liposomes (Part D)
Edited by NEJAT DÜZGÜNEŞ

VOLUME 388. Protein Engineering
Edited by DAN E. ROBERTSON AND JOSEPH P. NOEL

VOLUME 389. Regulators of G-Protein Signaling (Part A)
Edited by DAVID P. SIDEROVSKI

VOLUME 390. Regulators of G-Protein Signaling (Part B)
Edited by DAVID P. SIDEROVSKI

VOLUME 391. Liposomes (Part E)
Edited by NEJAT DÜZGÜNEŞ

VOLUME 392. RNA Interference
Edited by ENGELKE ROSSI

VOLUME 393. Circadian Rhythms
Edited by MICHAEL W. YOUNG

VOLUME 394. Nuclear Magnetic Resonance of Biological Macromolecules (Part C)
Edited by THOMAS L. JAMES

VOLUME 395. Producing the Biochemical Data (Part B)
Edited by ELIZABETH A. ZIMMER AND ERIC H. ROALSON

VOLUME 396. Nitric Oxide (Part E)
Edited by LESTER PACKER AND ENRIQUE CADENAS

VOLUME 397. Environmental Microbiology
Edited by JARED R. LEADBETTER

VOLUME 398. Ubiquitin and Protein Degradation (Part A)
Edited by RAYMOND J. DESHAIES

VOLUME 399. Ubiquitin and Protein Degradation (Part B)
Edited by RAYMOND J. DESHAIES

VOLUME 400. Phase II Conjugation Enzymes and Transport Systems
Edited by HELMUT SIES AND LESTER PACKER

VOLUME 401. Glutathione Transferases and Gamma Glutamyl Transpeptidases
Edited by HELMUT SIES AND LESTER PACKER

VOLUME 402. Biological Mass Spectrometry
Edited by A. L. BURLINGAME

VOLUME 403. GTPases Regulating Membrane Targeting and Fusion
Edited by WILLIAM E. BALCH, CHANNING J. DER, AND ALAN HALL

VOLUME 404. GTPases Regulating Membrane Dynamics
Edited by WILLIAM E. BALCH, CHANNING J. DER, AND ALAN HALL

VOLUME 405. Mass Spectrometry: Modified Proteins and Glycoconjugates
Edited by A. L. BURLINGAME

VOLUME 406. Regulators and Effectors of Small GTPases: Rho Family
Edited by WILLIAM E. BALCH, CHANNING J. DER, AND ALAN HALL

VOLUME 407. Regulators and Effectors of Small GTPases: Ras Family
Edited by WILLIAM E. BALCH, CHANNING J. DER, AND ALAN HALL

VOLUME 408. DNA Repair (Part A)
Edited by JUDITH L. CAMPBELL AND PAUL MODRICH

VOLUME 409. DNA Repair (Part B)
Edited by JUDITH L. CAMPBELL AND PAUL MODRICH

VOLUME 410. DNA Microarrays (Part A: Array Platforms and Web-Bench Protocols)
Edited by ALAN KIMMEL AND BRIAN OLIVER

VOLUME 411. DNA Microarrays (Part B: Databases and Statistics)
Edited by ALAN KIMMEL AND BRIAN OLIVER

VOLUME 412. Amyloid, Prions, and Other Protein Aggregates (Part B)
Edited by INDU KHETERPAL AND RONALD WETZEL

VOLUME 413. Amyloid, Prions, and Other Protein Aggregates (Part C)
Edited by INDU KHETERPAL AND RONALD WETZEL

VOLUME 414. Measuring Biological Responses with Automated Microscopy
Edited by JAMES INGLESE

VOLUME 415. Glycobiology
Edited by MINORU FUKUDA

VOLUME 416. Glycomics
Edited by MINORU FUKUDA

VOLUME 417. Functional Glycomics
Edited by MINORU FUKUDA

VOLUME 418. Embryonic Stem Cells
Edited by IRINA KLIMANSKAYA AND ROBERT LANZA

VOLUME 419. Adult Stem Cells
Edited by IRINA KLIMANSKAYA AND ROBERT LANZA

VOLUME 420. Stem Cell Tools and Other Experimental Protocols
Edited by IRINA KLIMANSKAYA AND ROBERT LANZA

VOLUME 421. Advanced Bacterial Genetics: Use of Transposons and Phage for Genomic Engineering
Edited by KELLY T. HUGHES

VOLUME 422. Two-Component Signaling Systems, Part A
Edited by MELVIN I. SIMON, BRIAN R. CRANE, AND ALEXANDRINE CRANE

VOLUME 423. Two-Component Signaling Systems, Part B
Edited by MELVIN I. SIMON, BRIAN R. CRANE, AND ALEXANDRINE CRANE

VOLUME 424. RNA Editing
Edited by JONATHA M. GOTT

VOLUME 425. RNA Modification
Edited by JONATHA M. GOTT

VOLUME 426. Integrins
Edited by DAVID CHERESH

VOLUME 427. MicroRNA Methods
Edited by JOHN J. ROSSI

VOLUME 428. Osmosensing and Osmosignaling
Edited by HELMUT SIES AND DIETER HAUSSINGER

VOLUME 429. Translation Initiation: Extract Systems and Molecular Genetics
Edited by JON LORSCH

VOLUME 430. Translation Initiation: Reconstituted Systems and Biophysical Methods
Edited by JON LORSCH

VOLUME 431. Translation Initiation: Cell Biology, High-Throughput and Chemical-Based Approaches
Edited by JON LORSCH

VOLUME 432. Lipidomics and Bioactive Lipids: Mass-Spectrometry–Based Lipid Analysis
Edited by H. ALEX BROWN

VOLUME 433. Lipidomics and Bioactive Lipids: Specialized Analytical Methods and Lipids in Disease
Edited by H. ALEX BROWN

VOLUME 434. Lipidomics and Bioactive Lipids: Lipids and Cell Signaling
Edited by H. ALEX BROWN

VOLUME 435. Oxygen Biology and Hypoxia
Edited by HELMUT SIES AND BERNHARD BRÜNE

VOLUME 436. Globins and Other Nitric Oxide-Reactive Protiens (Part A)
Edited by ROBERT K. POOLE

VOLUME 437. Globins and Other Nitric Oxide-Reactive Protiens (Part B)
Edited by ROBERT K. POOLE

VOLUME 438. Small GTPases in Disease (Part A)
Edited by WILLIAM E. BALCH, CHANNING J. DER, AND ALAN HALL

VOLUME 439. Small GTPases in Disease (Part B)
Edited by WILLIAM E. BALCH, CHANNING J. DER, AND ALAN HALL

VOLUME 440. Nitric Oxide, Part F Oxidative and Nitrosative Stress in Redox Regulation of Cell Signaling
Edited by ENRIQUE CADENAS AND LESTER PACKER

VOLUME 441. Nitric Oxide, Part G Oxidative and Nitrosative Stress in Redox Regulation of Cell Signaling
Edited by ENRIQUE CADENAS AND LESTER PACKER

VOLUME 442. Programmed Cell Death, General Principles for Studying Cell Death (Part A)
Edited by ROYA KHOSRAVI-FAR, ZAHRA ZAKERI, RICHARD A. LOCKSHIN, AND MAURO PIACENTINI

VOLUME 443. Angiogenesis: *In Vitro* Systems
Edited by DAVID A. CHERESH

VOLUME 444. Angiogenesis: *In Vivo* Systems (Part A)
Edited by DAVID A. CHERESH

VOLUME 445. Angiogenesis: *In Vivo* Systems (Part B)
Edited by DAVID A. CHERESH

VOLUME 446. Programmed Cell Death, The Biology and Therapeutic Implications of Cell Death (Part B)
Edited by ROYA KHOSRAVI-FAR, ZAHRA ZAKERI, RICHARD A. LOCKSHIN, AND MAURO PIACENTINI

VOLUME 447. RNA Turnover in Prokaryotes, Archaea and Organelles
Edited by LYNNE E. MAQUAT AND CECILIA M. ARRAIANO

SECTION ONE

DECAPPING ANALYSES

CHAPTER ONE

ANALYSIS OF mRNA DECAPPING

Shin-Wu Liu, Xinfu Jiao, Sarah Welch, *and* Megerditch Kiledjian

Contents

1. Introduction	4
2. Measuring Decapping Activities of Recombinant and Endogenous Dcp2	6
2.1. Preparation of cap-radiolabeled RNA substrates	7
2.2. Measuring Dcp2 decapping activity	9
2.3. Additional considerations	14
3. Measuring DcpS Activity	15
3.1. Preparation of radiolabeled cap substrate for DcpS decapping	16
3.2. Generation of bacterial-expressed DcpS protein	16
3.3. Generation of total-cell extract to detect endogenous DcpS activity	16
3.4. *In vitro* DcpS decapping assay	17
4. Migration of Cap Analogs with Various Thin-layer Chromatography Running Buffers	17
4.1. Preparation of various cap analog markers	18
Acknowledgment	19
References	19

Abstracts

The modulation of mRNA decay is a critical determinant in the regulation of gene expression. mRNAs in eukaryotes are primarily degraded by two major exonucleolytic pathways: the 5′ to 3′-and the 3′ to 5′-pathways, both of which are initiated by removal of the polyadenylated (poly(A)) tail. Hydrolysis of the 5′-cap structure, termed decapping, is a key step in the demise of mRNA. Two major decapping enzymes with distinct activities and substrate requirements have been identified. Dcp2 hydrolyzes the cap structure on an intact mRNA in the 5′ to 3′-decay pathway; Dcp2 scavenges the residual cap oligonucleotide resulting from the 3′ to 5′-decay pathway, as well as hydrolyzes the decapping product generated by Dcp2. In this chapter, we describe the methods for monitoring Dcp2 and DcpS decapping activities of bacterially expressed and endogenous human decapping enzymes.

Rutgers University, Department of Cell Biology and Neuroscience, Piscataway, New Jersey, USA

Methods in Enzymology, Volume 448   © 2008 Elsevier Inc.
ISSN 0076-6879, DOI: 10.1016/S0076-6879(08)02601-3   All rights reserved.

1. Introduction

The steady-state levels of cytoplasmic mRNAs depend on the combined rate of mRNA synthesis in the nucleus, mRNA transport from the nucleus to the cytoplasm, and mRNA degradation. The abundance of an mRNA is, therefore, determined by both its synthesis and decay. Thus, the regulation of mRNA stability becomes a critical determinant of gene expression.

In eukaryotes, mature mRNAs are protected against exonuclease attack by the presence of the 3′-poly (A) tail, as well as an m^{7}GpppN cap structure at the 5′-end. As shown in Fig. 1.1, normal polyadenylated mRNAs undergo degradation by the initial removal of the poly(A) tail, termed deadenylation (Decker and Parker, 1993; Muhlrad et al., 1995), followed by the degradation of the mRNA body in either a 5′ to 3′-manner or a 3′ to 5′-manner (Coller and Parker, 2004; Garneau et al., 2007; Liu and Kiledjian, 2006). In the 5′ to 3′-pathway, the 5′-cap is hydrolyzed by the catalytically active decapping enzyme Dcp2, which most likely functions within a larger decapping complex

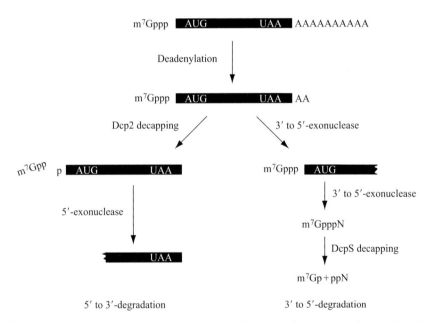

Figure 1.1 General eukaryotic mRNA decay pathways. In eukaryotes, the degradation of normal polyadenylated mRNAs initiates with deadenylation, followed by degradation of the mRNA body from either end. Decay from the 5′-end continues with the hydrolysis of the 5′-cap structure by a decapping enzyme Dcp2, followed by the 5′ to 3′-exonucleolytic digestion of the RNA body by Xrn1. Decay from the 3′-end continues with 3′ to 5′-exonucleolytic digestion of the RNA body by the exosome complex. The resulting cap dinucleotide or capped oligonucleotide is hydrolyzed by the scavenger-decapping enzyme, DcpS.

(Dunckley and Parker, 1999; Fenger-Gron et al., 2005; Steiger et al., 2003; van Dijk et al., 2002; Wang et al., 2002). The exposed mRNA 5′-end is subsequently degraded in the 5′ to 3′-direction by the Xrn1 exonuclease (Decker and Parker, 1994; Johnson, 1997). In the 3′ to 5′-pathway after deadenylation, the unprotected 3′-end is further degraded by the cytoplasmic exosome in a 3′ to 5′-manner (Anderson and Parker, 1998; Rodgers et al., 2002). The residual cap structure is then hydrolyzed by a scavenger decapping enzyme DcpS (Dcs1 in yeast) (Liu et al., 2002; Wang and Kiledjian, 2001). Regulation of decapping is an important process, because hydrolysis of the cap structure is irreversible and renders the mRNA susceptible to exonucleases in the case of 5′ to 3′-decay pathway, leading to the rapid decay of the remaining mRNA body (Coller and Parker, 2004; Cougot et al., 2004; Simon et al., 2006).

The substrate for Dcp2 is capped mRNA that is longer than 25 nucleotides (LaGrandeur and Parker, 1998; Steiger et al., 2003). Recognition and hydrolysis of the cap structure require an initial interaction with the RNA moiety on the substrate (Piccirillo et al., 2003). Hydrolysis of the 5′-cap by Dcp2 yields the products m^{7}GDP and a monophosphate-terminated RNA (van Dijk et al., 2002; Wang et al., 2002).

In *Saccharomyces cerevisiae*, Dcp2p and Dcp1p form a decapping complex, in which Dcp2p is the catalytic subunit, and Dcp1p serves as an enhancer of Dcp2p decapping activity (Coller and Parker, 2004; She et al., 2006; Steiger et al., 2003). In higher eukaryotes, there is a third adaptor component, Edc4 (Hedls or Ge-1), which stimulates Dcp2 decapping activity (Fenger-Gron et al., 2005; Yu et al., 2005). In addition to Dcp1 and Edc4, the Lsm1-7 protein complex, as well as Pat1p, Dhh1p, and the Edc1p, Edc2p, and Edc3p proteins, are also identified as activators of Dcp2 decapping *in vivo* (Coller and Parker, 2004). In addition to positive regulation, Dcp2 is also subject to negative regulation. PABP can inhibit Dcp2 activity *in vitro* (Khanna and Kiledjian, 2004; Wilusz et al., 2001). The cap-binding protein eIF4E inhibits Dcp2 activity both *in vitro* and *in vivo* (Caponigro and Parker, 1995; Khanna and Kiledjian, 2004; Ramirez et al., 2002; Schwartz and Parker, 1999; 2000). More recently, VCX-A, a protein implicated in X-linked mental retardation, has also been shown to be an inhibitor of Dcp2 (Jiao et al., 2006).

The Dcp2 proteins are members of a Nudix (nucleotide diphosphate linked to an X moiety) hydrolase family. Nudix hydrolases contain a conserved Nudix motif that consists of the 23 amino acids GX$_5$EX$_7$REUX-EEXGU, where X stands for any residue and U denotes hydrophobic residues (Koonin, 1993; Mejean et al., 1994). Other Nudix hydrolases have diverse substrates, including nucleoside triphosphates, nucleotide sugars, coenzymes such as NADH, FAD, or CoA, cell signaling molecules such as dinucleoside polyphosphates, and toxic metabolites such as ADP-ribose and toxic nucleotides (Bessman et al., 1996). Residues in the Dcp2 Nudix motif are required for catalysis, because mutations abolish the hydrolase activity (Dunckley and Parker, 1999; van Dijk et al., 2002; Wang et al., 2002). Similar to other Nudix proteins, Dcp2 requires cations, preferentially Mn^{2+}, as a cofactor for its decapping activity (Piccirillo et al., 2003).

In addition to the Nudix motif, there are at least two other conserved motifs in Dcp2 proteins (Wang et al., 2002). The first is Box A, which precedes the Nudix fold, and the second is Box B, which is at the carboxy terminus of the Nudix fold. Biochemical analysis revealed that the Nudix motif and Box B are required for Dcp2 activity in vitro; they have been proposed to be essential for RNA binding (Piccirillo et al., 2003; Shen and Kiledjian, 2006). Box A is not absolutely required for decapping activity in vitro (Piccirillo et al., 2003) but is necessary for the interaction of yeast Dcp1p with yeast Dcp2p (She et al., 2006).

Unlike Dcp2, after 3′ to 5′-degradation by the exosome, DcpS hydrolyzes the residual capped oligonucleotide that contains less than 9 nucleotides to generate m^7GMP (Liu et al., 2002). DcpS possesses a distinct hydrolase motif, termed the histidine triad (HIT) motif, and, thus, is a member of the HIT protein superfamily. Members of the HIT superfamily are nucleotide hydrolases and transferases that possess the conserved HUHUHU HIT motif, where U denotes a hydrophobic amino acid (Brenner, 2002; Brenner et al., 1999; Guranowski, 2000; Seraphin, 1992). The central histidine residue is critical for the hydrolase activity and has been proposed to serve as the nucleophile attacking the phosphate most proximal to the methylated guanosine (Lima et al., 1997).

Recently, the structural analysis of DcpS in complex with cap substrates has provided informative insight into its decapping mechanism. The crystallized structure revealed that DcpS is a dimeric enzyme with a distinct N terminus and C terminus linked by a flexible hinge (Chen et al., 2005; Gu et al., 2004). The DcpS dimer displays an asymmetric architecture having a simultaneous productive closed conformation formed at one protomer and a nonproductive open conformation formed at the second protomer, with each protomer bound by a cap substrate. On the basis of these structural results, a dynamic decapping model where the N terminus of DcpS flips back and forth for binding to and hydrolysis of the cap substrate was proposed (Chen et al., 2005; Gu et al., 2004) and subsequently confirmed (Liu et al., 2008).

In this chapter, we describe methods to monitor the decapping activities of recombinant and endogenous Dcp2 and DcpS, as well as general considerations in the experimental approaches.

2. Measuring Decapping Activities of Recombinant and Endogenous Dcp2

In the following, we describe decapping assays for recombinant human (h)Dcp2 expressed in bacteria or mammalian cells, as well as endogenous hDcp2 in cell extracts. The methods described focus on the biochemical activity of hDcp2. However, they can also be applied to Dcp2 from any organism, and with minor optimization of cation utilization, most likely will be useful to study any cellular or viral mRNA decapping enzyme. It should be noted that all solutions described in the following are generated with double-distilled diethylpyrocarbonate (DEPC)–treated H_2O.

2.1. Preparation of cap-radiolabeled RNA substrates

To prepare the cap-radiolabeled RNA substrate for hDcp2 decapping assays, uncapped RNA substrate is first generated by *in vitro* transcription. When testing a generic RNA for the assays, we generally use the polylinker region of the plasmid pcDNA3. A DNA template containing the polylinker sequence is PCR amplified with primers corresponding to the T7 promoter and SP6 promoter sequences that flank the polylinker region of pcDNA3. The DNA template is subsequently used for *in vitro* transcription to generate the RNA substrate. We recommend the use of SP6 polymerase for the RNA synthesis step, because we have found RNAs transcribed with SP6 polymerase are subsequently capped more efficiently than RNAs transcribed by T7 polymerase from the T7 promoter. If a specific RNA sequence rather than a generic sequence is to be tested for decapping, the DNA or cDNA of interest can similarly be PCR amplified with appropriate primers that insert a bacteriophage promoter sequence at the $5'$-end to enable *in vitro* transcription into RNA. Cap-labeled RNA is generated by use of the vaccinia virus capping enzyme, [α-^{32}P]GTP and S-adenosylmethionine (SAM) to label the $5'$-most phosphate within the cap relative to the methylated guanosine (m7G*pppG-, where the asterisk denotes radioactive phosphate). After the capping reaction, the [^{32}P]-cap–labeled RNA is gel purified to ensure the integrity of the RNA body and used as the substrate in hDcp2 decapping assays.

2.1.1. *In vitro* Transcription

1. Mix the following reagents for a 25 μl reaction: 5 μl 5× transcription buffer (Promega; Madison, WI), 5 μl 2.5 mM NTPs mixture (ATP, TTP, GTP, CTP), 2.5 μl 100 mM dithiothreitol (DTT), 2.5 μl 1 mg/ml bovine serum albumin, 1 μl (40 U) RNasin (RNase inhibitor, Promega), 1 μl (17 U) SP6 RNA polymerase (Promega), 200 ng PCR-amplified DNA template. Adjust the volume to 25 μl with H_2O. Although a T7 promoter–initiated transcript can also be used, as stated previously, we have found that, in general, SP6 RNA polymerase-transcribed RNAs are more efficiently cap-labeled than T7 RNA polymerase transcribed RNAs. The difference might be due to the presence of one guanosine at the $5'$-end of a SP6 RNA

polymerase-transcribed RNA compared with the presence of three guanosines at the 5′-end of a T7 RNA polymerase-transcribed RNA).
2. Incubate the reaction at 37 °C for 1 h.
3. Add 1 μl RQ1 DNase to digest the DNA template in the reaction. Incubate at 37 °C for 15 min.
4. Centrifuge a Microspin® G-50 columns (GE Healthcare, Piscataway, NJ) at 750g for 1 min. Discard the spun-down buffer in the collection tube.
5. Pass the RQ1 DNase-treated reaction through a precentrifuged Microspin® G-50 column by centrifuging the column at 750g for 2 min to separate the unincorporated NTPs from the *in vitro*–transcribed RNA.
6. Add 85 μl H$_2$O to adjust the volume to 100 μl for more convenient phenol/chloroform extraction.
7. Extract the RNA flow-through solution with phenol/chloroform once and chloroform twice.
8. Ethanol precipitate the RNA solution by adding 10 μg glycogen, 1/10 volume of 3 M sodium acetate (pH 5.2) and 2.5 volumes of 100% ethanol. Leave the solution at −80 °C for 10–30 min.
9. Spin the solution at full speed in a tabletop centrifuge for 20 min at 4 °C. Aspirate the supernatant.
10. Add 100 μl of 70% ethanol to the RNA pellet. Vortex briefly and centrifuge at 4 °C for 5 min. Aspirate the supernatant, and air-dry the RNA pellet for a few minutes.
11. Resuspend the RNA pellet in 15 μl H$_2$O.
12. Measure the RNA concentration by use of a spectrophotometer.

2.1.2. Generation of cap-labeled RNA

1. Mix the following reagents for a capping reaction: 1 μg of *in vitro*–transcribed RNA, 1 μl 15× capping buffer (750 mM Tris-HCl, pH 7.9, 18.75 mM MgCl$_2$, 90 mM KCl, 37.5 mM DTT, 1.5 mg/ml BSA), 1 μl 10 mM S-adenosylmethionine (SAM), 3.5 μl (20 U) vaccinia virus capping enzyme (this could be purchased from Ambion [Austin, TX] or purified according to Shuman (1990), 0.5 μl (20 U) RNasin, 2 μl (20 μCi) [α-^{32}P]GTP. Adjust the total volume to 15 μl by adding H$_2$O.
2. Incubate the reaction at 37 °C for 1 h.
3. Pass the reaction solution through a Microspin® G-50 column by centrifuging at 750g for 2 min to remove the unincorporated [α-^{32}P]GTP.
4. Ethanol precipitate as in steps 6 through 10 in section 2.1.1.
5. Resuspend the RNA pellet in 10 μl 80% formamide containing bromophenol blue and xylene cyanol dyes for the subsequent gel purification.

2.1.3. Gel purification of the cap-labeled RNA

1. Mix the following reagents: 21 g urea, 8.3 ml 30% Bis-acrylamide, 2.5 ml 10× TBE buffer. Add H$_2$O to adjust the volume to 50 ml.

Prepare the 10× TBE stock solution by dissolving 108 g of Tris base, 55 g of boric acid, and 40 ml of 0.5 M EDTA, pH 8.0, and adjust with H_2O to 1 L.
2. Incubate the solution in a 37 °C waterbath to dissolve the urea.
3. Add 460 μl 10% ammonium persulfate and 25 μl TEMED (N,N,N′,N′-tetramethyl ethylenediamine) to the gel solution. Pour the solution into an assembled gel cast. Leave the gel at room temperature for several minutes to solidify.
4. Prerun the gel at 15W for 15 min in 0.5× TBE.
5. Rinse the wells extensively with 0.5× TBE with a syringe to remove any particles or concentrated urea solution.
6. Load the cap-radiolabeled RNA sample after heating at 75 °C for 2 min directly onto the prerun gel, and run the gel at 15W for ∼40 min.
7. Disassemble the gel case, cover the gel with plastic wrap, and expose the gel to Kodak Biomax film for several minutes as necessary and develop the film. Include some form of luminescent tape on at least two edges of the gel to enable superimposition of the gel and the film. Locate the cap-labeled RNA in the gel by superimposing the developed film on the gel.
8. Excise the gel region that contains the cap-radiolabeled RNA with a clean razor blade, and place the gel slice in an Eppendorf tube. Grind the gel piece with a pipetman tip.
9. Add 400 μl elution buffer (20 mM Tris-HCl, pH 7.5, 0.5 M sodium acetate, 10 mM EDTA, 1% SDS) into the Eppendorf tube. Incubate at 37 °C overnight, or at 65 °C for 1 h.
10. Bore a hole into the bottom of a small Eppendorf tube (volume 0.5 ml) with a 23-gauge needle, and place the tube into a 1.5-ml Eppendorf tube for sample collection. Pack a small amount of glass wool at the bottom of the 0.5-ml tube to prevent gel pieces from exiting through the hole, and add the eluted RNA sample and the ground gel into the small tube. Centrifuge at 9500g for 2 min to separate the eluted cap-radiolabeled RNA from the ground gel.
11. Extract the RNA solution with phenol/chloroform once and with chloroform twice.
12. Ethanol precipitate the RNA as in steps 7 through 10 in section 2.1.1.
13. Dissolve the RNA pellet in 50 μl H_2O. Store the cap-labeled RNA sample at −20 °C.

2.2. Measuring Dcp2 decapping activity

In this section, we detail the various strategies to generate recombinant or endogenous hDcp2, the decapping assay parameters, and the pitfalls when detecting decapping activity.

2.2.1. Preparation of bacterial-expressed hDcp2 for decapping assays

Histidine-tagged Dcp2 can be prepared by expression of hDcp2 with the pET expression system (Novagen; San Diego, CA) with the plasmid pET28-hDcp2 (Wang *et al.*, 2002). BL21(DE3) bacterial cells are transformed with pET28-hDcp2, expression of histidine-tagged hDcp2 (His-hDcp2) is induced with isopropyl-β-D-thiogalactoside (IPTG), and the protein is affinity purified with a nickel column according to manufacturer's instructions.

Our purification deviates from the manufacturer's protocol in one important way. We include 300 to 500 mM urea and 0.5% Triton X-100 in the binding buffer and the wash buffer to minimize nonspecific protein interactions with the column. Otherwise, the preparation is contaminated with extensive bacterial exonuclease activity, as well as a pyrophosphatase that generates a similar decapping product as hDcp2 (see section 2.3.3 and Fig. 1.2). After the wash steps suggested by the manufacturer, wash the resin twice with 5 volumes of wash buffer containing no urea or Triton X-100. Elute the His-tagged Dcp2 with elution buffer as suggested by the manufacturer.

2.2.2. Preparation of recombinant flag-tagged hDcp2 expressed in human cells

1. A pcDNA3 plasmid-based expression construct, pcDNA3-Flag-hDcp2, which contains the 5′-Flag tag sequence upstream of the hDcp2 open-reading frame, is used to transfect HEK293T cells for the overexpression of Flag-hDcp2. The transfection is carried out with Lipofectamin 2000 (Invitrogen; Carlsbad, CA) according to the manufacturer's instructions.
2. Harvest cells 24 h after transfection by washing the cells with phosphate-buffered saline (0.14 M NaCl, 2.7 mM KCl, 1.5 mM KH$_2$PO$_4$, 8.1 mM Na$_2$HPO$_4$, pH 7.4). Resuspend cells in sonication buffer (150 mM KCl, 20 mM Tris-HCl pH 7.9, 0.2 mM EDTA), and sonicate for 30 sec twice with a 1-min interval with a Branson Sonifier 450 (Branson; Danbury, CT) with output set at 3. Place the tube on ice during sonication to prevent overheating the sample.
3. Centrifuge the cell debris at 20,000g for 5 min at 4 °C. Incubate the supernatant with 20 µl prewashed Anti-Flag M2 agarose beads (Sigma; Saint Louis, MO) at 4 °C for 3 h with mild rocking to allow Flag-hDcp2 to bind the beads.
4. Wash beads three times with 1 ml wash buffer (300 mM KCl, 20 mM Tris-HCl, pH 7.9, 0.2 mM EDTA).
5. Elute the bound Flag-hDcp2 with 100 µl elution buffer (wash buffer with 100 µg/ml Flag peptide [Sigma]). Add 10 µl glycerol to the eluted protein solution.

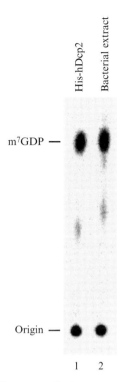

Figure 1.2 BL21 bacterial-cell extracts display an activity that can hydrolyze the cap structure of an RNA to release m^7GDP. BL21 cell extracts were prepared from untransformed cells similarly to the manner in which extracts were prepared from transformed BL21 cells for recombinant protein overexpression. Cells were lysed by sonication, and cell extract was serially diluted with decapping buffer. Decapping assays were carried out with 40 ng of recombinant Dcp2 and 2 μg of cell extract. Both reactions generated products that comigrate with m^7GDP.

6. Store the eluate at −80 °C. The protein samples should be stored in aliquots and should not be freeze-thawed more than three times, because repeated freeze-thawing reduces decapping activity.

2.2.3. Preparation of endogenous hDcp2 (P50 cytoplasmic fraction) for decapping assays

In general, endogenous hDcp2 activity is not readily detectable with cell extracts, in part caused by hDcp2 decapping inhibitory proteins in the extracts (Jiao *et al.*, 2006). A modest, yet detectable, activity can be achieved with centrifugal fractionation of cytoplasmic extract.

1. Resuspend 10^8 human erythroleukemia K562 cells in 1.5 ml of lysis buffer (10 mM Tris-HCl, pH 7.5, 10 mM potassium acetate, 1.5 mM

magnesium acetate, 2 mM DTT, 1× protease inhibitor [Roche; Indianapolis, IN]) and incubate on ice for 15 min.
2. Lyse the cells with 10 strokes of a type B Douce homogenizer on ice.
3. Remove the nuclei and debris with a 2000g centrifugation for 10 min at 4 °C.
4. Transfer the resulting supernatant to a new tube and centrifuge at 50,000g at 4 °C for 20 min.
5. Resuspend the resulting pellet in 200 to 300 µl protein buffer (10 mM Tris [pH 7.5], 100 mM potassium acetate, 2 mM magnesium acetate, 2 mM DTT) to allow the final protein concentration to be 8 to 10 µg/µl and used directly in decapping assays. The P50 fraction cannot be stored because it loses activity on freezing.

2.2.4. *In vitro* Dcp2 decapping assay

1. For the decapping assays with recombinant bacterial-expressed Dcp2 or Flag antibody immunoprecipitated Dcp2, incubate 100 ng or 10 ng of protein respectively with 1 to 10 fmol of [^{32}P] cap-labeled RNA substrate (∼3000 cpm/reaction) in decapping buffer (10 mM Tris, pH 7.5, 100 mM KOAc, 2 mM MgOAc, 2 mM DTT) supplemented with fresh 0.5 mM MnCl$_2$ for 30 min at 37 °C. The concentration of Flag-tagged Dcp2 can be determined by Western blot analysis compared with the signal obtained with known concentration of hDcp2. Adjust the decapping reaction volume to 20 µl with H$_2$O; 10 µl of 2× decapping buffer is added per 20 µl reaction. Terminate the reaction by incubating the reaction tube on ice.
2. For the decapping assays with cell extracts, incubate 50 µg of P50 with [^{32}P] cap-labeled RNA substrate in 1× decapping buffer supplemented with 0.5 mM MnCl$_2$ and 100 µM cold cap structure at 37 °C for 45 min. The purpose of including the cold cap structure, which is the substrate for DcpS, is to sequester the endogenous DcpS activity so as to prevent its hydrolyzing the Dcp2 decapping product m^{7}GDP, because m^{7}GDP can also serve as a substrate for DcpS (van Dijk *et al.*, 2003) After incubation, terminate the reaction with phenol/chloroform. Decapping reactions with 50 µg of cell extract have relatively higher amounts of protein, which interferes with the TLC migration. Therefore, phenol/chloroform extraction is necessary.
3. Spot 5 µl of each reaction onto a polyethylenimine-cellulose (PEI) thin-layer chromatography (TLC) plate (Sigma) that was prerun in H$_2$O to remove impurities and air dried for at least 30 min prior to use.
4. Develop the products on the TLC with 0.75 M LiCl (or any of the other solutions listed in Fig. 1.3 that optimally resolve the products of interest) at room temperature. Alternately, the decapping products can be

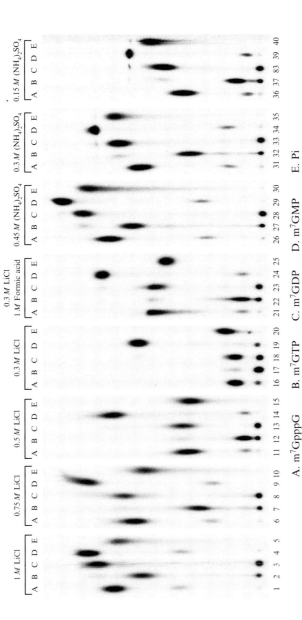

Figure 1.3 Cap analogs migrate differently under various buffer conditions. Four different cap analogs, m^{7}GpppG, m^{7}GTP, m^{7}GDP, and m^{7}GMP, as well as inorganic phosphate (Pi) were resolved with PEI-TLC developed in the indicated running buffers. The relative migration position of each cap analog and Pi differs, depending on the running buffer. For example, m^{7}GDP comigrates with Pi in 0.45 M (NH$_4$)$_2$SO$_4$ running buffer, but the two are more distinctly separated in 0.75 M LiCl running buffer. The identity of each cap analog or Pi spotted on the individual lanes is denoted at the bottom. On the basis of the information provided in this figure, one can select a buffer condition that serves the experimental needs for separation of specific cap analog combinations.

resolved by polyacrylamide gel electrophoresis as described in Chapter 8 by Sokoloski *et al.* in this volume of *Methods in Enzymology*.
5. Air-dry the TLC plates. Expose the TLC plates to Kodak BioMax film or use Phosphorimaging to quantitate the fraction of hydrolyzed capped-RNA. This fraction is calculated by dividing the signal of hydrolyzed substrate by the sum of the hydrolyzed substrate and unhydrolyzed substrate. All quantitations can be carried out with a Molecular Dynamics Phosphorimager (Storm860) with ImageQuant 5 software.

2.3. Additional considerations

2.3.1. Confirmation that the decapping product is m^{7}GDP

To ensure that the hDcp2 decapping product detected with TLC is m^{7}GDP and not a comigrating molecule, the reaction product can be treated with nucleoside diphosphate kinase (NDPK; Sigma). NDPK specifically converts nucleotide diphosphate, including a N7-methylated nucleotide diphosphate, into nucleotide triphosphate (Wang and Kiledjian, 2001). Therefore, m^{7}GDP will be phosphorylated to m^{7}GTP. To 10 μl of the decapping reaction product, add 1 unit nucleotide diphosphate kinase (NDPK) (Sigma) in decapping buffer in the presence of 1 mM ATP. After a 30-min incubation at 37 °C, resolve the sample by TLC.

2.3.2. The decapping buffer

The standard decapping buffer listed previously lacks an energy source (ATP, GTP, and creatine phosphate), which is not required for the decapping reaction but is required for exonucleolytic activity. Therefore, by use of the standard decapping buffer, we can detect decapping with minimal exonucleolytic decay of the RNA and, thereby, uncouple the two reactions (Song and Kiledjian, 2007). If a coupled decapping and decay reaction is desired, 1 mM ATP and 0.4 mM GTP, 10 mM creatine phosphate, and 0.1 mM spermine can be added to the decapping buffer. The resulting buffer on addition of an energy source is the same as our standard *in vitro* decay buffer that is normally used to follow RNA decay *in vitro* (Wang *et al.*, 1999).

2.3.3. Presence of a copurifying bacterial pyrophosphatase activity

A major concern with the use of bacterial-expressed and purified recombinant protein is the copurification of bacterial proteins. In addition to the presence of exoribonucleases in bacterial extract, we also detect pyrophosphatase activity that can hydrolyze cap-labeled RNA to generate a product that comigrates with m^{7}GDP (Fig. 1.2). Both these activities copurify with fusion proteins purified with either a glutathione S-transferase (GST) column for GST-fusion proteins or nickel columns for His-tagged fusion proteins (data not shown). As stated previously, both of these activities can be minimized by use of 300 mM urea and 0.5 % Triton X-100 in the binding buffer, as well as the column wash buffer. An additional approach to minimize copurifying contaminants is to elute the protein from the affinity column with a small volume such that the concentration of protein ranges from 1 μg/μl to 5 μg/μl. On dilution of the protein in the decapping reaction, copurifying contaminants will be minimized. If the original protein concentration is low and larger volumes of the protein are required, this will also increase the presence of contaminant proteins. The same principle holds for bacterial-expressed and purified proteins used to supplement the decapping reaction.

2.3.4. Strategies to improve decapping efficiency

As noted earlier, endogenous hDcp2 activity is not robust and is difficult to detect with unfractionated cell extract. We have uncovered two approaches that can be used to improve the efficiency of decapping. The first involves adding a uridine tract to the 3′-end of the RNA substrate. The normally low level of decapping (<1%) detected with cytoplasmic extract can be increased up to fourfold by the addition of five uridines at the 3′-end of an RNA (Song and Kiledjian, 2007). This was shown with a generic polylinker RNA derived from the pcDNA3 polylinker region transcribed with the SP6 promoter (for details see Song et al., 2007). Stimulation seems to be mediated by the binding of the LSm1-7 complex to the 3′-end and subsequent recruitment of an hDcp2-containing decapping complex (Song and Kiledjian, 2007). However, stimulation is context dependent, and not all RNAs with a uridine tract are competent to mediate increased decapping. Therefore, we suggest the use of the pcP RNA described in Song and Kiledjian (2007) as the substrate for Dcp2 decapping assays.

The second approach that can be used to increase the efficiency of decapping is to use a cap-labeled RNA substrate that is preferentially bound by hDcp2. We recently identified cellular mRNAs substrates of hDcp2 and demonstrated the 5′-most 60 nts of mRNA encoding Rrp41, which is a subunit of the exosome complex, is preferentially bound and selectively decapped by hDcp2 (Li et al., 2008). The increase in decapping was observed with both recombinant and endogenous hDcp2. Therefore, the use of an RNA substrate containing 60 nts of the Rrp41 mRNA 5′ UTR at its 5′-end will also improve the efficiency of detected decapping.

3. Measuring DcpS Activity

DcpS-like activity was first reported with HeLa cell extracts more than 30 years ago (Nuss et al., 1975). A DcpS-like activity was initially partially purified approximately 15 years later (Kumagai et al., 1992) and independently repurified and cloned 10 years after that (Liu et al., 2002). Unlike hDcp2 activity, DcpS activity is robust in cell extracts and can readily be detected in total-cell extract. In the following, we describe the assays for detecting DcpS activity.

3.1. Preparation of radiolabeled cap substrate for DcpS decapping

DcpS hydrolyzes cap dinucleotides or cap oligonucleotides that are smaller than 10 bases, but it is unable to hydrolyze longer capped RNAs (Liu et al., 2002). Therefore, decapping reactions are carried out with [^{32}P]-labeled cap dinucleotide (m7G★pppG) generated from cap-labeled RNA.

1. Generate cap-labeled RNA as described in section 2.1.2. Such an RNA will contain the [^{32}P] at the 5′-most phosphate of the cap(m7G★pppG-). After the capping reaction, cap-labeled RNA is ethanol precipitated and resuspended in 15 μl of H$_2$O.
2. Mix the following reagents for a 30 μl reaction: 1 μl (1 U) nuclease P1 (Sigma), 15 μl cap-labeled RNA, 11 μl H$_2$O, as well as 3 μl 10× nuclease P1 buffer (100 mM Tris-HCl, pH 7.5, 10 mM ZnCl$_2$). Incubate at 37 °C for 2 h. Nuclease P1 can hydrolyze the phosphodiester linkage of the RNA body. Because it does not hydrolyze the pyrophosphatase linkage within the cap, it leaves a cap dinucleotide (m7G★pppG).
3. Add 120 μl H$_2$O to adjust the volume to 150 μl for more convenient phenol/chloroform extraction.
4. Add equal volume of phenol/chloroform (150 μl), vortex vigorously, centrifuge at full speed in a tabletop centrifuge, and transfer the top aqueous phase containing the cap structure to a new tube.
5. Extract the [^{32}P]-labeled cap structure with chloroform once to remove the remaining phenol.
6. Transfer the aqueous layer containing the [^{32}P]-labeled cap structure substrate to a new tube.

3.2. Generation of bacterial-expressed DcpS protein

DcpS protein can be expressed and purified from bacterial BL21(DE3) cells transformed with the pET28-hDcpS (Liu et al., 2002) in a way that is similar to the protocol described in section 2.2.1 for hDcp2.

3.3. Generation of total-cell extract to detect endogenous DcpS activity

The robust nature of DcpS activity in extract eliminates the necessity to fractionate the extract to detect the activity. Therefore, total-cell extract can be prepared by ordinary sonication of the cells.

1. Harvest three plates (10 cm) of cells at approximately 80% confluence. Wash cells with phosphate buffer saline (0.14 M NaCl, 2.7 mM KCl, 1.5 mM KH$_2$PO$_4$, 8.1 mM Na$_2$HPO$_4$, pH 7.4). Resuspend cells in 1 ml sonication buffer (150 mM KCl, 20 mM Tris-HCl, pH 7.9, 0.2 mM EDTA). Carry out sonication of the cells as described in step 2 in 2.2.2.
2. Centrifuge the cell debris at full speed with a tabletop centrifuge for 10 min at 4 °C. Transfer the supernatant to a clean Eppendorf tube.
3. Determine the protein concentration.
4. Add glycerol to a final concentration of 10% and store in aliquots at −80 °C.

3.4. In vitro DcpS decapping assay

1. Set up a 20-μl reaction by incubating recombinant DcpS (1-10 ng) or total-cell extract (100 ng to 1μg) with 1 to 10 fmol [^{32}P]-labeled cap substrate (3000 cpm/reaction) in decapping buffer (10 mM Tris, pH 7.5, 100 mM KOAc, 2 mM MgOAc, 2 mM DTT) for 15 min at 37 °C. Total-cell extract can be diluted from the frozen aliquots above with 1× decapping buffer.
2. Place tubes on ice after the 15-min incubation. Alternately, reaction can be terminated by adding 4 μl 10 N formic acid.

If the reaction is carried out with cell extract, continue to step 3. If the reaction is carried out with recombinant protein, step 3 is not necessary, proceed to step 4.

3. Phenol/chloroform extract the reaction once and transfer the upper aqueous phase to a new tube.
4. Spot an aliquot of 5 μl of each reaction onto PEI-TLC plates (Sigma) that were prerun in H$_2$O and air dried as described in step 3 in 2.2.4.
5. Develop the products with 0.45 M (NH$_4$)$_2$SO$_4$ at room temperature.
6. Air-dry the TLC plates. Expose the TLC plates to Kodak BioMax film or Phosphorimager for quantitation of the fraction of hydrolyzed cap substrate. It is calculated by dividing the signal of hydrolyzed cap substrate (which is the produced decapping product m^{7}GMP) by the sum of the hydrolyzed product and unhydrolyzed substrate. All quantitations can be conducted with a Molecular Dynamics Phosphorimager (Storm860) and ImageQuant 5 software.

4. MIGRATION OF CAP ANALOGS WITH VARIOUS THIN-LAYER CHROMATOGRAPHY RUNNING BUFFERS

The various decapping products and reactants will migrate different distances on the PEI-TLC plate, depending on the running buffer used. Depending on the decapping reaction being carried out and the expected products, a running buffer that will maximally separate the reactants and various products must be used. In particular, it is important to note that both DcpS activity and phosphatase activity are very potent in extract. For example, when decapping assays are carried out with extract, the m^{7}GDP product of hDcp2 can further be hydrolyzed by DcpS to m^{7}GMP (van Dijk et al., 2003) and the m^{7}GMP in turn dephosphorylated to m^{7}G + radioactive Pi (Wang and Kiledjian, 2001). However, these decapping substrates and products might comigrate under certain buffer conditions, which would complicate interpretation of the data analysis. Fig. 1.3 lists the migration of four different cap analogs and Pi developed under eight different buffer conditions: 1 M LiCl; 0.75 M LiCl; 0.5 M LiCl; 0.3 M LiCl; 0.3 M LiCl with 1 M formic acid; 0.45 M (NH$_4$)$_2$SO$_4$; 0.3 M (NH$_4$)$_2$SO$_4$, and 0.15 M (NH$_4$)$_2$SO$_4$. As shown in Fig. 1.3, certain markers separate better under distinct buffer conditions. For example, Pi comigrates with m^{7}GDP in 0.45 M (NH$_4$)$_2$SO$_4$ but is more readily resolved in 0.75 M LiCl (compare lanes 8 and 10 to lanes 28 and 30). Another example is m^{7}GDP and m^{7}GpppG. They nearly comigrate in 0.75 M LiCl, but they more distinctly migrate in 0.45 M (NH$_4$)$_2$SO$_4$. Because each of the cap analogs and Pi listed in Fig. 1.3 can also serve as a marker, we list the conditions to generate each structure in the following.

4.1. Preparation of various cap analog markers

All cap analogs described in the following contain the [^{32}P] as the first phosphate after the methylated guanosine.

4.1.1. Preparation of m^{7}GpppG

1. Follow the procedure outlined in 2.1 for the preparation of m^{7}GpppG.

4.1.2. Preparation of m^{7}GDP

1. Carry out a decapping assay with bacterial-expressed recombinant hDcp2 and cap-labeled RNA substrate at 37 °C for 30 min as in 2.2.4 except use at least 0.5 μg hDcp2 to ensure complete digestion.
2. Phenol/chloroform extract the reaction once. The aqueous layer contains radiolabeled m^{7}GDP.

4.1.3. Preparation of m^{7}GTP

1. Incubate radiolabeled m^{7}GDP with 1 unit nucleoside diphosphate kinase (NDPK) (Sigma) in decapping buffer in the presence of 1 mM ATP.
2. Incubate the reaction at 37 °C for 30 min.
3. Phenol/chloroform extract the reaction once. The aqueous layer contains the radiolabeled m^{7}GTP.

4.1.4. Preparation of m⁷GMP

1. Carry out a DcpS decapping reaction with radiolabeled m7GpppG substrate at 37 °C for 15 min as in 2, except use 0.1 µg of DcpS per reaction for complete digestion.
2. The decapping reaction is phenol/chloroform extracted once. The aqueous layer contains the radiolabeled m⁷GMP.

4.1.5. The Preparation of Pi

1. Incubate 50 fmol of [γ-^{32}P]-ATP with calf intestinal alkaline phosphatase (CIP; New England Biolabs; Beverly, MA) in a buffer containing 100 mM NaCl, 50 mM Tris-HCl, 10 mM MgCl$_2$, 1 mM DTT, pH 7.9, for 1 h at 37 °C.
2. Phenol/chloroform extract the reaction once. The aqueous layer contains radioactive inorganic Pi.

ACKNOWLEDGMENT

Support for this work was provided by NIH grant GM67005 to M. K.

REFERENCES

Anderson, J. S. J., and Parker, R. P. (1998). The 3′ to 5′-degradation of yeast mRNAs is a general mechanism for mRNA turnover that requires the SKI2 DEVH box protein and 3′ to 5′-exonucleases of the exosome complex. *EMBO J.* **17,** 1497–1506.

Bessman, M. J., Frick, D. N., and O'Handley, S. F. (1996). The MutT proteins or "Nudix" hydrolases, a family of versatile, widely distributed, "housecleaning" enzymes. *J. Biol. Chem.* **271,** 25059–25062.

Brenner, C. (2002). Hint, Fhit, and GalT: Function, structure, evolution, and mechanism of three branches of the histidine triad superfamily of nucleotide hydrolases and transferases. *Biochemistry* **41,** 9003–9014.

Brenner, C., Bieganowski, P., Pace, H. C., and Huebner, K. (1999). The histidine triad superfamily of nucleotide-binding proteins. *J. Cell Physiol.* **181,** 179–187.

Caponigro, G., and Parker, R. (1995). Multiple functions for the poly(A)-binding protein in mRNA decapping and deadenylation in yeast. *Genes Dev.* **9,** 2421–2432.

Chen, N., Walsh, M. A., Liu, Y., Parker, R., and Song, H. (2005). Crystal structures of human DcpS in ligand-free and m7GDP-bound forms suggest a dynamic mechanism for scavenger mRNA decapping. *J. Mol. Biol.* **347,** 707–718.

Coller, J., and Parker, R. (2004). Eukaryotic mRNA decapping. *Annu. Rev. Biochem.* **73,** 861–890.

Cougot, N., Babajko, S., and Seraphin, B. (2004). Cytoplasmic foci are sites of mRNA decay in human cells. *J. Cell Biol.* **165,** 31–40.

Decker, C. J., and Parker, R. (1993). A turnover pathway for both stable and unstable mRNAs in yeast: Evidence for a requirement for deadenylation. *Genes Dev.* **7,** 1632–1643.

Decker, C. J., and Parker, R. (1994). Mechanisms of mRNA degradation in eukaryotes. *Trends Biochem. Sci.* **19**, 336–340.

Dunckley, T., and Parker, R. (1999). The DCP2 protein is required for mRNA decapping in *Saccharomyces cerevisiae* and contains a functional MutT motif. *EMBO J.* **18**, 5411–5422.

Fenger-Gron, M., Fillman, C., Norrild, B., and Lykke-Andersen, J. (2005). Multiple processing body factors and the ARE-binding protein TTP activate mRNA decapping. *Mol. Cell* **20**, 905–915.

Garneau, N. L., Wilusz, J., and Wilusz, C. J. (2007). The highways and byways of mRNA decay. *Nat. Rev. Mol. Cell Biol.* **8**, 113–126.

Gu, M., Fabrega, C., Liu, S. W., Liu, H., Kiledjian, M., and Lima, C. D. (2004). Insights into the structure, mechanism, and regulation of scavenger mRNA decapping activity. *Mol. Cell* **14**, 67–80.

Guranowski, A. (2000). Specific and nonspecific enzymes involved in the catabolism of mononucleoside and dinucleoside polyphosphates. *Pharmacol. Ther.* **87**, 117–139.

Jiao, X., Wang, Z., and Kiledjian, M. (2006). Identification of an mRNA-decapping regulator implicated in X-linked mental retardation. *Mol. Cell* **24**, 713–722.

Johnson, A. W. (1997). Rat1p and Xrn1p are functionally interchangeable exoribonucleases that are restricted to and required in the nucleus and cytoplasm, respectively. *Mol. Cell Biol.* **17**, 6122–6130.

Khanna, R., and Kiledjian, M. (2004). Poly(A)-binding-protein mediated regulation of hDcp2 decapping *in vitro*. *EMBO J.* **23**, 1968–1976.

Koonin, E. V. (1993). A highly conserved sequence motif defining the family of MutT-related proteins from eubacteria, eukaryotes and viruses. *Nucleic Acids Res.* **21**, 4847.

Kumagai, H., Kon, R., Hoshino, T., Aramaki, T., Nishikawa, M., Hirose, S., and Igarashi, K. (1992). Purification and properties of a decapping enzyme from rat liver cytosol. *Biochim. Biophys. Acta.* **1119**, 45–51.

LaGrandeur, T. E., and Parker, R. (1998). Isolation and characterization of Dcp1p, the yeast mRNA decapping enzyme. *EMBO J.* **17**, 1487–1496.

Li, Y., Song, M., and Kiledjian, M. (2008). Transcript-specific decapping and regulated stability by the human Dcp2 decapping protein. *Mol. Cell Biol.* **28**, 939–948.

Lima, C. D., Klein, M. G., and Hendrickson, W. A. (1997). Structure-based analysis of catalysis and substrate definition in the HIT protein family. *Science* **278**, 286–290.

Liu, H., and Kiledjian, M. (2006). Decapping the message: A beginning or an end. *Biochem. Soc. Trans.* **34**, 35–38.

Liu, H., Rodgers, N. D., Jiao, X., and Kiledjian, M. (2002). The scavenger mRNA decapping enzyme DcpS is a member of the HIT family of pyrophosphatases. *EMBO J.* **21**, 4699–4708.

Liu, S. W., Rajagopal, V., Patel, S. S., and Kiledjian, M. (2008). Mechanistic and kinetic analysis of the DcpS scavenger decapping enzyme. *J. Biol. Chem.* **283**, 16427-16436.

Mejean, V., Salles, C., Bullions, L. C., Bessman, M. J., and Claverys, J. P. (1994). Characterization of the mutX gene of *Streptococcus pneumoniae* as a homologue of *Escherichia coli* mutT, and tentative definition of a catalytic domain of the dGTP pyrophosphohydrolases. *Mol. Microbiol.* **11**, 323–330.

Muhlrad, D., Decker, C. J., and Parker, R. (1995). Turnover mechanisms of the stable yeast PGK1 mRNA. *Mol. Cell Biol.* **15**, 2145–2156.

Nuss, D. L., Furuichi, Y., Koch, G., and Shatkin, A. J. (1975). Detection in HeLa cell extracts of a 7-methyl guanosine specific enzyme activity that cleaves m7GpppNm. *Cell.* **6**, 21–27.

Piccirillo, C., Khanna, R., and Kiledjian, M. (2003). Functional characterization of the mammalian mRNA decapping enzyme hDcp2. *RNA* **9**, 1138–1147.

Ramirez, C. V., Vilela, C., Berthelot, K., and McCarthy, J. E. (2002). Modulation of eukaryotic mRNA stability via the cap-binding translation complex eIF4F. *J. Mol. Biol.* **318**, 951–962.

Rodgers, N. D., Wang, Z., and Kiledjian, M. (2002). Regulated alpha-globin mRNA decay is a cytoplasmic event proceeding through 3′ to 5′-exosome-dependent decapping. *RNA* **8,** 1526–1537.

Schwartz, D. C., and Parker, R. (1999). Mutations in translation initiation factors lead to increased rates of deadenylation and decapping of mRNAs in *Saccharomyces cerevisiae*. *Mol. Cell. Biol.* **19,** 5247–5256.

Schwartz, D. C., and Parker, R. (2000). mRNA decapping in yeast requires dissociation of the cap binding protein, eukaryotic translation initiation factor 4E. *Mol. Cell. Biol.* **20,** 7933–7942.

Seraphin, B. (1992). The HIT protein family: A new family of proteins present in prokaryotes, yeast and mammals. *DNA Seq.* **3,** 177–179.

She, M., Decker, C. J., Chen, N., Tumati, S., Parker, R., and Song, H. (2006). Crystal structure and functional analysis of Dcp2p from *Schizosaccharomyces pombe*. *Nat. Struct. Mol. Biol.* **13,** 63–70.

Shen, V., and Kiledjian, M. (2006). Decapper comes into focus. *Structure.* **14,** 171–172.

Shuman, S. (1990). Catalytic activity of vaccinia mRNA capping enzyme subunits coexpressed in *Escherichia coli*. *J. Biol. Chem.* **265,** 11960–11966.

Simon, E., Camier, S., and Seraphin, B. (2006). New insights into the control of mRNA decapping. *Trends Biochem. Sci.* **31,** 241–243.

Song, M., and Kiledjian, M. (2007). 3′-Terminal oligo U-tract-mediated stimulation of decapping. *RNA* **13,** 2356–2365.

Steiger, M., Carr-Schmid, A., Schwartz, D. C., Kiledjian, M., and Parker, R. (2003). Analysis of recombinant yeast decapping enzyme. *RNA* **9,** 231–238.

van Dijk, E., Cougot, N., Meyer, S., Babajko, S., Wahle, E., and Seraphin, B. (2002). Human Dcp2: A catalytically active mRNA decapping enzyme located in specific cytoplasmic structures. *EMBO J.* **21,** 6915–6924.

van Dijk, E., Le Hir, H., and Séraphin, B. (2003). DcpS can act in the 5′ to 3′-mRNA decay pathway in addition to the 3′ to 5′-pathway. *Proc. Natl. Acad. Sci. USA* **100,** 12081–12086.

Wang, Z., Day, N., Trifillis, P., and Kiledjian, M. (1999). An mRNA stability complex functions with poly(A)-binding protein to stabilize mRNA *in vitro*. *Mol. Cell. Biol.* **19,** 4552–4560.

Wang, Z., Jiao, X., Carr-Schmid, A., and Kiledjian, M. (2002). The hDcp2 protein is a mammalian mRNA decapping enzyme. *Proc. Natl. Acad. Sci. USA* **99,** 12663–12668.

Wang, Z., and Kiledjian, M. (2001). Functional link between the mammalian exosome and mRNA decapping. *Cell* **107,** 751–762.

Wilusz, C. J., Gao, M., Jones, C. L., Wilusz, J., and Peltz, S. W. (2001). Poly(A)-binding proteins regulate both mRNA deadenylation and decapping in yeast cytoplasmic extracts. *RNA* **7,** 1416–1424.

Yu, J. H., Yang, W. H., Gulick, T., Bloch, K. D., and Bloch, D. B. (2005). Ge-1 is a central component of the mammalian cytoplasmic mRNA processing body. *RNA* **11,** 1795–1802.

CHAPTER TWO

A Kinetic Assay to Monitor RNA Decapping Under Single-Turnover Conditions

Brittnee N. Jones,* Duc-Uy Quang-Dang,[†] Yuko Oku,[‡] and John D. Gross[†]

Contents

1. Introduction	24
2. Kinetic Equations	25
3. Materials	28
3.1. *In vitro* transcription	28
3.2. Cap-labeled RNA preparation	28
3.3. Substrate characterization	29
3.4. Protein expression and purification	30
4. Kinetic Assay	30
4.1. Overview	30
4.2. Enzyme quantification and dilution	32
4.3. Substrate mix	32
4.4. TLC analysis	32
4.5. Curve fitting	33
4.6. Metal activation	34
5. Summary	36
Acknowledgments	37
References	38

Abstract

The stability of all RNA polymerase II transcripts depends on the 5′-terminal cap structure. Removal of the cap is a prerequisite for 5′ to 3′-decay and is catalyzed by distinct cellular and viral decapping activities. Over the past decade, several decapping enzymes have been characterized through functional and structural studies. An emerging theme is that function is regulated by protein interactions; however, *in vitro* assays to dissect the effects on enzyme activity are unavailable. Here we present a kinetic assay to monitor decapping by the

---

* Program in Chemistry and Chemical Biology, University of California, San Francisco, California, USA
[†] Department of Pharmaceutical Chemistry, University of California, San Francisco, California, USA
[‡] Department of Neuroscience, Johns Hopkins University School of Medicine, Baltimore, Maryland, USA

heterodimeric yeast Dcp1/Dcp2 complex. Kinetic constants related to RNA binding and the rate of the catalytic step can be determined with recombinant enzyme and cap-radiolabeled RNA substrate, allowing substrate specificity and the role of activating factors to be firmly established.

1. INTRODUCTION

All RNA polymerase II transcripts contain an m^7GpppN (N is any nucleotide) cap structure (Shatkin, 1976). Cap recognition drives nearly every stage of mRNA metabolism, whereas hypermethylation of the cap, $m^{2,2,7}GpppN$, promotes nuclear localization of snoRNAs and snRNAs (Cougot et al., 2004; Moore, 2005; Reddy et al., 1992). Removal of the cap by decapping enzymes can trigger degradation by conserved 5′ to 3′-exonucleases (Rat1 or Xrn1) that recognize a 5′-monophosphate substrate (Johnson, 1997; Muhlrad et al., 1994; Stevens and Maupin, 1987).

Several decapping enzymes have been identified that act in different RNA decay pathways. The mRNA decapping enzyme, Dcp2, is part of a large mRNP responsible for deadenylation-dependent 5′ to 3′-decay, AU rich element (ARE)–mediated decay, nonsense-mediated decay (NMD), deadenylation-independent decapping, and decay of some miRNA targets (Coller and Parker, 2004; Fenger-Gron et al., 2005; Maquat, 2004; Parker and Sheth, 2007; Parker and Song, 2004; Rehwinkel et al., 2005; Weischenfeldt et al., 2005). The scavenger decapping enzyme, DcpS, functions with the exosome in 3′ to 5′-decay possibly to remove residual cap structures that may inhibit cap-binding proteins (Liu et al., 2002; van Dijk et al., 2003; Wang and Kiledjian, 2001). The nuclear decapping enzyme X29 functions in the turnover of U8 snoRNA, which is essential for vertebrate ribosome biogenesis (Ghosh et al., 2004; Peculis and Steitz, 1993; Tomasevic and Peculis, 1999). The Poxvirus family of decapping enzymes, D9 and D10, downregulate both host and viral m^7GpppN-capped RNA, thereby aiding with progression through the viral life cycle (Parrish and Moss, 2007; Parrish et al., 2007; Shors et al., 1999).

Structural and functional studies of decapping enzymes indicate that substrate binding and activity are regulated by protein interactions. Crystallographic analysis of S. pombe Dcp2 reveals a dumbbell-shaped structure composed of an α-helical domain, which binds the essential activator Dcp1, followed by a C-terminal catalytic domain containing a classic Nudix fold (Beelman, 1996; She et al., 2006). Dcp1 enhances the activity of Dcp2 in vitro and is required for function in yeast; however, the Nudix domain is necessary and sufficient for substrate binding and pyrophosphate chemistry (Deshmukh et al., 2008; Dunckley and Parker, 1999; Piccirillo et al., 2003; Steiger et al., 2003). X29 also contains a Nudix fold with additional insertions that mediate an extensive dimer interface (Scarsdale et al., 2006). The ribose of m7G is contacted by an aromatic residue from a neighboring molecule in the crystal, suggesting dimerization is coupled with substrate recognition and regulation of catalysis. The crystal structure of DcpS in complex with cap-dinucleotide reveals a homodimer containing a β-strand swapped regulatory domain followed by a catalytic domain related to the HIT protein family (Chen et al., 2005; Gu et al., 2004; Han et al., 2005). A structure-based model of catalysis by DcpS posits that decapping by one protomer could be coupled to product release in the other through binding of cap-dinucleotide (Gu et al., 2004). The HIT and Nudix folds each define separate superfamilies, where members can function as single domain enzymes (Bessman et al., 1996; Lima et al., 1997; Mildvan et al., 2005; Séraphin, 1992). Therefore, an emerging principle for decapping enzymes is that protein interactions provide a mechanism for modulating activity through insertions and domains that extend beyond a defined catalytic core.

The substrate preferences of decapping enzymes are dictated by geometry and sequence-specific contacts. Dcp2 has strong preferences for long mRNA substrates, unlike DcpS, which has no activity on long mRNAs but efficiently processes capped oligonucleotides (Liu et al., 2004; van Dijk et al., 2002). Similarly, the Poxvirus Nudix enzymes D9 and D10 preferentially decap long and short RNAs, respectively (Parrish and Moss, 2007; Parrish et al., 2007). Structural and biochemical studies indicate that Dcp2 and DcpS form specific contacts with m^7G, whereas X29 does not (Deshmukh et al., 2008; Gu et al., 2004; Scarsdale et al., 2006). Instead, X29 forms specific contacts with the RNA body that restrict decapping to the U8 snoRNA in a metal-dependent manner (Ghosh et al., 2004; Peculis et al., 2007; Tomasevic and Peculis, 1999). The length requirements of mRNA decapping enzymes may safeguard translating messages from precocious decapping. The differences in specificity among decapping enzymes are consistent with their biologic role to act on either an exclusive or broad set of substrates.

Regulation of decapping activity almost certainly extends beyond the binary protein interactions surveyed here. For example, X29 forms part of a large U8 snoRNP, DcpS interacts with the multisubunit exosome, and Dcp2 function is controlled by a dense network of protein interactions in different mRNA decay pathways (Eulalio et al., 2007; Ghosh et al., 2004; Krogan et al., 2006; Parker and Sheth, 2007). To better understand how protein interactions regulate decapping activity and substrate specificity, we have developed a simple in vitro kinetic assay to measure substrate binding and the rate of the catalytic step of decapping by the yeast Dcp1/2 complex.

2. Kinetic Equations

Single-turnover kinetics allows the evaluation of rate constants for formation of reaction intermediates (Johnson, 1992). When the concentration of enzyme [E] is in excess of substrate [S], the rate of the catalytic step is measured directly. This approach has been used extensively by researchers studying the mechanism of ribozymes (Herschlag and Cech, 1990a; Herschlag and Cech, 1990b; McConnell et al., 1993).

The distinction between single- and multiple-turnover kinetics is best illustrated by considering the mechanism:

$$E + S \underset{k_{\text{off}}}{\overset{k_{\text{on}}}{\leftrightarrow}} E \bullet S \overset{k_{\text{max}}}{\to} E \bullet P \overset{k_{\text{rel}}}{\to} E + P \qquad \text{(Scheme 1)}$$

where k_{max} is the rate of the catalytic step, k_{rel} is the rate of product release, and k_{on} and k_{off} are the rate for enzyme substrate association and dissociation, respectively.

Under multiple-turnover conditions, k_{cat} is obtained from the maximal velocity of the reaction under conditions of saturating substrate (Fersht, 1999). For scheme (1) above

$$k_{\text{cat}} = \frac{k_{\text{max}} * k_{\text{rel}}}{k_{\text{max}} + k_{\text{rel}}} \qquad (2.1)$$

Therefore, k_{cat} is not merely the rate of the catalytic step (k_{max}), because it also depends on the rate of product release. In this case, the interpretation of k_{cat} relies heavily on knowledge of whether the catalytic step or product release is rate limiting.

In single-turnover kinetics, the observed rates do not depend on product release, and the maximal velocity of the reaction corresponds to the rate of the catalytic step, k_{max}. The rate equations for single-turnover kinetics follow from the condition of excess enzyme over substrate ([E] > [S]). This ensures a nearly constant enzyme concentration throughout the course of the reaction, so that the concentration of enzyme at any time is approximately equal to the total enzyme concentration (E_T). Rate equations for substrate and product follow from a steady-state assumption on [E]:

$$\frac{\partial}{\partial t}[S] = -k_{\text{obs}}[S] \qquad (2.2)$$

where the observed first order rate constant, k_{obs}, describes the rate of conversion of substrate to product. The amount of substrate present in the system is given by the equation

$$[S] = [S]_0 e^{-k_{\text{obs}} t} \qquad (2.3)$$

Similarly, the amount of product produced at that time is also proportional to the observed rate of the reaction and can be found:

$$[P] = [S]_0 \left[1 - e^{-k_{obs}t}\right] \tag{2.4}$$

By monitoring product formation as a function of time, the observed first-order rate constant k_{obs} can be obtained by curve fitting. The kinetic constants K'_m and k_{max} can be obtained by measuring k_{obs} at several enzyme concentrations through the relation:

$$k_{obs} = \frac{k_{max} * E_T}{E_T + K'_m} \tag{2.5}$$

with k_{max} the rate of the catalytic step, E_T the total enzyme concentration and

$$K'_m = \frac{k_{off} + k_{max}}{k_{on}} \tag{2.6}$$

The prime symbol is used to distinguish the K_m measured under single-turnover conditions from that obtained under multiple-turnover conditions because both depend on k_{max}, whereas the latter also depends on k_{rel}. In the limit where $k_{max} \ll k_{off}$, or when the catalytic step is slow compared with the rate at which the enzyme substrate complex dissociates, K'_m is K_d. This point can be tested by comparing K'_m with K_d measured by direct binding of nonhydrolyzable substrate. If nonhydrolyzable substrate is not available, a chase experiment can be performed to determine whether k_{off} is fast relative to the catalytic step (Dixon, 1953). Hereafter the prime symbol for K_m will be omitted for simplicity.

It should be noted that the rate of the catalytic step reports on all substeps that occur after substrate binding and before the formation of product, including the rate of the chemical step. This would include a scenario in which an enzyme undergoes a conformational change from an open, inactive state to a closed, catalytically active form.

Although single-turnover experiments are valuable for directly measuring the rate of the catalytic step, they do not address whether this step is rate limiting in the biological context of multiple turnovers. The rate of product release relative to the catalytic step may be ascertained in one of three ways. First, if k_{cat} measured under multiple-turnover conditions is equal to the k_{max} measured in the single-turnover experiment, then product release is not rate limiting. Second, if K_m measured under multiple turnovers conditions is K_d for nonhydrolyzable substrate, then product release is not rate limiting. Third, the rate of product release may be measured directly and if $k_{rel} > k_{max}$, product release is not rate limiting.

3. Materials

3.1. In vitro transcription

Short RNAs (10 to 40 nt) are transcribed *in vitro* with 1 μM oligonucleotide template DNA, 10 μl 10× T7 buffer, 5 mM each NTP, 2 mM DTT, and 2 μl of recombinant T7 RNA polymerase (20U/ul) in the presence of 0.1 U RNasin in a total reaction volume of 100 μl (Milligan *et al.*, 1987) (see Appendix A for buffer recipes). The reaction is incubated at 37 °C for 3 h or until a considerable amount of magnesium pyrophosphate precipitate is visible. The product is purified using denaturing polyacrylamide gel electrophoresis (PAGE); visualized by UV shadowing; and subsequently excised, diced, and eluted overnight in gel extraction buffer (Appendix A). Eluted RNA undergoes phenol/chloroform extraction and ethanol precipitation, with the resulting pellet dissolved in water and quantified with UV spectrophotometry. The 5′-triphosphate RNA is stored at −20 °C in 5μl aliquots at a concentration of 20 μM.

3.2. Cap-labeled RNA preparation

Capping is carried out with 10 μl α-[^{32}P]-GTP (MP Biomedicals, 6000 Ci/mmol), 4 μl recombinant Vaccinia virus guanylyltransferase (Ambion 5 U/μL) in the presence of 1 mM S-adenosyl methionine (SAM, NEB) at a final volume of 20 μl as described (Haley *et al.*, 2003; also see Liu *et al.*, Chapter 1 in this issue). Quickspin columns (G25, Roche) are used to separate unincorporated [^{32}P] GTP. The capped RNA is separated from minor degradation products with denaturing PAGE. Cap-radiolabeled RNA is radioimaged (Kodak Biomax XAR) and subsequently excised, diced, and eluted overnight in gel extraction buffer (Appendix A). The aqueous product is phenol/chloroform extracted and ethanol precipitated. The precipitate is washed with 70% ethanol, resuspended in water, quantified by scintillation counting, and stored at 4 °C.

The aforementioned procedure can be adapted to prepare capped, radiolabeled RNA lacking the N^7-methyl modification (GpppG RNA) as described previously except 1 mM S-adenosyl homocysteine (SAH, Sigma) is used in the capping reaction instead of SAM. The integrity of the capped RNA is analyzed with denaturing PAGE against standards (Ambion).

Three measures were taken to ensure consistent, highly efficient preparation of cap-radiolabeled RNA. First, because SAM degrades rapidly once thawed, we use a fresh aliquot for each capping reaction. Second, we use commercial 10× capping reaction buffer (Ambion) to ensure RNase-free assay conditions. Third, a fresh aliquot of 5′-triphosphate RNA is used for each capping reaction to prevent multiple freeze–thaw cycles, which compromise the integrity of the 5′-triphosphate.

3.3. Substrate characterization

Before proceeding with kinetic assays, we evaluate the extent of N^7-methylation with nuclease P1. Typically, cap-radiolabeled RNA (50,000 cpm) is added to 10 μl nuclease P1 in buffer (Ambion) in a total volume of 20 μl, and incubated at 37 °C for 1 h as described (Wang and Kiledjian, 2001). Four microliters of the reaction are spotted onto a PEI-F thin-layer chromatography (TLC) plate (Baker), and TLC analysis is performed as described below. Migration of cap structures is visualized by phosphorimaging and compared with cold m^{7}GpppG and GpppG standards (NEB), loaded in separate lanes, and visualized by UV shadowing.

Nuclease P1 treatment of a 29-nt cap–radiolabeled RNA results in two species that comigrate with m^{7}GpppG and GpppG (Fig. 2.1). The fractions m^{7}GpppG RNA and GpppG RNA that compose total radiolabeled RNA (i.e., m^{7}GpppG, GpppG, and Origin), are 60 and 40% respectively. This information is used for later analysis of decapping data as an endpoint for fitting.

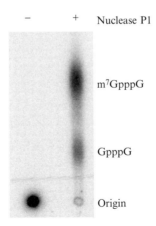

Figure 2.1 A capped RNA preparation before and after nuclease P1 treatment, visualized by TLC analysis. α-[^{32}P] labeled methylated and unmethylated GpppG caps run as indicated. The Origin represents unreacted α-[^{32}P] cap radiolabeled substrate prepared as described in the text.

The rates extracted with single-turnover kinetics do not depend on substrate concentration, unlike multiple-turnover experiments, which have an explicit substrate dependence (Eq. (2.4)). The presence of unmethylated GpppG RNA does not affect the observed kinetic constants given that, as with m^{7}GpppG RNA, unmethylated GpppG RNA is present in trace amounts relative to the enzyme.

3.4. Protein expression and purification

Budding yeast Dcp2 is essentially insoluble on its own, so we have developed a coexpression system for Dcp1/Dcp2 (Deshmukh et al.). The solubility of the Dcp1/Dcp2 complex is enhanced by fusing Dcp1 to the C-terminus of the B1 domain of streptococcal protein G. His-tagged GB1-Dcp1/Dcp2 (1 to 245) was expressed and purified by immobilized metal affinity chromatography with Talon resin (Clontech). Enzyme eluted with varying concentrations of imidazole in 10-ml fractions was pooled, concentrated, and purified by size-exclusion chromatography with a 10/30 GL Superdex 200 column (GE Healthcare) (Appendix A).

The benefit of the size exclusion analysis is threefold. First, it ensures a monodisperse species. Second, a singular peak for the Dcp1/2 complex is indicative of a 1:1 molar ratio of complex components. Size-exclusion chromatography is critical because single turnover kinetics relies heavily on accurate determinations of the enzyme concentration, and an excess of one subunit over the other will skew subsequent concentration measurements. Third, gel filtration into storage buffer ensures the removal of any inhibitors or contaminants in solution, ensuring maximally active enzyme. For example, we have observed that sulfate and phosphate ions, high salt, or polyamines, such as spermine, inhibit decapping. In addition, because we quantify the enzyme by UV spectrophotometry, we avoid detergents and other compounds that absorb at 280 nm. If chemicals that inhibit decapping or absorb at 280 nm are for storage, it is worth considering desalting columns (PD-10 Sephadex G-25M, Amersham) for removing these contaminants before quantification.

4. Kinetic Assay

4.1. Overview

The general workflow of the assay is presented in Fig. 2.2. To extract k_{max} and K_m from Eq. (2.5), we measure k_{obs} for m^{7}GDP formation at a series of enzyme concentrations.

For each enzyme concentration, a typical decapping reaction is initiated by adding 40 μl of enzyme mix to 80 μl of RNA substrate mix at 4 °C in an

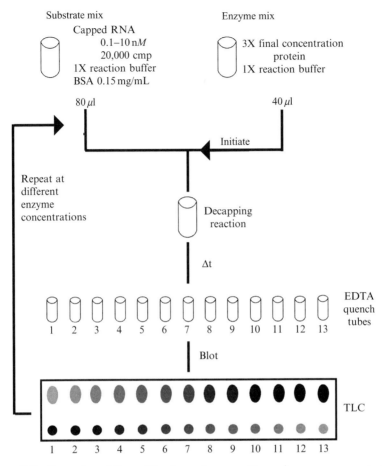

Figure 2.2 General workflow of the decapping assay. For each enzyme concentration, we initiate the decapping reaction by adding enzyme to substrate. Kinetics are followed by removing aliquots from the decapping reaction at different times and rapidly quenching with EDTA.

ultrathin-walled low retention PCR tube (Island Scientific). Kinetics is followed by manual pipetting with 6 μl of the decapping reaction withdrawn at variable times and rapidly quenched by addition to tubes containing 1 μl 0.5 M EDTA on ice. Four microliters of each quenched reaction are blotted onto TLC plates, developed, exposed, and analyzed. Reactions obeyed first-order kinetics and were typically followed for times out to $3/k_{obs}$. This ensures adequate data for fitting Eq. (2.4).

Kinetic constants are measured in the course of a single day with a single enzyme aliquot to ensure consistency and minimize reductions in enzyme activity because of aggregation.

4.2. Enzyme quantification and dilution

It is useful to measure rates (k_{obs}) at a range of enzyme concentrations above and below the K_m so that kinetic constants may be determined accurately through Eq. (2.5). For the yeast Dcp1/Dcp2 complex, with a K_m of 4 μM, we typically measure rates at enzyme concentrations of 1, 2, 3, 4, 5, 7, and 9 μM. Enzyme is prepared at three times the desired final concentration in 1× Dcp reaction buffer in low-retention microcentrifuge tubes (Fisher) (Appendix A). Serial protein dilutions are required to reduce variations between decapping time courses, reducing scatter in plots of Eq. (2.5), and increasing accuracy of k_{max} and K_m.

4.3. Substrate mix

Two general considerations must be made when preparing the substrate mix. To ensure single-turnover conditions, all assays are carried out with enzyme in excess of substrate. In our assay, this condition is strictly adhered to, because our lowest enzyme concentration is 1 μM, whereas cap-radiolabeled RNA varies between 0.1 to 10 nM. In practice, a 10-fold excess of enzyme is adequate. Second, to achieve sufficient signal to noise during TLC analysis, the decapping reaction must contain at least 160 cpm/ μl radiolabeled substrate.

We typically measure k_{obs} at seven different enzyme concentrations to obtain kinetic constants through Eq. (2.5). To ensure consistency in reagents, we prepare the substrate mix as a single batch and then aliquot it into separate reaction tubes for each enzyme concentration. Substrate mix is prepared in low-retention microcentrifuge tubes (Fisher) by adding cap-radiolabeled RNA (150,000 cpm), 8 μl RNasin (Promega, 40 U/μl), 96 μl RNase-free BSA (Promega, 1 mg/ml), 320 μl 2× Dcp reaction buffer and water for a final volume of 640 μl (Appendix A).

The inclusion of BSA and use of low-retention tubes decrease variability in rates from enzyme adherence to the tube surface. This adherence effect is especially pronounced at low concentrations and causes a lower effective enzyme concentration that skews calculated results. For the yeast Dcp1/Dcp2 complex, omission of BSA or use of standard microcentrifuge tubes gives rise to a sigmoidal plot of k_{obs} versus E instead of the hyperbolic behavior predicted from Eq. (2.5). However, a sigmoidal plot can also result from cooperative activation of enzyme through multimerization. This result may be corroborated by an assay to monitor multimerization, such as light-scattering or analytical ultracentrifugation.

4.4. TLC analysis

Liberation of m^7GDP product is visualized by spotting onto PEI-F cellulose TLC plates (J.T. Baker). TLC plates are prerun the entire length in ddH$_2$O. The plates are allowed to dry, and pencil marks are made every 1.5 cm, allowing 1 cm from the sides and bottom of the plate. Aliquots of each quenched reaction are spotted onto the marked TLC plate that is subsequently dried and developed in 0.75 M LiCl until solvent spans three-fourths the height of the plate. Dry plates are exposed to phosphorimaging screens (Amersham) overnight and scanned with a PhosphorImager (Typhoon, GE Healthscience Systems).

The spacing between TLC spots is important to ensure reaction products do not overlap during development. TLC plates may be prerun in advance and stored at 4 °C in a dry area or plastic bag. Migration of m^7GDP and GDP products are verified by UV shadowing with cold standards or treatment of products with NDPK as described (Wang and Kiledjian, 2001; also see Liu *et al.*, Chapter 1 in this volume).

4.5. Curve fitting

Exposed plates are scanned and processed with Imagequant (Molecular Dynamics) to calculate the fraction of m^7GDP released as cmp m^7GDP divided by cpm for total radiolabeled RNA (m^7GDP, GDP, and Origin) (Fig. 2.3).

The data are transferred to SigmaPlot (Systat Software), and the fraction m^7GDP (F) released is plotted versus time (minutes). The time courses are fit by nonlinear regression with the following equation (exponential rise to the max, three parameter, where a = F_{ep} and b = k_{obs}):

$$F(t) = F_{ep}\left[1 - e^{-k_{obs}t}\right] \quad (2.7)$$

with the endpoint (F_{ep}) and first-order rate constant for product formation (k_{obs}) as fit-parameters (Fig. 2.4).

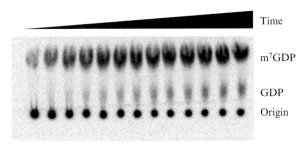

Figure 2.3 Time course for decapping as analyzed using TLC for 2 μM wild-type Dcp1/2 complex. Methylated and unmethylated products migrate as indicated. The fraction of m^7GDP product is calculated as m^7GDP product divided by total radiolabeled RNA (m^7GDP + GDP + Origin).

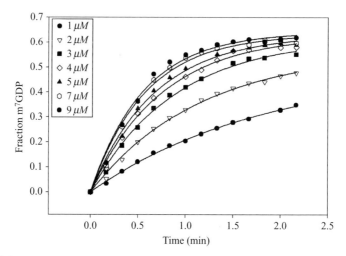

Figure 2.4 Time course for decapping at variable wild-type Dcp1/2 complex concentrations (1, 2, 3, 4, 5, 7, and 9 μM). Time points were taken every 10 sec, and m^{7}GDP released was plotted as a function of time. Data were fit to Eq. (2.7).

For very slow reactions that are linear throughout the time course, the observed rate constant is obtained from the initial rate, with the equation (standard curves, 2 parameter):

$$F(t) = y_0 + a * t \qquad (2.8)$$

where

$$k_{obs} = \frac{a}{F_{ep}}$$

with the endpoint (F_{ep}) obtained by nuclease P1 treatment. We obtain good first-order fits with endpoints between 0.5 and 0.6 for the fraction of m^{7}GDP (Fig. 2.5).

Rates determined are then plotted at different enzyme concentrations and fit to Eq. (2.5) (Hyperbola, three parameter, where $a = k_{max}$ and $b = K_m$) with k_{max} the apparent rate of the catalytic step and K_m corresponding to $\frac{k_{off} + k_{max}}{k_{on}}$ (Fig. 2.6).

4.6. Metal activation

Assays involving the Dcp1/2 complex are performed with 5 mM Mg^{2+}. This is an important variable in the reaction as the concentration, as well as type, of metal varies depending on enzyme. Optimum metal concentration

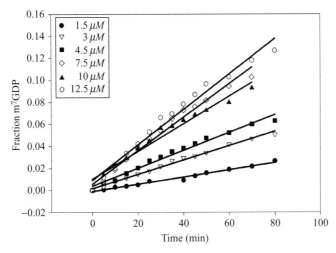

Figure 2.5 Linear time course for decapping at variable Dcp1/K135A Dcp2 complex concentrations (1.5, 3, 4.5, 7.5, 10, and 12.5 μM) obtained by plotting m^{7}GDP released as a function of time. Data were fit to Eq. (2.8).

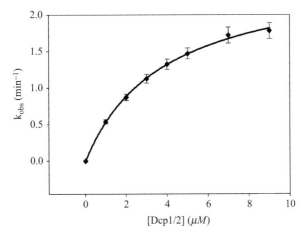

Figure 2.6 Graph of k_{obs}, versus wild-type Dcp1/2 complex concentration. Data were fit to Eq. (2.5) to obtain parameters of k_{max} and K_m. Error bars represent the standard error in k_{obs}.

is determined with the assay described with minor modifications: a single enzyme concentration is chosen, and the amount of metal is varied. Rates (k_{obs}) are plotted as a function of metal concentration, and the optimum metal concentration is determined.

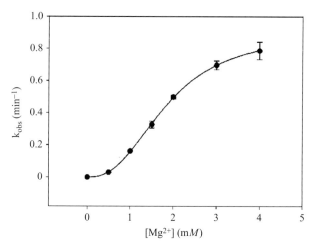

Figure 2.7 Graph of k_{obs} for a single concentration of wild-type Dcp1/2 complex at variable concentrations of metal. Data are fit to Eq. (2.9). Error bars represent the standard error in k_{obs}. For the wild-type Dcp1/Dcp2 complex, the concentration of metal needed for maximal enzyme activity is 5 mM. The Hill coefficient is 2.4, indicating at least two metals bind the enzyme.

In addition, the metal occupancy is determined by fitting the metal activation data to the following equation (Sigmoidal, Hill 3 parameter):

$$F(x) = \frac{ax^b}{(c^b + x^b)} \qquad (2.9)$$

where, a is F_{ep}, c corresponds to the concentration of metal that gives half-maximal activation, and b is the Hill coefficient, or minimum number of metals bound by enzyme (Fig. 2.7).

5. SUMMARY

Removal of the 5′-cap sentences an RNA to destruction and is a highly regulated step in 5′ to 3′-RNA decay. We have developed a simple *in vitro* assay for monitoring substrate binding and the catalytic step of decapping using recombinant yeast decapping enzyme complex (Dcp1/Dcp2) and cap-radiolabeled RNA. We anticipate that this procedure will be useful for dissecting the role of protein interactions at the level of regulating RNA specificity and enzyme activity. A powerful application of this assay will entail examining how decapping enzymes are regulated by co-activators, added as either recombinant purified proteins or in crude cell extracts. The combination of kinetic assays and structural studies will allow the molecular mechanisms of RNA decapping to be clearly defined.

ACKNOWLEDGMENTS

This work was funded by the Sandler Family Foundation in Basic Science. B. N. J. was supported by an NSF predoctoral fellowship. We thank Dr. Geeta Narlikar of UCSF for useful suggestions during the development of this assay and Dr. Roy Parker of the University of Arizona for sharing Dcp1 and Dcp2 expression vectors.

APPENDIX

Buffers

10× T7 Transcription Buffer

380 mM MgCl$_2$
400 mM Tris-HCl (pH 8.0)
25 mM Spermidine
0.1% Triton-X

Gel Extraction Buffer, pH 8.3

200 mM Tris-Cl
25 mM EDTA
300 mM NaCl
2% w/v SDS)

Dcp1/2 Lysis Buffer, pH 7.5

50 mM sodium phosphate (38.7mM Na$_2$HPO$_4$, 11.3 mM NaH$_2$PO$_4$)
300 mM NaCl
0.5% NP-40
5 mM 2-mercaptoethanol (βMe)
5% glycerol
Complete EDTA-free protease inhibitors (Roche)

Dcp1/2 Wash Buffer, pH 7.5

50 mM sodium phosphate (38.7 mM Na$_2$HPO$_4$, 11.3 mM NaH$_2$PO$_4$)
300 mM NaCl
5 mM 2-mercaptoethanol (βMe)
5% glycerol
20 mM Imidazole

Dcp1/2 Elution buffer, pH 7.5

50 mM sodium phosphate (38.7 mM Na$_2$HPO$_4$, 11.3 mM NaH$_2$PO$_4$)
300 mM NaCl
4 mM 2-mercaptoethanol (βMe)
5% glycerol
50, 100, 150, 200, 250 mM Imidazole Fractions

Gel Filtration Buffer, pH 7.5

50 mM HEPES
100 mM NaCl
5% glycerol
5 mM DTT

Storage Buffer, pH 7.5

50 mM HEPES
100 mM NaCl
20% glycerol
5 mM DTT

Substrate Mix

0.1–10 nM Cap-Radiolabeled RNA
8 U RNasin
0.15 mg/ml BSA
1X Dcp Reaction Buffer

Dcp Reaction Buffer (1×)

50 mM Tris-Cl (pH 8 at 25 °C)
50 mM NH$_4$Cl
0.01% NP-40
1 mM DTT
5 mM MgCl$_2$

REFERENCES

Beelman, C. A., Stevens, A., Caponigro, G., LaGrandeur, T. E., Hatfield, L., Fortner, D. M., and Parker, R. (1996). An essential component of the decapping enzyme required for normal rates of mRNA turnover. *Nature* **382,** 642–646.

Bessman, M. J., *et al.* (1996). The MutT proteins or "Nudix" hydrolases, a family of versatile, widely distributed, "housecleaning" enzymes. *J. Biol. Chem.* **271,** 25059–25062.

Chen, N., *et al.* (2005). Crystal structures of human DcpS in ligand-free and m7GDP-bound forms suggest a dynamic mechanism for scavenger mRNA decapping. *J. Mol. Biol.* **347,** 707–718.

Coller, J., and Parker, R. (2004). Eukaryotic mRNA decapping. *Annu. Rev. Biochem.* **73**, 861–890.

Cougot, N., et al. (2004). 'Cap-tabolism'. *Trends Biochem. Sci.* **29**, 436–444.

Deshmukh, M. V., et al. (2008). mRNA decapping is promoted by an RNA binding channel in Dcp2. *Mol. Cell.* In press.

Dixon, M. (1953). The determination of enzyme inhibitor constants. *Biochem. J.* **55**, 170–171.

Dunckley, T., and Parker, R. (1999). The DCP2 protein is required for mRNA decapping in *Saccharomyces cerevisiae* and contains a functional MutT motif. *EMBO J.* **18**, 5411–5522.

Eulalio, A., et al. (2007). P bodies: At the crossroads of post-transcriptional pathways. *Nat. Rev. Mol. Cell Biol.* **8**, 9–22.

Fenger-Gron, M., et al. (2005). Multiple processing body factors and the ARE binding protein TTP activate mRNA decapping. *Mol. Cell.* **20**, 905–915.

Fersht, A. R. (1999). Enzyme structure and mechanism. John Wiley & Sons, Chichester, UK (2nd ed.).

Ghosh, T., et al. (2004). *Xenopus* U8 snoRNA binding protein is a conserved nuclear decapping enzyme. *Mol. Cell.* **13**, 817–828.

Gu, M., et al. (2004). Insights into the structure, mechanism, and regulation of scavenger mRNA decapping activity. *Mol. Cell.* **14**, 67–80.

Han, G. W., et al. (2005). Crystal structure of an Apo mRNA decapping enzyme (DcpS) from Mouse at 1.83 A resolution. *Proteins* **60**, 797–802.

Herschlag, D., and Cech, T. R. (1990a). Catalysis of RNA cleavage by the Tetrahymena thermophila ribozyme. 1. Kinetic description of the reaction of an RNA substrate complementary to the active site. *Biochemistry* **29**, 10159–10171.

Herschlag, D., and Cech, T. R. (1990b). DNA cleavage catalysed by the ribozyme from *Tetrahymena*. *Nature* **344**, 405–409.

Johnson, A. W. (1997). Rat1p and Xrn1p are functionally interchangeable exoribonucleases that are restricted to and required in the nucleus and cytoplasm, respectively. *Mol. Cell Biol.* **17**, 6122–6130.

Johnson, K. A. (1992). Transient-state kinetic analysis of enzyme reaction pathways. *Enzymes* **XX**, 1–61.

Krogan, N. J., et al. (2006). Global landscape of protein complexes in the yeast *Saccharomyces cerevisiae*. *Nature* **440**, 637–643.

Lima, C. D., et al. (1997). Structure-based analysis of catalysis and substrate definition in the HIT protein family. *Science* **278**, 286–290.

Liu, H., et al. (2002). The scavenger mRNA decapping enzyme DcpS is a member of the HIT family of pyrophosphatases. *EMBO J.* **21**, 4699–4708.

Liu, S. W., et al. (2004). Functional analysis of mRNA scavenger decapping enzymes. *RNA* **10**, 1412–1422.

Maquat, L. E. (2004). Nonsense-mediated mRNA decay: Splicing, translation and mRNP dynamics. *Nat. Rev. Mol. Cell. Biol.* **5**, 89–99.

McConnell, T. S., et al. (1993). Guanosine binding to the *Tetrahymena* ribozyme: Thermodynamic coupling with oligonucleotide binding. *Proc. Natl. Acad. Sci. USA* **90**, 8362–8366.

Mildvan, A. S., et al. (2005). Structures and mechanisms of Nudix hydrolases. *Arch. Biochem. Biophys.* **433**, 129–143.

Moore, M. J. (2005). From birth to death: The complex lives of eukaryotic mRNAs. *Science* **309**, 1514–1518.

Muhlrad, D., et al. (1994). Deadenylation of the unstable mRNA encoded by the yeast MFA2 gene leads to decapping followed by 5′ to 3′-digestion of the transcript. *Genes Dev.* **8**, 855–866.

Parker, R., and Sheth, U. (2007). P bodies and the control of mRNA translation and degradation. *Mol. Cell* **25,** 635–646.
Parker, R., and Song, H. (2004). The enzymes and control of eukaryotic mRNA turnover. *Nat. Struct. Mol. Biol.* **11,** 121–127.
Parrish, S., and Moss, B. (2007). Characterization of a second vaccinia virus mRNA-decapping enzyme conserved in poxviruses. *J. Virol.* **81,** 12973–12978.
Parrish, S., et al. (2007). Vaccinia virus D10 protein has mRNA decapping activity, providing a mechanism for control of host and viral gene expression. *Proc. Natl. Acad. Sci. USA* **104,** 2139–2144.
Peculis, B. A., et al. (2007). Metal determines efficiency and substrate specificity of the nuclear NUDIX decapping proteins X29 and H29K (Nudt16). *J. Biol. Chem.* **282,** 24792–24805.
Peculis, B. A., and Steitz, J. A. (1993). Disruption of U8 nucleolar snRNA inhibits 5.8S and 28S rRNA processing in the *Xenopus* oocyte. *Cell* **73,** 1233–1245.
Piccirillo, C., et al. (2003). Functional characterization of the mammalian mRNA decapping enzyme hDcp2. *RNA* **9,** 1138–1147.
Reddy, R., et al. (1992). Methylated cap structures in eukaryotic RNAs: structure, synthesis and functions. *Pharmacol. Ther.* **54,** 249–267.
Rehwinkel, J., et al. (2005). A crucial role for GW182 and the DCP1:DCP2 decapping complex in miRNA-mediated gene silencing. *RNA* **11,** 1640–1647.
Scarsdale, J. N., et al. (2006). Crystal structures of U8 snoRNA decapping nudix hydrolase, X29, and its metal and cap complexes. *Structure* **14,** 331–343.
Séraphin, B. (1992). The HIT protein family: A new family of proteins present in prokaryotes, yeast and mammals. *DNA Seq.* **3,** 177–179.
Shatkin, A. J. (1976). Capping of eucaryotic mRNAs. *Cell* **9,** 645–653.
She, M., et al. (2006). Crystal structure and functional analysis of Dcp2p from *Schizosaccharomyces pombe*. *Nat. Struct. Mol. Biol.* **13,** 63–70.
Shors, T., et al. (1999). Down regulation of gene expression by the vaccinia virus D10 protein. *J. Virol.* **73,** 791–796.
Steiger, M., et al. (2003). Analysis of recombinant yeast decapping enzyme. *RNA* **9,** 231–238.
Stevens, A., and Maupin, M. K. (1987). A 5′ to 3′-exoribonuclease of *Saccharomyces cerevisiae*: Size and novel substrate specificity. *Arch. Biochem. Biophys.* **252,** 339–347.
Tomasevic, N., and Peculis, B. (1999). Identification of a U8 snoRNA-specific binding protein. *J. Biol. Chem.* **274,** 35914–35920.
van Dijk, E., et al. (2002). Human Dcp2: A catalytically active mRNA decapping enzyme located in specific cytoplasmic structures. *EMBO J.* **21,** 6915–6924.
van Dijk, E., et al. (2003). DcpS can act in the 5′ to 3′-mRNA decay pathway in addition to the 3′ to 5′-pathway. *Proc. Natl. Acad. Sci. USA* **100,** 12081–12086.
Wang, Z., and Kiledjian, M. (2001). Functional link between the mammalian exosome and mRNA decapping. *Cell* **107,** 751–762.
Weischenfeldt, J., et al. (2005). Messenger RNA surveillance: Neutralizing natural nonsense. *Curr. Biol.* **15,** R559–R562.

CHAPTER THREE

Purification and Analysis of the Decapping Activator Lsm1p-7p-Pat1p Complex from Yeast

Sundaresan Tharun

Contents

1. Introduction	42
2. Purification of the Lsm1p-7p-Pat1p Complex	43
2.1. Protocol	45
3. Analysis of RNA Binding by the Lsm1p-7p-Pat1p Complex	50
4. Concluding Remarks	53
Acknowledgments	53
References	53

Abstract

Biochemical analysis of the components of the mRNA decay machinery is crucial to understand the mechanisms of mRNA decay. The Lsm1p-7p-Pat1p complex is a key activator of decapping in the $5'$ to $3'$-mRNA decay pathway that is highly conserved in all eukaryotes. The first step in this pathway is poly(A) shortening that is followed by the selective decapping and subsequent $5'$ to $3'$-exonucleolytic degradation of the oligoadenylated mRNAs. Earlier studies suggested that the Lsm1p-7p-Pat1p complex preferentially associates with oligoadenylated mRNAs and facilitates their decapping *in vivo* (Tharun and Parker, 2001a; Tharun et al., 2000). They also showed that the Lsm1p through Lsm7p and Pat1p are involved in protecting the $3'$-ends of mRNAs *in vivo* from trimming (He and Parker, 2001). Therefore, to gain better insight into the biologic function of the Lsm1p-7p-Pat1p complex, it is important to determine its *in vitro* RNA binding properties. Here I describe the methods we use in my laboratory for the purification and *in vitro* RNA binding analysis of this complex from the budding yeast *Saccharomyces cerevisiae*. Purification was achieved with tandem affinity chromatography using a split-tag strategy. This involved use of a

Department of Biochemistry, Uniformed Services University of the Health Sciences (USUHS), Bethesda, Maryland, USA strain expressing FLAG-tagged Lsm1p and 6×His-tagged Lsm5p and purification by a two-step procedure with an anti-FLAG antibody matrix followed by a Ni–NTA matrix. The purified complex was analyzed for its RNA binding properties with gel mobility shift assays. Such analyses showed that this complex has the intrinsic ability to distinguish between oligoadenylated and polyadenylated RNAs and that it binds near the 3′-ends of RNAs (Chowdhury *et al.*, 2007). These observations, therefore, highlighted the importance of the intrinsic RNA binding properties of this complex as key determinants of its *in vivo* functions.

1. Introduction

Turnover of mRNA is an important control point in gene expression, and misregulation of mRNA stability can result in a variety of diseases (Cheadle *et al.*, 2005; Fraser *et al.*, 2005; Hollams *et al.*, 2002; Seko *et al.*, 2006; Steinman, 2007; Tharun and Parker, 2001b). Two major decay pathways conserved in eukaryotes are the 5′ to 3′-decay pathway and the 3′ to 5′-decay pathway. Poly(A) shortening is the first step in both pathways, leading to the generation of oligoadenylated mRNA from polyadenylated mRNA. The oligoadenylated mRNA is then degraded in a 3′ to 5′-exonucleolytic fashion by the exosome (3′ to 5′-pathway), or is decapped and then degraded by 5′ to 3′-exonucleolysis (5′ to 3′-pathway) (Coller and Parker, 2004; Meyer *et al.*, 2004).

Studies determining the characteristics of the components of the cellular mRNA decay machinery are pivotal to the understanding of the mechanism of mRNA decay. Such studies necessitate the use of both genetic and biochemical approaches, because each of these approaches by itself has certain limitations that can be complemented by one another. Biochemical studies on several mRNA decay factors have provided valuable insight in the past on the mechanism of mRNA decay, especially with regard to deadenylation and decapping (Dehlin *et al.*, 2000; Jiao *et al.*, 2006; Khanna and Kiledjian, 2004; Korner and Wahle, 1997; Tucker *et al.*, 2002).

In the 5′ to 3′-pathway, decapping is a crucial rate determining step that permits the degradation of the body of the message. The highly conserved Lsm1p-7p-Pat1p complex is required for normal rates of decapping *in vivo* (Boeck *et al.*, 1998; Bouveret *et al.*, 2000; Tharun *et al.*, 2000). Earlier *in vivo* studies suggesting that this complex selectively associates with oligoadenylated mRNPs and facilitates their decapping called for a thorough *in vitro* biochemical analysis of the purified complex to determine its RNA binding characteristics (Tharun and Parker, 2001a; Tharun *et al.*, 2000). Therefore, we purified this complex from the yeast *Saccharomyces cerevisiae* and carried out RNA binding studies (Chowdhury *et al.*, 2007).

2. PURIFICATION OF THE LSM1P-7P-PAT1P COMPLEX

In principle, the purified eukaryotic protein needed for an *in vitro* biochemical study can be obtained by purification from the original eukaryotic source or from *Escherichia coli* engineered to express that protein. If no critical posttranslational modifications are required for the protein activity, often the preferred method is to express the protein in *E. coli*, because it is easy to obtain large quantities of highly purified protein that way. Simple protein complexes composed of a few subunits can also be made by expressing the individual polypeptides in *E. coli* followed by purification of the complex from *E. coli* as exemplified by the studies on the yeast-decapping enzyme, which is a heterodimer of Dcp1p and Dcp2p (Steiger *et al.*, 2003). However, when the number of different polypeptides that make up the complex is large, its reconstitution with proteins expressed and purified from *E. coli* becomes more complicated. This is an important issue with regard to the Lsm1p-7p-Pat1p complex, because it is made of eight different polypeptides, namely, Lsm1p through Lsm7p and Pat1p. The Lsm1p through Lsm7p proteins are homologous to the Sm-proteins and belong to the family of Sm-like proteins (Anantharaman and Aravind, 2004; Hermann *et al.*, 1995; Salgado-Garrido *et al.*, 1999; Séraphin, 1995). On the basis of this homology, the crystal structure of the Sm-complex (Kambach *et al.*, 1999) and additional protein interaction data (Fromont-Racine *et al.*, 2000; Lehner and Sanderson, 2004), the relative arrangement of the Lsm protein subunits in the Lsm1p-7p complex is thought to be analogous to that of the Sm-complex (Bouveret *et al.*, 2000). However, given that the Sm-like proteins are known to exist naturally as complexes of multiple combinations, including homomeric complexes (Mura *et al.*, 2001; Schumacher *et al.*, 2002; Toro *et al.*, 2001; Wilusz and Wilusz, 2005), it is likely that the Lsm proteins have the potential to form complexes of nonnative quaternary structures under some experimentally contrived conditions. Indeed, it has been shown that the yeast Lsm and Sm proteins are able to form homomeric complexes when overexpressed in *E. coli* (Collins *et al.*, 2003). Therefore, purification from yeast is the best way to obtain the Lsm1p-7p-Pat1p complex in the native form.

Epitope tagging followed by purification with affinity chromatography provides a commonly used efficient protein purification strategy. Highly specific interactions between the epitope and the immobilized antibody (or other types of ligand) allow the purification of the protein under relatively mild conditions. The tandem affinity purification (TAP) strategy, which involves tagging with two different epitopes followed by sequential use of the two affinity chromatographic procedures that target each of the two epitope tags, is especially useful to achieve good purity (Puig *et al.*, 2001).

When purifying a multi-subunit complex with this strategy, one can either use a bipartite tag attached to one of the subunit proteins or tag two different subunits differently (split-tag strategy) (Caspary et al., 1999; Puig et al., 2001). To purify the Lsm1p-7p-Pat1p complex, we used the split-tag strategy for two reasons. (1) When only one subunit is tagged with a bipartite tag, one cannot distinguish between purification of the tagged subunit in uncomplexed form or in the entire complex. (2) Sm-like proteins are also known to exist in nature as hexameric complexes (Schumacher et al., 2002). Therefore, formation of "partial" versions of the Lsm1p-7p-Pat1p complex as a result of possible loss of one or more Lsm subunits during cell lysis and further purification steps could occur and, hence, cannot be completely ruled out. Given that, the split-tag strategy should greatly minimize the purification of such partial complexes if they were to form so that the final material is enriched for the intact complex.

There are eight Lsm proteins that are conserved in all eukaryotes (Lsm1p through Lsm8p), and they are known to associate in at least two combinations, namely, Lsm1p through Lsm7p and Lsm2p through Lsm8p, forming complexes of distinct functions (Bouveret et al., 2000; Mayes et al., 1999; Tharun et al., 2000). Thus, Lsm2p through Lsm7p are six subunits common to both the Lsm1p-7p and Lsm2p-8p complexes, whereas Lsm1p and Lsm8p, respectively, form the distinguishing subunits of these two complexes. So, while using the split-tag strategy for purification of the Lsm1p-7p-Pat1p complex, Lsm1p should obviously be one of the tagged subunits. Therefore, we used a yeast strain expressing epitope-tagged versions of Lsm1p and Lsm5p to purify the complex (Chowdhury et al., 2007).

To ensure that the tag has minimal effect on the function or folding of the tagged protein, we chose to use small tags, namely, the FLAG tag (sequence: DYKDDDDK) and the His-tag (sequence: HHHHHH) on the Lsm1p and Lsm5p proteins, respectively. This allowed us to use a purification procedure with anti-FLAG antibody matrix and Ni–NTA matrix in the first and second affinity purification steps, respectively (Chowdhury et al., 2007). Lsm1p was tagged at the N-terminus because we and others had shown that C-terminal tagging of this protein affects its function (Bouveret et al., 2000; Tharun et al., 2005), whereas we had shown earlier that the N-terminal portion of this protein was dispensable for its functions (Tharun et al., 2005). Lsm5p was tagged at the C-terminus, because our studies had shown that C-terminal tagging did not affect the function of this protein (Chowdhury et al., 2007).

Given the potential ability of the Lsm proteins to form nonnative complexes (see earlier), it is important to ensure that the strain used for the purification of the Lsm1p-7p-Pat1p complex expresses all the subunits of that complex at wild-type levels. This was ensured in the following ways. (1) The epitope-tagged genes *FLAG-LSM1* and *LSM5-6xHis* were made to be the only sources of the Lsm1p and Lsm5p proteins in this strain by deleting the chromosomal *LSM1* gene and expressing the *FLAG-LSM1* gene from a *CEN* vector and by tagging the chromosomal *LSM5* gene with 6× His tag. (2) Epitope tagging of the *LSM1* and *LSM5* genes was done without altering any of their flanking sequences, including the promoter and the UTRs (Chowdhury *et al.*, 2007).

The Sm-like protein complexes, in general, are RNA-binding complexes, and we had shown earlier that the Lsm1p-7p-Pat1p complex also associates with mRNAs *in vivo* (Tharun and Parker, 2001a; Tharun *et al.*, 2000). Therefore, while purifying this complex for RNA binding studies, it is important to ensure that the endogenously bound RNAs do not copurify with it. We had observed that coimmunoprecipitation of mRNA with this complex is abolished at salt concentrations greater than 100 mM. Therefore, our Lsm1p-7p-Pat1p complex purification procedure was set up such that it is carried out at a salt concentration of approximately 300 mM.

It is known that the 5' to 3'-exonuclease Xrn1p interacts with the Lsm1p-7p-Pat1p complex *in vivo* (Bouveret *et al.*, 2000). However, its copurification with this complex is not desirable for our studies, because it could interfere with the subsequent RNA binding analyses of the purified complex by causing degradation of the substrate RNAs used for such analyses. With the purification procedure that we use, Xrn1p does not detectably copurify with this complex, thus eliminating this problem.

2.1. Protocol

2.1.1. Stock solutions

1. 0.5 M NaH$_2$PO$_4$; pH 7.5.
2. 4 M NaCl.
3. 10% (v/v) Nonidet P-40 (NP-40).
4. 2.5 M Imidazole; pH 7.0.

Note: The imidazole stock is adjusted for pH with concentrated HCl and sterilized by filtration. Both 10% NP-40 and 2.5 M imidazole are stored at 4 °C.

2.1.2. Buffers (prepared with the stock solutions listed above)

1. F-Eq Buffer (50 mM NaH$_2$PO$_4$, pH 7.5; 100 mM NaCl; 0.05% NP40).
2. F-W Buffer (50 mM NaH$_2$PO$_4$, pH 7.5; 300 mM NaCl; 0.05% NP40).
3. H-W Buffer (50 mM NaH$_2$PO$_4$, pH 8.0; 300 mM NaCl; 0.1% NP40; 20 mM imidazole).
4. H-Elu Buffer (50 mM NaH$_2$PO$_4$, pH 8.0; 300 mM NaCl; 0.1% NP40; 250 mM imidazole).
5. FLAG peptide (5 μg/μl) solution. Dissolve 4 mg of FLAG peptide (Sigma) in the manufacturer's vial with 800 μl of F-W buffer. Store at −20 °C.

6. Lysis buffer. The lysis buffer is made by adding the appropriate volume of a 20× stock of "COMPLETE" protease inhibitor mix (Roche) to the F-Eq buffer to get a final 1× concentration. The 20× stock of "COMPLETE" was made and stored following the manufacturer's directions. The lysis buffer is made just before use.

Note: While making the H-W and H-Elu buffers, the pH should be adjusted to 8.0 with NaOH after combining the stock solutions before making up the final volume. These buffers are then sterilized by filtration. The F-Eq, F-W, H-W, and H-Elu buffers are all stored at 4 °C and used cold for the purification.

The protocol given in the following is for purification from a 200-ml culture of yeast. The protocol can be scaled accordingly for larger cultures.

2.1.3. Preparation of cell lysate

1. Grow a 200-ml culture of the *FLAG-LSM1, LSM5-6× His* yeast strain (Chowdhury *et al.*, 2007) until the culture reaches 1.0 A_{600}/ml.
2. Harvest the cells with a tabletop centrifuge. Wash the cells in sterile water and spin them down as two equal halves in two, 15-ml Falcon tubes so that each cell pellet represents ~100 ml of culture.

Note: If the cell pellets are not going to be used for purification immediately, they can be stored at −80 °C until further use.

3. Suspend each cell pellet in 1 ml of lysis buffer and add approximately 1½ Eppendorf (1.5 ml) tubefuls of acid-washed glass beads (Sigma) to it.
4. Close the Falcon tubes containing the cell suspension and vortex them vigorously for 1 min and then transfer them to ice bath for 1 min. Repeat this vortexing and chilling cycle 5 to 7 times.
5. Pierce the bottom of each of the two Falcon tubes with an 18-guage needle and collect the crude lysate into a larger tube. Pool the crude lysate obtained from both Falcon tubes.
6. Distribute the crude lysate into two 2-ml Eppendorf tubes and spin them at 10,000*g* for 20 min at 4 °C in a microfuge. Save the supernate. This is the lysate. Typically, approximately 2.3 ml of lysate is obtained from a 200-ml culture of yeast cells. Freeze ~10 μl of the lysate in −80 °C for later analysis.

2.1.4. Preclearing of lysate

Often cell lysates contain proteins that nonspecifically bind to various matrixes used for purification. In the preclearing step, such proteins are removed by exposing the lysate to an unrelated matrix. In our protocol we use the IgG Sepharose (Amersham) for that purpose.

7. Equilibrate the IgG Sepharose matrix with F-Eq buffer as follows. Transfer 50 μl of the IgG Sepharose matrix suspension into each of two 1.5-ml Eppendorf tubes. Spin in the microfuge at 10,000g for 30 sec and discard supernate. Wash the pelleted matrix in each tube three times with F-Eq buffer (with 1 ml each time) spinning down and discarding supernate as previously.
8. Distribute the lysate equally to the two tubes containing the IgG matrix. Mix to suspend the matrix and set the tubes mixing on the nutator (BD diagnostics) at 4 °C for 30 min.
9. Spin down the matrix at 10,000g for 30 sec and save the supernate. This is the precleared lysate.

2.1.5. Anti-FLAG M2 agarose matrix binding and elution

10. Transfer 50 μl of the anti-FLAG M2 agarose matrix (Sigma) suspension into each of two, 1.5-ml Eppendorf tubes. Equilibrate the two matrix aliquots with F-Eq buffer as described in step 7 for the IgG Sepharose matrix.
11. Distribute the precleared lysate equally to the two tubes containing the anti-FLAG M2 agarose matrix. Mix to suspend the matrix and set the tubes mixing on the nutator at 4 °C for 3 h.
12. Spin down the matrix in each tube at 10,000g for 30 sec. Remove supernate (the unbound fraction) and store at −80 °C for later analysis.
13. Wash the pelleted matrix in each tube three times with F-W buffer with 1 ml of buffer each time. Spin and discard supernate as previously.
14. To elute the proteins bound to the anti-FLAG M2 agarose matrix, add 200 μl of F-W buffer and 6 μl of FLAG peptide (5 μg/μl) solution to the washed matrix in each of the two tubes. Mix to suspend the matrix and set the tubes mixing on the nutator at 4 °C for 30 min.
15. Spin down the matrix and collect the supernate. This is the eluate. Pool the eluate fractions from the two tubes (total volume of ∼400 μl). Freeze ∼10 μl of the eluate for later analysis and use the remaining eluate to bind to the Ni–NTA matrix (see later).

2.1.6. Ni–NTA matrix binding and elution

16. Transfer 50 μl of the Ni–NTA agarose matrix (Invitrogen) suspension into a single 1.5-ml Eppendorf tube. Equilibrate the matrix as described in step 7 (for the IgG Sepharose matrix) but with H-W buffer and spin at 1000g.
17. Add the eluate obtained in step 15 to the equilibrated Ni–NTA matrix. Mix to suspend the matrix and set the tubes mixing on the nutator at 4 °C for 2 h.

18. Spin down beads at 1000g for 30 sec. Remove the supernate (unbound fraction) and store at −80 °C for later analysis. Wash the pelleted matrix three times with H-W buffer.
19. To elute the proteins bound to the Ni–NTA matrix, suspend the washed matrix thoroughly in 50 μl of H-Elu buffer. Spin down the matrix at 1000g for 30 sec and save the supernate. This is the final eluate fraction I. Repeat elution two more times and collect fractions II and III.

Approximately 7 μl of each of the final eluate fractions is usually sufficient to visualize the bands by SDS-polyacrylamide gel electrophoresis (SDS-PAGE) with a minigel followed by silver staining. Typically, most of the purified complex appears in the first two eluate fractions, with small amounts appearing in the third fraction. The fractions can then be pooled appropriately and used for RNA binding analyses after dialysis into 50 mM Tris, pH 7.5, 50 mM NaCl, 20% glycerol. The dialyzed samples are stored at −20 °C. Typically, approximately 100 picomoles of the complex are obtained from a 200-ml culture. When stored properly, we find the purified complex to be quite stable at least during the first few months after purification as assessed from its RNA binding activity. However, the activity may decrease with extended time. Nevertheless, even after several months considerable activity is retained. Figure 3.1 shows the band pattern of the purified complex after SDS-PAGE and silver staining.

Notes: (1) In case the purification is not successful or the yield is poor, the saved lysate, unbound and bound fractions of anti-FLAG M2 agarose matrix, and the unbound fraction of the Ni-NTA matrix can be analyzed by SDS-PAGE or Western analysis (with anti-FLAG antibodies or anti-His tag antibodies) to find out at which step the complex was lost during the procedure. Using this information one can determine whether binding with any of the two matrices is inefficient. (2) Because the anti-FLAG M2 agarose matrix uses immobilized anti-FLAG antibodies, reducing agents like β-mercaptoethanol and dithiothreitol should be avoided in the buffers used with this matrix, since they may inactivate the antibodies by reduction of their disulfide bonds. (3) The presence of strong reducing agents like dithiothreitol and chelating agents like EDTA at high concentrations in buffers used with the Ni–NTA matrix will impair the binding efficiency of the Ni–NTA matrix by causing reduction and chelation, respectively, of the Nickel ions. (4) pH of the buffers used with Ni-NTA matrix is very important, because binding of the His-tag by the Ni-NTA matrix is optimal at slightly basic pH (7.2 to 7.8) and poorer at acidic conditions. (5) The H-W buffer used with the Ni-NTA matrix in our procedure contains low concentration (20 mM) of imidazole to minimize nonspecific binding of proteins to this matrix. Such nonspecific binding is also minimized if the amount of Ni–NTA matrix used is such that the binding capacity of the

Purification of the Yeast Lsm1p-7p-Pat1p Complex

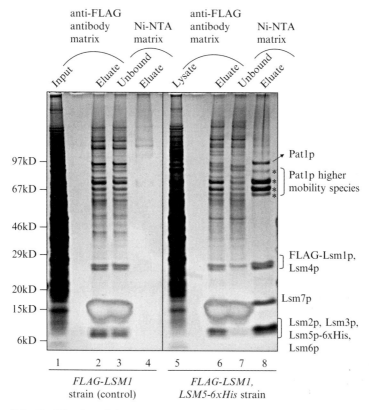

Figure 3.1 Purification of the Lsm1p-7p-Pat1p complex. Purification with anti-FLAG antibody matrix followed by Ni-NTA matrix as described in the text was performed with the lysate prepared from the *FLAG-LSM1, LSM5- 6× His* strain (lanes 5 through 8) or a control strain lacking the 6×His tag (lanes 1 through 4). Proteins present in each sample at different stages of purification (indicated on top) were visualized with SDS-PAGE analysis followed by silver staining. Mass spectrometry analysis of the gel slices containing the different bands observed with the final purified material (shown in lane 8) revealed the identities of the proteins present in those bands (indicated on the right). Pat1p migrates like a 97-kDa protein, although its predicted size is 88 kDa, and it is present in additional bands of higher mobility that seem to represent full-length forms of the protein having anomalous mobilities because of some modifications, truncated forms of the protein, or both. Positions of the size markers are indicated to the left. (Reproduced with permission from Chowdhury *et al.* [2007]).

matrix closely matches the amount of the His-tagged protein in the input material. Therefore, use of excess of Ni–NTA matrix should be avoided. (6) As mentioned earlier, the purification procedure presented here can be scaled up for purification of the complex from larger cultures like a 1-L culture. In such cases, binding to the matrixes is conveniently carried out in 15-ml Falcon tubes instead of Eppendorf tubes.

3. Analysis of RNA Binding by the Lsm1p-7p-Pat1p Complex

The method we use for determining the RNA binding activity of the Lsm1p-7p-Pat1p complex is the gel mobility shift assay which is a sensitive and widely used method for studying nucleic acid–protein interactions (Black *et al.*, 1998). Here the labeled nucleic acid is incubated with the nucleic acid binding protein, and the binding reaction is allowed to reach the equilibrium. The protein-bound nucleic acid molecules in the reaction mixture are then resolved from the unbound nucleic acid molecules on the basis of their lower electrophoretic mobility by native gel electrophoresis carried out under mild conditions that minimize the disruption of the nucleic acid–protein complexes. Quantitation of the signal in the bound and free nucleic acid bands after the electrophoresis reveals the fraction of nucleic acid that is bound at equilibrium.

As substrate for our gel shift assays, we routinely use small *in vitro* transcribed RNA molecules containing sequences derived from the 3' UTRs of different yeast mRNAs. We chose to use such RNAs, because earlier studies on the Lsm1p-7p-Pat1p complex had suggested that it associates with several mRNAs *in vivo* and that its binding site is near the 3'-end of the mRNA (He and Parker, 2001; Tharun and Parker, 2001a; Tharun *et al.*, 2000). Furthermore, it is known that Sm-like protein complexes in general prefer binding to U-rich sequences (Achsel *et al.*, 1999, 2001; Raker *et al.*, 1999), and such sequences are present near the 3'-end of a large fraction of yeast mRNAs (Graber *et al.*, 1999a,b). We typically use 40 to 45 nucleotides-long RNAs for the assays, because we found that size to be optimal for binding to the Lsm1p-7p-Pat1p complex. With shorter RNAs binding gets weaker (Chowdhury *et al.*, 2007), whereas with longer RNAs we did not find significant further enhancement of binding. The RNAs are made by *in vitro* transcription with T7 RNA polymerase. After *in vitro* transcription, the reaction mix is cleaned up with the NucAway spin column (Ambion). This is followed by gel purification of the RNA that is done with the materials and protocol of the "mirVana" kit (Ambion). The gel purification involves visualizing the products of transcription by denaturing polyacrylamide–urea gel electrophoresis and autoradiography followed by excision of the gel slice containing the RNA band of interest and extraction of the RNA from the gel slice. We use the radiolabeled "decade markers" (Ambion) as size markers in the denaturing gel. When plasmid constructs are available that express the RNA of interest under T7 promoter, the linearized plasmid is used as the template and *in vitro* transcription protocol provided by the supplier of T7 polymerase (Promega) was used. When such constructs are not available, oligonucleotide templates are used. For *in vitro* transcription with oligonucleotide templates we use the "mirVana" kit supplied by Ambion.

The RNA binding reaction of the radiolabeled RNA substrate with the purified Lsm1p-7p-Pat1p complex is done in the presence of ribonuclease inhibitor and excess of *E. coli* tRNA. We carry out 10-μl reactions containing 50 mM Tris (pH 7.5), 50 mM NaCl, 0.5 mM MgCl$_2$, 10% glycerol, 0.1% NP-40, 0.1 μg/μl *E. coli* tRNA, and 4 u/μl ribonuclease inhibitor (Promega). All of these components and the purified Lsm1p-7p-Pat1p complex are combined in 9-μl volume at room temperature. The reaction is started by the addition of 1 μl of radiolabeled RNA substrate, which is heat denatured just before use. After incubation at 30 °C for 45 min, the samples are quickly loaded onto a 6% native polyacrylamide gel that is run with 0.5× Tris-borate-EDTA (TBE) buffer in the cold room with a voltage gradient of 13 volts/cm. The gel and running buffer are made ahead of time and cooled to 4 °C before loading the samples. In a separate well, approximately 10 μl of 80% glycerol containing bromophenol blue is loaded to track the progress of the electrophoresis. Once the bromophenol blue band has crossed approximately 6 cm from the well, electrophoresis is stopped, and the gel is transferred to a Whatman paper. It is then covered with plastic wrap, dried in the gel dryer, and exposed to the phosphor screen.

Figure 3.2 shows the binding of *MFA2*(u) RNA that carries the 3′-most 42 nucleotides of the yeast *MFA2* mRNA (followed by three U-residues derived from the plasmid template) (Chowdhury *et al.*, 2007), with increasing concentrations of the Lsm1p-7p-Pat1p complex. These experiments revealed an apparent K_D of ~200 nM for the interaction of this RNA with the Lsm1p-7p-Pat1p complex (Chowdhury *et al.*, 2007). Given that the purified preparation of the complex used here is only approximately 50% active, this binding affinity is an underestimate (Chowdhury *et al.*, 2007). Consistent with this, with another preparation of the complex we observed higher affinity for this RNA (Chowdhury and Tharun, 2008). To allow the estimation of apparent K_D for the interaction, we generally use the radiolabeled substrate RNA in our gel shift assays at a low final concentration of approximately 0.05 nM. This RNA concentration is several-fold below the K_D for the interaction of the Lsm1p-7p-Pat1p complex with even its most preferred substrates we have studied.

The RNA binding studies carried out with a variety of different RNA substrates revealed that the Lsm1p-7p-Pat1p complex has a much higher affinity for RNAs with a 3′-oligo(A) tail (optimal length of approximately 5 A's) than unadenylated or polyadenylated RNAs. Moreover, in the case of unadenylated RNAs, the presence of a stretch of 6 or more U's near the 3′-end facilitates binding. Consistent with this, additional studies showed that this complex binds near the 3′-end of the RNA (Chowdhury *et al.*, 2007).

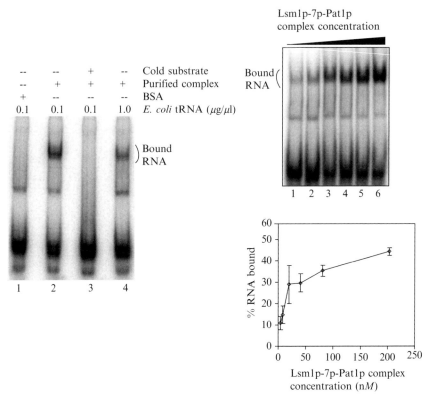

Figure 3.2 RNA binding analysis of the purified Lsm1p-7p-Pat1p complex with gel the mobility shift assay. The *MFA2*(u) RNA was synthesized *in vitro* in the presence of radiolabeled UTP and used at 0.05 nM in all the binding reactions shown. Left panel, Observed gel mobility shift of the *MFA2*(u) RNA is not due to nonspecific interactions. Binding reactions were carried out with the purified complex (at a final concentration of 40.6 nM; lanes 2, 3, and 4) or BSA (lane 1) in the presence of *E. coli* tRNA at a final concentration of 0.1 μg/μl (lanes 1, 2, and 3) or 1.0 μg/μl (lane 4) and in the presence (lane 3) or absence (lanes 1, 2, and 4) of cold *MFA2*(u) RNA (used at 8-fold molar excess over the hot *MFA2*(u) RNA). Right panel, Analysis of *MFA2*(u) RNA binding at increasing concentrations of the purified Lsm1p-7p-Pat1p complex. Purified Lsm1p-7p-Pat1p complex was present at final concentrations ranging from 4.06 to 203 nM (upper panel, lanes 1 through 6). A plot of the percentage of RNA bound (quantitated from the gel with phosphorimaging) vs the concentration of the Lsm1p-7p-Pat1p complex used is shown in the lower panel. Binding reactions and visualization of mobility shift of the RNA were carried out as described in the text. (Reproduced with permission from Chowdhury *et al.* [2007]).

These features that define the unique RNA binding properties of this complex provide useful assays to ensure the quality of the purified complex and enable the design of appropriate controls for the RNA binding studies with this complex.

4. Concluding Remarks

Purification of the yeast Lsm1p-7p-Pat1p complex and analysis of its RNA binding properties with the procedures described previously revealed that it has the intrinsic ability to distinguish between oligoadenylated and polyadenylated RNAs such that the former are bound with much higher affinity than the latter (and unadenylated RNA) (Chowdhury et al., 2007). This defines an important RNA binding characteristic of this complex and is consistent with earlier studies, suggesting that it preferentially associates with oligoadenylated mRNAs *in vivo* (Tharun and Parker, 2001a). These data support the notion that the selective association of this complex with oligoadenylated mRNAs *in vivo* is by virtue of its intrinsic ability to recognize such RNAs and suggest that the targeting of mRNAs by this complex is one of the key events that is critical for the deadenylation dependence of decapping in the 5′ to 3′-decay pathway. The *in vitro* studies with the purified complex also showed that this complex binds near the 3′-ends of RNAs. This binding property is consistent with the fact that the Lsm1p-7p-Pat1p complex protects the 3′-ends of mRNAs from trimming *in vivo*. The binding of this complex at the 3′-end of mRNAs suggests that such a protection is the result of steric inhibition of the trimming nuclease. Thus, the *in vitro* studies were useful to gain insight into the *in vivo* functions of this complex.

ACKNOWLEDGMENTS

This work was supported by NIH grant (GM072718) and USUHS intramural grant (C071HJ).

REFERENCES

Achsel, T., Brahms, H., Kastner, B., Bachi, A., Wilm, M., and Luhrmann, R. (1999). A doughnut-shaped heteromer of human Sm-like proteins binds to the 3′-end of U6 snRNA, thereby facilitating U4/U6 duplex formation *in vitro*. *EMBO J.* **18,** 5789–5802.

Achsel, T., Stark, H., and Luhrmann, R. (2001). The Sm domain is an ancient RNA-binding motif with oligo(U) specificity. *Proc. Natl. Acad. Sci. USA* **98,** 3685–3689.

Anantharaman, V., and Aravind, L. (2004). Novel conserved domains in proteins with predicted roles in eukaryotic cell-cycle regulation, decapping and RNA stability. *BMC Genomics* **5,** 45.

Black, D. L., Chan, R., Min, H., Wang, J., and Bell, L. (1998). "RNA: Protein interactions—A practical approach." Oxford University Press, New York.

Boeck, R., Lapeyre, B., Brown, C. E., and Sachs, A. B. (1998). Capped mRNA degradation intermediates accumulate in the yeast spb8-2 mutant. *Mol. Cell. Biol.* **18,** 5062–5072.

Bouveret, E., Rigaut, G., Shevchenko, A., Wilm, M., and Séraphin, B. (2000). A Sm-like protein complex that participates in mRNA degradation. *EMBO J.* **19,** 1661–1671.

Caspary, F., Shevchenko, A., Wilm, M., and Séraphin, B. (1999). Partial purification of the yeast U2 snRNP reveals a novel yeast pre-mRNA splicing factor required for prespliceosome assembly. *EMBO J.* **18,** 3463–3474.

Cheadle, C., Fan, J., Cho-Chung, Y. S., Werner, T., Ray, J., Do, L., Gorospe, M., and Becker, K. G. (2005). Stability regulation of mRNA and the control of gene expression. *Ann. N. Y. Acad. Sci.* **1058,** 196–204.

Chowdhury, A., Mukhopadhyay, J., and Tharun, S. (2007). The decapping activator Lsm1p-7p-Pat1p complex has the intrinsic ability to distinguish between oligoadenylated and polyadenylated RNAs. *RNA* **13,** 998–1016.

Chowdhury, A., and Tharun, S. (2008). *lsm1* mutations impairing the ability of the Lsm1p-7p-Pat1p complex to preferentially bind to oligoadenylated RNA affect mRNA decay *in vivo*. *RNA* **14,** 2149–2158.

Coller, J., and Parker, R. (2004). Eukaryotic mRNA decapping. *Annu. Rev. Biochem.* **73,** 861–890.

Collins, B. M., Cubeddu, L., Naidoo, N., Harrop, S. J., Kornfeld, G. D., Dawes, I. W., Curmi, P. M., and Mabbutt, B. C. (2003). Homomeric ring assemblies of eukaryotic Sm proteins have affinity for both RNA and DNA. Crystal structure of an oligomeric complex of yeast SmF. *J. Biol. Chem.* **278,** 17291–17298.

Dehlin, E., Wormington, M., Korner, C. G., and Wahle, E. (2000). Cap-dependent deadenylation of mRNA. *EMBO J.* **19,** 1079–1086.

Fraser, M. M., Watson, P. M., Fraig, M. M., Kelley, J. R., Nelson, P. S., Boylan, A. M., Cole, D. J., and Watson, D. K. (2005). CaSm-mediated cellular transformation is associated with altered gene expression and messenger RNA stability. *Cancer Res.* **65,** 6228–6236.

Fromont-Racine, M., Mayes, A. E., Brunet-Simon, A., Rain, J. C., Colley, A., Dix, I., Decourty, L., Joly, N., Ricard, F., Beggs, J. D., and Legrain, P. (2000). Genome-wide protein interaction screens reveal functional networks involving Sm-like proteins. *Yeast* **17,** 95–110.

Graber, J. H., Cantor, C. R., Mohr, S. C., and Smith, T. F. (1999a). Genomic detection of new yeast pre-mRNA 3′-end-processing signals. *Nucl. Acids Res.* **27,** 888–894.

Graber, J. H., Cantor, C. R., Mohr, S. C., and Smith, T. F. (1999b). In silico detection of control signals: mRNA 3′-end-processing sequences in diverse species. *Proc. Natl. Acad. Sci. USA* **96,** 14055–14060.

He, W., and Parker, R. (2001). The yeast cytoplasmic LsmI/Pat1p complex protects mRNA 3′-termini from partial degradation. *Genetics* **158,** 1445–1455.

Hermann, H., Fabrizio, P., Raker, V. A., Foulaki, K., Hornig, H., Brahms, H., and Luhrmann, R. (1995). snRNP Sm proteins share two evolutionarily conserved sequence motifs which are involved in Sm protein–protein interactions. *EMBO J.* **14,** 2076–2088.

Hollams, E. M., Giles, K. M., Thomson, A. M., and Leedman, P. J. (2002). MRNA stability and the control of gene expression: Implications for human disease. *Neurochem. Res.* **27,** 957–980.

Jiao, X., Wang, Z., and Kiledjian, M. (2006). Identification of an mRNA-decapping regulator implicated in X-linked mental retardation. *Mol. Cell* **24,** 713–722.

Kambach, C., Walke, S., Young, R., Avis, J. M., de la Fortelle, E., Raker, V. A., Luhrmann, R., Li, J., and Nagai, K. (1999). Crystal structures of two Sm protein complexes and their implications for the assembly of the spliceosomal snRNPs. *Cell* **96,** 375–387.

Khanna, R., and Kiledjian, M. (2004). Poly(A)-binding-protein–mediated regulation of hDcp2 decapping *in vitro*. *EMBO J.* **23,** 1968–1976.

Korner, C. G., and Wahle, E. (1997). Poly(A) tail shortening by a mammalian poly(A)-specific 3′-exoribonuclease. *J. Biol. Chem.* **272,** 10448–10456.

Lehner, B., and Sanderson, C. M. (2004). A protein interaction framework for human mRNA degradation. *Genome Res.* **14,** 1315-1323.

Mayes, A. E., Verdone, L., Legrain, P., and Beggs, J. D. (1999). Characterization of Sm-like proteins in yeast and their association with U6 snRNA. *EMBO J.* **18,** 4321–4331.

Meyer, S., Temme, C., and Wahle, E. (2004). Messenger RNA turnover in eukaryotes: pathways and enzymes. *Crit. Rev. Biochem. Mol. Biol.* **39,** 197–216.

Mura, C., Cascio, D., Sawaya, M. R., and Eisenberg, D. S. (2001). The crystal structure of a heptameric archaeal Sm protein: Implications for the eukaryotic snRNP core. *Proc. Natl. Acad. Sci. USA* **98,** 5532–5537.

Puig, O., Caspary, F., Rigaut, G., Rutz, B., Bouveret, E., Bragado-Nilsson, E., Wilm, M., and Séraphin, B. (2001). The tandem affinity purification (TAP) method: A general procedure of protein complex purification. *Methods* **24,** 218–229.

Raker, V. A., Hartmuth, K., Kastner, B., and Luhrmann, R. (1999). Spliceosomal U snRNP core assembly: Sm proteins assemble onto an Sm site RNA nona nucleotide in a specific and thermodynamically stable manner. *Mol. Cell. Biol.* **19,** 6554–6565.

Salgado-Garrido, J., Bragado-Nilsson, E., Kandels-Lewis, S., and Séraphin, B. (1999). Sm and Sm-like proteins assemble in two related complexes of deep evolutionary origin. *EMBO J.* **18,** 3451–3462.

Schumacher, M. A., Pearson, R. F., Moller, T., Valentin-Hansen, P., and Brennan, R. G. (2002). Structures of the pleiotropic translational regulator Hfq and an Hfq-RNA complex: A bacterial Sm-like protein. *EMBO J.* **21,** 3546–3556.

Seko, Y., Cole, S., Kasprzak, W., Shapiro, B. A., and Ragheb, J. A. (2006). The role of cytokine mRNA stability in the pathogenesis of autoimmune disease. *Autoimmun. Rev.* **5,** 299–305.

Séraphin, B. (1995). Sm and Sm-like proteins belong to a large family: Identification of proteins of the U6 as well as the U1, U2, U4 and U5 snRNPs. *EMBO J.* **14,** 2089–2098.

Steiger, M., Carr-Schmid, A., Schwartz, D. C., Kiledjian, M., and Parker, R. (2003). Analysis of recombinant yeast decapping enzyme. *RNA* **9,** 231–238.

Steinman, R. A. (2007). mRNA stability control: A clandestine force in normal and malignant hematopoiesis. *Leukemia* **21,** 1158–1171.

Tharun, S., He, W., Mayes, A. E., Lennertz, P., Beggs, J. D., and Parker, R. (2000). Yeast Sm-like proteins function in mRNA decapping and decay. *Nature* **404,** 515–518.

Tharun, S., Muhlrad, D., Chowdhury, A., and Parker, R. (2005). Mutations in the *Saccharomyces cerevisiae* LSM1 gene that affect mRNA decapping and 3′-end protection. *Genetics* **170,** 33–46.

Tharun, S., and Parker, R. (2001a). Targeting an mRNA for decapping: displacement of translation factors and association of the Lsm1p-7p complex on deadenylated yeast mRNAs. *Mol. Cell* **8,** 1075–1083.

Tharun, S., and Parker, R. (2001b). Turnover of mRNA in eukaryotic cells. *In* "RNA", (D. Soll, S. Nishimura, and P. Moore, eds.), pp. 245–257. Pergamon, Oxford.

Toro, I., Thore, S., Mayer, C., Basquin, J., Séraphin, B., and Suck, D. (2001). RNA binding in an Sm core domain: X-ray structure and functional analysis of an archaeal Sm protein complex. *EMBO J.* **20,** 2293–2303.

Tucker, M., Staples, R. R., Valencia-Sanchez, M. A., Muhlrad, D., and Parker, R. (2002). Ccr4p is the catalytic subunit of a Ccr4p/Pop2p/Notp mRNA deadenylase complex in *Saccharomyces cerevisiae*. *EMBO J.* **21,** 1427–1436.

Wilusz, C. J., and Wilusz, J. (2005). Eukaryotic Lsm proteins: Lessons from bacteria. *Nat. Struct. Mol. Biol.* **12,** 1031–1036.

CHAPTER FOUR

Reconstitution of Recombinant Human LSm Complexes for Biochemical, Biophysical, and Cell Biological Studies

Bozidarka L. Zaric[*] and Christian Kambach[†]

Contents

1. Introduction	58
2. Cloning	61
3. Protein Expression and Purification	65
4. LSm Complex Reconstitution	68
5. LSm Complex Functional Assays	70
6. Outlook	72
References	72

Abstract

Sm and Sm-like (LSm) proteins are an ancient family of proteins present in all branches of life. Having originally arisen as RNA chaperones and stabilizers, the family has diversified greatly and fulfills a number of central tasks in various RNA processing events, ranging from pre-mRNA splicing to histone mRNA processing to mRNA degradation. Defects in Sm/LSm protein-containing ribonucleoprotein assembly and function lead to severe medical disorders like spinal muscular atrophy. Sm and LSm proteins always assemble into and function in the form of ringlike hexameric or heptameric complexes whose composition and architecture determine their intracellular location and RNA and effector protein binding specificity and function Sm/LSm complexes that have been assembled *in vitro* from recombinant components provide a flexible and invaluable tool for detailed cell biological, biochemical, and biophysical studies on these biologically and medically important proteins. We describe here protocols for the construction of bacterial LSm coexpression vectors, expression and purification of LSm proteins and subcomplexes, and the *in vitro* reconstitution of fully functional human LSm1-7 and LSm2-8 heptameric complexes.

[*] Institut Curie, UMR 7147, Equipe: Recombinaison et Instabilité Génétique, Paris, France
[†] Biomolecular Research, Paul Scherrer Institut, Villigen PSI, Switzerland

Methods in Enzymology, Volume 448 © 2008 Elsevier Inc.
ISSN 0076-6879, DOI: 10.1016/S0076-6879(08)02604-9 All rights reserved.

1. Introduction

Sm and Sm-like (LSm) proteins bind to a large variety of RNAs and are thought to have arisen very early in evolution as RNA chaperones and stabilizers (Beggs, 2005). Their central role in RNA maturation and metabolism is underscored by their widespread distribution; LSm genes are found in all three branches of life (Moller et al., 2002; Salgado-Garrido et al., 1999). Where bacteria have only one LSm gene (Hfq), Archaebacteria have, in general, two, and Crenarchaeota three (Khusial et al., 2005). In contrast, eukaryotic genomes contain at least 15 (in the case of yeast) or 18 (higher eukaryotes) LSm paralogs, most of which are essential.

LSm proteins are characterized by a bipartite sequence motif of approximately 80 amino acids (Hermann et al., 1995) that forms a common fold, the Sm fold. The Sm fold consists of a five-stranded, strongly bent antiparallel ß-sheet with an N-terminal α-helical extension (Kambach et al., 1999b). In more recently discovered LSm family members, the Sm motifs are fused to distinct C-terminal domains that include a methyltransferase domain (Albrecht and Lengauer, 2004) or, domains of unknown function (Anantharaman and Aravind, 2004)

Sm/LSm proteins always appear as homomeric (in the case of prokaryotes) or heteromeric (in eukaryotes) ringlike multimers. These ring-shaped complexes, generally containing either six or seven subunits, are the functional LSm protein unit.

LSm gene diversification in eukaryotes has led to the appearance of several heptameric rings with discrete intracellular distributions, RNA binding specificities, and functions. The canonical Sm core domain, composed of the Sm proteins B/B′, D1, D2, D3, E, F, and G, binds to a conserved, single-stranded stretch of uridine-rich residues in the spliceosomal U1, U2, U4, and U5 snRNAs, which forms the Sm site. The resulting core snRNPs are then assembled with specific proteins to form the mature snRNPs active in nuclear pre-mRNA splicing (Beggs, 2005). The LSm2-8 complex binds to the 3′-end of U6 snRNA in the cell nucleus. It has been shown to be required for the nuclear accumulation and retention of U6 snRNA (Spiller et al., 2007). The closely related cytoplasmic LSm1-7 complex binds to the 3′ UTR of mRNAs destined for degradation (Bouveret et al., 2000; Tharun et al., 2000). Remarkably, LSm1-7 differs from LSm2-8 only by the exchange of one single subunit, LSm1 for LSm8. This difference then changes the complex location (cytoplasmic vs nuclear), RNA binding specificity (mRNA vs U6 snRNA), and function (mRNA degradation vs splicing). In the U7 snRNP-specific Sm complex, the canonical Sm proteins D1 and D2 are replaced by LSm10 and 11, respectively (Pillai et al., 2003). The U7 snRNP-specific Sm complex functions in histone mRNA processing. In *Trypanosoma brucei* U2 snRNP, SmD3 and SmB are replaced by two novel LSm proteins, LSm15K and LSm16.5K, leading to a core snRNP particle with altered RNA-binding specificity (Wang et al., 2006). In conclusion, Sm/LSm proteins appear as exchangeable building blocks, forming complexes whose composition and architecture determine their intracellular distribution, RNA interaction specificity, and function.

Assembly of sets of proteins and RNAs into functional RNPs is an imposing task for cells. Most often, assembly is a multistep process with defined intermediates in several subcellular compartments. For spliceosomal snRNP biogenesis, a large assembly machine encompassing several multisubunit effector protein complexes has evolved. The most important of these is the SMN complex, composed of the SMN protein and Gemins 2 to 7 (Meister et al., 2002; Yong et al., 2004). By binding both to U snRNAs (Yong et al., 2004) and Sm proteins (Brahms et al., 2001; Friesen et al., 2001), SMN provides a structural platform for Sm protein–U snRNA interactions. Gemins 6 and 7 have been found to adopt an Sm fold even though they do not possess any significant sequence similarity to Sm proteins. The Gemin 6/7 heterodimer arranges in an Sm-like architecture (Ma et al., 2005). These findings suggest a mechanism for the chaperone-like function of the SMN complex, which ensures both the efficiency and specificity of Sm core domain assembly (Meister et al., 2002). This role extends to the mixed Sm/LSm U7 snRNP core domain (Pillai et al., 2003). However, whether the SMN complex is also involved in LSm1-7 or LSm2-8 assembly remains an open question. On the other hand, the similarity in composition and architecture of the different LSm complexes together with their clearly orthogonal functions argues for the presence of discrete assembly pathways for each complex encompassing discrimination and quality/specificity control steps: Incorrectly assembled RNPs could potentially lead to disastrous consequences. The biochemistry of these pathways is as yet almost completely unknown, but the existence of an assembly machinery can safely be assumed. In view of the similarity in composition and architecture, it is possible that LSm1-7 and LSm2-8 share certain intermediates in partially overlapping assembly pathways. Alternatively, the two tracks could be completely independent of one another, both spatially and in terms of components involved.

A major difference between the canonical Sm core domain and the LSm complexes lies in their respective stabilization by cognate RNA: The canonical Sm proteins bind to their target snRNAs in the cytoplasm, which leads to snRNA cap hypermethylation and subsequent reimport of the immature pre-snRNP into the nucleus (Kambach et al., 1999a). Sm proteins absolutely require their cognate snRNA to assemble into a complex, a process that can be carried out *in vitro* with high efficiency (Raker et al., 1996). In contrast, the LSm2-8 complex is stable in the absence of U6 snRNA (Achsel et al., 1999). However, it is not known whether LSm2 to 8 assembles in the cytoplasm without RNA substrate and migrates to the nucleus as one entity. Alternatively, LSm proteins could be transported to the nucleus individually and there assemble onto the U6 snRNA. The recent discovery that in yeast importin β is required for LSm8 nuclear accumulation (Spiller et al., 2007) strengthens the latter scenario, although, in contrast to a subset of the canonical Sm proteins (Bordonne, 2000), no nuclear localization signals have been identified in LSm proteins. Biogenesis of the LSm1 to 7 complex, which accumulates in cytoplasmic foci, or processing (P)-bodies (Ingelfinger et al., 2002; Sheth and Parker, 2003), is even less well characterized. In particular, it has not been established whether LSm1 to 7 is stable in the absence of its RNA substrate in vivo. However, the broad range of 3′ UTR sequences bound by LSm1-7 makes it unlikely that such substrates play a decisive role in LSm1-7 biogenesis.

In order to gain a thorough functional understanding of the LSm complexes, it is necessary to understand the processes involved in LSm complex biogenesis. Common ways to investigate LSm protein intracellular migration, distribution, and association with other factors are fluorescence labeling by GFP fusion or immunostaining, and subsequent colocalization or fluorescence resonance energy transfer (FRET), or coimmunoprecipitation. Although such studies can reveal LSm steady-state accumulation and, in some instances, pairwise LSm interactions (Ingelfinger et al., 2002) or binding of effector proteins (Tharun and Parker, 2001), they do not address where the LSm complexes assemble, the existence or nature of intermediates and/or assembly factors, or how LSm proteins or (sub)complexes are transported through the cell. A nuclear import receptor for LSm2-8 has not yet been identified. Such studies would profit greatly from the availability of fully defined, homogeneous, preassembled and, in most instances, specifically labeled LSm complexes (i.e., recombinant, in vitro reconstituted particles). Although in principle, at least LSm2 to 8 can be purified by dissociation from native snRNPs (Achsel et al., 1999), the milligram levels of highly pure, homogeneous samples required for biophysical studies and, in particular, X-ray crystallography can only be obtained as recombinant material. LSm complexes containing specific isotope-labeled subunits would function as a tool to map LSm–LSm or LSm–effector protein interfaces by NMR in cases in which crystallization attempts fail.

Furthermore, such particles could be used both for in vivo cell biologic studies and for in vitro biochemistry experiments (e.g., with immobilized complexes and cellular extracts to identify interacting components under specific conditions). LSm protein complexes bind to specific effector proteins and carry out specific RNA processing functions, depending on their composition. Because LSm1 and LSm8 represent the only differing components in LSm1-7 and LSm2-8, these disparate effects are likely due to the differences in LSm1 and LSm8 sequences. The Sm motif is presumably engaged in LSm–LSm (β-strands 4 and 5) and LSm–RNA (loops L3 and L5 [Mura *et al.*, 2003; Urlaub *et al.*, 2001]) contacts. Therefore, domains for protein–protein interactions with additional non-LSm proteins are expected to reside in the C-termini. Recombinant chimeric LSm1 and LSm8 proteins incorporated *in vitro* could help to address the precise functional role of these segments. Similarly, the biochemical significance of the symmetric dimethyl-arginine residues in the C-terminal tails of LSm4 (Brahms *et al.*, 2001) could be studied in more detail.

To obtain such defined LSm complex samples also suitable for biophysical and structural characterization, we set out to adapt the *in vitro* reconstitution scheme established for the canonical Sm core domain (Sumpter *et al.*, 1992; Stefan Walke, 2000) to the human LSm1-7 and LSm2-8 complexes. Starting with heterodimeric and trimeric subcomplexes chosen on the basis of pairwise closest homology to a given Sm protein and the known order of subunits in the Sm heptameric ring (see later), we assembled both complexes from exclusively recombinant material in the absence of RNA. We could show the purified complexes to be functional in terms of RNA-binding specificity and intracellular migration (Zaric *et al.*, 2005). In this chapter, we describe the cloning scheme, the purification of the subunits and subcomplexes, the *in vitro* assembly protocol, the subsequent isolation of essentially pure heteroheptameric species, and biophysical characterization of subcomplexes. Functional analyses of recombinant LSm1 to 7 and LSm2 to 8 have been described in Zaric *et al.* (2005) and will briefly be referenced.

2. Cloning

The following description of the Sm/LSm complex reconstitution protocol has evolved from initial attempts for well over a decade, and some aspects of both the cloning scheme and the *in vitro* assembly are historically determined and may not in all cases reflect the current experimental state of the art. Possible alternatives that could lead to a faster and more efficient expression protocol, improved yields, and a more homogeneous complex preparation will be discussed in the last section of this chapter.

The key to successful expression and purification of the first canonical Sm proteins was the discovery that they exist as defined subcomplexes in the cell (Raker *et al.*, 1996). Coexpressing SmD1D2, SmD3B and SmEFG subcomplexes led to high yields of soluble proteins, whereas individual expression was either unsuccessful or gave only insoluble material that could not be refolded (Kambach *et al.*, 1999b; C. Kambach, unpublished observations). On the basis of a reconstitution protocol developed for native Sm proteins from HeLa cell nuclear extract (Sumpter *et al.*, 1992), *in vitro* assembly of a complete snRNP core domain was then readily achieved by mixing the subcomplexes with U snRNA in a 1:1 molar ratio, incubating, and purifying the resulting preparation by anion exchange chromatography (C. Kambach and S. Walke, unpublished).

Sequence comparisons of the yeast LSm protein family indicate that each canonical Sm protein has a corresponding LSm protein (Fromont-Racine et al., 1997; Salgado-Garrido et al., 1999), with the exception of SmB, which aligns almost equally well with LSm1 and LSm8. On the basis of these homologies, an arrangement for the 7 LSm proteins in the LSm1 to 7 and LSm2 to 8 rings can be formulated that is partially supported by genetic studies in yeast (Pannone et al., 2001) and fluorescence experiments in human cells (Ingelfinger et al., 2002) (Fig. 4.1). Although the full complement of pairwise interactions in the LSm complexes remains to be established, we chose this model as a guideline to construct coexpression vectors encoding the homologs of SmD1D2, LSm23, of SmD3B, LSm48, and of SmEFG, LSm567 (Zaric et al., 2005). An LSm41 expression plasmid could not be constructed because of cellular toxicity, and we had to express LSm4 and LSm1 singly for the reconstitution of LSm1 to 7.

In the course of establishing a reconstitution protocol for the two LSm complexes, we noted that akin to canonical Sm proteins, yield and solubility of individual LSm proteins were greatly increased by coexpression. However, this effect seems to occur irrespective of the combinations, not just those representing presumed nearest neighbors in the final heptameric complex (see Table 1 in Zaric et al. [2005]). In addition to the constructs listed in the referenced table, the following dicistrons were tested for expression: LSm56, LSm47, and LSm42. In these cases, supernatant/pellet ratios were not quantified. It should be noted that physiologic LSm paralogs of the Sm subcomplexes D1D2, D3B, and EFG have never been described, and no one has carried out a systematic Sm coexpression study assessing a potential correlation between neighborhood in native heteromultimers and

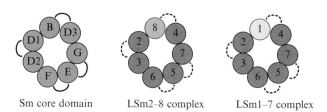

Figure 4.1 Scheme for canonical Sm core domain (left), LSm2 to 8 (center), and LSm1 to 7 (right) complexes. Paralogs are shown in corresponding positions in the three rings. Proven Sm subcomplexes are indicated by black arcs, their potential LSm counterparts by broken arcs. The latter have not been identified *in vivo* but can serve to design working coexpression constructs. For details, see text.

solubility as in the case of TFIID (Fribourg *et al.*, 2001). This LSm promiscuity likely represents a severe obstacle for thermodynamic control over an unassisted *in vitro* reconstitution and is probably the main reason for the overall low efficiency of the process (see later). Concerning the *in vivo* situation, it emphasizes even more strongly the need for an assembly machinery that ensures specificity.

The layout for a combinatorial LSm coexpression vector construction scheme was fraught with two main unknowns: The number and identity of LSm proteins sensibly coexpressed in a single coexpression plasmid could only be guessed from analogy to the canonical Sm protein overexpression strategy developed earlier (see above and Kambach *et al.* [1999b]). However, the *order* of cistrons in a given vector can have a huge impact on subcomplex yield. For example, the expression vector SmD1D2:pQE30 produces roughly 100-fold more heterodimer than SmD2D1:pQE30 (C. Kambach, unpublished observation). Furthermore, the efficiency of protease cleavage differs by several orders of magnitude between homologous constructs (see later). Neither aspect is predictable from sequence alone. Ideally, the cloning scheme should be flexible enough to accommodate any combination of subunits in any order. However, with the restriction-ligation–based cloning method we adopted for our project, this demand could not be accomplished with the resources available. We therefore chose to arrange the LSm cistrons in the same way as in the homologous Sm constructs. The expression tests proved this approach viable, possibly because of the rather indiscriminate enhancement of LSm protein production by coexpressing any other LSm protein.

Although in the case of the canonical Sm proteins, some coexpression constructs bearing a T7 promoter gave better yields than their T5 promoter counterparts (SmD1D2:pET15b vs SmD1D2:pQE30), in particular, SmD3B could only be produced using the pQE30 vector. We therefore chose to produce all our LSm proteins and subcomplexes in pQE30.

Having decided on a cloning strategy, we needed to implement it experimentally. Initial subcloning trials revealed that carrying out a complex cloning scheme like the one intended required consistently high yields of good purity plasmid DNA at all stages. This was more easily obtainable from a promoterless (to avoid potential gene toxicity effects), high copy number vector like pUC. We therefore proceeded to first construct a pUC19 derivative that suited our needs.

The first step was to remove the pUC19 Nde1 site, because Nde1 was to be used for expression cassette insertion. We then introduced an intergenic region (IGR) adopted from the mini-cistron constructs developed by Schoner *et al.* (1990), between the pUC19 multiple cloning site (MCS) sites *Bam*HI and *Hind*III creating pUC19.0. This segment contained *Bam*HI/*Xba*I–IGR–

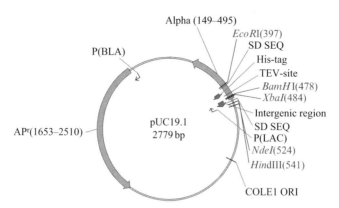

Figure 4.2 Vector map of pUC19.1, the cloning vector used to construct all LSm coexpression cassettes.

NdeI/HindIII. The EcoRI–BamHI segment encoding a Shine–Dalgarno sequence and translation start including the Met-Arg-Gly-Ser-His$_6$-tag was then transferred from pQE30 (Qiagen AG) to this vector. Finally, a tobacco etch virus (TEV) protease cleavage site linker oligonucleotide (sequences <u>GATCTGAAAACCTGTATTTCCAGG</u> and <u>GATCCCTGGAAATA</u>-CAGGTT-TTCA, 5′-BglII and 3′-BamHI sites underlined) was subcloned into the vector's BamHI site, creating pUC19.1 (Fig. 4.2). Correct insertion of the linker was verified by PCR with the upper linker strand primer and the generic M13 reverse primer (M13r, CAGGAAACAGCTATGACC).

Human LSm genes were cloned from a cDNA library or EST clones as described (Zaric et al., 2005). The preparation of the LSm expression cassettes depended on the chosen order of cistrons in the subcomplex expression vectors: Designated first cistrons were amplified as BamHI–cDNA–XbaI fragments, second and third cistrons as NdeI–cDNA–BglII–HindIII fragments. All were then cloned into pUC19. LSm internal restriction sites interfering with the cloning scheme were removed by site-directed mutagenesis, such as BamHI from LSm4 and EcoRI from LSm5. After sequence confirmation, first cistrons were subcloned into pUC19.1 as BamHI–XbaI fragments, followed by second cistrons introduced as NdeI–HindIII fragments. In the case of LSm567, LSm7 was subcloned as a NdeI–HindIII fragment into pUC19.1, then the BamHI–HindIII fragment retrieved from this construct was inserted into BglII–HindIII cut LSm56:pUC19.1. The final polycistronic expression cassettes were transferred in a final step into EcoRI–HindIII cut pQE30. As a representative example, Fig. 4.3 shows the map of LSm567:pQE30.

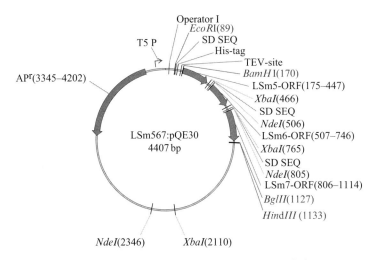

Figure 4.3 Vector map of LSm567:pQE30, with genetic control elements, expression features, and all restriction sites relevant for LSm subcloning. The vector NdeI and XbaI sites are not present in pUC19.1.

3. PROTEIN EXPRESSION AND PURIFICATION

Expression protocols had to be adjusted to the behavior of individual subcomplexes. All expression cultures were grown to an OD_{600} of ~0.8 at 37 °C, then the temperature was adjusted to the appropriate value (25 to 37 °C), cells were induced with 1 mM IPTG, and grown for the prescribed induction time (4 to 48 h). LSm1 and LSm23 were expressed for 20 h at 25 °C in SG13009[pREP4]. LSm567 could only be produced in satisfactory yields and stoichiometric subunit ratios by 48 h of expression at 37 °C in the strain SG13009[pREP4], whereas LSm4 and in particular LSm48 were best expressed in BLR[pREP4] for 4 h at 25 °C. After induction, cells were harvested by centrifugation (20 min at 5000 rpm in a Sorvall RC3B Plus centrifuge), and cell pellets were frozen at −20 °C until further processing.

Nonstoichiometric production of subunits is sometimes observed with polycistronic expression vectors (Kim *et al.*, 2004). In the case of LSm subcomplexes, it seems that the significantly reduced solubility of individual subunits relative to the heteromultimers "automatically" adjusts the stoichiometry of the subcomplex by precipitating out excess protein. We did not conduct a systematic analysis of insoluble material from our expression trials, but according to SDS–PAGE assessment and reversed-phase chromatography, subunits were generally produced in equal amounts. The exception was LSm567, in which LSm6 was produced at a lower level than LSm5 and LSm7. Trimers containing substoichiometric amounts of LSm6 could be separated from species that are more homogeneous by anion exchange chromatography (see later).

Cell pellets were thawed and resuspended in 5 ml of lysis buffer per g cell wet weight. Lysis buffer was 20 mM HEPES-Na, pH 7.5, 0.5 or 1 M NaCl, 5 mM β-mercaptoethanol, 10 mM imidazole-Cl, pH 7.5, and one tablet of Complete^{TM} protease inhibitor–EDTA per 50 ml. LSm4 and LSm48 are low-salt intolerant and had to be kept at $\geq$1 M NaCl at all times. In particular, LSm4 and LSm48 benefited from the presence of 4 M urea in the lysis and subsequent chromatography buffers to minimize nucleic acid contamination and aggregation. Lysis was carried out with a Constant Systems Pressure cell (model Basic Z) at 1.5 kbar, followed by sonication (Sonics Vibracell, 5-min total process time, 60% amplitude, 0.5-sec on/0.5-sec off), and DNase 1 treatment (adding 10 mM MgCl$_2$, 1 mM MnCl$_2$, and 0.01 mg/ml DNase 1 from Boehringer Mannheim, incubation on ice for 20').

The first purification step was immobilized metal ion affinity chromatography (IMAC). The lysate was loaded on a 5 ml Ni^{2+}-charged HiTrap-Chelating Sepharose column (GE Healthcare) at 2 to 5 ml/min. The column was washed with 60 mM and the target protein/subcomplex eluted in 250 mM imidazole (step gradient 10%, then 50% IMAC buffer B). Because of the lower stability of the LSm48 subcomplex, loading and washing speed had to be reduced to 0.5 ml/min for this target. Insufficiently pure samples were dialyzed against low salt buffer (20 mM HEPES-Na, pH 7.5, 150 mM NaCl, 5 mM β-mercaptoethanol, corresponding to AIEX buffer A) and purified using anion-exchange chromatography (Source 15Q, GE Healthcare, linear gradient 150 to 1000 mM NaCl). Because of low salt intolerance, both LSm4 and LSm48 could not be purified this way; and the IMAC eluate had to be used directly for reconstitution. In the case of LSm567, anion-exchange chromatography proved essential to remove subcomplexes containing substoichiometric amounts of LSm6.

All N-terminal His$_6$-tags were constructed to be cleavable by means of a TEV protease cleavage site. Cleavage of purified LSm proteins and subcomplexes was carried out at room temperature. Cleavage efficiencies varied greatly, from almost complete cleavage overnight for LSm567 to less than 90% cleavage after 7 days for LSm48 and complete lack of cleavage for LSm23. One possible reason for these differences is the formation of higher order structures of varying stability that could obstruct the cleavage sites. Uncleaved species, the cleaved-off tag and the (His$_6$-tagged) TEV protease, were subsequently removed by IMAC. We used a starting buffer without imidazole and a linear gradient 0 to 500 mM imidazole for post-cleavage IMAC (other components identical to precleavage IMAC buffers). IMAC eluates were generally used directly for reconstitution. Because of the overall rather low efficiency of the reconstitution process (see later), we generally decided to forego cleavage and subsequent purification to maximize complex yields. Hence, in most reconstitution trials, the final heptamer contained either three (LSm2 to 8) or four (LSm1 to 7) His$_6$-tags, from LSm1, LSm2, LSm4, and LSm5.

Biophysical characterization of the subcomplexes did not give fully internally consistent results: LSm48 aggregated too heavily to yield interpretable data using static light scattering, analytical ultracentrifugation runs, or electron microscopy. The LSm23 dimer (with one His$_6$-tag) has a nominal MW of 25 kDa. By reverse-phase chromatography, we could verify that both subunits were present in equimolar ratio in the purified subcomplex. Calibrated gel filtration yielded a Gaussian peak indicating a MW of approximately 70 kDa. In sedimentation equilibrium centrifugation, LSm23 formed a mixture of higher-order structures, with the octamer as the predominant species (see Table II in Zaric et al. [2005]). In the current discussion, the basis for oligomer nomenclature is the respective subcomplex, not the subunit. Static light scattering yielded a MW of 86 kDa, between a hexamer and an octamer. Similarly, the LSm567 subcomplex (trimer MW 33 kDa) was found to sediment as a mixture of monomeric subunits, trimer, and higher-order structures, whereas static light scattering yielded a MW of 77 kDa, between hexamer and nonamer, with low polydispersity. Negative-stain electron microscopy showed a homogeneous population of ringlike structures similar to those observed for native SmEFG (Plessel et al., 1997), and LSm2-8 (Achsel et al., 1999) for each subcomplex (Zaric et al., 2005).

With the assumption that subcomplex higher-order structures will have to be disrupted for an efficient LSm1-7 and LSm2-8 reconstitution, we analyzed subcomplex stability by gel filtration chromatography in urea, varying its concentration from 4 to 8 M. Both LSm23 and LSm567 elute as a mixture of hexamers and higher oligomers (LSm23: hexadecamer, LSm567: dodecamer) whose composition shifts toward the hexameric species as the urea concentration increases. In the presence of 4 M urea, LSm23 falls apart into monomers to a certain extent, which was more pronounced at higher urea concentrations. Both LSm23 and LSm567 elute at higher volumes in urea than under fully native buffer conditions, indicating that they possibly adopt a less globular, more open, conformation potentially more suitable for association with other LSm proteins. LSm48 forms high MW aggregates under native buffer conditions, eluting close to the exclusion volume of the Superdex 200 column; 4 M urea disrupts most of these aggregates to yield a main broad peak at 15.4 ml elution volume. This corresponds to a MW of ∼40 kDa (the nominal MW of the heterodimer is 28.1 kDa).

In summary, the higher-order structures formed by these subcomplexes prove to be surprisingly stable, with hexamers as the predominant species. In terms of heteroheptamer reconstitution, we had to rely on assay conditions being compatible with a dynamic exchange of subunits between the different higher order structures, and with the equilibrium being dominated by the most stable species, namely the full heptameric complex.

4. LSm Complex Reconstitution

Reconstitution was initiated by adjusting the buffer conditions for the individual subunits/subcomplex samples to 4 M urea, 1 M NaCl, 20 mM HEPES-NaOH, pH 7.5, and 5 mM β-mercaptoethanol. In the case of LSm2-8, these were LSm23, LSm48, and LSm567. For LSm1-7, LSm48 was replaced by the two single proteins LSm1 and LSm4. Individual protein/subcomplex concentration was between 2 and 6 mg/ml. The absolute amount per assay was between 0.5 and 1 μmol (~40 to 80 mg). The subunits/subcomplexes were incubated in the above buffer for 1 to 2 h at 37 °C. After mixing equimolar amounts of the chosen components, incubation was continued for a further 2 to 3 h at 37 °C. The samples were then dialyzed against the same buffer minus urea for 6 h at 4 °C. For the second dialysis step, the NaCl concentration was reduced to 0.5 M. At this stage, we generally observed substantial turbidity or precipitation, presumably because of formation of nonphysiologic species with reduced solubility. The sample was then concentrated to 1 ml and purified with a Superdex 200 HR 10/30 gel filtration column (GE Healthcare), in a buffer containing 150 mM NaCl. The main peak eluted at an elution volume corresponding to roughly twice the calculated MW of LSm2-8 (160 kDa) and from SDS-PAGE seemed to contain all seven proteins, although not in stoichiometric amounts. As a next step, we purified the main gel filtration peak using an anion exchange column (1 ml Resource 15Q, GE Healthcare), in a gradient 150 to 1000 mM NaCl. Several peaks containing differing subunit compositions eluted from this column. The main peak eluting at approximately 250 mM NaCl contained essentially pure LSm2 to 8. On reapplication of this sample to the anion exchange column, it eluted as one single peak and did not split up, which is an indication it represented a stable, homogeneous species. As a final step, the sample was resubjected to gel filtration chromatography, now yielding a Gaussian peak eluting at 12.6 ml (Fig. 4.4). This corresponds to a MW slightly larger than the calculated MW for LSm2 to 8 (110 vs 86 kDa). The SDS gel shows the presence of equal amounts of individual subunits. The overall complex yield after all purification steps was generally on the order of 5 mg (i.e., between 5 and 10%).

In a modified reconstitution scheme, we used TEV-cleaved LSm567 and LSm48 together with His$_6$-tagged LSm23. The reconstitution protocol itself was the same as previously. The samples were then purified with

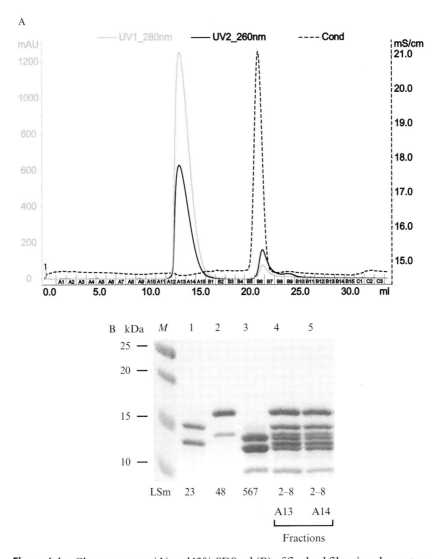

Figure 4.4 Chromatogram (A) and 12% SDS gel (B) of final gel filtration chromatography for the reconstituted LSm2-8 complex. The gel shows individual subcomplexes LSm23 (lane 1), LSm48 (lane 2), and LSm567 (lane 3), as well as fraction A13 and A14 from the gel filtration peak at 14.6-ml elution volume.

Ni-IMAC, followed by two anion exchange and one gel filtration chromatography. In this case, the final sample eluted at 12.96 ml on the Superdex 200 HR10/30 column, corresponding to ~100 kDa (the nominal MW is 81 kDa).

LSm1-7 reconstitution was carried out in essentially the same way as LSm2-8. Here, a single anion exchange chromatography on a Resource 15Q column (GE Healthcare) was sufficient to produce a stoichiometric sample homogeneous enough for final purification by gel filtration chromatography (Fig. 4.5).

5. LSm Complex Functional Assays

To verify the functionality of our reconstituted LSm complexes, we first tested them in an electrophoretic mobility shift assay with U6 snRNA as the substrate. Fully assembled LSm2-8 led to a band shift, whereas LSm1-7 or reconstituted particles missing one or more subunits did not. In a tissue culture cell microinjection study with fluorescently labeled LSm1 to 7 and LSm2 to 8, the former stayed in the cytoplasm; the latter migrated to the nucleus through an active, inhibitable process (Zaric et al., 2005).

Recombinant LSm1 to 7 and LSm2 to 8 complexes were also used in a study investigating the stabilization of poxvirus mRNA through binding of LSm1-7 to poly-A tracts in its $5'$ UTR (Bergman et al., 2007). In the course of this study, the activity of fully purified samples was compared with the activity of preparations directly taken from the second dialysis step after reconstitution (see section 4). Although both samples were active, the amount of the unpurified sample required to shift the same ratio of RNA was 5 to $10\times$ higher (K. Moraes, personal communication). This indicates a large proportion of soluble, yet inactive, molecules in the unpurified sample, presumably consisting of multimers with the wrong composition and/or architecture.

The purified heptamers were visualized by negative-stain electron microscopy in the same manner as the subcomplexes (Zaric et al., 2005). Both yield ring-shaped assemblies with similar sizes to the one observed in an electron microscopic study of native LSm2 to 8 isolated from HeLa cell extract (Achsel et al., 1999). This result indicates that LSm1 to 7 can also assemble into a complex and remains stable in the absence of cognate RNA. Whether both complexes assemble in this RNA-free form in the cell, to be transported intact to their respective site of action, remains to be investigated.

Although LSm6 and the LSm567 subcomplex yielded (weakly diffracting) crystals, all attempts to crystallize LSm1 to 7 and LSm2 to 8 where unsuccessful (Zaric et al., 2005). One likely cause could be the presence of incorrectly assembled species even in the purified heptamers at levels undetectable by functional assays, yet high enough to prevent crystallization. High-resolution structural studies on LSm complexes will have to await improved production/reconstitution protocols.

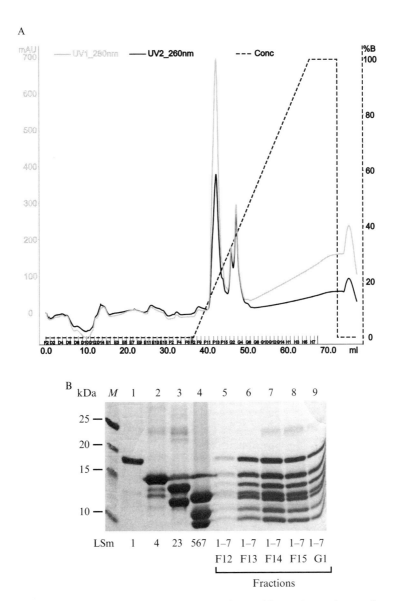

Figure 4.5 Chromatogram (A) and 12% SDS gel (B) of first anion exchange chromatography for reconstituted LSm1 to 7, after initial dialysis and gel filtration chromatography. LSm subunits LSm1 (lane 1), LSm4 (lane 2), LSm23 (lane 3), and LSm567 (lane 4) are shown in the gel, as are fractions F12 to G1 from the main peak eluting in 22% buffer B.

6. Outlook

In summary, we have presented an *in vitro* reconstitution scheme for human heteroheptameric LSm complexes that yields homogeneous, functional LSm1 to 7 and LSm2 to 8 complexes. The resulting samples are of high quality that can be used in a variety of cell biologic and biochemical studies. Nevertheless, the complex protocol and relatively low reconstitution efficiencies leave room for potential improvements. *In vivo*, the SMN complex is instrumental in assembling the Sm core domain and likely participates in LSm complex formation (Meister *et al.*, 2002) as well. The presence of the (homohexameric) Hfq Sm homolog in *E. coli* (Schumacher *et al.*, 2002) argues that LSm complex assembly mechanisms exist also in eubacteria; however, correct positioning of several different subunits in the same ring may not be supported. In any case, it can be assumed carrying out the reconstitution process in a living cell may lead to higher yields and more homogeneous molecule populations. A prerequisite for implementing this concept experimentally is a reliable, efficient, yet flexible, cloning scheme for the construction of coexpression vectors encoding the whole LSm complex. Such schemes have made very substantial progress in recent years. In particular, the MultiBac cloning scheme suits the requirements of LSm coexpression vector build up (Berger *et al.*, 2004). Originally designed for baculovirus expression, it has now been adapted to other hosts, like *E. coli* and cultured mammalian cells (I. Berger, personal communication). Its inherent flexibility presents the option to test expression in different hosts, in particular higher eukaryotes possessing the full Sm/LSm complex assembly machinery.

Apart from simplifying the reconstitution and purification processes, polycistronic or multigenic coexpression vectors for LSm complexes would very likely increase the yield and homogeneity of complexes produced. The resulting particles could be pure enough for structural studies, a goal that has as yet eluded experimental efforts. A recombination-based, reversible cloning scheme would allow the simple exchange of one subunit in a given heptameric complex against another, very suitable for the variety in subunit composition found *in vivo*. Combining such a scheme with carefully chosen in-frame restriction sites would allow the insertion and exchange of single or multiple complementary epitope tags for biochemical and cell biologic studies.

REFERENCES

Achsel, T., Brahms, H., Kastner, B., Bachi, A., Wilm, M., and Lührmann, R. (1999). A doughnut-shaped heteromer of human Sm-like proteins binds to the 3′-end of U6 snRNA, thereby facilitating U4/U6 duplex formation in vitro. *EMBO J.* **18,** 5789–5802.

Albrecht, M., and Lengauer, T. (2004). Novel Sm-like proteins with long C-terminal tails and associated methyltransferases. *FEBS Lett.* **569,** 18–26.

Anantharaman, V., and Aravind, L. (2004). Novel conserved domains in proteins with predicted roles in eukaryotic cell-cycle regulation, decapping and RNA stability. *BMC. Genomics* **5,** 45.

Beggs, J. D. (2005). LSm proteins and RNA processing. *Biochem. Soc. Trans.* **33,** 433–438.

Berger, I., Fitzgerald, D. J., and Richmond, T. J. (2004). Baculovirus expression system for heterologous multiprotein complexes. *Nat. Biotechnol.* **22,** 1583–1587.

Bergman, N., Moraes, K. C., Anderson, J. R., Zaric, B., Kambach, C., Schneider, R. J., Wilusz, C. J., and Wilusz, J. (2007). LSm proteins bind and stabilize RNAs containing 5'-poly(A) tracts. *Nat. Struct. Mol. Biol.* **14,** 824–831.

Bordonne, R. (2000). Functional characterization of nuclear localization signals in yeast Sm proteins. *Mol. Cell Biol.* **20,** 7943–7954.

Bouveret, E., Rigaut, G., Shevchenko, A., Wilm, M., and Séraphin, B. (2000). A Sm-like protein complex that participates in mRNA degradation. *EMBO J.* **19,** 1661–1671.

Brahms, H., Meheus, L., de, B., V, Fischer, U., and Lührmann, R. (2001). Symmetrical dimethylation of arginine residues in spliceosomal Sm protein B/B' and the Sm-like protein LSm4, and their interaction with the SMN protein. *RNA* **7,** 1531–1542.

Fribourg, S., Romier, C., Werten, S., Gangloff, Y. G., Poterszman, A., and Moras, D. (2001). Dissecting the interaction network of multiprotein complexes by pairwise coexpression of subunits in *E. coli*. *J. Mol. Biol.* **306,** 363–373.

Friesen, W. J., Paushkin, S., Wyce, A., Massenet, S., Pesiridis, G. S., Van, D. G., Rappsilber, J., Mann, M., and Dreyfuss, G. (2001). The methylosome, a 20S complex containing JBP1 and pICln, produces dimethylarginine-modified Sm proteins. *Mol. Cell Biol.* **21,** 8289–8300.

Fromont-Racine, M., Rain, J. C., and Legrain, P. (1997). Toward a functional analysis of the yeast genome through exhaustive two-hybrid screens. *Nat. Genet.* **16,** 277–282.

Hermann, H., Fabrizio, P., Raker, V. A., Foulaki, K., Hornig, H., Brahms, H., and Lührmann, R. (1995). snRNP Sm proteins share two evolutionarily conserved sequence motifs which are involved in Sm protein-protein interactions. *EMBO J.* **14,** 2076–2088.

Ingelfinger, D., Arndt-Jovin, D. J., Lührmann, R., and Achsel, T. (2002). The human LSm1-7 proteins colocalize with the mRNA-degrading enzymes Dcp1/2 and Xrn1 in distinct cytoplasmic foci. *RNA* **8,** 1489–1501.

Kambach, C., Walke, S., and Nagai, K. (1999a). Structure and assembly of the spliceosomal small nuclear ribonucleoprotein particles. *Curr. Opin. Struct. Biol.* **9,** 222–230.

Kambach, C., Walke, S., Young, R., Avis, J. M., de la, F. E., Raker, V. A., Lührmann, R., Li, J., and Nagai, K. (1999b). Crystal structures of two Sm protein complexes and their implications for the assembly of the spliceosomal snRNPs. *Cell* **96,** 375–387.

Khusial, P., Plaag, R., and Zieve, G. W. (2005). LSm proteins form heptameric rings that bind to RNA via repeating motifs. *Trends Biochem. Sci.* **30,** 522–528.

Kim, K. J., Kim, H. E., Lee, K. H., Han, W., Yi, M. J., Jeong, J., and Oh, B. H. (2004). Two-promoter vector is highly efficient for overproduction of protein complexes. *Protein Sci.* **13,** 1698–1703.

Ma, Y., Dostie, J., Dreyfuss, G., and Van Duyne, G. D. (2005). The Gemin6-Gemin7 heterodimer from the survival of motor neurons complex has an Sm protein-like structure. *Structure* **13,** 883–892.

Meister, G., Eggert, C., and Fischer, U. (2002). SMN-mediated assembly of RNPs: A complex story. *Trends Cell Biol.* **12,** 472–478.

Moller, T., Franch, T., Hojrup, P., Keene, D. R., Bachinger, H. P., Brennan, R. G., and Valentin-Hansen, P. (2002). Hfq: A bacterial Sm-like protein that mediates RNA-RNA interaction. *Mol. Cell* **9,** 23–30.

Mura, C., Kozhukhovsky, A., Gingery, M., Phillips, M., and Eisenberg, D. (2003). The oligomerization and ligand-binding properties of Sm-like archaeal proteins (SmAPs). *Protein Sci.* **12,** 832–847.

Pannone, B. K., Kim, S. D., Noe, D. A., and Wolin, S. L. (2001). Multiple functional interactions between components of the Lsm2-Lsm8 complex, U6 snRNA, and the yeast La protein. *Genetics* **158,** 187–196.

Pillai, R. S., Grimmler, M., Meister, G., Will, C. L., Lührmann, R., Fischer, U., and Schumperli, D. (2003). Unique Sm core structure of U7 snRNPs: Assembly by a specialized SMN complex and the role of a new component, Lsm11, in histone RNA processing. *Genes Dev.* **17,** 2321–2333.

Plessel, G., Lührmann, R., and Kastner, B. (1997). Electron microscopy of assembly intermediates of the snRNP core: Morphological similarities between the RNA-free (E.F.G) protein heteromer and the intact snRNP core. *J. Mol. Biol.* **265,** 87–94.

Raker, V. A., Plessel, G., and Lührmann, R. (1996). The snRNP core assembly pathway: Identification of stable core protein heteromeric complexes and an snRNP subcore particle in vitro. *EMBO J.* **15,** 2256–2269.

Salgado-Garrido, J., Bragado-Nilsson, E., Kandels-Lewis, S., and Séraphin, B. (1999). Sm and Sm-like proteins assemble in two related complexes of deep evolutionary origin. *EMBO J.* **18,** 3451–3462.

Schoner, B. E., Belagaje, R. M., and Schoner, R. G. (1990). Enhanced translational efficiency with two-cistron expression system. *Methods Enzymol.* **185,** 94–103.

Schumacher, M. A., Pearson, R. F., Moller, T., Valentin-Hansen, P., and Brennan, R. G. (2002). Structures of the pleiotropic translational regulator Hfq and an Hfq-RNA complex: A bacterial Sm-like protein. *EMBO J.* **21,** 3546–3556.

Sheth, U., and Parker, R. (2003). Decapping and decay of messenger RNA occur in cytoplasmic processing bodies. *Science* **300,** 805–808.

Spiller, M. P., Boon, K. L., Reijns, M. A., and Beggs, J. D. (2007). The Lsm2-8 complex determines nuclear localization of the spliceosomal U6 snRNA. *Nucleic Acids Res.* **35,** 923–929.

Stefan Walke, Ph.D. thesis. (2000). University of Cambridge, UK.

Sumpter, V., Kahrs, A., Fischer, U., Kornstadt, U., and Lührmann, R. (1992). In vitro reconstitution of U1 and U2 snRNPs from isolated proteins and snRNA. *Mol. Biol. Rep.* **16,** 229–240.

Tharun, S., He, W., Mayes, A. E., Lennertz, P., Beggs, J. D., and Parker, R. (2000). Yeast Sm-like proteins function in mRNA decapping and decay. *Nature* **404,** 515–518.

Tharun, S., and Parker, R. (2001). Targeting an mRNA for decapping: Displacement of translation factors and association of the Lsm1p-7p complex on deadenylated yeast mRNAs. *Mol. Cell* **8,** 1075–1083.

Urlaub, H., Raker, V. A., Kostka, S., and Lührmann, R. (2001). Sm protein-Sm site RNA interactions within the inner ring of the spliceosomal snRNP core structure. *EMBO J.* **20,** 187–196.

Wang, P., Palfi, Z., Preusser, C., Lucke, S., Lane, W. S., Kambach, C., and Bindereif, A. (2006). Sm core variation in spliceosomal small nuclear ribonucleoproteins from *Trypanosoma brucei*. *EMBO J.* **25,** 4513–4523.

Yong, J., Wan, L., and Dreyfuss, G. (2004). Why do cells need an assembly machine for RNA-protein complexes. *Trends Cell Biol.* **14,** 226–232.

Zaric, B., Chami, M., Remigy, H., Engel, A., Ballmer-Hofer, K., Winkler, F. K., and Kambach, C. (2005). Reconstitution of two recombinant LSm protein complexes reveals aspects of their architecture, assembly, and function. *J. Biol. Chem.* **280,** 16066–16075.

SECTION TWO

POLYADENYLATION/ DEADENYLATION ANALYSES

CHAPTER FIVE

REGULATED DEADENYLATION *IN VITRO*

Aaron C. Goldstrohm,* Brad A. Hook,[†] *and* Marvin Wickens*

Contents

1. Introduction	78
2. *In vitro* Deadenylation Systems	80
2.1. Advantages	80
2.2. A brief history	81
2.3. Optimizing deadenylation conditions	82
2.4. RNA substrates	84
2.5. Protocol: 5′-labeling of synthetic substrate RNA	86
2.6. Regulated deadenylation *in vitro*	87
2.7. Purification of Pop2p deadenylase complexes	88
2.8. Test enzymatic activity of purified complex	91
2.9. Purify recombinant regulator	94
2.10. Confirm protein interactions	97
2.11. Assay-regulated deadenylation *in vitro*	97
3. Troubleshooting	100
3.1. Enzyme concentration	100
3.2. Regulator concentration	100
3.3. Contaminating ribonuclease	101
3.4. Buffer conditions	101
References	101

Abstract

The 3′-poly(A) tail, found on virtually all mRNAs, is enzymatically shortened by a process referred to as "deadenylation." Deadenylation is a widespread means of controlling mRNA stability and translation. The enzymes involved—so-called deadenylases—are surprisingly diverse. They are controlled by RNA sequences commonly found in 3′-untranslated regions (UTRs), which bind regulatory factors.

   Both RNA-binding proteins and microRNAs accelerate deadenylation of specific mRNAs. In some cases, regulators enhance deadenylation by binding to and recruiting specific deadenylases to the target mRNA. The many hundreds of potential regulators encoded in mammalian genomes (both RNA-binding

---

* Department of Biochemistry, University of Wisconsin—Madison, Madison, Wisconsin, USA
[†] Promega Corporation, Madison, Wisconsin, USA

proteins and microRNAs) and the numerous deadenylases, coupled with the many potential regulatory sites represented in 3′ UTRs of mRNAs, provide fertile ground for regulated deadenylation. Recent global studies of poly(A) regulation support this conclusion. Biochemical and genetic approaches will be essential for exploring regulated deadenylation.

The methods we describe focus on the reconstruction *in vitro* of regulated deadenylation with purified components from yeast. We discuss broadly the strategies, problems, and history of *in vitro* deadenylation systems. We combine this with a more detailed discussion of the purification, activity, and regulation of the *Saccharomyces cerevisiae* Ccr4p-Pop2p deadenylase complex and its regulation by PUF (Pumilio and Fem-3 binding factor) RNA-binding proteins.

1. Introduction

Gene expression is extensively controlled by posttranscriptional mechanisms, including mRNA stability and translation (Garneau *et al.*, 2007, 2007b). Indeed, mRNA stabilities vary by four orders of magnitude, and protein levels vary by as much as six orders of magnitude (Beyer *et al.*, 2004; Ghaemmaghami *et al.*, 2003; Huh *et al.*, 2003; Lackner *et al.*, 2007; Vasudevan *et al.*, 2006; Yang *et al.*, 2003). These types of regulation permeate biology, contributing to quantitative, spatial, and temporal control of protein production. They are required for development and differentiation, cell growth, immunity, and memory formation (Bramham and Wells, 2007; Colegrove-Otero *et al.*, 2005, 2007b; Garneau *et al.*, 2007; Kimble and Crittenden, 2007; Khabar, 2007; Khabar and Young, 2007).

The poly(adenosine) (poly(A)) tail plays a central role in mRNA regulation (Gray and Wickens, 1998; Kuhn and Wahle, 2004). Poly(A) is added to the 3′-end of nascent mRNAs by the nuclear polyadenylation machinery (Zhao *et al.*, 1999). Once the mRNA is in the cytoplasm, poly(A) has at least two duties: to promote translation and to stabilize the mRNA. Both of these effects are mediated by poly(A) binding protein (PABP), which coats the poly(A) tail (Gorgoni and Gray, 2004; Kuhn and Wahle, 2004). PABP interacts with translation initiation factors, including the eIF4F complex that recognizes the 5′-7-methyl guanosine cap structure and promotes translation initiation (Gorgoni and Gray, 2004; Kuhn and Wahle, 2004). PABP wards off attack by cellular exoribonucleases (with some exceptions) by blocking their access to the mRNA (Goldstrohm and Wickens, 2008; Gorgoni and Gray, 2004; Kuhn and Wahle, 2004).

"Deadenylation" is the enzymatic shortening of the poly(A) tail over time. This process typically initiates, and is the rate-limiting step of, general mRNA decay (Parker and Song, 2004). Enzymes progressively degrade the poly(A) tail at a slow, basal rate, beginning at the 3′-end (Fig. 5.1). Once poly(A) is shortened to a certain length (typically 10 to 15 A's in yeast;

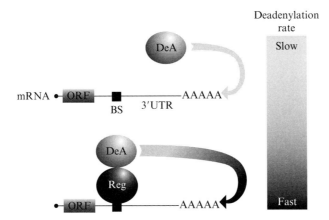

Figure 5.1 Regulators enhance deadenylation by recruiting a deadenylase to the mRNA. A deadenylase (DeA) slowly degrades the poly(A) tail of an mRNA that contains an open-reading frame (ORF) and a 3'-untranslated region (3' UTR) with a binding site (BS) for a sequence-specific, RNA-binding regulatory factor (Reg). The regulator interacts with the deadenylase and recruits it to the mRNA, thereby accelerating the rate of deadenylation.

slightly longer in mammals), the mRNA is degraded by either of two pathways. The first is a 5' to 3'-pathway wherein the triphosphate linkage of the 5'-cap is cleaved by a "decapping" enzyme (DCP2), and the mRNA body is subsequently degraded from the 5'-end by XRN1 exonuclease. The second pathway acts 3' to 5' and is catalyzed by a complex of exonucleases, the exosome (Garneau et al., 2007; Parker and Song, 2004).

"Deadenylases" are defined as exoribonucleases that degrade poly(A) in a 3' to 5'-direction (Goldstrohm and Wickens, 2008; Parker and Song, 2004). In doing so, they release 5'-AMP (Astrom et al., 1991, 1992; Lowell et al., 1992). Many such enzymes exist (Goldstrohm and Wickens, 2008). In mammals, informatics suggest as many as 12 deadenylases (Dupressoir et al., 2001; Goldstrohm and Wickens, 2008; Wagner et al., 2007). Deadenylases have preference for 3'-poly(A), although in some cases can degrade nonadenosine homopolymers with reduced efficiency (Astrom et al., 1991; Bianchin et al., 2005; Chen et al., 2002; Lowell et al., 1992; Thore et al., 2003; Tucker et al., 2001; Uchida et al., 2004). All known deadenylases are Mg^{2+}-dependent enzymes that belong to either of two superfamilies, defined by conserved nuclease sequence motifs that are necessary for catalysis (Goldstrohm and Wickens, 2008). The DEDD superfamily contains relatives of the POP2, PARN, and PAN2 deadenylases (Goldstrohm and Wickens, 2008; Thore et al., 2003; Zuo and Deutscher, 2001). The second superfamily includes members related to a class of exonucleases, endonucleases, and phosphatases, known as the EEP superfamily (Dlakic, 2000;

Dupressoir et al., 2001; Goldstrohm and Wickens, 2008). CCR4 and Nocturnin deadenylases belong to the EEP superfamily (Goldstrohm and Wickens, 2008).

Deadenylases function in general mRNA turnover, but also have specialized regulatory roles (Goldstrohm and Wickens, 2008; Parker and Song, 2004). Sequence-specific, RNA-binding regulators can target individual deadenylases to specific mRNAs (Fig. 5.1) (Chicoine et al., 2007; Goldstrohm and Wickens, 2008; Goldstrohm et al., 2006; Kim and Richter, 2006; Lykke-Andersen and Wagner, 2005; Moraes et al., 2006). For example, the PUF family of RNA-binding regulatory proteins can repress specific mRNAs by interacting with and recruiting a deadenylase complex (Goldstrohm et al., 2006, 2007; Hook et al., 2007; Kadyrova et al., 2007). Yeast PUF proteins directly bind the Pop2p subunit of the Ccr4p-Pop2p deadenylase complex and thereby accelerate deadenylation and decay and/or repress translation of target mRNAs. Regulation of the mRNA encoding *HO* mating-type switching endonuclease by two yeast PUF proteins is exemplary (Goldstrohm et al., 2006; Hook et al., 2007).

Recruitment of deadenylases by regulators is likely widespread (Beilharz and Preiss, 2007; Goldstrohm and Wickens, 2008; Grigull et al., 2004; Lackner et al., 2007; Meijer et al., 2007). The potential for mRNA control is enormous, with hundreds of protein and microRNA regulators, a multitude of deadenylases, and the abundance of 3' UTR controls (Chen and Rajewsky, 2007; Keene, 2007; Lander et al., 2001). Biochemical approaches will be essential for dissecting mechanisms and control of deadenylation. Here we focus on *in vitro* methods for analyzing deadenylation, concentrating on the yeast, *S. cerevisiae*. With the PUF regulation of the Ccr4p-Pop2p deadenylase complex as an example, we describe an *in vitro* assay system that recapitulates regulated deadenylation, allowing mechanistic, enzymatic, and molecular genetic analysis.

2. IN VITRO DEADENYLATION SYSTEMS

In vitro systems have been indispensable for identifying deadenylases; indeed, the first deadenylases were discovered with *in vitro* assays (Astrom et al., 1991, 1992; Lowell et al., 1992; Sachs and Deardorff, 1992). Moreover, *in vitro* biochemical systems are important for identifying enzymatic properties and dissecting control mechanisms of mRNA deadenylation.

2.1. Advantages

The benefits of *in vitro* deadenylation analysis are manifold. First, this approach permits study of isolated, individual deadenylases out of the many in eukaryotic cells (Goldstrohm and Wickens, 2008). Several deadenylase mutants are nonviable, complicating genetic analysis (Chiba et al., 2004; Molin and Puisieux, 2005; Reverdatto et al., 2004). Also, the interconnections and feedback among RNA-related steps, particularly translation and mRNA decay pathways, are circumvented by turning to the test tube (Jacobson and Peltz, 1996; Parker and Sheth, 2007; Schwartz and Parker, 1999, 2000).

In vitro analysis is particularly useful for examining the enzymatic properties of deadenylases, including reaction kinetics and cofactor requirements (Astrom et al., 1991, 1992; Chen et al., 2002; Korner and Wahle, 1997; Martinez et al., 2000; Viswanathan et al., 2003). The effects on substrate structures and RNA sequences on substrate recognition and catalysis can also be examined (Bianchin et al., 2005; Dehlin et al., 2000; Gao et al., 2000; Martinez et al., 2001; Viswanathan et al., 2003, 2004). Importantly, these properties may influence regulatory outcomes: targeting an mRNA with a highly active, processive enzyme elicits more rapid repression than an enzyme with low activity.

Mechanistic analysis of deadenylation, and in particular of its regulation, also can be accomplished *in vitro*. For instance, the effects of regulatory factors on enzymatic activity can be determined (Balatsos et al., 2006; Goldstrohm et al., 2006; Moraes et al., 2006). The detailed mechanisms of the interaction between regulator and deadenylase can be dissected.

Although the *in vitro* studies are powerful, molecular genetics and *in vivo* analysis clearly are a necessary complement (Chen et al., 2002; Goldstrohm et al., 2006, 2007; Hook et al., 2007; Tucker et al., 2001, 2002; Viswanathan et al., 2004). Without them, it is difficult to know how faithfully an *in vitro* system recapitulates biologic regulation. The fact that many enzymes are capable of removing poly(A) is a serious complication and can send *in vitro* analysis down a misleading path.

2.2. A brief history

Multiple *in vitro* systems have been established over the past 15 years, beginning with the use of cell extracts (Astrom et al., 1991; Sachs and Deardorff, 1992). Among the first deadenylases purified, Pan2p was initially identified as a PAB-stimulated deadenylase present in yeast extracts (Lowell et al., 1992; Sachs and Deardorff, 1992). Vertebrate PARN was purified from HeLa cell nuclear and cytoplasmic extracts (Astrom et al., 1991, 1992). These enzymes were subsequently purified by conventional chromatography, studied in reconstituted assays, identified, and cloned (Boeck et al., 1996; Brown et al., 1996; Copeland and Wormington, 2001; Dehlin et al., 2000; Korner and Wahle, 1997; Korner et al., 1998; Martinez et al., 2000; Ren et al., 2002, 2004; Sachs, 2000).

Extract systems also have revealed PARN-dependent deadenylation followed by ATP-dependent degradation of the mRNA body (Ford and

Wilusz, 1999; Ford et al., 1999; Fritz et al., 2000). The addition of poly(A) competitor, which likely alleviates PABP-mediated stabilization, was critical (Ford and Wilusz, 1999). Similar methods were later applied to insect cells and trypanosomes (Milone et al., 2004; Opyrchal et al., 2005). Regulated deadenylation in *Drosophila* extracts revealed ATP-dependent deadenylation and has only been observed in this system (Jeske et al., 2006; Temme et al., 2004).

Genetic analysis identified the heterodimeric Ccr4p and Pop2p complex as responsible for deadenylation activity in yeast, with Ccr4p predominating *in vivo* (Daugeron et al., 2001; Tucker et al., 2001, 2002). *In vitro* analysis with affinity tag–purified complexes, and later recombinant purified enzymes, verified their activities (Chen et al., 2002; Tucker et al., 2001; 2002). Orthologous proteins in mammals and flies were subsequently shown to be active in deadenylases *in vitro* (Baggs and Green, 2003; Bianchin et al., 2005; Chen et al., 2002; Morita et al., 2007; Viswanathan et al., 2004; Wagner et al., 2007).

We studied the regulation of deadenylation by yeast PUF proteins, a family of sequence-specific repressor proteins found throughout eukarya (Wickens et al., 2002). PUF repression correlates with deadenylation, and we found that PUFs make a direct protein contact with the Pop2p subunit of the Ccr4p-Pop2p complex (Goldstrohm et al., 2006, 2007; Hook et al., 2007).

Here we describe the *in vitro* system that we used to reconstitute regulated deadenylation with purified components. We consider each of the steps in developing an *in vitro* deadenylation system in detail. These are diagrammed in Fig. 5.2, which serves as a guide to the remainder of the chapter.

2.3. Optimizing deadenylation conditions

Developing optimal conditions to analyze the properties of a deadenylase is relatively straightforward. However, in reconstituting *regulated* deadenylation, the situation is complicated by the fact that at least three different biochemical events are required: the regulator needs to bind the RNA, the regulator needs to bind the deadenylase or deadenylase complex, and the enzyme has to be catalytically active (Fig. 5.2). Each of these biochemical events has its own distinct optima. With this in mind, we recommend first establishing functional deadenylation conditions, followed by analysis of regulator binding to the substrate under those conditions. If the RNA-binding regulator is functional, one can proceed to test regulated deadenylation *in vitro*. Otherwise, it may be necessary to adjust the deadenylation buffer conditions to suit both deadenylation and RNA binding.

A range of conditions has been used for *in vitro* studies and merits consideration here. These conditions vary with the enzymes and regulators.

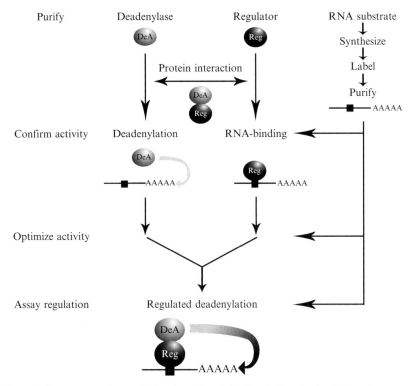

Figure 5.2 Strategy for analysis of regulated deadenylation *in vitro*. To analyze regulated deadenylation *in vitro*, begin by preparing purified deadenylase (DeA), regulator (Reg), and radioactively labeled RNA substrate. Confirm protein interactions between the regulator and deadenylase. Measure the deadenylase activity and RNA-binding activity of deadenylase and regulator, respectively. Titrate each component and optimize concentrations and reaction conditions. Finally, combine deadenylase and regulator with substrate RNA to observe enhancement of deadenylation.

Optimized *in vitro* conditions have been reported for vertebrate PARN (Astrom et al., 1992; Korner and Wahle, 1997; Martinez et al., 2000), yeast Pan2p (Lowell et al., 1992), and yeast Ccr4p (Viswanathan et al., 2003).

Magnesium ions are essential for all known deadenylases, and both magnesium chloride and acetate have been used successfully over a range of 0.01 to 3 mM (Astrom et al., 1991; Baggs and Green, 2003; Chen et al., 2002; Daugeron et al., 2001; Lowell et al., 1992; Tucker et al., 2001; Wagner et al., 2007). Excessive magnesium can be inhibitory (Astrom et al., 1992; Viswanathan et al., 2003). In general, other divalent cations cannot substitute (Astrom et al., 1992; Korner and Wahle, 1997).

Common buffers include Tris-HCl, HEPES-KOH, and KPO$_4$, with a pH range of 6.8 to 8.5. Monovalent cations, sodium, or potassium in the form of NaCl, KCl, or potassium acetate, have been used, typically ranging from 0 to 200 nM. In some instances, either no salt or excessive salt concentrations were inhibitory (Astrom et al., 1992; Viswanathan et al., 2003).

Detergents such as Nonidet P40/IGEPAL and Tween-20 have been included in some cases, with concentrations up to 0.2 and 0.02% being tolerated, respectively (Copeland and Wormington, 2001; Daugeron et al., 2001; Tucker et al., 2001; Wagner et al., 2007). We recommend avoiding detergents unless they are necessary for protein solubility or for reducing spurious interactions.

Other common reaction additives include DTT, spermidine, BSA, polyvinyl alcohol, or glycerol. DTT is commonly used as a reducing agent, with concentrations up to 1 mM. Under certain conditions, the polyamine spermidine (up to 2 mM) can stimulate PARN or PAN2 (Korner and Wahle, 1997; Lowell et al., 1992; Uchida et al., 2004). BSA may help stabilize proteins and acts as a molecular crowding agent, with concentrations as high as 0.2 mg/ml. Other molecular crowding agents such as polyvinyl alcohol (MW 10000) have been used with concentrations up to 2.5% (Astrom et al., 1991; Ford and Wilusz, 1999). Glycerol can be included up to at least 10%. The RNase A inhibitor RNasin has also been included to minimize spurious contamination.

Several extract systems have included ATP regeneration systems, such as 20 mM creatine phosphate, 80 ng/μl creatine kinase, and 800 μM ATP (Ford and Wilusz, 1999; Temme et al., 2004). As a cautionary note, high levels of nucleotides can be inhibitory to several deadenylases, possibly by chelating Mg^{2+} (Astrom et al., 1992). Likewise, inclusion of EDTA is not advisable, although very low levels have been tolerated in some systems (Astrom et al., 1991; Martinez et al., 2000).

Reactions are commonly performed at temperatures ranging from 25 to 37 °C. For instance, yeast Ccr4p activity is optimal at 37 °C, and nearly inactive at 0 °C and above 55 °C (Viswanathan et al., 2003).

2.4. RNA substrates

Several methods can be used to create RNA substrates for deadenylation analysis. Minimally, RNA substrates must contain a high-affinity binding site for the sequence-specific regulator and a 3'-poly(A) tail. We have successfully observed regulated deadenylation in the yeast system with regulator RNA dissociation constants between 10 and 250 nM, as measured with purified protein and RNA (Goldstrohm et al., 2006; Hook et al., 2007). Poly(A) tails may be as long as several hundred nucleotides or as short as a single adenosine (Sachs and Deardorff, 1992; Viswanathan et al., 2003).

Substrates can be transcribed in vitro or chemically synthesized. Benefits of synthetic RNAs include a homogeneous starting material, high purity and quality, and precise amount. Purity is commonly quite good, and gel purification is typically not necessary; however, costs may be considerable and lengths are limited. In our own experience, synthetic substrates as long as 54 nucleotides, including 14 nucleotide poly(A) tails, have yielded excellent results (Goldstrohm et al., 2006, 2007; Hook et al., 2007).

Transcripts can be produced by bacteriophage RNA polymerases, such as T7, from plasmid templates encoding poly(A) tracts (Kreig and Johnson, 1996). The plasmid template is linearized with a specific restriction enzyme to produce "runoff transcripts" with a defined 3′-end. Common restriction sites used to linearize the template produce RNAs with nonadenosine 3′-ends, arising from the restriction site 3′-overhangs. Because some deadenylases are inhibited by nonadenosine 3′-terminal residues, it is advisable to linearize the template with a restriction enzyme, such as Nsi1, which produces a 3′-terminus encoding only on a single additional adenosine nucleotide (Ford and Wilusz, 1999). Another common method is to generate PCR-amplified or double-stranded oligonucleotide templates that encode 3′-poly(A) tails as transcription templates (Ford and Wilusz, 1999). Longer poly(A) tails can also be added to RNAs by means of a tailing reaction with poly(A) polymerases, although it is difficult to control the length of the often heterogeneous tails (Brown et al., 1996; Daugeron et al., 2001; Ford et al., 1999; Lowell et al., 1992; Martin and Keller, 1998; Temme et al., 2004).

2.4.1. Uniform labeling

Radioactive nucleotides, typically [α-^{32}P]-UTP, can be incorporated into the body of the RNA during transcription to produce substrates with high specific radioactivity (Kreig and Johnson, 1996). Transcription systems are available commercially for transcribing labeled RNAs *in vitro* (for example, from Promega or Ambion). Unincorporated nucleotides and truncated transcripts must be removed by size exclusion chromatography and/or gel purification (see later).

2.4.2. 5′-End labeling

RNA substrates can be labeled at the 5′-end by T4 polynucleotide kinase with [γ-^{32}P]-ATP. Deadenylation is then observed by progressive shortening from the 3′-end of the labeled RNA, in which all degradation products retain identical signal levels, thereby facilitating quantitation. Synthetic RNAs with a 5′-hydroxyl terminus, are ready to be phosphorylated, whereas transcribed RNAs, possessing 5′-triphosphate terminus, should be dephosphorylated with calf intestinal phosphatase before 5′-end labeling. Synthetic RNAs with 5′-fluorescent labels (for example, fluorescein derivatives) have also been used successfully to analyze deadenylation *in vitro* (Morita et al., 2007).

2.4.3. 3′-End labeling

Labeled poly(A) tail substrates may also be created with poly(A) polymerases and [α-^{32}P]-ATP (Brown et al., 1996; Daugeron et al., 2001; Ford et al., 1999; Lowell et al., 1992; Martin and Keller, 1998; Temme et al., 2004). This approach can be used to assay deadenylase activity by release of TCA-soluble AMP (Sachs and Deardorff, 1992).

2.4.4. 5′-Capping

A 5′-7-methyl guanosine cap structure can be added to RNA substrates if necessary (Kreig and Johnson, 1996). The 5′-cap can stabilize substrates in extract systems, presumably by protecting the RNA from 5′-exonucleolytic attack (Parker and Song, 2004). Furthermore, the 5′-cap activates PARN deadenylase activity and so may be desirable when studying PARN-like activities (Dehlin et al., 2000; Gao et al., 2000; Martinez et al., 2001).

Cap analog can be incorporated by RNA polymerases during *in vitro* transcription; however, the incorporation is typically incomplete, and the cap analog can be incorporated in reverse orientation. To increase capping efficiency and maintain proper orientation, Vaccinia virus capping enzyme can be used. Capping enzyme can label RNAs with a [α-^{32}P]-GTP and S-adenosyl methionine (capping enzyme is available from Ambion) (See chapter 1 by Liu *et al* for details).

2.4.5. Gel purification

For transcribed substrates, gel purification of full-length product is recommended. Free nucleotides from the transcription reaction can inhibit deadenylases. Free cap analog inhibits PARN. Purification is easily accomplished by electrophoresis through denaturing urea-polyacrylamide gels (Gilman, 2000). For unlabeled RNAs, UV shadowing is used to visualize the RNA; for radioactively labeled RNAs, autoradiography can be used. A gel slice containing the RNA is excised, and the RNA is eluted and precipitated by standard methods (Gilman, 2000). For transcribed RNA substrates, we prefer to transcribe them without the radiolabel, gel purify the RNA, and measure its final concentration. Specific quantities of substrate RNA can then be 5′-labeled as needed.

2.5. Protocol: 5′-labeling of synthetic substrate RNA

Synthetic substrate RNAs (from Dharmacon or Integrated DNA Technologies) can be labeled at the 5′-end with the following protocol.

Assemble labeling reaction:

9 μl ddH$_2$O, sterile, RNase free
2 μl 10× kinase buffer (Promega)
4 μl 10 μM RNA (40 pmol total per reaction)

3 μl [γ-^{32}P]-ATP, 6000 Ci/mmol, 10 μCi/ml (Perkin-Elmer)
2 μL T4 polynucleotide kinase (Promega)
20 μl final volume, final concentration is 2 pmol/μl
Incubate at 37 °C for 1.5 h.
Heat inactivate at 70 °C for 20 min.
Remove free nucleotides and salts by size exclusion spin chromatography.
   Apply sample to spin columns, Nucaway (Ambion) or G25 Sepharose (GE Healthcare), according to the manufacturers' recommendations.
Store labeled RNA at −20 °C until ready to use.

2.6. Regulated deadenylation *in vitro*

2.6.1. Purified components

The general strategy requires multiple steps (Fig. 5.2). First, the deadenylase and sequence-specific regulator are purified, and their ability to interact confirmed. Second, the radioactively labeled, polyadenylated substrate RNAs are prepared. Third, the enzyme is titrated, and the RNA-binding activity of the regulator assayed. Finally, the effect of the regulator on the deadenylase is measured by combining the two assays.

Deadenylase complexes have been successfully purified for *in vitro* analysis from natural sources by conventional chromatography (Boeck *et al.*, 1996; Korner and Wahle, 1997; Korner *et al.*, 1998; Lowell *et al.*, 1992; Martinez *et al.*, 2000). Now, affinity purification techniques often are used, including epitope tags or tandem affinity purification (TAP) methods (Chen *et al.*, 2002; Tucker *et al.*, 2001, 2002; Viswanathan *et al.*, 2003). We purified Ccr4p-Pop2p deadenylase complexes from yeast strains expressing either chromosomally integrated or episomal, affinity-tagged proteins (Goldstrohm *et al.*, 2006, 2007; Hook *et al.*, 2007). The tags were engineered with tobacco etch virus (TEV) protease sites to allow specific elution of the bead bound, purified complexes by TEV cleavage (Puig *et al.*, 2001; Rigaut *et al.*, 1999).

Purification from natural sources offers several advantages. Native deadenylase complexes are isolated, maintaining potentially important posttranslational modifications and protein subunits. For instance, several subunits of the CCR4–POP2 complex function as adaptors that bridge regulators and the deadenylase subunits by means of protein interactions (TOB, NOT subunits) (Brown *et al.*, 1996; Chicoine *et al.*, 2007; Ezzeddine *et al.*, 2007; Goldstrohm *et al.*, 2007; Kadyrova *et al.*, 2007). However, there are disadvantages as well; native complexes are not completely defined without additional analysis and are likely heterogeneous in composition *in vivo*. (For instance, multiple forms of CCR4–POP2 complexes have been detected in yeast and mammals [Denis and Chen, 2003; Morel *et al.*, 2003]).

Deadenylation can also be reconstituted with recombinant deadenylases isolated from bacteria (Baggs and Green, 2003; Balatsos *et al.*, 2006;

Bianchin et al., 2005; Korner et al., 1998; Martinez et al., 2001; Ren et al., 2002, 2004; Viswanathan et al., 2004; Wagner et al., 2007). The obvious advantage is that recombinant enzymes are fully defined. However, it has proven difficult to obtain full-length forms of several deadenylases from bacterial expression systems. Active, truncated derivatives have successfully been purified as an alternative (Daugeron et al., 2001; Simon and Seraphin, 2007). The purified protein may require other proteins to interact with the regulator (Brown et al., 1996; Chicoine et al., 2007; Kadyrova et al., 2007).

Molecular genetics can be combined with biochemical purification to create and isolate complexes with specific mutations or with missing specific subunits. This powerful approach permits the structure–function analysis of enzymatic properties and the evaluation of protein domains and specific protein interactions (Chen et al., 2002; Goldstrohm et al., 2007; Hook et al., 2007; Kim and Richter, 2006; Tucker et al., 2002).

2.7. Purification of Pop2p deadenylase complexes

In the following protocol, we describe a method for purification of TAP-tagged Pop2p deadenylase complexes from yeast. We recommend collecting samples at each step of the purification to monitor purification efficiency (Fig. 5.3).

2.7.1. Yeast strain

The TAP-tagged POP2 haploid *Saccharomyces cerevisiae* strain (Open Biosystems) contains a C-terminal TAP tag integrated into the POP2 chromosomal locus by PCR-mediated gene modification. Expression of the POP2-TAP protein is driven by the natural POP2 promoter. For yeast culture techniques and media, refer to Guthrie and Fink (2002).

2.7.2. Grow yeast cultures

Grow a 50-ml starter culture in YPAD media overnight at 30 °C. The following morning, dilute the culture into 4 L of YPAD and grow to an optical density at 660 nm of 1.5 (4.5×10^7 cells/ml).

2.7.3. Harvest cells

Harvest cells at $4000g$ in a JLA8.1 rotor (Beckman-Coulter) at 4 °C. Resuspend cell pellets in ice-cold ddH$_2$O. Transfer the suspension to a 50-ml plastic conical tube. Collect cells by centrifugation in a refrigerated tabletop centrifuge at $4000g$. Decant the supernatant from the cell pellets, which will be approximately 10 to 15 ml of packed cell volume. Cell pellets can be frozen and stored at −80 °C at this stage.

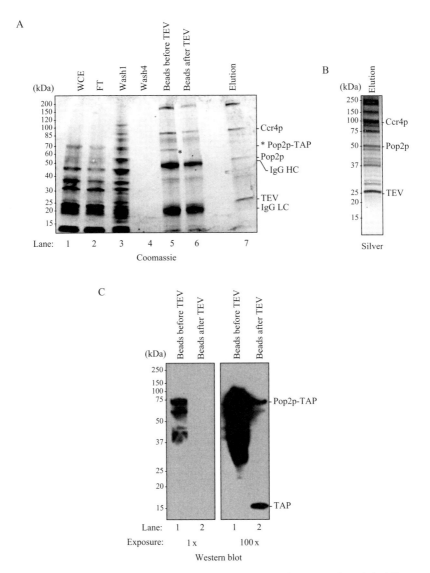

Figure 5.3 Purification of TAP-tagged Pop2p deadenylase complex. (A) Aliquots from the indicated purification steps of TAP-Pop2p were analyzed by SDS-PAGE and Coomassie blue protein stain, including: whole cell extract (WCE, 0.1% of the total), the flow through (FT, 0.1% of the total) material, and the first (Wash 1) and last (Wash 4) wash steps (0.3% of total of each). Purified material bound to the IgG agarose beads before and after elution with TEV protease (0.5% of total of each), as well as the eluted Pop2p complex (370 ng, representing approximately 8% of the total) were also analyzed. Molecular weights (in kilodaltons) are indicated on the left. On the right, the mobility of Pop2p-TAP, and Pop2p after TEV cleavage, in addition to TEV protease, is indicated. Also, the likely mobility of the Ccr4p deadenylase subunit is shown.

2.7.4. Lyse cells

Resuspend cell pellets, on ice, in 1 packed cell volume of TNEMN150 buffer (50 mM Tris-HCl (pH 8), 0.5% (v/v) Nonidet P40, 1 mM EDTA, 2 mM MgCl$_2$ and 150 mM NaCl) supplemented with 2× protease inhibitor cocktail (Complete, Roche). Dispense suspension into prechilled microfuge tubes containing 500 µl dry volume of sterile acid-washed glass beads (500 µm in diameter, Sigma). Mechanically lyse cells by "bead-bashing" for 15 min with a vortex genie with a bead-basher attachment (Disruptor Genie, Scientific Industries, Inc.). Clarify lysates at 16,000g at 4 °C for 10 min and then combine supernatants (the whole cell extract) in an ice-cold 15-ml conical tube.

2.7.5. Bind deadenylase to beads

Add rabbit IgG agarose beads (120 µl of a 1:1 slurry, binding capacity of 5 mg/ml of resin, Sigma) to the whole cell extract and allow binding to occur for 2 h at 4 °C on a rocking platform (for example, a Nutator).

2.7.6. Wash beads

Collect beads by centrifugation in a refrigerated tabletop centrifuge at 1000g for 5 min. Wash beads four times with (15 ml) 100-bed volumes of TNEMN150. The first wash will contain cellular protein, similar in pattern to that in the whole-cell extract or "flow-through" material (Fig 5.3A, lanes 1-3); the fourth should be virtually devoid of protein (Fig. 5.3A, lane 4).

2.7.7. Exchange buffer

To remove detergents and equilibrate purified protein in the downstream deadenylation buffer, wash the beads two times with 15 ml of deadenylation buffer (50 mM Tris-HCl, pH 8, 20 mM NaCl, 0.1% MgCl$_2$, 10% glycerol).

Major bands from IgG heavy and light chains, present on the IgG agarose beads, are indicated. (B) To further assess purity and composition, the TAP-purified Pop2p deadenylase complex (370 ng), after TEV elution, was analyzed by SDS-PAGE and silver staining. Approximately a dozen protein bands are abundant, with major bands of 50 kDa and 95 kDa corresponding to Pop2p and Ccr4p, respectively. (C) Western blot analysis, with peroxidase/antiperoxidase of purified Pop2-TAP material bound to IgG agarose beads before and after TEV cleave and elution (0.5% of total for each). Two exposures of the same blot are shown (a short exposure is on the right while a 100-fold longer exposure is on the right). TAP-tagged Pop2p migrates at approximately 71 kDa. Several minor degradation products are seen under the major 71-kDa band. TEV cleavage liberates the Pop2p protein from the IgG binding domains of the TAP tag. The TAP tag fragment remains bound to the IgG beads, seen in the longer exposure (right panel). A minor amount of uncleaved Pop2-TAP remains on the beads after TEV cleavage.

2.7.8. Elution of deadenylase complex

Elute purified deadenylase complex from the beads with TEV protease (AcTEV, Invitrogen). Cleavage can be performed for 12 h at 4 °C with 8 U of AcTEV in 1-bed volume (60 µl) of deadenylation buffer.

2.7.9. Analyze purified deadenylase complex

Measure the protein concentration of the eluted deadenylase by Bradford assay (Bio-Rad) or similar methods. Assess the purity and composition of the deadenylase complex by SDS-PAGE followed by Coomassie and silver staining (Fig. 5.3A and B). Western blotting of the deadenylase subunits is advisable (Fig. 5.3C). For TAP-tagged deadenylase subunits, a mixture of peroxidase/ antiperoxidase (Sigma, at 1 to 5000-fold dilution), which recognizes the IgG binding domains of the TAP tag, is used to detect the purified Pop2p-TAP protein. Western analysis of a portion of the bead-bound material should reveal a strong signal at MW of 71 kDa (Fig. 5.3C). TEV cleavage efficiency can be monitored by Coomassie staining and Western blotting of the beads before and after cleavage (Fig. 5.3A, lanes 5 and 6, and Fig. 5.3C, lanes 1 and 2). After TEV treatment, the IgG binding domain of the TAP tag is detected on the beads (Fig. 5.3C, lane 2). Because the IgG binding domain is removed from Pop2p, the eluted Pop2p deadenylase is no longer detectable by peroxidase/antiperoxidase. When purifying Pop2p-Ccr4p complexes, we also coexpressed T7-tagged Ccr4p with the TAP-tagged Pop2p. Copurification of both deadenylase subunits was confirmed by Western blotting with anti-T7 (Novagen) and peroxidase/antiperoxidase antibodies, respectively (Goldstrohm *et al.*, 2006, 2007).

2.8. Test enzymatic activity of purified complex

The next step is to test the enzymatic activity of the purified deadenylase and determine the optimal amount of enzyme. To measure deadenylation rates, collect samples from the reactions over a time course, typically between 0 to 120 min. The relative rate can be adjusted by altering the enzyme concentration. An example of such a time course analysis of deadenylation activity versus protein concentration is shown in Fig. 5.4A for the yeast Pop2p deadenylase complex. At low concentrations or short times, very little activity is observed (Fig 5.4A, lanes 9 to 14); longer times and high concentrations result in degradation of all RNA molecules (Fig. 5.4A, lanes 3 to 8).

2.8.1. Control reactions

To control for potential ribonuclease contamination of reaction components, include a reaction in which substrate RNA is incubated with deadenylase buffer (Fig. 5.4B, lanes 3 to 8). To control for nonspecific ribonuclease contamination of the purified deadenylase preparation,

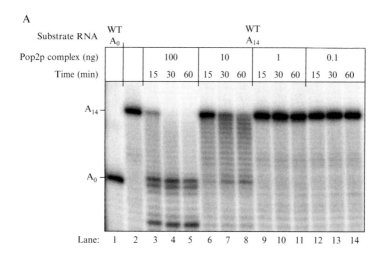
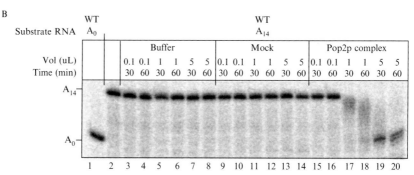

Figure 5.4 Assaying the enzymatic activity of the purified deadenylase complex. (A) The enzymatic activity of the TAP tag–purified Pop2p deadenylase complex was measured with *in vitro* deadenylation assays. Deadenylation was time and concentration dependent. Decreasing amounts of the Pop2p complex (indicated at the top) were incubated with radioactively labeled substrate RNA with 14 adenosine residues at its 3′ end (WT A_{14}). Reaction times are indicated at the top. Reaction products were then separated by denaturing polyacrylamide gel electrophoresis. The migration of marker RNAs without (WT A_0) or with poly(A) tail (WT A_{14}) are indicated on the left side; 100 ng of Pop2p complex was sufficient to rapidly degrade the poly(A) tail, and, in fact, degradation proceeded into the body of the RNA substrate. At 10 ng, the deadenylase complex slowly degraded the poly(A) tail, with fully deadenylated product accumulating at later time points. Lower concentrations of deadenylase were inactive. (B) Control deadenylation reactions were performed by incubating labeled substrate RNA with deadenylation buffer alone to control for ribonuclease contamination or with eluate from a mock TAP purification. As a positive control, purified Pop2p complex was included, resulting in deadenylation of the substrate.

perform a mock TAP purification from a wild-type yeast strain that lacks a TAP-tagged protein. No degradation of substrate should be observed with the mock-purified material (Fig. 5.4B, lanes 9 to 14), whereas an equal volume of purified deadenylase is fully active (Fig. 5.4B, lanes 15 to 20). Another excellent control is to purify mutated, catalytically inactive versions of the deadenylase and test this material for the absence of deadenylation, thereby proving that the observed activity originates from the deadenylase enzyme and not a contaminant (Goldstrohm et al., 2007).

2.8.2. Assemble deadenylation reactions

Use the following general method to test the activity of the purified deadenylase complex. Results for analysis of the Pop2p complex isolated from yeast are presented in Fig. 5.4.

Dilute substrate RNA to 200 fmol/μl concentration with deadenylation buffer.

Boil for 5 min; place on ice.

Assemble reactions on ice. Titrate the purified deadenylase complex over a wide range of concentrations (for instance, a 1000-fold range was tested in Fig. 5.4).

Purified deadenylase complex (0 to 100 ng of Pop2p-TAP complex)
200 fmol of substrate RNA (final concentration, 10 nM or 10 fmol/μl).
20 μl final reaction volume in deadenylation buffer.
Collect 5 μl from each sample. This is the zero time point sample. Transfer sample to ice-cold microfuge tube containing 5 μl of denaturing gel loading dye [95% (v/v) formamide, 18 mM EDTA, 0.025% (w/v) SDS, 0.025% (w/v) xylene cyanol, 0.025% (w/v) bromophenol blue].
Incubate at 30 °C in a waterbath.
Collect 5-μl samples at 15-, 30-, and 60-min time points. Place samples in ice-cold microfuge tubes with 5 μl of denaturing gel loading dye, which effectively terminates the reaction.
Place on ice or -20 °C until ready to load gel.
Samples can be stored at -80 °C.
Boil the samples for 5 min before loading onto gel.

2.8.3. Analyze products on a denaturing polyacrylamide gel

Prepare 12% (w/v) acrylamide gel with 29:1 acrylamide/bis-acrylamide ratio, 7 M urea and 1× TBE buffer with a size of 15 × 25 cm × 0.5 mm. Note that the gel percentage will vary depending on the size range of RNAs being analyzed.

Prerun gel for 30 min at 1100 volts.
Flush urea and unpolymerized acrylamide from gel wells with running buffer before loading samples.
Load samples into gel wells.

Electrophorese gel at 1100 volts for approximately 1 h. Stop current when the bromophenol blue dye is approximately 2 inches from the bottom of the gel.

Carefully transfer the gel to 3MM Whatman paper and wrap in plastic.

Expose gel to a PhosphorImager screen.

Visualize data with a PhosphorImager.

2.8.4. Interpret data

Analysis of the *in vitro* deadenylation data should demonstrate concentration- and time-dependent shortening of the substrate RNA by the deadenylase (Fig. 5.4A). Deadenylation should halt, or substantially pause, once the $3'$-adenosines are removed. Endonucleolytic cleavage products, with no evidence of progressive shortening, are telltale signs of *Exoribonucleolytic* contamination. No degradation should be observed in the control reactions, either mock or deadenylase buffer alone (Fig. 5.4B).

To detect enhancement of deadenylation by a regulator, the purified deadenylase must be carefully titrated so that its activity is rate limiting (for instance, see Fig. 5.4A, lanes 6, 7, and 8). For analysis of proteins that inhibit deadenylation, it will be desirable to use more enzyme, although not a vast excess, so that deadenylation is efficient in the absence the regulator (Fig. 5.4A, lanes 3 to 5). Very high amounts of enzyme will cause very rapid deadenylation, and in some cases, the enzyme will begin to degrade the body of the substrate RNA (Fig. 5.4A, lanes 3 to 5).

2.9. Purify recombinant regulator

Purify regulators, fused to an affinity tag such as GST, from *Escherichia coli* with standard methods (2007a). We purified PUF-GST fusion proteins for use in deadenylation assays (Goldstrohm et al., 2006). After binding to glutathione-agarose beads, wash the bead-bound regulator three times with high salt (500 mM NaCl) buffer and detergent (0.5% Nonidet P40). These stringent washes help remove contaminating ribonucleases. Next, wash the beads twice with deadenylation buffer to remove high salt and detergent. Finally, elute GST fusion protein in deadenylation buffer containing 10 mM glutathione. We elute in deadenylation buffer to match buffer conditions with downstream deadenylation and RNA-binding assays. Elution conditions should maintain solubility and activity of the regulator. For PUF proteins, like GST-Puf4p, we typically include up to 50% glycerol in the eluted stock to maintain protein solubility.

2.9.1. Analyze purified regulator

Measure the protein concentration of the purified regulator by Bradford assay (Bio-Rad) or similar methods. Assess the purity by Coomassie blue staining of protein separated by SDS-PAGE gel (Fig. 5.5A).

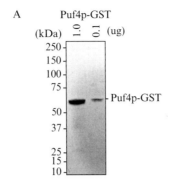
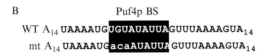
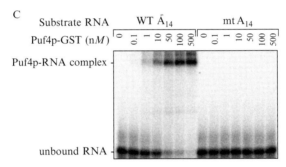

Figure 5.5 Measuring the RNA-binding affinity and specificity of purified RNA-binding protein, Puf4p. (A) Purified, recombinant yeast Puf4p, fused to the GST affinity tag, was purified from *E. coli* and analyzed by SDS-PAGE and Coomassie blue stain (1.0 and 0.1 μg of Puf4p-GST). (B) RNA substrates used in RNA binding and deadenylase assays. The wild-type synthetic substrate contains the Puf4p binding site from the yeast HO mRNA and 14 adenosines at its 3′-end (WT A_{14}). In the mutant RNA version (mt A_{14}), the UGU trinucleotide that is critical for Puf4p binding was changed to ACA. (C) The RNA-binding affinity and the specificity of the purified Puf4p-GST was determined by electrophoretic mobility shift assay. Increasing concentrations, indicated at the top of the gel, of Puf4p-GST were incubated with either the WT A_{14} or mt A_{14} radioactively labeled substrate RNAs. Puf4p-GST bound wild-type RNA with an approximate dissociation constraint of 10 ± 5 nM while no binding was detected with the mutant RNA (Hook *et al.*, 2007).

2.9.2. Test activity of purified regulator

Assess the RNA-binding activity of the regulator before proceeding to deadenylation assays. To measure RNA-binding activity and specificity, we perform electrophoretic mobility shift assays (EMSA) with radioactively labeled RNA substrates, both wild-type RNAs and versions with mutations in the regulator's binding site (Hellman and Fried, 2007). For Puf4p, we used two substrate RNAs; the first RNA had a wild-type Puf4p binding site from *HO* mRNA, and the second substrate contained mutations in that binding site (Fig. 5.5B) (Hook *et al.*, 2007). It is important that the EMSAs be performed in deadenylation reaction buffer with a range of regulator concentrations. Results from this experiment are used to determine the amount of regulator used in subsequent deadenylation reactions.

2.9.3. Electrophoretic mobility shift assay (EMSA)

Determine the RNA-binding activity of purified regulatory protein with the following protocol. As an example, EMSA analysis of Puf4p binding to either wild-type or mutant RNA substrates is shown in Fig. 5.5.

Prepare native polyacrylamide gel. Typically we use a 15 × 15 × 1-mm gel composed of 6% polyacrylamide with 29:1 acrylamide/bis-acrylamide and 0.5× TBE running buffer.
Boil RNA for 5 min; place on ice.
Assemble RNA-binding reaction.
200 fmol of radioactively labeled substrate RNA (wild-type or mutant).
Purified regulator protein (0 to 500 nM GST-Puf4p).
20 μl final volume in deadenylation buffer.
Incubate the reactions at 25 °C for 30 min.
During the binding reaction, prerun gel at 300 volts for 30 min at 4 °C with 0.5× TBE running buffer.
Add 4 μl of 5× EMSA loading buffer (10% Ficoll 400, 5% DMSO, 0.1% bromophenol blue).
Flush gel wells thoroughly with running buffer.
Load samples onto gel.
Electrophorese for 3 h at 300 volts and 4 °C.
Transfer gel to Whatmann 3MM paper, wrap with plastic, expose to PhosphorImager screen.
Visualize data with a PhosphorImager.

2.9.4. Interpret data

Determine apparent binding constants (K_d) from the RNA-binding data. Binding must be specific, as judged by a comparison of apparent K_d for mutant and wild-type RNAs. For instance, we observed high-affinity, specific binding of Puf4p-GST to wild-type RNA substrate, but not a mutant version (Fig. 5.5B and C). If nonspecific binding to the mutant target is observed, alter conditions to establish specificity while maintaining high affinity. Titration of competing nucleic acids, such as tRNA or synthetic ribopolymers, is a common practice for optimizing binding specificity (Hellman and Fried, 2007).

2.10. Confirm protein interactions

Recruitment of the deadenylase obviously is dependent on protein–protein interactions between the regulator and deadenylase complex (Fig. 5.1). For example, Puf4p binds directly to the Pop2p subunit of the Ccr4p-Pop2p deadenylase complex, as do other PUF proteins (Goldstrohm et al., 2006; Hook et al., 2007). It is essential that the reaction conditions used to assay deadenylation are compatible with those needed for the protein–protein contacts. These contacts can be assayed with purified regulator and deadenylase complex by means of assays such as coimmunoprecipitation or GST pull down (Goldstrohm et al., 2006; Hook et al., 2007).

2.11. Assay-regulated deadenylation *in vitro*

Measure the effect of a regulator on deadenylase activity by combining the regulator, a limiting amount of enzyme, and a substrate RNA with a high-affinity binding site for the regulator (Fig. 5.6). Regulation by recruitment depends on several factors, including the RNA-binding activity of the regulator, protein interactions between regulator and deadenylase, and the enzymatic activity of the deadenylase. Recruitment results in the acceleration of the deadenylation rate (Fig. 5.6).

To measure sequence-specific deadenylase recruitment by a regulator, both wild-type and mutant RNA substrates should be tested. The reaction rate in the presence of the regulator is compared with that of the deadenylase alone. Titration of the regulator may be necessary to observe maximum effect. Ideally, we use the minimal amount of regulator that binds all of the substrate RNA, as determined in the EMSA analysis. Note that excess regulator can potentially inhibit deadenylation, essentially squelching the recruitment of deadenylase. Finally, the order of addition of reaction components may affect the observations. We have found that preincubation of the regulator and deadenylase on ice, although not essential, can facilitate formation of the regulator–deadenylase complex. Subsequently, the substrate RNA is added, and the reactions are immediately shifted 30 °C to allow deadenylation to occur.

Control reactions should include substrate RNA incubated with deadenylase buffer alone and regulator alone (as controls for nuclease contamination of reaction components) and deadenylase alone (as a baseline level of activity). As mobility markers for the denaturing gel analysis, include labeled substrate RNA with and without a poly(A) tail (Fig. 5.6A, lanes 1 and 2).

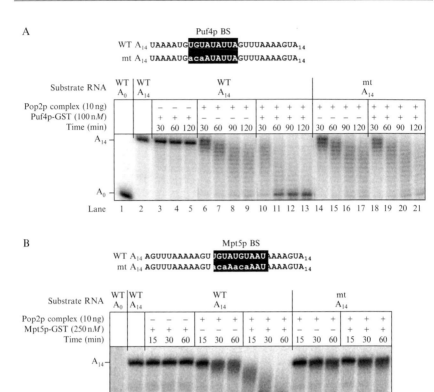

Figure 5.6 PUF proteins recruit and enhance deadenylation by the Pop2p deadenylase complex. (A) Puf4p enhances deadenylation. Deadenylation of wild-type RNA (WT A_{14}) containing a Puf4p binding site and, at the 3′-end, 14 adenosines, was analyzed in the presence of purified Pop2p deadenylase complex (10 ng) without (lanes 6 to 9) or with Puf4p-GST (100 nM, lanes 10 to 13) for the indicated times. As a negative control, Puf4p-GST was incubated alone with the substrate RNA (lanes 3 to 5). The Pop2p complex slowly deadenylated wild-type RNA, and when Puf4p was combined with the Pop2p complex, deadenylation was enhanced, resulting in the accumulation of fully deadenylated RNA (A_0). Mutant substrate RNA (mt A_{14}), which Puf4p could no longer bind, was deadenylated by the Pop2p complex at the same rate as wild-type RNA (lanes 14 to 17); however, addition of Puf4p had no effect on deadenylation of the mutant RNA (lanes 18 to 21). Sequences of the synthetic substrate RNAs are shown at the top. The migration of marker RNAs without (WT A_0) or with poly(A) tail (WT A_{14}) are indicated on the left side. (B) Mpt5p accelerates deadenylation. A similar, but separate, analysis with the PUF protein Mpt5p and RNA substrates (shown at the top) with either wild-type or mutant Mpt5p binding sites (BS) demonstrates regulated deadenylation with a limiting amount of Pop2p complex deadenylase activity.

2.11.1. Assemble deadenylation reactions

Dilute substrate RNA to 200 fmol/μl concentration with deadenylation buffer.
Boil for 5 min; place on ice.
Assemble reactions on ice. First add deadenylase buffer, then enzyme, then regulator protein.
Purified deadenylase complex (10 ng of Pop2p complex).
Purified regulator (100 nM of GST-Puf4p).
20 μl final reaction volume in deadenylation buffer.
Preincubate on ice up to 15 min.
Add 200 fmol of substrate RNA (final concentration, 10 nM or 10 fmol/μl).
Collect 5 μl from each sample. This is the zero time point sample. Transfer sample to ice cold microfuge tube containing 5 μl of denaturing gel loading dye.
Incubate at 30 °C.
Collect 5-μl samples at 15-, 30-, and 60-min time points. Place samples in ice-cold microfuge tubes and add 5 μl of denaturing gel loading dye.
Store samples on ice or −20 °C until ready to load gel.
Boil the samples for 5 min before loading onto gel.
Analyze samples by denaturing polyacrylamide gel as described earlier.

2.11.2. Interpret data

As an example of data analysis, consider Fig. 5.6A, in which we analyzed regulation of Ccr4p–Pop2p deadenylase complex by Puf4p. The RNA substrate in the control reaction, containing only the regulator (Puf4p), was unaffected (Fig. 5.6A, lanes 3 to 5). If degradation is observed, this is indicative of nuclease contamination (see "Troubleshooting" section). RNA incubated with the limiting amount of purified deadenylase was slowly deadenylated, with very little fully deadenylated product seen at the longest time point (Fig. 5.6A, lanes 6 to 9). If the regulator enhances deadenylation by recruiting the enzyme to the substrate, deadenylation of wild-type substrate will be accelerated when both are combined, resulting in the accumulation of fully deadenylated RNA over time (Fig. 5.6A, lanes 10 to 13). In some instances, the regulator may hyperactivate the deadenylase, causing degradation to proceed beyond the poly(A) tail and into the body of the RNA substrate (Hook et al., 2007). With deadenylase alone, the rate of poly(A) shortening of the mutant substrate should be identical to wild-type RNA (Fig. 5.6A, lanes 14 to 17); however, in the presence of both regulator and deadenylase, deadenylation should not be enhanced, because RNA-binding and recruitment cannot occur (Fig. 5.6A, lanes 18 to 21). Mutations that disrupt the interaction between the two proteins are ideal additional controls. An equivalent analysis with a different PUF protein, Mpt5p, is included for comparison (Fig 5.6B). Like Puf4p,

Mpt5p enhances deadenylation by the Ccr4p-Pop2p complex (Goldstrohm *et al.*, 2006). In the example shown in Fig. 5.6B, the activity of Pop2p complex added was lower than Fig. 5.6A, and, hence, very little deadenylation was observed in the absence of the PUF protein or its binding site (Fig. 5.6B, lanes 6 to 8 and 12 to 14). The addition of Mpt5p greatly accelerated deadenylation of the wild-type RNA substrate, with completely deadenylated RNA observed at 60 min (Fig. 5.6B, lanes 9 to 11), whereas the RNA substrate with mutated Mpt5p binding site was unaffected (Fig. 5.6B, lanes 15 to 17).

3. Troubleshooting

3.1. Enzyme concentration

Establishing the right amount of purified deadenylase complex is essential to the success of these assays. Too much enzyme leads to rapid degradation of the substrate, likely obscuring any enhancement. Too little enzyme, or enzyme with poor activity, obviously also prevents analysis. For this reason, proper storage of the enzyme is necessary. For short-term storage, we maintain small aliquots of enzyme in deadenylation buffer at $-20\,°C$. Long-term storage is done at $-80\,°C$. After a month or so, we prefer to prepare freshly purified deadenylase. Because the specific activity of purified enzyme can vary over time and between preparations, it is essential to titrate the deadenylase before assaying regulated deadenylation.

Comparison of Fig. 5.6 A and B, both of which lead to readily interpretable results, reveals the effect of differences in Pop2p complex concentration and activity. The higher activity used in Fig. 5.6A leads to more deadenylation in the absence of the PUF protein (compare Fig. 5.6A lanes 6 to 8 with 5.6B lanes 6 to 8).

3.2. Regulator concentration

The amount of RNA-binding regulator used in the *in vitro* deadenylation reactions is critical. Carefully determine the RNA-binding activity by titration, and use this information to guide the amount of regulator used in deadenylation assays. Still, it may be necessary to also titrate the regulator in deadenylation assays. If too little regulator is used, then enhanced deadenylation cannot occur because most of the RNA is not bound. Too much regulator inhibits sequence-specific deadenylation, either because most regulator-deadenylase complexes are not engaged with the RNA or because the RNA-binding protein nonspecifically binds the RNA, blocking deadenylase access.

3.3. Contaminating ribonuclease

These assays are extremely sensitive to contaminating nucleases, and all precautions should be taken at all steps to ensure RNase-free procedures, reagents, glassware, and plastic ware. Addition of RNase A inhibitors, such as RNAsin (Promega), may help diminish common sources of contamination, but we recommend avoiding other chemical RNase inhibitors or chelating agents that may inhibit deadenylases. The addition of competing nucleic acids may help diminish the effect of low levels of contaminants but may require subsequent adjustment of deadenylase concentration. In our experience, recombinant proteins from bacterial sources are a common source of contamination. To reduce this problem, increase the stringency of washing steps during the purification procedure, including with high salt (1 M NaCl), more detergent, additional washes with more volume, and longer duration.

3.4. Buffer conditions

High demands are placed on the buffer conditions used in these assays. The buffer must be compatible with deadenylase enzymatic activity, regulator-deadenylase protein interaction and, high affinity, specific RNA-binding by the regulator. Arriving at these conditions can require considerable effort. In some cases, compromising one or more activities may be necessary. Conditions that are suboptimal for one or more parameters may be required.

REFERENCES

Protein expression. (2007a). In "Current Protocols in Molecular Biology" Bob Kingston and Roger Brent (F. M. Ausable, R. Brent, R. E. Kingston, D. M. Moore, J. G. Seidman, J. A. Smith, and K. Struchl. eds.), Chapter 16, pp. 16.1–16.8. John Wiley and Sons, Inc, New York.

"Translational Control in Biology and Medicine." (2007b) Cold Spring Harbor Laboratory Press, Cold Spring Harbor, NY.

Astrom, J., Astrom, A., and Virtanen, A. (1991). In vitro deadenylation of mammalian mRNA by a HeLa cell 3′-exonuclease. *EMBO J.* **10**, 3067–3071.

Astrom, J., Astrom, A., and Virtanen, A. (1992). Properties of a HeLa cell 3′-exonuclease specific for degrading poly(A) tails of mammalian mRNA. *J. Biol. Chem.* **267**, 18154–18159.

Baggs, J. E., and Green, C. B. (2003). Nocturnin, a deadenylase in Xenopus laevis retina: A mechanism for posttranscriptional control of circadian-related mRNA. *Curr. Biol.* **13**, 189–198.

Balatsos, N. A., Nilsson, P., Mazza, C., Cusack, S., and Virtanen, A. (2006). Inhibition of mRNA deadenylation by the nuclear cap binding complex (CBC). *J. Biol. Chem.* **281**, 4517–4522.

Beilharz, T. H., and Preiss, T. (2007). Widespread use of poly(A) tail length control to accentuate expression of the yeast transcriptome. *RNA* **13**, 982–997.

Beyer, A., Hollunder, J., Nasheuer, H. P., and Wilhelm, T. (2004). Post-transcriptional expression regulation in the yeast Saccharomyces cerevisiae on a genomic scale. *Mol. Cell Proteomics* **3**, 1083–1092.

Bianchin, C., Mauxion, F., Sentis, S., Seraphin, B., and Corbo, L. (2005). Conservation of the deadenylase activity of proteins of the Caf1 family in human. *RNA* **11**, 487–494.

Boeck, R., Tarun, S., Jr., Rieger, M., Deardorff, J. A., Muller-Auer, S., and Sachs, A. B. (1996). The yeast Pan2 protein is required for poly(A)-binding protein-stimulated poly(A)-nuclease activity. *J. Biol. Chem.* **271**, 432–438.

Bramham, C. R., and Wells, D. G. (2007). Dendritic mRNA: Transport, translation and function. *Nat. Rev. Neurosci.* **8**, 776–789.

Brown, C. E., Tarun, S. Z., Jr., Boeck, R., and Sachs, A. B. (1996). PAN3 encodes a subunit of the Pab1p-dependent poly(A) nuclease in Saccharomyces cerevisiae. *Mol. Cell. Biol.* **16**, 5744–5753.

Chen, J., Chiang, Y. C., and Denis, C. L. (2002). CCR4, a $3'$-$5'$-poly(A) RNA and ssDNA exonuclease, is the catalytic component of the cytoplasmic deadenylase. *EMBO J.* **21**, 1414–1426.

Chen, K., and Rajewsky, N. (2007). The evolution of gene regulation by transcription factors and microRNAs. *Nat. Rev. Genet.* **8**, 93–103.

Chiba, Y., Johnson, M. A., Lidder, P., Vogel, J. T., van Erp, H., and Green, P. J. (2004). AtPARN is an essential poly(A) ribonuclease in *Arabidopsis*. *Gene* **328**, 95–102.

Chicoine, J., Benoit, P., Gamberi, C., Paliouras, M., Simonelig, M., and Lasko, P. (2007). Bicaudal-C recruits CCR4-NOT deadenylase to target mRNAs and regulates oogenesis, cytoskeletal organization, and its own expression. *Dev. Cell* **13**, 691–704.

Colegrove-Otero, L. J., Minshall, N., and Standart, N. (2005). RNA-binding proteins in early development. *Crit. Rev. Biochem. Mol. Biol.* **40**, 21–73.

Copeland, P. R., and Wormington, M. (2001). The mechanism and regulation of deadenylation: Identification and characterization of Xenopus PARN. *RNA* **7**, 875–886.

Daugeron, M. C., Mauxion, F., and Seraphin, B. (2001). The yeast POP2 gene encodes a nuclease involved in mRNA deadenylation. *Nucl. Acids Res.* **29**, 2448–2455.

Dehlin, E., Wormington, M., Korner, C. G., and Wahle, E. (2000). Cap-dependent deadenylation of mRNA. *EMBO J.* **19**, 1079–1086.

Denis, C. L., and Chen, J. (2003). The CCR4-NOT complex plays diverse roles in mRNA metabolism. *Progr. Nucl. Acid Res. Mol. Biol.* **73**, 221–250.

Dlakic, M. (2000). Functionally unrelated signalling proteins contain a fold similar to Mg^{2+}-dependent endonucleases. *Trends Biochem. Sci.* **25**, 272–273.

Dupressoir, A., Morel, A. P., Barbot, W., Loireau, M. P., Corbo, L., and Heidmann, T. (2001). Identification of four families of yCCR4- and Mg^{2+}-dependent endonuclease-related proteins in higher eukaryotes, and characterization of orthologs of yCCR4 with a conserved leucine-rich repeat essential for hCAF1/hPOP2 binding. *BMC Genomics* **2**, 9.

Ezzeddine, N., Chang, T. C., Zhu, W., Yamashita, A., Chen, C. Y., Zhong, Z., Yamashita, Y., Zheng, D., and Shyu, A. B. (2007). Human TOB, an anti-proliferative transcription factor, is a PABP-dependent positive regulator of cytoplasmic mRNA deadenylation. *Mol. Cell. Biol.* **27**, 219–227.

Ford, L. P., Watson, J., Keene, J. D., and Wilusz, J. (1999). ELAV proteins stabilize deadenylated intermediates in a novel *in vitro* mRNA deadenylation/degradation system. *Genes Dev.* **13**, 188–201.

Ford, L. P., and Wilusz, J. (1999). An *in vitro* system with HeLa cytoplasmic extracts that reproduces regulated mRNA stability. *Methods-A Companion to Methods in Enzymology* **17**, 21–17.

Fritz, D. T., Ford, L. P., and Wilusz, J. (2000). An *in vitro* assay to study regulated mRNA stability. *Sci. STKE* **2000**, PL1.

Gao, M., Fritz, D. T., Ford, L. P., and Wilusz, J. (2000). Interaction between a poly(A) specific ribonuclease and the $5'$-cap influences mRNA deadenylation rates *in vitro*. *Mol. Cell* **5**, 479–488.

Garneau, N. L., Wilusz, J., and Wilusz, C. J. (2007). The highways and byways of mRNA decay. *Nat. Rev. Mol. Cell. Biol.* **8,** 113–126.
Ghaemmaghami, S., Huh, W. K., Bower, K., Howson, R. W., Belle, A., Dephoure, N., O'Shea, E. K., and Weissman, J. S. (2003). Global analysis of protein expression in yeast. *Nature* **425,** 737–741.
Gilman, M. (2000). Ribonuclease protection assay. *In* "Current Protocols in Molecular Biology" (F.M Ausable, R Brent, R. E. Kingston, D. M. Moore, J. G. Seidman, J. A. Smith, and K. Struhl, eds.), pp. 4.7.1–4.7.8. John Wiley and Sons, Inc., New York.
Goldstrohm, A. C., Seay, D. J., Hook, B. A., and Wickens, M. (2006). PUF proteins bind Pop2p to regulate messenger RNAs. *Nat. Struct. Mol. Biol.* **13,** 533–539.
Goldstrohm, A. C., Seay, D. J., Hook, B. A., and Wickens, M. (2007). PUF protein-mediated deadenylation is catalyzed by Ccr4p. *J. Biol. Chem.* **282,** 109–114.
Goldstrohm, A. C., and Wickens, M. (2008). Multifunctional mRNA deadenylase complexes. *Nat. Rev. Mol. Cell Biol.* **9,** 337–344.
Gorgoni, B., and Gray, N. K. (2004). The roles of cytoplasmic poly(A)-binding proteins in regulating gene expression: a developmental perspective. *Brief Funct. Genomic Proteomic* **3,** 125–141.
Gray, N. K., and Wickens, M. (1998). Control of translation initiation in animals. *Annu. Rev. Cell Dev. Biol.* **14,** 399–458.
Grigull, J., Mnaimneh, S., Pootoolal, J., Robinson, M. D., and Hughes, T. R. (2004). Genome-wide analysis of mRNA stability using transcription inhibitors and microarrays reveals posttranscriptional control of ribosome biogenesis factors. *Mol. Cell Biol.* **24,** 5534–5547.
Guthrie, C., and Fink, G. R. (2002). Guide to yeast genetics and molecular and cell biology. *In* "Methods in Enzymology," Volume B p. 609. Academic Press, San Diego, CA.
Hellman, L. M., and Fried, M. G. (2007). Electrophoretic mobility shift assay (EMSA) for detecting protein-nucleic acid interactions. *Nat. Protoc.* **2,** 1849–1861.
Hook, B. A., Goldstrohm, A. C., Seay, D. J., and Wickens, M. (2007). Two yeast PUF proteins negatively regulate a single mRNA. *J. Biol. Chem.* **282,** 15430–15438.
Huh, W. K., Falvo, J. V., Gerke, L. C., Carroll, A. S., Howson, R. W., Weissman, J. S., and O'Shea, E. K. (2003). Global analysis of protein localization in budding yeast. *Nature* **425,** 686–691.
Jacobson, A., and Peltz, S. W. (1996). Interrelationships of the pathways of mRNA decay and translation in eukaryotic cells. *Annu. Rev. Biochem.* **65,** 693–739.
Jeske, M., Meyer, S., Temme, C., Freudenreich, D., and Wahle, E. (2006). Rapid ATP-dependent deadenylation of nanos mRNA in a cell-free system from *Drosophila* embryos. *J. Biol. Chem.* **281,** 25124–25133.
Kadyrova, L. Y., Habara, Y., Lee, T. H., and Wharton, R. P. (2007). Translational control of maternal cyclin B mRNA by nanos in the *Drosophila* germline. *Development* **134,** 1519–1527.
Keene, J. D. (2007). RNA regulons: Coordination of post-transcriptional events. *Nat. Rev. Genet.* **8,** 533–543.
Khabar, K. S. (2007). Rapid transit in the immune cells: The role of mRNA turnover regulation. *J. Leukoc. Biol.* **81,** 1335–1344.
Khabar, K. S., and Young, H. A. (2007). Post-transcriptional control of the interferon system. *Biochimie.* **89,** 761–769.
Kim, J. H., and Richter, J. D. (2006). Opposing polymerase-deadenylase activities regulate cytoplasmic polyadenylation. *Mol. Cell* **24,** 173–183.
Kimble, J., and Crittenden, S. L. (2007). Controls of germline stem cells, entry into meiosis, and the sperm/oocyte decision in Caenorhabditis elegans. *Annu. Rev. Cell. Dev. Biol.* **23,** 405–433.

Korner, C. G., and Wahle, E. (1997). Poly(A) tail shortening by a mammalian poly(A)-specific 3′-exoribonuclease. *J. Biol. Chem.* **272,** 10448–10456.

Korner, C. G., Wormington, M., Muckenthaler, M., Schneider, S., Dehlin, E., and Wahle, E. (1998). The deadenylating nuclease (DAN) is involved in poly(A) tail removal during the meiotic maturation of Xenopus oocytes. *EMBO J.* **17,** 5427–5437.

Kreig, P. A., and Johnson, A. D. (1996). In vitro synthesis of mRNA. *In* "A Laboratory Guide to RNA: Isolation, Analysis, and Synthesis." (P. A. Kreig, ed.), pp. 141–153. Wiley-Liss Inc., New York, NY.

Kuhn, U., and Wahle, E. (2004). Structure and function of poly(A) binding proteins. *Biochim. Biophys. Acta* **1678,** 67–84.

Lackner, D. H., Beilharz, T. H., Marguerat, S., Mata, J., Watt, S., Schubert, F., Preiss, T., and Bahler, J. (2007). A network of multiple regulatory layers shapes gene expression in fission yeast. *Mol. Cell* **26,** 145–155.

Lander, E. S., Linton, L. M., Birren, B., Nusbaum, C., Zody, M. C., Baldwin, J., Devon, K., Dewar, K., Doyle, M., FitzHugh, W., Funke, R., Gage, D., Harris, K., *et al.* (2001). Initial sequencing and analysis of the human genome. *Nature* **409,** 860–921.

Lowell, J. E., Rudner, D. Z., and Sachs, A. B. (1992). 3′ UTR-dependent deadenylation by the yeast poly(A) nuclease. *Genes Dev.* **6,** 2088–2099.

Lykke-Andersen, J., and Wagner, E. (2005). Recruitment and activation of mRNA decay enzymes by two ARE-mediated decay activation domains in the proteins TTP and BRF-1. *Genes Dev.* **19,** 351–361.

Martin, G., and Keller, W. (1998). Tailing and 3′-end labeling of RNA with yeast poly(A) polymerase and various nucleotides. *RNA* **4,** 226–230.

Martinez, J., Ren, Y. G., Nilsson, P., Ehrenberg, M., and Virtanen, A. (2001). The mRNA cap structure stimulates rate of poly(A) removal and amplifies processivity of degradation. *J. Biol. Chem.* **276,** 27923–27929.

Martinez, J., Ren, Y. G., Thuresson, A. C., Hellman, U., Astrom, J., and Virtanen, A. (2000). A 54-kDa fragment of the Poly(A)-specific ribonuclease is an oligomeric, processive, and cap-interacting Poly(A)-specific 3′-exonuclease. *J. Biol. Chem.* **275,** 24222–24230.

Meijer, H. A., Bushell, M., Hill, K., Gant, T. W., Willis, A. E., Jones, P., and de Moor, C. H. (2007). A novel method for poly(A) fractionation reveals a large population of mRNAs with a short poly(A) tail in mammalian cells. *Nucl. Acids Res.* **35,** e132.

Milone, J., Wilusz, J., and Bellofatto, V. (2004). Characterization of deadenylation in trypanosome extracts and its inhibition by poly(A)-binding protein Pab1p. *RNA* **10,** 448–457.

Molin, L., and Puisieux, A. (2005). C. elegans homologue of the Caf1 gene, which encodes a subunit of the CCR4-NOT complex, is essential for embryonic and larval development and for meiotic progression. *Gene* **358,** 73–81.

Moraes, K. C., Wilusz, C. J., and Wilusz, J. (2006). CUG-BP binds to RNA substrates and recruits PARN deadenylase. *RNA* **12,** 1084–1091.

Morel, A. P., Sentis, S., Bianchin, C., Le Romancer, M., Jonard, L., Rostan, M. C., Rimokh, R., and Corbo, L. (2003). BTG2 antiproliferative protein interacts with the human CCR4 complex existing *in vivo* in three cell-cycle-regulated forms. *J. Cell Sci.* **116,** 2929–2936.

Morita, M., Suzuki, T., Nakamura, T., Yokoyama, K., Miyasaka, T., and Yamamoto, T. (2007). Depletion of mammalian CCR4b deadenylase triggers elevation of the p27Kip1 mRNA level and impairs cell growth. *Mol. Cell Biol.* **27,** 4980–4990.

Opyrchal, M., Anderson, J. R., Sokoloski, K. J., Wilusz, C. J., and Wilusz, J. (2005). A cell-free mRNA stability assay reveals conservation of the enzymes and mechanisms of mRNA decay between mosquito and mammalian cell lines. *Insect Biochem. Mol. Biol.* **35,** 1321–1334.

Parker, R., and Sheth, U. (2007). P bodies and the control of mRNA translation and degradation. *Mol. Cell* **25,** 635–646.

Parker, R., and Song, H. (2004). The enzymes and control of eukaryotic mRNA turnover. *Nat. Struct. Mol. Biol.* **11,** 121–127.

Puig, O., Caspary, F., Rigaut, G., Rutz, B., Bouveret, E., Bragado-Nilsson, E., Wilm, M., and Seraphin, B. (2001). The tandem affinity purification (TAP) method: A general procedure of protein complex purification. *Methods* **24,** 218–229.

Ren, Y. G., Kirsebom, L. A., and Virtanen, A. (2004). Coordination of divalent metal ions in the active site of poly(A)-specific ribonuclease. *J. Biol. Chem.* **279,** 48702–48706.

Ren, Y. G., Martinez, J., and Virtanen, A. (2002). Identification of the active site of poly(A)-specific ribonuclease by site-directed mutagenesis and Fe(2+)-mediated cleavage. *J. Biol. Chem.* **277,** 5982–5987.

Reverdatto, S. V., Dutko, J. A., Chekanova, J. A., Hamilton, D. A., and Belostotsky, D. A. (2004). mRNA deadenylation by PARN is essential for embryogenesis in higher plants. *RNA* **10,** 1200–1214.

Rigaut, G., Shevchenko, A., Rutz, B., Wilm, M., Mann, M., and Seraphin, B. (1999). A generic protein purification method for protein complex characterization and proteome exploration. *Nat. Biotechnol.* **17,** 1030–1032.

Sachs, A. (2000). Physical and functional interactions between the mRNA cap structure and the poly(A) tail. *In* "Translational Control of Gene Expression" (N. Sonenberg, J. Hershey, and M. B. Mathews, eds.), pp. 447–465. Cold Spring Harbor Laboratory Press, Cold Spring Harbor, NY.

Sachs, A. B., and Deardorff, J. A. (1992). Translation initiation requires the PAB-dependent poly(A) ribonuclease in yeast. *Cell* **70,** 961–973.

Schwartz, D. C., and Parker, R. (1999). Mutations in translation initiation factors lead to increased rates of deadenylation and decapping of mRNAs in Saccharomyces cerevisiae. *Mol. Cell Biol.* **19,** 5247–5256.

Schwartz, D. C., and Parker, R. (2000). mRNA decapping in yeast requires dissociation of the cap binding protein, eukaryotic translation initiation factor 4E. *Mol. Cell Biol.* **20,** 7933–7942.

Simon, E., and Seraphin, B. (2007). A specific role for the C-terminal region of the Poly(A)-binding protein in mRNA decay. *Nucl. Acids Res.* **35,** 6017–6028.

Temme, C., Zaessinger, S., Meyer, S., Simonelig, M., and Wahle, E. (2004). A complex containing the CCR4 and CAF1 proteins is involved in mRNA deadenylation in *Drosophila*. *EMBO J.* **23,** 2862–2871.

Thore, S., Mauxion, F., Seraphin, B., and Suck, D. (2003). X-ray structure and activity of the yeast Pop2 protein: A nuclease subunit of the mRNA deadenylase complex. *EMBO Rep.* **4,** 1150–1155.

Tucker, M., Staples, R. R., Valencia-Sanchez, M. A., Muhlrad, D., and Parker, R. (2002). Ccr4p is the catalytic subunit of a Ccr4p/Pop2p/Notp mRNA deadenylase complex in Saccharomyces cerevisiae. *EMBO J.* **21,** 1427–1436.

Tucker, M., Valencia-Sanchez, M. A., Staples, R. R., Chen, J., Denis, C. L., and Parker, R. (2001). The transcription factor associated Ccr4 and Caf1 proteins are components of the major cytoplasmic mRNA deadenylase in Saccharomyces cerevisiae. *Cell* **104,** 377–386.

Uchida, N., Hoshino, S., and Katada, T. (2004). Identification of a human cytoplasmic poly(A) nuclease complex stimulated by poly(A)-binding protein. *J. Biol. Chem.* **279,** 1383–1391.

Vasudevan, S., Seli, E., and Steitz, J. A. (2006). Metazoan oocyte and early embryo development program: A progression through translation regulatory cascades. *Genes Dev.* **20,** 138–146.

Viswanathan, P., Chen, J., Chiang, Y. C., and Denis, C. L. (2003). Identification of multiple RNA features that influence CCR4 deadenylation activity. *J. Biol. Chem.* **278,** 14949–14955.

Viswanathan, P., Ohn, T., Chiang, Y. C., Chen, J., and Denis, C. L. (2004). Mouse CAF1 can function as a processive deadenylase/3′ to 5′-exonuclease *in vitro* but in yeast the deadenylase function of CAF1 is not required for mRNA poly(A) removal. *J. Biol. Chem.* **279,** 23988–23995.

Wagner, E., Clement, S. L., and Lykke-Andersen, J. (2007). An unconventional human Ccr4-Caf1 deadenylase complex in nuclear Cajal bodies. *Mol. Cell Biol.* **27,** 1686–1695.

Wickens, M., Bernstein, D. S., Kimble, J., and Parker, R. (2002). A PUF family portrait: 3′UTR regulation as a way of life. *Trends Genet.* **18,** 150–157.

Yang, E., van Nimwegen, E., Zavolan, M., Rajewsky, N., Schroeder, M., Magnasco, M., and Darnell, J. E., Jr. (2003). Decay rates of human mRNAs: Correlation with functional characteristics and sequence attributes. *Genome Res.* **13,** 1863–1872.

Zhao, J., Hyman, L., and Moore, C. (1999). Formation of mRNA 3′-ends in eukaryotes: Mechanism, regulation, and interrelationships with other steps in mRNA synthesis. *Microbiol. Mol. Biol. Rev.* **63,** 405–445.

Zuo, Y., and Deutscher, M. P. (2001). Exoribonuclease superfamilies: Structural analysis and phylogenetic distribution. *Nucl. Acids Res.* **29,** 1017–1026.

CHAPTER SIX

CELL-FREE DEADENYLATION ASSAYS WITH *DROSOPHILA* EMBRYO EXTRACTS

Mandy Jeske *and* Elmar Wahle

Contents

1. Introduction	107
2. Preparation of *Drosophila* Embryo Extracts	110
3. Preparation of Substrate RNA	112
4. Deadenylation Assay with *Drosophila* Embryo Extracts	113
5. Characterization of Sequence-Dependent Deadenylation in *Drosophila* Embryo Extracts	115
Acknowledgments	116
References	116

Abstract

Deadenylation initiates degradation of most mRNAs in eukaryotes. Regulated deadenylation of an mRNA plays an important role in translation control as well, especially during animal oogenesis and early embryonic development. To investigate the mechanism of sequence-dependent deadenylation, we established an *in vitro* system derived from 0- to 2-h-old *Drosophila* embryos. These extracts faithfully reproduce several aspects of the regulation of *nanos* mRNA: They display translation repression and deadenylation both mediated by the same sequences within the *nanos* 3' UTR. Here, we describe detailed protocols for preparing *Drosophila* embryo extracts, and their use in deadenylation assays exemplified with exogenous RNA substrates containing the *nanos* 3' UTR.

1. INTRODUCTION

Poly(A) tails are almost universal modifications of eukaryotic mRNAs. They are added during posttranscriptional processing of the pre-mRNA in the cell nucleus in a reaction consisting of two steps, endonucleolytic cleavage of the primary transcript and addition of a poly(A) tail (Wahle and

Institute of Biochemistry and Biotechnology, Martin-Luther-University Halle-Wittenberg, Halle, Germany

Rüegsegger, 1999). Nuclear polyadenylation normally generates poly(A) tails of a more or less uniform but species-specific length, approximately 80 nucleotides in yeast and 250 nucleotides in mammalian cells. Poly(A) tails serve multiple important functions in the cell, including enhancing translation initiation (Gorgoni and Gray, 2004).

After export to the cytoplasm, poly(A) tails are gradually shortened. This deadenylation reaction is the first step of decay for most mRNAs (Meyer et al., 2004; Parker and Song, 2004). Deadenylation is considered the rate-limiting step in mRNA turnover for three reasons: First, in general, subsequent steps do not occur until the poly(A) tail has been shortened beyond some critical limit. Second, although deadenylation is a relatively slow process even for unstable mRNAs, subsequent steps are usually fast and often proceed without detectable intermediates. Third, unstable mRNAs are deadenylated rapidly, whereas stable mRNAs are deadenylated slowly. Rapid deadenylation and degradation are often mediated by sequences within the 3′ UTR of an mRNA, for example the so-called AU-rich elements (AREs) found in many unstable mammalian mRNAs (Barreau et al., 2005). Several ARE-binding proteins have been identified that mediate both accelerated poly(A) shortening and subsequent RNA degradation (Barreau et al., 2005). MicroRNAs can also promote deadenylation of mRNAs by targeting their 3′ UTRs (Giraldez et al., 2006; Jing et al., 2005; Wakiyama et al., 2007; Wu et al., 2006).

Control of mRNA degradation is biologically important: The rate of mRNA decay, together with the rates of transcription and RNA processing, determines the steady-state level of the mRNA in the cytoplasm which, in turn, is the main determinant of the rate of protein synthesis. Rapid turnover of an mRNA species not only decreases its steady-state level but also accelerates the adjustment to a new steady-state level on either upregulation or downregulation of the rate of synthesis (Ross, 1995).

Oocytes and early embryos are transcriptionally quiescent; thus, oocyte maturation and the first stages of development depend on maternally provided mRNA, and regulation of protein synthesis must rely on posttranscriptional control until the onset of zygotic transcription. Many maternal transcripts remain silent until their translation is required at specific stages of early development. In this biologic context, controlled changes in poly(A) tail length are used to regulate translation: Poly(A) tail shortening causes translational repression, and poly(A) tail elongation plays a role in translational activation (Gorgoni and Gray, 2004; Richter, 2000). At these early developmental stages, deadenylation does not necessarily result in immediate degradation of the RNA.

Posttranscriptional regulation has been studied extensively in early Drosophila embryos (Hentze et al., 2007; Johnstone and Lasko, 2001; Thompson et al., 2007). One well-understood example is the localization-dependent regulation of nanos mRNA, which is crucial for proper development of the posterior body part of the *Drosophila* embryo (Gavis and Lehmann, 1992, 1994). During oogenesis, approximately 5% of the *nanos* mRNA population is localized to the posterior pole of the oocyte, whereas the rest of the RNA is uniformly distributed (Bergsten and Gavis, 1999; Wang and Lehmann, 1991). After fertilization, only the localized *nanos* mRNA is translationally activated, whereas most of the RNA within the early embryo stays dormant (Gavis and Lehmann, 1994). This spatially restricted translation leads to a Nanos protein gradient emanating from the posterior pole. The function of Nanos protein is consequently restricted to the posterior part of the embryo (Gavis and Lehmann, 1992). Repression of translation of nonlocalized *nanos* mRNA is mediated by the protein Smaug, which binds to 3′ UTR sequences referred to as Smaug recognition elements (SRE) (Dahanukar *et al.*, 1999; Smibert *et al.*, 1996, 1999). Bulk *nanos* mRNA has a very short poly(A) tail (Sallés *et al.*, 1994). The nonlocalized *nanos* mRNA, together with other maternal mRNAs, is gradually degraded during the first 3 h of embryonic development, and this also depends on the SREs (Bashirullah *et al.*, 1999; Smibert *et al.*, 1996; Tadros *et al.*, 2007). Smaug has been shown to interact with the CCR4-NOT deadenylase and to be required for CCR4-dependent deadenylation of *nanos in vivo* (Semotok *et al.*, 2005; Zaessinger *et al.*, 2006).

Cell-free systems derived from various sources have been used to study distinct aspects of mRNA turnover. Accelerated deadenylation mediated by specific RNA sequences could be observed in extracts from cultured mammalian cells, trypanosomes, and *Xenopus* eggs (Brewer and Ross, 1988; Ford *et al.*, 1999; Legagneux *et al.*, 1995; Milone *et al.*, 2002; Voeltz *et al.*, 2001; Wakiyama *et al.*, 2007). Cell-free systems are essential for detailed investigations of the mechanism of sequence-dependent mRNA deadenylation. Most of all, they serve as the starting point for attempts to reconstitute the reaction from purified components.

We have established a cell-free system from *Drosophila* embryos that reproduces the SRE-dependent translational repression of *nanos* mRNA and promotes rapid deadenylation of SRE-containing reporter RNAs (Jeske *et al.*, 2006) (Fig. 6.1). Therefore, the cell-free system described here should be an excellent tool to study not only the mechanism of sequence-dependent deadenylation but also the interplay of deadenylation and translation regulation. Deadenylation of exogenous RNA substrates in this extract is not followed by degradation and, therefore, can be analyzed independently of subsequent mRNA decay steps. Furthermore, the embryo extract exhibits very little unspecific background degradation activity, which facilitates data interpretation and quantification. Deadenylation in *Drosophila* embryo extract is sequence-dependent, because control RNAs that carry a single point mutation in each SRE are resistant to deadenylase activity. The sequence-specific deadenylation is robust and highly reproducible from batch to batch.

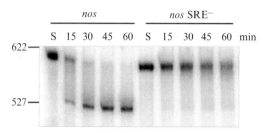

Figure 6.1 Sequence-specific deadenylation of *nanos* mRNA. The [^{32}P]-labeled RNA substrates (10 nM) were incubated in deadenylation reactions with 20% (v/v) *Drosophila* embryo extract. The *nos* RNA carries a segment of nucleotides 7 to 161 from the *nanos* 3′ UTR containing the two SREs. The *nos* SRE⁻ RNA differs from the *nos* RNA by two point mutations changing the CUGGC loop sequence to CUCGC in each of the two SREs. At the times indicated, the RNA was isolated and separated in a denaturing 5% polyacrylamide gel. S, Unreacted substrate RNA. The migration of DNA markers is indicated on the left. (For a more detailed description of the RNA substrates, refer to Jeske *et al.* [2006].)

2. Preparation of *Drosophila* Embryo Extracts

Drosophila embryo extract is prepared from a mixed population of 0- to 2-h-old embryos. Here, we describe a method for extract preparation from approximately 50 mg to 3-g embryos per 2-h collection. For large-scale extract preparation (>5-g embryos per 2-h collection) use the equipment described in Moritz (2000) or Gebauer and Hentze (2007).

For propagation of flies, refer to standard procedures (e.g., Sission, 2000). We maintain flies at 25 °C in 3 to 4 egg-laying cages with a volume of 14 l each. Follow published protocols to prepare embryo collection trays and yeast paste (e.g., Gebauer and Hentze, 2007; Moritz, 2000). Prewarm the yeast-smeared collection trays for approximately 15 min at 37 °C before use. Change the trays in the fly cages every 2 h. Prepare extract immediately after each 2-h embryo collection to obtain both high translation and deadenylation activity. It is possible to collect embryos every 2 h, keep them at 4 °C, and prepare extracts once at the end of the day, but activity is generally lower and less reproducible.

1. Rinse embryos off all trays with a stream of cold tapwater with a soft paint brush and collect them into a set of three stapled sieves differing in their mesh size (Retsch, Haan, Germany). The upper sieve (710 μm mesh) retains flies, the middle sieve (355 μm) retains fly body parts, and the lower sieve (125 μm) retains embryos, but yeast can pass through them all. To handle a small amount of embryos (<200 mg), use a self-made sieve. For this purpose, cut a 50-ml Falcon tube in half and discard

the bottom. Also cut the cap of the Falcon tube, discard the flat top, and retain the threaded ring. Screw the ring on the top half of the tube over a piece of Mira cloth (28-μm pore size; Calbiochem) that thus covers one opening of the tube. When this self-made sieve is used, remove flies and body parts from the collected embryos with forceps.

2. Remove the chorion of the embryo by bleaching at room temperature: Rinse embryos out of the last sieve into a beaker with distilled water and add 1 volume of sodium hypochlorite solution (Roth; stock solution contains 12% chlorine). Swirl the beaker occasionally. If the self-made collection sieve is being used, submerge embryos in this sieve in a beaker containing 1:2 diluted sodium hypochlorite solution. Dechorionation is a critical step during extract preparation, because insufficient dechorionation results in incomplete lysis of embryos and excessive bleaching harms the embryos. The efficiency of a sodium hypochlorite solution depends on its age and storage conditions. Therefore, follow the dechorionation process by eye. Removal of the chorion is complete when embryos have lost their dorsal appendages and tend to stick to the glass and to each other. Expect dechorionation to take approximately 1 min.

3. Pour embryos back into the sieve and wash extensively with cold tap-water until they no longer smell of chlorine. It is important to remove sodium hypochlorite completely, because it can be detrimental to the activity of extract proteins.

4. Dry the embryos by blotting with paper towels from the bottom of the sieve. Use a spatula to scrape the embryos into a chilled dounce homogenizer (Wheaton) of known weight. Weigh again to determine the mass of embryos. Use a 1-ml, 2-ml, or 7-ml dounce homogenizer for the lysis of up to 500-mg, 1-g, or 3.5-g embryos, respectively.

5. Add 1-ml of ice-cold lysis buffer consisting of 30 mM HEPES-KOH, pH 7.4, 100 mM potassium acetate, 2 mM magnesium acetate, 5 mM dithiothreitol (DTT), and 1 mg/ml Pefabloc SC (Roche) for 1 g of damp embryos (Tuschl *et al.*, 1999). Manually homogenize the embryos on ice by moving the tight-fitting pestle up and down until the pestle can be easily moved within the glass tube (approximately 5 to 10 strokes).

6. Transfer the lysate to 1.5-ml Eppendorf tubes and spin in a microcentrifuge for 20 min at 14,500g and 4 °C. Sedimentation results in a fatty upper layer, a pellet containing nuclei and cell debris, and the soluble cytoplasmic phase in between.

7. Remove the fatty upper layer by touching and lifting it off with a cut pipet tip (layer sticks to tip). Separate the cytoplasmic phase from the pellet and flash freeze it in aliquots in liquid nitrogen. Aliquot size should be such that refreezing after use is not necessary. On storage at -80 °C, translation and deadenylation activity of the extract is stable for at least 3 and 6 months, respectively. However, we recommend storing the extracts in liquid nitrogen, because stability of both translation and

deadenylation activity can be extended to more than 1 year. The extract protein concentration should be approximately 25 to 35 mg/ml (Bio Rad assay).

Even if translational regulation is not the subject of interest, it may be useful to test the extract for translational activity, because translation activity is a very sensitive indicator of general extract quality: Every extract batch active in translation had a good deadenylation activity, but not every batch of extract active in deadenylation showed translation. We also observe a tenfold decrease of translational activity after every thaw–freeze cycle, whereas deadenylation is not strongly affected. Finally, translation activity is more sensitive during prolonged storage. In general, the translation activity of the embryo extract should be at least as high as that of micrococcal nuclease-treated rabbit reticulocyte lysate (Promega).

3. Preparation of Substrate RNA

To study deadenylation, *in vitro*-synthesized radiolabeled mRNA substrates are used. RNA substrates should be short enough to allow good separation of deadenylated product from polyadenylated substrate within a polyacrylamide gel.

Perform runoff transcription of a linearized template DNA under standard conditions with the appropriate bacteriophage RNA polymerase. Use of 10 μCi of [α-^{32}P]-UTP in a 25-μl reaction in the presence of 1 mM nonlabeled UTP results in RNA of suitable specific radioactivity. To obtain higher specific activities and more efficient incorporation of the label, in particular for the synthesis of shorter RNAs, one can reduce the concentration of nonlabeled UTP to 0.1 mM. After incubation, remove the plasmid DNA by a 20-min digestion with 1.5 units of RNase-free DNase I (Roche). Phenol-extract, ethanol-precipitate, and dissolve the RNA in 30 to 50 μl diethylpyrocarbonate-treated, double distilled water. Gel purification of the RNA is necessary only if the products of the runoff transcription are not uniform. Nonincorporated nucleotides are usually removed if ethanol precipitation is carried out from 2.5 M ammonium acetate. The absence of free nucleotides can be verified easily by thin-layer chromatography on polyethylenimine cellulose plates (Jeske *et al.*, 2006). If necessary, the precipitation can be repeated, or the RNA can be purified by a Sephadex G50 spin column.

To calculate the concentration of the RNA, remove a small aliquot of the reaction mixture at any time during the transcription reaction and count in a scintillation counter. Calculate the specific radioactivity of UTP from the known quantity of the nonlabeled nucleotide and the amount of radioactivity. Determine the radioactivity of an aliquot of the purified RNA by scintillation counting. Calculate the amount of RNA from the incorporated radioactivity, the specific activity of the UTP, and the RNA's nucleotide composition.

For convenience, the mRNA's poly(A) tail should be encoded within the plasmid used for runoff transcription. The vector we use, originally constructed by Fátima Gebauer, is linearized with *Hind*III, which leads to RNAs containing 72 adenylate residues and 3 nonadenylate nucleotides (GCU) downstream of the poly(A) tract (Jeske *et al.*, 2006). The three non-A nucleotides do not prevent deadenylation, whereas 11 or more additional nonadenylate nucleotides inhibit the reaction (data not shown). To avoid nonadenylate residues at the 3′-end, one can also design a vector that is linearized with an enzyme cutting outside its recognition sequence (e.g., *Bbs*I): If the recognition sequence is placed at the appropriate distance from the end of the poly(A) tail, cleavage will be within the homopolymeric sequence (Kerwitz *et al.*, 2003). Alternatively, poly(A) tails can be added after transcription with the help of commercial poly(A) polymerase. Although this takes additional time, it has the advantage that tails of any length can be added by variations of enzyme concentration or incubation time. At least with some poly(A) polymerases, elongation products are more homogeneous if the starting RNA already has a few A's at the 3′-end. Nevertheless, gel purification will usually be necessary to generate a deadenylation substrate with a uniform tail length.

The rate of deadenylation of the *nanos* RNA is independent of a 5′-cap. Also, capped and uncapped RNAs do not differ in their stabilities during incubation in extract (Jeske *et al.*, 2006). Therefore, we omit incorporation of cap analogs during *in vitro* transcription. However, a cap can easily be incorporated, if desired, by the addition of 7 mM m^{7}GpppG (New England Biolabs or KEDAR, Warsaw, Poland) to the reaction.

4. Deadenylation Assay with *Drosophila* Embryo Extracts

To study the timecourse of sequence-dependent deadenylation, set up a reaction with a volume of 10 μl for each intended time point.

1. Assemble the deadenylation reaction on ice: 5 to 20 nM labeled poly(A)$^+$ RNA (concentration refers to RNA molecules, not mononucleotides), 20 to 40% (v/v) extract, 16 mM HEPES-KOH, pH 7.4, 50 mM potassium acetate, 2.5 mM magnesium acetate, 100 μM spermidine, 250 μg/ml tRNA from yeast (Roche), 80 μg/ml creatine kinase (Roche), 20 mM creatine phosphate (Sigma), and 800 μM ATP. (These

concentrations refer to the entire reaction volume and ignore components introduced with the extract.) Mix the components gently. If you make up your own ATP stock solution, adjust the pH to ~7 with NaOH and determine the concentration from the extinction coefficient ($\varepsilon = 15{,}400$ at 259 nm). Store creatine kinase as a 10 mg/ml stock solution in 50% (v/v) glycerol and 20 mM HEPES-KOH, pH 7.4. The enzyme is active for at least 1 year. Creatine phosphate is stored as a 1 M stock solution. Use of a premix containing HEPES, potassium acetate, magnesium acetate, spermidine, and tRNA facilitates assembly of the reaction. All these reagents should be stored at $-20\,^\circ\text{C}$. Spermidine can precipitate RNA, especially at low salt concentration and low temperature. Be sure not to expose RNA (tRNA or substrate) to higher concentrations of spermidine in the absence of salt. We do not find it necessary to add RNase inhibitors to the deadenylation reaction. No reaction takes place as long as the mixture is left on ice.

2. Incubate the reaction mixture at $25\,^\circ\text{C}$. For every time point, transfer 10 to 190 μl stop solution (25 mM EDTA, pH 8).
3. Add 200 μl phenol/chloroform (1:1) and vortex thoroughly. Separate the aqueous phase from the organic phase by spinning for 30 min at maximum speed at room temperature in a microcentrifuge.
4. Transfer 120 μl of the upper aqueous phase into a fresh 1.5-ml tube. Be sure not to touch the protein interphase with the pipette tip. Precipitate the RNA by addition of 40 μl of 10 M ammonium acetate supplemented with 0.2 μg/μl glycogen (precipitation carrier; Roche) and 400 μl ethanol. Invert the tube several times and sediment the precipitated RNA by spinning for 30 min at maximum speed at room temperature in a microcentrifuge.
5. Remove the entire supernatant from the pellet and let the pellet air-dry completely. Dissolve the RNA in 4 μl of formamide loading buffer (deionized formamide supplemented with 0.1% [w/v] bromphenol blue, 0.1% [w/v] xylene cyanol, 10 mM EDTA pH 8) by vortexing and denature the sample for 3 min at $90\,^\circ\text{C}$. Prepare an aliquot of unreacted substrate RNA for comparison.
6. Separate the RNA in an appropriate polyacrylamide gel containing 8.3 M urea according to standard protocols (Sambrook and Russell, 2001). For RNAs migrating more slowly than the dyes, it is often not necessary to run the gel at high temperature. Long gels can, therefore, conveniently be run overnight at room temperature. Analyze the gel by phosphoimaging. A representative example of *nanos* RNA deadenylation is shown in Fig. 6.1.

To confirm that shortening of the RNA is caused by deadenylation, perform a digestion of the reaction product with RNase H and oligo(dT) according to Jeske et al. (2006).

5. CHARACTERIZATION OF SEQUENCE-DEPENDENT DEADENYLATION IN *DROSOPHILA* EMBRYO EXTRACTS

Deadenylation of the *nanos* mRNA strongly depends on an ATP-regenerating system consisting of ATP, creatine kinase, and creatine phosphate (Jeske *et al.*, 2006). $3'$-dATP can substitute for ATP. This can be technically useful, because the analog acts as a chain terminator in polyadenylation; if the pattern of deadenylation is not changed by the substitution, any influence of polyadenylation can be excluded. Divalent cations are essential to deadenylation; addition of 6 mM EDTA completely inhibits the activity (data not shown). This observation is consistent with the fact that the activity of all known deadenylases depends on Mg^{2+} ions (Meyer *et al.*, 2004). A second reason for the Mg^{2+}-dependence of the reaction is the requirement for an ATP-regenerating system; ATP-dependent enzymes require Mg^{2+} ions. Hence, EDTA solution is used as stopping reagent.

The deadenylation system is sensitive to detergents: The presence of 0.05% of the nonionic detergent Nonidet-P40 (NP-40; Merck) within either the embryo lysis buffer or the assay buffer reduces the deadenylation activity. Even high amounts of NP-40 (tested up to 1%) do not block the activity completely, but, remarkably, the sequence-specificity is lost; under theses conditions, an otherwise stable control RNA is deadenylated to the same (low) extent as the regulated RNA (data not shown).

Deadenylation of *nanos* was performed at different temperatures ranging from 10 to 40 °C in 5 °C steps. Sequence-specific deadenylation was observed between 10 °C and 30 °C with the maximum rate between 25 °C and 30 °C. Incubation at 35 °C resulted in unspecific degradation of the RNA, and incubation at 40 °C or 0 °C gave no activity at all (data not shown).

The SRE-specific deadenylation can tolerate as much as 100 mM potassium chloride or potassium acetate beyond the standard salt concentration without significant decrease in the rate. Higher salt amounts gradually reduce the deadenylation rate without affecting the sequence specificity (data not shown).

Other *in vitro* systems need to be supplemented with poly(A), presumably as a competitor, to uncover their deadenylation activity on exogenous RNA substrates (Ford *et al.*, Milone *et al.*, 2002; 1999; Wilusz *et al.*, 2001). In contrast, deadenylation in *Drosophila* embryo extract does not depend on competitor poly(A). Instead, addition of high amounts of poly(A) (3.9 μg/ 10 μl reaction) slightly decreases the deadenylation rate (data not shown).

The same embryo extracts also deadenylate constructs containing the $3'$ UTR of the *hunchback* RNA (data not shown). In contrast, the *Hsp70* and *Hsp83* mRNAs, which are rapidly deadenylated *in vivo* (Bönisch et al., 2007,

Semotok *et al.*, 2005), remain stable in the extract (data not shown). One might expect that micrococcal nuclease treatment of the extract to remove potentially competing endogenous mRNAs should improve sequence-dependent deadenylation of exogenous RNA. However, our studies on both *nanos* and *hunchback* 3′ UTRs revealed that micrococcal nuclease treatment of the extract decreases the deadenylation rate in both cases (data not shown).

ACKNOWLEDGMENTS

We are grateful to Sylke Meyer and Dorian Freudenreich for their contributions to the development of the *in vitro* system and Martine Simonelig for many helpful discussions. We thank Gunter Reuter, Rainer Dorn, Peter Becker, and Fátima Gebauer for help and advice on fly work. Fátima Gebauer also donated the transcription vector containing the DNA-encoded poly(A) tail.

REFERENCES

Barreau, C., Paillard, L., and Osborne, H. B. (2005). AU-rich elements and associated factors: Are there unifying principles? *Nucl. Acids Res.* **33**, 7138–7150.

Bashirullah, A., Halsell, S. R., Cooperstock, R. L., Kloc, M., Karaiskakis, A., Fisher, W. W., Fu, W., Hamilton, J. K., Etkin, L. D., and Lipshitz, H. D. (1999). Joint action of two RNA degradation pathways controls the timing of maternal transcript elimination at the midblastula transition in *Drosophila melanogaster*. *EMBO J.* **18**, 2610–2620.

Bergsten, S. E., and Gavis, E. R. (1999). Role for mRNA localization in translational activation but not spatial restriction of *nanos* RNA. *Development* **126**, 659–669.

Bönisch, C., Temme, C., Moritz, B., and Wahle, E. (2007). Degradation of *hsp70* and other mRNAs in *Drosophila* via the 5′ to 3′-pathway and its regulation by heat shock. *J. Biol. Chem.* **282**, 21818–21828.

Brewer, G., and Ross, J. (1988). Poly(A) shortening and degradation of the 3′- A+U-rich sequences of human c-myc mRNA in a cell-free system. *Mol. Cell Biol.* **8**, 1697–1708.

Dahanukar, A., Walker, J. A., and Wharton, R. P. (1999). Smaug, a novel RNA-binding protein that operates a translational switch in *Drosophila*. *Mol. Cell* **4**, 209–218.

Ford, L. P., Watson, J., Keene, J. D., and Wilusz, J. (1999). ELAV proteins stabilize deadenylated intermediates in a novel *in vitro* mRNA deadenylation/degradation system. *Genes Dev.* **13**, 188–201.

Gavis, E. R., and Lehmann, R. (1992). Localization of *nanos* RNA controls embryonic polarity. *Cell* **71**, 301–313.

Gavis, E. R., and Lehmann, R. (1994). Translational regulation of *nanos* by RNA localization. *Nature* **369**, 315–318.

Gebauer, F., and Hentze, M. W. (2007). Studying translational control in *Drosophila* cell-free systems. *Methods Enzymol.* **429**, 23–33.

Giraldez, A. J., Mishima, Y., Rihel, J., Grocock, R. J., Van, D. S., Inoue, K., Enright, A. J., and Schier, A. F. (2006). Zebrafish MiR-430 promotes deadenylation and clearance of maternal mRNAs. *Science* **312**, 75–79.

Gorgoni, B., and Gray, N. K. (2004). The roles of cytoplasmic poly(A)-binding proteins in regulating gene expression: A developmental perspective. *Brief. Funct. Genomic. Proteomic.* **3**, 125–141.

Hentze, M. W., Gebauer, F., and Preiss, T. (2007). *Cis*-regulatory sequences and *trans*-acting factors in translational control. In "Translational Control in Biology and Medicine" (M. B. Mathews, N. Sonenberg, and J. W. B. Hershey, eds.), pp. 269–295. Cold Spring Harbor Laboratory Press, Cold Spring Harbor, NY.

Jeske, M., Meyer, S., Temme, C., Freudenreich, D., and Wahle, E. (2006). Rapid ATP-dependent deadenylation of *nanos* mRNA in a cell-free system from *Drosophila* embryos. *J. Biol. Chem.* **281**, 25124–25133.

Jing, Q., Huang, S., Guth, S., Zarubin, T., Motoyama, A., Chen, J., Di, P. F., Lin, S. C., Gram, H., and Han, J. (2005). Involvement of microRNA in AU-rich element-mediated mRNA instability. *Cell* **120**, 623–634.

Johnstone, O., and Lasko, P. (2001). Translational regulation and RNA localization in *Drosophila* oocytes and embryos. *Annu. Rev. Genet.* **35**, 365–406.

Kerwitz, Y., Kühn, U., Lilie, H., Knoth, A., Scheuermann, T., Friedrich, H., Schwarz, E., and Wahle, E. (2003). Stimulation of poly(A) polymerase through a direct interaction with the nuclear poly(A) binding protein allosterically regulated by RNA. *EMBO J.* **22**, 3705–3714.

Legagneux, V., Omilli, F., and Osborne, H. B. (1995). Substrate-specific regulation of RNA deadenylation in *Xenopus* embryo and activated egg extracts. *RNA* **1**, 1001–1008.

Meyer, S., Temme, C., and Wahle, E. (2004). Messenger RNA turnover in eukaryotes: Pathways and enzymes. *Crit Rev. Biochem. Mol. Biol.* **39**, 197–216.

Milone, J., Wilusz, J., and Bellofatto, V. (2002). Identification of mRNA decapping activities and an ARE-regulated 3' to 5'-exonuclease activity in trypanosome extracts. *Nucl. Acids Res.* **30**, 4040–4050.

Moritz, M. (2000). Preparing cytoplasmic extracts from *Drosophila* embryos. In "*Drosophila* protocols" (W. Sullivan, M. Ashburner, and R. S. Hawley, eds.), pp. 570–575. Cold Spring Harbor Laboratory Press, Cold Spring Harbor, NY.

Parker, R., and Song, H. (2004). The enzymes and control of eukaryotic mRNA turnover. *Nat. Struct. Mol. Biol.* **11**, 121–127.

Richter, J. D. (2000). Influence of polyadenylation-induced translation on metazoan development and neuronal synaptic function. In "Translational Control of Gene Expression" (N. Sonenberg, J. W. B. Hershey, and M. B. Mathews, eds.), pp. 785–805. Cold Spring Harbor Laboratory Press, Cold Spring Harbor, NY.

Ross, J. (1995). mRNA stability in mammalian cells. *Microbiol. Rev.* **59**, 423–450.

Sallés, F. J., Lieberfarb, M. E., Wreden, C., Gergen, J. P., and Strickland, S. (1994). Coordinate initiation of *Drosophila* development by regulated polyadenylation of maternal messenger RNAs. *Science* **266**, 1996–1999.

Sambrook, J., and Russell, D. W. (2001). "Molecular Cloning." Cold Spring Harbor Laboratory Press, Cold Spring Harbor, NY.

Semotok, J. L., Cooperstock, R. L., Pinder, B. D., Vari, H. K., Lipshitz, H. D., and Smibert, C. A. (2005). Smaug recruits the CCR4/POP2/NOT deadenylase complex to trigger maternal transcript localization in the early *Drosophila* embryo. *Curr. Biol.* **15**, 284–294.

Sission, J. C. (2000). Culturing large populations of *Drosophila* for protein biochemistry. In "*Drosophila* Protocols." (W. Sullivan, M. Ashburner, and R. S. Hawley, eds.), pp. 540–551. Cold Spring Harbor Laboratory Press, Cold Spring Harbor, NY.

Smibert, C. A., Wilson, J. E., Kerr, K., and Macdonald, P. M. (1996). Smaug protein represses translation of unlocalized *nanos* mRNA in the *Drosophila* embryo. *Genes Dev.* **10**, 2600–2609.

Smibert, C. A., Lie, Y. S., Shillinglaw, W., Henzel, W. J., and Macdonald, P. M. (1999). Smaug, a novel and conserved protein, contributes to repression of *nanos* mRNA translation *in vitro*. *RNA* **5,** 1535–1547.

Tadros, W., Goldman, A. L., Babak, T., Menzies, F., Vardy, L., Orr-Weaver, T., Hughes, T. R., Westwood, J. T., Smibert, C. A., and Lipshitz, H. D. (2007). SMAUG is a major regulator of maternal mRNA destabilization in *Drosophila* and its translation is activated by the PAN GU kinase. *Dev. Cell* **12,** 143–155.

Thompson, B., Wickens, M., and Kimble, J. (2007). Translational control in development. *In* "Translational Control in Biology and Medicine." (M. B. Mathews, N. Sonenberg, and J. W. B. Hershey, eds.), pp. 507–544. Cold Spring Harbor Laboratory Press, Cold Spring Harbor, NY.

Voeltz, G. K., Ongkasuwan, J., Standart, N., and Steitz, J. A. (2001). A novel embryonic poly(A) binding protein, ePAB, regulates mRNA deadenylation in *Xenopus* egg extracts. *Genes Dev.* **15,** 774–788.

Wahle, E., and Ruegsegger, U. (1999). 3′-End processing of pre-mRNA in eukaryotes. *FEMS Microbiol. Rev.* **23,** 277–295.

Wakiyama, M., Takimoto, K., Ohara, O., and Yokoyama, S. (2007). Let-7 microRNA-mediated mRNA deadenylation and translational repression in a mammalian cell-free system. *Genes Dev.* **21,** 1857–1862.

Wang, C., and Lehmann, R. (1991). Nanos is the localized posterior determinant in *Drosophila. Cell* **66,** 637–647.

Wilusz, C. J., Gao, M., Jones, C. L., Wilusz, J., and Peltz, S. W. (2001). Poly(A)-binding proteins regulate both mRNA deadenylation and decapping in yeast cytoplasmic extracts. *RNA* **7,** 1416–1424.

Wu, L., Fan, J., and Belasco, J. G. (2006). MicroRNAs direct rapid deadenylation of mRNA. *Proc. Natl. Acad. Sci. USA* **103,** 4034–4039.

Zaessinger, S., Busseau, I., and Simonelig, M. (2006). Oskar allows *nanos* mRNA translation in *Drosophila* embryos by preventing its deadenylation by Smaug/CCR4. *Development* **133,** 4573–4583.

CHAPTER SEVEN

Measuring CPEB-Mediated Cytoplasmic Polyadenylation-Deadenylation in *Xenopus laevis* Oocytes and Egg Extracts

Jong Heon Kim* and Joel D. Richter*,†

Contents

1. Introduction	120
2. Monitoring Polyadenylation of Cyclin B1 RNA in *Xenopus laevis* Oocytes	121
2.1. Principle	121
2.2. Materials	122
2.3. Methods	123
2.4. Notes	125
3. Detection of Nuclear and Cytoplasmic Polyadenylation of *X. laevis* Cyclin B1 mRNA	126
3.1. Principle	126
3.2. Materials	126
3.3. Methods	126
3.4. Notes	128
4. Monitoring Deadenylation and Re-adenylation of *X. laevis* Cyclin B1 RNA	128
4.1. Principle	128
4.2. Materials	129
4.3. Methods	129
4.4. Notes	129
5. Detection of Endogenous Cytoplasmic Poly(A) Polymerase Activity	130
5.1. Principle	130
5.2. Materials	130
5.3. Methods	130
5.4. Notes	131

* Research Institute, National Cancer Center, Goyang, Gyeonggi, South Korea
† Program in Molecular Medicine, University of Massachusetts Medical School, Worcester, Massachusetts, USA

Methods in Enzymology, Volume 448 © 2008 Elsevier Inc.
ISSN 0076-6879, DOI: 10.1016/S0076-6879(08)02607-4 All rights reserved.

6. Overexpression of Exogenous mRNAs and Measuring
   Polyadenylation-Deadenylation   132
   6.1. Methods   132
7. Preparation of *X. laevis* Egg Extracts and Measuring
   Polyadenylation-Deadenylation with Cyclin B1 RNA   133
   7.1. Principle   133
   7.2. Materials   133
   7.3. Methods   134
Acknowledgments   137
References   137

Abstract

The regulation of poly(A) tail length is one important mechanism for controlling gene expression during early animal development. In *Xenopus* oocytes, the polyadenylation-deadenylation of several essential dormant mRNAs, including cyclin B1 mRNA, are controlled by the *cis*-acting cytoplasmic polyadenylation element (CPE) and the hexanucleotide AAUAAA through their associations with protein factors CPEB and CPSF, respectively. CPE-containing, as well as CPE-lacking, pre-mRNAs acquire long poly(A) tails in the nucleus; after their export to the cytoplasm, there is subsequent deadenylation of CPE-containing mRNAs that is controlled by the CPEB-associated factor PARN, a poly(A)-specific ribonuclease. In general, re-adenylation after meiotic maturation of CPE-containing mRNAs is mediated by Gld2, a poly(A) polymerase. Moreover, embryonic poly(A)-binding protein, ePAB, is required for the subsequent elongation and stabilization of the poly(A) tail against PARN and other deadenylating enzymes. In this chapter, we present detailed information for measuring CPEB-mediated cytoplasmic polyadenylation-deadenylation in *Xenopus laevis* oocytes and egg extracts.

1. INTRODUCTION

Polyadenylation control of translation plays an essential role in vertebrate germ cells, embryos, and neurons (Mendez and Richter, 2001; Richter, 2007). The regulation of cytoplasmic polyadenylation of a set of mRNAs requires two 3′-untranslated region (UTR) *cis*-determinants: cytoplasmic polyadenylation element (CPE) and hexanucleotide AAUAAA. These key sequences are bound by the protein factors CPEB (Hake and Richter, 1994; Paris *et al.*, 1991) and cleavage and polyadenylation specificity factor (CPSF) (Bilger *et al.*, 1994), respectively. Generally, poly(A) tail length has a close relationship with the rate of translation. In *Xenopus laevis* oocytes responding to the progesterone stimulation of the meiotic divisions, poly(A) tail elongation is tightly coupled to translational activation.

Irrespective of the presence of the CPE in their 3′ UTR sequences, mRNAs acquire long poly(A) tails in the nucleus, whereas after transport to the cytoplasm, only the CPE-containing RNAs undergo deadenylation, which is commensurate with translational repression (Huarte *et al.*, 1992). This deadenylation process is controlled by the CPEB-interacting factor poly(A)-specific ribonuclease (PARN) (Kim and Richter, 2006). In oocytes, the cytoplasmic poly(A) polymerase Gld2 (Kwak *et al.*, 2004; Wang *et al.*, 2002), which mediates polyadenylation, is also active (Barnard *et al.*, 2004). However, PARN activity is particularly strong, thus the poly(A) tails are shortened and are maintained in that state (Kim and Richter, 2006). Progesterone-stimulation of oocyte maturation leads to the phosphorylation of CPEB at serine 174 (Mendez *et al.*, 2000), causing the dissociation of PARN from the CPEB-mediated cytoplasmic polyadenylation complex (Kim and Richter, 2006). This event allows the poly(A) polymerase Gld2 to catalyze polyadenylation by default (Kim and Richter, 2006).

Another protein factor, ePAB [embryonic poly(A)-binding protein] (Voeltz *et al.*, 2001), is necessary for the cytoplasmic polyadenylation processes. Initially, ePAB is associated with CPEB in oocytes, but during maturation, ePAB dissociates from this factor and interacts with newly elongated poly(A) tails (Kim and Richter, 2007). ePAB release from CPEB occurs through the phosphorylation of CPEB on six residues by the kinase complex RINGO/cdk1 (Ferby *et al.*, 1999; Kim and Richter, 2007; Mendez *et al.*, 2002; Padmanabhan and Richter, 2006). ePAB binding to the newly elongated poly(A) tail protects the poly(A) tail from nuclease attack and promotes translation initiation through the association with eIF4G (Kim and Richter, 2007).

To analyze polyadenylation-deadenylation and the closely coupled translational activation by CPEB and its associated factors, we have used *X. laevis* oocytes and egg extracts, which are powerful tools for analyzing these biochemical processes. In this chapter, we present detailed information for measuring cytoplasmic polyadenylation and deadenylation in oocytes and egg extracts.

2. Monitoring Polyadenylation of Cyclin B1 RNA in *Xenopus laevis* Oocytes

2.1. Principle

After meiotic maturation, CPE-containing mRNAs in *X. laevis* oocytes are robustly polyadenylated (e.g., cyclin B1 mRNA acquires a poly(A) length of ~200 to 250 nts). Oocytes are a convenient experimental system for measuring polyadenylation of CPE-containing RNAs through the injection of exogenous radiolabeled short RNA probes.

2.2. Materials

2.2.1. Generation of radiolabeled probe

1. Transcription optimized 5× buffer (Promega).
2. 100 mM dithiothreitol (DTT).
3. 10× NTPs for "hot and capped" RNA (10 mM ATP, CTP, GTP, and 1 mM UTP).
4. RNaseOUT (Invitrogen) (40 units/μl).
5. T3, T7, or SP6 RNA polymerase (Promega; 17 units/μl).
6. G(5')ppp(5')G RNA cap structure analog (New England Biolabs).
7. Linearized DNA template with upstream T3, T7, or SP6 phage promoter sequences.
8. [α-^{32}P]UTP (3000 Ci/mmol).
9. 0.5 M EDTA (pH 8.0).
10. QIAquick Nucleotide Removal Kit (Qiagen).
11. SequaGel Sequencing System (National Diagnostics).
12. Sterile, nuclease free water.
13. 10× TBE: 1.78 M Tris base, 1.78 M boric acid, 40 mM EDTA (pH 8.0); autoclave, and store at room temperature.
14. 98% Formamide loading dye for denaturing polyacrylamide gels: mix 98 ml of formamide, 5 mg of bromophenol blue, 5 mg of xylene cyanol FF, and 2 ml 0.5 M EDTA (pH 8.0) for a 100 ml stock solution; store at $-20\,^{\circ}$C.
15. Ammonium persulfate: prepare 10% solution in sterile water and store at 4°C.
16. N,N,N,N'-tetramethyl-ethylenediamine (TEMED).
17. PelletPaint (Novagen): coprecipitant.
18. Phenol/chloroform/isoamyl alcohol (25:24:1, v/v) and chloroform/isoamyl alcohol (24:1, v/v).
19. Absolute and 70% ethanol.
20. 3 M sodium acetate (pH 5.2).

2.2.2. Oocyte collection, injection, incubation, and retrieval of radiolabeled RNA probes

1. Modified Barth's saline (MBS): prepare 0.1 M CaCl$_2$, and 10× MBS (880 mM NaCl, 10 mM KCl, 10 mM MgSO$_4$, 50 mM HEPES [pH 7.8], 25 mM NaHCO$_3$); autoclave, and store at room temperature. Prepare the final 1× MBS solution (0.7 mM CaCl$_2$, 88 mM NaCl, 1 mM KCl, 1 mM MgSO$_4$, 5 mM HEPES [pH 7.8], 2.5 mM NaHCO$_3$) by mixing 100 ml of 10× MBS plus 7 ml of 0.1 M CaCl$_2$ and adjust the volume up to 1 L with sterile, distilled water. Do not autoclave the mixed buffer again, because this usually causes precipitation.

2. 4× Collagenase/Dispase (4× CD): mix 50 mg/L of streptomycin (optional), 80 mg of collagenase (type II: Sigma-Aldrich), and 48 mg of Dispase (Roche) in 10 ml 1× MBS.
3. Progesterone (Sigma-Aldrich).
4. Guanidium thiocyanate-phenol–based RNA extraction reagents (e.g., TRIzol [Invitrogen]), isopropanol, and absolute chloroform.
5. Equipment: Nanoject II (Drummond Scientific Company), micropipette puller, fine point tweezers (Drummond Scientific Company), glass petri dishes, and plastic disposable pipette droppers (BD Falcon).

2.3. Methods

1. Linearize 1 to 2 μg of plasmid DNA containing the minimal 3' UTR region of a CPE-containing gene (e.g., the cyclin B1 gene) with the proper restriction enzyme.
2. Extract reaction mixture with phenol/chloroform/isoamyl alcohol (25:24:1, v/v) followed by chloroform: isoamyl alcohol (24:1, v/v) extraction. Precipitate the DNA with absolute ethanol, 3 M sodium acetate (pH 5.2), and PelletPaint (Novagen). PelletPaint is useful for the visualization (pink color) of precipitated DNA to minimize loss of the precipitate.
3. Synthesis of "hot and capped" RNAs: dissolve DNA pellet in 2 μl of nuclease-free water.
4. Add the following to a sterile and nuclease-free 1.5-ml microcentrifuge tube to make final 20 μl of *in vitro* transcription mixture.
   1 μl 100 mM DTT
   1 μl RNaseOUT (40 units/μl)
   3 μl RNA polymerase (17 units/μl)
   4 μl G(5')ppp(5')G RNA cap structure analog
   4 μl Transcription optimized 5× buffer
   3 μl [α-^{32}P]UTP (3000 Ci/mmol)
   2 μl 10× NTPs for "hot and capped" RNAs
   2 μl Linearized DNA template (0.5 to 1.0 μg/μl)
5. Incubate the *in vitro* transcription mixture for 15 min at 37 °C and then add 1 μl of 20 mM GTP in a 20-μl reaction volume. After 75 min of further incubation, stop the reaction by adding 1 μl of Turbo DNase (2 units/μl; Ambion) and incubate for 15 min at 37 °C.
6. After the reaction steps, unincorporated free [α-^{32}P]UTPs are easily removed with a nucleotide removal kit (Qiagen). To obtain a sufficiently high concentration of radiolabeled RNA, elute the RNA with a minimal volume of sterile and nuclease-free water. Usually eluting with ~30 μl gives the best results.
7. Adjust radioactivity to approximately 500,000 to 1,000,000 cpm/μl with sterile and nuclease-free water after scintillation counting.

8. Check the integrity of the radiolabeled RNA with urea-containing denaturing polyacrylamide electrophoresis (PAGE) (usually 5% acrylamide monomer). National diagnostics supplies the SequaGel Sequencing System, which is a useful premade denaturing gel.
9. Once the gel has polymerized, carefully remove the comb and wash the wells with water. Place the gel plate in the electrophoresis apparatus. Add 1× TBE to the upper and lower parts of the gel unit and preelectrophorese for 10 to 15 min at 100 to 150 V. Turn off the power supply.
10. Aliquot of RNAs are mixed with 1 to 2× formamide loading dye, incubated for 5 to 10 min at 60 to 70 °C, and then quickly chill on ice for at least 5 min. After brief centrifugation and gentle mixing, put the RNAs on the ice bath before loading onto the gel.
11. Wash the wells to remove accumulated urea, load the samples, and start the electrophoresis at constant voltage (100 to 150 V). (The *X. laevis* cyclin B1 RNA containing a 125-nt minimal CPE exactly comigrates with xylene cyanol FF in a 5% acrylamide gel).
12. Keep the radioactive RNA at −80 °C before microinjection or *in vitro* translation.
13. Oocyte preparation: A sexually mature animal has more than 10,000 oocytes of all stages, but only a few hundred oocytes are required for most experiments. Therefore, we often obtain oocytes from one animal multiple times.
14. Measure the volume of isolated pieces of *Xenopus* ovary and add one volume of 4× CD plus three volumes of 1× MBS buffer.
15. Incubate oocytes with 1× MBS containing collagenase/dispase on a slowly shaking rocker at approximately 20 to 23 °C for 2.5 to 3 h.
16. After incubation, remove excess collagenase/dispase with fresh 1× MBS. Usually, dead and small-stage oocytes (stage I, II, and III) float to the top of 50-ml conical tubes during the washing step.
17. After collagenase/dispase treatment and the washing steps, we usually check oocyte quality under the light microscope. Do not inject the oocytes for at least 1 to 2 h to provide sufficient amount of time to detect dead oocytes. Dead oocytes can usually be identified by the lack of smooth or uniform pigmentation at the animal pole and should be discarded. Oocytes that are undergoing atresia have a mottled appearance at the animal pole.
18. Collect healthy, large (∼1.4-mm diameter) and "white-banded" stage VI (SVI) oocytes and transfer to fresh 1× MBS in a petri dish.
19. Cover the oocytes to protect from drying. They are usually stable for at least 2 days (48 h) in 1× MBS at 20 to 23 °C.
20. Injection equipment consists of a Nanoject II (Drummond Scientific Company), micropipette puller, fine point tweezers (Drummond Scientific Company), and plastic disposable pipette droppers (BD Falcon).

Make a fine point micropipette with a micropipette puller. We usually inject 50 nl of RNA into each oocyte in a single injection.
21. Monitor the health of the oocytes for at least 1 to 2 h because the injection can cause damage to the oocytes. Inject 5 to 10 oocytes for monitoring cytoplasmic polyadenylation. A CPE-disrupting mutation [CPE(−)] can be used as a polyadenylation control. Remove the dead oocytes after injection. After 1 to 2 h, add progesterone (final concentration 20 nM) in the incubation buffer and mix gently.
22. Progesterone treatment of oocytes induces a white spot ("germinal vesicle break down" or GVBD) on top of the animal pole within 5 to 12 h. The time at which GVBD occurs after progesterone exposure can vary widely. Collect five oocytes showing GVBD and transfer to a 1.5-ml centrifuge tube. Remove the incubation buffer as much as possible and quickly freeze the oocytes with dry ice (−80 °C).
23. To retrieve total RNA from the oocytes, we usually use TRIzol (Invitrogen) or TRI reagent (Molecular Research Center). It is not necessary to add a carrier for the precipitation of the radiolabeled RNA probe, because there is an enormous amount of rRNA in the oocytes that promotes precipitation.
24. Add 50 μl of sterile, nuclease-free 1× MBS plus RNaseOUT (final 1 unit/μl) to the 5 oocytes. Vortex and homogenize the oocytes first, because they are usually difficult to homogenize in TRIzol directly. Add 250 μl of TRIzol and follow the protocol of the manufacturer for the RNA extraction.
25. Do not dry the pellet completely following an isopropanol precipitation and 70% ethanol wash. Completely dried RNA pellets are difficult to solubilize. Semidried total RNA pellets are dissolved with 20 to 50 μl of sterile, nuclease-free water; 1 μl of dissolved sample is used for scintillation counting.
26. Load the samples in the wells and start the electrophoresis at constant voltage (100 to 150 V). When the xylene cyanol FF (slower migrating dye) reaches the bottom of the gel, turn off the power supply. A low-percentage acrylamide gel (usually less than 4%) is not easy to handle. Be careful when transferring it to Whatman paper for drying. Usually, urea-containing gels are very difficult to dry completely. Radioactivity can be monitored by X-ray film or PhosphorImager.

2.4. Notes

All plastic ware should be sterile and nuclease free. To make sterile and nuclease-free water, 0.01% diethylpyrocarbonate (DEPC) is added to water and stirred at least 1 h with protection from light. DEPC is easily removed by double-autoclaving. In addition, a very long enzymatic treatment of the ovary usually causes lysis of the oocytes. The procedure for the *X. laevis* oocyte isolation and incubation is described (Smith *et al.*, 1991).

3. DETECTION OF NUCLEAR AND CYTOPLASMIC POLYADENYLATION OF *X. LAEVIS* CYCLIN B1 mRNA

3.1. Principle

Pre-mRNAs, irrespective of whether they contain CPEs, usually acquire long poly(A) tails. After splicing and export to the cytoplasm, CPE-containing RNAs undergo deadenylation (Huarte *et al.*, 1992). This deadenylation process is controlled by the CPEB-interacting factor poly(A)-specific ribonuclease PARN (Kim and Richter, 2006). Instead of a Northern blot for the analysis of poly(A) of specific mRNAs. A simple polymerase chain reaction referred to as the PAT assay [poly(A) tail assay] (Salles *et al.*, 1999) can be used to detect the poly(A) status of a specific mRNA.

3.2. Materials

1. GenElute Mammalian Genomic DNA Miniprep Kit (Sigma-Aldrich).
2. TRIzol (Invitrogen).
2. $d(T)_{18}$-linker (500 µg/ml; 5'-GCGAGCTCCGCGGCCGCGT$_{18}$-3').
3. SuperScript II (Invitrogen).
4. Platinum Taq (Invitrogen).
5. Intron-specific oligomers.
6. Turbo DNase (2 units/µl; Ambion).
7. 5× First strand buffer (Invitrogen) (250 mM Tris-HCl [pH 8.3], 375 mM KCl, 15 mM MgCl$_2$).

3.3. Methods

1. Prepare genomic DNA from oocytes by use of GenElute Mammalian Genomic DNA Miniprep Kit (Sigma-Aldrich). The starting oocyte number should be more than 100.
2. On the basis of the genomic structure of *Xenopus tropicalis* cyclin B1 intron 8 (intron 8 to 9; 178-nt long), intron 8 of cyclin B1 from *X. laevis* was amplified and sequenced a forward primer in exon 8 and reverse primer in exon 9. Information of *X. tropicalis* cyclin B1 genomic sequence was obtained from Ensembl Genome Browser (http://www.ensembl.org/index.html). Exon 9 is the last exon of *X. laevis* cyclin B1 gene. The entire sequence of *X. laevis* intron 8 is noted as follows (underlined and italicized sequence).

GCATATGGCCAAGAACATCATCAAGGTGAACAAAGGACTA
ACCAAGCATCTG*GTAAGCTTTTTAGCCATTCAAGACAAGTT*
GTTAATTACTATATATGCATAACCCTTGCTTGAGGGGGGGG
GGAAATGTGTGAAGCTTGCTTCCATATTTCCATGGCAGT
TAGTCTGGTGGGTAAACTAGGTTAGCATGAAATAGATTAC
*CTTGTGACCTAAATATAACTTGCTTTCCAG*ACTGTTAAGAAC
AAGTATGCTAGCAGCAAACAAATGAAGATCAGCACGATTCC
ACAGCTGAGGTCAGATGTTGTTGTGGAAATGGCCCGCCCA
CTCATG*TG*AAGGACTACGTGGCATTCCAATTGTGTATTGTT
GGCACCATGTGCTTCTGTAATAGTGTATTGTGTTTTTAATG
TTTTACTGGTTTTAATAAAGCTCATTTTAACATG

Exon 8 specific forward primer: 5′-GCATATGGCCAAGAACATCATCAAGG-3′

Exon 9 specific reverse primer: 5′-CATGTTAAAATGAGCTTTATTAAAACCAG-3′

3. For the analysis of poly(A) of cyclin B1 pre-mRNA, 5 μg of total oocyte RNA is prepared and treated with Turbo DNase. The DNase reaction is noted as follows:
   4 μl 5× First strand buffer
   5 μl Total oocyte RNA (concentration 1 μg/μl)
   2.5 μl Turbo DNase (2 units/μl)
   8.5 μl Sterile, nuclease-free water
4. Incubate reaction mixture at 37 °C for 20 min and then heat inactivate DNase at 65 °C for 10 min.
5. After a brief centrifugation, 2 μl oligo d(T)$_{18}$-linker and 2 μl of 10 mM dNTPs are added, followed by a further incubation at 65 °C for 5 min.
6. After quick chilling on ice, briefly centrifuge the reaction mixture, collect the contents of the tube, and add the following:
   24 μl Reaction mixture
   4 μl 5× First strand buffer
   4 μl 100 mM DTT
   2 μl RNaseOUT
   4 μl Sterile, nuclease-free water
7. Incubate reaction mixture at 42 °C for 2 min.
8. Add 2 μl (400 units) of SuperScript II RT and mix by pipetting gently up and down and incubate at 42 °C for 1 h.
9. Inactivate the reaction by heating at 70 °C for 15 min.
10. 1 μl of cDNA template was used in a 25 μl PCR reaction with Platinum Taq (Invitrogen). Perform PCR reaction as follows:
    1 μl Template (cDNA)
    2.5 μl 10× PCR buffer (MgCl$_2$)
    0.5 μl 10 mM dNTPs
    0.5 μl Forward oligomer (25 pmol/μl; intron- or exon-specific primers)

0.5 μl Reverse oligomer (25 pmol/μl; $d(T)_{18}$-linker or exon-specific oligomers)
0.75 μl 50 mM $MgCl_2$
0.1μl Platinum Taq
19.15 μl Sterile, distilled water

Oligomers for PCR are used as follows:

Intron 8–specific forward primers
    a. 5'-AGGTTAGCATGAAATAGATTACCTT-3'
    b. 5'-GTGAAGCTTGCTTCCCATATTTCC-3'
Exon-specific primers (c), (d), and oligo $d(T)_{18}$-linker (e)
    c. 5'-GCATATGGCCAAGAACATCATCAAGG-3'
    d. 5'-CATGTTAAAATGAGCTTTATTAA AACCAG-3'
    e. 5'-GCGAGCTCCGCGGCCGCGT$_{18}$-3'

11. After an initial denaturation step at 94 °C for 2 min, the PCR condition is as follows: 94 °C for 0.5 min, 56 °C for 1 min, 72 °C for 1.5 min for 40 cycles.
12. The amplified products are analyzed in a 1.5 % agarose gel or 5% non-denaturing polyacrylamide gel and visualized by ethidium bromide staining.

3.4. Notes

$MgCl_2$ is critical for the DNase activity. A minimum of 10 mM is required for full DNase activity.

4. Monitoring Deadenylation and Re-adenylation of *X. laevis* Cyclin B1 RNA

mRNAs acquire long poly(A) tails in the nucleus; after transport to the cytoplasm, only the CPE-containing RNAs undergo deadenylation and translational silencing (Huarte *et al.*, 1992). This deadenylation process is controlled by the CPEB-interacting factor poly(A)-specific ribonuclease PARN (Copeland and Wormington, 2001; Kim and Richter, 2006; Korner *et al.*, 1998). In general, re-adenylation after meiotic maturation of CPE-containing mRNAs is mediated by Gld2 (Barnard *et al.*, 2004; Kwak *et al.*, 2004; Wang *et al.*, 2002).

4.1. Principle

Post-transcriptional polyadenylation of synthetic RNAs with *Escherichia coli* poly(A) polymerase (E-PAP) is useful for increasing the translational potential of capped mRNA in injected oocytes. An oligo(dA-dT) track can be inserted after the CPE-containing 3' UTR sequence within the plasmid.

However, 50 to 100 nt oligo(dA-dT) tracts within the vector can cause instability of the insert when propagated in bacteria, which may restrict use of these vectors. Alternately, E-PAP easily adds a poly(A) tail of more than 200 nucleotides to the 3′-termini of RNA. Currently, a convenient polyadenylation kit is available from Ambion. We use this kit to generate suitable poly(A) tails on synthetic mRNAs.

4.2. Materials

1. Poly(A) Tailing kit (Ambion).
2. 0.5 M EDTA (pH 8.0).

4.3. Methods

1. Save a 5 l aliquot of the final transcription mixture from "hot and capped" RNA (section 2.3) to be used as a nonadenylated control.
2. To a sterile and nuclease free 1.5 ml microcentrifuge tube add the following:
   15 μl DNase-treated *in vitro* transcribed mixture
   10 μl 5× E-PAP reaction buffer (Ambion)
   5 μl 25 mM MnCl$_2$
   5 μl 10 mM ATP
   13 μl Sterile, nuclease-free water
   2 μl E-PAP (Ambion)
3. Incubate the reaction mixture at 37 °C. Perform a time course for the *in vitro* polyadenylation reaction, generally 5, 10, and 20 min. Take approximately 15 μl reaction mixture at each time point.
4. Aliquot 1.25 μl of 0.5 M EDTA (pH 8.0) into the microcentrifuge tube before transfer of 15 μl of polyadenylation mixture to avoid unexpected delay. The reaction is completely terminated by additional heating at 70 °C for 10 min.
5. Free nucleotides are removed with a nucleotide removal kit (Qiagen), and the extent of polyadenylation is monitored by 5% denaturing PAGE. An example of the results obtained is shown in Fig. 7.1.

4.4. Notes

The 10-min time point usually yields a 200 to 250 nt poly(A) tail of >95% RNA; >20 min incubation yields much longer and hetero-disperse poly(A) species. The extent of polyadenylation critically depends on the quality of E-PAP enzyme; hence, the investigator should check the quality of each lot of enzyme.

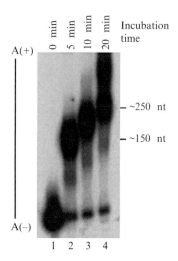

Figure 7.1 Time course for *in vitro* polyadenylation. Aliquots from each time point were analyzed with 5% denaturing PAGE. See "Methods" 4.3. for the detailed experimental scheme. The spectrum of A(+) to A(−) represents the polyadenylation status of cyclin B1 RNA.

5. Detection of Endogenous Cytoplasmic Poly(A) Polymerase Activity

5.1. Principle

Gld2 is an unconventional cytoplasmic poly(A) polymerase (Kwak *et al.*, 2004; Wang *et al.*, 2002) that catalyzes the cytoplasmic polyadenylation of CPE-containing mRNAs after meiotic maturation of *X. laevis* oocytes (Barnard *et al.*, 2004). Through the coinjection of radioinert CPE-containing RNA plus [α-^{32}P]NTPs, endogenous Gld2 activity can easily be monitored.

5.2. Materials

1. Actinomycin D (Sigma-Aldrich).
2. [α-^{32}P]NTPs (usually [α-^{32}P]ATP and [α-^{32}P]UTP)].
See additional materials in section 2.2.1.

5.3. Methods

1. Radioinert RNA: unlabeled RNAs containing a CPE (e.g., cyclin B1) can be generated by *in vitro* transcription. See "Methods" in section 2.3.

2. Add actinomycin D (5 μg/ml concentration) in 1× MBS and incubate oocytes for 3 h to block transcription. Cover petri dish during actinomycin D incubation because it is light sensitive.
3. Prepare unlabeled CPE-containing RNA at a 0.5 μg/μl concentration.
4. Mix ([α-^{32}P]NTP: unlabeled RNA, or sterile nuclease free water = 1:1). Mix each component at equal volume (1:1). (e.g., [α-^{32}P] NTP + water or [α-^{32}P]NTP + RNA).
5. Inject 50 nl of mixture into 10 to 20 oocytes.
6. After approximately 1 to 2 h incubation, add progesterone to induce meiotic maturation.
7. Collect oocytes that have undergone GVBD. Extract total RNA and analyze radioactivity as described in "Methods" 2.3. An example of cyclin B1 RNA is shown in Fig. 7.2.

5.4. Notes

Actinomycin D does not affect oocyte maturation. Therefore, do not remove once the experiment has begun.

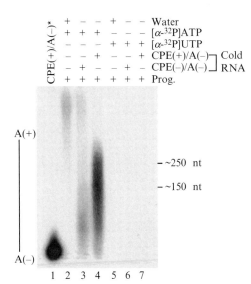

Figure 7.2 Detection of endogenous cytoplasmic poly(A) polymerase activity (i.e., Gld2 activity). See "Methods" 5.3. for the detailed experimental scheme. Radioactive cyclin B1 mRNA was analyzed with 5% denaturing PAGE. CPE(+)/A(−)*, CPE (+)/A(−), and CPE(−)/A(−) represent radiolabeled CPE-containing cyclin B1 3′ UTR RNA, cold CPE-containing cyclin B1 3′ UTR RNA, and cold CPE-disrupted cyclin B1 3′ UTR RNA, respectively.

6. OVEREXPRESSION OF EXOGENOUS mRNAs AND MEASURING POLYADENYLATION-DEADENYLATION

Convenient kits for the generation of capped and polyadenylated mRNA are available from Ambion [mMESSAGE mMACHINE and Poly(A) Tailing Kit]. Epitope-tagged proteins are generally useful for the detection of recombinant protein from mRNA that is generated by *in vitro* transcription. The polyadenylation-deadenylation pattern can be monitored by a simple stepwise double-injection of capped-polyadenylated mRNA that expresses a heterologous protein and followed radiolabeled RNA.

6.1. Methods

1. Capped and polyadenylated mRNAs are easily generated by mMESSAGE mMACHINE and Poly(A) Tailing Kit (Ambion). Plasmids encoding regulatory proteins [usually poly(A) polymerase or deadenylase] are linearized by suitable restriction enzymes and then transcribed *in vitro* with suitable phage promoters (e.g., T3, T7, and SP6).
2. Epitope tags such as Myc, FLAG, and HA are useful for checking the expression profile of injected mRNA.
3. Generally, the addition of poly(A) tail by E-PAP enhances expression of target mRNA.
4. To overexpress epitope-tagged proteins, oocytes are injected with 50 nl solutions of capped-polyadenylated mRNA in water at concentrations of 0.5 to 1.0 $\mu g/\mu l$.
5. Radiolabeled RNA probes are generated as described in section 2.3. RNAs are polyadenylated *in vitro* with Poly(A) Tailing Kit (Ambion); free nucleotides are removed with a nucleotide removal kit (Qiagen). The integrity of the probes is verified by denaturing 5% PAGE.
6. The translation of injected mRNA usually is maximal by 12 to 16 h.
7. After an incubation of 12 to 18 h before the second injection, 5 injected oocytes per time point are collected and used for immunoblotting.
8. Inject either polyadenylated or nonadenylated radiolabeled RNA into oocytes that express the heterologous protein from the injected mRNA.
9. After incubating the double-injected oocytes for approximately 0.5 to 1 h, add progesterone if the induction of meiotic maturation is required.
10. Incubate oocytes for varying times and then extract RNA with TRIzol (Invitrogen).
11. Radioactivity is analyzed in a 3.5 to 6% denaturing polyacrylamide gel.

7. Preparation of *X. laevis* Egg Extracts and Measuring Polyadenylation-Deadenylation with Cyclin B1 RNA

7.1. Principle

X. laevis egg extracts are useful for the biochemical analysis of polyadenylation. However, the extracts are very sensitive to dilution and, therefore, residual buffer must be removed before the immunodepletion of protein. For this purpose, the depletions are best performed with magnetic bead–based affinity columns.

7.2. Materials

7.2.1. Egg extract preparation

1. 250× frog salt (5 M NaCl).
2. 10× MMR (50 mM HEPES [pH to 7.8 with NaOH], 1 mM EDTA, 1 M NaCl, 20 mM KCl, 10 mM MgCl$_2$, 20 mM CaCl$_2$); autoclave and store at room temperature.
3. 20× Extraction buffer (XB) salt stock (2 M KCl, 20 mM MgCl$_2$, 2 mM CaCl$_2$; autoclave or filter sterilize).
4. L-cysteine (Sigma-Aldrich).
5. 10 N NaOH.
6. Cytochalasin B.
7. Protease inhibitors (leupeptin, chymostatin, and pepstatin; Sigma-Aldrich).
8. Human chorionic gonadotropin (hCG; Sigma-Aldrich).
9. Proper gauge needles.
10. Top oil: Nyosil M-25; filter sterilize.
11. 20× energy mix (150 mM creatine phosphate, 20 mM ATP, 20 mM MgCl$_2$).
12. De-jellying solution (2% L-cystine in 1× XB); 100 ml.
13. Beckman 50 Ultra-clear tubes; small (13 × 51 mm) or large (14 × 89 mm).
14. 1× XB in nuclease and protease-free, sterile water.

7.2.2. Depletion of protein with specific antibody and polyadenylation assay

1. Protein A Dynal beads (Invitrogen).
2. MagneSphere magnetic separation stand (Promega).
3. Suitable antibodies.

7.2.3. Depletion of protein with specific interacting protein and polyadenylation assay

1. MagneGST glutathione particles (Promega).
2. GST-fusion protein induced *Escherichia coli* crude soluble fraction.

7.2.4. *In vitro* polyadenylation-translation assay

1. *Xenopus* egg extract.
2. 4× poly(A) polymerase buffer without ATP (80 mM Tris-HCl [pH 7.5], 200 mM KCl, 10 mM MnCl$_2$, 200 µg/ml bovine serum albumin [BSA], 40% glycerol).
3. 20× energy mix (150 mM creatine phosphate, 20 mM ATP, 20 mM MgCl$_2$).
4. RNaseOUT (Invitrogen).
5. Nuclease-free water.
6. Formamide loading dye.
7. Dual-luciferase reporter assay kit (Promega).
8. Passive lysis buffer (Promega).

7.3. Methods

7.3.1. Egg extract preparation

1. Extracts from unfertilized *X. laevis* eggs are prepared according to published methods (Desai *et al.*, 1999; Murray and Kirschner, 1989) with slight modifications. Primed frogs are usually purchased from NASCO.
2. 12 to 18 h before egg extract preparation, usually 2 to 3 frogs are induced to ovulate by injection with 500 U of hCG into each animal. Inject the hCG at the border between the leg and the back of the animal.
3. After injection, place frogs in individual tanks containing 1 to 2 L of 1× frog salt (20 mM NaCl). Monitor animals at least 1 h and clean up all excretion.
4. Keep animals at 16 to 20 °C.
5. After 12 to 18 h, the frogs usually start to lay eggs. If the laying speed is very slow, gently squeeze the frogs to obtain enough eggs. Transfer the eggs to the 1× MMR containing beakers with end-trimmed disposable pipette dropper.
6. Prepare 100 ml of de-jellying solution for each frog. Usually ∼10 to 30 ml of eggs are obtained from each animal.
7. Wash eggs with 1× XB several times. Remove dead, activated, puffy, or irregular eggs with a disposable pipetter dropper.

8. De-jellying step: 100 ml of 2% cysteine in 1× XB (pH 8.0); freshly prepared; adjust to pH 8.0 with ~0.7 ml of 10 N NaOH per 100 ml of buffer.
9. Add ~30 ml of 2% cysteine containing 1× XB slowly to the eggs and gently swirl. Repeat this step three times. However, do not exceed more than 5 min for all steps. The eggs will become very dense and packed at the bottom of the beaker.
10. Neutralize 2% cysteine with 1× XB and rinse eggs with fresh 1× XB several times. Pack the eggs into Beckman 50 Ultra-clear tubes and remove all excess buffer.
11. Add 2.5 to 3 ml of 1× XB in sterile water containing protease inhibitors (leupeptin, chymostatin, and pepstatin; 10 μg/ml each) plus cytochalasin B (50 μg/ml). Rinse eggs with buffer and remove as much of the buffer as possible. Finally, add 1 ml of buffer on top of the tube and add 1 ml of Nyosil M-25 sequentially.
12. Centrifuge the eggs in a SW55 swinging bucket 1 to 2 min at 1000 to 2000 rpm. The temperature should be approximately 16 to 18 °C.
13. After centrifugation, most of the buffer is located on top of the Nyosil M-25; remove excess buffer to minimize dilution effect of egg extracts.
14. Crush the eggs by centrifugation at 10,000 rpm for 15 min at 16 to 18 °C. Store tubes with crushed eggs on ice.
15. The multiple layers are composed of (from top to bottom): yolk, oil, cytoplasmic extract, and pigment.
16. Puncture the tube near the bottom of the cytoplasmic layer with an 18-gauge needle on a 1-ml syringe and gently withdraw the extract. Add 1/20 volume of energy mix, protease inhibitors (final 10 μg/ml each), and cytochalasin B (final 50 μg/ml). Keep the aliquot at the −80 °C.

7.3.2. Depletion of protein with specific antibody and polyadenylation assay

1. Prepare 60 μl of Protein A Dynal (Invitrogen) beads washed 1 to 2 times with 1× PBS.
2. Remove residual PBS and add 750 μl of 1× PBS plus 2 μg of control or target-specific antibody (e.g., anti-PARN antibody). Incubate mixture 2 to 2.5 h at 4 °C to make antibody-protein A-magnetic bead conjugate.
3. Wash the conjugate three times with 1× PBS. After the last wash, remove buffer as much as possible to minimize dilution effect of the egg extract.
4. Add 25 μl of egg extract plus 1.25 μl of RNaseOUT to the antibody-magnetic beads pellet. Gently vortex tubes on top of a vortexing platform and incubate 1.25 h at 4 °C. Repeat this step twice.
5. Separate depleted egg extracts and antibody-magnetic beads with MagneSphere Magnetic Separation Stand (Promega). Use 4 μl of depleted extracts for polyadenylation assay and 1 μl for immunoblot.

6. Polyadenylation assay with depleted extracts are assessed as follows:
   4 μl Egg extract
   0.5 μl 20× energy mix
   2.5 μl 4× poly (A) polymerase buffer (80 mM Tris [pH 7.0], 200 mM KCl, 2.8 mM MnCl$_2$, 400 μg/ml BSA, 40% glycerol)
   0.5 μl 10 mM ATP
   0.5 μl RNaseOUT
   1 μl Radioactive RNA (25,000 cpm)
   Up to 10 μl nuclease-free water
7. Incubate mixture at 23 °C for 1.5 h and terminate reaction by adding 40 μl sterile, nuclease-free water plus 100 μl of phenol/chloroform/isoamyl alcohol (25:24:1, v/v).
8. After extraction, take 40 μl of supernatant and add 40 μl of formamide loading dye and load 12 μl on a denaturing PAGE to analyze polyadenylation status. Do not touch the protein-enriched layer during the pipetting step.

7.3.3. Depletion of protein with specific interacting protein and polyadenylation assay

This method is essentially that described by Svitkin and Sonenberg for poly (A) binding protein depletion (Svitkin and Sonenberg, 2004). We slightly modified this method to optimize it for *Xenopus* egg extracts.

1. Prepare 10 μl of MagneGST (Promega) beads and wash beads 1 to 2 times with 1× PBS.
2. A GST-fusion protein produced in *E. coli* soluble fraction is generally used for the protein-conjugated affinity column. The amount of GST-fusion protein should be quantified empirically with a simple resin-binding assay. Alternately, purified GST-fusion protein can be used for making an affinity column.
3. Wash the MagneGST beads with lysis buffer (20 mM Na-phosphate [pH 7.6], 300 mM NaCl, 0.5 mM phenylmethanesulphonylfluoride, 10 mM imidazole, 1 mM β-mercaptoethanol, and 10% glycerol [v/v]). Usually 2 to 2.5 μg of GST or GST-fusion protein is used for making the protein affinity column.
4. After extensive washing, incubate the GST-fusion protein-conjugated beads with 25 μl of egg extracts (contains 7.5 mM creatine phosphate [Roche], 1 mM ATP, and 1 mM MgCl$_2$) plus 1 unit/μl RNaseOUT. Gently shake tubes on top of a vortexing platform and incubate 30 min at 4 °C. Repeat this step three times followed by one time with unconjugated beads.
5. Polyadenylation assays are carried out at 23 °C for 1.5 h with 0.4 volume of egg extract, 25,000 cpm of [^{32}P]-lableled RNA, and 0.6 volumes of

reaction mixture containing 20 mM Tris–HCl (pH 7.5), 50 mM KCl, 2.5 mM MnCl$_2$, 50 µg/ml BSA, 10% glycerol (v/v) final concentration.

6. After a single phenol/chloroform extraction, the RNA probe is analyzed on 3.5 or 5% denaturing PAGE. Check the polyadenylation pattern by PhosphorImager or autoradiography.

7.3.4. In vitro polyadenylation-translation assay

For the *in vitro* translation experiments, a previous method (Patrick *et al.*, 1989) was adapted and modified. The reaction assay is as follows:

6 µl Egg extract

2 µl 5× Egg extract translation buffer (100 mM creatine phosphate [Roche], 2.5 mM spermidine (Sigma-Aldrich), 5 mM MnCl$_2$, 200 mM KCl, 50 mM Tris–HCl [pH 7.5])

1 µl 10% Supplement (composed of 1 µg of calf liver tRNA [Novagen], 0.2 µl of 1 mM complete amino acid mixture (Promega), and 2 µg of creatine phosphokinase [Roche])

0.2 µl RNaseOUT (Invitrogen)

1 fmol [^{32}P]-labeled RlucB1-CPE(+) mRNA (Kim and Richter, 2007)

Up to 10 µl nuclease-free water

Incubate the reaction mixture at 23 °C for 1.5 h and dilute appropriately with 1× passive lysis buffer (Promega) and subject to a luciferase assay.

ACKNOWLEDGMENTS

We are grateful to the Richter lab members for helpful comments. J. H. K. and J. D. R. were supported by grants from the National Cancer Center, Korea (0810040-1) and the US National Institutes of Health (GM46779), respectively.

REFERENCES

Barnard, D. C., Ryan, K., Manley, J. L., and Richter, J. D. (2004). Symplekin and xGLD-2 are required for CPEB-mediated cytoplasmic polyadenylation. *Cell* **119,** 641–651.

Bilger, A., Fox, C. A., Wahle, E., and Wickens, M. (1994). Nuclear polyadenylation factors recognize cytoplasmic polyadenylation elements. *Genes Dev.* **8,** 1106–1116.

Copeland, P. R., and Wormington, M. (2001). The mechanism and regulation of deadenylation: identification and characterization of *Xenopus* PARN. *RNA* **7,** 875–886.

Desai, A., Murray, A., Mitchison, T. J., and Walczak, C. E. (1999). The use of *Xenopus* egg extracts to study mitotic spindle assembly and function *in vitro*. *Methods Cell Biol.* **61,** 385–412.

Ferby, I., Blazquez, M., Palmer, A., Eritja, R., and Nebreda, A. R. (1999). A novel p34^{cdc2}-binding and activating protein that is necessary and sufficient to trigger G$_2$/M progression in *Xenopus* oocytes. *Genes Dev.* **13,** 2177–2189.

Hake, L. E., and Richter, J. D. (1994). CPEB is a specificity factor that mediates cytoplasmic polyadenylation during *Xenopus* oocyte maturation. *Cell* **79,** 617–627.

Huarte, J., Stutz, A., O'Connell, M. L., Gubler, P., Belin, D., Darrow, A. L., Strickland, S., and Vassalli, J. D. (1992). Transient translational silencing by reversible mRNA deadenylation. *Cell* **69,** 1021–1030.

Kim, J. H., and Richter, J. D. (2006). Opposing polymerase-deadenylase activities regulate cytoplasmic polyadenylation. *Mol. Cell* **24,** 173–183.

Kim, J. H., and Richter, J. D. (2007). RINGO/cdk1 and CPEB mediate poly(A) tail stabilization and translational regulation by ePAB. *Genes Dev.* **21,** 2571–2579.

Korner, C. G., Wormington, M., Muckenthaler, M., Schneider, S., Dehlin, E., and Wahle, E. (1998). The deadenylating nuclease (DAN) is involved in poly(A) tail removal during the meiotic maturation of *Xenopus* oocytes. *EMBO J.* **17,** 5427–5437.

Kwak, J. E., Wang, L., Ballantyne, S., Kimble, J., and Wickens, M. (2004). Mammalian GLD-2 homologs are poly(A) polymerases. *Proc. Natl. Acad. Sci. USA* **101,** 4407–4412.

Mendez, R., Barnard, D., and Richter, J. D. (2002). Differential mRNA translation and meiotic progression require Cdc2-mediated CPEB destruction. *EMBO J.* **21,** 1833–1844.

Mendez, R., Hake, L. E., Andresson, T., Littlepage, L. E., Ruderman, J. V., and Richter, J. D. (2000). Phosphorylation of CPE binding factor by Eg2 regulates translation of c-*mos* mRNA. *Nature* **404,** 302–307.

Mendez, R., and Richter, J. D. (2001). Translational control by CPEB: a means to the end. *Nat. Rev. Mol. Cell. Biol.* **2,** 521–529.

Murray, A. W., and Kirschner, M. W. (1989). Cyclin synthesis drives the early embryonic cell cycle. *Nature* **339,** 275–280.

Padmanabhan, K., and Richter, J. D. (2006). Regulated Pumilio-2 binding controls RINGO/Spy mRNA translation and CPEB activation. *Genes Dev.* **20,** 199–209.

Paris, J., Swenson, K., Piwnica-Worms, H., and Richter, J. D. (1991). Maturation-specific polyadenylation: *In vitro* activation by p34cdc2 and phosphorylation of a 58-kD CPE-binding protein. *Genes Dev.* **5,** 1697–1708.

Patrick, T. D., Lewer, C. E., and Pain, V. M. (1989). Preparation and characterization of cell-free protein synthesis systems from oocytes and eggs of *Xenopus laevis*. *Development* **106,** 1–9.

Richter, J. D. (2007). CPEB: a life in translation. *Trends Biochem. Sci.* **32,** 279–285.

Salles, F. J., Richards, W. G., and Strickland, S. (1999). Assaying the polyadenylation state of mRNAs. *Methods* **17,** 38–45.

Sive, H. L., Grainger, R. M., and Harland, R. M. (2000). Early Development of *Xenopus laevis*: A Laboratory Manual. pp. 288–289. Cold Spring Harbor Laboratory Press, New York.

Smith, L. D., Xu, W. L., and Varnold, R. L. (1991). Oogenesis and oocyte isolation. *Methods Cell Biol.* **36,** 45–60.

Svitkin, Y. V., and Sonenberg, N. (2004). An efficient system for cap- and poly(A)-dependent translation *in vitro*. *Methods Mol. Biol.* **257,** 155–170.

Voeltz, G. K., Ongkasuwan, J., Standart, N., and Steitz, J. A. (2001). A novel embryonic poly(A) binding protein, ePAB, regulates mRNA deadenylation in *Xenopus* egg extracts. *Genes Dev.* **15,** 774–788.

Wang, L., Eckmann, C. R., Kadyk, L. C., Wickens, M., and Kimble, J. (2002). A regulatory cytoplasmic poly(A) polymerase in *Caenorhabditis elegans*. *Nature* **419,** 312–316.

CHAPTER EIGHT

The Preparation and Applications of Cytoplasmic Extracts from Mammalian Cells for Studying Aspects of mRNA Decay

Kevin J. Sokoloski, Jeffrey Wilusz, *and* Carol J. Wilusz

Contents

1. Introduction	140
2. Preparation of HeLa-Cell Cytoplasmic Extracts	142
2.1. Preparation of HeLa-Cell S100 cytoplasmic extracts	143
2.2. Standardization of cytoplasmic extract activity	145
2.3. Protocol	147
3. Preparation of RNA Substrates	148
3.1. Transcription of polyadenylated and nonadenylated RNA substrates	148
3.2. Production of cap-labeled RNA substrate	151
4. Evaluating mRNA Decay with Cytoplasmic Extracts	152
4.1. Assaying deadenylation rates	152
4.2. Protocol	152
4.3. Determining exonuclease activity	153
4.4. Protocol	155
4.5. Assaying decapping activity	156
4.6. Protocol	156
4.7. Analysis of *trans*-acting factors with ultraviolet crosslinking	157
4.8. Protocol	158
4.9. Immunoprecipitation protocol	160
4.10. Immunodepletion of factors from cytoplasmic extracts	161
4.11. Protocol	161
5. Concluding Remarks	161
Acknowledgments	162
References	162

Department of Microbiology, Immunology and Pathology, Colorado State University, Ft. Collins, Colorado, USA

Methods in Enzymology, Volume 448   © 2008 Elsevier Inc.
ISSN 0076-6879, DOI: 10.1016/S0076-6879(08)02608-6   All rights reserved.

Abstract

HeLa S100 cytoplasmic extracts have been shown to effectively recapitulate many aspects of mRNA decay. Given their flexibility and the variety of applications readily amenable to extracts, the use of such systems to probe questions relating to the field of RNA turnover has steadily increased over time. Cytoplasmic extract systems have contributed greatly to the field of RNA decay by allowing valuable insight into RNA–protein interactions involving both the decay machinery and stability/instability factors. A significant advantage of these systems is the ability to assess the behaviors of several transcripts within an identical static environment, reducing errors within experimental replications. The impact of the cytoplasmic extract/*in vitro* RNA decay technology may be further advanced through manipulations of the extract conditions or the environment of the cells from which it is made. For instance, an extract may be produced from cells after depletion of a specific factor by RNAi, giving insight into the role of that factor in a particular process.

The goals of this chapter are threefold. First, we will familiarize the reader with the process of producing high-quality, reliable HeLa-Cell cytoplasmic extracts. Second, a method for the standardization of independent extracts is described in detail to allow for dependable extract-to-extract comparisons. Finally, the use and application of cytoplasmic extracts with regard to assaying several aspects of mRNA turnover are presented. Collectively these procedures represent an important tool for the mechanistic analysis of RNA decay in mammalian cells.

1. INTRODUCTION

The biochemical reconstitution of cellular processes has become an invaluable tool for the explanation of many features of RNA biology. This is due, in part, to the relative simplicity and reliability of extract production, as well as the ability to address focused mechanistic and biochemical questions using cell-free systems that are often difficult to assess using living cells. Because of their high rate of division coupled with the ability to grow in suspension, HeLa S3 cells make an excellent source for cytoplasmic extracts. To date, nuclear and cytoplasmic extracts from HeLa and other cell lines have been an asset for evaluating mRNA transcription (Handa *et al.*, 1981; Manley *et al.*, 1980), splicing (Handa *et al.*, 1981; Hernandez and Keller, 1983), polyadenylation (Mifflin and Kellems, 1991; Moore and Sharp, 1985), translation (Brown and Ehrenfeld, 1979), and stability/decay (Ford and Wilusz, 1999; Opyrchal *et al.*, 2005), the latter being the focus of this chapter.

Most cytoplasmic mRNA decay is initiated by shortening of the poly(A) tail by one or more members of a cadre of cellular deadenylases (Garneau et al., 2007). The body of the transcript is then degraded by two exonucleolytic pathways, perhaps acting in unison (Garneau et al., 2007; Murray and Schoenberg, 2007). In one pathway, the 5′-cap of the mRNA is removed after deadenylation. This decapping step is mediated by DCP2 in conjunction with DCP1 and several auxiliary factors, including the Lsm protein complex (Chowdhury et al., 2007; Cohen et al., 2005; Piccirillo et al., 2003; Wang et al., 2002). The decapped mRNA becomes a substrate for rapid 5′ to 3′-exonuclease digestion by XRN1 (Garneau et al., 2007). Many aspects of this pathway are associated with cellular cytoplasmic structures called P bodies (Parker and Sheth, 2007). In the second pathway, the mRNA is degraded after deadenylation in a 3′ to 5′-direction by a multicomponent complex called the exosome (Houseley et al., 2006). After extensive degradation of the body of the mRNA in this pathway, the 5′-cap is removed and recycled by a scavenger decapping activity (Gu et al., 2004; Liu and Kiledjian, 2005). Alternate pathways of decay do exist, including variations of the ones outlined earlier, as well as endonuclease-initiated pathways (e.g., Orban and Izaurralde, 2005). Collectively, mRNA decay pathways play a significant role in both the regulation and quality control of gene expression in a eukaryotic cell (Cheadle et al., 2005; Garcia-Martinez et al., 2004).

HeLa cytoplasmic extracts have been shown to faithfully recapitulate many general and regulated aspects of the cellular decay machinery. To date, HeLa cytoplasmic extracts have been used to determine the substrate requirements and rates of deadenylation by the PARN deadenylase (Ford and Wilusz, 1999), characterize aspects of both the 5′ to 3′ and 3′ to 5′-exonuclease activities (Chen et al., 2001; Ford and Wilusz, 1999; Mukherjee et al., 2002), study DCP2 and scavenger decapping activities (Bergman et al., 2002; Liu et al., 2002), and explore the identities and roles of numerous regulatory proteins (Chen et al., 2001; Ford et al., 1999; Gherzi et al., 2004; Zhang et al., 1993). This work has made a substantial contribution to our mechanistic understanding of mRNA decay. A comprehensive approach to recapitulating and analyzing these activities is described below. In addition, a means to standardize extracts to allow for more accurate comparisons of independent extracts is presented. Standardization, for example, will prove invaluable for mechanistic analyses involving extracts made from independent cell lines exhibiting RNAi-mediated knockdown of decay factors. Finally, a UV cross-linking approach is described to aid in the rapid visualization of protein factors that may be involved in the regulation of mRNA decay by novel or message-specific RNA elements. Collectively, these methods should be useful for the in-depth mechanistic analysis of decay factors and the identification of new regulatory proteins.

2. Preparation of HeLa-Cell Cytoplasmic Extracts

HeLa-Cell extracts have consistently proven to be valuable tools for determining mechanistic aspects of physiologic processes. It is important to note that extracts can be readily produced from HeLa-Cells grown under a variety of circumstances. Altering the cellular environment may yield extracts that can be used to directly address many novel questions in the field. For instance, extracts may be produced from HeLa-Cell pretreated with small molecule inhibitors of cellular processes or infected with a given virus before harvesting. Perhaps of most general use is the ability to produce extracts after si/shRNA knockdown of a targeted gene of interest. Because most of these modifications to the cellular environment are rather specialized and occur before the actual creation of the cytoplasmic extract, they will not be described in great detail here. Section 2.1 provides a detailed description of cytoplasmic extract preparation.

Section 2.2 of this chapter deals with the issue of standardization when comparing results obtained with independently generated extract preparations. The production of a single HeLa-Cell extract is rarely, if ever, sufficient to accomplish a goal. It is important to first stress that the *in vitro* RNA decay assays outlined in the following protocols are highly reproducible from extract to extract. This is largely based on the fact that extracts are often made with large volumes (liters) of cultured cells, minimizing population effects, thereby reducing the overall variation between preparations. However, despite this, individual extract preparations can occasionally behave differently from one another, even when equal concentrations of protein are used. This discrepancy between preparations likely arises because of a loss of activity or differential extraction of select components. Although one can adjust incubation times to get similar results between individual extract preparations, these empirical adjustments are not possible, for example, if one is attempting to compare the relative activities of extracts made from wild-type HeLa-Cells and a line that is knocked-down for the expression of a suspected regulatory decay factor. In this case, it is useful to make extracts from knockdown cells expressing a variant of the factor in question that is immune to the si/shRNA-mediated knockdown to rescue any defect, further demonstrating the role of the targeted factor. To a similar end, recombinant protein may be added back to a deficient extract to confirm the roles performed by the factor of interest.

Independent extracts should first be normalized on the basis of protein concentration. Determination of protein concentration is often approached through spectrophotometric means (for example, a Bradford assay), in which a linear curve of absorbance is determined through serial dilution of a known protein standard. A sample of unknown concentration is then titrated, and protein concentration is extrapolated from the standard curve. Although this method allows for the determination of gross protein concentration, protein concentration alone is not necessarily implicitly informative of the quality of the extract. This observation is likely due to the possibility of contaminating serum, high levels of cellular lipids within the extract, and/or the presence of large amounts of cellular RNA. An approach to provide a more consistent means of standardizing individual extracts is outlined in Section 2.2.

2.1. Preparation of HeLa-Cell S100 cytoplasmic extracts

Before producing a HeLa-Cell cytoplasmic extract, one must first obtain a sufficient quantity of cells. To this end, one can use a 6-L spinner flask to culture large volumes of HeLa-Cells in JMEM (Hyclone) supplemented with 10% horse serum (Hyclone) and appropriate antibiotics. Spinner culture is performed at 37 °C without CO_2 and, therefore, does not require a specialized incubator. The use of horse serum is preferred over fetal bovine serum for its lower cost and because it produces a cleaner extract because of less membrane blebbing. Once a cell culture has been expanded to a relatively large volume (3 to 6 L), we find it is most economical to make several independent extracts on consecutive days by removing 50% of the culture for extraction and replenishing with fresh medium. This saves considerable time, considering that multiple independent extracts are needed for results to be statistically significant.

To produce a large-scale extract, several liters of cells grown to a density of $\sim$0.5 to 1.0 × 10^6 cells/ml will be needed. The cells should be healthy and actively dividing before harvest to ensure high-quality extract. Clumping of the cells should be minimal; this is prevented by increasing the rate of spin within the flask.

Harvest the desired volume of HeLa-Cells by means of centrifugation at 300g for 5 min at 4 °C. After centrifugation, completely decant and dispose of the supernatant in a manner appropriate for potentially hazardous biologic materials. To the HeLa-Cell pellets, which should be tight and uniform in color, add 50 ml of ice-cold phosphate buffered saline (PBS; Hyclone) and gently resuspend until homogenous in appearance. Transfer the cell suspension to a 50-ml centrifuge tube.

Wash the cells by pelleting the HeLa-Cell suspension with centrifugation at 300g for 5 min at 4 °C. Decant the supernatant as previously described and, once, more resuspend the cell pellet in 50 ml of PBS. Pellet the cell suspension with centrifugation and reserve HeLa cell pellets for hypotonic swelling in the next step. Note that because of the presence of excess medium, it is not unusual for the first wash to be slightly pink. The supernatant from the second wash should be clear and colorless. If not, additional washing is necessary. Take note of the approximate volume of the final cell pellet, because this will be used as a reference later. In general, from a liter of HeLa S3 cells one should obtain a cell pellet of ~0.5 to 1 ml in volume.

To swell the cells for subsequent lysis, resuspend the washed HeLa-Cell pellet with 5× the cell pellet volume (CPV) of chilled Buffer A (10 mM HEPES, pH 7.9,/1.5 mM MgCl$_2$/10 mM KCl/1 mM DTT). Gently resuspend the HeLa-Cells by inversion, and incubate on ice for 10 min. After osmotic swelling, centrifuge the cells at 300g for 5 min at 4 °C. Decant and discard the supernatant carefully, because cell pellets will not be as densely packed as they were when pelleted out of a PBS wash.

After a successful osmotic swelling, the cell pellet volume should have approximately doubled in size, and the supernatant should be generally clear. A murky supernatant is indicative of premature lysis, which can adversely affect the quality of the extract. If significant premature lysis is noted, the osmotic pressure may be modulated by modification of the buffer A salt concentrations. A simple titration to determine the proper concentration of the potassium chloride in buffer A is usually sufficient to determine the appropriate range for swelling for a given cell type.

To the swollen cell pellet, add 2× the CPV of buffer A. Gently resuspend the cell pellet until homogenous by slow inversion. Transfer the swollen cell suspension to an appropriate-sized dounce (Kontes) with a loose-fitting pestle that has been prechilled on ice. Swiftly dounce the swollen cells with 10 strokes of the pestle to rupture the cell membrane. Transfer the whole cell lysate to a fresh, prechilled 50-ml centrifuge tube. If desired, the cells may be visually inspected to assess lysis of the plasma membrane and integrity of the nuclear membrane by conventional microscopy.

To clear the dounced extract of intact nuclei, centrifuge the lysate for 10 min at ~850g. The resulting loosely pelleted nuclei should be approximately one half the original cell pellet volume and may be ruddy in color. Carefully pipette the supernatant into a fresh tube, leaving behind the nuclei and a small volume of the supernatant. Avoiding nuclear contamination will result in a higher quality extract, with a noticeably lower nonspecific exonuclease activity. Estimate and note the volume of the cytoplasmic fraction, because this will be used as a reference in the next step.

Add 0.09× cytoplasmic fraction volumes (CFV) of cold buffer B (300 mM HEPES, pH 7.9/30 mM MgCl$_2$/1.4 M KCl) to the cytoplasmic fraction and gently mix by inversion. Evenly aliquot the cytoplasmic fraction into several balanced ultracentrifuge tubes. Clarify the cytoplasmic fraction by centrifugation at 4 °C for 1 h at 100,000g in an ultracentrifuge.

After centrifugation, the cytoplasmic fraction will have clarified, and a large organelle/debris pellet will be visible. Remove the supernatant, which is the S100 cytoplasmic extract, to a fresh tube. At the surface of the cytoplasmic extract, there will often be a lipid "slick" that is white in color and creamy in consistency. In general, we try to avoid this layer during removal of the cytoplasmic extract. In our experience, inclusion of traces of this top lipid layer does not appreciably affect the activity of the extract; however, it may affect the spectrophotometric determination of the extracts' concentration. Estimate and take note of the volume of the cytoplasmic extract.

To the cytoplasmic extract, add 0.2 volumes of 80% glycerol. Mix by inversion. Aliquot the finished cytoplasmic extract and store in a $-80\,^{\circ}\mathrm{C}$ freezer. Flash-freezing in liquid nitrogen before storage is not essential. In our hands, cytoplasmic extracts when stored in an ultralow freezer have remained active for several years. Although, in general, cytoplasmic extracts have been observed to be tolerant to freeze–thaw cycling, it is advisable to aliquot extracts according to use to avoid numerous freeze–thaw cycles that could affect activity. Finally, before use, always be sure to thaw cytoplasmic extracts on ice to avoid any possible decreases in activity from mistreatment.

2.2. Standardization of cytoplasmic extract activity

When comparing independent cytoplasmic extract preparations, it is important to normalize the extracts to afford confidence in analyzing the results of your assays. First, the concentration of total protein must be determined with a Bradford or similar spectrophotometric assay. Although this determination provides a rough guide for the amount of extract to use for a given assay, it is not a determination of enzyme levels or activity. It is for this reason that we advocate further measuring the levels of the components responsible for activity. To minimize, if not eliminate, the variation between extracts, one can calibrate the activity of the different extracts by determining relative levels of the enzyme responsible for the observed phenomenon (for example, a deadenylase). The relationship between the extracts may then be applied when analyzing the result to detect alterations of stability because of biologically relevant phenomena rather than extract-to-extract differences.

As an example, suppose that one wishes to determine whether an RNA-binding factor results in altered deadenylation on association (Fig. 8.1A). A cell line deficient in this hypothetical factor (protein X) has been generated through application of RNAi technology. Furthermore, the protein X-knockdown cells have been used to produce an extract with the protocol described previously and adjusted to the same overall protein concentration as an extract made from wild-type/control cells. The rates of deadenylation for a given substrate are then determined in independent extracts from

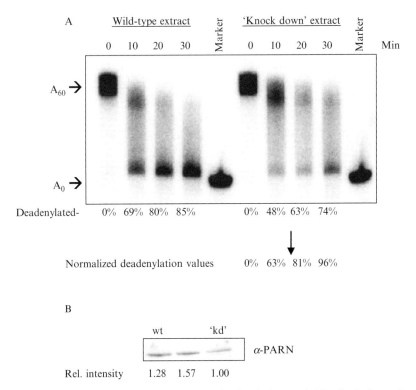

Figure 8.1 A method for normalizating deadenylation activities in independent extracts. (A) Representative deadenylation reactions performed in wild-type (wt) extract and a hypothetical extract deficient in Protein "X" ('knock down', kd) as outlined in the text. The lanes labeled "marker" contain an unadenylated version of the input RNA. Listed immediately below the gel is the percent of RNA substrate that has undergone deadenylation at each time point. The normalized deadenylation values for the time points in the "knockdown" extract are listed below the downward-facing arrow. (B) Western blotting analysis to determine the relative levels of PARN in individual extracts. "wt" and "kd" refer to the wild-type and "knocked-down" extracts, respectively, used in A. The middle lane is from a third independent extract not related to the experiment in A. The relative intensities of the PARN band as determined by densitometry are listed at the bottom of the panel. α, anti.

wild-type cells and the protein X knockdown cell line with the protocol described in section 4.1. As depicted in Fig. 8.1A, the rates of deadenylation are ~1.4× different between wild-type/control and protein X knockdown extracts. At first glance, it seems that the factor does, indeed, influence deadenylation of the substrate. However, before conclusions as to the putative roles of factor X are made, simple extract-to-extract variation must be ruled out as the cause of this observed difference in deadenylation rates.

Obviously, multiple extracts from the knockdown cell line should be compared with multiple wild-type extracts. However, it is also possible to minimize the effects of extract-to-extract variation in an individual experiment. One helpful test that we have used is to evaluate the relative levels of Poly(A) ribonuclease (PARN) in independent extracts by Western blotting (Fig. 8.1B). By taking into account the relative intensity of PARN band detected between independent extracts, a calibration factor can be determined. After adjusting the observed rates of deadenylation to reflect the levels of PARN present in individual extracts/assays by multiplying the observed values by the calibration factor, the researcher ultimately finds that the original differences in deadenylation rates can be largely explained by minor variations of PARN levels in the extract rather than a biologically relevant difference between the cells. Clearly, application of this type of extract normalization will give increased confidence in the interpretation of extract-to-extract comparisons in *in vitro* RNA decay assays.

There is one additional caveat to the hypothetical scenario just presented. Although the fact that PARN levels are reduced in the knockdown cell line may in and of itself be biologically relevant, the interaction between the RNA substrate and PARN does not seem at this level of analysis to have been significantly altered as determined after calibration. Should this downregulation of PARN levels prove to be reproducible, further experimentation is necessary to determine whether the depletion of protein "X" could potentially alter the overall rates of deadenylation within the cell by affecting deadenylase levels.

2.3. Protocol

To calibrate independent extracts as described in the preceding example, we routinely add 5 μg of extract to a microfuge tube for each independent extract and adjust each sample to an equal volume with buffer D (20 mM HEPES, pH 7.9/100 mM KCl/20% (v/v) glycerol/1 mM DTT). Add a sufficient amount of 2× SDS protein loading dye (500 mM Tris-Cl, pH 6.8/2% (w/v) SDS/100 mM DTT/5% (v/v) glycerol/0.004% (w/v) bromophenol blue) and mix completely. Heat the samples at 90 °C for a minimum of 3 min before electrophoresis. Load an equal amount of each sample onto a SDS-containing 10% polyacrylamide gel. Electrophorese samples in the gel at constant current (30 mA) until the desired resolution has been obtained. Transfer the proteins to a membrane with a conventional Western blotting protocol. Once transfer is complete, block the membrane and probe it to detect PARN with available antibodies (e.g., BIOO Scientific) according to the supplier's specifications. With image analysis software, evaluate the intensity of the bands relative to each other and calculate the ratios of extract needed to attain similar levels of PARN protein. As outlined in Fig. 8.1, these ratios are often an excellent indication of relative extract activity and can be used to normalize results and provide additional confidence in conclusions drawn from independent extracts. In principle, a similar approach can be used to normalize decapping and exosome protein levels, assuming appropriate antibodies are available.

3. Preparation of RNA Substrates

To use cytoplasmic extracts to assay aspects of mRNA decay as described later, one must first produce the required RNA substrates. Depending on the assay, different features must be present (or absent) from the RNA substrate. These features include a 3′-poly(A) tail, a 5′-^{7me}GpppN cap, as well as selected [^{32}P]-labeled nucleotides.

Because the yield of an *in vitro* transcription is largely reliant on the DNA template, the selection of an appropriate transcription vector is the first step toward producing a high-quality transcript. To prepare RNAs that possess a 60-base poly(A) tail located precisely at their 3′-end, we routinely use a pGEM-4 (Promega) derivative termed pGEM A60 as the parent construct for our transcription templates. As depicted in Fig. 8.2, this vector has been modified to include a polyadenylate tail of 60 residues immediately after the *Hind*III site of the multiple cloning site (MCS). Importantly, following the 60-base adenylate tract is an *Nsi*I site. Cleavage with *Nsi*I, therefore, produces a template that will place the poly(A) tract at the precise 3′-end of the resulting transcript driven from the SP6 polymerase promoter. Individual sequences/elements to be assayed for their effects on mRNA stability are cloned into sites within the MCS. It is important to ensure that your insert does not contain an *Nsi*1 site. In the event that this cannot be avoided, a pGEM-A60 variant with an *Ssp*1 site following the poly(A) tract can be used.

3.1. Transcription of polyadenylated and nonadenylated RNA substrates

It is often desirable to produce DNA templates specific to either the adenylated or nonadenylated form of the sequence to be assayed. To this end, the pGEM A60 transcription vector must be independently restriction digested according to the manufacturer's instructions with *Hind*III (to produce the nonadenylated form) and *Nsi*I (for the adenylated form). After digestion of the vector, reaction mixtures are diluted with 350 μl of high salt column buffer (HSCB: 25 mM Tris-HCl, pH 7.6/400 mM NaCl/ 0.1% sodium dodecyl sulfate) supplemented with 10 μg of proteinase K and incubated at 37 °C for 15 min to remove any possible contaminating RNases. After proteinase treatment, the template DNA is extracted with phenol chloroform/isoamyl alcohol (PCI; (v/v/v) 25:24:1) and precipitated in ethanol to recover the template DNA.

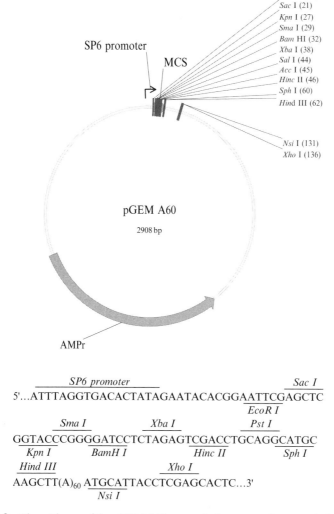

Figure 8.2 Plasmid map of the pGEM A60 transcription vector. Inset text is descriptive of the multiple cloning sites located upstream of the poly(A) encoding stretch.

To PCI extract the DNA, adjust the reaction volume to 100 μl and add an equal volume of PCI. Mix the solutions to milky homogeneity by vortexing and centrifuge the samples at ∼16,000g for 2 min to separate the organic and aqueous phases. After centrifugation, carefully remove the upper (aqueous) layer to a fresh microfuge tube. Adjust the salt, if necessary, by adding 0.1 volumes of 3 M sodium acetate, pH 5.0, and add 2.5 volumes of 100% ethanol, inverting the tube several times to ensure complete mixing. Precipitate the sample on dry ice for a minimum of 10 min.

Centrifuge the samples at $\sim$16,000g for 10 min to pellet the DNA templates. Remove the supernatant carefully to avoid dislodging the DNA pellet and wash the pellet by adding 200 μl of 80% ethanol to the tube. Briefly centrifuge the sample and remove the supernatant. Vacuum-dry the pellet to remove excess ethanol, and resuspend the DNA plasmid template to a final concentration of 1 μg/μl.

In a 1.5-ml microfuge tube, mix 1 μg of DNA template, 1 μl of 10× SP6 polymerase buffer (NEB), 1 μl of a 10× U-labeling/capping rNTP mix (10 mM ATP/5 mM CTP/0.5 mM GTP/0.5 mM UTP), 1 μl of 5 m$M^{7\mathrm{me}}$GpppG cap analog (GE Healthcare), 4.5 μl of [γ-^{32}P]-UTP (800 mCi/mmol, MP Biomedicals), 0.5 μl RNase Inhibitor (Roche) and 1 μl of SP6 RNA Polymerase (NEB). Note that radiolabeling with alternate nucleotides can be achieved simply by use of a similar amount of radiolabeled NTP and adjusting the concentration of the unlabeled version of the nucleotide in the reaction to 0.5 mM. It is important to assemble these reagents at room temperature to avoid the possible precipitation of the DNA template by the spermidine present in the reaction mix. Gently mix by pipetting the solution and incubate the complete transcription reaction at 37 °C for at least 1 h. To increase the yield from a transcription, extending the transcription time is advised, because it increases the number of transcription initiation events; this is particularly useful when transcribing smaller substrates.

After the incubation, adjust the final volume to a total of 160 μl with RNase-free water. PCI extract the mixture as performed previously for the preparation of the DNA template, disposing of radioactive wastes appropriately. Precipitate the RNA in a fresh tube by the addition of 40 μl of 10 M ammonium acetate and 2.5 volumes of 100% ethanol accompanied by incubating on dry ice for at least 10 min. Centrifuge the precipitated transcriptions at $\sim$16,000g for 10 min. If necessary, the samples may be stored in ethanol for prolonged periods until sufficient time to complete the protocol is available. The use of ammonium ions rather than sodium ions to precipitate the RNA significantly reduces the precipitation of unincorporated nucleotides.

Carefully (but as completely as possible) remove the radioactive supernatant and dispose of appropriately. Wash the pellets by adding 200 μl of 80% ethanol. Remove the supernatant after a quick spin and vacuum-dry the pellet to remove all traces of ethanol. Resuspend the RNA pellet in 10 μl of RNA loading buffer (20 mM Tris-HCl, pH 7.5/8 M urea/1 mM EDTA/ 0.002% (w/v) bromophenol blue/0.002% (w/v) xylene cyanol). Heat the samples at 90 °C for 30 sec and chill on ice to relax any secondary structures.

Gel purify the transcriptions by resolving them in a 5% denaturing (7 M urea) polyacrylamide gel. After electrophoresis, gently separate the glass plates and cover the gel with plastic wrap. Expose the radioactive gel to autoradiograph film (Kodak) for a short time (usually less than 1 min) and develop the film to determine the position of the RNA in the gel. Excise the desired transcription products with a sterile razor blade, minimizing the amount of gel around the band, and immerse the gel slices in 400 µl of HSCB supplemented with 10 µg of proteinase K. The protease is added to ensure the integrity of the RNA during elution by degrading any trace RNases that may have contaminated the sample during processing. Elute the RNA from the gel slice by incubating overnight at room temperature. After elution, PCI extract and ethanol precipitate as previously without the addition of excess salt, because the HSCB elution buffer has sufficient amounts for precipitation. Resuspend the RNA samples in ~21 µl of RNase-free distilled water. Analyze 1 µl of the final volume with a liquid scintillation counter and adjust the final counts per minute (cpm) to 100,000 cpm. A typical transcription reaction will yield approximately 10 to 20 million total counts. This value can be used to calculate the specific activity of the transcript, which is important when comparing different RNA substrates (note that Ambion, Inc. has several resources available in its technical library to assist in these calculations). Occasionally, higher yields of RNA may be attained; these variations are largely dictated by the relative specific activity of the nucleotide, the number of residues capable of incorporating the radiolabeled nucleotide, and, to a lesser extent, the quality of the DNA template.

3.2. Production of cap-labeled RNA substrate

The procedure for producing the cap-labeled RNA substrate for the decapping assay in section 4.3 is rather straightforward. First, a source of unlabeled RNA substrate with a 5′-triphosphate needs to be generated. This may be produced with the protocol listed as 3.1 with minor modifications. To produce an unlabeled substrate, use equal molar concentrations of rNTPs (5 mM) and use 1 µl of dH$_2$O in lieu of cap analog. Proceed as described in section 3.1 of this chapter, stopping before addition of gel loading dye.

After precipitation of the transcription reactions, resuspend the RNA directly in a mixture containing 0.5 µl 10 mM S-adenosylmethionine (SAM; Sigma), 0.5 µl 15× capping buffer (750 mM Tris-HCl, pH 7.9/60 M MgCl$_2$/20 mM spermidine/100 mM DTT), and 4.5 µl [γ-^{32}P]-GTP (800 mCi/mmol). Mix thoroughly and add 1 µl RNase inhibitor (Roche), as well as 1 µl of vaccinia capping enzyme (guanylyltransferase; Ambion). Incubate the reaction at 37°C for a minimum of 1 h. Proceed with PCI extraction, ethanol precipitation, and gel purification as described in section 3.1. Because the RNA is only labeled at the 5′-end, the specific activity of cap-labeled transcripts is lower than that of internally labeled RNAs. Normally, approximately 1 to 5 million cpm is obtained from an average capping reaction.

4. Evaluating mRNA Decay with Cytoplasmic Extracts

Extracts prepared as described previously have been successfully used to reproduce many aspects of the cytoplasmic mRNA decay pathways, including deadenylation, exonucleolytic decay (both 5′ to 3′ and 3′ to 5′), and decapping. Outlined in the following are the standard assays that we use to assay each of these enzymatic activities.

4.1. Assaying deadenylation rates

The predominant pathway of cellular mRNA decay initiates through removal of the 3′-poly(A) tail (Garneau *et al.*, 2007). The interaction of the RNA transcript with the cellular deadenylation machinery plays several important physiologic roles within the cell. For example, the translational efficiency of many mRNAs depends on the presence of a poly(A) tail (Kuhn and Wahle, 2004). In addition, deadenylation also plays regulatory roles in the cell as observed with nocturnin, a circadian deadenylase responsible for the regulation of many mRNAs (Garbarino-Pico *et al.*, 2007; Green and Besharse, 1996). Aside from increasing the translational activity, the poly(A) tail is responsible for protecting the 3′-end of the transcript from the activities of the mRNA decay machinery. Finally, deadenylation is the first and often the rate-limiting step in the decay of most mRNAs (Garneau *et al.*, 2007). Therefore, it is important to determine how the cellular deadenylation machinery interacts with a given RNA transcript. To date, PARN has been the predominant deadenylase in the cytoplasmic extracts from the cell types that we have examined from both vertebrate and invertebrate cell lines (Ford and Wilusz, 1999; Opyrchal *et al.*, 2005). Therefore, the following protocol describes an effective *in vitro* assay for measuring the activity of this major cellular deadenylase.

4.2. Protocol

Into a fresh 1.5-ml microfuge tube, combine 6.5 μl of 10% (w/v) polyvinyl alcohol (PVA), 2 μl of phosphocreatine/adenosine triphosphate (PC/ATP, 250 mM phosphocreatine/12.5 mM ATP) and 2 μl of poly-adenylic acid (Poly(A), 500 ng/μl; Sigma) on ice. Add 16 μl of HeLa cytoplasmic extract ($\sim$5 mg/ml), mix gently by pipetting, and return to ice. Add 2 μl of 5′-capped, radiolabeled polyadenylated RNA substrate ($\sim$100 K cpm/μl). Mix the reaction by gentle pipetting. The addition of exogenous poly(A) is necessary for activation of deadenylation, because it competes off cellular poly(A) binding proteins (Ford and Wilusz, 1999).

Assessing mRNA Decay in Cytoplasmic Extracts

To produce an initial or "zero" time point, immediately remove 5 µl of the reaction mixture and pipette it into 400 µl of HSCB buffer for future processing. Incubate the rest of the reaction tube at 30 °C, removing 5 µl, and processing as earlier at the desired time points. The sample volume removed can be adjusted depending on the number of time points. Depending on the sensitivity of detection, as little as 2 µl of sample may be accurately removed and processed as a time point.

On completion of the time course (or between time points if there is sufficient time), PCI extract the samples that have been mixed with HSCB buffer. To each sample tube, add an equal volume of PCI and vortex until homogenous. Separate the organic and aqueous phases by centrifugation of the samples at $\sim$16,000g for 2 min. Transfer the aqueous (upper) layer, taking special care to not contaminate the sample with the organic phase, to a fresh 1.5-ml microfuge tube. It is advisable (but not absolutely necessary) to add 10 µg of tRNA to the mixtures to aid in quantitative precipitation of the RNA. However, if this potentially could interfere with the resolution of the RNA substrate during gel electrophoresis, glycogen can be used instead as a carrier. Add 1 ml of cold 100% ethanol. Invert the tube several times and vortex to mix completely and precipitate on dry ice for at least 10 min. If desired, the samples may be stored at $-80\,°C$ until all samples in the time course protocol are brought to this stage.

Centrifuge the samples at $\sim$16,000g for 10 min to pellet the precipitated RNA. Remove the supernatant completely, taking care to retain the pelleted RNA. Wash the pellet by the addition of 200 µl of 80% ethanol, quick spin, and remove the supernatant. Vacuum-dry the pellets for 30 sec. Resuspend the RNA samples in 5 µl of RNA loading buffer. To ensure complete resuspension of the pellet, vortex the sample tubes vigorously. Heat the samples to 90 °C for 30 sec and snap-chill on ice to relax any secondary structures present in the RNA substrate. Load samples onto a 5% denaturing (7 M urea) polyacrylamide gel, and run at 600 V/60 mA/15 W until sufficient resolution has been obtained. Dry the gel with a slab dryer and expose to a PhosphorImager screen.

Rates of deadenylation may be determined by performing densitometric analysis of adenylated and deadenylated species. A typical pattern of deadenylation is exhibited in Fig. 8.3A.

4.3. Determining exonuclease activity

After removal of the 3′-poly(A) tail, the body of the mRNA can be degraded in one of two ways. Because the 3′-end of the transcript is no longer protected from exonuclease activity by the poly(A) tail, the transcript can be decayed in a 3′ to 5′-direction. This event occurs as result of the enzymatic activity of the cellular exosome, a large complex of proteins, many of which contain signature motifs associated with exonuclease activity

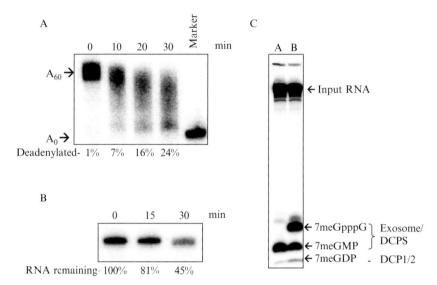

Figure 8.3 Examples of assays to evaluate specific aspects of RNA decay. (A) A representative deadenylation reaction involving a capped and polyadenylated GEM A_{60} transcript. Reactions were stopped at the time points indicated and products analyzed in a 5% denaturing acrylamide gel. The marker lane represents the fully deadenylated version of the input transcript. The percent of fully deadenylated product at each time point is presented immediately below the gel. (B) Representative 3′ to 5′-exosome decay assay with the unadenylated form of the Gem A_{60} transcript (GemA_0). Reactions were stopped at the time points indicated and products analyzed in a 5% denaturing acrylamide gel. The percent of input RNA remaining at each time point is denoted below the gel. (C) A representative decapping assay. GemA_0 RNA labeled exclusively at its 5′-cap structure was incubated under decapping conditions for 30 min in the absence (lane A) or presence of cap analog (lane B). Excess cap analog blocks the action of DCPS and allows the terminal product of exosome decay to be observed. Reaction products were purified and analyzed in a 20% denaturing acrylamide gel as described in the text. The position of the relevant reaction products is indicated on the right.

(Chen et al., 2001; Mukherjee et al., 2002). Alternately, the 5′-cap structure may be removed by DCP2 in a process termed decapping. After removal of the cap, the body of the transcript is decayed by XRN1, a 5′ to 3′-exonuclease (Garneau et al., 2007). Described in the following is a protocol for assaying the activity of both of these exonucleolytic pathways *in vitro*.

Often it can be advantageous to determine the directionality of RNA decay. To ascertain which pathway is predominantly responsible for the decay of a transcript in an *in vitro* decay assay, one can alter the RNA substrate to produce decay products whose size allows one to discern the responsible pathway. In brief, this involves the inclusion of modified bases within the body of the transcript that will effectively block progress of an exonuclease and result in trapped intermediates. A modification that we have had significant success with as an exonucleolytic block is the inclusion of tandem, site-specific phosphorothioate linkages that prevent degradation past the modified point in the transcript. Incubation of phosphorothioate-containing transcripts results in the formation exonucleolytic decay intermediates whose size will allow the predominant pathway of decay (Mukherjee et al., 2002). An alternative to the phosphorothioate strategy is the inclusion of a homopolymeric guanosine tract in the interior of the transcript. This results in the formation of decay intermediates, because G-rich stretches within RNAs have been shown to block exonucleolytic progress (Decker and Parker, 1993). In our experience, however, the performance of internal poly(G) tracts has been variable, and they seem be more effective at blocking $5'$ to $3'$-rather than $3'$ to $5'$-decay in extracts. Further details on these procedures involving phosphorothioate modifications can be found in Mukherjee et al. (2002).

4.4. Protocol

To assay for XRN1 activity, prepare a polyadenylated RNA substrate that contains a $5'$-monophosphate. This can be readily done by adding 5 mM $5'$-GMP instead of cap analog to a standard transcription reaction described in section 3.1. As the poly(A) competitor RNA is absent from our exonuclease assays, the presence of PABP on the poly(A) tail of the RNA substrate will effectively inhibit $3'$ to $5'$-decay of these substrates (Ford and Wilusz, 1999). Assaying the $3'$ to $5'$-decay activity of the exosome simply requires a source of capped, nonadenylated RNA substrate.

In a 1.5-ml microfuge tube on ice, prepare a reaction mixture containing 2 μl of PC/ATP (250 mM phosphocreatine/12.5 mM ATP), 6.5 μl of 10% PVA, 2 μl of RNase-free distilled water, 16 μl of HeLa cytoplasmic extract ($\sim$5 mg/ml) and gently mix by pipetting. Introduce into the reaction tube 2 μl of $5'$-capped, internally radiolabeled, unadenylated RNA substrate ($\sim$100,000 cpm/μl) for an exosome assay or a $5'$-monophosphate, internally radiolabeled and polyadenylated RNA substrate for a $5'$ to $3'$-exonuclease assay. Mix thoroughly and return to ice; the reaction mixture is now complete.

Immediately remove 5 μl of the reaction mixture and transfer the sample into 400 μl of HSCB. This is your zero time point. Incubate the reaction tube at 30 °C. Remove 5 μl at each desired time point, transferring the samples into 400 μl of HSCB to immediately halt the reaction. If more than four total time points are desired, the sample volume may be adjusted accordingly.

Process the time point samples by vortexing until homogenous after the addition of an equal volume of PCI. Centrifuge the mixture at $\sim$16,000g for 2 min to separate the organic and aqueous phases. Transfer the upper layer (aqueous phase) to a fresh tube, and precipitate through the addition of

1 ml of 100% ethanol. Carrier tRNA should be added to ensure efficient precipitation of all samples. Precipitate the RNA by incubating the sample tubes on dry ice for at least 10 min. Centrifuge at $\sim$16,000g for 10 min to pellet the RNA samples. Carefully remove the ethanol, discarding it in a manner appropriate for radioactive materials. Wash the RNA pellet by adding 200 μl of 80% ethanol. Briefly centrifuge the samples and remove the ethanol wash. Dry the resulting pellets for 30 sec under a vacuum.

Resuspend the pellets in 5 μl of RNA loading buffer. For quantitative results, it is important to ensure that the pellets are completely resuspended before gel loading. After heating the samples to 90 °C for 30 sec, incubate on ice and load the samples onto a 5% polyacrylamide gel containing 7 M urea. Electrophorese at 600 V/60 mA/15 W until the desired resolution has been obtained. Visualize the data by scanning with a PhosphorImager (e.g., Typhoon, GE Lifesciences). The rates of decay may be determined with densitometry and regression analysis of the remaining RNA substrate with respect to time. A representative exosome assay is depicted in Fig. 8.3B.

4.5. Assaying decapping activity

After the deadenylation of a cellular mRNA, the 5'-cap can be removed as the next step in the major 5' to 3'-pathway of mRNA decay. Removal of the 5'-cap structure leaves the 5'-end of the transcript unprotected from 5' to 3'-exonuclease activity. Two distinct decapping activities can be observed in cytoplasmic extracts. First, the DCP2 enzyme in concert with auxiliary factors cleaves long transcripts to produce a ^{7me}GDP product (Cohen *et al.*, 2005; Piccirillo *et al.*, 2003; Wang *et al.*, 2002). Alternately, after decay of most of the body of the transcript by the exosome, DCPS, a cap-salvage enzyme, will cleave the resulting short oligomer and produce ^{7me}GMP (Gu *et al.*, 2004; Liu *et al.*, 2002).

4.6. Protocol

In a fresh microfuge tube, combine the following on ice: 1 μl 10× CE buffer (500 mM Tris-HCl, pH 7.9/300 mM (NH$_4$)$_2$SO$_4$/1 mM MgCl$_2$), 5 μl RNase-free distilled water, and 4 μl of HeLa cytoplasmic extract ($\sim$5 mg/ml). To this mixture, add 1 μl of cap-labeled RNA substrate ($\sim$100,000 cpm/μl), gently pipetting to mix. Production of RNA labeled exclusively at the cap structure is outlined in section 3.2 of this chapter. (See also Chapter 1 by Liu *et al.* in this issue).

Incubate the reaction tube 30 °C for the desired time. On completion of incubation, immediately add 10 μl of ddH$_2$O. Add 20 μl of phenol to the diluted reaction, and mix thoroughly by vortexing. Chloroform is not added to the extraction to prevent the formation of a substantial amount of precipitate at the interface of this small volume extraction. Centrifuge the sample at ∼16,000g for 3 min. Carefully remove 15 μl of the upper (aqueous) layer to a fresh microfuge tube. Add 4 μl of RNA loading buffer to the sample, heat it to 90 °C for 30 sec, and snap-chill on ice. Load the samples onto a 20% denaturing polyacrylamide gel containing 7 M urea. Run the gel at 750 V/60 mA/20 W for approximately 30 min, or until sufficient resolution has been achieved. After electrophoresis, dry the gel with a heated slab dryer and expose to a PhosphorImager screen for visualization. As depicted in Fig. 8.3C, this single gel-based system effectively resolves free phosphate, ^{7m}GMP, ^{7m}GDP, and ^{7m}GpppG (the major product of exosome activity in extracts). As a marker for the expected products, unradiolabeled ^{7me}-G mono and diphosphates, as well as ^{7me}GpppG, can be obtained commercially and may be resolved alongside the radiolabeled species. The markers can be visualized with UV shadowing. Alternately, radiolabeled methylated nucleobases can be generated as described in Chapter 1 by Lui et al.

4.7. Analysis of *trans*-acting factors with ultraviolet crosslinking

Protein–RNA interactions can also be readily visualized with HeLa cytoplasmic extracts by a variety of assays. One powerful aspect of the *in vitro* mRNA stability system described in this chapter is that protein–RNA interaction cannot only be detected but also directly correlated with function when used in combination with competition assays or immunodepletion. Brief incubation of the RNA substrate in the presence of HeLa cytoplasmic extract in one of the decay systems discussed previously results in the formation of mRNP complexes. Subsequent exposure to ultraviolet (UV) radiation results in the formation of covalently linked RNA–protein complexes (Wilson and Brewer, 1999; Wilusz and Shenk, 1988). Although the overall efficiency of this cross-linking process is relatively poor (∼1%), the ability to detect RNA–protein complexes is made possible by the sensitivity afforded by the use of radiolabeled RNA substrates. RNase digestion of the cross-linked RNA–protein complexes yields proteins covalently bound to short radioactive RNA fragments. Separation by use of standard SDS-PAGE results in visualization of radiolabeled RNA–protein complexes whose migration in the gel is similar to the migration of the uncross-linked protein, allowing for a reasonable estimation of molecular weight.

Despite the power of this UV cross-linking approach, there are several caveats in its application. First, UV crosslinking relies on the production of very short-lived, free radical intermediates that require an angstrom or less distance to form a covalent bond between the RNA substrate and the *trans*-acting factor. This technique cannot, therefore, detect *trans*-acting factors that are not intimately associated with a given radiolabeled base. However, it should be emphasized that not all proteins intimately associated with an RNA are bound in a conformation that is accessible to covalent bond formation on photoactivation. In these cases, perhaps a photoactivatable nucleotide can be used as described elsewhere in this volume. Second, identification of a protein requires that it cross-link to a region of the RNA containing radiolabeled nucleotides. If a protein is binding to a poly (G) tract, for example, and the RNA substrate is radiolabeled at U residues, the technique will not detect the poly(G) binding protein. Radiolabeling the RNA substrate at alternative bases (e.g., labeling at guanosine residues rather than uridine as in the preceding example) can overcome this potential pitfall. Third, although the UV cross-linking procedure will generally give you very few bands on a gel, giving the appearance of a highly purified protein, it must be stressed that it is not by any means a purification procedure. Identification of the cross-linked factor itself will rely on further purification and mass spectrometry or the use of antibody detection to follow up on insightful hypotheses on the basis of approximate molecular weights.

Described in the following is a standard protocol for the detection of mRNP constituents after UV irradiation. Following this are basic protocols for evaluating detected *trans*-acting factors by immunoprecipitation and also immunodepletion of factors from extracts. This protocol is very useful for assessing the relative contribution of candidate factors to RNA decay processes observed in the *in vitro* system.

4.8. Protocol

Place a microtiter plate (Nunc) on ice for precooling. Combine in a fresh 1.5-ml microfuge tube (on ice) 1 μl of PC/ATP (250 mM phosphocreatine/12.5 mM ATP), 1 μl of poly(A) (500 ng/μl), 8 μl of HeLa cytoplasmic extract, and 1 μl of RNA substrate ($\sim$100,000 cpm/μl). Note that this is the same reaction as the deadenylation assay described previously. If you are specifically interested in pursuing factors regulating decapping, it is recommended that you set up the corresponding decapping reaction conditions for your cross-linking experiment.

Mix gently and incubate at 30 °C for 5 min. Transfer all 12 μl of the reaction to a single well on the microtiter plate. Avoid creating bubbles on the surface of the drop to ensure even exposure to the UV radiation. Transfer the entire microtiter plate (still on ice) to a UV Stratalinker 2400 (Stratagene). Irradiate the uncovered microtiter plate containing the RNA samples with 1.8 mJ of UV radiation on ice. Because the distance between the ultraviolet source and the sample can affect the efficiency of the cross linking, it is important to always use a standardized platform height. In the preceding instance, a Stratalinker with a UV sensor is the source of the UV radiation. The microtiter plate should be virtually equidistant from the bulbs as the sensor to ensure proper irradiation.

Immediately transfer the samples from the microtiter plate to a set of fresh 1.5-ml microfuge tubes containing 25 μg RNase A. If desired, 0.5 μl RNase one (Roche Biomedical), which cuts at every RNA base rather than just pyrimidines, can be added in addition to (or instead of) RNase A. Gently mix and incubate for 15 min at 37 °C.

After completion of the RNase digestion, add an equal volume of 2× SDS protein dye. Mix the samples by vortexing, and before SDS-PAGE boil the samples for a minimum of 3 min. Centrifuge the samples for 2 min at ~16,000g to remove any precipitated proteins. Load the samples on a 10 or 12% polyacrylamide gel containing SDS, and run the gel until sufficient resolution has been obtained. Dry the gel with a slab dryer and expose to a PhosphorImager plate for visualization. An example of a typical UV cross-linking assay is shown in Fig. 8.4.

If the RNAs being evaluated are known to be unstable in the extract (and you are searching for regulatory factors that may promote the instability), the reaction may be supplemented with 1 μl of 50 mM EDTA to block exonuclease activity but permit mRNP assembly. It should be noted that although this will inhibit most exonuclease activity, including that of the cellular decay machinery, it could theoretically also affect the binding of

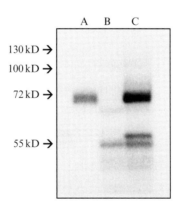

Figure 8.4 Identification of cellular proteins observed by means of ultraviolet cross-linking of radiolabeled RNAs in the *in vitro* RNA decay system. Lanes A through C represent three independent radiolabeled RNA substrates that were incubated in the *in vitro* RNA decay system, irradiated with UV light, RNase treated, and analyzed on a 12% SDS acrylamide gel followed by phosphorimaging. The positions of molecular weight markers are noted on the left. Note that a ~70-kDa factor associates with transcripts A and C but is not associated with transcript B. Likewise a ~55-kDa factor associates only with transcripts B and C, whereas a ~58-kDa protein uniquely crosslinks to transcript C.

trans-acting factors. Given the affinities of EDTA for divalent cations other than magnesium, for example zinc, one must consider the unfortunate consequences of adding EDTA to the reaction on RNA-binding proteins that require divalent cations to bind nucleic acid. Thus, it is always advisable to perform reactions in the presence and absence of EDTA to ensure that the pattern of bands (rather than the intensity) does not change in the presence of EDTA. If EDTA cannot be added to stabilize the substrate, alternative RNA stabilizers can be directly incorporated into the RNA substrate such as 3′-terminal hairpins or alternative cap analogs.

4.9. Immunoprecipitation protocol

As stated in the introduction to this section, if the identity of an associated *trans*-acting factor is known, or suspected, its identity can be further confirmed through the use of immunoprecipitation after UV cross linking.

Because cross-linked bands are sometimes faint and immunoprecipitation is not always efficient, it is advisable to perform three to four cross-linking reactions and pool them before RNase digestion and subsequent immunoprecipitation. After RNase digestion of the cross-linked samples in the protocol described previously, remove 5 to 10 μl of the reaction and reserve it as the "input." Add 400 μl of NET2 buffer (40 mM Tris-HCl, pH 7.5/200 mM NaCl/0.05% Nonidet P40) to the remainder of the sample. Briefly vortex the samples and centrifuge at ~16,000g for 3 min to remove any protein that precipitated during the cross-linking step. Transfer the supernatant to a fresh 1.5-ml microfuge tube. Add the desired antibody (amount as dictated by supplier's recommendations) to the clarified supernatant. Incubate at 4 °C on a nutating mixer for 1 h. Add 20 μl of formalin-fixed, protein-A–positive *Staphylococcus aureus* cells (Roche Biomedical) that have been washed with NET2 buffer and resuspended to produce a 50% slurry. Incubate the mixture on ice for a minimum of 15 min. It is always advisable to run parallel experiments with preimmune serum as a control to ensure that precipitated bands are indeed specific.

Pellet the *S. aureus* cells by a brief centrifugation step. Remove and discard the supernatant and add 500 μl of NET2 buffer to the pellet. Completely resuspend the pellet by vigorous vortexing—the entire pellet needs to be resuspended, not simply washed. Repeat the washing step at least four more times. After the final washing step, resuspend the cell pellet in 20 μl of 6× SDS protein dye. Analyze the samples, including the "input" control in the manner described previously for a conventional cross-linking analysis. The bands detected in this manner can be extremely faint, thus longer exposures to a phosphor screen may be necessary. It is also advisable to load the input sample on a separate region of the gel, because this lane will be considerably darker than the immunoprecipitated lanes.

4.10. Immunodepletion of factors from cytoplasmic extracts

To determine the role of a given cellular factor with respect to a cellular process, immunodepletion may be used in lieu of, or to confirm, the results obtained through an RNAi-based knockdown.

4.11. Protocol

Begin the procedure by thawing an aliquot of HeLa cytoplasmic extract on ice. Although the following protocol is designed to yield 200 μl of depleted cytoplasmic extract, the volumes may be adjusted accordingly to produce larger or smaller quantities.

Once completely thawed, add to the extract 3 to 5 μl of the antibody (depending on concentration and quality of the antibody, according to the supplier's instructions). As a control, preimmune sera must be used in parallel depletion reactions to ensure that any observed effects are not a result of handling during the depletion treatment. Mix thoroughly and incubate for 1 h at 4 °C. After antibody binding, immediately add 20 μl of a 50% solution of formalin-fixed, protein-A–positive *S. aureus* suspension in buffer D. Alternately, protein-A sepharose beads or equivalent may be readily substituted. Incubate the mixture for ~30 min at 4 °C to allow the antibody–protein A complexes to form. After complex formation, centrifuge the samples briefly at full speed to pellet the *S. aureus* cells. Remove the supernatant and further aliquot for later use in the preceding assays. Depletion of the target protein can be quantitatively observed through the use of conventional Western blotting, although this may not be useful if your protein migrates close to the Ig heavy or light chains. Immunodepletion can be repeated several times to ensure adequate depletion of the target protein is obtained.

5. Concluding Remarks

As shown previously, cytoplasmic extracts are capable of recapitulating many aspects of mRNA decay and represent a valuable technology to include in the repertoire of any molecular biologist. Furthermore, these protocols are applicable to any conceivable RNA substrate, greatly expanding the possible applications of this particular technology.

Any single experimental approach has its limitations in the study of mRNA decay. *In vitro* studies allow the assessment of biochemical parameters and significant insight into mechanisms, but may not always fully recapitulate cellular events. Assays of relative RNA stability in tissue culture cells allow cell biology parameters to be evaluated but are limited by the fact that these are largely tumor cells with altered regulation that may impact on the biologic relevance of the results that are obtained. Transgenic animals are undoubtedly the most biologically relevant system in which to study decay, but they present inherent technical difficulties in analyses because of their complexity. Furthermore, it is often difficult to fully establish mechanisms from *in vivo* analyses alone. To determine precisely the mechanisms by which the stability of a given transcript is regulated, a combination of approaches clearly must be applied.

ACKNOWLEDGMENTS

We thank all the members of the Wilusz laboratory, past and present, for their contributions to this work. This work was funded by NIH grant #GM072481 to J. W.

REFERENCES

Bergman, N., Opyrchal, M., Bates, E. J., and Wilusz, J. (2002). Analysis of the products of mRNA decapping and 3′ to 5′-decay by denaturing gel electrophoresis. *RNA* **8,** 959–965.

Brown, B. A., and Ehrenfeld, E. (1979). Translation of poliovirus RNA *in vitro*: Changes in cleavage pattern and initiation sites by ribosomal salt wash. *Virology* **97,** 396–405.

Cheadle, C., Fan, J., Cho-Chung, Y. S., Werner, T., Ray, J., Do, L., Gorospe, M., and Becker, K. G. (2005). Stability regulation of mRNA and the control of gene expression. *Ann. N. Y. Acad. Sci. USA* **1058,** 196–204.

Chen, C. Y., Gherzi, R., Ong, S. E., Chan, E. L., Raijmakers, R., Pruijn, G. J., Stoecklin, G., Moroni, C., Mann, M., and Karin, M. (2001). AU binding proteins recruit the exosome to degrade ARE-containing mRNAs. *Cell* **107,** 451–464.

Chowdhury, A., Mukhopadhyay, J., and Tharun, S. (2007). The decapping activator Lsm1p-7p-Pat1p complex has the intrinsic ability to distinguish between oligoadenylated and polyadenylated RNAs. *RNA* **13,** 998–1016.

Cohen, L. S., Mikhli, C., Jiao, X., Kiledjian, M., Kunkel, G., and Davis, R. E. (2005). Dcp2 Decaps m2,2,7GpppN-capped RNAs, and its activity is sequence and context dependent. *Mol. Cell Biol.* **25,** 8779–8791.

Decker, C. J., and Parker, R. (1993). A turnover pathway for both stable and unstable mRNAs in yeast: Evidence for a requirement for deadenylation. *Genes Dev.* **7,** 1632–1643.

Ford, L. P., Watson, J., Keene, J. D., and Wilusz, J. (1999). ELAV proteins stabilize deadenylated intermediates in a novel *in vitro* mRNA deadenylation/degradation system. *Genes Dev.* **13,** 188–201.

Ford, L. P., and Wilusz, J. (1999). An *in vitro* system using HeLa cytoplasmic extracts that reproduces regulated mRNA stability. *Methods* **17,** 21–27.

Garbarino-Pico, E., Niu, S., Rollag, M. D., Strayer, C. A., Besharse, J. C., and Green, C. B. (2007). Immediate early response of the circadian polyA ribonuclease nocturnin to two extracellular stimuli. *RNA* **13,** 745–755.

Garcia-Martinez, J., Aranda, A., and Perez-Ortin, J. E. (2004). Genomic run-on evaluates transcription rates for all yeast genes and identifies gene regulatory mechanisms. *Mol. Cell* **15,** 303–313.

Garneau, N. L., Wilusz, J., and Wilusz, C. J. (2007). The highways and byways of mRNA decay. *Nat. Rev. Mol. Cell Biol.* **8**, 113–126.
Gherzi, R., Lee, K. Y., Briata, P., Wegmuller, D., Moroni, C., Karin, M., and Chen, C. Y. (2004). A KH domain RNA binding protein, KSRP, promotes ARE-directed mRNA turnover by recruiting the degradation machinery. *Mol. Cell* **14**, 571–583.
Green, C. B., and Besharse, J. C. (1996). Identification of a novel vertebrate circadian clock-regulated gene encoding the protein nocturnin. *Proc. Natl. Acad. Sci. USA* **93**, 14884–14888.
Gu, M., Fabrega, C., Liu, S. W., Liu, H., Kiledjian, M., and Lima, C. D. (2004). Insights into the structure, mechanism, and regulation of scavenger mRNA decapping activity. *Mol. Cell* **14**, 67–80.
Handa, H., Kaufman, R. J., Manley, J., Gefter, M., and Sharp, P. A. (1981). Transcription of Simian virus 40 DNA in a HeLa whole cell extract. *J. Biol. Chem.* **256**, 478–482.
Hernandez, N., and Keller, W. (1983). Splicing of *in vitro* synthesized messenger RNA precursors in HeLa-Cell extracts. *Cell* **35**, 89–99.
Houseley, J., LaCava, J., and Tollervey, D. (2006). RNA-quality control by the exosome. *Nat. Rev. Mol. Cell Biol.* **7**, 529–539.
Kuhn, U., and Wahle, E. (2004). Structure and function of poly(A) binding proteins. *Biochim. Biophys. Acta* **1678**, 67–84.
Liu, H., *et al.* (2002). Scavenger decapping activity facilitates 5′ to 3′-mRNA decay. *EMBO J.* **21**, 4699–4708.
Manley, J. L., Fire, A., Cano, A., Sharp, P. A., and Gefter, M. L. (1980). DNA-dependent transcription of adenovirus genes in a soluble whole-cell extract. *Proc. Natl. Acad. Sci. USA* **77**, 3855–3859.
Mifflin, R. C., and Kellems, R. E. (1991). Coupled transcription-polyadenylation in a cell-free system. *J. Biol. Chem.* **266**, 19593–19598.
Moore, C. L., and Sharp, P. A. (1985). Accurate cleavage and polyadenylation of exogenous RNA substrate. *Cell* **41**, 845–855.
Mukherjee, D., Gao, M., O'Connor, J. P., Raijmakers, R., Pruijn, G., Lutz, C. S., and Wilusz, J. (2002). The mammalian exosome mediates the efficient degradation of mRNAs that contain AU-rich elements. *EMBO J.* **21**, 165–174.
Murray, E. L., and Schoenberg, D. R. (2007). A+U-rich instability elements differentially activate 5′ to 3′ and 3′ to 5′-mRNA decay. *Mol. Cell Biol.* **27**, 2791–2799.
Opyrchal, M., Anderson, J. R., Sokoloski, K. J., Wilusz, C. J., and Wilusz, J. (2005). A cell-free mRNA stability assay reveals conservation of the enzymes and mechanisms of mRNA decay between mosquito and mammalian cell lines. *Insect Biochem. Mol. Biol.* **35**, 1321–1334.
Orban, T. I., and Izaurralde, E. (2005). Decay of mRNAs targeted by RISC requires XRN1, the Ski complex, and the exosome. *RNA* **11**, 459–469.
Parker, R., and Sheth, U. (2007). P bodies and the control of mRNA translation and degradation. *Mol. Cell* **25**, 635–646.
Piccirillo, C., Khanna, R., and Kiledjian, M. (2003). Functional characterization of the mammalian mRNA decapping enzyme hDcp2. *RNA* **9**, 1138–1147.
Wang, Z., Jiao, X., Carr-Schmid, A., and Kiledjian, M. (2002). The hDcp2 protein is a mammalian mRNA decapping enzyme. *Proc. Natl. Acad. Sci. USA* **99**, 12663–12668.
Wilson, G. M., and Brewer, G. (1999). Identification and characterization of proteins binding A + U-rich elements. *Methods* **17**, 74–83.
Wilusz, J., and Shenk, T. (1988). A 64 kd nuclear protein binds to RNA segments that include the AAUAAA polyadenylation motif. *Cell* **52**, 221–228.
Zhang, W., Wagner, B. J., Ehrenman, K., Schaefer, A. W., DeMaria, C. T., Crater, D., DeHaven, K., Long, L., and Brewer, G. (1993). Purification, characterization, and cDNA cloning of an AU–rich element RNA-binding protein, AUF1. *Mol. Cell Biol.* **13**, 7652–7665.

SECTION THREE

NUCLEASES IN mRNA DECAY

CHAPTER NINE

IN VITRO ASSAYS OF 5′ TO 3′-EXORIBONUCLEASE ACTIVITY

Olivier Pellegrini, Nathalie Mathy, Ciarán Condon, *and* Lionel Bénard

Contents

1. Introduction	168
2. Purification of *Xrn*1	170
3. *In Vitro* RNA Substrate Synthesis	171
4. Degradation of RNA by XRN1 Depends on the Nature of the 5′-End	173
4.1. TCA-based exoribonuclease assays	173
4.2. Gel-based exoribonuclease assays	174
4.3. Fluorescence-based exoribonuclease assays	176
5. Determining the Directionality of Decay	176
5.1. Degradation of doubly labeled RNA by *Xrn*1	176
5.2. Degradation of 5′- and 3′-labeled synthetic RNAs by *Xrn*1	179
6. Conclusions and Prospects	180
References	181

Abstract

The major cytoplasmic 5′ to 3′-exoribonuclease activity is carried out by the *Xrn*1 protein in eukaryotic cells. A number of different approaches can be used to study multifunctional *Xrn*1 protein activity *in vitro*. In this chapter, we concentrate on methods used in our laboratory to analyze *Xrn*1 5′ to 3′-exoribonuclease activity. Some of these techniques may also be suitable for detecting 3′ to 5′-exoribonuclease or endoribonuclease activity. For these reasons, these assays can be used to isolate new proteins with ribonuclease activity and, when performed in combination with *in vivo* experiments, will contribute to a new level of understanding of the function of these factors.

CNRS UPR 9073 (affiliated with Université de Paris 7—Denis Diderot), Institut de Biologie Physico-Chimique, Paris, France

1. Introduction

In eukaryotic cells, most cytoplasmic messenger RNAs (mRNAs) are degraded through two alternative pathways, each of which is initiated by the removal of the poly(A) tail by deadenylases. Subsequently, the cap (5′-m7GpppN) structure is removed by the decapping complex, and the mRNA is degraded by the major cytoplasmic enzyme *Xrn1* (Caponigro and Parker, 1996). Alternatively, deadenylated mRNAs can be degraded from their 3′-ends by the exosome, a multimeric complex possessing 3′ to 5′-exoribonuclease activity (see for review Parker and Song [2004]).

The 5′ to 3′-exoribonuclease *Xrn1* is evolutionarily conserved, and orthologs have been systematically identified in all eukaryotes investigated (Bashkirov *et al.*, 1997; Kastenmayer and Green, 2000; Sato *et al.*, 1998; Souret *et al.*, 2004; Szankasi and Smith, 1996). The highly conserved N-terminal domain, which participates in the 5′ to 3′-exonuclease activity and is also found in the nuclear counterpart of *Xrn1*, Rat1 (Amberg *et al.*, 1992), clearly defines the prototypical branch of the 5′ to 3′-exoribonucleases in eukaryotes (Solinger *et al.*, 1999).

The XRN1 gene is not essential to *Saccharomyces cerevisiae*, because mutations in the XRN1 gene are viable, although they result in defects in the turnover of pre-rRNA (Henry *et al.*, 1994; Stevens *et al.*, 1991) and mRNA (reviewed in Caponigro and Parker [1996]). Because *Xrn1* protein functions subsequent to decapping of the mRNA by the Dcp1/Dcp2 complex (Coller and Parker, 2004), xrn1 mutant *S. cerevisiae* strains accumulate 5′-monophosphorylated intermediates (Caponigro *et al.*, 1993). In accordance with this observation *in vivo*, the *Xrn1*-mediated hydrolysis of RNAs bearing a 5′-cap is almost undetectable *in vitro*, whereas robust *Xrn1* activity is observed on 5′-monophosphorylated substrates (Stevens and Poole, 1995). Interestingly, RNAs with a 5′-triphosphate or a 5′-hydroxyl (5′-OH) group are also refractory to degradation by *Xrn1 in vitro* (Mathy *et al.*, 2007; Stevens and Poole, 1995).

The role of *Xrn1* is also evident in the mRNA quality control mechanism that degrades aberrant mRNAs having premature translation termination codons. This mechanism is known as nonsense-mediated mRNA decay (NMD) (reviewed in Amrani *et al.* [2006]; NMD substrates can be decapped in yeast and human cells (Hagan *et al.*, 1995; Lejeune *et al.*, 2003), whereas NMD in *Drosophila* has been shown to be initiated by endonucleolytic cleavage. Degradation of the decapped species in yeast and humans, and the 3′-product of the endonucleolytic cleavage in *Drosophila*, is catalyzed by *Xrn1* (Gatfield and Izaurralde [2004] and reviewed in Conti and Izaurralde [2005]). Thus, NMD in *Drosophila* is reminiscent of the *Xrn1*p-catalyzed hydrolysis of mRNAs cleaved by the endonuclease Rnt1 in *S. cerevisiae* (Zer and Chanfreau, 2005). Similar to this observation for the

NMD pathway, it has been demonstrated that depletion of Xrn1 leads to the accumulation of the 3'-decay intermediate generated by the multimeric RNA–induced silencing complexes (RISC) (Orban and Izaurralde, 2005; Souret et al., 2004). It seems, therefore, that both the NMD and RNA interference machineries use Xrn1 to degrade targeted RNAs.

Xrn1 orthologs can be found associated within evolutionary conserved foci in the cytoplasm colocalized with other conserved proteins required for the 5' to 3'-mRNA degradation pathway. These specialized cytoplasmic foci are known as mRNA processing bodies (P bodies) or GW bodies (Ingelfinger et al., 2002; Sheth and Parker, 2003; see chapter by Nissan and Parker). However, Xrn1 is dispensable for the assembly of these foci. Thus, the role of Xrn1 in these structures can be viewed as a house-keeping function that limits the formation of unwanted RNA/protein aggregates in the cytoplasm (Sheth and Parker, 2003).

In addition to defects in RNA turnover, Xrn1 mutants exhibit pleiotropic phenotypes, including slow growth, loss of viability on nitrogen starvation, meiotic arrest, defective sporulation, and defects in microtubule-related processes (Heyer et al., 1995; Page et al., 1998; Solinger et al., 1999; Tishkoff et al., 1995). Deletion of the XRN1 gene is also known to produce significant changes in protein accumulation and expression patterns. Proteome analysis has revealed that many of the proteins up-or down regulated in an Xrn1-deficient strain are involved in amino acid biosynthesis, amine and nitrogen metabolism, or produce other perturbations in sucrose, purine, and pyrimidine metabolism (Ross et al., 2004). Caenorhabditis elegans embryos depleted of Xrn1 protein die because of failure to undergo epithelium closure (Newbury and Woollard, 2004). In Drosophila, mutations in the XRN1 gene result in defects in gastrulation and thorax closure (Newbury, 2006). A deficiency in the expression of the human ortholog, SEP1, has been correlated with the appearance of osteosarcoma (Zhang et al., 1998). It has yet to be shown that these phenotypes are directly related to a deficiency in exoribonuclease activity. However, consistent with these observations, Xrn1 is capable of binding and cleaving a variety of nucleic acid substrates besides RNA in vitro, such as single-stranded DNA and double-stranded DNA (Johnson and Kolodner, 1991). For DNA substrates, Xrn1 was found to have a preference for G4 tetraplex-containing DNAs, a structure that may form at telomeres (Chernukhin et al., 2001; Liu and Gilbert, 1994). Xrn1 has been identified as a homologous DNA pairing protein (Kolodner et al., 1987; Tishkoff et al., 1995). Xrn1 has also been shown to promote in vitro assembly of tubulin into microtubules with high efficiency (Solinger et al., 1999).

A number of different properties of Xrn1 can be studied in vitro (Johnson and Kolodner, 1991; Solinger et al., 1999). In this chapter, we concentrate on the methods used to purify Xrn1 and analyze its 5' to 3'-exoribonuclease activity. These different approaches may be used to define the catalytic domain of Xrn1 or to provide information on other functions not related to exonuclease activity. Site-specific mutations in *Xrn1* can separate *Xrn1* function in sporulation, microtubule assembly, RNA binding, and as an exonuclease (Page *et al.*, 1998; Solinger *et al.*, 1999). Therefore, these types of assays can be performed on similar proteins and carried out in combination with genetic experiments to provide new insights into protein function.

2. Purification of *XRN1*

Our protocol for *Xrn1* purification is adapted from different protocols (Johnson and Kolodner, 1991; Stevens, 1980; 2001). *S. cerevisiae* *Xrn1* is purified from the haploid protease–deficient strain C131BYS86/pRDK249 grown in HC-Ura medium containing the nonfermentable carbon sources lactic acid (2%) and glycerol (3%). The plasmid pRDK249 was constructed by A. W. Johnson (Johnson and Kolodner, 1991). The cells are collected 16 h after the addition of 2% galactose, pelleted, and frozen at −80 °C until further use.

All procedures are performed at 4 °C. *Xrn1* purification is monitored by SDS-PAGE and TCA-based exoribonuclease assays (Fig. 9.1 and see the second part following for assay details). In a typical preparation, 6.5 g of cells are resuspended in 20 ml buffer B (20 mM Tris-HCl, pH 7.5, 10% glycerol, 0.5 mM DTT, 1 mM EDTA, 0.1 mM PMSF) containing 150 mM NaCl,

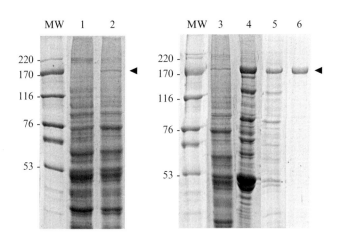

Figure 9.1 Electrophoretic analysis by Coomassie blue–stained SDS-PAGE (7.5%) showing *Xrn1* overproduction and purification. Full-length *Xrn1* protein (175 kDa) is indicated with an arrowhead. MW, protein size standard in kDa. Lane 1, extract from uninduced culture. Lane 2, extract from culture induced with 2% galactose. Loaded samples correspond to 42.5 μg of protein. Lane 3, DE52 flow through. Lane 4, HiTrap heparin HP pool. Lane 5, Mono Q HR 5/5 pool. Lane 6, purified *Xrn1* eluted from Superdex 200 HR 26/60.

one protease inhibitor cocktail tablet (Roche), and 5 μg/ml DNase I. The cells are disrupted by two passages through a French press (1200 bars, 20,000 psi) and then centrifuged at 10,000 rpm in an SS34 rotor (Sorvall) for 30 min at 4 °C.

The resulting supernatant is applied at a rate of 1 ml/min to a DE52 (Whatman) column (2 cm^2 × 8 cm) equilibrated in buffer B containing 150 mM NaCl and washed with the same buffer. The flow-through fraction (90 ml) is loaded on a 5-ml HiTrap heparin (GE Healthcare) column also equilibrated in buffer B containing 150 mM NaCl. The column is subsequently washed extensively in buffer B containing 300 mM NaCl and eluted with a linear gradient (40 ml) from 300 mM to 1 M NaCl in buffer B. The fractions containing *Xrn1* (elution at approximately 580 mM NaCl) are pooled, extensively dialyzed against buffer B containing 100 mM NaCl, and applied to Mono Q HR 5/5 column (GE Healthcare) equilibrated in the same buffer. A linear gradient (20 ml) of increasing ionic strength from 100 mM to 500 mM NaCl in buffer B is used for the elution. *Xrn1* elutes at approximately 250 mM NaCl. The peak fractions are pooled and loaded on a Superdex 200 HR 26/60 column (GE Healthcare) equilibrated in buffer B containing 250 mM NaCl. The elution volume of the protein corresponds to an apparent molecular mass of 340 kDa, suggesting that *Xrn1* (175 kDa) is a dimer. Peak fractions containing *Xrn1* are pooled and concentrated to 0.2 mg/ml with an Amicon Ultra-4 filter (30-kDa molecular weight cutoff). The purified protein shows no loss of activity after 2 years at −80 °C in 30% glycerol.

3. *IN VITRO* RNA SUBSTRATE SYNTHESIS

Degradation of RNA by *Xrn1* depends on the nature of the 5′-end, with 5′-monophosphorylated RNAs being the best substrates for hydrolysis. Uniformly labeled RNAs, such as 16S ribosomal RNAs or poly(A), can be used to assay *Xrn1* and can be prepared by labeling cultures or synthesized with polynucleotide phosphorylase, respectively (Stevens, 2001). Our standard procedure is to synthesize specific RNAs by transcribing linearized plasmids or PCR products containing phage T7, T3, or SP6 polymerase promoters (see Ambion's MEGAscript instruction manual). Nonradioactive and uniformly labeled transcripts, 5′-triphosphate (5′-triP) and capped (5′-cap) RNAs are synthesized according to the manufacturer's instructions (Ambion, MEGAscript). RNAs bearing 5′-cap groups are easily obtained by adding a 3.2-fold excess of cap analog (Ambion) over GTP to the *in vitro* transcription reaction (Table 9.1). On the basis of minor modifications of this procedure, we can also produce 5′-monophosphorylated (5′-monoP) or 5′-hydroxylated (5′-OH) RNAs, by adding a 6.25-fold excess of GMP or guanosine over GTP, respectively, to the reaction mixture (Table 9.1).

Table 9.1 *In vitro* transcription assembly

	Synthesized RNA			
	5′-triP	5′-MonoP	5′-OH	5′-capped
Nuclease-free water	14	11	11	11
T7 Reaction buffer 10×	4	4	4	4
ATP 75 mM	4	4	4	4
UTP 75 mM	4	4	4	4
CTP 75 mM	4	4	4	4
GTP 75 mM	4	1	1	1
GMP 100 mM	0	6	0	0
Guanosine 100 mM	0	0	6	0
Cap analog 40 mM	0	0	0	6
DNA template 1µg/µl (0.25 to 1 µM)	2	2	2	2
[^{32}P] or [^{3}H] nucleotide				
T7 RNA polymerase	4	4	4	4
Total volume	40 µl	40 µl	40 µl	40 µl

Guanosine, GMP are from Sigma-Aldrich; labeled nucleotides from GE Healthcare; other components are from Ambion. DNA template can be made from PCR products or linearized plasmids.
Abbreviations: RNAs with 5′-triphosphate (5′-triP), 5′-monophosphate (5′-MonoP), 5′-hydroxyl (5′-OH), and 5′-m7GpppG (5′-CAP) groups.

During the transcription reaction preparation, it is highly recommended to assemble the reaction components at room temperature, as shown in Table 9.1, to avoid coprecipitation of the DNA template with the spermidine present in the reaction buffer. In a 40-µl reaction, uniformly [^{32}P]-labeled transcripts are produced by adding 5 µl of [α^{32}P]-UTP (3000 Ci/mmol, 10 mCi/ml). It is not necessary to reduce the quantity of cold UTP.

The synthesis of capped, 5′-OH or 5′-monoP RNAs is expected to give a twofold to threefold lower RNA yield than synthesis of 5′-triP RNAs. The time of incubation at 37 °C depends on the length of the RNA to be synthesized, 6 h for RNAs above 0.5 kb in length and overnight for shorter RNAs. After incubation, we systematically degrade the template DNA by DNase I treatment (Ambion). Unincorporated nucleotides are removed with G50 spin columns (GE-Amersham). Synthesized RNAs are quantified by measurement of the optical density at 260 nm, and RNA integrity is checked by separation and visualization of the RNA samples on agarose gels. After the G50 column, we systematically take advantage of working with [^{32}P]-labeled RNAs to purify them in 5% polyacrylamide/7 M urea gels (for RNAs less than 100 nucleotides, we recommend the use of 14%

polyacrylamide/7 M urea gels for purification). The RNA band position is revealed by autoradiography (film exposure between 5 to 15 min). The corresponding polyacrylamide slice is cut out and eluted in 400 μl of buffer 10 mM Tris, pH7.5, 300 mM NaCl, and 1 mM EDTA in a 1.5-ml Eppendorf tube (incubation for 3 to 6 h at room temperature, or overnight at 4 °C) and precipitated with 1 ml of cold ethanol. The RNA pellet is resuspended in nuclease-free water at the desired concentration, and an aliquot is counted for 5 min in a scintillation counter (Packard Model Tri-Carb 2100TR, the [^{32}P] channel is 18 to 2000 keV). For example, the synthesis yield of a [^{32}P]-labeled 5′-monophosphorylated RNA of 300 bases is approximately 40 μg; 1 μl in 2 ml scintillation fluid of a 2.5 μM solution of this purified RNA will correspond to approximately 150,000 cpm (i.e., approximately 60,000 cpm/nmol).

4. DEGRADATION OF RNA BY XRN1 DEPENDS ON THE NATURE OF THE 5′-END

4.1. TCA-based exoribonuclease assays

$Xrn1$ is usually assayed in a 300- or 600-μl reaction volume containing 1 to 5 pmol RNA in buffer containing 30 mM Tris-HCl, pH 8, 2 mM MgCl$_2$, 50 mM NH$_4$Cl, 0.5 mM DTT, 20 μg/ml acetylated bovine serum albumin (BSA) (Stevens, 2001). The amounts of enzyme can vary from 0.002 to 0.2 μg. Reactions are incubated at 30 °C or 37 °C. At the desired times, a 60-μl aliquot is removed to an Eppendorf tube containing 60 μl of stop solution (ice-cold 7% w/v trichloroacetic acid (TCA) containing 100 μg/ml acetylated BSA). Tubes are allowed to stand on ice for 10 min. Reactions are centrifuged in a refrigerated benchtop centrifuge (Eppendorf) at 13,200 rpm for 10 min at 4 °C. A volume of 40 μl, carefully taken from the top of the supernatant, is removed to 2 ml scintillation fluid, vortexed, and counted for 5 min in a scintillation counter. The intact RNA is precipitated while the hydrolyzed mononucleotides remain in the solution, and the amount of released nucleotide can be quantitated with a scintillation counter. This procedure gives the lowest [^{32}P] background when purified RNAs exceeding 1 kb in length are used for the assay. TCA-based assays with shorter, 250 base or 600 base RNAs, sometimes have a high [^{32}P] background. As shown in Fig. 9.2, [^{32}P] release by $Xrn1$ is almost undetectable for RNAs bearing a 5′-triphosphate group, a 5′-cap, or a 5′-hydroxyl group, whereas hydrolysis of 5′-monophosphorylated RNA is almost complete in the first 10 min.

During the purification of $Xrn1$, we use TCA assays to determine the fractions containing 5′ to 3′-exonuclease activity. To accurately select the fractions containing peaks of 5′ to 3′-exonuclease activity, we recommend diluting (1/10 or 1/50) aliquots of fractions exhibiting the 175-kDa band on SDS-PAGE gels to avoid a complete release of [^{32}P] in the first minute

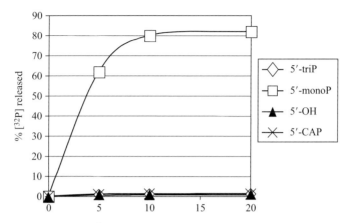

Figure 9.2 Xrn1 releases [^{32}P] from uniformly labeled 5′-monophosphorylated RNA. Plasmid pT7luc (Gallie, 1991) linearized by BamHI is the DNA template used for *in vitro* synthesis of [^{32}P]-labeled luciferase mRNAs (1.8 kb) with different 5′-end features; 2.5 pmol of RNA was incubated with 0.2 μg Xrn1 at 37 °C for the times indicated. RNAs with 5′-hydroxyl (5′-OH), 5′-m7GpppG (5′-CAP), 5′ triphosphate (5′-triP), 5′-monophosphate (5′-monoP) groups are indicated in the legend. Samples were precipitated with TCA, centrifuged, and the amount of [^{32}P] label released measured by scintillation counting and reported as a percentage of the input.

(Fig. 9.1); 1 μl of these dilutions tested for 5 min at 30 °C under the conditions described previously is sufficient to observe nonsaturating enzyme kinetics. Triphosphorylated RNAs are not degraded when peak fractions of 5′ to 3′-exonuclease activity are assayed. A comparison of the degradation of 5′-monophosphorylated and 5′-triphosphorylated RNA substrates is, therefore, the best method for selecting fractions containing active Xrn1.

4.2. Gel-based exoribonuclease assays

Consistent with its exonucleolytic activity, degradation of labeled RNAs by Xrn1 generates [α^{32}P]-UMP products. However, TCA-based assays are inadequate to demonstrate that the [^{32}P] label released corresponds to the liberation of mononucleotides, because small RNA fragments are also precipitated by this method. One way to detect mononucleotide products is to load aliquots of the reaction in 20% denaturing polyacrylamide gels (Oussenko *et al.*, 2002). Uniformly labeled RNAs are synthesized with [α^{32}P]-UTP, and a kinetic analysis of RNA degradation by Xrn1 is performed. Xrn1 is assayed in a 60-μl reaction volume containing the assay buffer described previously. The amounts of enzyme can vary from 0.002 to 0.2 μg, and the amount of substrate, from 0.25 pmol to 5 pmol. The reaction is incubated at 30 °C or 37 °C, and at desired times, the reaction is stopped by removing a 5-μl aliquot to an Eppendorf tube containing 5 μl 95% formamide, 20 m*M* EDTA, 0.05%

bromophenol blue, and 0.05% xylene cyanol. The mixtures are loaded directly on 20% denaturing polyacrylamide gels (Fig. 9.3A), and the radioactivity in the UMP band is quantified using a PhosphorImager and plotted on a graph (Fig. 9.3B). A typical reaction contains labeled [α^{32}P]-UTP 16S rRNA substrates that are either monophosphorylated, triphosphorylated, 5'-capped or bear a 5'-hydroxyl group. We have also used a 5'-monophosphorylated RNA with an 18-nt polyG sequence inserted 5 nucleotides from the 5'-end. Such polyG tracts are known to inhibit yeast *Xrn1* activity *in vivo* (Muhlrad *et al.*, 1994) and *in vitro* (Stevens, 2001). A typical experiment with these substrates is shown in Fig. 9.3B. An analysis of the release of mononucleotides confirms that RNA hydrolysis by *Xrn1* is completely inhibited by the 5'-triphosphate group, the 5'-hydroxyl, and the 5'-cap structure and is partially inhibited by the presence of a polyG sequence (Fig. 9.3B). Release of mononucleotides is clearly predominant when a 5'-monophosphorylated RNA is used as a substrate.

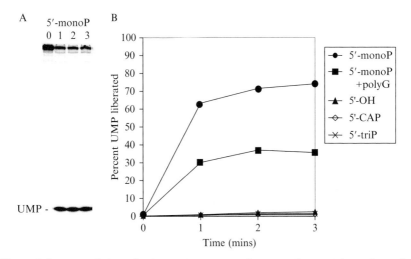

Figure 9.3 Degradation of 16S rRNA precursor fragments by *Xrn1* depends on the nature of the 5'-end. (A) Kinetic analysis of the degradation of short pre-16S precursor rRNAs (Mathy *et al.*, 2007) bearing 5'-monophosphate (5'-monoP) groups by *Xrn1*. The mixture was loaded on 20% denaturing polyacrylamide gels; 0.25 pmol of uniformly labeled [α^{32}P]-UTP was incubated with 0.1 μg native *Xrn1* for the time indicated. The migration position of nonradioactive UMP revealed by UV shadowing at 254 nm is indicated to the right of the autoradiogram. (B) Graph of kinetic analysis of short pre-16S rRNA degradation bearing different 5'-groups by *Xrn1*. RNAs with 5'-hydroxyl (5'-OH), 5'-m7GpppG (5'-CAP), 5'-triphosphate (5'-triP), 5'-monophosphate (5'-monoP) groups or a 5'-monophosphate group and polyG sequence (5'-monoP + polyG) are indicated in the legend. [α^{32}P]-UMP bands visualized in an acrylamide gel and quantified using a PhosphorImager, are plotted on the graph; 0.25 pmol of RNA was incubated with 0.1 μg native *Xrn1* at 37 °C for the times indicated.

4.3. Fluorescence-based exoribonuclease assays

TCA-based assays are useful in the purification of *Xrn1*, because RNA substrates differing in their 5'-extremities can be exploited to detect fractions containing active *Xrn1* and can be used to distinguish 5' to 3'-exoribonuclease activity from endonuclease or 3' to 5'-exonuclease activities. However, these assays are time consuming and require handling of radioactivity, centrifugation, and scintillation counting. We, therefore, recently developed a fluorescence-based exoribonuclease assay that exploits the ability of SYBR green to bind to RNA. The affinity of SYBR Green II (Molecular Probes) for single-stranded RNAs is higher than that of ethidium bromide, and its fluorescence is highest when bound to RNA molecules and practically undetectable in the presence of mononucleotides. It is, therefore, possible to follow the hydrolysis of SYBR Green-bound RNA through the decrease in fluorescence. RNAs are mixed in the standard *Xrn1* reaction buffer with SYBR Green II (Stock solution 10,000× in DMSO). We recommend the use of a 2500-fold dilution of SYBR Green II stock solution (SYBR Green II 4×) in reaction buffer. We have also assayed SYBR Green II 10× in a standard gel-based exonuclease assay and observed no inhibitory effect on *Xrn1* activity. The concentration of RNA substrates has to be adapted to maintain a maximum fluorescence yield of the RNA/SYBR Green II complex. A 4× SYBR Green solution is optimal for 5 to 10 μM of polymerized nucleotides. Thus, we typically assay a 20-nt RNA at 500 nM (10 μM nt) or a 128-nt RNA at 50 nM (6.4 μM nt) concentration. Diluted aliquots from fractions or purified *Xrn1* (0.2 μg) is mixed in 20 μl of standard reaction buffer containing SYBR Green II 4× in tubes adapted for a real-time PCR instrument (Rotor Gene) (Fig. 9.4). As with the TCA-based assay, we can follow the presence of an exoribonuclease activity dependent on the monophosphorylated nature of the 5'-end of RNA, this time through the decrease in SYBR Green fluorescence bound to intact RNA. The advantage is that results are obtained in a non-radioactive real-time assay in less than 10 min, using a spectrofluorometer or a real-time PCR instrument configured for the detection of SYBR Green fluorescence (excitation wavelength 470 nm, detection wavelength 510 nm) (Fig. 9.4).

5. Determining the Directionality of Decay

5.1. Degradation of doubly labeled RNA by *Xrn1*

Confirmation of a 5' to 3'-progression by *Xrn1* on a single RNA substrate can be obtained from experiments where the RNA is labeled both at its 5'-end with [γ^{32}P]-ATP and internally with [^{3}H]-ATP. 5'-Hydroxylated RNA is synthesized in the presence of 12 μl of [2, 5', 8-[^{3}H]]-ATP (60 Ci/mmol, 1 mCi/ml, GE-Amersham) according to Table 9.1. The amount of cold

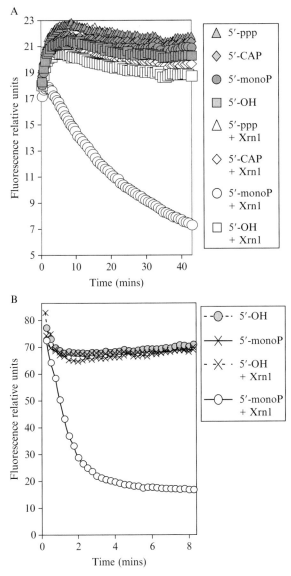

Figure 9.4 Real-time fluorescence-based analysis of RNA hydrolysis by Xrn1. (A) Fluorescence of 128 nt RNAs (50 nM) complexed to SYBR green II (4×) decreases in the presence of Xrn1 (0.2 μg) when the 5′-group is a monophosphate. Abbreviations: RNAs with 5′-hydroxyl groups (5′-OH), 5′-m7GpppG (5′-CAP), 5′-triphosphate (5′-triP), 5′-monophosphate (5′-monoP). (B) Degradation of a 20-nt synthetic RNA (5′-UGGUGGUGGAUCCCGGGAUC-3′) with 5′-monophosphorylated (5′-monoP) or 5′-hydroxyl (5′-OH) group. Fluorescence of RNAs (500 nM) complexed to SYBR green II (4×) decreases in the presence of Xrn1 (0.2 μg) when the 5′-group is a monophosphate. The fluorescence decrease was analyzed on real-time PCR instrument (Rotor Gene, set up on GAIN 5, 30 °C).

ATP is reduced twofold in the synthesis mixture. After DNase I treatment, the RNA is precipitated in the presence of LiCl (Ambion) or in the presence of 0.3 M Na acetate (pH 5.3) and 3 volumes of ethanol. Fifty picomoles (or less) of ^{3}H-RNA is labeled with $(\gamma^{32}$P)-ATP (3000 Ci, 1 mCi/ml, GE-Amersham) and 20 units of T4 polynucleotide kinase (Biolabs), in a total volume of 50 µl, for 30 min at 37 °C. Unincorporated nucleotides are removed with G50 spin columns (GE-Amersham). A TCA-based assay is carried out in the presence of an excess of *Xrn1* (1 µg) and 250 nmol of RNA (Fig. 9.5). The experiment is performed at 0 °C to allow better resolution of the liberation of [^{32}P] from the 5′-end and ^{3}H from the internal portion of the RNA substrate. *Xrn1* seems more distributive at this temperature. Reactions are precipitated with TCA, centrifuged, and the release of labeled mononucleotides into the supernatant over time is measured by scintillation counting (Counter Packard Model Tri-Carb 2100TR). The ^{3}H channel is 1 to 18 keV, and the [^{32}P] channel is 18 to 2000 keV. Tritium values are corrected for contamination of the ^{3}H channel by [^{32}P], determined by calibration of the

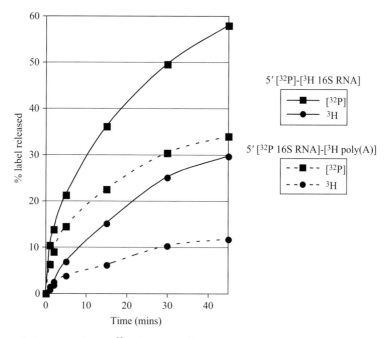

Figure 9.5 *Xrn1* releases [^{32}P] from the 5′-end of doubly labeled 16S rRNA before ^{3}H from the body of the transcript; 0.25 pmol of [^{32}P]/[^{3}H]-labeled 16S rRNA was incubated with 1 µg *Xrn1* at 0 °C for the times indicated. Full-length tritiated RNA with a [^{32}P] label at the 5′-end: 5′-[^{32}P]-[^{3}H-16S RNA] and uniformly [^{32}P]-labeled RNA with a (^{3}H)-poly(A) tail at the 3′-end: 5′-[^{32}P]-16S RNA]-[^{3}H poly(A)]. Samples were precipitated with TCA, centrifuged, and the amount of [^{32}P] and [^{3}H] label released measured by scintillation counting and are reported as a percentage of the input.

counter (in our case 1.6%). The [^{32}P] label at the 5′-extremity is the primary product released initially at 0 °C, whereas tritium is released later, confirming the 5′ to 3′-directionality of RNA decay by Xrn1 (Fig. 9.5).

An alternate approach to generating a double-labeled RNA is by the addition of a ^{3}H-poly(A) tail to a uniformly [^{32}P]-labeled RNA. 5′-monoP [^{32}P]-labeled RNAs are synthesized according to Table 9.1. After DNase I treatment and precipitation, 25 to 50 pmol of RNAs are incubated for 30 min at 37 °C in a final volume of 50 μl containing yeast Poly(A) polymerase buffer (USB), 0.5 μl of 10 mM ATP, 20 μl of [2, 5′, 8-[^{3}H]]-ATP (60 Ci/mmol, 1 mCi/ml, GE-Amersham), and 1200 units of yeast Poly(A) polymerase (USB). Unincorporated nucleotides are removed with G50 spin columns (GE-Amersham), and doubly labeled RNAs are assayed as described previously (Fig. 9.5).

5.2. Degradation of 5′- and 3′-labeled synthetic RNAs by Xrn1

The 5′ to 3′-orientation of exonucleolytic degradation by Xrn1 can also be demonstrated in a gel-based assay. Here we will describe degradation assays with a 20-nt synthetic RNA (5′-UGGUGGUGGAUCCCGGGAUC-3′) [^{32}P]-labeled at either its 5′ or 3′-extremity. 5′-Labeling of 50 pmol of this RNA is performed with 20 units of T4 polynucleotide kinase (Biolabs) and 5 μl of [γ^{32}P]-ATP (3000 Ci, 10 mCi/ml, GE-Amersham) in a final volume of 50 μl, for 30 min at 37 °C. 3′-Labeling of 50 pmol of this RNA is performed with RNA ligase (Biolabs) and 5 μl [γ^{32}P]-pCp (3000 Ci, 10 mCi/ml, GE-Amersham) in a final volume of 50 μl and in the presence of 10% DMSO, incubated overnight at 4 °C. 3′-Labeled RNAs are precipitated in 0.3 M Na acetate/ethanol as previously, resuspended in nuclease-free water, and are 5′-monophosphorylated with 20 units of T4 polynucleotide kinase and 1 mM of cold ATP. For both RNA preparations, unincorporated nucleotides are removed with G50 spin columns (GE-Amersham), and RNAs are gel purified.

An RNA amount corresponding to 100,000 cpm is assayed according to the protocol described for gel-based exoribonuclease assay in the presence of 0.2 μg Xrn1. To get a better snapshot of the production of processing intermediates progressively shortened from the 5′-end, we recommend performing the degradation assays at 0 °C, although similar results have been obtained with Xrn1 at 30 °C (data not shown).

Digestion of the 5′-labeled species by Xrn1 results in the accumulation of the first nucleotide, UMP (Fig. 9.6). Because the RNA is labeled at the 5′-end, no intermediate-length RNA products are detected. Digestion of the 3′-labeled species, on the other hand, results in the production of many intermediate species differing in length by only one nucleotide, consistent with exonucleolytic attack from the 5′-end. These experiments provide conclusive proof of exonucleolytic degradation of RNAs in the 5′ to 3′-direction by Xrn1.

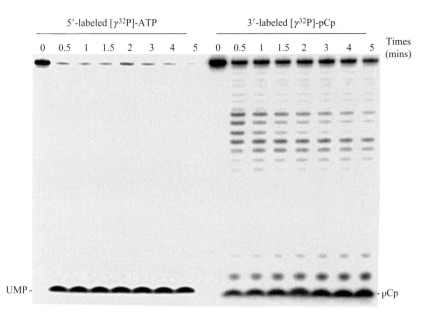

Figure 9.6 Degradation of a 5′-labeled or 3′-labeled 20-nt synthetic RNA by *Xrn1* at 0 °C; 10 pmol RNA (100,000 cpm) labeled at either the 5′- or 3′-end was incubated with 0.2 μg *Xrn1* at 0 °C for the times indicated. Aliquots from the mixture are loaded on 20% denaturing polyacrylamide gels. The migration position of mononucleotide products (UMP/pCp) is indicated.

6. Conclusions and Prospects

The different techniques presented here are important in determining the existence of exoribonuclease activity. They are not limited to the demonstration of 5′ to 3′ or 3′ to 5′-directionality of this activity and can also be helpful in detecting endonuclease activity. Until recently, only eukaryotic proteins belonging to the *Xrn1* family of proteins were thought to have 5′ to 3′-exoribonuclease activity. However, we have used these approaches to show that the recently identified *Bacillus subtilis* endoribonuclease, RNase J1, also possesses 5′ to 3′-exoribonuclease activity (Mathy *et al.*, 2007). RNase J1 is a ubiquitous essential enzyme that contains an N-terminal metallo-β–lactamase domain, interrupted by the so-called β-CASP domain (Callebaut *et al.*, 2002) and is very highly conserved throughout the bacterial and archacal kingdoms (Even *et al.*, 2005). The β-lactamase domain is responsible for the 5′ to 3′-exoribonuclease activity. This discovery paves the way for the discovery of new RNases with similar activity. For example, the mRNA cleavage and polyadenylation specificity factor, CPSF-73, another of the metallo-β–lactamases, is also suspected of having both endonuclease and 5′ to 3′-exonuclease activity (Dominski et al., 2005), although this latter activity has not yet been directly demonstrated.

In vitro exonuclease assays with purified Xrn1 have also been performed to determine the existence of inhibitors of 5′ to 3′-exoribonuclease, such as the metabolite 3′-phospho-adenosine-5′-phosphate (pAp) (Dichtl et al., 1997). Inhibition of Xrn1 by pAp has been shown to occur in the presence of toxic ions such as sodium or lithium, which inhibit Met22, the enzyme metabolizing pAp to AMP (Murguia et al., 1996). This biochemical approach led to the demonstration that cells maintain Xrn1 activity by limiting the accumulation of this inhibitor during sodium stress by means of a Gcn4-dependent regulatory mechanism that triggers additional production of Met22 (Todeschini et al., 2006).

Recent work by Kiledjian and colleagues raised the possibility that other inhibitors of Xrn1 activity also exist. The 3′ to 5′-mRNA degradation pathway generates a 5′-capped oligoribonucleotide that is hydrolyzed by the scavenger decapping activity of Dcs1p (Liu et al., 2002). This decapping activity has been shown to facilitate Xrn1 5′ to 3′-exoribonuclease activity (Liu and Kiledjian, 2005) and has led to the hypothesis that accumulation of m7GpppN species functions as a negative effector of Xrn1 activity. Assays to investigate ribonucleases combined with site-specific mutations in the ribonucleases, their potential inhibitors, and ribonuclease-associated factors is, therefore, a way to improve our understanding of the control of mRNA degradation. The discovery of new potential effectors of 5′ to 3′-mRNA degradation with these assays will provide important information with regard to the pathways involved in the regulation of mRNA degradation.

REFERENCES

Amrani, N., Dong, S., He, F., Ganesan, R., Ghosh, S., Kervestin, S., Li, C., Mangus, D. A., Spatrick, P., and Jacobson, A. (2006). Aberrant termination triggers nonsense-mediated mRNA decay. Biochem. Soc. Trans. **34**, 39–42.

Bashkirov, V. I., Scherthan, H., Solinger, J. A., Buerstedde, J. M., and Heyer, W. D. (1997). A mouse cytoplasmic exoribonuclease (mXRN1p) with preference for G4 tetraplex substrates. J. Cell. Biol. **136**, 761–773.

Callebaut, I., Moshous, D., Mornon, J. P., and de Villartay, J. P. (2002). Metallo-beta-lactamase fold within nucleic acids processing enzymes: The beta-CASP family. Nucleic Acids Res. **30**, 3592–3601.

Caponigro, G., Muhlrad, D., and Parker, R. (1993). A small segment of the MAT alpha 1 transcript promotes mRNA decay in Saccharomyces cerevisiae: A stimulatory role for rare codons. Mol. Cell. Biol. **13**, 5141–5148.

Caponigro, G., and Parker, R. (1996). Mechanisms and control of mRNA turnover in Saccharomyces cerevisiae. Microbiol. Rev. **60**, 233–249.

Chen, N., Walsh, M. A., Liu, Y., Parker, R., and Song, H. (2005). Crystal structures of human DcpS in ligand-free and m7GDP-bound forms suggest a dynamic mechanism for scavenger mRNA decapping. J. Mol. Biol. **347**, 707–718.

Chernukhin, I. V., Seago, J. E., and Newbury, S. F. (2001). Drosophila 5′ to 3′-exoribonuclease Pacman. Methods Enzymol. **342**, 293–302.

Coller, J., and Parker, R. (2004). Eukaryotic mRNA decapping. *Annu. Rev. Biochem.* **73**, 861–890.
Conti, E., and Izaurralde, E. (2005). Nonsense-mediated mRNA decay: Molecular insights and mechanistic variations across species. *Curr. Opin. Cell. Biol.* **17**, 316–325.
Dichtl, B., Stevens, A., and Tollervey, D. (1997). Lithium toxicity in yeast is due to the inhibition of RNA processing enzymes. *EMBO J.* **16**, 7184–7195.
Dominski, Z., Yang, X. C., and Marzluff, W. F. (2005). The polyadenylation factor CPSF-73 is involved in histone-pre-mRNA processing. *Cell* **123**, 37–48.
Even, S., Pellegrini, O., Zig, L., Labas, V., Vinh, J., Brechemmier-Baey, D., and Putzer, H. (2005). Ribonucleases J1 and J2: Two novel endoribonucleases in *B. subtilis* with functional homology to *E. coli* RNase E. *Nucleic Acids Res.* **33**, 2141–2152.
Gallie, D. R. (1991). The cap and poly(A) tail function synergistically to regulate mRNA translational efficiency. *Genes Dev.* **5**, 2108–2116.
Gatfield, D., and Izaurralde, E. (2004). Nonsense-mediated messenger RNA decay is initiated by endonucleolytic cleavage in Drosophila. *Nature* **429**, 575–578.
Hagan, K. W., Ruiz-Echevarria, M. J., Quan, Y., and Peltz, S. W. (1995). Characterization of cis-acting sequences and decay intermediates involved in nonsense-mediated mRNA turnover. *Mol. Cell. Biol.* **15**, 809–823.
Henry, Y., Wood, H., Morrissey, J. P., Petfalski, E., Kearsey, S., and Tollervey, D. (1994). The 5′-end of yeast 5.8S rRNA is generated by exonucleases from an upstream cleavage site. *EMBO J.* **13**, 2452–2463.
Heyer, W. D., Johnson, A. W., Reinhart, U., and Kolodner, R. D. (1995). Regulation and intracellular localization of *Saccharomyces cerevisiae* strand exchange protein 1 (Sep1/Xrn1/Kem1), a multifunctional exonuclease. *Mol. Cell. Biol.* **15**, 2728–2736.
Ingelfinger, D., Arndt-Jovin, D. J., Luhrmann, R., and Achsel, T. (2002). The human LSm1-7 proteins colocalize with the mRNA-degrading enzymes Dcp1/2 and Xrn1 in distinct cytoplasmic foci. *RNA* **8**, 1489–1501.
Johnson, A. W., and Kolodner, R. D. (1991). Strand exchange protein 1 from Saccharomyces cerevisiae. A novel multifunctional protein that contains DNA strand exchange and exonuclease activities. *J. Biol. Chem.* **266**, 14046–14054.
Kastenmayer, J. P., and Green, P. J. (2000). Novel features of the XRN-family in Arabidopsis: Evidence that AtXRN4, one of several orthologs of nuclear Xrn2p/Rat1p, functions in the cytoplasm. *Proc. Natl. Acad. Sci. USA* **97**, 13985–13990.
Kolodner, R., Evans, D. H., and Morrison, P. T. (1987). Purification and characterization of an activity from *Saccharomyces cerevisiae* that catalyzes homologous pairing and strand exchange. *Proc. Natl. Acad. Sci. USA* **84**, 5560–5564.
Lejeune, F., Li, X., and Maquat, L. E. (2003). Nonsense-mediated mRNA decay in mammalian cells involves decapping, deadenylating, and exonucleolytic activities. *Mol. Cell.* **12**, 675–687.
Liu, H., and Kiledjian, M. (2005). Scavenger decapping activity facilitates 5′ to 3′-mRNA decay. *Mol. Cell. Biol.* **25**, 9764–9772.
Liu, H., Rodgers, N. D., Jiao, X., and Kiledjian, M. (2002). The scavenger mRNA decapping enzyme DcpS is a member of the HIT family of pyrophosphatases. *EMBO J.* **21**, 4699–46708.
Liu, Z., and Gilbert, W. (1994). The yeast KEM1 gene encodes a nuclease specific for G4 tetraplex DNA: Implication of *in vivo* functions for this novel DNA structure. *Cell.* **77**, 1083–1092.
Mathy, N., Benard, L., Pellegrini, O., Daou, R., Wen, T., and Condon, C. (2007). 5′ to 3′-exoribonuclease activity in bacteria: Role of RNase J1 in rRNA maturation and 5′-stability of mRNA. *Cell* **129**, 681–692.
Murguia, J. R., Belles, J. M., and Serrano, R. (1996). The yeast HAL2 nucleotidase is an *in vivo* target of salt toxicity. *J. Biol. Chem.* **271**, 29029–29033.

Newbury, S., and Woollard, A. (2004). The 5′ to 3′-exoribonuclease xrn-1 is essential for ventral epithelial enclosure during *C. elegans* embryogenesis. *RNA.* **10,** 59–65.

Newbury, S. F. (2006). Control of mRNA stability in eukaryotes. *Biochem. Soc. Trans.* **34,** 30–34.

Orban, T. I., and Izaurralde, E. (2005). Decay of mRNAs targeted by RISC requires XRN1, the Ski complex, and the exosome. *RNA* **11,** 459–469.

Oussenko, I. A., Sanchez, R., and Bechhofer, D. H. (2002). *Bacillus subtilis* YhaM, a member of a new family of 3′ to 5′-exonucleases in gram-positive bacteria. *J. Bacteriol.* **184,** 6250–6259.

Page, A. M., Davis, K., Molineux, C., Kolodner, R. D., and Johnson, A. W. (1998). Mutational analysis of exoribonuclease I from *Saccharomyces cerevisiae*. *Nucleic Acids Res.* **26,** 3707–3716.

Parker, R., and Song, H. (2004). The enzymes and control of eukaryotic mRNA turnover. *Nat. Struct. Mol. Biol.* **11,** 121–127.

Ross, P. L., Huang, Y. N., Marchese, J. N., Williamson, B., Parker, K., Hattan, S., Khainovski, N., Pillai, S., Dey, S., Daniels, S., Purkayastha, S., Jushaz, P., *et al.* (2004). Multiplexed protein quantitation in *Saccharomyces cerevisiae* using amine-reactive isobaric tagging reagents. *Mol. Cell Proteomics* **3,** 1154–1169.

Sato, Y., Shimamoto, A., Shobuike, T., Sugimoto, M., Ikeda, H., Kuroda, S., and Furuichi, Y. (1998). Cloning and characterization of human Sep1 (hSEP1) gene and cytoplasmic localization of its product. *DNA Res.* **5,** 241–246.

Sheth, U., and Parker, R. (2003). Decapping and decay of messenger RNA occur in cytoplasmic processing bodies. *Science.* **300,** 805–808.

Solinger, J. A., Pascolini, D., and Heyer, W. D. (1999). Active-site mutations in the *Xrn*1p exoribonuclease of *Saccharomyces cerevisiae* reveal a specific role in meiosis. *Mol. Cell. Biol.* **19,** 5930–5942.

Souret, F. F., Kastenmayer, J. P., and Green, P. J. (2004). AtXRN4 degrades mRNA in Arabidopsis and its substrates include selected miRNA targets. *Mol. Cell.* **15,** 173–183.

Stevens, A. (1980). Purification and characterization of a *Saccharomyces cerevisiae* exoribonuclease which yields 5′-mononucleotides by a 5′-leads to 3′-mode of hydrolysis. *J. Biol. Chem.* **255,** 3080–3085.

Stevens, A. (2001). 5′-exoribonuclease 1: *Xrn1*. *Methods Enzymol.* **342,** 251–259.

Stevens, A., Hsu, C. L., Isham, K. R., and Larimer, F. W. (1991). Fragments of the internal transcribed spacer 1 of pre-rRNA accumulate in *Saccharomyces cerevisiae* lacking 5′ to 3′-exoribonuclease 1. *J. Bacteriol.* **173,** 7024–7028.

Stevens, A., and Poole, T. L. (1995). 5′-exonuclease-2 of *Saccharomyces cerevisiae*. Purification and features of ribonuclease activity with comparison to 5′-exonuclease-1. *J. Biol. Chem.* **270,** 16063–16069.

Szankasi, P., and Smith, G. R. (1996). Requirement of *S. pombe* exonuclease II, a homologue of S. cerevisiae Sep1, for normal mitotic growth and viability. *Curr. Genet.* **30,** 284–293.

Tishkoff, D. X., Rockmill, B., Roeder, G. S., and Kolodner, R. D. (1995). The sep1 mutant of *Saccharomyces cerevisiae* arrests in pachytene and is deficient in meiotic recombination. *Genetics.* **139,** 495–509.

Todeschini, A. L., Condon, C., and Benard, L. (2006). Sodium-induced GCN4 expression controls the accumulation of the 5′ to 3′-RNA degradation inhibitor, 3′-phosphoadenosine 5′-phosphate. *J. Biol. Chem.* **281,** 3276–3282.

Zer, C., and Chanfreau, G. (2005). Regulation and surveillance of normal and 3′-extended forms of the yeast aci-reductone dioxygenase mRNA by RNase III cleavage and exonucleolytic degradation. *J. Biol. Chem.* **280,** 28997–29003.

Zhang, Z., Simons, A. M., Prabhu, V. P., and Chen, J. (1998). Strand exchange protein 1 (Sep1) from *Saccharomyces cerevisiae* does not promote branch migration *in vitro*. *J. Biol. Chem.* **273,** 4950–4956.

CHAPTER TEN

RECONSTITUTION OF RNA EXOSOMES FROM HUMAN AND *SACCHAROMYCES CEREVISIAE*: CLONING, EXPRESSION, PURIFICATION, AND ACTIVITY ASSAYS

Jaclyn C. Greimann *and* Christopher D. Lima

Contents

1. Introduction	186
2. Cloning Strategies for Recombinant Protein Expression	190
3. PCR and Subcloning Protocols	191
4. PCR and Subcloning Protocols for Yeast cDNA	191
4.1. Yeast *RRP41/RRP45* cDNA	191
4.2. Yeast *MTR3/RRP42* cDNA	193
4.3. Yeast *RRP46/RRP43* cDNA	193
4.4. Yeast *RRP4, RRP40, CSL4, RRP44, RRP6* cDNA	193
4.5. Yeast *RRP6* cDNA fused to *SMT3* cDNA	193
5. PCR and Subcloning Protocols for Human cDNA	194
5.1. Human *RRP45/RRP41* cDNA	194
5.2. Human *RRP42/MTR3* cDNA	194
5.3. Human *RRP43* and *RRP46* cDNA fused to *SMT3* cDNA	194
5.4. Human RRP4, RRP40, and CSL4 cDNA	194
6. Expression and Purification of Yeast Exosome Proteins	195
6.1. Yeast Rrp41/Rrp45	196
6.2. Yeast Mtr3/Rrp42	196
6.3. Yeast Rrp46/Rrp43	197
6.4. Yeast Rrp4, Rrp40, and Csl4	197
6.5. Yeast Rrp44	197
6.6. Yeast Rrp6	198
7. Expression and Purification of Human Exosome Proteins	198
7.1. Human Rrp45/Rrp41	199
7.2. Human Rrp42/Mtr3	200

Structural Biology Program, Sloan-Kettering Institute, New York, NY, USA

7.3. Human Rrp43	200
7.4. Purification of human Rrp42/Mtr3/Rrp43	200
7.5. Human Rrp46	200
7.6. Human Rrp4, Rrp40, and Csl4	201
8. Reconstitution and Purification of Human and Yeast Exosomes	201
8.1. The yeast nine-subunit exosome	202
8.2. The yeast ten-subunit exosome	202
8.3. The yeast eleven-subunit exosome	203
8.4. Further purification of yeast exosomes	203
8.5. The human nine-subunit exosome	205
9. Exoribonuclease Assays	206
10. Comparative Exoribonuclease Assays with Different RNA Substrates	207
11. Conclusions	208
Acknowledgments	208
References	209

Abstract

Eukaryotic RNA exosomes participate in 3′ to 5′-processing and degradation of RNA in the nucleus and cytoplasm. RNA exosomes are multisubunit complexes composed of at least nine distinct proteins that form the exosome core. Although the eukaryotic exosome core shares structural and sequence similarity to phosphorolytic archaeal exosomes and bacterial PNPase, the eukaryotic exosome core has diverged from its archaeal and bacterial cousins and appears devoid of phosphorolytic activity. In yeast, the processive hydrolytic 3′ to 5′-exoribonuclease Rrp44 associates with exosomes in the nucleus and cytoplasm. Although human Rrp44 appears homologous to yeast Rrp44, it has not yet been shown to associate with human exosomes. In the nucleus, eukaryotic exosomes interact with Rrp6, a distributive hydrolytic 3′ to 5′-exoribonuclease. To facilitate analysis of eukaryotic RNA exosomes, we will describe procedures used to clone, express, purify, and reconstitute the nine-subunit human exosome and nine-, ten-, and eleven-subunit yeast exosomes. We will also discuss procedures to assess exoribonuclease activity for reconstituted exosomes.

1. INTRODUCTION

Classical genetics identified one of the first proteins involved in ribosome biosynthesis, the ribosomal RNA processing protein 4 (Rrp4; Mitchell et al., 1996). Protein A–tagged-Rrp4 copurified with a 300-kDa complex, including at least four additional proteins Rrp41, Rrp42, Rrp43, and Rrp44 and complementation of yeast *rrp4-1* with human Rrp4 facilitated purification of the human exosome and identification of additional exosome subunits in both yeast and human exosomes (Allmang et al., 1999; Mitchell et al., 1997). These and subsequent studies led to the discovery that human and yeast exosomes include a core of nine to eleven proteins, six that share similarity to bacterial RNase PH (Rrp41, Rrp45, Rrp42, Rrp43, Mtr3, and Rrp46), three that share similarity to S1/KH RNA binding domains (Rrp4, Rrp40, and Csl4), and two that share similarity to bacterial hydrolytic enzymes, RNase D (Rrp6) and RNase II/R (Rrp44). Ten of the eleven exosome proteins are essential for yeast viability, except Rrp6, and depletion of any elicited RNA-processing defects (Allmang et al., 1999; Schneider et al., 2007).

Homologs for each exosome gene and respective protein are found in human and yeast, but their association in the cell may differ (Graham et al., 2006). In yeast, affinity-purified exosomes from both nuclear and cytoplasmic fractions contain yeast Rrp44, whereas human exosomes have not been copurified with human Rrp44 (Allmang et al., 1999; Raijmakers et al., 2004). Consequently, it is currently believed that nine-subunit (Exo9) or ten-subunit (Exo10) exosomes compose core cytoplasmic exosomes in human and yeast, respectively, whereas exosomes in the nucleus associate with the additional subunit Rrp6 (Allmang et al., 1999).

The proteins involved in $3'$ to $5'$-RNA decay share evolutionary relationships among prokaryotic, archaeal, and eukaryotic organisms. Once initiated, $3'$ to $5'$-RNA decay in bacteria proceeds by means of two distinct $3'$ to $5'$-exoribonuclease activities, one catalyzed by processive hydrolytic exoribonucleases (RNase II/R), and the other catalyzed by a processive phosphorylase (PNPase). Bacterial PNPase is a core constituent of an RNA decay complex termed the degradosome, a complex composed of PNPase, the endonuclease RNase E, the helicase RhlB, and enolase (Fig. 10.1; Carpousis, 2002). Each PNPase protomer includes four domains, two C-terminal RNA binding domains (S1 and KH domains) and two core domains that bear sequence and structural homology to RNase PH (a phosphate-dependent exoribonuclease). The N-terminal RNase PH-like core domain is noncatalytic, whereas the second RNase PH-like core domain contains the catalytic residues required to activate phosphate for nucleophilic attack at the RNA phosphodiester backbone. Both RNase PH-like core domains contribute to formation of a composite surface for substrate binding within the central pore (Fig. 10.1; Carpousis, 2002; Symmons et al., 2000).

Archaeal exosomes were uncovered through genomic analysis and by affinity purification, revealing that archaeal exosomes are composed of at least four proteins (Rrp41, Rrp42, Csl4, and Rrp4), so named for their relationship to the analogous eukaryotic proteins (Evguenieve-Hackenberg et al., 2003; Koonin et al., 2001). Archaeal Rrp42 and Rrp41 share sequence and structural similarity to the first and second RNase PH-like domains in PNPase, respectively (Lorentzen et al., 2005), whereas Csl4 and Rrp4 share similarity to the S1 and KH domains within PNPase (Buttner et al., 2005).

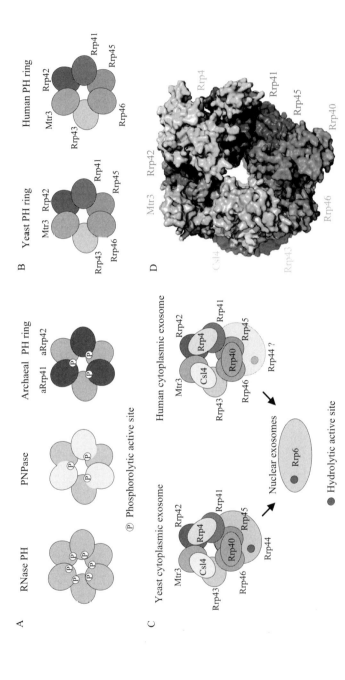

Figure 10.1 Schematic of bacterial, archaeal, and eukaryotic RNase PH rings and excsomes. (A) The RNase PH ring from RNase PH (left) indicating phosphorolytic active sites (circles containing a P) within the six subunits. The PNPase RNase PH-like ring (middle) indicating that PNPase encompasses a fusion between N-terminal and C-terminal RNase PH-like domains. The C-terminal PH-like domain contains the phosphorolytic active site, whereas the N-terminal and C-terminal PH-like domains encompass a composite substrate-binding surface. Active sites are indicated. Archaeal RNase PH-like ring (right) indicating that archaeal exosomes include an RNase PH-like ring composed of a

Structural studies have revealed the architecture of six- and nine-subunit archaeal exosomes. In the former, structures showed that archaeal exosomes form stable six-subunit RNase PH-like rings through oligomerization of three Rrp41/Rrp42 heterodimers (Fig. 10.1; Buttner et al., 2005; Lorentzen and Conti, 2005; Lorentzen et al., 2005). In the latter, structures revealed that nine-subunit archaeal exosomes are formed by capping the PH-like ring with either three Csl4 or three Rrp4 subunits (Buttner et al., 2005; Lorentzen et al., 2007). In the cell, archaeal exosomes are presumed to form complexes with mixtures of Csl4 and Rrp4, although the stoichiometry of this complex is not known. Subsequent studies revealed that RNA substrates must penetrate the pore to gain access to the phosphorolytic active sites, although it is not currently understood whether RNA is recruited to these complexes by means of direct interactions with Rrp41, Rrp42, Rrp4, or Cls4 (Lorentzen et al., 2007).

The architecture of the nine-subunit human exosome core was recently determined (Liu et al., 2006), revealing structural relationships among the eukaryotic exosome, PNPase, and archaeal exosomes. In this case, the eukaryotic RNase PH-like ring is composed of six different gene products, and single copies of Csl4, Rrp4, and Rrp40 cap the RNase PH-like ring (Fig.10.1; Liu et al., 2006). Although the RNase PH-like domains resemble the phosphorolytic PH-like domains in PNPase and archaeal exosomes, recent studies suggest that eukaryotic exosomes have diverged mechanistically from their bacterial and archaeal cousins (Dziembowski et al., 2007; Liu et al., 2006; 2007). In these studies, phosphorolytic activity was not observed for nine-subunit PH-ring complexes from either yeast or humans.

---

trimer of Rrp41/Rrp42 heterodimers. In this instance, the Rrp41 subunit contains the active site analogous to the active PNPase PH-like domain, whereas Rrp42 is analogous to the PNPase N-terminal PH-like domain. (B) RNase PH-like rings for yeast and human Rrp41, Rrp45, Rrp42, Mtr3, Rrp43, Rrp46. Although Rrp41, Rrp46, and Mtr3 appear more similar to archaeal Rrp41 and the C-terminal PNPase PH domain, none has been shown to contain a functional phosphorolytic catalytic site. Rrp42, Rrp43, and Rrp45 appear more similar to archaeal Rrp42 and the N-terminal PNPase PH domain. The position of labels for individual subunits indicate differences between human and yeast with respect to the subcomplexes purified. In the case of yeast, Rrp41/Rrp45, Mtr3/Rrp42, Rrp46/Rrp43 copurified as subcomplexes, whereas in humans, Rrp43/Mtr3/Rrp42 and Rrp41/Rrp45 were purified as subcomplexes and Rrp46 was purified as a protomer. (C) Organization of the cytoplasmic and nuclear exosomes from humans and yeast. The S1/KH domain proteins Csl4, Rrp4, and Rrp40 cap the RNase PH-like ring. Rrp44 is known to associate with the yeast nine-protein core, although Rrp44 has not been shown to associate with the human exosome core (indicated by "Rrp44?"). Arrows indicate that respective cytoplasmic exosomes associate with Rrp6 in nuclear fractions. Filled circles indicate the location of hydrolytic active sites. (D) The x-ray structure of the human exosome in a similar orientation to that observed in (C), indicating the positions of respective subunits and oriented to highlight the central pore common to RNase PH-like rings from all kingdoms of life.

Although phosphorolytic activities were initially reported for the human nine-subunit exosome (Liu et al., 2006), these activities were later determined to be due to contamination by bacterial PNPase (Liu et al., 2007). It remains unclear why eukaryotic nine-subunit exosome cores have lost their phosphorolytic activity. We posit that phosphorolytic activity may have been lost to facilitate evolution of binding sites within the nine-subunit core, either for interaction with RNA substrates or for interaction with protein cofactors, including the hydrolytic exoribonucleases Rrp44 and Rrp6, as well as TRAMP and Ski complexes (Houseley et al., 2006).

In this chapter, we will describe procedures used to clone and express eukaryotic exosome subunits and protocols for reconstitution of exosomes that include the nine-subunit human exosome, and nine-, ten-, and eleven-subunit exosomes from the budding yeast *Saccharomyces cerevisiae*. We will also describe biochemical assays used to assess the catalytic activities of these exosome preparations.

2. Cloning Strategies for Recombinant Protein Expression

To prepare individual exosome proteins in the absence of potential contaminating cofactors or exosome-associated subunits derived from a eukaryotic host, we embarked on expression and purification of eukaryotic exosome proteins from recombinant bacterial hosts. Some human and yeast exosome proteins could be obtained as single homogeneous polypeptides; however, other exosome proteins required obligate coexpression partners to maintain solubility and to facilitate copurification of stoichiometric subcomplexes. In some instances, coexpression did not suffice, and protein expression required the assistance of fusion partners.

Primer sets were designed for exosome genes from yeast and human for subcloning into Duet vectors (Novagen), pSMT3-Topo, or pSMT3 (Mossessova and Lima, 2000; Custom-Topo adapted by Invitrogen). Duet vectors facilitate coexpression by offering the opportunity to subclone genes into two distinct multiple cloning sites (MCS1 and MCS2) within a single plasmid. Engineered primers for Duet or pSMT3 vectors contained sequences for unique restriction endonuclease sites to facilitate ligation into respective MCS. In contrast, the pSMT3-Topo vector used directional Topo cloning by means of flap ligation (Invitrogen). In this instance, the 5′-primer required no additional engineered sequences, whereas the 3′-primer required addition of a 5′-CACC-3′ to facilitate flap ligation, which enabled ligation of the 3′-recessed end to a blunt ended acceptor site (Cheng and Shuman, 2000). Both pSMT3 and pSMT3-Topo vectors are based on pET-28b (Novagen) and encode fusions between the respective polypeptide and an

N-terminal Hexa-histidine Smt3 polypeptide. Fusions to the yeast SUMO ortholog Smt3 can increase expression levels and enhance solubility (Mossessova and Lima, 2000). The His6-Smt3 fusion can be liberated from the protein of interest by incubation in the presence of the Smt3 protease Ulp1. The following protocols are the result of a number of coexpression combinations tested, the most efficient of which will be presented.

3. PCR AND SUBCLONING PROTOCOLS

Coding DNA for yeast exosome genes could be amplified by PCR from *S. cerevisiae* genomic DNA (*W303-1A*), because none of the yeast genes included introns. Coding DNA for human exosome genes was amplified by PCR from human placental cDNA (Ambion, Inc.). Touchdown PCR methods were used in conjunction with a high-fidelity polymerase such as Pfu (Stratagene). Native stop codons were maintained in all respective coding sequences. PCR products designed for Topo ligation were gel-purified (Qiagen) and subsequently ligated into pSMT3-Topo vector with directional TOPO-ligation protocols (Invitrogen). Vectors and PCR products designated for pRSF-Duet, pDuet, or pSMT3 were digested with appropriate restriction enzymes (NEB), and the vectors were treated with calf intestinal phosphatase (NEB) to dephosphorylate 5′-ends. Vectors and PCR products were gel-purified (Qiagen) and ligated with T4 DNA ligase (NEB). Many RNase PH-like proteins were insoluble when expressed alone (discussed later; Liu *et al.*, 2006; Lorentzen *et al.*, 2005), so exosome RNase PH-like genes were cloned in pairs into pDuet-1 MCS1 and MCS2 in sequential fashion. Cloning into MCS1 encodes an N-terminal fusion to a Hexa-histidine polypeptide (His$_6$), whereas an untagged polypeptide was encoded through the use of MCS2. Ligated products were transformed into chemically competent One Shot *E. coli* TOP10 cells (Invitrogen). Plasmid DNA was obtained from a single colony, sequence-verified, and used to transform *E. coli* expression strains.

4. PCR AND SUBCLONING PROTOCOLS FOR YEAST CDNA

4.1. Yeast *RRP41/RRP45* cDNA

Yeast *RRP41* and *RRP45* cDNA were inserted into MCS1 and MCS2 of plasmid pRSF-Duet-1(Novagen), respectively, by use of restriction sites engineered into positions flanking the coding sequence. The plasmid pRSF-*RRP41*(MCS1)/*RRP45*(MCS2) encodes two polypeptides, an

Table 10.1 Cloning and expression

Human	Vector	MCS	5′-site	3′-site	Nonnative N-terminal amino acids
RRP45	pETDuet-1	MCS I	EcoRI	SalI	MGSSHHHHHHSQDPNSH
RRP41	pRSFDuet-1	MCS II	BglII/BamHI[a]	SalI/XhoI[a]	MADP
RRP42	pRSFDuet-1	MCS I	BamHI	XhoI/SalI[a]	MGSSHHHHHHSQDP
MTR3	pSMT3	MCS II	NdeI	SalI/XhoI[a]	—
RRP43	pSMT3	MCS	BamHI	SalI/XhoI[a]	DP
RRP46	pSMT3-Topo	—	Blunt	Flap	SL
CSL4	pRSFDuet-1	MCS I	BamHI	HindIII	MGSSHHHHHHSQDP
RRP4	pRSFDuet-1	MCS I	BamHI	SacI	MGSSHHHHHHSQDPH
RRP40	pRSFDuet-1	MCS I	BamHI	PstI	MGSSHHHHHHSQDP

Yeast	Vector	MCS	5′-site	3′-site	Nonnative N-terminal amino acids
RRP41	pRSFDuet-1	MCS I	BamHI	SalI	MGSSHHHHHHSQDPH
RRP45	pRSFDuet-1	MCS II	BglII/BamHI[a]	SalI/XhoI[a]	MADPH
MTR3	pRSFDuet-1	MCS I	EcoRI	XhoI/SalI[a]	MGSSHHHHHHSQDPNSH
RRP42	pRSFDuet-1	MCS II	BglII/BamHI[a]	SalI/XhoI[a]	MADPH
RRP46	pRSFDuet-1	MCS I	BamHI	SalI	MGSSHHHHHHSQDPH
RRP43	pRSFDuet-1	MCS II	BglII/BamHI[a]	SalI/XhoI[a]	MADP
CSL4	pRSFDuet-1	MCS I	BamHI	SalI	MGSSHHHHHHSQDP
RRP4	pRSFDuet-1	MCS I	BamHI	SalI	MGSSHHHHHHSQDP
RRP40	pRSFDuet-1	MCS I	BamHI	SalI	MGSSHHHHHHSQDPH
RRP44	pRSFDuet-1	MCS I	EcoRI	XhcI/SalI[a]	MGSSHHHHHHSQDPNS
RRP6	pRSFDuet-1	MCS I	EcoRI	SalI	MGSSHHHHHHSQDPNS
RRP6	pSMT3-Topo	—	Blunt	Flap	SL

[a] Denotes loss of restriction site after ligation.

N-terminal His-tagged Rrp41 fusion and untagged Rrp45 (Table 10.1). This plasmid was used to transform *E. coli* BL21 (DE3) Codon Plus RIL (Stratagene).

4.2. Yeast *MTR3/RRP42* cDNA

Yeast *MTR3* and *RRP42* cDNA were inserted into MCS1 and MCS2 of plasmid pRSF-Duet-1, respectively, by use of restriction sites engineered into positions flanking the coding sequence. The plasmid pRSF-MTR3 (MCS1)/*RRP42*(MCS2) encodes two polypeptides, an N-terminal His-tagged Mtr3 fusion and untagged Rrp42 (Table 10.1). This plasmid was used to transform *E. coli* BL21 (DE3) Codon Plus RIL.

4.3. Yeast *RRP46/RRP43* cDNA

Yeast *RRP46* and *RRP43* cDNA were inserted into MCS1 and MCS2 of plasmid pRSF-Duet-1, respectively, by use of restriction sites engineered into positions flanking the coding sequence. The plasmid pRSF-*RRP46* (MCS1)/*RRP43*(MCS2) encodes two polypeptides, an N-terminal His-tagged RRP46 fusion and untagged RRP43 (Table 10.1). This plasmid was used to transform *E. coli* BL21 (DE3) Codon Plus RIL.

4.4. Yeast *RRP4, RRP40, CSL4, RRP44, RRP6* cDNA

Yeast *RRP4, RRP40, CSL4, RRP44*, or *RRP6* cDNA was individually inserted into MCS1 within plasmid pRSF-Duet-1 to generate plasmids pRSF-*RRP4*, pRSF-*RRP40*, pRSF-*CSL4*, pRSF-*RRP44*, and pRSF-*RRP6*, respectively. These constructs encoded in-frame fusions of the respective exosome polypeptides to N-terminal Hexa-histidine polypeptides (Table 10.1). These plasmids were used to transform *E. coli* BL21 (DE3) Codon Plus RIL. To facilitate cloning within pRSF-Duet-1 MCS1, deliberate point mutations were inserted within the 5′-primers to encode proteins with the following amino acid substitutions: Rrp4 (S2A), RRP6 (T2A), and RRP44 (S2A). *RRP40* cDNA contained a point mutation that resulted in substitution of leucine for phenylalanine at position 160. This mutation was inherent to the parental yeast strain *(W303-1A)* that was used to prepare genomic DNA.

4.5. Yeast *RRP6* cDNA fused to *SMT3* cDNA

Yeast *RRP6* cDNA was also ligated into p*SMT3*-Topo (Mossessova and Lima, 2000; custom plasmid; Invitrogen) resulting in plasmid p*SMT3-RRP6* that encodes a fusion between an N-terminal His_6-Smt3 polypeptide and Rrp6. The His_6-Smt3 polypeptide can be liberated from Rrp6 by digestion with the Smt3 protease Ulp1 to generate Rrp6 with a nonnative N-terminal Ser-Leu polypeptide (Table 10.1). This plasmid was used to transform *E. coli* BL21 (DE3) Codon Plus RIL.

5. PCR AND SUBCLONING PROTOCOLS FOR HUMAN cDNA

5.1. Human *RRP45/RRP41* cDNA

Human *RRP45* and *RRP41* cDNA were inserted into MCS1 and MCS2 of plasmid pDuet-1 (Novagen), respectively (Table 10.1). Plasmid pDuet-*RRP45*(MCS1)/*RRP41*(MCS2) encodes two polypeptides, an N-terminal His-tagged Rrp45 fusion and untagged Rrp41. This plasmid was used to transform *E. coli* BL21(DE3) Codon Plus RIL.

5.2. Human *RRP42/MTR3* cDNA

Human *RRP42* and *MTR3* cDNA were inserted into pRSF-Duet-1. *RRP42* was placed into MCS1, whereas *MTR3* was integrated into MCS2 (Table 10.1). Plasmid pRSF-*RRP42*(MCS1)/*MTR3*(MCS2) encodes two polypeptides, an N-terminal His-tagged Rrp42 fusion and untagged Mtr3. This plasmid was used to transform *E. coli* BL21 (DE3) STAR (Invitrogen).

5.3. Human *RRP43* and *RRP46* cDNA fused to *SMT3* cDNA

Human *RRP43* or *RRP46* cDNA was subcloned into p*SMT3*-Topo (Mossessova and Lima, 2000; custom plasmid; Invitrogen), resulting in plasmids p*SMT3*-Rrp43 and p*SMT3-RRP46*, respectively (Table 10.1). These plasmids generate fusions between an N-terminal His$_6$-tagged Smt3 polypeptide and the respective exosome polypeptide. These plasmids were used to transform *E. coli* BL21 (DE3) Codon Plus RIL.

5.4. Human RRP4, RRP40, and CSL4 cDNA

Human *RRP4*, *RRP40*, or *CSL4* cDNA was inserted into MCS1 of plasmid pRSF-Duet-1 to generate plasmids pRSF-*RRP4*, pRSF-*RRP40*, and pRSF-CSL4, respectively (Table 10.1). These expression constructs encode in-frame polypeptide fusions that contain an N-terminal Hexahistidine tag followed by the respective exosome polypeptides. These plasmids were used to transform *E. coli* BL21 (DE3) Codon Plus RIL.

6. EXPRESSION AND PURIFICATION OF YEAST EXOSOME PROTEINS

Large-scale expression cultures were obtained by fermentation in a BioFlo 3000 bioreactor (New Brunswick); 2 ml of LB culture was inoculated with 50 µl glycerol stock and incubated for 5 to 6 h at 37 °C. This starter culture was used to inoculate 500 ml Superbroth (SB) media (Teknova), which was then incubated at 37 °C overnight, but for no more than 12 h. The overnight culture was used to inoculate the 10 L of SB in the BioFlo 3000. Cultures were maintained at 37 °C and infused with air by mixing at 800 rpm. When the OD_{600} reached 1.5, cultures were cooled to 30 °C, and expression was induced by addition of IPTG to a final concentration of 0.75 mM before the culture reached OD_{600} of 2. Cultures were fermented for an additional 3 to 4 h at 30 °C. Cell cultures for expression of recombinant yeast exosome proteins were harvested by centrifugation at 4 °C at 9000g (Beckman JLA-8.1). Supernatant was discarded and cell pellets were suspended in 20% sucrose and 50 mM Tris-HCl, pH 8.0, at a concentration of 0.5 mg/ml, snap-frozen in liquid nitrogen, and stored at −80 °C. Unless otherwise noted, cell suspensions from 3.5 L of the 10 L fermenter culture were thawed and prepared for sonication in buffer containing 20% sucrose, 50 mM Tris-HCl, pH 8.0, 0.5 M NaCl, 20 mM imidazole, 0.1% IGEPAL, 1 mM β-mercaptoethanol (BME), 1 mM phenyl-methane-sulfonyl fluoride (PMSF), and 10 µg/ml DNase (Sigma). Over 20 min, cells were disrupted by sonication with 8× 30-sec pulses at 80% power (Branson Digital Sonifier). Insoluble cell debris was removed by centrifugation at 39,000g (Beckman JA-20).

All protein purification was conducted at 4 °C. Metal-affinity chromatography (Ni-NTA SuperFlow; Qiagen) was used to affinity purify each recombinant yeast protein that contained a fusion to an N-terminal Hexahistidine polypeptide. Lysate was applied to a chromatography column that contained 5 to 10 ml Ni–NTA resin equilibrated in Ni-wash buffer (20 mM imidazole, 350 mM NaCl, 10 mM Tris-HCl, pH 8.0, 1 mM BME) and washed with an additional 5 column volumes of Ni-wash buffer. Proteins were eluted from the resin in batch by application of Ni-elution buffer (250 mM imidazole, 350 mM NaCl, 20 mM Tris-HCl. pH 8.0, 1 mM BME). Fractions containing the protein or proteins of interest were pooled and passed through a 0.2-µm filter before application of the sample to a HiLoad Superdex 200 26/60 (GE Healthcare) gel filtration column equilibrated with gel filtration buffer (350 mM NaCl, 20 mM Tris-HCl, pH 8.0, 1 mM BME). Fractions were analyzed by SDS-PAGE (4 to 12% Bis-Tris polyacrylamide, MES buffer; Invitrogen), and those containing the protein or proteins of interest were pooled and concentrated with Centriplus or Centricon filtration devices (Amicon). DTT was added to the concentrated sample to a final concentration of 1 mM. Samples were snap-frozen in liquid nitrogen and stored at $-80\,^\circ$C.

Most of the yeast RNase PH-like proteins did not behave well when expressed as individual polypeptides. Yeast Rrp46 and Rrp43 were insoluble when expressed individually. When coexpressed and copurified, both were obtained in the soluble fraction. Yeast Rrp41 was obtained as a soluble protein after small-scale expression (40 ml), but it was insoluble when expressed at preparative scales (10 L). Yeast Mtr3 was soluble, but this protein was prone to aggregation during size exclusion chromatography. With the exception that Rrp45 expression was not attempted at large scale, Rrp42 was the only yeast RNase PH-like protein that we could express, purify, and obtain as a monodisperse sample after size exclusion chromatography. As a result of these studies, the six yeast RNase PH-like proteins were expressed in pairwise manner as described later. In each case, one of the respective RNase PH-like proteins contained an N-terminal His$_6$-tag for affinity purification by Ni–NTA chromatography, whereas the other untagged PH-like protein copurified by means of interactions with its His$_6$–PH–like protein partner.

6.1. Yeast Rrp41/Rrp45

A 40-g cell pellet (wet weight) was suspended in lysis buffer, sonicated, and cleared by centrifugation. Yeast Rrp41/Rrp45 was purified by metal-affinity (Ni–NTA) chromatography. Fractions containing Rrp41/Rrp45 were pooled and applied to a gel filtration column (Superdex 200). Fractions containing Rrp41/Rrp45 were pooled, concentrated to 10 mg/ml, snap-frozen in liquid nitrogen, and stored at $-80\,^\circ$C. We typically obtain 30 mg of Rrp41/Rrp45 from a 40-g cell pellet.

6.2. Yeast Mtr3/Rrp42

A 50-g cell pellet (wet weight) was suspended in lysis buffer, sonicated, and cleared by centrifugation. Yeast Mtr3/Rrp42 was purified by metal-affinity chromatography. Fractions containing Mtr3/Rrp42 were pooled and applied to a gel filtration column (Superdex 200). This sample contained the Mtr3/Rrp42 heterodimer and excess Mtr3. Although the heterodimer appeared monodisperse during gel filtration, Mtr3 was prone to aggregation and eluted in the void volume of the column. Fractions containing Mtr3/Rrp42 were pooled, concentrated to 7 mg/ml, snap-frozen in liquid nitrogen, and stored at $-80\,^\circ$C. We typically obtain 15 mg of Mtr3/Rrp42 from a 50-g cell pellet.

6.3. Yeast Rrp46/Rrp43

Coexpression and purification of Rrp46/Rrp43 initially failed with SB media and large-scale fermentation insofar as Rrp46 underwent limited degradation. Although Rrp46 degradation products maintained the ability to interact with Rrp43, these subcomplexes were not competent for reconstitution into exosomes. As such, we sought alternate methods to obtain a more homogenous sample. Strains were cultured in ten 2-L baffled flasks with each flask containing 1 L of SB media. Cultures were inoculated with 5 ml overnight culture (SB) and grown at 37 °C to an OD_{600} of 2. Cultures were cooled on ice for 30 min. Ethanol was added to a final concentration of 2%, and IPTG was added to a final concentration of 0.25 mM. Cultures were then incubated at 18 °C for 18 h. The 10-L culture was harvested, and the cell pellet (80 g; wet weight) was suspended in lysis buffer, sonicated, and cleared by centrifugation. Yeast Rrp46/Rrp43 was purified by metal-affinity chromatography. Fractions containing Rrp46/Rrp43 were pooled, concentrated to 8 mg/ml, snap-frozen in liquid nitrogen, and stored at −80 °C. We typically obtain 10 mg of Rrp46/Rrp43 from an 80-g cell pellet. We attempted additional purification steps to improve the quality of these preparations, including ion exchange chromatography and fusion to His_6-Smt3, but none have resulted in improved yields or purity for this heterodimer.

6.4. Yeast Rrp4, Rrp40, and Csl4

These three proteins were purified independently, but because their purification schemes are nearly identical, we will discuss them within a single paragraph. A 40-g cell pellet (wet weight) was suspended in lysis buffer, sonicated, and cleared by centrifugation. Yeast Rrp4, Rrp40, and Csl4 were purified by metal-affinity chromatography. Fractions containing the respective protein were pooled and applied to a gel filtration column (Superdex 200). Yeast Rrp4, Rrp40, and Csl4 could each be isolated as monodisperse proteins by gel filtration. Proteins were concentrated to 10 mg/ml, snap frozen in liquid nitrogen, and stored at -80°C. We typically obtain 40 mg, 15 mg, and 30 mg of Rrp4, Rrp40, and Csl4, respectively, from a 40-g cell pellet.

6.5. Yeast Rrp44

A 40-g cell pellet (wet weight) was obtained, suspended in lysis buffer, sonicated, and cleared by centrifugation. Yeast Rrp44 was purified by metal-affinity (Ni–NTA) chromatography, and fractions containing Rrp44 were pooled and applied to a gel filtration column (Superdex 200). Fractions containing Rrp44 were pooled, concentrated to 10 mg/ml, snap-frozen in liquid nitrogen, and stored at −80 °C. We typically obtain 40 mg of Rrp44 from a 40-g cell pellet.

6.6. Yeast Rrp6

A 25-g cell pellet (wet weight) was obtained, suspended in lysis buffer, sonicated, and cleared by centrifugation. Rrp6 was purified by metal-affinity chromatography. Fractions containing Rrp6 were pooled and applied to a gel filtration column (Superdex 200). Fractions containing Rrp6 were pooled, concentrated to 3 mg/ml, snap-frozen in liquid nitrogen, and stored at −80 °C. We typically obtain 2 mg of His_6-Rrp6 from a 25-g cell pellet. The yield for Rrp6 was approximately one order of magnitude below that achieved for other yeast exosome proteins; however, large-scale expression of His_6-Smt3-Rrp6 improved the purity of this protein. Cultures were grown and processed as previously destribed. After gel filtration, His_6-Smt3-Rrp6 was incubated with the Ulp1 SUMO protease at a ratio of 1000:1 at 4 °C for 6 to 8 h to liberate Rrp6 from His_6-Smt3 (Mossessova and Lima, 2000). Rrp6 was separated from Smt3 by a second gel filtration step (Superdex 200). Fractions containing Rrp6 were pooled, concentrated to 3 mg/ml, snap-frozen in liquid nitrogen, and stored at −80 °C. We typically obtain 2 mg of Rrp6 from a 40-g cell pellet. The overall yield was lower when compared with His_6-Rrp6, but the purity was superior.

7. EXPRESSION AND PURIFICATION OF HUMAN EXOSOME PROTEINS

For strains containing plasmids encoding human exosome proteins, 500 ml LB cultures containing appropriate antibiotics were inoculated with 200 µl from the respective glycerol stock and grown overnight at 37 °C. The 500-ml overnight culture was used to inoculate 10 L of SuperBroth (SB). Cultures were grown by fermentation with a BioFlo 3000 reactor (New Brunswick) at 37 °C to an OD_{600} of 2 to 3, cooled to 30 °C, induced for expression by addition of 0.75 mM IPTG, and grown for 4 h at 30 °C. Strains containing Rrp42/Mtr3 were induced for 6 h. Cells were harvested by centrifugation at 6000g (Beckman JLA-8.1) for 15 min at 4 °C. The supernatant was discarded, and cell pellets were suspended in 50 mM Tris-HCl, pH 8.0, and 20% sucrose at a concentration of 0.5 g cell wet weight per ml. Suspended pellets were distributed into 50-ml conical tubes followed by snap-freezing in liquid nitrogen before storage at −80 °C. Cell suspensions were equilibrated in lysis buffer (50 mM Tris-HCl, pH 8.0, 20% sucrose, 350 mM NaCl, 10 mM imidazole, 1 mM BME, 0.1% IGEPAL, 10 µg/ml DNase, and 1 mM PMSF). Cells were disrupted by sonication over 20 min with 8 × 30-sec pulses at 80% power (Branson Digital Sonifier). Insoluble cell debris was removed by centrifugation at 44,000g (Beckman JA-20) for 1 h at 4 °C.

All protein purification was conducted at 4 °C. Metal-affinity chromatography (Ni–NTA SuperFlow; Qiagen) was used to affinity-purify each recombinant human protein that contained N-terminal Hexa-histidine or His$_6$-Smt3 polypeptide fusions. Human proteins were isolated from the soluble fraction of *E. coli* lysate by mixing the supernatant with 10 ml of Ni–NTA Superflow resin with stirring for 1 to 2 h in Ni-wash buffer (20 m*M* Tris-HCl, pH 8.0, 350 m*M* NaCl, 10 m*M* imidazole, and 1 m*M* BME). The slurry was loaded into a chromatography column and washed with 100 to 200 ml Ni-wash buffer. Proteins were eluted by batch by means of application of Ni-elution buffer (20 m*M* Tris-HCl, pH 8.0, 350 m*M* NaCl, 250 m*M* imidazole, and 1 m*M* BME). Fractions containing the protein or proteins of interest were pooled and the protein concentration determined by Bradford method with Bio-Rad Protein Assay reagent. Where noted, samples were applied to a gel filtration column (Superdex 200 26/60 or Superdex 75 26/60) equilibrated with gel filtration buffer (20 m*M* Tris-HCl, pH 8.0, 350 m*M* NaCl, and 1 m*M* BME). Before any steps involving FPLC chromatography columns, samples were passed through a 0.2-μm filter. After each purification step, fractions were analyzed by SDS-PAGE with 4 to 12% polyacrylamide gradient gels (NuPAGE, MES buffer; Invitrogen), and fractions containing the protein or proteins of interest were pooled and concentrated with Centriplus or Centricon filtration devices (Amicon). DTT was added to the concentrated samples to a final concentration of 1 m*M*. Samples were snap-frozen in liquid nitrogen and stored at −80 °C.

As observed for the yeast system, several of the human RNase PH-like proteins did not behave well when expressed as individual polypeptides, with notable exception. Unlike yeast Rrp46/Rrp43, human Rrp46 did not copurify with human Rrp43, so Rrp43 and Rrp46 were expressed as individual polypeptides. As a result of these studies, we expressed four of the human RNase PH-like proteins in a pairwise manner as described later. In these two cases, one RNase PH-like protein contained an N-terminal His$_6$-tag for affinity purification by Ni–NTA chromatography, whereas the other native PH-like protein copurified by means of interactions with its His$_6$-PH-like protein partner.

7.1. Human Rrp45/Rrp41

A 20-g cell pellet (wet weight) was suspended in lysis buffer, sonicated, and the lysate cleared by centrifugation. Rrp45/Rrp41 was purified by metal-affinity chromatography. Fractions containing Rrp45/Rrp41 were pooled and applied to a gel filtration column (Superdex 200), and fractions containing Rrp45/Rrp41 were pooled, concentrated to 5 to 15 mg/ml, snap-frozen in liquid nitrogen, and stored at −80 °C. We typically obtain 30 mg of Rrp45/Rrp41 from a 20-g cell pellet.

7.2. Human Rrp42/Mtr3

A 40-g cell pellet (wet weight) was suspended in lysis buffer, sonicated, and cleared by centrifugation. Rrp42/Mtr3 was purified by metal-affinity chromatography, and fractions containing Rrp42/Mtr3 were pooled and concentrated. Typical yields were 10 to 15 ml at 5 to 10 mg/ml, although this is an overestimate of Rrp42/Mtr3 yield because of contaminating proteins after elution from Ni–NTA. Samples were snap-frozen in liquid nitrogen and stored at $-80\,^\circ\text{C}$. This sample will be used in reconstitution of the Rrp42/Mtr3/Rrp43 complex (see later).

7.3. Human Rrp43

A 20-g cell pellet (wet weight) was suspended in lysis buffer, sonicated, and cleared by centrifugation. His_6-Smt3-Rrp43 was purified by metal-affinity chromatography. Fractions containing Smt3-Rrp43 were pooled and concentrated. Typical yields were 10 to 15 ml at 10 mg/ml, although this is an overestimate of Rrp43 yield because of contaminating proteins after elution from Ni-NTA and because the Smt3 tag was not liberated from Rrp43 until it was mixed with Rrp42/Mtr3 (see later). Samples containing Smt3-Rrp43 were snap-frozen in liquid nitrogen and stored at $-80\,^\circ\text{C}$.

7.4. Purification of human Rrp42/Mtr3/Rrp43

This ternary complex was obtained by reconstituting Rrp42/Mtr3 with His_6-Smt3-Rrp43 (see earlier) by mixing in equimolar ratio in the presence of Ulp1 at a ratio of 1000:1 Smt3-Rrp43/Ulp1 (Mossessova and Lima, 2000). This mixture was incubated at 4° overnight or until Smt3 cleavage was complete. The sample was applied to a gel filtration column (Superdex 200). Fractions containing Rrp42/Mtr3/Rrp43 were pooled, concentrated, and buffer-exchanged into 100 mM NaCl, Tris-HCl, pH 8.0, 1 mM BME. The sample was then applied to anion exchange resin (Mono Q 10/10) and eluted with a NaCl gradient from 100 mM to 450 mM over 15 column volumes. Fractions containing Rrp42/Mtr3/Rrp43 were pooled and concentrated to 5 mg/ml. Typical yields for Rrp42/Mtr3/Rrp43 were 10 to 20 mg from a 40-g cell pellet for Rrp42/Mtr3 and a 20-g cell pellet for Rrp43. Samples were snap-frozen in liquid nitrogen and stored at $-80\,^\circ\text{C}$.

7.5. Human Rrp46

A 20-g cell pellet (wet weight) was suspended in lysis buffer, sonicated, and the lysate cleared by centrifugation. His_6-Smt3-Rrp46 was purified by metal-affinity chromatography. Smt3 was liberated from Rrp46 by incubating the mixture overnight at 4 °C at a ratio of 1000:1 for Smt3-Rrp46/

Ulp1. The sample was applied to a gel filtration column (Superdex 75). Fractions containing Rrp46 were pooled, concentrated to 10 mg/ml, snap-frozen in liquid nitrogen, and stored at $-80\,°C$. We typically obtain 20 to 30 mg of Rrp46 from a 20-g cell pellet.

7.6. Human Rrp4, Rrp40, and Csl4

For each strain, a 20-g cell pellet (wet weight) was suspended in lysis buffer, sonicated, and the lysate cleared by centrifugation. Human Rrp4, Rrp40, or Csl4 was purified by metal-affinity chromatography, and fractions containing the respective protein were applied to a gel filtration column (Superdex 75). Fractions containing the respective protein were pooled, concentrated to 10 to 15 mg/ml, snap-frozen in liquid nitrogen, and stored at $-80\,°C$. We typically obtain 20 to 30 mg for Csl4, Rrp4, or Rrp40 from a 20-g cell pellet.

8. Reconstitution and Purification of Human and Yeast Exosomes

Eukaryotic exosomes are reconstituted by combining individual proteins or protein subcomplexes (see earlier; Liu et al., 2006). Unlike archaeal exosomes, six-subunit RNase PH-like rings do not assemble spontaneously with the six human or yeast RNase PH-like proteins, although nine-subunit complexes could be reconstituted by combining the eukaryotic RNase PH-like proteins (Rrp41, Rrp42, Rrp43, Rrp45, Rrp46, and Mtr3) along with the S1/KH-domain proteins (Rrp4, Csl4, and Rrp40). To form the yeast ten- or eleven-subunit exosomes, Rrp44 and Rrp6 were included in the reconstitution. Equimolar ratios for respective subunits were ensured before reconstitution by determining protein concentrations on the day of reconstitution with the Bradford method. We conducted reconstitution experiments at two scales. Small-scale reconstitutions used $\sim$2.5 mg of total protein in 1 ml (7 nM per subunit), whereas large-scale reconstitutions have used as much as 30 mg of total protein in 10 ml (100 nM per subunit). In each instance, reconstitution began by mixing the proteins in high salt buffer (350 mM NaCl, 20 mM Tris-HCl, pH 8.0, 1 mM BME). The samples are placed into dialysis tubing with a molecular cutoff no greater than 15 kDa (Spectra-Por). Reconstitution takes place in two steps. The samples are first dialyzed against reconstitution buffer containing 100 mM NaCl, 20 mM Tris-HCl, pH 8.0, 1 mM BME and then transferred for dialysis against low salt reconstitution buffer (50 mM NaCl, 20 mM Tris-HCl, pH 8.0, 1 mM BME).

8.1. The yeast nine-subunit exosome

An equimolar ratio of subunits was combined in high salt reconstitution buffer and placed into dialysis tubing. The sample was dialyzed against 1 L reconstitution buffer containing 100 mM NaCl for 2 to 4 h at 4 °C. The sample was transferred and dialyzed against 1 L low salt reconstitution buffer for 6 to 8 h at 4 °C. After dialysis, the sample was filtered and applied to a gel filtration column (Superdex 200) equilibrated with low salt reconstitution buffer. If reconstitution was successful, the resulting complex eluted as a monodisperse peak with an apparent molecular weight of 400 kDa. Fractions containing the complex were pooled, concentrated to 5 to 6 mg/ml in 20 mM Tris-HCl, pH 8.0, 50 mM NaCl, 1 mM TCEP (Tris(2-carboxyethyl) phosphine), snap-frozen in liquid nitrogen, and stored at -80° C.

Incorrect estimation of concentrations for the respective subunits often led to unsuccessful reconstitution experiments. In these instances, exosome complexes eluted as two distinct but overlapping peaks during gel filtration. The observed peaks differed from peak positions observed for individual protomers or subcomplexes, suggesting the presence of partially assembled complexes. In most cases, poor reconstitution profiles for the yeast exosome core were due to substoichiometric quantities of Rrp46, presumably caused by Rrp46 degradation products acquired during coexpression with Rrp43 (see earlier). Although Rrp46 degradation products interact with Rrp43, these products do not integrate into nine-subunit complexes. By adding 2.5-fold higher molar ratios of Rrp46/Rrp43 to the reconstitution mixture, reconstitutions resulted in a single monodisperse peak on gel filtration which corresponded to a nine-subunit complex as assessed by SDS-PAGE and Coomassie staining (10% polyacrylamide, MOPS buffer, Invitrogen).

8.2. The yeast ten-subunit exosome

As with the yeast nine-protein complex, stoichiometric quantities of the ten yeast exosome proteins were combined in high salt reconstitution buffer, including Rrp44 at 1.5× molar excess while maintaining Rrp46/Rrp43 at 2.5× molar excess. Once mixed, the sample was placed into dialysis tubing and dialyzed in two steps against reconstitution buffer containing 100 mM NaCl and 50 mM NaCl, respectively (see earlier). After dialysis, the sample was filtered and applied to a gel filtration column (Superdex 200) equilibrated with low salt reconstitution buffer. If reconstitution was successful, the complex eluted from gel filtration as a monodisperse peak with an apparent molecular weight of 500 kDa. Fractions containing the complex were pooled, concentrated to 6 mg/ml in 20 mM Tris-HCl, pH 8.0, 50 mM NaCl, 1 mM TCEP, snap-frozen in liquid nitrogen, and stored at -80 °C.

8.3. The yeast eleven-subunit exosome

The eleven yeast proteins were mixed in a molar ratio (Rrp46/Rrp43 at 2.5×, Rrp44 at 1×, Rrp6 at 1.5×) in high salt reconstitution buffer. The sample was dialyzed as earlier in two steps against reconstitution buffer containing 100 mM NaCl and 50 mM NaCl, respectively. The mixture was filtered and applied to a gel filtration column (Superdex 200). The complex eluted from gel filtration as a monodisperse peak with an apparent molecular weight of 600 kDa. SDS-PAGE and Coomassie staining revealed that fractions from this peak contained each of the eleven proteins. Fractions containing the complex were pooled, concentrated to 6 mg/ml in 20 mM Tris-HCl, pH 8.0, 50 mM NaCl, 1 mM TCEP, snap-frozen in liquid nitrogen, and stored at $-80\,°C$.

8.4. Further purification of yeast exosomes

The quality of exosome preparations can be significantly improved through purification of the assembled complexes by ion-exchange chromatography. The yeast nine-subunit exosome does not bind to cation exchange resin, but it does interact with anion exchange resin (Mono Q, GE Healthcare). After gel filtration, the complex was applied to a Mono Q 5/5 column equilibrated with 50 mM NaCl, 20 mM Tris-HCl, pH 8.0, 1 mM BME and eluted by a NaCl gradient from 50 mM NaCl to 1 M NaCl over 20 column volumes (Fig. 10.2A, left). The nine-subunit complex eluted in two peaks, and SDS-PAGE analysis revealed that the minor peak lacked Rrp40 and Rrp43, whereas the major peak contained each of the nine subunits. The major peak eluted at approximately 300 mM NaCl. To ensure that the complex did not disassemble after anion exchange, fractions containing the peak were subjected to analytical gel filtration (Superose 6 10/30 GL; GE Healthcare; equilibrated in low salt reconstitution buffer). The sample eluted as a monodisperse peak that contained all nine proteins (Fig. 10.2A, right).

To improve purification, the ten-subunit complex was applied to a Mono Q 5/5 column and eluted as described for the nine-subunit complex. As before, the complex eluted in two peaks, one that included a substoichiometric complex (apparent absence of Rrp40 and Rrp43) and one that included a stoichiometric complex. To further improve reconstitution of the ten-protein complex, we mixed nine-subunit yeast exosomes purified by gel filtration and anion exchange with Rrp44 in high salt reconstitution buffer. The sample was placed into dialysis tubing and dialyzed in two steps into low salt reconstitution buffer (see earlier). When the mixture was applied to gel filtration (Superdex 200), the resulting peak contained all ten proteins as observed previously. This ten-protein complex was then applied to Mono Q 5/5 and eluted with a NaCl gradient, resulting in one major peak that eluted at approximately 300 mM NaCl and contained all ten

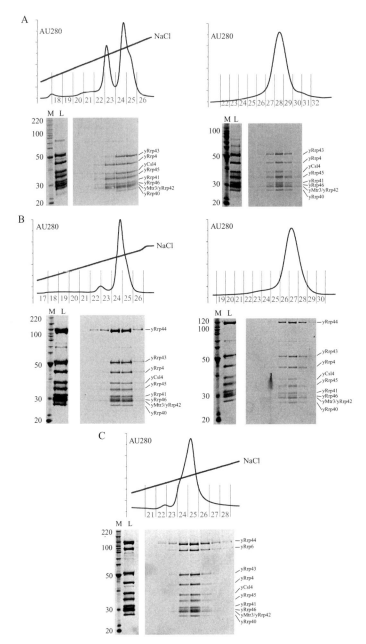

Figure 10.2 Ion exchange and gel filtration for yeast exosomes. (A) The nine-subunit yeast exosome. The left panel indicates the chromatogram and SDS-PAGE gel for purification of the yeast nine-subunit complex by anion exchange (Mono Q 5/5). The right panel depicts the gel filtration profile (Superose 6 10/30) and SDS-PAGE analysis for the yeast nine-protein complex after purification by anion exchange. (B) The ten-subunit

proteins in apparent molar ratios (Fig. 10.2B, left). To ensure that the complex did not disassemble after anion exchange, fractions containing the peak were subjected to analytical gel filtration (Superose 6 10/30 GL; equilibrated in low salt reconstitution buffer). The sample eluted as a monodisperse peak that contained all ten proteins (Fig. 10.2B, right).

This procedure was also used to improve the quality of the eleven-subunit exosome. The eleven-subunit complex was applied to anion exchange resin (Mono Q 5/5) and eluted as described for the nine- and ten-subunit complexes. In this instance, the eleven-protein complex eluted in a major peak at approximately 300 mM NaCl, although a shoulder is apparent on the left side of the peak. SDS-PAGE analysis of the resulting fractions revealed a complex that contained each of the eleven subunits (Fig. 10.2C).

8.5. The human nine-subunit exosome

Reconstitution of the human nine-subunit exosome was performed by mixing stoichiometric quantities of Rrp41/Rrp45 with Rrp42/Mtr3/Rrp43, Rrp46, Csl4, Rrp4, and Rrp40 in high salt reconstitution buffer. This mixture (10 ml) was placed into dialysis tubing and dialyzed for 4 to 6 h at 4 °C against 1 L 20 mM Tris-HCl, pH 8.0, 100 mM NaCl, and 1 mM BME. The sample was then dialyzed overnight at 4 °C against 1 L low salt reconstitution buffer containing 10 mM Tris-HCl, pH 8.0, 50 mM NaCl, and 1 mM BME. Insoluble material was removed by centrifugation and by passing the mixture through a 0.2-μm filter. The sample was then loaded onto a gel filtration column (Superdex 200) equilibrated with low salt reconstitution buffer. The complex eluted as a monodisperse peak and fractions containing the desired complex were analyzed by SDS-PAGE (4 to 12% Nu-PAGE gels and Coomassie-staining). Fractions containing the complex were pooled, concentrated to 6 to 9 mg/ml in 10 mM Tris-HCl, 50 mM NaCl, 1 mM DTT, snap-frozen in liquid nitrogen, and stored at -80 °C.

To improve the quality and purity of the nine-subunit exosome, the sample was applied to either cation- or anion-exchange resin (Mono S or Mono Q, respectively). Unlike the yeast complex, the human exosome

---

yeast exosome. The left panel indicates the chromatogram and SDS-PAGE gel for purification of the yeast ten-subunit complex by anion exchange (Mono Q 5/5). The right panel depicts the gel filtration profile (Superose 6 10/30) and SDS-PAGE analysis for the yeast ten-protein complex after purification by anion exchange. (C) The yeast eleven-subunit exosome. The panel indicates the chromatogram and SDS-PAGE gel for purification of the yeast nine-subunit complex by anion exchange (Mono Q 5/5). We have not yet obtained material after anion exchange for purification by gel filtration. Subunit positions are labeled adjacent to the respective gels. Fractions and lanes are aligned to assist in relating chromatograms to respective SDS-PAGE gels. Fraction numbers in chromatograms indicate 0.5 ml volumes after injection. SDS-PAGE gels are stained with Sypro-Ruby (Bio-Rad).

core interacts with both cation- and anion-exchange resins. In each instance, the complex was eluted from the ion exchange column with a gradient of NaCl from 50 mM NaCl to 400 mM NaCl, eluting at approximately 300 mM NaCl from either Mono Q or Mono S (Liu et al., 2007). As discussed previously, this step was critical to purify the human exosome away from contaminating exoribonucleases that are carried over from the recombinant E. coli host. For the human exosome core, separation was best achieved by cation exchange chromatography.

9. Exoribonuclease Assays

Archaeal exosomes and bacterial PNPase catalyze phosphate-dependent (phosphorolytic) exoribonuclease activity (Lorentzen et al., 2005; Symmons et al., 2000). Because of similarities observed among PNPase, archaeal Rrp41/Rrp42, and eukaryotic PH-like subunits, it was hypothesized that the eukaryotic exosome might possess phosphate-dependent activity. The eukaryotic exosome also possesses hydrolytic exoribonuclease activities, namely through association with Rrp44 and Rrp6, two subunits that share sequence similarity to bacterial RNase II/R and RNase D, respectively. Previous biochemical analyses on a few exosome subunits supported the hypothesis that most eukaryotic exosome proteins possessed 3' to 5'-exoribonuclease activity (reviewed in Raijmakers et al., 2004); however, more recent analysis suggests this is not true (Dziembowski et al., 2007; Liu et al., 2006; 2007). To determine the biochemical activities of the exosome, we tested each reconstituted exosome, as well as each individual exosome protein or subcomplex, in assays for hydrolytic and phosphorolytic 3' to 5'-exoribonuclease activity (Liu et al., 2006).

All RNA substrates were synthetically derived (Invitrogen), and each included a fluorescein attached to the 5'-end of the substrate to enable detection of reaction products by exciting fluorescein at 473 nm and detecting fluorescence at 520 nm. To detect reaction products, samples were separated by denaturing polyacrylamide gel electrophoresis (PAGE) and detected by scanning gels on a flat transparent surface with a Fuji FLA-5000 equipped with a FITC filter.

Preliminary biochemical analysis used end point assays in combination with protein titration from 100 pM to 1 μM. Each protein or subcomplex was prepared as a 10× stock (1 nM to 10 μM) by diluting proteins in protein dilution buffer (10 mM Tris-HCl, pH 8.0, 10 mM DTT, 50 mM KCl, 5 mM MgCl$_2$, 1 U/μl RNase Inhibitor (NEB) with and without 10 mM sodium phosphate buffer [pH 8.0]). The reaction buffer was analogous to protein dilution buffer, but included 11.12 nM RNA. Reactions were initiated by mixing 2 μl 10× protein stock and 18 μl reaction buffer to yield samples that contained 1× protein and 10 nM RNA. Reactions were incubated at 37 °C and quenched after 45 min by addition of 20 μl TBE loading buffer (Invitrogen). Reaction products were separated with 15% TBE-urea polyacrylamide gels (Invitrogen). RNA products were detected by exciting fluorescein at 473 nm and detecting fluorescence at 520 nm with a Fuji FLA-5000 equipped with a FITC filter. Images were processed with a linear scale with Fuji Film Multi-Gauge V2.02.

Two RNA substrates were initially assessed, an AU-rich RNA (ARE) substrate that contained two tandem repeats of the AU-rich domain derived from the 3′ UTR of the TNFα gene and a generic RNA sequence derived from previous studies (Liu et al., 2006; Wang and Kiledjian, 2001). These assays revealed that none of the yeast exosome proteins contain phosphate-dependent exoribonuclease activity, and only two of the subunits were capable of catalyzing hydrolytic exoribonuclease activity, Rrp44 and Rrp6 (Burkard and Butler, 2000; Liu et al., 2006; Mitchell et al., 1997). For the human exosome, we initially reported phosphorolytic activities for the Rrp41/Rrp45 heterodimer and for the nine-subunit human exosome (Liu et al., 2006), but these activities were later determined to result from contamination by E. coli polynucleotide phosphorylase (PNPase; Liu et al., 2007). This discovery necessitated further purification of reconstituted exosome complexes from yeast and human by ion-exchange chromatography (described earlier) to remove PNPase from our samples. Our results now suggest that neither human nor yeast nine-subunit exosomes are capable of catalyzing phosphorolytic activity, distinguishing eukaryotic exosomes from their archaeal and bacterial counterparts.

10. Comparative Exoribonuclease Assays with Different RNA Substrates

We have assessed 10 different 49 nucleotide RNA substrates in comparative biochemical assays with our exosome preparations, including the aforementioned AU-rich RNA and generic RNA. In addition, we used a poly-adenylate RNA (Poly(A)) and three RNA chimeras that included generic RNA sequences followed by Poly(A), generic RNA sequences followed by AU-rich RNA, and a substrate containing AU-rich RNA followed by Poly(A) (Liu et al., 2006). Four 49-nucleotide AU-rich RNA substrates were also synthesized, which contained a 20-nucleotide GNRA stem loop (eight GC base pairs and GCAA tetraloop; Heus and Pardi, 1991). The GNRA substrates differed from each other with respect to the position of the stem loop within the 49 nucleotide substrate such that the AU-rich sequence elements 3′ to the GNRA stem loop varied from 5, 10, 20, to 29 nucleotides in length with commensurate shortening of AU-rich sequences 5′ to the GNRA stem loop. These substrates were used to assess the effect of stable RNA secondary structure on exosome activities and to test the hypothesis that RNA substrates must pass through the pore to gain access to exosome catalytic sites as suggested previously for archaeal exosomes (Lorentzen and Conti, 2005, Lorentzen et al., 2007).

Exosome proteins and complexes were analyzed for activity over time with fixed concentrations of substrate and protein. For these experiments, 100-μl reactions contained 10 mM Tris-HCl, pH 8.0, 10 mM DTT, 50 mM KCl, 5 mM MgCl$_2$, 1 U/μl RNase Inhibitor (NEB), 10 nM RNA, and 10 nM protein. Reactions were initiated by addition of the 10× protein stock (as earlier). Reactions were incubated at 37 °C, and 10 μl were removed at discrete time points, usually after 1, 2, 4, 8, or 16 min. Samples were quenched by addition of 10 μl TBE loading buffer followed by snap-freezing in liquid nitrogen. Reaction products were separated by electrophoresis with 15% TBE-urea gels (Invitrogen) and detected by fluorescence with the FLA-5000. Images were processed with a linear scale with Fuji Film Multi-Gauge V2.02.

11. Conclusions

In this chapter, we described methods to clone, express, purify, and reconstitute eukaryotic exosomes from human and yeast. To date, biochemical characterization of exosomes has been mainly accomplished by purifying exosomes from their endogenous source by affinity techniques. Although it seems possible to purify exosomes from yeast to near homogeneity, these preparations suffer from potential contamination by endogenous cofactors, and there are still problems associated with purification of stoichiometric complexes (Dziembowski et al., 2007; Wang et al., 2007). Purification of human exosomes from tissue or cell culture seems even more daunting. The protocols presented herein should enable analysis of exosome function through reconstitution of mutant subunit isoforms and by analysis of substrate dependencies on pure components obtained before and after exosome reconstitution. These methods, combined with biochemical and genetic analysis, should facilitate functional studies of RNA exosomes during RNA processing and decay.

ACKNOWLEDGMENTS

We thank Quansheng Liu for his contributions to this chapter, particularly methods used to purify and reconstitute human exosomes. J. C. G. is a trainee in the Tri-Institutional Program in Chemical Biology. J. C. G. and C. D. L. are supported in part by a grant from the National Institutes of Health (GM079196). C. D. L. acknowledges additional support from the Rita Allen Foundation.

REFERENCES

Allmang, C., Kufel, J., Chanfreau, G., Mitchell, P., Petfalski, E., and Tollervey, D. (1999). Functions of the exosome in rRNA, snoRNA and snRNA synthesis. *EMBO J.* **18**, 5399-5410.

Allmang, C., Petfalski, E., Podtelejnikov, A., Mann, M., Tollervey, D., and Mitchell, P. (1999). The yeast exosome and human PM-Scl are related complexes of $3'$ to $5'$-exonucleases. *Genes Dev.* **13**, 2148-2158.

Burkard, K. T., and Butler, J. S. (2000). A nuclear $3'$ to $5'$-exonuclease involved in mRNA degradation interacts with Poly(A) polymerase and the hnRNA protein Npl3p. *Mol. Cell Biol.* **20**, 604-616.

Buttner, K., Wenig, K., and Hopfner, K.-P. (2005). Structural framework for the mechanism of Archaeal exosomes in RNA processing. *Mol. Cell* **20**, 461-471.

Carpousis, A. J. (2002). The *Escherichia coli* RNA degradosome: Structure, function and relationship in other ribonucleolytic multienzyme complexes. *Biochem. Soc. Trans.* **30**, 150-155.

Cheng, C., and Shuman, S. (2000). Recombinogenic flap ligation pathway for intrinsic repair of topoisomerase IB-induced double-strand breaks. *Mol. Cell Biol.* **20**, 8059-8068.

Dziembowski, A., Lorentzen, E., Conti, E., and Séraphin, B. (2007). A single subunit, Dis3, is essentially responsible for yeast exosome core activity. *Nat. Struc. Molec. Biol.* **14**, 15-22.

Evguenieva-Hackenberg, E., Walter, P., Hochleitner, E., Lottspeich, F., and Klug, G. (2003). An exosome-like complex in *Sulfolobus solfataricus*. *EMBO Rep.* **4**, 889-893.

Graham, A. C., Kiss, D. L., Andrulis, E. D. (2006). Differential distribution of exosome subunits at the nuclear lamina and in cytoplasmic foci. *Mol. Biol. Cell.* **17**, 1399-1409.

Heus, H. A., and Pardi, A. (1991). Structural features that give rise to the unusual stability of RNA hairpins containing GNRA loops. *Science.* **253**, 191-194.

Houseley, J., LaCava, J., and Tollervey, D. (2006). RNA-quality control by the exosome. *Nat. Rev. Mol. Cell Biol.* **7**, 529-539.

Koonin, E. V., Wolf, Y. I., and Aravind, L. (2001). Prediction of the archaeal exosome and its connections with the proteasome and the translation and transcription machineries by a comparative-genomic approach. *Gen. Res.* **11**, 240-252.

Liu, Q., Greimann, J. C., and Lima, C. D. (2006). Reconstitution, activities, and structure of the eukaryotic RNA exosome. *Cell* **127**, 1223-1237.

Liu, Q., Greimann, J. C., and Lima, C. D. (2007). Reconstitution, activities, and structure of the eukaryotic RNA exosome. *Cell* **131**, 188-189.

Lorentzen, E., Walter, P., Fribourg, S., Evguenieva-Hackenberg, E., Klug, G., and Conti, E. (2005). The archaeal exosome core is a hexameric ring structure with three catalytic subunits. *Nat Struct. Molec. Biol.* **12**, 575-581.

Lorentzen, E., and Conti, E. (2005). Structural basis of $3'$-end RNA recognition and exoribonucleolytic cleavage by an exosome RNase PH core. *Mol. Cell* **20**, 473-481.

Lorentzen, E., Dziembowski, A., Lindner, D., Seraphin, B., and Conti, E. (2007). RNA channelling by the archaeal exosome. *EMBO Rep.* **8**, 470-476.

Mitchell, P., Petfalski, E., and Tollervey, D. (1996). The $3'$-end of yeast 5.8S rRNA is generated by an exonuclease processing mechanism. *Genes Dev.* **10**, 502-513.

Mitchell, P., Petfalski, E., Shevchenko, A., Mann, M., and Tollervey, D. (1997). The exosome: A conserved eukaryotic RNA processing complex containing multiple $3'$ to $5'$-exoribonucleases. *Cell* **91**, 457-466.

Mossessova, E., and Lima, C. D. (2000). Ulp1-SUMO crystal structure and genetic analysis reveal conserved interactions and a regulatory element essential for cell growth in yeast. *Mol. Cell* **5**, 865-876.

Raijmakers, R., Schilders, G., and Pruijn, G. J. (2004). The exosome, a molecular machine for controlled RNA degradation in both nucleus and cytoplasm. *Eur. J. Cell Biol.* **83**, 175–183.

Schneider, C., Anderson, J. T., and Tollervey, D. (2007). The exosome subunit Rrp44 plays a direct role in RNA substrate recognition. *Mol. Cell* **27**, 324–331.

Symmons, M. F., Jones, G. H., and Luisi, B. F. (2000). A duplicated fold is the structural basis for polynucleotide phosphorylase catalytic activity, processivity, and regulation. *Structure* **8**, 1215–1226.

Wang, Z., and Kiledjian, M. (2001). Functional link between the mammalian exosome and mRNA decapping. *Cell* **107**, 751–762.

Wang, H.-W., Wang, J., Ding, F., Callahan, K., Bratkowski, M. A., Butler, J. S., Nogales, E., and Ke, A. (2007). Architecture of the yeast Rrp44 exosome complex suggests routes of RNA recruitment for 3'-end processing. *Proc. Nat. Acad. Sci. USA* **104**, 16844–16849.

CHAPTER ELEVEN

BIOCHEMICAL STUDIES OF THE MAMMALIAN EXOSOME WITH INTACT CELLS

Geurt Schilders *and* Ger J. M. Pruijn

Contents

1. Introduction	212
2. Identifying Protein–Protein Interactions by the Mammalian Two-Hybrid System	213
2.1. Protocol	214
2.2. Comments	216
3. Characterization of Different Exosome Subsets by Glycerol Sedimentation	218
3.1. Protocol	219
3.2. Comments	220
4. Studying Exosome Function with RNAi	222
4.1. Protocol	222
4.2. Comments	223
References	224

Abstract

A key component responsible for $3'$- to $5'$-RNA turnover in eukaryotic cells is the exosome, a multisubunit complex present in both the nucleus and cytoplasm of the cell. Here we describe several methods that can be applied to study the structure and function of the exosome in mammalian cell lines. The mammalian two-hybrid system has been successfully used to identify protein–protein interactions between exosome core components. Cell and glycerol gradient fractionation procedures are described that allow the identification and characterization of different exosome subsets. Finally, a protocol to study the function of the exosome in RNA turnover with RNA interference is presented.

Department of Biomolecular Chemistry, Nijmegen Center for Molecular Life Sciences, Institute for Molecules and Materials, Radboud University Nijmegen, Nijmegen, The Netherlands

1. INTRODUCTION

The exosome is a multisubunit complex containing $3'$- to $5'$-exoribonuclease activity and has been shown to be vital for several cellular processes, such as mRNA turnover (Bousquet-Antonelli et al., 2000; Das et al., 2003; Jacobs Anderson et al., 1998; van Dijk et al., 2007), ribosomal and small nucle(ol)ar RNA processing (Allmang et al., 1999a; Schilders et al., 2005; Stoecklin et al., 2005; van-Hoof et al., 2000) as well as several RNA surveillance pathways (Houseley et al., 2006; Kadaba et al., 2006; LaCava et al., 2005; Vanacova et al., 2005; Wyers et al., 2005). The exosome and its associated proteins can be divided into several types of subunits. The human exosome contains nine components that are shared by the cytoplasmic and nuclear complex and are, therefore, termed "core" components (Chen et al., 2001). Accessory components are defined as active $3'$- to $5'$-exoribonucleases, which contribute to the catalytic activity of the core exosome. Examples are PM/Scl-100 (Rrp6p in yeast), which belongs to the RNase D family of exoribonucleases, and Dis3p/Rrp44p, a member of the RNase R family of exoribonucleases, which so far has been found to be associated with the exosome in several eukaryotes, but not in humans. Auxiliary proteins can be defined as proteins that either directly or indirectly (e.g., via accessory proteins) interact with the core complex and aid in the function of the exosome, such as RNA helicases and RNA-binding proteins.

Six of the human core components (hRrp41, hRrp42, hMtr3, hRrp46, PM/Scl-75 [Rrp45 in yeast], and OIP2 [Rrp43 in yeast]) contain an RNase PH domain (RPD), whereas the other three components (hCsl4, hRrp4, hRrp40) contain an S1 domain in addition to a KH or zinc-ribbon domain. The assembly of the complex formed by the nine core components was found to be similar to the eubacterial polynucleotide phosphorylase (PNPase) complex, suggesting that their three-dimensional structure might also be similar. The crystal structure of PNPase revealed that the S1 domains form a crown of RNA-binding domains around the space in the center of the trimer (Symmons et al., 2000). On the basis of the PNPase structure and taking into account the mutual subunit interactions observed in different two-hybrid systems, a model for the human exosome was generated in which the RNase PH domain containing subunits form a hexameric ring, and the three proteins with the S1 (and KH or zinc-ribbon) RNA binding domains are positioned at the outer surface of this ring (Lehner et al., 2004; Raijmakers et al., 2002). Biophysical support for this ringlike structure was provided by the crystal structure of the reconstituted human exosome. The structure that revealed that the arrangement of the exosome components is consistent with the model of the human exosome, although the S1 domain proteins were found to be associated with one side of the ring rather than with the periphery of the ring (Liu *et al.*, 2006) (see also Chapter 10 by Greimann and Lima in this volume). Nevertheless, these data showed that it is possible to derive a reliable model of a multisubunit complex from the results obtained in a mammalian two-hybrid system.

In yeast, the exosome was reported to be present in differently sized complexes (Mitchell *et al.*, 1997). Also, in humans, the core exosome has been reported to be associated with different complexes sedimenting at ∼10S and 60 to 80S in glycerol gradients (Schilders *et al.*, 2005; van Dijk *et al.*, 2007). For the U3 snoRNP, the high-molecular-weight (60 to 80S) complexes were shown to represent preribosomes (Granneman *et al.*, 2004; Lukowiak *et al.*, 2000). Therefore, the sedimentation of the human exosome in similarly sized complexes suggested that the 60 to 80S complexes correspond to exosomes associated with preribosomal complexes. Interestingly, a subset of exosome-auxiliary proteins have been identified that only cosediment with the 60 to 80S complexes and not with the 10S complexes, suggesting that these proteins are involved in ribosome synthesis. Indeed, both the human exosome and the exosome-auxiliary proteins cosedimenting with the 60 to 80S exosome complexes have been implicated in the 3'-end processing of the 5.8S rRNA (Schilders *et al.*, 2005; Stoecklin *et al.*, 2005). These data demonstrated that by biochemical fractionation experiments, different exosome subsets containing additional auxiliary proteins could be identified and characterized.

Most of the biochemical and genetic approaches that have been applied in the past to identify the functions of the exosome and its accessory and auxiliary components have used the yeast *Saccharomyces cerevisiae* or *in vitro* systems with individual components. With the possibility of RNA-interference (RNAi) technology in mammalian cells (Elbashir *et al.*, 2001), the study of exosome function has been extended to identify the role of the mammalian exosome in rRNA processing (Schilders *et al.*, 2005; Stoecklin *et al.*, 2005), mRNA turnover (Stoecklin *et al.*, 2005; van Dijk *et al.*, 2007), and mRNA surveillance pathways, such as for mRNAs containing a premature translation termination codon (Lejeune *et al.*, 2003). It is anticipated that RNAi will probably also be applied in combination with the microarray technology to identify the role of the exosome in the expression of protein coding genes and small noncoding RNAs (sn(o)RNAs, miRNAs) on a genome-wide scale. Here we describe several protocols that have been successfully applied in our laboratory to study the human exosome.

2. IDENTIFYING PROTEIN–PROTEIN INTERACTIONS BY THE MAMMALIAN TWO-HYBRID SYSTEM

One of the major advantages of studying the interaction between two mammalian proteins in the mammalian two-hybrid system is that these proteins are expressed in their "natural" environment (i.e., the most optimal conditions for proper folding, posttranslational modification, and oligomerization). For studying protein–protein interactions with the Checkmate mammalian two-hybrid system (Promega), three plasmids are cotransfected into mammalian cells. The pACT plasmid contains a VP16 transcriptional activation domain upstream of the cDNA of interest, whereas the pBIND plasmid contains a GAL4 DNA-binding domain sequence upstream of the second cDNA of interest. The third pG5luc plasmid contains five GAL4 binding sites upstream of a firefly luciferase gene that acts as a reporter for the interaction between the proteins to be studied (Fig. 11.1). It should be noted that the pBIND plasmid also expresses the *Renilla* luciferase, which allows for normalization of differences in transfection efficiency. The detailed protocol described in the following is optimized for COS-1 cells and FuGENE transfection reagent. However, other mammalian cell lines and transfection reagents can be used as well, although this may require minor optimization.

2.1. Protocol

1. 3 to 4×10^5 actively growing COS-1 cells are seeded into a single well of a 6-well plate (Greiner Bio-One) and incubated overnight in 2 ml culture medium (DMEM containing 10% heat-inactivated fetal calf serum [FCS]) at 37 °C in a CO_2 incubator.
2. Per transfection 1 μg of pACT and pBIND, either with or without insert, and 1 μg of the pG5luc reporter plasmid (Promega) are added to 100 μl of DMEM.
3. To the DNA/DMEM mixture, 4 μl of FuGENE (Roche) is added and, after gentle mixing, incubated for 20 min at room temperature.
4. Meanwhile, the COS-1 cells are washed twice with PBS and incubated with 4 ml DMEM containing 10% FCS.
5. The FuGENE/DNA complexes are added to the cells and gently mixed.
6. After incubation for 48 h at 37 °C in a CO_2 incubator, cells are washed twice with PBS and lysed by the addition of 500 μl of passive lysis buffer (Dual Luciferase Reporter (DLR)-kit [Promega]).
7. After lysing the cells for 20 min on a shaking incubator at room temperature, cell extracts are transferred to a tube and placed on ice.
8. The expression levels of both the firefly luciferase and the control *Renilla* luciferase are determined by activity measurements with the Dual Luciferase Reporter Assay System (Promega) and a Lumat LB 9507 Luminometer (Berthold).
9. For the firefly luciferase activity, 100 μl of firefly luciferase substrate is added to 20 μl of cell extract, and the luminescence is measured to detect the interaction between the proteins of interest.

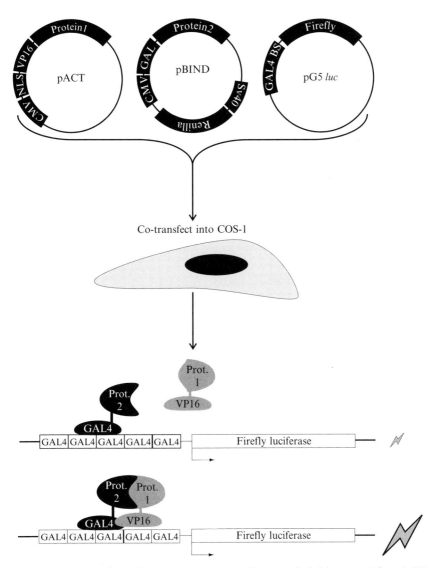

Figure 11.1 Principle of the checkmate mammalian two-hybrid system. The pACT, pBIND, and pG5*luc* vectors are introduced into COS-1 cells with FuGENE. After culturing for 48 h, cells are lysed and extracts analyzed for luciferase activity. The pG5luc vector contains five GAL4 binding sites upstream of the firefly luciferase gene. When the fusion partners of the GAL4 DNA-binding domain (from pBIND) and the VP16 activation domain (from pACT) do not interact, luciferase expression will not be induced, resulting in little firefly luciferase activity. An interaction between the two fusion proteins on the other hand will result in an increase of firefly luciferase expression.

10. To monitor transfection efficiency, *Renilla* luciferase activity is measured by the addition of 100 μl of *Renilla* luciferase substrate solution to the mixture of step 8, and luminescence is measured.
11. After obtaining the raw data, the following calculations can be performed. The observed firefly luciferase activity can be normalized for differences in the transfection efficiency by dividing the firefly luciferase activity values by the *Renilla* luciferase activity values observed with the same cell extract and multiplied by 1000. Subsequently, the luciferase activity can be expressed in relative luminescence units (RLU). In all experiments we included hRrp42 in pBIND and hCsl4 in pACT as a positive control and set the normalized luminescence resulting from the interaction between these two exosome subunits at 100 RLU. This approach allowed a direct comparison of the results obtained in the distinct experiments. In addition, it is crucial to include background controls in every experiment (i.e., luciferase measurements with extracts from cells transfected with either the "empty" pACT or the "empty" pBIND vector, which replaces the cDNA containing plasmid). An example of a mammalian two-hybrid experiment is shown in Table 11.1.

2.2. Comments

It is important to note that the two-hybrid system can result in false-positive interactions, which might result from activation of the reporter independent of the protein interaction or because of (weak) nonspecific binding. Therefore, it is important to determine a cutoff value. In the two-hybrid experiments we have carried out, an interaction is considered positive when the RLU exceeds the sum of the corresponding "empty-vector" control measurements. For the experiment documented in Table 11.1, this results in the sum of 48 + 45 RLU, which is well below the observed 292 RLU for the interaction of C1D with PM/Scl-100.

Note that false-negative results can also occur. Several common problems could be: the protein is not properly modified; the protein is not expressed to sufficiently high levels; the protein is insoluble; fusion of the protein to the GAL4 or VP16 domain interferes with the proper folding of the protein or shields the interaction surface. Expression levels can easily be tested by Western blot analysis with specific antibodies directed against the GAL4 and VP16 domains. If Western blot analysis shows that a protein is poorly expressed, expression can be enhanced by the addition of sodium butyrate, which is known to stimulate expression from the CMV promoter (Wilkinson *et al.*, 1992). If sodium butyrate is to be used, it should be added 24 h after transfection to a final concentration of 5 mM.

Another factor that may lead to false-negative results is the subcellular localization of the fusion protein. The interaction between the proteins of interest has to occur in the nucleus to induce transcription of the firefly

Table 11.1 Example of a single two-hybrid interaction between C1D and PM/Scl-100

Plasmids	Firefly luciferase activity[a]	Renilla luciferase activity[a]	Normalized luciferase activity	Relative luminescence units (RLU)
pACT–hCsl4 pBIND–hRrp42	26135	3308947	(26135/3308947) × 1000 = 7.90	7.90 × (100/7.90) = 100
pACT–empty pBIND–C1D	15745	4085106	(15745/4085106) × 1000 = 3.85	3.85 × (100 / 7.90) = 48
pACT–PM/Scl-100 pBIND–empty	4354	1229814	(4354/1229814) × 1000 = 3.54	3.54 × (100/7.90) = 45
pACT–PM/Scl-100 pBIND–C1D[b]	47934	2079513	(47934/2079513) × 1000 = 23.05	23.05 × (100/7.90) = 292

[a] Arbitrary units.
[b] Interaction should be corrected for background levels: 292 − 48 − 45 = 199.

luciferase gene. The subcellular localization can be tested by indirect immunofluorescence and by use of the antibodies directed to the GAL4 or VP16 domains. Note that the pACT vector contains a nuclear localization sequence (NLS) in front of the VP16 fusion protein that might circumvent the problem in most cases. In contrast the pBIND vector does not contain a NLS, but presumably the proteins will already interact in the cytoplasm and may be exported into the nucleus as a heterodimer by means of the NLS present in the pACT vector.

3. CHARACTERIZATION OF DIFFERENT EXOSOME SUBSETS BY GLYCEROL SEDIMENTATION

As described previously, the human exosome is composed of a core consisting of nine subunits conserved in both the nuclear and cytoplasmic exosomes, as well as auxiliary proteins that are only associated with a subset of exosome complexes. To identify different exosome-containing macromolecular complexes, biochemical fractionation experiments can be performed in which total or compartment-specific cell extracts can be fractionated by glycerol gradient sedimentation (Fig. 11.2). Subsequently, distribution of the exosome complexes in the gradient can be monitored by

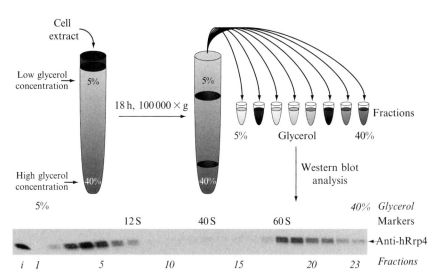

Figure 11.2 Schematic representation of glycerol gradient sedimentation analysis. A HEp-2 cell lysate is loaded onto a 5 to 40% glycerol gradient and after centrifugation for 18 h at 100,000g, fractions are collected manually, separated with SDS-PAGE, and analyzed by Western blot analysis. The sedimentation of hRrp4 is visualized by immunoblotting with a mouse monoclonal anti-hRrp4 antibody.

Western blot and immunoprecipitation analysis. In addition, activity assays can be performed with either the gradient fractions or the immunoprecipitated material to functionally characterize the exosome subsets. In the following section a method is described for the fractionation and characterization of different exosome subsets in mammalian cells.

3.1. Protocol

3.1.1. Preparation of cytoplasmic and nuclear extracts with digitonin

1. 5×10^6 HEp-2 cells are grown to 70% confluence, harvested, and resuspended in 500 μl lysis buffer (20 mM HEPES/KOH at pH 7.6, 150 mM NaCl, 0.5 mM DTE, 0.5 mM PMSF).
2. Permeabilization of the plasma membrane is achieved by the addition of digitonin to a final concentration of 0.025%. After mixing gently, the cells are incubated for 10 min at room temperature. Note that lysis by digitonin is selective for cholesterol-containing membranes like the plasma membrane, with minimal nuclear permeabilization and leakage of nuclear proteins.
3. The lysate is centrifuged at 1000g for 5 min at room temperature.
4. The supernatant, which contains the cytoplasmic material, is transferred to a new tube.
5. The pellet is resuspended in 500 μl lysis buffer and is used to prepare the nuclear extract.
6. Both the cytoplasmic and the nuclear material are homogenized by sonication for 3×20 sec with a Branson microtip.
7. Triton X-100 is added to the homogenates to a final concentration of 0.2% (v/v).
8. Lastly, insoluble material is removed by centrifugation at 12,000g for 10 min at 4 °C. The purity of the nuclear and cytoplasmic fractions can be determined by Western blot analysis. As a nuclear marker, we use a polyclonal serum (from an autoimmune patient) that recognizes topoisomerase I, whereas a mouse monoclonal antibody directed against eIF-2α is used as a cytoplasmic marker. It is particularly important to check for leakage of proteins from the nucleus during cell permeabilization, and a relatively soluble nuclear protein should be used to monitor proper cell fractionation, because proteins stably associated with chromatin, the nucleolus, or other higher order nuclear substructures will not easily leak from a permeabilized nucleus.

3.1.2. Preparation of 5 to 40% glycerol gradients

9. For each gradient, two times 8 ml of lysis buffer supplemented with 0.2% Triton X-100 is prepared containing 5 or 40% glycerol, respectively.

10. Add 6 ml of 5% glycerol lysis buffer to the bottom of an ultracentrifuge tube (the tube should be compatible with the rotor required for the centrifugation step) (see step 6).
11. Subsequently, 6 ml of 40% glycerol lysis buffer is added slowly to the bottom of the tube with a long needle, while not disturbing the 5% glycerol layer. Note that the order of steps is important for creating an optimal gradient.
12. The 5 to 40% glycerol gradient is generated with a Biocomp Gradient Master 107. Note that depending on the size of the complexes to be analyzed and the resolution of complexes in a particular region of the gradient, glycerol gradients with different glycerol concentration ranges can also be used.
13. The cell lysate (0.5 ml), prepared as described previously, is layered on top of the 12-ml gradient.
14. The gradients are centrifuged in a Sorvall TH641 (or equivalent swing-out) rotor for 18 h at 100,000g at 4 °C.
15. After centrifugation, 500-μl fractions are collected manually by a micropipet from top to bottom and stored at −20 °C until further analysis.

3.2. Comments

The collected fractions can be separated with SDS-PAGE, and the proteins of interest can be visualized by Western blotting. Note that when the pellet fraction from the gradient is also analyzed, it can contain large protein aggregates resulting in a false-negative signal in this fraction.

An overview of available anti-exosome subunit antibodies is given in Table 11.2. In our experiments, core exosome components seemed to be slightly more abundant in the nuclear fraction of HEp-2 cells (Brouwer et al., 2001). In contrast to yeast, where Rrp6p is restricted to the nucleus, PM/Scl-100 is found in substantial amounts in the cytoplasm, whereas the auxiliary proteins MPP6 and C1D are restricted to the nuclear fraction. With regard to the distribution of exosome complexes in glycerol gradients, the nuclear exosome seems to be equally divided between high-molecular-weight fractions and low-molecular-weight exosome complexes. The latter most likely represent core exosomes associated with PM/Scl-100. In the cytoplasm, most exosome complexes are found as low-molecular-weight complexes, and only a small percentage seems to be stably associated with larger complexes (van Dijk et al., 2007).

RNA can be isolated from the gradient fractions with TRIzol (Invitrogen) and analyzed with denaturing gel electrophoresis followed by Northern blot hybridization. As markers for 40S and 60S complexes, fractionation of the 18S and 28S rRNAs can be monitored with agarose gel electrophoresis and ethidium bromide staining. As an additional marker

Table 11.2 Antibodies and siRNAs to study the mammalian (human) exosome

Protein	Antibody	siRNA (sequence of sense strand (5′ to 3′))	Reference(s)
hRrp4	Rabbit polyclonal Mouse monoclonal	AGCUUUCACACAGAUCAACdTdT	(Allmang et al., 1999b; Schilders et al., 2005; van Dijk et al., 2007)
hRrp40	Rabbit polyclonal	GAAUAUGGGUUAAGGCAAA	(Brouwer et al., 2001; Stoecklin et al., 2005)
OIP2	Rabbit polyclonal	—	(Jiang et al., 2002)
PM/Scl-75	Rabbit polyclonal	GCGUGAUCCUGUACCAUUA GCCAAGAUGCUCCCAUAAUdTdT	(Mukherjee et al., 2002; Stoecklin et al., 2005; van Dijk et al., 2007)
hRrp41	Rabbit polyclonal	UGUGCAGGUGCUACAGGCAdTdT	(Brouwer et al., 2001; Lejeune et al., 2003; Schilders et al., 2005)
hRrp46	Rabbit polyclonal	GCAAAGAGAUUUUCAACAA CAACACGUCUUCCGUUUCU	(Brouwer et al., 2001; Stoecklin et al., 2005)
PM/Scl-100	Rabbit polyclonal	GUACAACCCAGGAUAUGUGdTdT	(Brouwer et al., 2001; Lejeune et al., 2003)
hMtr4	—	GCCUAUGCACUUCAAAUGAdTdT	(Schilders et al., 2007)
C1D	—	UUGUUCAAGUGGAUCCAACdTdT	(Schilders et al., 2007)
MPP6	Rabbit polyclonal	GAGCACUGGUACUUGGAUUdTdT CAGUAGAGCUUGAUGUGUCdTdT GAUAUGAGACCUUGGUGGdTdT	(Schilders et al., 2005)
hSki3	Rabbit polyclonal	—	(Zhu et al., 2005)
hSki8	Rabbit polyclonal	UGACCAACCAGUACGGUAUdTdT	(Zhu et al., 2005)

for 12S and 60 to 80S complexes in the gradient, a U3 snoRNA probe can be used.

Immunoprecipitations can be performed directly with the collected fractions, and the precipitated complexes can be further characterized. For example, the activity of the complex can be determined with an *in vitro* exonuclease activity assay (Brouwer *et al.*, 2001; Mitchell *et al.*, 1997).

4. Studying Exosome Function with RNAi

The identification of small interfering RNAs (siRNAs) that are able to suppress the expression of genes in a sequence-specific manner in cultured mammalian cells has made it possible to study the function of the exosome in human cells and has revealed that the exosome plays a key role in several $3'$- to $5'$-RNA turnover pathways. Here, we describe a general procedure for the downregulation of exosome components with RNAi (RNAi) in human cell lines.

4.1. Protocol

1. HEp-2 cells, grown to 70% confluent monolayers are seeded in a 6-well plate (approximately 1.5×10^5 cells/well) and cultured overnight at 37 °C in a humidified 5% CO_2 incubator.
2. For each well in a 6-well plate, 2 μl of a 20 μM siRNA stock solution is diluted in 183 μl OptiMEM (Gibco) and gently mixed.
3. 3 μl of Oligofectamine (Invitrogen) is diluted in OptiMEM to a final volume of 15 μl and after gentle mixing incubated for 10 min at room temperature.
4. After the 10-min incubation, the diluted Oligofectamine is added to the diluted siRNA, gently mixed, and incubated for 20 min at room temperature.
5. Meanwhile, the HEp-2 cells are washed twice with PBS and incubated with 800 μl of DMEM (+10% FCS). For most cell lines it is best to use serum-free culture medium during transfection, although for HEp-2 cells, FCS containing medium can be used without loss of transfection efficiency.
6. The 200 μl of siRNA-Oligofectamine mixture is added to each well, and the 6-well plate is gently swirled. Note that it is important to mix the transfection components just before its addition to the cells to achieve a high and reproducible transfection efficiency.
7. The cells are incubated for 4 h at 37 °C in a 5% CO_2 incubator.
8. Two ml of DMEM containing 10% FCS is added.

9. If the cells are cultured for periods longer than 48 h, it is necessary to retransfect the cells, because we have observed a reduction in knockdown efficiency 48 h post-transfection.

4.2. Comments

The optimal incubation times, amount of siRNA, and transfection reagents should be determined empirically for each cell type. However, the protocol described above should be a good starting point for optimization for most mammalian cell lines, for example, by replacing Oligofectamine with an equal amount of an alternate transfection reagent.

With regard to efficient and successful exosome knockdown experiments, there are some specific points to bear in mind. We observed that knockdown of core exosome components, such as hRrp41 and hRrp4, by RNAi is more efficient in the cytoplasm than in the nucleus (Fig. 11.3). Therefore, RNAi effects observed until 48 h after transfection are likely to be due primarily to knockdown of the cytoplasmic exosome. For studying the function of core exosome components in the nucleus, it will be required to elongate the transfection time to at least 72 h. However, it should be clear that the effects observed at that time point might as well be indirectly caused by the impairment of cytoplasmic exosome function. Moreover, knockdown of core exosome components inhibits cell growth, which could lead

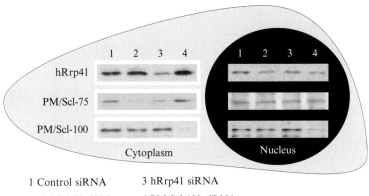

Figure 11.3 Exosome subunit knockdown efficiency in different cellular compartments. Knockdown of hRrp41, PM/Scl-75, and PM/Scl-100 by RNAi was analyzed with Western blotting and nuclear and cytoplasmic extracts prepared from HEp-2 cells. In the cytoplasmic fractions, significantly reduced levels of hRrp41, PM/Scl-75, and PM/Scl-100 were observed, whereas in the nuclear fractions only an efficient knockdown of PM/Scl-100 was detected. Importantly, this demonstrates that the cytoplasmic exosome is more efficiently downregulated than the nuclear exosome (see also Van Dijk *et al.* [2007]).

to nonspecific side effects. PM/Scl-100 on the other hand is efficiently downregulated in the nucleus and, thus, might be a more attractive target to downregulate the function of the nuclear exosome (see Fig. 11.3).

If reporter constructs are used, such as β-globin mRNA expression constructs to study 3′- to 5′-mRNA turnover on exosome knockdown, approximately 8×10^6 HEp-2 cells are grown to 70% confluent monolayers and are transfected by electroporation with 20 μg of plasmid DNA in 900 μl of DMEM containing 10% FCS. HEp-2 cells are electroporated at 260 V and 950 μF with a Gene-Pulser II (Bio-Rad), subsequently seeded in a 6-well plate (approximately 1.5×10^5 cells/well), and cultured overnight at 37 °C in a humidified 5% CO_2 incubator. Then the siRNA transfection protocol just described can be applied to study the effects on the expression of the reporter mRNA. Note that it is important to discriminate between the effects of exosome depletion on mRNA degradation and on other processes that influence steady-state mRNA levels, such as transcription. To do so, we use HEp-2 cells stably transfected with the plasmid pTet-tTAk, which expresses a transcriptional activator in the absence of tetracycline. The addition of tetracycline results in the inhibition of transcription of the reporter constructs, which eliminates the possibility that the effects are due to transcriptional events.

It has been reported for the *Trypanosoma brucei* exosome that knockdown of a single core exosome component results in codepletion of other exosome components (Estevez *et al.*, 2003). Similarly, it should be noted that knockdown of the human core exosome components hRrp41 and hRrp4 by RNAi also leads to destabilization of other exosome components, which makes it difficult to study the function of a single exosome component by this technology. On the basis of the crystal structure of the human exosome, it is expected that knockdown of other exosome components will also result in the codepletion of the other core exosome components, which makes RNAi unsuitable to study the role of an individual exosome component. However, PM/Scl-75 seems to be an exception, because no codepletion of other core components is observed upon knockdown of PM/Scl-75 (Van Dijk *et al.*, 2007). Also PM/Scl-100 can be downregulated without reducing the levels of core subunits. SiRNAs that have been successfully applied to downregulate the levels of exosome subunits in human cells are also listed in Table 11.2.

REFERENCES

Allmang, C., Kufel, J., Chanfreau, G., Mitchell, P., Petfalski, E., and Tollervey, D. (1999a). Functions of the exosome in rRNA, snoRNA and snRNA synthesis. *EMBO J.* **18**, 5399–5410.

Allmang, C., Petfalski, E., Podtelejnikov, A., Mann, M., Tollervey, D., and Mitchell, P. (1999b). The yeast exosome and human PM-Scl are related complexes of 3′- to 5′-exonucleases. *Genes Dev.* **13**, 2148–2158.

Bousquet-Antonelli, C., Presutti, C., and Tollervey, D. (2000). Identification of a regulated pathway for nuclear pre-mRNA turnover. *Cell* **102,** 765–775.

Brouwer, R., Allmang, C., Raijmakers, R., van Aarssen, Y., Egberts, W. V., Petfalski, E., van Venrooij, W. J., Tollervey, D., and Pruijn, G. J. (2001). Three novel components of the human exosome. *J. Biol. Chem.* **276,** 6177–6184.

Chen, C. Y., Gherzi, R., Ong, S. E., Chan, E. L., Raijmakers, R., Pruijn, G. J., Stoecklin, G., Moroni, C., Mann, M., and Karin, M. (2001). AU binding proteins recruit the exosome to degrade ARE-containing mRNAs. *Cell* **107,** 451–464.

Das, B., Butler, J. S., and Sherman, F. (2003). Degradation of normal mRNA in the nucleus of *Saccharomyces cerevisiae*. *Mol. Cell Biol.* **23,** 5502–5515.

Elbashir, S. M., Harborth, J., Lendeckel, W., Yalcin, A., Weber, K., and Tuschl, T. (2001). Duplexes of 21-nucleotide RNAs mediate RNA interference in cultured mammalian cells. *Nature* **411,** 494–498.

Estevez, A. M., Lehner, B., Sanderson, C. M., Ruppert, T., and Clayton, C. (2003). The roles of intersubunit interactions in exosome stability. *J. Biol. Chem.* **278,** 34943–34951.

Granneman, S., Vogelzangs, J., Luhrmann, R., van Venrooij, W. J., Pruijn, G. J., and Watkins, N. J. (2004). Role of pre-rRNA base pairing and 80S complex formation in subnucleolar localization of the U3 snoRNP. *Mol. Cell Biol.* **24,** 8600–8610.

Houseley, J., LaCava, J., and Tollervey, D. (2006). RNA-quality control by the exosome. *Nat. Rev. Mol. Cell Biol.* **7,** 529–539.

Jacobs Anderson, J. S., and Parker, R. P. (1998). The $3'$- to $5'$-degradation of yeast mRNAs is a general mechanism for mRNA turnover that requires the SKI2 DEVH box protein and $3'$- to $5'$-exonucleases of the exosome complex. *EMBO J.* **17,** 1497–1506.

Jiang, T., and Altman, S. (2002). A protein subunit of human RNase P, Rpp14, and its interacting partner, OIP2, have $3'$- to $5'$-exoribonuclease activity. *Proc. Natl. Acad. Sci. USA* **99,** 5295–5300.

Kadaba, S., Wang, X., and Anderson, J. T. (2006). Nuclear RNA surveillance in *Saccharomyces cerevisiae*: Trf4p-dependent polyadenylation of nascent hypomethylated tRNA and an aberrant form of 5S rRNA. *RNA* **12,** 508–521.

LaCava, J., Houseley, J., Saveanu, C., Petfalski, E., Thompson, E., Jacquier, A., and Tollervey, D. (2005). RNA degradation by the exosome is promoted by a nuclear polyadenylation complex. *Cell* **121,** 713–724.

Lehner, B., and Sanderson, C. M. (2004). A protein interaction framework for human mRNA degradation. *Genome Res.* **14,** 1315–1323.

Lejeune, F., Li, X., and Maquat, L. E. (2003). Nonsense-mediated mRNA decay in mammalian cells involves decapping, deadenylating, and exonucleolytic activities. *Mol. Cell* **12,** 675–687.

Liu, Q., Greimann, J. C., and Lima, C. D. (2006). Reconstitution, activities, and structure of the eukaryotic RNA exosome. *Cell* **127,** 1223–1237.

Lukowiak, A. A., Granneman, S., Mattox, S. A., Speckmann, W. A., Jones, K., Pluk, H., Venrooij, W. J., Terns, R. M., and Terns, M. P. (2000). Interaction of the U3-55k protein with U3 snoRNA is mediated by the box B/C motif of U3 and the WD repeats of U3-55k. *Nucleic Acids Res.* **28,** 3462–3471.

Mitchell, P., Petfalski, E., Shevchenko, A., Mann, M., and Tollervey, D. (1997). The exosome: A conserved eukaryotic RNA processing complex containing multiple $3'$- to $5'$-exoribonucleases. *Cell* **91,** 457–466.

Mukherjee, D., Gao, M., O'Connor, J. P., Raijmakers, R., Pruijn, G., Lutz, C. S., and Wilusz, J. (2002). The mammalian exosome mediates the efficient degradation of mRNAs that contain AU-rich elements. *EMBO J.* **21,** 165–174.

Raijmakers, R., Egberts, W. V., van Venrooij, W. J., and Pruijn, G. J. (2002). Protein protein interactions between human exosome components support the assembly of

RNase PH-type subunits into a six-membered PNPase-like ring. *J. Mol. Biol.* **323**, 653–663.

Schilders, G., van Dijk, E., and Pruijn, G. J. (2007). C1D and hMtr4p associate with the human exosome subunit PM/Scl-100 and are involved in pre-rRNA processing. *Nucleic Acids Res.* **35**, 2564–2572.

Schilders, G., Raijmakers, R., Raats, J. M. H., and Pruijn, G. J. M. (2005). MPP6 is an exosome-associated RNA-binding protein involved in 5.8S rRNA maturation. *Nucleic Acids Res.* **33**, 6795–6804.

Stoecklin, G., Mayo, T., and Anderson, P. (2005). ARE-mRNA degradation requires the 5′- to 3′-decay pathway. *EMBO Rep.* **7**, 72–77.

Symmons, M. F., Jones, G. H., and Luisi, B. F. (2000). A duplicated fold is the structural basis for polynucleotide phosphorylase catalytic activity, processivity, and regulation. *Structure Fold Des.* **8**, 1215–1226.

van Dijk, E. L., Schilders, G., and Pruijn, G. J. (2007). Human cell growth requires a functional cytoplasmic exosome, which is involved in various mRNA decay pathways. *RNA* **13**, 1027–1035.

van Hoof, A., Lennertz, P., and Parker, R. (2000). Yeast exosome mutants accumulate 3′-extended polyadenylated forms of U4 small nuclear RNA and small nucleolar RNAs. *Mol. Cell. Biol.* **20**, 441–452.

Vanacova, S., Wolf, J., Martin, G., Blank, D., Dettwiler, S., Friedlein, A., Langen, H., Keith, G., and Keller, W. (2005). A new yeast poly(A) polymerase complex involved in RNA quality control. *PLoS. Biol.* **3**, e189.

Wilkinson, G. W., and Akrigg, A. (1992). Constitutive and enhanced expression from the CMV major IE promoter in a defective adenovirus vector. *Nucleic Acids Res.* **20**, 2233–2239.

Wyers, F., Rougemaille, M., Badis, G., Rousselle, J. C., Dufour, M. E., Boulay, J., Regnault, B., Devaux, F., Namane, A., Seraphin, B., Libri, D., and Jacquier, A. (2005). Cryptic pol II transcripts are degraded by a nuclear quality control pathway involving a new poly(A) polymerase. *Cell* **121**, 725–737.

Zhu, B., Mandal, S. S., Pham, A. D., Zheng, Y., Erdjument-Bromage, H., Batra, S. K., Tempst, P., and Reinberg, D. (2005). The human PAF complex coordinates transcription with events downstream of RNA synthesis. *Genes Dev.* **19**, 1668–1673.

CHAPTER TWELVE

Determining *In Vivo* Activity of the Yeast Cytoplasmic Exosome

Daneen Schaeffer, Stacie Meaux, Amanda Clark, *and* Ambro van Hoof

Contents

1. Introduction	228
2. Is My Favorite RNA Degraded and/or Processed by the Exosome?	229
2.1. Core exosome mutants	229
2.2. Mutants in cytoplasmic exosome cofactors	231
2.3. Nuclear exosome cofactors mutants	231
3. Is the Cytoplasmic Exosome Active in my Mutant or Under my Conditions?	232
3.1. The use of growth to analyze the degradation of aberrant transcripts by the cytoplasmic exosome	232
3.2. The use of synthetic lethality to analyze the degradation of normal transcripts by the cytoplasmic exosome	234
3.3. Assessing cytoplasmic exosome activity by measuring the stability of aberrant mRNAs	235
3.4. The use of RNA stability to analyze the degradation of normal transcripts by the cytoplasmic exosome	236
3.5. The use of the killer assay to analyze the activity of the cytoplasmic exosome	236
References	238

Abstract

A 3′-exoribonuclease complex, termed the exosome, has important functions in the cytoplasm, as well as in the nucleus, and is involved in 3′-processing and/or decay of many RNAs. This chapter will discuss methods to study cytoplasmic exosome function in yeast with *in vivo* approaches. The first section will describe mutants that are available to study the processing or decay of a specific RNA by the nuclear or cytoplasmic exosome. The second section will discuss methods to determine whether the cytoplasmic exosome is functional

University of Texas Health Science Center-Houston, Department of Microbiology and Molecular Genetics, Houston, Texas, USA under a specific condition(s) with reporter mRNAs that are known substrates of this complex.

1. Introduction

The yeast RNA exosome is composed of 10 essential subunits. Six of the subunits (Rrp41p, Rrp42p, Rrp43p, Rrp45p, Rrp46p, and Mtr3p) share sequence similarity to *E. coli* RNase PH and PNPase, which are two processive, phosphorolytic 3′ to 5′-exoribonucleases (Liu *et al.*, 2006; Symmons *et al.*, 2000). Despite their similarity, the phosphorolytic active site is not conserved in the yeast proteins and, therefore, they do not have RNase activity (Allmang *et al.*, 1999b and Mitchell *et al.*, 1997). Three additional subunits, Csl4p, Rrp4p, and Rrp40p, contain nucleic acid binding domains, either S1 or KH, suggesting that these subunits bind RNA. The tenth subunit, Rrp44p, shares homology with *E. coli* RNase II and RNase R, which are two processive, hydrolytic 3′ to 5′-exoribonucleases (Mitchell *et al.*, 1996). *In vitro* and *in vivo* data suggests that Rrp44 is the only subunit responsible for core exosome activity (Dziembowski *et al.*, 2007; Liu *et al.*, 2006).

The eukaryotic exosome is present in both the cytoplasm and the nucleus, and it requires different cofactors depending on its cellular location. The cytoplasmic exosome associates with Ski7p and the Ski2p/Ski3p/Ski8p complex (Araki *et al.*, 2001; van Hoof *et al.*, 2002). Deletion of the *SKI2, SKI3, SKI7,* or *SKI8* genes disrupts all known functions of the cytoplasmic exosome but does not affect cell viability (Jacobs Anderson and Parker, 1998; van Hoof *et al.*, 2000b). Thus, none of the cytoplasmic functions of the exosome are required for cell viability. The nuclear exosome associates with Rrp6p, Rrp47p, Mtr4p, and other proteins. Although the *MTR4* gene is essential, the *RRP6* and *RRP47* genes are not (Briggs *et al.*, 1998; de la Cruz *et al.*, 1998; Mitchell *et al.*, 2003). For a more complete discussion of nuclear exosome cofactors refer to Houseley *et al.* (2006).

The cytoplasmic exosome degrades mRNAs from the 3′-end. This pathway is not essential because it is redundant with the Xrn1p-mediated 5′ to 3′-degradation pathway (Muhlrad *et al.*, 1995). Interestingly, the cytoplasmic exosome preferentially degrades aberrant transcripts, including (1) those lacking all in-frame termination codons, termed nonstop transcripts (van Hoof *et al.*, 2002), (2) unadenylated mRNAs (Meaux and van Hoof, 2006), (3) 5′-mRNA fragments generated by endonucleolytic cleavage in no-go decay (Doma and Parker, 2006), and (4) RNAs from the L and M viruses in yeast (Brown and Johnson, 2001; Meaux and van Hoof, 2006; Toh and Wickner, 1980).

In addition to functioning in RNA degradation in the cytoplasm, the exosome is also required for RNA processing and degradation in the nucleus. Specifically, the nuclear exosome functions in processing the 3′-ends of stable, structured RNAs, including 5.8S rRNA and many snoRNAs (Allmang *et al.*, 1999a; Mitchell *et al.*, 1996; van Hoof *et al.*, 2000a). The nuclear exosome also degrades aberrant precursor RNAs, including pre-tRNA (Kadaba *et al.*, 2004; LaCava *et al.*, 2005; Vanacova *et al.*, 2005) and pre-rRNA (Allmang *et al.*, 2000; LaCava *et al.*, 2005).

This chapter will cover methods used to study RNA degradation by the cytoplasmic exosome with intact yeast. The first section is for those who are studying the degradation or processing of their favorite RNA and are curious as to whether the cytoplasmic exosome is involved in this process. The second section addresses methods to test whether the cytoplasmic exosome is active under specific conditions. If you have isolated a yeast strain that you think might be defective in exosome function, or you suspect the cytoplasmic exosome is inactive under certain conditions, the methods in the second section should be helpful.

2. Is My Favorite RNA Degraded and/or Processed by the Exosome?

The known functions of the exosome were discovered because exosome mutants display defects in RNA processing and degradation. This idea can be extended to any RNA of interest. If the exosome processes an RNA, exosome mutants will accumulate processing intermediates. Similarly, if an RNA is degraded by the exosome, its half-life will be stabilized in exosome mutants. The exosome process of interest (i.e., degradation or processing) dictates which type of assays one can use to determine whether the cytoplasmic exosome is active on an RNA of interest. This section will describe available exosome mutants.

2.1. Core exosome mutants

All subunits of the exosome are essential; therefore, conditional mutants must be used to study this complex. Two types of conditional exosome mutants are available. Various groups have generated temperature-sensitive alleles of many of the exosome subunits (Table 12.1). Each of these temperature-sensitive alleles seems to inactivate the entire exosome complex; therefore, they can be used interchangeably. Another set of mutants has the exosome subunits under the control of a regulatable promoter. The Tollervey laboratory has generated strains that express exosome subunits from a galactose–regulatable (*GAL*) promoter, some of which are also epitope tagged (Allmang *et al.*, 1999b; Mitchell *et al.*, 1997). Each of these alleles is expressed in the presence of galactose but repressed when cells are

Table 12.1 Temperature-sensitive mutations affecting the exosome

Exosome factor	Temperature-sensitive allele	Reference
Rrp4p	rrp4-1	Mitchell et al., 1996
Rrp40p	mtr14-1	Kadowaki et al., 1994
Rrp41p	ski6-100	Jacobs, Anderson, and Parker, 1998
Rrp42p		
Rrp43p	rrp43-1, rrp43-2, rrp43-3	Oliveira et al., 2002
Rrp44p	mtr17-1; dis3-1 to dis3-14	Kadowaki et al., 1994; Suzuki et al., 2001
Rrp45p		
Rrp46p		
Csl4p		
Mtr3p	mtr3-1	Kadowaki et al., 1994
Mtr4p	mtr4-1, dob1-1	Kadowaki et al., 1994; de la Cruz et al., 1998

exposed to glucose. An alternative to the *GAL*-regulated exosome subunits is a collection of strains that have exosome subunits expressed from a tetracycline (TET)-repressible promoter. These were made as part of a genome-wide effort to put each of the essential yeast genes under control of the TET promoter (Mnaimneh et al., 2004) and are commercially available through Open Biosystems (http://www.openbiosystems.com/GeneExpression/Yeast/Tet%2DPromoters/).

As indicated in the introduction, Rrp44p seems to be the only active $3'$ to $5'$-exoribonuclease in the exosome. A point mutation that inactivates this activity has been characterized and is useful to determine whether this activity is required for various exosome functions (Dziembowski et al., 2007). Although complete deletion of the *RRP44* gene is lethal, this point mutation reduces growth but is not lethal. In addition to the activity of Rrp44p, the nuclear exosome cofactor Rrp6p is also a $3'$ to $5'$-exoribonuclease, and a point mutation that eliminates Rrp6p activity has been characterized (Phillips and Butler, 2003).

Core cytoplasmic exosome mutants are useful because they are defective in both cytoplasmic and nuclear exosome functions. However, these mutants should be used cautiously because the usual caveats of working with conditional alleles apply. The galactose-inducible or tetracycline-repressible alleles require long incubation times (>12 h) in glucose or tetracycline to fully repress expression (see also chapter 14 by Coller and Chapter 20 Passos and Parker in this volume of MIE). This prolonged incubation could lead to secondary effects that may be responsible for some of the observed phenotypes. In contrast, the use of a temperature-sensitive mutant allows for quick inactivation of the protein, which may limit, but not completely eliminate, such secondary effects. Galactose-inducible or tetracycline-repressible alleles are relatively easy to construct and use; however, the expression is not completely repressed in the presence of glucose (for galactose-inducible genes) or tetracycline (for tetracycline-repressible genes). Similarly, temperature-sensitive alleles may not be completely inactivated at the restrictive temperature and/or completely active at the permissive temperature. Thus, for a thorough analysis, the use of a combination of temperature-sensitive and regulatable-promoter alleles might be advisable.

2.2. Mutants in cytoplasmic exosome cofactors

The cytoplasmic functions of the exosome can be studied with strains expressing mutated cytoplasmic exosome cofactors (i.e., Ski2p, Ski3p, Ski7p, or Ski8p). Mutations in these cofactors disrupt all known functions of the cytoplasmic exosome (Jacobs Anderson and Parker, 1998; van Hoof et al., 2000b). Thus, each of the *ski* mutants can be used to measure the stability and/or level of an RNA of interest. The exosome mutant *ski4-1* can also be used. Although the *ski4-1* mutation is in the core exosome, it only affects activity of the cytoplasmic exosome, presumably because it disrupts the interaction between the core exosome and its cytoplasmic cofactors (van Hoof et al., 2000b).

The advantage of using mutations in the *SKI2, 3, 7,* or *8* gene is that none affects cell growth; therefore, indirect effects resulting from the cells being sick or dying do not occur. However, it is theoretically possible that the cytoplasmic exosome functions independently of the Ski cofactors in yet-to-be appreciated ways. In addition, Ski8p has a role in meiotic recombination that is independent of the exosome (Arora et al., 2004), and other Ski proteins may have similar exosome-independent functions. To fully determine the role of the cytoplasmic exosome, additional analyses should be performed with strains harboring mutations in core exosome subunits.

2.3. Nuclear exosome cofactors mutants

As mentioned previously, the nuclear exosome processes and degrades various RNA substrates. To analyze the role of the nuclear exosome, strains expressing mutated Rrp6p, Rrp47p, or Mtr4p can be used. Although the nuclear exosome is essential, Rrp6p and Rrp47p are not. This suggests that either deletion of the *RRP6* or *47* gene does not completely inactivate the nuclear exosome and/or that Mtr4p has some functions that are independent of the exosome. For analysis of nuclear exosome functions, temperature-sensitive alleles of the *MTR4* gene are available (see Table 12.1).

3. Is the Cytoplasmic Exosome Active in my Mutant or Under my Conditions?

The second approach to studying RNA degradation by the cytoplasmic exosome is to test whether the exosome is active in a mutant of interest. A variety of assays can be used to test exosome activity because the cytoplasmic exosome is involved in degrading normal, unadenylated, and nonstop transcripts. In general, if the cytoplasmic exosome is active in a specific mutant, these transcripts will be degraded. However, if a mutant of interest is defective for cytoplasmic exosome activity, these transcripts will be stabilized. Reporter constructs are available that allow the functionality of the exosome to be determined, either indirectly by a growth assay or directly by mRNA decay analysis.

3.1. The use of growth to analyze the degradation of aberrant transcripts by the cytoplasmic exosome

Unadenylated mRNAs and nonstop mRNAs are rapidly degraded by the cytoplasmic exosome; therefore, reporters that generate such transcripts can be used to test for cytoplasmic exosome function. A nonstop *his3* reporter mRNA (*his3-nonstop*) was created by van Hoof *et al.* (2002), and it is possible to generate an unadenylated *his3* mRNA by ribozyme cleavage (*his3-RZ*) (Meaux and van Hoof, 2006). In a wild-type background, these reporter mRNAs result in low *his3* transcript levels and slow growth in medium lacking histidine. However, in a mutant with a defective cytoplasmic exosome, these *his3* reporter transcripts are stabilized, resulting in increased growth in medium lacking histidine. The His3 growth assay is performed by transforming one of these reporter constructs into a strain of interest. If a factor of interest is involved in degrading aberrant transcripts, a mutation in that gene will stabilize the *his3* reporter transcripts and allow for better growth than a wild-type strain on medium lacking histidine. An example of the expected results from the His3p growth assay is shown in Fig. 12.1, and a detailed protocol of this assay is outlined in the following.

3.1.1. Protocol

1. Transform a mutant strain of interest with a plasmid-borne *his3-RZ* or *his3-nonstop* reporter. Also introduce the plasmid into isogenic wild-type and *ski* mutant strains to serve as negative and positive controls, respectively. The available reporter plasmids also contain a *URA3* marker, which allows for selection of transformants. Thus, for this assay, the starting strain should be a *his3, ura3* mutant.
2. Grow each strain in 5-ml of SC-URA medium to mid-log phase at 30 °C.

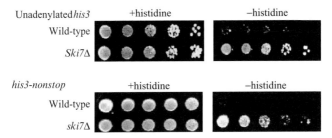

Figure 12.1 The *in vivo* activity of the cytoplasmic exosome can be conveniently monitored in a growth assay. An unadenylated *his3* mRNA (top panel) or a nonstop *HIS3* mRNA (lower panel) is normally rapidly degraded by the cytoplasmic exosome. This rapid degradation limits the production of the His3 enzyme and histidine, causing slow growth in the absence of added histidine (right panels). In a mutant strain that has an inactivated cytoplasmic exosome (e.g., *ski7Δ*), the reporter mRNAs are stabilized, resulting in higher levels of histidine biosynthesis and increased growth. The indicated strains were serially diluted, and a decreasing number of cells were spotted on plates that either contained histidine (+, left) or lacked histidine (−, right).

3. Make fivefold serial dilutions of each strain in a 96-well plate and spot 5 μl onto each of the following plates: (1) complete medium, (2) SC-URA, (3) SC-HIS, (4) SC-URA-HIS. See Burke *et al.* (2000) for media recipes. Although spotting can be done manually, a 48 pin or 96 pin "pronger" (e.g. Sigma Aldrich catalog number R2383) greatly facilitates doing this accurately. Streaking cells is another alternative, but it is less quantitative because of the reduced control of how many cells are plated.
4. Incubate plates at 30 °C for 3 or more days, depending on the growth rate of the strains tested.

Note: A *ski* mutant containing a *his3-nonstop* reporter will not grow as quickly as a HIS+ strain. Thus, 3 to 5 days of growth may be needed to see a clear difference between the growth of the wild-type and *ski* control strains.

Note: The *his3* plasmids used for this assay express His3p from its normal promoter. The *GAL* promoter-driven reporters described in the following section allow for the direct measurement of mRNA half-lives but do not work well in this growth assay.

This assay provides a quick and simple way to analyze the degradation of aberrant transcripts by the cytoplasmic exosome and is especially useful for high-throughput screening of potential mutants (Wilson *et al.*, 2007). A disadvantage of this method is that it uses growth as an indirect measure of mRNA levels. Therefore, exosome mutants are indistinguishable from mutants that increase expression by some other way (i.e., increased transcription or protein stabilization). Although with pronging the assay is semiquantitative, direct measurements of mRNA levels are more appropriate for fully quantitative analyses (see Section 3.3.1.).

3.2. The use of synthetic lethality to analyze the degradation of normal transcripts by the cytoplasmic exosome

Normal mRNAs can be degraded by two redundant pathways (Muhlrad et al., 1995; see also chapter 14 by Coller and chapter 20 by Passos and Parker in this volume of MIE). Therefore, mutants in one or the other pathway are viable, but a double mutant that is defective in both pathways cannot survive (Jacobs Anderson and Parker, 1998). An assay to study cytoplasmic exosome function is performed by combining a mutant of interest with a mutant in the 5′ to 3′-degradation pathway and testing for growth of the double mutant. Survival indicates that the cytoplasmic exosome in the mutant of interest is functional. Conversely, a mutant that is defective in 3′ to 5′-degradation will be synthetically lethal when combined with the 5′ to 3′-degradation mutant.

Deactivation of the 5′ to 3′-decay pathway can be accomplished by mutating any of the factors involved, including those involved in decapping and/or the 5′ to 3′-exoribonuclease, Xrn1p. These factors can be deleted because they are not essential in yeast. However, a deletion of these factors would prohibit a double mutant, defective in both the 5′ to 3′- and 3′ to 5′-degradation pathways, from growing. Therefore, a more convenient 5′ to 3′-degradation mutant to use in this assay is one that is conditional (i.e., temperature-sensitive). Two temperature-sensitive decapping mutants, *dcp1-2* and *dcp2-7*, each allow for a functional 5′ to 3′-decay pathway at the permissive temperature but not at the nonpermissive temperature. The permissive and nonpermissive temperatures for *dcp1-2* are 23 °C and 30 °C and for *dcp2-7* 30 °C and 37 °C. An example of the expected results from a synthetic lethality growth assay is shown in Fig. 12.2.

An advantage of the growth assay is that it can be used to screen a large number of mutants. However, a disadvantage is that the mutant being tested could be synthetically lethal with the 5′ to 3′-degradation machinery for reasons independent of the cytoplasmic exosome. To correct for this, additional assays that directly analyze the decay kinetics of a transcript should be performed to confirm the results obtained in the growth assay (see Section 3.4).

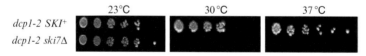

Figure 12.2 Yeast has two general pathways for mRNA degradation. Because these two pathways are redundant, mutant yeast defective in one pathway are viable, but simultaneous inactivation of both pathways is lethal. The *dcp1-2* strain has a conditionally inactive decapping pathway at 30 °C and above. Thus, *dcp1-2* strains lack cytoplasmic exosome activity and fail to grow at temperatures above 30 °C. The indicated strains were serially diluted, and a decreasing number of cells were spotted on plates and incubated at the indicated temperatures.

3.3. Assessing cytoplasmic exosome activity by measuring the stability of aberrant mRNAs

As mentioned previously, the cytoplasmic exosome is the main enzyme responsible for degrading unadenylated and nonstop mRNAs. Although the growth assays described previously are convenient, they are not quantitative and only indirectly test for cytoplasmic exosome activity. A more direct way to determine whether a mutant of interest has a functional cytoplasmic exosome is to measure the stability of an unadenylated or nonstop reporter mRNA. To accurately determine the half-lives of these reporters, their synthesis must be repressed. Although this can be accomplished in various ways (see also chapter 14 by Coller and chapter 20 Passos and Parker in this volume of MIE), we routinely use reporter constructs containing the *GAL1* promoter, which allows for induction of transcription in the presence of galactose but rapid repression in the presence of glucose. Several *GAL* promoter–driven reporters that produce unadenylated mRNAs have been made by creating a reporter mRNA harboring a hammerhead ribozyme (Dower *et al.*, 2004; Meaux and van Hoof, 2006). The hammerhead ribozyme cleaves the primary transcript, resulting in an unadenylated mRNA that normally is rapidly degraded by the cytoplasmic exosome. Similarly, *GAL* promoter-driven nonstop reporters generate nonstop mRNAs that are rapidly degraded by the cytoplasmic exosome. For the hammerhead reporters, versions based on HIS3 and the ZZ domain of Protein A have proven useful (Meaux and van Hoof, 2006), whereas useful nonstop mRNAs have been created based on *HIS3*, *PGK1*, or the ZZ domain of Protein A (van Hoof *et al.*, 2002; Wilson *et al.*, 2007). Each of these can be used interchangeably to measure cytoplasmic exosome function. The advantage of the use of the *his3* or *pgk1pG-nonstop* reporters is that the cognate genes are biologically relevant to yeast. A standard protocol is outlined in the following.

3.3.1. Protocol

1. Transform a plasmid encoding the reporter mRNA into the strain of interest. Most available plasmids have a *URA3* marker to facilitate this. Also transform the plasmid into isogenic wild-type and *ski2*, *ski3*, *ski7* or *ski8* control strains.
2. Grow cells in 50 ml of SC-URA medium containing 2% galactose to mid-log phase at 30 °C.
3. Harvest the cells by centrifugation and resuspend in 20 ml of SC-URA.
4. Immediately remove a 2-ml cell aliquot (t_0 time point). Centrifuge this and all subsequent aliquots for 10 sec in a picofuge or microcentrifuge. Remove medium/supernatant and freeze cell pellet in dry ice.
5. To the remaining culture, add glucose to a final concentration of 2%.

6. Continue incubation in a shaking waterbath at 30 °C. Remove and harvest additional 2-ml aliquots of cells 1, 2, 3, 4, 6, 8, 10, and 15 min after the addition of glucose.
7. Isolate RNA by a standard method (e.g., Caponigro et al., 1993).

Perform Northern blotting to quantitate the amount of reporter mRNA at each time point. After correction for differences in loading, a logarithmic graph of the percentage of mRNA versus time allows for the assessment of the half-life.

The half-life of the unadenylated or nonstop mRNA should be larger in the *ski* strain than in the wild-type strain. As a control, the stability of endogenous normal mRNAs should not differ. As controls, *GAL1, 7*, or *10* are convenient normal mRNAs, because addition of glucose also turns off their synthesis.

3.4. The use of RNA stability to analyze the degradation of normal transcripts by the cytoplasmic exosome

As explained previously, in strains that have a defect in decapping, the cytoplasmic exosome is the main activity that degrades normal mRNAs. This can be exploited by measuring the stability of *GAL1, 7*, or *10* mRNA. The stability assay is performed in a mutant with a defect in decapping (e.g., *dcp1-2* or *dcp2-7*). A standard protocol is outlined as follows.

3.4.1. Protocol

1. Grow 50 ml of cells in YEP medium containing 2% galactose to mid-log phase at room temperature.
2. Shift the cells to 37 °C for 1 h to inactivate the decapping enzyme.
3. Harvest the cells by centrifugation and resuspend in 20 ml YEP medium prewarmed to 37 °C.
4. Subsequent steps are identical to step 4 etc of protocol 3.3.1.

3.5. The use of the killer assay to analyze the activity of the cytoplasmic exosome

Most mutants affecting the cytoplasmic exosome were initially isolated in a screen for "superkiller" mutants (Toh and Wickner, 1980; Wickner, 1978). Many yeast strains contain the L-A virus and M satellite virus. The M satellite virus encodes a protein toxin that is secreted from infected cells. Mutant strains with defects in cytoplasmic exosome function secrete more of this killer toxin and, thus, are superkillers. The molecular mechanism of the Ski-complex antiviral function is incompletely understood, but it seems likely that the cytoplasmic exosome limits killer secretion by degrading unadenylated viral RNA (Brown et al., 2000; Meaux and van Hoof, 2006).

The killer assay, which can be used to analyze the antiviral effects of the cytoplasmic exosome, is performed by spotting infected, toxin-producing cells that contain a mutation of interest on a lawn of uninfected, non-killer cells. The premise of this assay is that the infected cells secrete the killer toxin that then kills the surrounding sensitive, uninfected cells, thus creating a halo, or zone of clearance. If the halo surrounding the toxin-producing cells is larger than wild type, the mutant of interest has enhanced toxin production. A detailed protocol is outlined as follows (adapted from Somers and Bevan, 1969).

3.5.1. Protocol

1. Grow individual colonies of the killer strain, which contains both dsRNA viruses and a mutation of interest, and a sensitive, uninfected strain on YEP medium containing 2% glucose at 15 to 20 °C.

   Note: Both the L-A virus and the killer toxin are thermolabile (Wickner, 1978); therefore, all cultures and incubations should be performed within 15 to 20 °C.

2. Using a freshly grown overnight liquid (YPD medium) culture, spread 0.5 ml of the sensitive, non-killer strain onto dried YPD-medium plates containing methylene blue (see recipe following). This will ensure an evenly distributed lawn.

3. Replica plate or spot killer strain onto the lawn of sensitive cells. To establish defined halos, spotted cells should be highly concentrated.

4. Incubate the plates at 15 to 20 °C for 4 to 6 days.

   Note: Methylene blue stains dead cells and facilitates detection of a zone of clearance, which represents death of the sensitive, non-killer cells. Methylene blue plates are made by adding 5.0 g of Bacto yeast extract, 10.0 g of peptone, and 10.0 g of Bacto agar to 450 ml of double-distilled H_2O. Autoclave for 30 min on the liquid cycle. After slight cooling, add dextrose to a final concentration of 2%, sodium citrate to a final molarity of 0.1 M, and adjust the pH to 4.7. Finally, add methylene blue to a final concentration of 0.003% (Wickner and Leibowitz, 1976).

   The advantage of this assay is that it provides easily detectable phenotypes and is a quick way to screen a large number of mutants for their effects on viral RNA production. A disadvantage, however, is that this analysis is not quantitative.

REFERENCES

Allmang, C., Kufel, J., Chanfreau, G., Mitchell, P., Petfalski, E., and Tollervey, D. (1999a). Functions of the exosome in rRNA, snoRNA and snRNA synthesis. *EMBO J.* **18**, 5399–5410.

Allmang, C., Mitchell, P., Petfalski, E., and Tollervey, D. (2000). Degradation of ribosomal RNA precursors by the exosome. *Nucleic Acids Res.* **28**, 1684–1691.

Allmang, C., Petfalski, E., Podtelejnikov, A., Mann, M., Tollervey, D., and Mitchell, P. (1999b). The yeast exosome and human PM-Scl are related complexes of 3′ to 5′-exonucleases. *Genes Dev.* **13**, 2148–2158.

Araki, Y., Takahashi, S., Kobayashi, T., Kajiho, H., Hoshino, S., and Katada, T. (2001). Ski7p G protein interacts with the exosome and the Ski complex for 3′ to 5′-mRNA decay in yeast. *EMBO J.* **20**, 4684–4693.

Arora, C., Kee, K., Maleki, S., and Keeney, S. (2004). Antiviral protein Ski8 is a direct partner of Spo11 in meiotic DNA break formation, independent of its cytoplasmic role in RNA metabolism. *Mol. Cell* **13**, 549–559.

Briggs, M. W., Burkard, K. T., and Butler, J. S. (1998). Rrp6p, the yeast homologue of the human PM-Scl 100-kDa autoantigen, is essential for efficient 5.8 S rRNA 3′-end formation. *J. Biol. Chem.* **273**, 13255–13263.

Brown, J. T., Bai, X., and Johnson, A. W. (2000). The yeast antiviral proteins Ski2p, Ski3p, and Ski8p exist as a complex *in vivo*. *RNA* **6**, 449–457.

Brown, J. T., and Johnson, A. W. (2001). A cis-acting element known to block 3′-mRNA degradation enhances expression of polyA-minus mRNA in wild-type yeast cells and phenocopies a ski mutant. *RNA* **7**, 1566–1577.

Burke, D., Dawson, D., and Stearns, T. (2000). "Methods in Yeast Genetics: A Cold Spring Harbor Laboratory Course Manual." Cold Spring Harbor Laboratory Press, Woodbury, New York.

Caponigro, G., Muhlrad, D., and Parker, R. (1993). A small segment of the MAT alpha 1 transcript promotes mRNA decay in *Saccharomyces cerevisiae*: A stimulatory role for rare codons. *Mol. Cell Biol.* **13**, 5141–5148.

de la Cruz, J., Kressler, D., Tollervey, D., and Linder, P. (1998). Dob1p (Mtr4p) is a putative ATP-dependent RNA helicase required for the 3′-end formation of 5.8S rRNA in *Saccharomyces cerevisiae*. *EMBO J.* **17**, 1128–1140.

Doma, M. K., and Parker, R. (2006). Endonucleolytic cleavage of eukaryotic mRNAs with stalls in translation elongation. *Nature* **440**, 561–564.

Dower, K., Kuperwasser, N., Merrikh, H., and Rosbash, M. (2004). A synthetic A tail rescues yeast nuclear accumulation of a ribozyme-terminated transcript. *RNA* **10**, 1888–1899.

Dziembowski, A., Lorentzen, E., Conti, E., and Seraphin, B. (2007). A single subunit, Dis3, is essentially responsible for yeast exosome core activity. *Nat. Struct. Mol. Biol.* **14**, 15–22.

Houseley, J., LaCava, J., and Tollervey, D. (2006). RNA-quality control by the exosome. *Nat. Rev. Mol. Cell Biol.* **7**, 529–539.

Jacobs Anderson, J. S., and Parker, R. (1998). The 3′ to 5′-degradation of yeast mRNAs is a general mechanism for mRNA turnover that requires the SKI2 DEVH box protein and 3′ to 5′-exonucleases of the exosome complex. *EMBO J.* **17**, 1497–1506.

Kadaba, S., Krueger, A., Trice, T., Krecic, A. M., Hinnebusch, A. G., and Anderson, J. (2004). Nuclear surveillance and degradation of hypomodified initiator tRNAMet in *S. cerevisiae*. *Genes Dev.* **18**, 1227–1240.

Kadowaki, T., Chen, S., Hitomi, M., Jacobs, E., Kumagai, C., Liang, S., Schneiter, R., Singleton, D., Wisniewska, J., and Tartakoff, A. M. (1994). Isolation and characterization of *Saccharomyces cerevisiae* mRNA transport-defective (mtr) mutants. *J. Cell Biol.* **126**, 649–659.

LaCava, J., Houseley, J., Saveanu, C., Petfalski, E., Thompson, E., Jacquier, A., and Tollervey, D. (2005). RNA degradation by the exosome is promoted by a nuclear polyadenylation complex. *Cell* **121**, 713–724.

Liu, Q., Greimann, J. C., and Lima, C. D. (2006). Reconstitution, activities, and structure of the eukaryotic RNA exosome. *Cell* **127**, 1223–1237.

Meaux, S., and van Hoof, A. (2006). Yeast transcripts cleaved by an internal ribozyme provide new insight into the role of the cap and poly(A) tail in translation and mRNA decay. *RNA* **12**, 1323–1337.

Mitchell, P., Petfalski, E., Houalla, R., Podtelejnikov, A., Mann, M., and Tollervey, D. (2003). Rrp47p is an exosome-associated protein required for the 3'-processing of stable RNAs. *Mol. Cell Biol.* **23**, 6982–6992.

Mitchell, P., Petfalski, E., Shevchenko, A., Mann, M., and Tollervey, D. (1997). The exosome: A conserved eukaryotic RNA processing complex containing multiple 3' to 5'-exoribonucleases. *Cell* **91**, 457–466.

Mitchell, P., Petfalski, E., and Tollervey, D. (1996). The 3'-end of yeast 5.8S rRNA is generated by an exonuclease processing mechanism. *Genes Dev.* **10**, 502–513.

Mnaimneh, S., Davierwala, A. P., Haynes, J., Moffat, J., Peng, W. T., Zhang, W., Yang, X., Pootoolal, J., Chua, G., Lopez, A., Trochesset, M., Morse, D., *et al.* (2004). Exploration of essential gene functions via titratable promoter alleles. *Cell* **118**, 31–44.

Muhlrad, D., Decker, C. J., and Parker, R. (1995). Turnover mechanisms of the stable yeast PGK1 mRNA. *Mol. Cell Biol.* **15**, 2145–2156.

Oliveira, C. C., Gonzales, F. A., and Zanchin, N. I. (2002). Temperature-sensitive mutants of the exosome subunit Rrp43p show a deficiency in mRNA degradation and no longer interact with the exosome. *Nucleic Acids Res.* **30**, 4186–4198.

Phillips, S., and Butler, J. S. (2003). Contribution of domain structure to the RNA 3'-end processing and degradation functions of the nuclear exosome subunit Rrp6p. *RNA* **9**, 1098–1107.

Somers, J. M., and Bevan, E. A. (1969). The inheritance of the killer character in yeast. *Genet. Res.* **13**, 71–83.

Symmons, M. F., Jones, G. H., and Luisi, B. F. (2000). A duplicated fold is the structural basis for polynucleotide phosphorylase catalytic activity, processivity, and regulation. *Structure* **8**, 1215–1226.

Toh, E. A., and Wickner, R. B. (1980). "Superkiller" mutations suppress chromosomal mutations affecting double-stranded RNA killer plasmid replication in *Saccharomyces cerevisiae*. *Proc. Natl. Acad. Sci. USA* **77**, 527–530.

van Hoof, A., Frischmeyer, P. A., Dietz, H. C., and Parker, R. (2002). Exosome-mediated recognition and degradation of mRNAs lacking a termination codon. *Science* **295**, 2262–2264.

van Hoof, A., Lennertz, P., and Parker, R. (2000a). Yeast exosome mutants accumulate 3'-extended polyadenylated forms of U4 small nuclear RNA and small nucleolar RNAs. *Mol. Cell Biol.* **20**, 441–452.

van Hoof, A., Staples, R. R., Baker, R. E., and Parker, R. (2000b). Function of the ski4p (Csl4p) and Ski7p proteins in 3' to 5'-degradation of mRNA. *Mol. Cell Biol.* **20**, 8230–8243.

Vanacova, S., Wolf, J., Martin, G., Blank, D., Dettwiler, S., Friedlein, A., Langen, H., Keith, G., and Keller, W. (2005). A new yeast poly(A) polymerase complex involved in RNA quality control. *PLoS Biol.* **3**, e189.

Wickner, R. B. (1978). Twenty-six chromosomal genes needed to maintain the killer double-stranded RNA plasmid of *Saccharomyces cerevisiae*. *Genetics* **88**, 419–425.

Wickner, R. B., and Leibowitz, M. J. (1976). Two chromosomal genes required for killing expression in killer strains of *Saccharomyces cerevisiae*. *Genetics* **82**, 429–442.

Wilson, M. A., Meaux, S., and van Hoof, A. (2007). A genomic screen in yeast reveals novel aspects of nonstop mRNA metabolism. *Genetics* **177**, 773–784.

CHAPTER THIRTEEN

Approaches for Studying PMR1 Endonuclease-Mediated mRNA Decay

Yuichi Otsuka *and* Daniel R. Schoenberg

Contents

1. Introduction	242
2. Identification of Endonuclease Cleavage Sites within mRNA	244
2.1. Identification of 3′-products generated by endonuclease cleavage *in vivo*	244
2.2. Identification of 5′-cleavage products generated by endonuclease *in vivo*	249
3. Analysis of PMR1-Containing Complexes	251
3.1. Expression of PMR1 in transfected mammalian cells	252
3.2. Preparation of postmitochondrial extracts	252
3.3. Sucrose density gradient analysis of PMR1-containing complexes	253
3.4. Glycerol gradient analysis of PMR1 containing complexes	256
4. Affinity Recovery of PMR1-Containing Complexes	257
4.1. IgG-Sepharose selection of PMR1-TAP complexes	258
4.2. Immunoprecipitation of PMR1 with immobilized anti-myc antibody	259
5. Analysis of PMR1 Activity *In Vivo* and *In Vitro*	260
5.1. *In vivo* analysis of PMR1 activity	260
5.2. *In vitro* analysis of PMR1 activity	261
6. Summary	262
Acknowledgments	262
References	262

Abstract

Although most eukaryotic mRNAs are degraded by exonucleases acting on either end of the molecule, a subset of mRNAs undergo endonuclease cleavage within the mRNA body. Endonuclease cleavage can be activated by cellular stress, extracellular signals, or by ribosome stalling, as might occur at a

Department of Molecular and Cellular Biochemistry and The RNA Group, The Ohio State University, Columbus, Ohio, USA premature termination codon. Only a few eukaryotic mRNA endonucleases have been identified, and of these, polysomal ribonuclease 1 (PMR1) is the best characterized. A notable feature of PMR1-mediated mRNA decay is that it acts on specific mRNAs while they are engaged by translating ribosomes. This chapter begins with several procedures used to characterize *in vivo* endonuclease cleavage of any mRNA by any endonuclease. These include approaches for identifying the 5′-end(s) downstream of an endonuclease cleavage site (S1 nuclease protection and primer extension), and a ligation-mediated RT-PCR approach developed in our laboratory for identifying the 3′-ends upstream of a cleavage site. We then describe a number of approaches used to characterize PMR1-mediated mRNA decay in cultured cells. PMR1 participates in a number of different complexes. We show several approaches for studying these complexes, and we describe techniques for isolating and characterizing PMR1-interacting proteins and its target mRNAs. Although the various techniques described here have proven their usefulness in studying PMR1, they can be generalized to studying decay by any other mRNA endonuclease.

1. Introduction

Most mRNAs are generally thought to be degraded by exonuclease-catalyzed mechanisms in which the mRNA is degraded either in a 3′ to 5′-manner by the cytoplasmic exosome or in a 5′ to 3′-manner by enzymes that remove the 5′-cap and degrade the mRNA body with 5′ to 3′-polarity (reviewed in Garneau *et al.* [2007]). In addition to these pathways, a subset of mRNAs undergoes endonuclease cleavage within the mRNA body followed by degradation of the upstream and downstream decay products by exonucleases. Although it was generally thought that endonuclease-mediated mRNA decay was limited to a few specialized mRNAs, the overall process is functionally analogous to RNA interference, where endonuclease cleavage results in the rapid disappearance of the mRNA body. The mRNA endonucleases that have been identified to date include PMR1, G3BP-1 (Gallouzi *et al.*, 1998), IRE-1 (Hollien and Weissman, 2006), and Dom34 (Lee *et al.*, 2007). Although Argonaute 2 can be thought of as an mRNA endonuclease, it is distinguished from these group protein ribonucleases by its requirement for a short antisense RNA. PMR1 was originally identified as an estrogen-induced endonuclease activity that appeared concomitantly with the estrogen-induced destabilization of serum protein mRNAs in *Xenopus* liver (Pastori *et al.*, 1991b). It differs structurally from most ribonucleases in that it is closely related to the peroxidase gene family, and the active form is processed from a larger precursor (Chernokalskaya *et al.*, 1998). PMR1-mediated mRNA decay also differs fundamentally from other types of mRNA decay in that it forms a specific complex with its translating substrate mRNA, and it is in this context that cleavage initiates mRNA decay. Interestingly, a similar mechanism was recently reported for IRE-1, which is activated by unfolded protein to cleave endoplasmic reticulum associated mRNAs (Hollien and Weissman, 2006).

The ability of PMR1 to catalyze mRNA decay depends on its participation in a number of different complexes. PMR1 is inherently unstable, and alterations in its binding to Hsp90 result in rapid degradation by the 26S proteasome (Peng et al., 2007). When examined on sucrose density gradients, PMR1 sediments with polysomes and in a lighter mRNP complex at the top of the gradient (Yang and Schoenberg, 2004). Dissociating ribosomes with EDTA releases polysome-bound PMR1 as a complex with its substrate mRNA (termed complex I) that sediments at ~680 kDa in glycerol gradients (Yang and Schoenberg, 2004). This treatment has no impact on the smaller complex (termed complex II), which sediments at ~140 kDa in glycerol gradients.

Each of these complexes plays a functional role in PMR1-mediated mRNA decay. PMR1 must be phosphorylated on Y650 to join complex I on polysomes and catalyze mRNA decay (Yang et al., 2004), and this process occurs in complex II. A key component of the latter complex is the oncogenic tyrosine kinase c-Src. Tyrosine phosphorylation of PMR1 "licenses" PMR1 for binding to polysomes and, as such, is a requisite step in mRNA decay (Peng and Schoenberg, 2007). PMR1 is also a stress-responsive protein, and in stressed cells the direct interaction of its N-terminal domain with TIA-1 recruits complex I containing PMR1 and its substrate mRNA to stress granules (Yang et al., 2006), where mRNA decay is stalled. Thus, understanding the function of PMR1 in mRNA decay depends on understanding how its participation in a number of macromolecules guides this process.

Our early work used whole animals and primary frog hepatocyte cultures to characterize much of the biology of PMR1. However, technical difficulties led us to the use of mammalian cell lines that recapitulate the process of PMR1 targeting to polysomes and PMR1-mediated mRNA decay. Although the work described here focuses on the frog protein expressed in mammalian cells, its mammalian ortholog behaves similarly. The latter is somewhat larger but, like frog PMR1, is found on polysomes, is tyrosine phosphorylated, and forms a complex with its substrate mRNA. Whereas most of our work on frog PMR1 uses albumin as a typical substrate mRNA, the mammalian ortholog is best characterized for its role in degrading nonsense-containing β-globin mRNA in erythroid cells (Bremer et al., 2003; Stevens et al., 2002).

This chapter is organized roughly in the order of experiments one would follow when first determining whether a particular mRNA undergoes endonucleolytic cleavage, followed by experiments used to characterize the behavior and complexes involved in PMR1-mediated mRNA decay. Although these approaches are described in the context of this particular enzyme, they are generally applicable to studying any other mRNA endonuclease. On a final note, throughout the text we refer to both the frog and mammalian enzyme generically as PMR1. In this context, it denotes the fully processed mature protein that participates in the process of endonuclease-mediated mRNA decay. In our published work, the frog protein is referred to as PMR60 to distinguish the active, processed 60 kDa form of the protein from the 80 kDa precursor (PMR80) from which it is derived.

2. Identification of Endonuclease Cleavage Sites within mRNA

In general, endonuclease cleavage products, including those generated by PMR1, cannot be detected, because they are rapidly degraded by 5′ to 3′ or 3′ to 5′-exonucleases. This makes it difficult to determine how many mRNAs are targeted by endonuclease-mediated decay. If one knows that a particular mRNA is targeted by endonuclease decay, it is at least theoretically possible to use techniques like Northern blotting to detect the downstream cleavage product by knocking down Xrn1 or the upstream cleavage product by knocking down one or more exosome proteins (Gatfield and Izaurralde, 2004). The protocols described in the following use S1 nuclease protection and primer extension assays to detect the downstream products of endonuclease cleavage and a sensitive ligation-mediated PCR assay to detect the upstream products. All three of these have been used successfully in our laboratory to detect endonuclease-generated decay intermediates in RNA from cultured cells without resorting to knocking down of 5′ or 3′-decay and products generated by *in vitro* endonuclease cleavage.

2.1. Identification of 3′-products generated by endonuclease cleavage *in vivo*

2.1.1. S1 nuclease protection assay

2.1.1.1. Overview S1 nuclease protection assay is a highly sensitive method for the detection, mapping, and quantification of specific RNA fragments in total cellular RNAs. The basis of S1 nuclease protection assay is the hybridization of a 5′-end-labeled single-stranded antisense DNA probe to a target RNA and then digesting unhybridized probe and cellular RNAs with single strand–specific S1 nuclease. Once the reaction is done, S1 nuclease is inactivated, and the remaining hybridized probe and RNA are precipitated. Finally, the protected probe is electrophoresed in a denaturing polyacrylamide gel and visualized by autoradiography, or more commonly, phosphorimaging. With proper controls, this can be used to quantify the degree of endonuclease cleavage.

Although double-stranded nucleic acids are generally resistant to S1, cleavage can occur under conditions of excess enzyme and high salt (Vogt, 1973). Also, because it does not effectively cleave at single base mismatches, S1 nuclease protection can be used to survey the effects of changing a single base across a particular RNA, such as one might encounter when studying the impact of changing the location of a premature termination codon. The susceptibility of mismatches to cleavage by S1 increases with the length of the mismatched sequence but this varies depending on the mismatching sequences (Brookes and Solomon, 1989). Thus, if the potential exists for mismatches between the target mRNA and the probe, one needs to optimize the reaction conditions to ensure that the assay detects genuine cleavage products. Use of dioxane in the S1 nuclease reaction enhances the ability of S1 nuclease to cleave at single base mismatches (Howard *et al.*, 1999). It is imperative that there is some additional sequence at the 3′-end of the probe that does not hybridize to the target mRNA so that undigested probe can be differentiated from probe that is protected by hybridization to intact mRNA. We have found it particularly useful to make probes by asymmetric PCR, although one can also use a cDNA clone that is linearized by cleaving within a multiple cloning site to give a 3′-end that does not hybridize to the target mRNA.

2.1.1.2. Protocol
2.1.1.2.1. End-labeling of DNA primer

1. 50 pmol of an antisense oligonucleotide (18 to 30 nt long) is combined with 10 μl of [γ-^{32}P]-ATP (3000 Ci/mmol) and 2 μl of polynucleotide kinase (Roche, 10 U/μl) in 50 μl of kinase buffer (50 mM Tris-HCl, pH 8.2, 10 mM MgCl$_2$, 5 mM dithiothreitol (DTT), 0.1 mM spermidine, 0.1 mM EDTA).
2. The mixture is incubated at 37 °C for 1 h, and the reaction is terminated by heating at 65 °C for 10 min. If you wish to determine the degree of incorporation at this step, dilute 1 μl into 100 μl of PBS and dot 1 μl of this onto a glass fiber filter. This is placed into 10% TCA, rinsed with ethanol, dried, and counted.
3. 1 μl of glycogen (Roche, 20 mg/ml), 13 μl of 7.5 M ammonium acetate, and 160 μl of ethanol are added to the mixture and labeled oligonucleotides are recovered by precipitation at −20 °C.
4. The precipitated DNA is recovered by centrifuging for 15 min at 12,000g in a refrigerated microcentrifuge, washed with 70% ethanol, and the pellet is dissolved in 21 μl of water.

2.1.1.2.2. Generation of single-stranded probes by asymmetric PCR

1. A flexible approach is to use asymmetric PCR, because the probe can be prepared from anywhere within a cloned DNA without regard to the

location relative to a multiple cloning site; 7 µl of the radiolabeled primer is added to a reaction containing 2.5 µl of 50 mM of MgCl$_2$, 1 µl of 10 mM dNTPs, 0.3 µl of Taq DNA polymerase (Invitrogen), 1.6 pmol of a sense primer that contains additional noncomplementary sequence at 5'-end and 100 ng of template DNA for PCR in 25 µl of PCR buffer (20 mM Tris-HCl, pH 8.4, 50 mM KCl). Note that the ratio of sense to antisense primer in the reaction is 1:10. The DNA template used for probe preparation is a PCR fragment that has been amplified with the same sense primer as earlier and an antisense primer that is located downstream of the previously labeled antisense primer.
2. The mixture is heated at 94 °C for 3 min, and PCR amplification is performed for 30 cycles at 94 °C for 30 sec, 54 °C for 30 sec, and 72 °C for 1 min, followed by extension for 5 min at 72 °C.
3. PCR products are separated on 6% polyacrylamide/urea gel and exposed to X-ray film to detect radiolabeled probes.
4. Radiolabeled probes are eluted from gel slices in 300 µl of probe elution buffer (0.5 M ammonium acetate, 1 mM EDTA, 0.2% SDS) by incubation for 3 to 4 h at 37 °C or overnight at 4 °C for maximal recovery.
5. The amount of radioactivity in the recovered probe is determined by scintillation counting.

2.1.1.2.3. S1 nuclease protection

1. 5×10^5 cpm of labeled DNA probe is ethanol-precipitated together with 5 to 50 µg of RNA, or yeast tRNA as a control. The pellet is rinsed with 70% ethanol and dried for 5 min on the bench.
2. The pellet is dissolved with 30 µl of hybridization buffer (40 mM 1,4-piperazine-diethanesulfonic acid (PIPES), pH 6.4, 1 mM EDTA, 0.4 M NaCl, 80% formamide).
3. The solution is heated at 95 °C for 4 min, transferred immediately to 52 °C, and incubated overnight.
4. S1 nuclease solution is prepared by adding 250 units of S1 nuclease (Promega) to 1 ml of S1 nuclease buffer (0.28 M NaCl, 0.05 M sodium acetate (pH 4.6), 4.5 mM ZnSO$_4$, 20 µg/ml sheared salmon sperm DNA).
5. 300 µl of ice-cold S1 nuclease solution is added to each hybridization mixture.
6. 300 µl of ice-cold S1 nuclease buffer without S1 nuclease is added to yeast RNA control tube for probe integrity.
7. The reaction mixture is incubated at 28 °C for 2 h followed by addition of 80 µl of stop solution (4 M ammonium acetate, 0.1 M EDTA, 50 µg/ml yeast tRNA).
8. Because the preceding step removes most of the nucleic acid in the reaction, 0.5 µl of glycogen carrier (Roche, 20 mg/ml) is added next, followed by 60 µl of 7.5 M ammonium acetate and 400 µl of isopropanol. This is then stored at −20 °C for at least 1 h.

9. The radiolabeled products are recovered by centrifuging at 12,000g for 15 min in a refrigerated microcentrifuge, and the pellet is rinsed with 70% ethanol. This is then air-dried for 5 min and dissolved in 5 μl of gel loading buffer (95% formamide, 18 mM EDTA, 0.025% [w/v] SDS, 0.025% [w/v] xylene cyanol, 0.025% [w/v] bromophenol blue).
10. Samples are heated at 95 °C for 3 min, chilled on ice immediately, and loaded onto a polyacrylamide/urea gel. Because of the greater amount of radioactivity of the undigested probe, only 5% of this is loaded onto the gel.
11. To pinpoint the exact location of endonuclease cleavage, the adjacent lanes should contain a dideoxy sequencing ladder that is prepared with the radiolabeled primer used for generating the S1 probe.
12. After electrophoresis, the gel is dried, and protected fragments are visualized by the PhosphorImager. An example of the use of S1 protection to identify an endonuclease cleavage product is shown in Fig. 13.1. In this experiment, S1 protection was used to map a previously unidentified endonuclease cleavage site in nonsense-containing β-globin mRNA expressed in murine erythroleukemia cells.

2.1.2. Primer extension assay

2.1.2.1. Overview The primer extension assay is an alternate method for detecting and mapping the 5′-end of endonuclease cleavage products. In this approach, a 5′-end-labeled antisense primer is hybridized to the complementary region of a target RNA and extended by reverse transcriptase to generate cDNAs extending to the sites of endonuclease cleavage and/or the 5′-end of the mRNA. The products are analyzed in a denaturing polyacrylamide gel and visualized by phosphoimaging, commonly with a dideoxy sequencing ladder, as described previously, to map precisely the location of the cleavage site. Because secondary structures within an mRNA can result in pausing of reverse transcriptase, one must include a control of an *in vitro* synthesized transcript that is treated at the same time and electrophoresed in a parallel lane of the gel. Effective hybridization of the primer is the most critical step for success in primer extension analysis, and the amount of template RNA, primer, and the hybridization temperature need to be optimized. The particular reverse transcriptase used may also affect the outcome. AMV reverse transcriptase is more robust than MMLV and less sensitive to the effect of secondary structure. However, MMLV reverse transcriptase generates longer products than AMV reverse transcriptase.

2.1.2.2. Protocol

1. 1 to 5×10^5 cpm of a 5′-end-labeled antisense oligonucleotide is ethanol-precipitated with 10 to 25 μg of total cellular RNA. In a separate tube, 20 ng of *in vitro* transcript corresponding to the region of your target

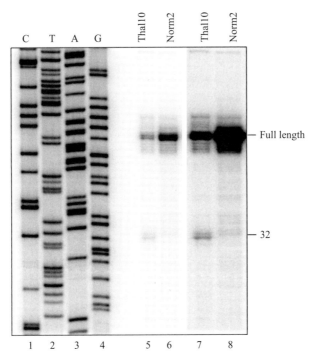

Figure 13.1 Identification of a PTC-induced endonuclease cleavage site at position 32 in human β-globin mRNA. Norm2 are murine erythroleukemia cells stably expressing normal human β-globin mRNA, and Thal10 are a matched line stably expressing the same mRNA with a premature termination codon (PTC) at codon 60/61 (Stevens et al., 2002). RNA from each of these cell lines was analyzed by S1 nuclease protection with a probe spanning the first 250 nt of β-globin mRNA. A dideoxy sequencing ladder in lanes 1 to 4 was used to identify a new endonuclease cleavage site at position 32 relative to the 5′-end of the mRNA. Lanes 7 and 8 are darker exposures of lanes 5 and 6.

mRNA and 10 μg of yeast tRNA as a carrier are ethanol-precipitated with the labeled oligonucleotide. This will serve as a control for reverse transcriptase pausing sites. The recovered pellet is washed with 70% of ethanol and dried for 5 min on the bench.
2. The precipitated RNA and oligonucleotide are dissolved in 10 μl of annealing buffer (50 mM Tris-HCl, pH 8.3, 75 mM KCl), heated at 65 °C for 10 min, followed by cooling slowly to 37 °C, and then put on ice.
3. 4 μl of buffer (125 mM Tris-HCl, pH 8.3, 187.5 mM KCl, 15 mM MgCl$_2$), 1 μl of 10 mM dNTPs, 2 μl of 0.1 M DTT, 1 μl of water, and 1 μl of RNaseOUT (Invitrogen, 40 U/μl) are added to the annealing mixture and incubated at 42 °C for 2 min.
4. 1 μl of MMLV reverse transcriptase (Invitrogen, 200 U/μl) is added, and the mixture is incubated for 50 min at 42 °C.

5. The reaction is stopped by adding 20 µl of loading buffer directly or by ethanol precipitation and dissolving the pellet in loading buffer.
6. The experimental samples and the *in vitro* transcript control reaction are loaded onto a polyacrylamide/urea gel and visualized using a PhosphorImager. Any bands seen in both experimental and control samples correspond to reverse transcriptase pausing sites and should be excluded from the list of *in vivo* cleavage sites.
7. It is also essential to include a dideoxy sequencing ladder in adjacent lanes to pinpoint the exact site of endonuclease cleavage. Note that the sequencing ladder should use the same radiolabeled primer.

2.2. Identification of 5′-cleavage products generated by endonuclease *in vivo*

S1 nuclease protection can be used to detect and map the 3′-end of upstream endonuclease cleavage products; however, this requires use of a 3′-end-labeled antisense probe, which is more difficult to prepare than the 5′-end-labeled probes described previously. Here, we describe an alternate approach that uses ligation-mediated RT-PCR that we used successfully to identify *in vivo* decay intermediates of albumin and c-myc mRNA (Hanson and Schoenberg, 2001).

2.2.1. Ligation-mediated reverse transcription–polymerase chain reaction (LM-RT-PCR)

2.2.1.1. Overview Protein endonucleases and RISC generate products with a 5′-monophosphate on the downstream fragment and a 3′-hydroxyl on the upstream fragment. 3′-end mapping by LM-RT-PCR takes advantage of the latter. In the first step, total-cell RNA is ligated to a DNA oligonucleotide bearing a 5′-phosphate and 3′-amino group. The 5′-phosphate allows this to be ligated to the 3′-hydroxyl of the upstream cleavage product and the 3′-amino group blocks concatamerization of the primer during ligation. A primer complementary to the ligated oligonucleotide is used for reverse transcription and PCR is then performed with a 5′-[^{32}P]–labeled sense primer that is specific to the mRNA of interest and the primer used for reverse transcription. The PCR products are then separated in a denaturing polyacrylamide/urea gel and visualized by a PhosphorImager. This application is straightforward and extremely sensitive.

2.2.1.2. Protocol
  2.2.1.2.1. Primer ligation

1. 2 µg of total or cytoplasmic RNA is combined with 1 µg of the ligation primer (5′-P-CCAGGTGGATAGTGCTCAATCTCTAGATCG-NH$_3$) in 15 µl of ligation buffer (50 mM Tris-HCl, pH 8.0, 10 mM

MgCl$_2$, 20 mM ATP, 2 mM DTT, 10 µg/ml BSA, 1 mM hexamine cobalt chloride, 25% [w/v] polyethylene glycol 8000, 20 units RNaseOUT).
2. The mixture is incubated overnight at 4 °C with 15 units of T4 RNA ligase (New England BioLabs). The conditions described here are optimized to add the primer onto approximately 50% of the 3′-ends.
3. The reaction is terminated by extracting twice with an equal volume of phenol/chloroform/isoamyl alcohol (25:24:1), and the products are recovered by adding 1 µl of glycogen, 1/10 volume of 3 M sodium acetate (pH 5.4) and 2.5 volumes of cold ethanol. This is kept at −20 °C for at least 2 h to maximize product recovery.
4. Nucleic acid is recovered by centrifuging at 12,000g for 15 min in a refrigerated microcentrifuge, and the resulting pellet is dissolved in 10 µl of DEPC-treated water.

2.2.1.2.2. Labeling of gene-specific oligonucleotide primer

1. 500 ng of the desired sense primer is combined with 3 µl of γ-[^{32}P]-ATP (3000 Ci/mmol) and 1 µl of polynucleotide kinase (Roche, 10 U/µl) and 2.5 µl of 10× kinase buffer (500 mM Tris-HCl, pH 8.2, 100 mM MgCl$_2$, 50 mM DTT, 1 mM spermidine, 1 mM EDTA) in final volume of 25 µl.
2. The mixture is incubated at 37 °C for 1 h, and the reaction is terminated by heating at 65 °C for 10 min.

2.2.1.2.3. RT-PCR

1. 10 µl of the preceding reaction mixture is mixed with 1 µl of 10 mM dNTPs and 250 ng of antisense primer (5′-CGATCTAGAGATT-GAGCAC-3′) in 1 µl of DEPC-treated water, which is complementary to the last 19 nt of ligation primer.
2. The solution is heated to 70 °C for 10 min and quickly cooled on ice to anneal the primer, followed by addition of 4 µl of 5× RT buffer (250 mM Tris-HCl, pH 8.3, 375 mM KCl, 15 mM MgCl$_2$), 2 µl of 0.1 M DTT and 1 µl of RNaseOUT to yield a total volume of 19 µl.
3. This is heated at 42 °C for 2 min followed by the addition of 1 µl of Superscript II reverse transcriptase (Invitrogen, 200 U/µl) and incubated for an additional 50 min.
4. The reaction is terminated by heating at 70 °C for 15 min.
5. 3 µl of the preceding reaction mixture is transferred to a Hot Start tube (Molecular BioProducts) and mixed with 2.5 µl of 10× PCR buffer (200 mM Tris-HCl, pH 8.4, 500 mM KCl), 3 µl of 25 mM of MgCl$_2$, and 3 µl of 10 mM dNTPs.
6. The wax bead of the Hot Start tube is melted by heating at 75 °C for 30 sec followed by cooling quickly on ice.

7. 2 µl of 5′-end-labeled gene specific primer, 0.5 µl of Taq DNA polymerase (Invitrogen), and 11 µl of DEPC-treated water are then added to the mixture in the final volume of 25 µl.
8. The mixture is heated at 94 °C for 2 min and PCR amplification is performed for 25 cycles at 94 °C for 1 min, 58 °C for 30 sec, and 72 °C for 1 min, followed by extension for 3 min at 72 °C.
9. The reaction mixtures are then extracted with phenol/chloroform/isoamyl alcohol, and the amplified products are recovered by ethanol precipitation.
10. The pellet is dissolved in 6 µl of loading buffer, heated at 95 °C for 4 min, and electrophoresed on a polyacrylamide/urea gel. After electrophoresis, gel is dried and visualized using a PhosphorImager.
11. Two approaches can be used to pinpoint the site of endonuclease cleavage. The easiest is to prepare a sequencing ladder with cDNA of the target mRNA and subtract from this the size (19 nt) of the ligated primer. If one wishes to obtain precise positional information or look for nontemplated nucleotides that might be added after cleavage, the gel can be visualized by film autoradiography after which the bands are excised as described previously for purifying radiolabeled probes, reamplified, cloned, and sequenced (Hanson and Schoenberg, 2004).

3. Analysis of PMR1-Containing Complexes

PMR1-mediated mRNA decay depends on its interaction with a number of different complexes, and we generally study these by sedimentation in sucrose and glycerol gradients. Polysome profile analysis with sucrose density gradients is the method of choice for studying the interaction of PMR1 with its translating substrate mRNA. As noted previously, decay depends on both the binding of PMR1 to polysomes and ongoing translation of its substrate mRNA. Tyrosine phosphorylation at position 650 by c-Src is required for the targeting of PMR1 to polysomes, so it is important to know the relative expression and activity of c-Src in cells used for analyzing PMR1-mediated mRNA decay. Dissociating polysomes by adding EDTA to extracts or adding puromycin to cells before lysis releases PMR1 in two main complexes; a ∼680 kDa complex (termed complex I) that contains PMR1 and its substrate mRNA, and a ∼140 kDa complex (complex II), which contains PMR1 and c-Src. As one might expect, complex II is the precursor to complex I and the site where tyrosine phosphorylation occurs. The following approaches describe methods used to study PMR1 in transfected mammalian cells and their application to studying PMR1-mediated mRNA decay.

3.1. Expression of PMR1 in transfected mammalian cells

Our early work used primary frog hepatocyte cultures to characterize the properties of PMR1. However, these cells are difficult to culture and transfect, so we investigated mammalian cell lines for their ability to recapitulate PMR1 targeting to polysomes and PMR1-specific endonuclease activity. As a general rule of thumb, the more differentiated epithelial cell lines such as MEL, MCF7, and Jurkat express the cross-reacting ortholog of frog PMR1, whereas other cell lines (e.g., Cos-1, U2OS, HeLa, CHO, 3T3) do not. Much of our work has been done with Cos-1 and U2OS osteosarcoma cells, because the sedimentation of PMR1 complexes in these cells closely matches that seen in frog liver and hepatocytes cultures. Most work studying the various PMR1-containing complexes uses a catalytically inactive form of PMR1, termed PMR60°. PMR60° lacks an endonuclease activity but still participates in all of the relevant complexes. As noted above, PMR60 also corresponds to the form that is processed from an 80 kDa precursor. All of our expression plasmids include an N-terminal myc epitope tag for ease of detection. In most experiments, 1.5 to 2×10^6 cells are plated on 100 mm dishes the afternoon before transfection. The next morning, cells are transfected with 6 μg of the plasmid expressing epitope-tagged PMR60 with commercial lipid-based transfection reagents (e.g., Fugene, Lipofectamine), and cells are harvested 40 h after transfection.

3.2. Preparation of postmitochondrial extracts

The protocol described in the following was developed for use with a 100 mm dish and should be adjusted for the amount of sample.

1. After aspirating the medium cells are washed twice with ice-cold PBS and harvested into 10 ml of PBS with a cell scraper. They are then collected by centrifugation at 500g for 5 min.
2. The cell pellet is gently resuspended in 0.5 ml of lysis buffer (10 mM HEPES-KOH, pH 7.5, 10 mM KCl, 10 mM MgCl$_2$, 50 mM NaF, 0.5% NP-40, 2 mM DTT, 0.5 mM phenylmethylsulfonyl fluoride (PMSF), 25 μl/ml protease inhibitor cocktail (Sigma-Aldrich), 10 μl/ml phosphatase inhibitor cocktails I and II (Sigma-Aldrich) 400 U/ml RNaseOUT), and the tube is placed on ice for 10 min. Note that the volume of lysis buffer can be adjusted as needed for subsequent steps.
3. The cells are then homogenized with 15 to 30 strokes with a chilled Douce homogenizer (a pestle). Note that this must be done gently, and care must be taken to avoid foaming, particularly during the upstroke. It is a good idea to look at a drop of extract under the microscope after every 5 strokes to monitor the extent of lysis.

4. The homogenate is centrifuged at 4 °C for 10 min at 15,000g to remove cell debris and nuclei, and the postmitochondrial supernatant is either used directly (e.g., for gradient analysis) or quickly frozen and stored at −80 °C. Note that if problems with polysome dissociation are encountered, 100 µg/ml of cycloheximide can be added to the PBS used for washing the cells and to the lysis buffer.

3.3. Sucrose density gradient analysis of PMR1-containing complexes

3.3.1. Linear sucrose density gradients

1. Place a gradient maker, peristaltic pump, and solutions into a cold room or cold box. Each linear 11 ml 10 to 40% sucrose gradient is prepared in a 14 × 89 mm ultracentrifuge tube (Beckman) with solutions of 10 and 40% sucrose in 10 mM HEPES-KOH, pH 7.5, 10 mM KCl, 10 mM MgCl$_2$, 0.5% NP-40, 0.5 mM PMSF, 25 µl/ml of protease inhibitor cocktail, 100 U/ml RNaseOUT, 2 mM DTT, 100 µg/ml cycloheximide. Sucrose gradients are quite stable and can be made the day before harvesting cells.
2. To ensure there is a sufficient amount of sample in each fraction, we use postmitochondrial extracts prepared from a 150 mm culture dish containing cells at the density described previously.
3. 0.5 ml is layered carefully on top of each gradient, and they are centrifuged at 225,000g for 2 to 3.5 h at 4 °C in a Sorvall TH641 or Beckman SW41 rotor with slow start and a slow stop. The length of centrifugation time depends on the desired distribution of polysomes across the gradient, with a broader distribution seen at shorter times.
4. 0.5 ml fractions are collected with a peristaltic pump to drive the solution through a UV monitor into the fraction collector. Alternately, if a monitor is not available, one can use a spectrophotometer or Nanodrop instrument to read the absorbance of each fraction at OD$_{260}$.
5. Protein samples from each fraction are concentrated by TCA precipitation. Each fraction is diluted with an equal volume of water to reduce the concentration of sucrose to <0.5 M; 100% TCA solution is added to the sample to a final volume of 10 to 20%, followed by incubation on ice for at least 1 h. The precipitated protein is recovered by centrifuging for 20 min at 12,000g in a refrigerated microcentrifuge, washed with 1 ml of ice-cold acetone, and recovered by centrifugation again. The pellet is dried by speed vacuum centrifugation for approximately 5 min and dissolved in SDS loading buffer.
6. The distribution of PMR1 is analyzed by Western blotting with antibody to the N-terminal myc tag. In addition, we commonly use

antibody to ribosomal protein S6 (Santa Cruz Biotechnology) to confirm the distribution of polysomes on the gradient. An example of this is shown in Fig. 13.2. In this particular experiment (from Yang *et al.*, 2004), PMR1-containing fractions were recovered by binding of TAP-tagged protein to IgG-Sepharose, and material recovered after Tobacco Etch Virus (TEV) protease cleavage was analyzed by Western blotting with antibody to the myc tag (upper panel) and S6 (bottom panel). The use of TAP-tagged PMR1 for characterizing complexes is described later. If one coexpresses a PMR1 target, such as albumin mRNA, its distribution can also be analyzed by Northern blotting, RPA, or semiquantitative RT-PCR of RNA isolated from gradient fractions.

7. As an additional control, polysomes can be dissociated by adjusting the postmitochondrial extract to 50 mM EDTA before gradient centrifugation or by adding 200 μg/ml of puromycin to cells 30 min before harvesting.

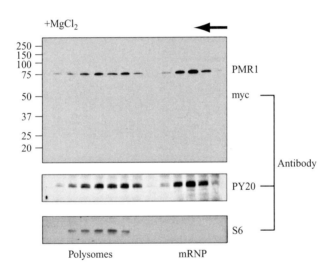

Figure 13.2 Sedimentation of PMR1-containing complexes in a linear sucrose density gradient. Postmitochondrial extract from Cos-1 cells expressing PMR1-TAP were separated in a linear 10 to 40% sucrose gradient. In the upper panels, odd-numbered fractions were recovered on IgG-Sepharose, eluted by TEV-protease cleavage, and analyzed by Western blotting with antibody to the myc tag on PMR1 (upper panel) and the phosphotyrosine monoclonal antibody PY20 (middle panel). In the bottom panel, even-numbered fractions were TCA precipitated and analyzed by Western blotting with antibody to ribosomal protein S6. Adapted from Yang *et al.* (2004).

3.3.2. Sucrose step gradients

Sucrose step gradients are a simpler method to determine the distribution of PMR1 between polysomes and smaller complexes, such as complex II. In this approach, postmitochondrial extract is centrifuged through a 2 ml gradient with a tabletop ultracentrifuge. The polysomes pellet to the bottom of the gradient and the smaller mRNP complexes accumulate at the interface between two sucrose pads.

1. 1 ml of 35% sucrose prepared as described previously is placed into the bottom of a 11 × 34 mm ultracentrifuge tube (Beckman) and subsequently overlaid with 1 ml of 10% sucrose, taking care not to disturb the interface. To facilitate sample recovery, the interface and the top of the gradient should be marked on the tube.
2. 0.2 ml of postmitochondrial extract isolated from cells in a 100 mm dish is layered carefully on top, and the tube is centrifuged for 1 h at 147,000g in a Sorvall Discovery M120 SE ultracentrifuge at 4 °C with an S55S-1123 rotor.
3. 200 µl are removed from the top of the gradient and from the interface and saved for analysis, and any remaining solution is removed carefully, taking care not to disturb the polysome pellet on the bottom of the tube. This is then dissolved in 200 µl of lysis buffer.
4. The distribution of PMR1 is analyzed by Western blotting with antibody that reacts with the myc tag. As described previously, the sedimentation of polysomes to the bottom can be confirmed by Western blotting with S6 antibody. An example of the usefulness of this quick and easy technique is shown in Fig. 13.3, where step gradients were used to show

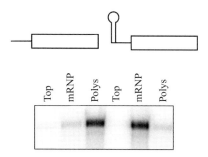

Figure 13.3 Separation of polysomes and mRNP complexes through discontinuous sucrose step gradients. Postmitochondrial extracts from Cos-1 cells expressing native albumin mRNA or albumin mRNA with a strong stem-loop in the 5′ UTR (SL-albumin) were applied to step gradients consisting of 1 ml each of 10 and 35% sucrose. The top, middle, and pellet fractions were collected after centrifugation and adjusted to equal volumes, and RNA recovered from each was analyzed by RPA (adapted from Yang and Schoenberg, 2004).

that native albumin mRNA sediments to the bottom of a step gradient with polysomes. In contrast, albumin mRNA that harbors a strong stem-loop structure, which that prevents translation initiation, sediments with mRNPs (Yang and Schoenberg, 2004).

3.4. Glycerol gradient analysis of PMR1 containing complexes

As noted earlier, the functional unit of PMR1-mediated mRNA decay is a ~680 kDa complex containing PMR1 and its translating substrate mRNA. This is released from polysomes by treating with EDTA or adding puromycin to cells before harvest, and the resulting complexes are analyzed by sedimentation on glycerol gradients, which are more amenable to studying these smaller complexes than sucrose density gradients.

1. An 11 ml linear density gradient of 10 to 40% (v/v) glycerol in 10 mM HEPES-KOH, pH 7.5, 10 mM KCl, 10 mM MgCl$_2$, 0.5 % NP-40, 0.5 mM PMSF, 25 μl/ml of protease inhibitor cocktail, 100 U/ml RNase-OUT, 2 mM DTT is prepared in a 14 × 89 mm ultracentrifuge tube as described previously.
2. Cells are treated with 200 μg/ml of puromycin for 30 min before harvesting or postmitochondrial extract is adjusted to 50 mM EDTA to dissociate polysomes; 0.5 ml of the postmitochondrial extract prepared in a single 150 mm dish of cells is layered carefully on top of each gradient and centrifuged for 20 h at 83,000g in a Sorvall TH641 rotor at 4 °C. A molecular size marker (GE Healthcare) containing thyroglobulin (MW, 669,000), ferritin (MW, 440,000), catalase (MW, 232,000), lactate dehydrogenase (MW, 140,000), and bovine serum albumin (MW, 67,000) is fractionated on a parallel gradient.
3. 0.5 ml of fractions are collected, and individual fractions from the gradient containing molecular size markers are assayed with the Bio-Rad (Bradford) protein assay to identify fractions containing each of the markers.
4. As with sucrose gradient fractions, these are diluted with an equal volume of buffer before TCA precipitation (see section 3.3.1.5).
5. The sedimentation of PMR1 is analyzed as described previously by Western blotting with antibody to the N-terminal myc tag. An example of the usefulness of glycerol gradients for analyzing PMR1-containing complexes is shown in Fig. 13.4 (modified from Yang et al., 2004). The upper panels show the gradient distribution of native PMR1 or PMR1 containing the Y650F mutation, which prevents phosphorylation by c-Src. Complexes I and II are seen in gradients performed with native PMR1 but not the mutated form, and RT-PCR of RNA recovered with the pooled fractions of each show that target (albumin) mRNA is only recovered with native PMR in complex I (bottom panels).

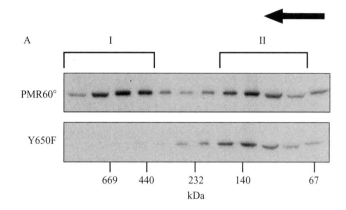

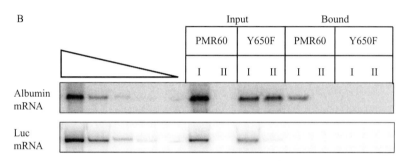

Figure 13.4 Glycerol gradient sedimentation of PMR1-containing complexes. Cos-1 cells were transfected with plasmids expressing albumin and luciferase mRNAs together with TAP-tagged catalytically inactive PMR1 or the Y650F form of the protein. Cells were treated with puromycin before harvesting to dissociate polysomes, and post-mitochondrial extracts were separated in 10 to 40% glycerol gradients. The collected fractions were analyzed by Western blotting with antibody to the myc-tag (upper panel), and the two major complexes are indicated (complex I: ~680 kDa, complex II: ~140 kDa). Fractions making up each of these complexes were pooled. In the bottom panels, the recovered RNA was analyzed by RT-PCR for albumin and luciferase mRNAs (adapted from Yang et al., 2004).

4. Affinity Recovery of PMR1-Containing Complexes

Numerous approaches are available for identifying and characterizing macromolecular complexes. For maximal flexibility, we have engineered plasmids expressing PMR1 with an N-terminal myc epitope tag and a C-terminal tandem affinity purification (TAP) (Puig et al., 2001), and used both tags to systematically identify PMR1-interacting proteins and target mRNAs. Both approaches can be used for recovering complexes from transiently transfected cells, but a more powerful approach is to generate stable cell lines expressing tagged PMR1 under a regulated promoter. The following describes approaches for recovering PMR1-containing complexes that take advantage of the gentle recovery afforded by eluting TAP-tagged proteins from IgG-Sepharose by cleaving within the tag with TEV protease, and approaches with immobilized myc monoclonal antibody to immunoprecipitate PMR1 and its associated proteins.

4.1. IgG-Sepharose selection of PMR1-TAP complexes

The protocol described in the following was created for use with transient transfected cells in a 100 mm dish or stable cell lines in a 150 mm dish and can be varied depending on the sample size.

4.1.1. Equilibration and blocking of IgG Sepharose beads

1. IgG Sepharose 6 Fast Flow (GE Healthcare) is washed twice with at least 5 bed volumes of water by end-over-end rotation at 4 °C.
2. The beads are treated for 5 min with 5 bed volumes of 0.5 M of acetic acid (pH 3.4) and washed four times with at least 5 bed volumes of TST buffer (50 mM Tris-HCl, pH 7.5, 150 mM NaCl, 0.05% Tween 20).
3. Step 2 is repeated once.
4. The beads are washed twice with at least 5 bed volumes of IPP150 buffer (10 mM Tris-HCl, pH 7.5, 150 mM NaCl, 0.1% NP-40) and the pH of the wash buffer is monitored to ensure complete removal of the acetic acid.
5. To reduce and/or prevent nonspecific binding, the beads are incubated for 2 h at 4 °C with 50 μg/ml of BSA or postmitochondrial extract from nontransfected cells.
6. Unbound protein is removed by washing the beads 5 times with at least 5 bed volumes of IPP150 buffer.

4.1.2. Sample application, washing, and elution with TEV protease

1. 0.5 ml of postmitochondrial extract prepared as described earlier is either used directly or treated with 50 mM EDTA to dissociate polysomes; 50 μl of extract is kept as an input fraction.
2. 50 μl of preblocked IgG-Sepharose beads is added to 450 μl of extract, and this is incubated for 2 h at 4 °C with end-over-end rotation.
3. The beads are then washed 4 times with 0.5 ml of IPP150 buffer and twice with TEV protease buffer (50 mM Tris-HCl, pH 8.0, 0.5 mM EDTA, 1 mM DTT).
4. Bound complexes are eluted by adding 100 μl of TEV protease buffer containing 5 μl of TEV protease (Invitrogen, 10 U/μl) and incubating at 4 °C for 2 to 5 h. Alternately, one can perform this step for 1 h at 25 °C,

but initially both approaches should be compared to ensure the increased temperature does not result in dissociation of PMR1-containing complexes.
5. Complexes recovered in this manner can be subsequently analyzed for PMR1-interacting proteins or RNAs.
6. The TAP recovery approach is flexible and can be used for several downstream applications. One example of this is seen in Fig. 13.2, where individual sucrose gradient fractions from cells expressing PMR1-TAP were recovered on IgG-Sepharose, released with TEV protease cleavage, and analyzed by Western blotting with antibodies to the myc tag on PMR1 and the anti-phosphotyrosine monoclonal antibody PY20. Another useful application of the TAP tag is for studying the relative strength of protein interactions with PMR1, where IgG-Sepharose–bound complexes can be washed with increasing salt to identify proteins that bind with high-affinity versus low-affinity interactions. Because TEV protease cleavage generates PMR1 with a C-terminal calmodulin binding–peptide complexes recovered from IgG-Sepharose can be processed through a second round of selection on calmodulin agarose, followed by elution with EDTA (Puig et al., 2001). This protein is suitable for analysis of posttranslational modifications by mass spectrometry. We have also taken advantage of this flexible recovery procedure to look at the interaction of PMR1 with its substrate mRNA. An example of this is seen in the glycerol gradient analysis of Fig. 13.4, where the fractions corresponding to complex I and complex II were pooled, separated on IgG-Sepharose, and RNA recovered after TEV protease cleavage was analyzed by RT-PCR. We are currently combining this with microarrays to analyze the scope of PMR1-mediated mRNA decay.

4.2. Immunoprecipitation of PMR1 with immobilized anti-myc antibody

1. 600 μl of postmitochondrial extracts from transfected cells or stable cell lines are prepared as described in 3.2.4.1.
2. 60 μl of extracts is kept as an input fraction and 540 μl is incubated overnight at 4 °C with 20 μl of myc monoclonal antibody–coupled beads (Santa Cruz Biotechnology: Cat. No. sc-40AC) with end-over-end rotation.
3. The beads are recovered by centrifuging at 500g at 4 °C for 1 min, and the supernatant is kept as the unbound fraction.
4. The beads are washed at least five times with 600 μl ice-cold IPP150, each time recovering the beads by centrifugation.
5. 50 μl of SDS loading buffer is added to the beads, followed by heating for 5 min at 90 °C.

6. PMR1 and its associated proteins are detected by Western blotting with either a rabbit polyclonal antibody or a mouse monoclonal antibody to the myc tag. Because PMR1 must be phosphorylated by c-Src to bind to polysomes and catalyze endonuclease cleavage, it is important to determine its tyrosine phosphorylation state. Most commonly, this is done by probing Western blots of protein recovered as described previously with the anti-phosphotyrosine monoclonal antibody PY20 (BD Biosciences). Also, because c-Src binds directly to PMR1, its recovery can be monitored with an antibody to this protein (Santa Cruz Biotechnology B-12 Cat. No. sc-8056).

5. Analysis of PMR1 Activity *In Vivo* and *In Vitro*

5.1. *In vivo* analysis of PMR1 activity

The *in vivo* activity of PMR1 is monitored with plasmid vectors expressing a known substrate mRNA together with a known control. At present, the best-characterized substrates are *Xenopus* serum protein mRNAs (Pastori et al., 1991a) of which we commonly use an albumin minigene. To date, the best-characterized mammalian target mRNA for PMR1 is human β-globin (Bremer et al., 2003), and this is degraded by PMR1 in both erythroid and nonerythroid cells. Firefly luciferase mRNA is not degraded by PMR1, and it is commonly used as a cotransfected control for enzymatic activity.

RNase protection assay (RPA) and semiquantitative RT-PCR with radiolabeled primers are the assays used most commonly to monitor PMR1-induced changes in target mRNA. We most frequently use RPA, because it is quite sensitive and enables us to examine substrate and control mRNAs in the same sample. As noted previously, PMR1 must be phosphorylated by c-Src to form a complex with its translating substrate mRNA on polysomes, so the amount and activity of c-Src must be determined for any cell line used for analyzing PMR1-mediated mRNA decay.

5.1.1. Protocol

1. 1.5 to 2×10^6 cells in a 100 mm dish are transfected with 4 μg of a plasmid expressing the myc-tagged catalytically active form of PMR1 or, as a control, myc-GFP plus 1 μg each of a plasmid expressing full-length albumin, mRNA, or luciferase mRNA. Cells are harvested 24-to-40 h after transfection, and total-cell RNA and protein are isolated with TRIzol reagent (Invitrogen).
2. Antisense probes are prepared from cDNA plasmids for each of the target mRNAs. Each probe should have a specific activity of 3×10^8 cpm/μg.

3. 2 to 8 × 10⁴ cpm or 150 to 600 pg of a labeled target RNA probe and a control RNA probe are precipitated together with 5 to 50 μg of total or cytoplasmic RNA or yeast RNA as a control by ethanol; the pellet is washed with 70% ethanol and dried for 5 min on the bench. The amounts of albumin and luciferase mRNAs are then determined by RPA with the Ambion RPA III kit. Products separated in a denaturing polyacrylamide/urea gel are visualized by phosphorimaging to quantify the relative differences in albumin/luciferase mRNA. Furthermore, Western blotting should be performed with antibody to the myc tag on PMR1 or GFP to monitor their expression in each sample and with antibody to luciferase as a control for relative transfection efficiencies.

5.2. *In vitro* analysis of PMR1 activity

Although the ability to recapitulate PMR1-mediated mRNA decay in cultured cells allows for the analysis of functional domains, one must also keep in mind that this is an enzyme, and changes to the protein can have consequences for enzymatic activity. This becomes particularly important when studying mutations that alter the binding of PMR1 to polysomes, because these can modulate decay either by increasing or decreasing polysome targeting or by interfering with the folding of the active site. To study this, we take advantage of the gentle recovery afforded by the TAP tag approach to measure target mRNA degradation *in vitro* by protein recovered from transfected cells.

1. PMR1-TAP is recovered from postmitochondrial extracts or complexes recovered by gradient sedimentation by binding onto IgG-Sepharose (see section 4.1). As a specificity control, a parallel recovery is done from cells expressing myc-tagged GFP, also containing a TAP tag. Western blotting with antibody to the myc tag is used to assess protein recovery after TEV-protease cleavage.
2. A uniformly labeled transcript is prepared that contains the mapped PMR1 cleavage sites within the 5′-end of albumin mRNA (Chernokalskaya *et al.*, 1997) or another known target mRNA (e.g., human β-globin). It is best to use substrates with mapped endonuclease cleavage sites so that one can monitor both loss of input RNA and production of cleavage products. An alternate approach is to prepare a 5′-end-labeled transcript by treating with tobacco acid pyrophosphatase followed by shrimp alkaline phosphatase, and with T4 polynucleotide kinase to label the dephosphorylated RNA with γ-[^{32}P]-ATP. All radiolabeled RNAs should be gel purified before use in activity assays. Note that luciferase mRNA is not cleaved well *in vitro* by PMR1 and can be used as a specificity control for PMR1 activity.

5.2.1. Protocol

1. 2 to 10 × 10^4 cpm of a labeled RNA transcript is mixed with 5 to 10 μl of myc-tagged PMR1 or GFP that has been recovered by TEV-protease cleavage of protein bound onto IgG-Sepharose in 200 μl buffer containing 10 mM Tris-HCl, pH 7.5, 50 mM KCl, 2 mM MgCl$_2$, and 2 mM DTT.
2. The mixture is incubated at 25 °C for 0, 5, 10, and 30 min.
3. 20 μl of formamide gel loading buffer is added, and the reaction mixture is heated at 95 °C for 5 min to stop the reaction.
4. 5 to 10 μl of this is loaded onto a denaturing polyacrylamide/urea gel, and the dried gel is visualized using a PhosphorImager.
5. Generally, it is sufficient to use markers of the appropriate size to determine sites of endonuclease cleavage *in vitro*.

6. SUMMARY

The protocols described here provide a comprehensive approach to characterizing PMR1-mediated mRNA decay *in vivo* and *in vitro*. Although these were developed for studying this particular mRNA endonuclease, many of these can be used generally to study other endonuclease-catalyzed decay processes.

ACKNOWLEDGMENTS

Work in the Schoenberg laboratory on PMR1 is supported by PHS grant R01 GM38277. Y. O. is supported by a postdoctoral fellowship from Great Rivers Affiliate of the American Heart Association.

REFERENCES

Bremer, K. A., Stevens, A., and Schoenberg, D. R. (2003). An endonuclease activity similar to *Xenopus* PMR1 catalyzes the degradation of normal and nonsense-containing human β-globin mRNA in erythroid cells. *RNA* **9,** 1157–1167.

Brookes, A. J., and Solomon, E. (1989). Evaluation of the use of S1 nuclease to detect small length variations in genomic DNA. *Eur. J. Biochem.* **183,** 291–296.

Chernokalskaya, E., Dompenciel, R. E., and Schoenberg, D. R. (1997). Cleavage properties of a polysomal ribonuclease involved in the estrogen-regulated destabilization of albumin mRNA. *Nucleic Acids Res.* **25,** 735–742.

Chernokalskaya, E., Dubell, A. N., Cunningham, K. S., Hanson, M. N., Dompenciel, R. E., and Schoenberg, D. R. (1998). A polysomal ribonuclease involved in the destabilization of albumin mRNA is a novel member of the peroxidase gene family. *RNA* **4,** 1537–1548.

Gallouzi, I. E., Parker, F., Chebli, K., Maurier, F., Labourier, E., Barlat, I., Capony, J. P., Tocque, B., and Tazi, J. (1998). A novel phosphorylation-dependent RNase activity of GAP-SH3 binding protein: A potential link between signal transduction and RNA stability. *Mol. Cell. Biol.* **18,** 3956–3965.

Garneau, N. L., Wilusz, J., and Wilusz, C. J. (2007). The highways and byways of mRNA decay. *Nat. Rev. Mol. Cell. Biol.* **8,** 113–126.

Gatfield, D., and Izaurralde, E. (2004). Nonsense-mediated messenger RNA decay is initiated by endonucleolytic cleavage in *Drosophila*. *Nature* **429,** 575–578.

Hanson, M. N., and Schoenberg, D. R. (2001). Identification of *in vivo* mRNA decay intermediates corresponding to sites of in vitro cleavage by polysomal ribonuclease 1. *J. Biol. Chem.* **276,** 12331–12337.

Hanson, M. N., and Schoenberg, D. R. (2004). Application of ligation-mediated reverse transcription polymerase chain reaction to the identification of *in vivo* endonuclease-generated messenger RNA decay intermediates. *Methods Mol. Biol.* **257,** 213–222.

Hollien, J., and Weissman, J. S. (2006). Decay of endoplasmic reticulum-localized mRNAs during the unfolded protein response. *Science* **313,** 104–107.

Howard, J. T., Ward, J., Watson, J. N., and Roux, K. H. (1999). Heteroduplex cleavage analysis using S1 nuclease. *Biotechniques* **27,** 18–19.

Lee, H. H., Kim, Y. S., Kim, K. H., Heo, I., Kim, S. K., Kim, O., Kim, H. K., Yoon, J. Y., Kim, H. S., Kim Do, J., Lee, S. J., Yoon, H. J., *et al.* (2007). Structural and functional insights into Dom34, a key component of no-go mRNA decay. *Mol. Cell* **27,** 938–950.

Pastori, R. L., Moskaitis, J. E., Buzek, S. W., and Schoenberg, D. R. (1991a). Coordinate estrogen-regulated instability of serum protein coding messenger RNAs in *Xenopus laevis*. *Mol. Endocrinol.* **5,** 461–468.

Pastori, R. L., Moskaitis, J. E., and Schoenberg, D. R. (1991b). Estrogen-induced ribonuclease activity in *Xenopus* liver. *Biochemistry* **30,** 10490–10498.

Peng, Y., and Schoenberg, D. R. (2007). c-Src activates endonuclease-mediated mRNA decay. *Mol. Cell* **25,** 779–787.

Peng, Y., Liu, X., and Schoenberg, D. R. (2008). Hsp90 stabilizes the PMR1 mRNA endonuclease to degradation by the 26S proteasome. *Mol. Biol. Cell* **19,** 546-552.

Puig, O., Caspary, F., Rigaut, G., Rutz, B., Bouveret, E., Bragado-Nilsson, E., Wilm, M., and Seraphin, B. (2001). The tandem affinity purification (TAP) method: a general procedure of protein complex purification. *Methods: A Companion to Methods in Enzymology* **24,** 218–229.

Stevens, A., Wang, Y., Bremer, K., Zhang, J., Hoepfner, R., Antoniou, M., Schoenberg, D. R., and Maquat, L. E. (2002). Beta-globin mRNA decay in erythroid cells: UG site-preferred endonucleolytic cleavage that is augmented by a premature termination codon. *Proc. Natl. Acad. Sci. USA* **99,** 12741–12746.

Vogt, V. M. (1973). Purification and further properties of single-strand-specific nuclease from *Aspergillus oryzae*. *Eur. J. Biochem.* **33,** 192–200.

Yang, F., Peng, Y., Murray, E. L., Otsuka, Y., Kedersha, N., and Schoenberg, D. R. (2006). Polysome-bound endonuclease PMR1 Is targeted to stress granules via stress-specific binding to TIA-1. *Mol. Cell Biol.* **26,** 8803–8813.

Yang, F., and Schoenberg, D. R. (2004). Endonuclease-mediated mRNA decay involves the selective targeting of PMR1 to polyribosome-bound substrate mRNA. *Mol. Cell* **14,** 435–445.

Yang, F., Peng, Y., and Schoenberg, D. R. (2004). Endonuclease-mediated mRNA decay requires tyrosine phosphorylation of polysomal ribonuclease 1 (PMR1) for the targeting and degradation of polyribosome-bound substrate mRNA. *J. Biol. Chem.* **279,** 48993–49002.

SECTION FOUR

MEASURING mRNA HALF-LIFE *IN VIVO*

CHAPTER FOURTEEN

Methods to Determine mRNA Half-Life in *Saccharomyces cerevisiae*

Jeff Coller

Contents

1. Introduction	268
2. The Use of Inducible Promoters	269
2.1. The GAL1 UAS	269
2.2. A primer on galactose metabolism	270
2.3. Transcriptional shut-off	270
2.4. Transcriptional pulse-chase	272
2.5. Differences between transcriptional shut-off and transcriptional pulse-chase	275
2.6. A GAL promoter bonus	275
2.7. The TET-off system	275
3. Measuring mRNA Decay by Use of Thermally Labile Alleles of RNA Polymerase II	276
4. Measuring mRNA Decay with Thiolutin	277
5. RNA Extractions	277
5.1. Recipe for LET	278
5.2. RNase H cleavages of 3′ UTRs	278
5.3. 10× Hybridization mix	279
5.4. 2× RNase H buffer	279
5.5. Stop mix	280
6. Northern Blot Analysis	280
7. Loading Controls	280
8. Determination of mRNA Half-Lives	280
9. Concluding Remarks	282
Acknowledgments	282
References	282

Center for RNA Molecular Biology, Case Western Reserve University, School of Medicine, Cleveland Ohio, USA

Abstract

Much of our understanding of eukaryotic mRNA decay has come from studies in budding yeast, *Saccharomyces cerevisiae*. The facile nature of genetic and biochemical manipulations in yeast has allowed detailed investigations into the mRNA decay pathway and the identification of critical factors required for each step. Discussed herein are standard protocols for measuring mRNA half-lives in yeast. It should be noted, however, that variations of these assays are possible. In addition, a few *a priori* considerations are addressed.

1. Introduction

The degradation of mRNA is a vital aspect of gene regulation. Shutting off mRNA expression ensures that previously transcribed mRNAs do not translate *ad infinitum*. Moreover, mRNA degradation is used by the cell as a site of regulatory responses (reviewed in Coller and Parker, 2004). Last, specialized mRNA turnover systems exist that recognize and degrade aberrant mRNAs, thereby increasing the quality control of mRNA biogenesis (reviewed in Baker and Parker, 2004). Given these functions, it is important to understand how decay rates of different mRNAs are controlled and how aberrant mRNAs are targeted for destruction. Much can be learned, therefore, by determining the mechanism of how a specific transcript is degraded under various biologic conditions or in distinct genetic backgrounds.

Polyadenylated mRNAs can be degraded in eukaryotic cells by two general pathways (Fig. 14.1). In both cases, the degradation of the transcript begins with the shortening of the poly(A) tail at the 3′-end of the mRNA (reviewed in Coller and Parker, 2004). In yeast, shortening of the poly(A) tail primarily leads to removal of the 5′-cap structure (decapping), thereby exposing the transcript to digestion by a 5′ to 3′-exonuclease (Fig. 14.1A; Decker and Parker, 1993; Hsu and Stevens 1993; Muhlrad *et al.*, 1994).

mRNAs can also be degraded in a 3′ to 5′-direction after deadenylation (Fig. 14.1B) (Muhlrad *et al.*, 1995). 3′ to 5′-degradation of mRNAs is catalyzed by the exosome (Anderson and Parker 1998; Mukherjee *et al.*, 2002; Rodger *et al.*, 2002), which is a large complex of 3′ to 5′-exonucleases functioning in several RNA degradative and processing events (reviewed in van Hoof and Parker 2002). For the yeast mRNAs that have been studied, the process of 3′ to 5′-decay is slower than decapping and 5′ to 3′-decay (Cao and Parker, 2001). However, it is likely that for some yeast mRNAs, or in other eukaryotic cells, 3′ to 5′-degradation will be the primary mechanism of mRNA degradation after shortening of the poly(A) tail (e.g., Higgs and Colbert 1994).

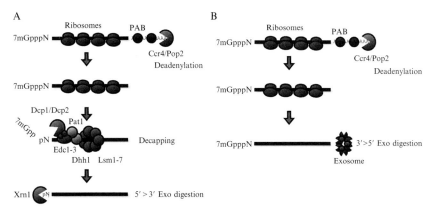

Figure 14.1 Pathways of mRNA decay in yeast. (A) Shows the major mRNA pathway that is initiated by deadenylation, followed by removal of the 5′-cap (decapping), and lastly exonucleolytic digestion of the mRNA body in a 5′ > 3′-direction. (B) Degradation of mRNA can also occur by a minor pathway which is initated by deadenylation, and then destruction of the mRNA body in a 3′ > 5′-direction by the exosome.

In this review, I do not focus on the mechanisms of mRNA decay, but rather, how half-life measurements are made within the cell. The techniques I describe have proven useful in determining the steps of message decay. It should be noted, however, that it has been necessary to couple these assays with genetic manipulations and other techniques to powerfully demonstrate how a particular transcript is degraded (Mulhrad et al., 1995). For the assays described here, the overarching theme is the same: RNA transcription is ceased in some manner, and the decay rates of specific transcripts are monitored. The method for stopping transcription depends on the questions asked and the information desired. I refer you to chapter 20 by Passos and Parker in this issue for details on how to determine the polarity of decay and questions related to the enzymatic activities of the decay machinery.

2. THE USE OF INDUCIBLE PROMOTERS

The use of reporter mRNAs that are expressed under the control of inducible promoters is the most common technique for measuring mRNA decay in yeast. The preferred choice is the galactose promoter; however, others systems have been recently exploited, especially the TET$_{off}$ repressible promoter.

2.1. The GAL1 UAS

The *GAL1* *u*pstream *a*ctivating *s*equence (UAS) is also known as the GAL promoter or galactose inducible promoter. The GAL promoter mediates the expression of galactose metabolism genes, and it can confer galactose inducibility to heterologous genes (reviewed in Lohr et al., 1995). Simply put, the GAL promoter induces mRNA expression in the presence of the sugar galactose and rapidly shuts off transcription in the presence of glucose. This provides a powerful method for transcriptional pulse-chase and shut-off experiments in which mRNA decay kinetics is analyzed.

2.2. A primer on galactose metabolism

Before designing an RNA decay analysis with the GAL promoter, it is important to consider how yeasts use galactose. The GAL promoter serves as a binding site for the transcriptional regulator, Gal4p (reviewed in Lohr et al., 1995). Gal4p has bipartite activities: a DNA binding domain, and a transcriptional activation domain. Gal4p binds strongly to the GAL UAS in the presence of galactose, inducing transcription through its activation domain. The addition of glucose causes a second gene, Gal80p, to bind to Gal4p and mask its activation domain, thereby immediately inhibiting further RNA synthesis.

The choice of yeast strain is important when the *GAL1* UAS is used to control gene expression. For example, the Gal4p protein is exploited for the yeast two-hybrid system (reviewed in Traven et al., 2006). For this reason, yeast strains used in two-hybrid assays lack both Gal4p and Gal80p (*gal4Δ* and *gal80Δ*). It is not possible, therefore, to measure mRNA decay with the GAL promoter in a two-hybrid strain, because it lacks galactose-dependent regulation. It is advisable to test laboratory strains for galactose-dependent regulation before the GAL promoter is used to measure mRNA decay. This can be accomplished by Northern blot analysis for the *GAL1*, *GAL7*, or *GAL10* genes (see following). The yeast strains that we commonly use are BY4741 and BY4742; the WT backgrounds used in the *Saccharomyces* Genome Deletion Project (Winzeler et al., 1999). Other strains are also commonly used, including S288C and W303 (He and Jacobson 1995; Muhlrad et al., 1994).

In theory, mRNA decay of any gene can be measured with the GAL promoter. This requires that the gene of interest be engineered to be transcribed under GAL control, either by integrating the GAL promoter into the chromosomal gene (Longtine et al., 1998) or placing the gene on a plasmid that contains the GAL promoter (Coller and Parker, 2005). Once established, the experiment can be performed in two distinct ways: either by transcriptional shut-off or by a transcriptional pulse-chase.

2.3. Transcriptional shut-off

Transcriptional shut-off by use of the GAL promoter is perhaps the simplest and most straightforward approach to determine decay kinetics of an mRNA. Cells are grown in the presence of galactose until reaching mid-log phase.

Because galactose is always present, the mRNA reporter is constitutively expressed and at steady state when analysis is performed, which has certain ramifications on the information that is gathered (Section 2.5). Shut-off is achieved by concentrating the cells and resuspending in medium that contains glucose. Time points are then quickly taken, RNA extracted, and analysis by Northern blot ensues. The following is a detailed protocol for this analysis:

- An overnight culture is grown in 20 ml of SGS medium (Section 2.3.1).
- The next day, 200 ml of SGS medium is inoculated overnight to an $OD_{600} = 0.05$ U/1 ml (this number is variable depending on spectrophotometer but reflects approximately 1×10^6 cells/mL). Grow the cells in a 1-L flask to ensure appropriate aeration. The temperature for growth depends on the experiment. WT strains grow best at 30 °C; mutant strains may only grow at 24 °C.
- Cells are harvested when reaching an $OD_{600} = 0.400$ U/1 ml (3×10^7 cells/mL), by centrifugation in appropriately sized conical tubes.
- Working quickly, the galactose medium is poured off, and the cell pellet is resuspended in 20 ml of S medium (Section 2.3.2). Place the culture in a small 50-ml flask in a shaking waterbath.
- 1 ml of 40% glucose (2% final concentration) is added, and time points immediately taken.
- Use the following time course: 0, 2, 4, 6, 8, 10, 15, 20, 25, 30, 40, 50, and 60 min.
- Harvest each time point quickly. We use a 5 ml pipet man to remove 2 ml of culture. Aliquots are placed in 2-ml tubes. The aliquot is quickly spun down in a small desktop picofuge, and the medium is removed by aspiration. The cell pellet is then quickly frozen on dry ice or liquid nitrogen. Ensure that the culture continues to shake in the waterbath between each time point.
- Once time course is complete, cell pellets can be stored at −80 °C indefinitely.
- Extract mRNA and analyze reporter by Northern blot analysis.

2.3.1. SGS Medium (1 L)

20 g D-galactose
10 g Sucrose
1.7 g Yeast nitrogen base without amino acids or ammonium acetate
5.0 g Ammonium acetate
2.0 g Amino acid dropout mix (variable dependent on selective requirements)[1]
pH to 6.5 with 10 N NaOH

---

[1] See Burke et al. (2000) for recipe.

2.3.2. S-Medium (1 L)

1.7 g Nitrogen base without amino acids or ammonium acetate
5.0 g Ammonium acetate
2.0 g Amino acid drop-out mix (variable dependent on selective requirements)[1]
pH to 6.5 with 10N NaOH

2.4. Transcriptional pulse-chase

Measuring mRNA decay with transcriptional pulse-chase is a bit more challenging, but the information obtained is exquisite. The use of a transcriptional pulse-chase allows for the analysis of the decay of a synchronized population of mRNA. This gives a wealth of information, because product-precursor relationships are visible (Fig. 14.2B). A synchronous pool of transcripts is made by keeping reporter expression off until just before analysis. The GAL promoter is repressed by growing cells in medium containing the nonfermentable carbon source raffinose. Once the appropriate cell density is reached, the mRNA is induced by adding galactose for a brief time, and then transcription is shut off by adding glucose. This procedure generates a burst of newly synthesized transcripts whose decay can be followed.

The following is a detailed protocol for transcriptional shut-offs:

- An overnight culture is grown in 20 ml of SR medium (Section 2.4.1).
- The next day, 200 ml of SR medium is inoculated overnight to an OD_{600} = 0.05 U/1 ml (1×10^6 cells/mL). Grow the cells in a 1-L flask to ensure appropriate aeration. The temperature for growth depends on the experiment.
- Cells are harvested when reaching an OD_{600} = 0.300 U/1 ml (2.5×10^7 cells/mL) by centrifugation in appropriately sized conicals. For this analysis, it is vital that the cell not be allowed to grow above an OD_{600} = 0.300 U/1 ml. Growth above this optical density can cause spontaneous induction of the GAL promoter (Section 2.4.1).
- Pour off medium and resuspend cell pellet in 20 ml of S medium (Section 2.3.2). Place the culture in a small 50-ml flask in a shaking waterbath.
- Harvest a 2-ml aliquot to use as a preinduction control. Treat this sample like a normal timepoint (i.e., remove medium by centrifugation and place tube on dry ice or in liquid nitrogen).
- Add 2 ml of 20% galactose (2% final concentration) and incubate culture for 8 min in shaking waterbath.
- Separate medium from cells by centrifugation (working quickly, a 1-min. spin at 4000 rpm is sufficient).

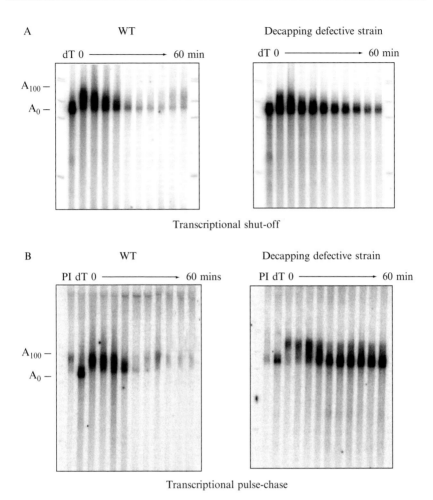

Figure 14.2 Comparison of a transcriptional shut-off (A). vs. a transcriptional pulse-chase (B). The specific reporter used is the MFA2 gene under control of the GAL1 UAS. dT represents a sample of the RNA treated with oligo dT and RNaseH in order to indicate the size of fully deadenylated RNA. PI indicates a pre-induction control. The left panels represent analysis in WT cells, while the right panel represents an experiment performed in a decapping defective strain (i.e. dcp2 mutant).

- Resuspend pellet in 20 ml of S medium, place in a 50-ml flask, and place in shaking waterbath.
- Add 2 ml of 40% glucose (2% final concentration) and take time points immediately.
- Use the following time course: 0, 2, 4, 6, 8, 10, 15, 20, 25, 30, 40, 50, and 60 min.
- Harvest each time point quickly. We use a 5-ml pipet man to remove 2 ml of culture. Aliquots are placed in 2-ml tubes. The aliquot is quickly

spun down in a small desktop picofuge, and the medium is removed by aspiration. The cell pellet is then quickly frozen on dry ice or liquid nitrogen. Ensure that the culture continues to shake in the waterbath between each time point.
- Once time course is complete, cell pellets can be stored at $-80\,°C$ almost indefinitely.
- Extract mRNA and analyze reporter mRNA by Northern blot analysis (see following).

2.4.1. SR Medium (1 L)

20 g Raffinose
10 g Sucrose
1.7 g Yeast nitrogen base without amino acids or ammonium acetate
5.0 g Ammonium acetate
2.0 g Amino acid dropout mix (variable dependent on selective requirements)[1]
pH to 6.5 with 10 N NaOH

The biggest problem in obtaining a good transcriptional pulse-chase is preventing preinduction of the reporter. Raffinose is a trisaccharide composed of galactose, fructose, and glucose. Raffinose is spontaneously hydrolyzed at low pH to generate galactose and sucrose. It is important, therefore, to ensure the pH of the culture does not become acidic, because this would generate galactose and induce the GAL promoter. We minimize the likelihood of spontaneous raffinose hydrolysis by adjusting our medium to pH = 6.5 before use. Notably, yeast synthetic medium usually has a pH of 5.0 or lower. Second, we never allow the cell density to get above an OD_{600} of greater than 0.3 U/ml (2.5×10^7 cells/mL). Yeast cells secrete the enzymes *invertase* and *α-galactosidase*. *Invertase* will hydrolyze raffinose into melibiose and fructose. Melibiose is hydrolyzed by *α-galactosidase* into galactose and glucose, and this induces Gal4p-mediated gene expression (Johnston and Hooper, 1982). The higher the cell density, the higher the secreted *invertase* and *α-galactosidase* levels within the culture, and, therefore, the higher the probability of spontaneous induction of the GAL promoter. This has several implications for a pulse-chase experiment. First, cell density must be kept low enough to avoid significant conversion of raffinose but high enough to be able to obtain significant amounts of RNA from samples. Second, extracellular *invertase* and *α-galactosidase* accumulates in medium. Therefore, if cultures overgrow when conducting a pulse-chase, they cannot be simply diluted back to a lower OD. If diluting cultures back, it is advisable to pellet cells by centrifugation, wash with fresh medium, and then resuspend in new medium to an appropriate OD. Last, the choice of yeast strain can also have an impact on *invertase* levels in the medium. The *S. cerevisiae* genome contains six unlinked loci that encode *invertase*, *SUC1-5*, and *SUC7* (Carlson and Botstein 1982). Yeast strains can carry any number or combination of *invertase* genes. For example, S288C strains only contain the *SUC2* gene (Carlson and Botsein, 1982). It is theoretically possible to avoid preinduction by use of sucrose as the carbon source rather than raffinose (Ambro Van Hoof 2008), although this has not been tested rigorously.

2.5. Differences between transcriptional shut-off and transcriptional pulse-chase

Both a transcriptional shut-off and a transcriptional pulse-chase will give important information about the half-life of an mRNA. The difference between these two analyses is that a shut-off is a decay from steady state, whereas a pulse-chase monitors the decay of a synchronized population (Fig. 14.2). There are times when one analysis is preferred over the other. Shut-off experiments are much easier than pulse-chase experiments, because cells grow better in galactose than in raffinose, and preinduction is not a concern. A transcriptional shut-off, however, gives only an approximation of how mRNA decay is impaired. In a transcriptional pulse-chase, the progression of a transcript through the steps of mRNA decay, (i.e., deadenylation, decapping, and exonucleolytic digestion) can be observed. This is especially important when trying to determine what step in decay is affected by an experimental condition (Fig. 14.2B).

2.6. A GAL promoter bonus

One particular advantage of choosing to use the GAL promoter for mRNA decay analysis is that, besides the reporter GAL UAS fusion, it is also possible to monitor the decay of three endogenous mRNAs: the *GAL1*, *GAL7*, and *GAL10* transcripts. All three mRNAs are naturally controlled by galactose and thus will be regulated transcriptionally like the reporter. These become powerful controls especially when determining transcript specific effects.

2.7. The TET-off system

Although not as common, mRNA decay has been successfully monitored by shutting off transcription with a tetracycline-dependent repressible promoter. Developed by Gari *et al*. (1997), the gene of interest is fused to the tetracycline operator (tetO). The plasmid must also express a second gene called the tTA transactivator. The tTA transactivator consists of the VP16 activator domain of herpes simplex virus fused to the tetracycline-inducible repressor (tetR) from the Tn10-encoded tetracycline-resistance operon (Gari *et al*., 1997). Under these conditions, genes under tetO control are expressed when tetracycline is absent, but transcription is quickly shut-off when the antibiotic is added to media. Commonly, doxycycline, which is a derivative of tetracycline, is used. This system is useful when trying to control expression of a single mRNA without altering carbons sources (Hilleren and Parker, 2003). However, this approach is limited, because only decay from the steady state can be performed.

3. MEASURING mRNA DECAY BY USE OF THERMALLY LABILE ALLELES OF RNA POLYMERASE II

Another useful method for analyzing mRNA decay takes advantage of the temperature-sensitive allele *rpb1-1* (Nonet *et al.*, 1987). The *rpb1-1* allele maps to the *RPO21* gene, which encodes the largest subunit of RNA polymerase II. At the permissive temperature, *rpb1-1* functions normally. Shifting to the restrictive temperature leads to a quick and synchronous disruption of all pol II transcription. The advantage of this approach is that the analysis of endogenous mRNAs can be measured. No prior cloning or alteration of promoter elements is required. In addition, decay analysis with *rpb1-1* is not dependent on nutrient conditions, allowing the effects of stress or starvation to be monitored (Hilgers *et al.*, 2006). Last, because *rpb1-1* stops all pol II transcription, decay of any mRNAs can be monitored. The disadvantage of *rpb1-1* is that the allele must be incorporated into different strains if distinct genetic backgrounds are desired. In addition, *rbp1-1* shut-off experiments analyze decay from the steady state. Therefore, product-precursor relationships cannot be determined as in transcriptional pulse-chase experiments. Nonetheless, it has proven advantageous to use *rpb1-1* to measure mRNA decay in some experimental settings. The following is a detailed protocol.

- An overnight culture is grown at 24 °C in 20 ml of medium (medium choice depends on need).
- The next day, 200 ml of medium is inoculated with the overnight, and cells are grown to an $OD_{600} = 0.05$ U/1 ml (1×10^6 cells/mL). In a 1-L flask to ensure appropriate aeration. Cultures must be grown at 24 °C.
- Ten milliliters of medium is preheated to 56 °C at least 1 h before performing the shut-off and, a 50-ml flask is preheated to 37 °C in a shaking waterbath.
- Cells are harvested when reaching an $OD_{600} = 0.400$ U/1 ml (3×10^7 cells/mL), by centrifugation in appropriately sized conicals. Resuspend cell pellet in 10 ml of room temperature medium.
- Working quickly, the preheated medium is poured into the freshly harvested 10-ml culture. The combination of 10 ml of medium at 56 °C and 10 ml of culture at 24 °C gives a rapid shift to 37 °C.
- Place the culture in the preheated 50-ml flask in a shaking waterbath set at 37 °C.
- Immediately take time points at 0, 2, 4, 6, 8, 10, 15, 20, 25, 30, 40, 50, and 60 min.

- Harvest each time point quickly as previously described.
- Cell pellets can be stored at −80 °C indefinitely.

Extract mRNA and analyze reporter by Northern blot analysis.

4. Measuring mRNA Decay with Thiolutin

Chemical means can also be used to measure mRNA decay in yeast (Herrick et al., 1990). Thiolutin is a sulfur-containing antibiotic from *Streptomyces laterosporus* that inhibits RNA synthesis directed by all three yeast RNA polymerase. Adding thiolutin at low concentrations is sufficient to completely block RNA synthesis, thereby allowing mRNA decay analysis. Like the use of the *rpb1-1* allele, this approach allows the analysis of endogenous mRNAs. Moreover, the effects of stress or nutrient starvation on mRNA decay can be monitored, because carbon source choice is irrelevant. Last, the decay of *any* RNA can be monitored because thiolutin blocks the activity of all three RNA polymerases. Unlike an *rpb1-1* mutant, however, a thiolutin shut-off can be conducted on any preexisting strain; genetic engineering is not required. The disadvantage of thiolutin, like *rpb1-1*, is that shut-off experiments limit the analysis to decay from the steady state. Last, new evidence suggests that some mRNAs may be stabilized by the response of cells to the drug (Pelechano and Pérez-Ortín, 2007). Nonetheless, for many mRNAs, thiolutin has proven useful in monitoring decay kinetics. The protocol is analogous to the protocols we have previously described for transcriptional shut-off experiments except that 3 μg/ml (final concentration) of thiolutin is added before taking time points.

5. RNA Extractions

Irrespective of how mRNA decay analysis is performed, once samples are collected, it is necessary to extract the mRNA and analyze decay rates. The following is our protocol for extracting total-cell mRNA from yeast. We analyze *total* RNA by Northern blotting. Because many mutations and conditions can affect the decapping step exclusively, it is important to not poly(A)-select mRNA by use of oligo(dT) columns before analysis. Alterations in decapping stabilize poly(A)-minus mRNAs. Therefore oligo(dT) selection would result in the loss of information.

- Resuspend frozen yeast cell pellets (in 2-ml tube) in 150 μl LET (Section 5.1).
- Add 150 μl phenol equilibrated with LET.

- Add equal volume (approximately 300 to 400 μl) of glass beads (bead size = 500 μm; available from Sigma-Aldrich; cat #G8772).
- Vortex in MultiMixer (VWR cat# 58816-115) for 5 min at top speed.
- Add 250 μl DEPC H_2O and 250 μl 1:1 phenol/chloroform equilibrated with LET.
- Vortex in MultiMixer for an additional 5 min followed by centrifugation for 5 min at 14,000 rpm.
- Transfer aqueous phase (approximately 450 μl, top layer) to new 1.5-ml tube.
- Add 450 μl of 1:1 phenol/chloroform equilibrated with LET. Vortex 60 sec and spin 3 min at 14,000 rpm.
- Transfer aqueous phase to new 1.5-ml tube.
- Add 400 μl of chloroform. Vortex 60 sec and spin 3 min at 14,000 rpm.
- Transfer aqueous phase to a new 1.5-ml tube. Add 40 μl of 3 M sodium acetate and 800 μl of 100% cold ethanol. Mix well and place at $-20\,°C$ for 1 h.
- Collect RNA by centrifugation at room temperature for 10 min at 14,000 rpm. Wash pellet with 500 μl 70% EtOH and recentrifuge for 5 min. Drain supernatant and dry pellet (either air-dry or dry in SpeedVac [no heat]).
- Resuspend pellet in 50 to 150 μl DEPC dH_2O.
- Quantify each RNA sample by measuring the absorbance at 260 nm with spectrophotometer. Assay 4 μl of RNA in 996 μl of dH_2O.
- Determine concentration of each sample on the basis of the extinction coefficient for RNA of 40 μg/ml. (*Note*: if the preceding dilution is used, simply multiply the OD_{260} by 10; this equals the concentration of each sample in μg/μl).
- Analyze 10 to 30 μg of each sample by Northern blot.

5.1. Recipe for LET

25 mM Tris, pH 8.0
100 mM LiCl
20 mM EDTA
(All reagents should be made in DEPC-treated distilled H_2O).

5.2. RNase H cleavages of 3′ UTRs

mRNA decay in yeast is initiated by deadenylation (reviewed in Coller and Parker, 2004). The rate of poly(A) shortening, therefore, is an important consideration in decay analysis. The poly(A) tail typically ranges in size from 10 to 100 adenosines. Because the tail is small, the size difference between adenylated and deadenylated transcripts is negligible for most large mRNAs. It is necessary, therefore, to first cleave an mRNA within its 3′ UTR with an antisense oligonucleotide and RNase H to generate a smaller RNA fragment that can be analyzed by high-resolution Northern blotting after electrophoresis in polyacrylamide gels. The antisense oligo that we typically use is 20 nucleotides in size and binds near the stop codon of the gene of interest. The following is the protocol for generating a smaller mRNA fragment from full-length mRNA.

- Dry down in SpeedVac 10 μg of total-cell RNA and 300 ng antisense oligo (can dry down as much as 40 μg of RNA and still maintain the 300 ng of oligo; do not overdry, because it will be difficult to resuspend). To control for poly(A) tail lengths, also treat a sample with antisense oligo and 300 ng of oligo d(T).
- Resuspend pellet in 10 μl of 1× hybridization mix (Section 5.3).
- Heat sample at 68 °C for 10 min. Cool slowly to 30 °C. Pulse spin down.
- Add 9.5 μl of 2× RNase H buffer and 0.5 μl RNase H. Mix well.
- Incubate sample at 30 °C for 60 min.
- Add 180 μl stop mix. Extract sample with 200 μl of phenol/chloroform. Remove aqueous phase and extract with 200 μl of chloroform.
- Precipitate RNA by adding 500 μl of 100% ethanol. Freeze at −20 °C for 60 min.
- Spin down sample for 10 min at room temperature. Wash RNA pellet with 300 μl of 70% ethanol. Dry either at room temperature or under vacuum in Speedvac (once again, do not overdry).
- Resuspend sample in 10 μl of DEPC-treated water and add 10 μl of loading dye. Heat sample at 100 °C for 5 min before loading in 6 to 8% denaturing acrylamide gel. After resolving by PAGE, perform a Northern Blot analysis and detect RNA fragment with an mRNA-specific probe.
- Prepare solutions with DEPC-treated distilled water and store in aliquots at −20 °C.

5.3. 10× Hybridization mix

0.25 M Tris-HCl, pH 7.5
10 mM EDTA
0.5 M NaCl

5.4. 2× RNase H buffer

40 mM Tris-HCl, pH 7.5
20 mM MgCl$_2$
100 mM NaCl
2 mM DTT
60 μg/ml BSA

5.5. Stop mix

0.04 mg/ml tRNA
20 mM EDTA
300 mM NaOAc

6. NORTHERN BLOT ANALYSIS

After performing an mRNA decay analysis, it is possible to obtain a half-life measurement by resolving mRNA in a formaldehyde agarose gel followed by Northern blot. To obtain information about the deadenylation rate, it is necessary to analyze mRNA or RNase H fragments by resolution in a 6% polyacrylamide gel followed by Northern blotting. Protocols for these techniques are described elsewhere (Sambrook *et al.*, 1989 and Chapter 17 by Chen *et al.* in this volume).

7. LOADING CONTROLS

It is important to consider several controls in an mRNA decay analysis. First, if monitoring deadenylation rate, a sample of mRNA (usually from the first time point) should be treated with RNase H and oligo (dT) to generate a nonadenylated marker that indicates the size of the fully deadenylated mRNA (Figs. 14.2 and 14.3). Second, the Northern blot should be reprobed with a loading control. *SCR1* RNA is ideal for this purpose. *SCR1* RNA is the 7S signal recognition particle RNA used by the cell for cotranslational protein targeting to the endoplasmic reticulum. Importantly, this small 522-nt RNA is noncoding and transcribed by RNA polymerase III and is unaffected by changes in *mRNA* transcription except in the case of thiolutin. Thiolutin inhibits all three RNA polymerases, including *SCR1* RNA levels, albeit only slightly within the recommended 1 h time course. *SCR1* RNA is abundant and can be easily detected with a gamma ^{32}P-ATP kinase-labeled oligonucleotide probe.

8. DETERMINATION OF MRNA HALF-LIVES

Yeast mRNAs generally decay with first-order kinetics (Herrick *et al.*, 1990). Half-lives, therefore, can be represented by the equation $t_{1/2} = 0.693/k$, where $k =$ the rate constant for mRNA decay (i.e., percent change over time). Half-life calculations for a particular mRNA transcript

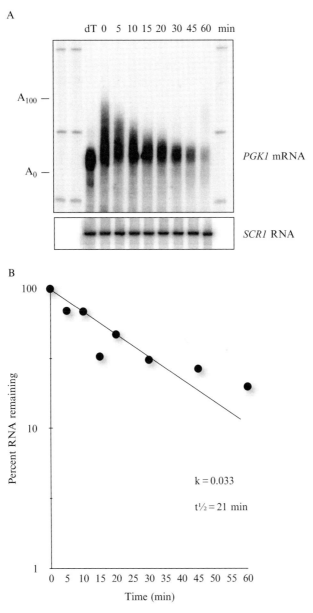

Figure 14.3 Transcriptional shut-off analyzes of the PGK1 mRNA (A) and calculation of PGK1 half-life (B). In (A) the mRNA reporter was cleaved by RNaseH using an antisense oligo and then resolved by PAGE. dT represents a sample that was also cleaved with oligo d(T) and RNaseH to indicate size of fully deadenylated mRNA. Below is a Northern of the SCR1 RNA which is used as a loading control. (B) is the quantitation of the experiment shown in (A) and representation of these data on a semi-log plot. This was used to calculate the half-life based on the protocol outlined in the text.

are made by quantitating the amount of mRNA present at each time point with a PhosphorImager. This value should be normalized for loading variations with a control like *SCR1* RNA (Fig. 14.3A). The value for k can be determined with a semilogarithmic plot of the concentration of mRNA over time and determining the slope of a best-fit line (slope=k; Fig. 14.3B). Once the rate constant is determined, the half-life can be calculated. It is strongly advised to measure mRNA half-lives at least three times to ensure reproducibility.

9. Concluding Remarks

In summary, the techniques described in this review provide useful methods for measuring mRNA decay rates in yeast. The most important consideration in choosing a particular technique is to determine the type of information that is desired. If simple half-life measurements are needed, then a transcriptional shut-off experiment will suffice. If information is sought about the sub-step of decay affected, then a transcriptional pulse-chase is required.

ACKNOWLEDGMENTS

The techniques described are amalgamated from the work of many laboratories and individuals, especially Drs. Roy Parker, Allan Jacobson, Stuart Peltz, Denise Muhlrad, Caroline Decker, David Herrick, Kristian Baker, Ambro Van Hoof, Pat Hilleren, and Richard Young. I also thank Drs. Wenqian Hu, Thomas Sweet, and Sarah Geisler for advice and criticism of the manuscript. Funding is provided by National Institutes of Health (GM080465).

REFERENCES

Anderson, J. S., and Parker, R. P. (1998). The 3′ to 5′-degradation of yeast mRNAs is a general mechanism for mRNA turnover that requires the SKI2 DEVH box protein and 3′ to 5′-exonucleases of the exosome complex. *EMBO J.* **17,** 1497–1506.

Baker, K. E., and Parker, R. (2004). Nonsense-mediated mRNA decay: Terminating erroneous gene expression. *Curr. Opin. Cell Biol.* **16,** 293–299.

Burke, D., Dawson, D., and Stearns, T. (2000). *In* "Methods in yeast genetics: A Cold Spring Harbor Laboratory Course Manual," Cold Spring Harbor Laboratory Press, Cold Spring Harbor, N.Y.

Cao, D., and Parker, R. (2001). Computational modeling of eukaryotic mRNA turnover. *RNA* **7,** 1192–1212.

Carlson, M., and Botstein, D. (1982). Two differentially regulated mRNAs with different 5′-ends encode secreted with intracellular forms of yeast invertase. *Cell* **28,** 145–154.

Coller, J., and Parker, R. (2004). Eukaryotic mRNA decapping. *Annu. Rev. Biochem.* **73,** 861–890.

Coller, J., and Parker, R. (2005). General translational repression by activators of mRNA decapping. *Cell* **122,** 875–886.

Decker, C. J., and Parker, R. (1993). A turnover pathway for both stable and unstable mRNAs in yeast: Evidence for a requirement for deadenylation. *Genes Dev.* **7,** 1632–1643.

Gari, E., Piedrafita, L., Aldea, M., and Herrero, E. (1997). A set of vectors with a tetracycline-regulatable promoter system for modulated gene expression in *Saccharomyces cerevisiae*. *Yeast* **13,** 837–848.

He, F., and Jacobson, A. (1995). Identification of a novel component of the nonsense-mediated mRNA decay pathway by use of an interacting protein screen. *Genes Dev.* **9,** 437–454.

Herrick, D., Parker, R., and Jacobson, A. (1990). Identification and comparison of stable and unstable mRNAs in *Saccharomyces cerevisiae*. *Mol. Cell. Biol.* **10,** 2269–2284.

Higgs, D. C., and Colbert, J. T. (1994). Oat phytochrome A mRNA degradation appears to occur via two distinct pathways. *Plant Cell* **6,** 1007–1019.

Hilgers, V., Teixeira, D., and Parker, R. (2006). Translation-independent inhibition of mRNA deadenylation during stress in *Saccharomyces cerevisiae*. *RNA* **12,** 1835–1845.

Hilleren, P. J., and Parker, R. (2003). Cytoplasmic degradation of splice-defective pre-mRNAs and intermediates. *Mol. Cell* **12,** 1453–1465.

Hsu, C. L., and Stevens, A. (1993). Yeast cells lacking $5'\rightarrow 3'$-exoribonuclease 1 contain mRNA species that are poly(A) deficient and partially lack the $5'$-cap structure. *Mol. Cell Biol.* **13,** 4826–4835.

Johnston, S. A., and Hopper, J. E. (1982). Isolation of the yeast regulatory gene GAL4 and analysis of its dosage effects on the galactose/melibiose regulon. *Proc. Natl. Acad. Sci. USA* **79,** 6971–6975.

Lohr, D., Venkov, P., and Zlatanova, J. (1995). Transcriptional regulation in the yeast GAL gene family: A complex genetic network. *FASEB J.* **9,** 777–787.

Longtine, M. S., McKenzie, A., 3rd, Demarini, D. J., Shah, N. G., Wach, A., Brachat, A., Philippsen, P., and Pringle, J. R. (1998). Additional modules for versatile and economical PCR-based gene deletion and modification in Saccharomyces cerevisiae. *Yeast* **14,** 953–961.

Muhlrad, D., Decker, C. J., and Parker, R. (1994). Deadenylation of the unstable mRNA encoded by the yeast MFA2 gene leads to decapping followed by $5'\rightarrow 3'$ digestion of the transcript. *Genes Dev.* **8,** 855–866.

Muhlrad, D., Decker, C. J., and Parker, R. (1995). Turnover mechanisms of the stable yeast PGK1 mRNA. *Mol. Cell Biol.* **15,** 2145–2156.

Mukherjee, D., Gao, M., O'Connor, J. P., Raijmakers, R., Pruijn, G., Lutz, C. S., and Wilusz, J. (2002). The mammalian exosome mediates the efficient degradation of mRNAs that contain AU-rich elements. *EMBO J.* **21,** 165–174.

Nonet, M., Scafe, C., Sexton, J., and Young, R. (1987). Eucaryotic RNA polymerase conditional mutant that rapidly ceases mRNA synthesis. *Mol. Cell Biol.* **7,** 1602–1611.

Pelechano, V., and Perez-Ortin, J. E. (2007). The transcriptional inhibitor thiolutin blocks mRNA degradation in yeast. *Yeast*.

Rodgers, N. D., Wang, Z., and Kiledjian, M. (2002). Characterization and purification of a mammalian endoribonuclease specific for the alpha-globin mRNA. *J. Biol. Chem.* **277,** 2597–2604.

Sambrook, J., Fritsch, E. F., and Maniatis, T. (1989). "Molecular Cloning: A Laboratory Manual," 2nd ed. Cold Spring Harbor Laboratory Press, Cold Spring Harbor, N.Y.

Traven, A., Jelicic, B., and Sopta, M. (2006). Yeast Gal4: A transcriptional paradigm revisited. *EMBO Rep.* **7,** 496–499.

van Hoof, A., and Parker, R. (2002). Messenger RNA degradation: Beginning at the end. *Curr. Biol.* **12,** R285–287.

Winzeler, E. A., Shoemaker, D. D., Astromoff, A., Liang, H., Anderson, K., Andre, B., Bangham, R., Benito, R., Boeke, J. D., Bussey, H., *et al.* (1999). Functional characterization of the *S. cerevisiae* genome by gene deletion and parallel analysis. *Science* **285,** 901–90.

CHAPTER FIFTEEN

MRNA DECAY ANALYSIS IN *DROSOPHILA MELANOGASTER*: DRUG-INDUCED CHANGES IN GLUTATHIONE *S*-TRANSFERASE D21 MRNA STABILITY

Bünyamin Akgül[*,†] and Chen-Pei D. Tu[*]

Contents

1. Introduction	286
2. Materials and Methods	287
2.1. Materials	287
2.2. Transgenic constructs and nomenclature	287
2.3. Pentobarbital and heat shock treatments	290
2.4. RNA isolation and RNase Protection Assays	291
2.5. Determining 5′- and/or 3′-ends and decay intermediates of *gst*D21 mRNAs	293
3. Concluding Remarks	295
Acknowledgment	296
References	296

Abstract

We have established an *in vivo* system to investigate mechanisms by which pentobarbital (PB), a psychoactive drug with a sedative effect, changes the rate of decay of *gst*D21 mRNA (encoding a *Drosophila* glutathione *S*-transferase). Here we describe methods for the use of *hsp*70 promoter-based transgenes and transgenic lines to determine mRNA half-lives by RNase protection assays in *Drosophila*. We are able to identify and map putative decay intermediates by cRT-PCR and DNA sequencing of the resulting clones. Our results indicate that the 3′-UTR of *gst*D21 mRNA is responsive to PB by regulating mRNA decay

[*] Department of Biochemistry and Molecular Biology, The Pennsylvania State University, University Park, Pennsylvania, USA
[†] Department of Molecular Biology and Genetics, Izmir Institute of Technology, Izmir, Turkey

and that the *cis*-acting element(s) responsible for the PB-mediated stabilization resides in a 59 nucleotide sequence in the 3′-UTR of the *gst*D21 mRNA (Akgül and Tu, 2007).

1. Introduction

Xenobiotics and drugs affect gene expression at different levels, including mRNA stability (Akgül and Tu, 2005; Köhle and Bock, 2007; Nakata *et al.*, 2006). We have reported that the sedative drug pentobarbital (PB) can stabilize the mRNA of an intronless gene encoding glutathione S-transferase (gst) D21 in *Drosophila melanogaster* (Akgül and Tu, 2004, 2007; Tang and Tu, 1995). The mechanism(s) by which PB may affect mRNA stability are not clear. Since cultured cells usually do not respond well, if at all, to psychoactive drugs like PB, we needed to develop a simple *in vivo* system to investigate the mechanism by which PB affects *gst*D21 mRNA stability. We chose the *Drosophila* system over the rodents because of the intronless nature of the *gst*D21 gene and the relative ease of maintenance and manipulation.

With the *Drosophila* system, we have at our disposal genetic and molecular tools such as transposition mutagenesis, siRNA technology, and an extensive collection of mutants generated by *p*-element insertions, classical mutagenesis, and other approaches (Rubin, 1988). On establishing an *in vivo* PB-responsive system for analyzing RNA decay, we can generate transformant flies with various designer (e.g., deletion and chimeric) constructs and, in these rendered contexts, identify in the mRNA sequences *cis*-acting elements important for RNA decay.

The major challenge to study RNA turnover in a whole organism is to devise method(s) that selectively and quickly activate transcription of the target gene to a very high level, followed by rapid shut-off of gene activation for monitoring the decay rate of the target mRNA under normal or well-controlled physiologic state of the organism. For the *Drosophila* system, we chose the *hsp*70 promoter to activate target transgene expression at 35 °C. When heat shock factor associates with the *hsp*70 promoter on heat shock (HS), the pausing RNA polymerase II complex at the promoter is immediately released for elongation. The promoter returns to the uninduced state within less than 60 min after removal of heat shock (Tang *et al.*, 2000), and the organism quickly returns to the normal physiologic state. These attributes make the HS-responsive *hsp*70 promoter suitable for use in studying RNA decay *in vivo*. In the following, we describe a system to investigate mRNA decay under the effects of PB using the *gst*D21 mRNA as the reporter. Furthermore, by incorporating the chimeric gene approach, we are able to dissect the D21 mRNA for the *cis*-acting elements responsible for drug-mediated stabilization of *gst*D21 mRNAs in *Drosophila* (Akgül and Tu, 2007).

2. MATERIALS AND METHODS

2.1. Materials

Bacteriologic media were purchased from Invitrogen/Life Technologies (Rockville, MD), and chemicals from ICN, Life Technologies or Sigma-Aldrich (St. Louis, MO). Sodium pentobarbital was purchased from Sigma-Aldrich. Oligonucleotides were obtained from Integrated DNA Technologies, Inc. (Coralville, IA). Radioactive nucleotides ([α-^{32}P] UTP, specific activity ~800 mCi/μmol) were purchased from ICN (Irvine, CA). RPA III kits were purchased from Ambion (Austin, TX). Restriction enzymes were products of New England Biolabs (NEB) (Beverly, MA) or American Allied Biochemicals (Aurora, CO). Pfu DNA polymerase and Quick-change® site-directed mutagenesis kit were purchased from Stratagene (San Diego, CA). T4 DNA ligase, MMLV-reverse transcriptase, SP6 RNA polymerase and the plasmid vectors pGEM T Easy® and pSP64(A) for in vitro transcription were purchased from Promega (Madison, WI). Tobacco acid phosphatase was from Epicentre Technologies (Madison, WI). T7 RNA polymerase was a generous gift from Bi-Cheng Wang (University of Georgia, Athens, GA). E. coli DH5α competent cells and Pfx DNA polymerase were products of Life Technologies. The plasmid vector pCaSpeR-hs-act for Drosophila transformation was obtained from C. S. Thummel of the University of Utah (Thummel et al., 1988). The Δ2-3 line {P[ry$^+$ 2-3](99B)} (Robertson et al., 1988) expressing transposase and the yw line were obtained from Susan Abmayr and David Gilmour, respectively, both of the Department of Biochemistry and Molecular Biology, The Pennsylvania State University (University Park, PA). All glassware used in RNA work is baked at 180 °C overnight. All solutions except for Tris buffers are treated with 0.1% DEPC (Sigma) overnight under a hood and autoclaved for 15 min and placed under the hood at least for 5 h to evaporate the residual DEPC in the solutions. Because DEPC changes the chemical structure of Tris, the stock solutions are prepared first with DEPC-treated water and then filtered through a 0.22-μm Nalgene filter.

2.2. Transgenic constructs and nomenclature

We use small plasmid vectors pBluescript or pGEM T Easy (Promega) for initial cloning of PCR products and manipulation of gene segments before transferring the chimeric DNA constructs into the transformation vector pCaSpeR-hs-act (Thummel et al., 1988). Site-directed mutagenesis and swapping of DNA fragments between homologous genes can be accomplished more easily with small plasmids. When we need a restriction site in the coding region to facilitate DNA fragment swapping, we introduce base changes that conserve the amino acid sequence to minimize any possible effect of mis-sense mutation(s) on the stability of transgenic mRNA(s) (e.g., the swapping at nucleotide 567 in Fig. 15.1, D21-170-D1). We then use site-directed mutagenesis to restore some mutations back to the wild-type sequence. A typical reaction contains 30 to 50 ng of double-stranded template DNA, 300 nM of each of the two primers, 0.2 mM dNTPs, 1× *Pfu* DNA polymerase buffer, and 2.5 U *Pfu* DNA polymerase (Stratagene). PCR conditions are as follows: 1 cycle of 94 °C for 1 min; 18 cycles of 94 °C for 30 sec, 55 °C for 1 min, and 68 °C for 11 min. The PCR product is then digested with 1 U of *Dpn* I to remove the methylated template DNA, and an aliquot is used for transformation (Sambrook and Russell, 2001).

Once the desired constructs in pBluescript or pGEM T Easy are generated and their sequences confirmed, a pair of primers with appropriate restriction sites at the 5′-ends is used to amplify the inserted DNA for cloning into the pCaSpeR-hs-act transformation vector (Thummel *et al.*, 1988). Inclusion of a few additional nucleotides 5′-to the restriction sites on the PCR primers facilitates complete digestion of the amplified products (Moreira and Noren, 1995). We use commercial fragment isolation kits to purify the restriction enzyme-digested PCR products for further manipulations. The pCaSpeR-hs-act transformation vector has eight unique restriction sites in the multiple cloning site (MCS) sequences. We typically use the *Bam*HI and *Eco*RI sites for DNA insertions so that the extraneous MCS

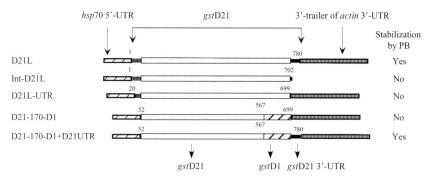

Figure 15.1 Transgenic constructs of *gst*D21. Nucleotides (nts) 1 to 780 represent the complete *gst*D21(L) mRNA sequence (5′-UTR, nts 1 to 52; coding region, nts 53 to 699; 3′-UTR, nts 700 to 762 for S form and nts 700 to 780 for the L form) (Akgül and Tu, 2004). Each transgenic RNA should have the same heterologous UTR context, containing the 5′-UTR of *hsp*70 and the 3′-trailer of *actin* 3′-UTR. Open boxes represent the *gst*D21 protein-coding region. The native 5′-UTR of *gst*D21 mRNA is represented by a checkered box preceding the protein coding region, and the native 3′-UTR is specified by a solid box. The *gst*D1 (nt 567 to 699) sequence in the chimeric genes D21-170-D1 and D21-170-D1 + D21UTR are shown in cross-hatched boxes. Pentobarbital-mediated stabilization of a given D21 transgenic mRNA is indicated by a Yes or No.

sequences are absent in the final constructs. All transgenes have the same heterologous sequence context by having the *hsp70* 5′-UTR and the trailer region of the *actin*5C's 3′-UTR. The 5′-UTR of the *hsp70* should afford optimum translation under heat shock conditions, and by doing so any potential effect of translation on the stability of the transgenic mRNAs may be avoided or minimized.

2.2.1. Microinjection and establishment of transgenic lines

We follow the procedure of Thummel *et al.* (1998) for *Drosophila* transformation, which is a modification of the initial approach demonstrated by Rubin and Spradling (1982 and 1983). We will limit the discussion of *Drosophila* transformation to those that apply to our results. Plasmid DNAs (chimeric genes on pCaSpeR-hs-act vector) should be of the highest possible purity for decent transformation efficiency. We use DNA preparations (Qiagen Midi Plasmid DNA preparation kit) at 0.5 mg/ml in ddH$_2$O and with an A_{260}/A_{280} ratio of ≥ 1.8 for transformation. We collect embryos at room temperature of 23 to 25 °C for 45 min, but microinjection is carried out in an 18 °C room, and the procedure is completed within 45 min after the embryo harvest. The glue used to line up the embryos on the microscope cover slip is prepared by vortex-mixing double-sided Scotch adhesive tape in n-hexane. The hexane supernatant is then empirically adjusted to the appropriate viscosity in 1.5-ml microcentrifuge tubes before use. We avoid the use of any dye marker in the DNA solution to optimize transformation efficiency; the injection can be easily visualized by a slight postinjection disturbance at the posterior end of the embryo. We minimize the number of nontransformants by use of the injection needle to smash the aged (cellularized) or uninjected embryos.

The newly eclosed G_0 flies are crossed individually to *yw* to remove any background transposase. All the female flies used in crosses are virgins. Yellow-eyed to red-eyed G_1 progeny with longer body bristles (Sb⁻) are recrossed with *yw*. Each stable line is established through sibling crosses of colored-eyed G_2 flies. On average, for each DNA construct, we inject 300 embryos, of which 10% reach adulthood with 25% being transformants. We typically maintain at least three separate lines for each transgene construct.

By our design, we use heat shock to induce the transgenes driven by the *hsp70* promoter, whereas the endogenous *gst*D21 gene is induced by PB (Akgül and Tu, 2004). Therefore, increases in the level of transgene-derived mRNAs upon simultaneous treatment with HS and PB (HSPB-treated) that are detected beyond that by HS alone, could be attributed to stabilization of the transgenic gstD21 mRNA by PB. When combined treatment is intended, it should be ensured that the endogenous gene is not induced by heat shock.

2.3. Pentobarbital and heat shock treatments

Pentobarbital (PB) regulates *gst* gene expression at both the transcriptional and posttranscriptional levels (Tang and Tu, 1995). Because of PB's water solubility, we chose this xenobiotic as a model drug to identify *cis*-acting determinants responsible for drug-induced changes of *gst*D21 mRNA stability. We discuss the procedure for PB and heat shock treatment of flies in this section.

Adult flies maintained on a standard cornmeal medium (Tang and Tu, 1994) (2- to 3-day-old, ~150 to 200 flies per bottle) are starved for 5 h in clean milk bottles with foam plugs at room temperature. We place a strip (3 × 10 cm) of 3MM paper saturated (~800 µl solution per strip) with either a 5% sucrose solution (control flies) or 200 mg/ml PB in 5% sucrose (PB-treated flies) in each bottle for the treatment for 2 h at room temperature. We avoid the use of dripping strips because the flies become stuck to the bottle wall (thus lower recovery). For a time-course analysis of RNA decay (e.g., calculation of the half-life of a given chimeric mRNA), flies are treated with PB for 2 h and then transferred to clean milk bottles containing a 3MM paper strip saturated with a 5% sucrose solution for 0 (control), 0.25, 0.5, 1, 2, 3, 4, 8, 10, and 12 h.

Heat shock of flies is carried out in milk bottles (~150 to 200 flies per bottle) at 35 °C for 1 h in a hybridization oven (Model 2000, Microhybridization incubator, Robbins Scientific Company, Sunnyvale, CA). A 3MM paper strip saturated with a 5% sucrose solution (~800 µl) is placed in the bottles during the HS treatment to prevent dehydration of the flies. Two strips are used per bottle when HS treatment is performed at elevated temperatures (e.g., 37 °C). (An empty milk bottle with a foam plug requires ~15 min to reach 35 °C from room temperature and takes ~6 min to drop to 31 °C after removal of the bottle from the 35 °C oven.) The extent of chimeric mRNA induction can be adjusted by controlling HS temperatures (e.g., elevated expression of transgenes at 37 °C relative to 35 °C). To minimize variations of temperature for HS treatment, different lines of transgenic flies to be compared in the same experiment are heat-shocked at the same time. To concurrently overexpress D21 transgene (heat shock–inducible) and endogenous *gst*D21 (PB-inducible) mRNAs, the flies are treated with PB at room temperature for 1 h and then at 35 °C for the second hour (Akgül and Tu, 2002). For a time-course study, the flies are transferred, after the treatment, into clean milk bottles containing 3MM paper strips saturated with a 5% sucrose solution kept at room temperature. Flies are collected in liquid nitrogen–treated 250-ml wide-mouth centrifuge bottles, snap-frozen with copious amounts of liquid nitrogen, and stored in 15-ml polypropylene tubes at −80 °C until use. We avoid the use of polystyrene tubes for storing flies as they easily crack in liquid nitrogen.

2.4. RNA isolation and RNase Protection Assays

Despite the availability of many rapid RNA isolation kits, we have consistently obtained excellent yield with high-quality RNA from both embryos and adults by following the procedure of Ullrich et al. (1977). Up to 2 mg of total-fly RNA can be isolated from 1 g of frozen flies. We prefer to use guanidine hydrochloride over guanidine isothiocyanate in the homogenization buffer because the former is easier to dissolve. This procedure is not appropriate for isolating RNAs smaller than 200 nt, because they do not efficiently cosediment with the high-molecular-weight RNAs (www.ambion.com). Typically, 0.5 to 1 g of frozen flies or embryos are homogenized in 7 ml of lysis buffer (4 M guanidinium hydrochloride, 1 M β-mercaptoethanol, 0.1 M NaOAc, pH 5, and 0.01 M EDTA) in a 15-ml Douce homogenizer first with Pestle B ($\sim$20 strokes) and then with Pestle A ($\sim$10 strokes). CsCl (1.05 g) is added per ml of the homogenate until it is fully dissolved at room temperature. The solution ($\sim$8 ml) is then transferred into a clear ultracentrifuge tube underlaid with 2 ml of DEPC-treated, cushion solution (saturated CsCl solution containing 10 mM EDTA and 50 mM NaOAc, pH 5). The mixture is then centrifuged at room temperature in a Beckman Ti70.1 rotor at 139,000g for 25 h. The flaky RNA visible to the naked eye below a non-UV–absorbing thick band is drawn with a 1-ml syringe with an 18-gauge needle. The RNA in CsCl solution $\sim$1 ml per tube is diluted with 3 volumes of DEPC-treated water before ethanol (2.5 volumes) precipitation in a Corex tube (15 ml or 30 ml). The RNA is collected by centrifugation at 13,000 rpm at 4 °C in a swinging bucket rotor (e.g., Sorvall HB4). The RNA dissolved in DEPC-treated water is aliquoted to prevent subsequent multiple freeze-and-thaw cycles. The quality of RNA prepared by this method survives long-term storage ($-$80 °C) without any degradation for up to 3 y.

To monitor changes in the expression of a specific chimeric *gst* mRNA, we have reliably used the ribonuclease protection assay (RPA) (Calzone et al., 1987). It is feasible to assess the quantity, as well as the ends of a transcript (e.g., alternatively polyadenylated mRNAs) (Fig. 15.2A) by this method. In addition, we can simultaneously analyze multiple RNA species with cRNA probes of different lengths (e.g., Akgül and Tu, 2002; 2004) by following Ambion's instructions for RPA assays. Riboprobes (cRNAs) are prepared from linearized plasmid DNAs containing the target sequences in the antisense orientation cloned downstream of a T7 or SP6 RNA polymerase promoter. We obtain excellent radiolabeled *in vitro* transcripts with 0.5 µg linearized plasmid DNA according to Promega's instructions. Only the mRNA sequences perfectly complementary to the radioactive riboprobe would be protected from the subsequent RNase digestions. We prepare riboprobe clones from DNAs of the chimeric plasmids used for

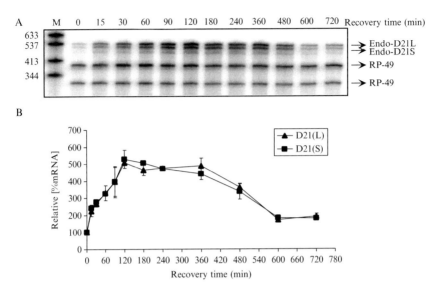

Figure 15.2 RNase protection assay analysis of pentobarbital-stabilized D21 transcripts. Pentobarbital treatment and RPA analysis were performed as indicated in "Pentobarbital Treatment" and "RNA Isolation and RPA Analyses" sections. The RPA analysis was carried out with 40 μg of total-fly RNA and a radiolabeled D21.AS.SmaI riboprobe. (A) The riboprotected D21 fragments were resolved in an 8% denaturing polyacrylamide gel. M, a radiolabeled *in vitro*–transcribed RNA markers, with sizes shown to the left of the panel. Endo-D21L and Endo-D21S, the L (long) and the S (short) forms of the endogenous *gst*D21 mRNAs, respectively; RP-49, the endogenous control ribosomal protein 49 mRNA. (B) mRNA amounts in (A) were normalized against the level of RP-49 mRNA (upper band) and expressed relative to the control level (100%).

establishing the transgenic lines. All plasmid DNAs used for riboprobe preparation must be correctly sequenced, because a single nucleotide difference would result in unhybridized bases that could be subjected to RNase cleavage and generation of partially ribo-protected fragments.

We typically use 20 to 40 μg of total-fly RNA in each RPA reaction to analyze *gst*D21 mRNA decay. However, less RNA would suffice if the expression level of the mRNA is relatively higher. The riboprobe-protected products are analyzed by denaturing polyacrylamide gel electrophoresis (0.75-mm thick, 8%, 19:1), followed by exposure in a PhosphorImager cassette for 1 to 5 h at room temperature depending on the intensity of the signal(s). The relative amount of each mRNA is then quantified. PhosphorImager signals are normalized against an endogenous reference mRNA (e.g., RP-49) (O'Connell and Rosbash, 1984). There is no need for drying the gel, because riboprobes with high specific activity are generated through *in vitro* transcription in the presence of $[\alpha^{-32}P]$ UTP (800 mCi/μmol). The cRNA probe is purified through a 6% denaturing polyacrylamide gel and extraction buffer (Ambion) and is included at levels threefold to fourfold (determined empirically) in excess to the target mRNA. Although one preparation of radioactive probes is sufficient for more than 100 reactions, prolonged storage at −80 °C over 2 days results in undesirable smears in RPA signals, presumably because of radiation-induced breakage of the probe.

2.5. Determining 5′- and/or 3′-ends and decay intermediates of *gst*D21 mRNAs

This method is based on circularization of decapped mRNAs followed by RT-PCR (Couttet et al, 1997) (Also see Chapter 22 by Grange). Although several methods are available to identify the 5′- or 3′-end of an mRNA individually, we have successfully used circular RT-PCR to simultaneously map both ends of various *gst*D21s at the nucleotide level (Akgül and Tu, 2007). It has been particularly useful in mapping multiple *gst*D21 decay intermediates at the nucleotide level. We first use a nested set of radioactive probes in RPA analyses to roughly estimate the ends of the intermediates (Akgül and Tu, 2007). These results are then used as a guide in designing the appropriate pair of primers (∼50 nt from the ends) for the subsequent RT-PCR experiment. This procedure can also be used to determine the length of the poly(A) tail of an mRNA, if desired (Couttet et al, 1997).

For cRT-PCR reactions, we treat 10 µg of total RNA with 1 U of RNase-free DNase (Promega) for 20 min at 37 °C in 50 µl of 50 mM Tris-HCl, pH 7.5, and 10 mM $MgCl_2$ to remove any residual genomic DNA contamination. After phenol-chloroform (Ambion) extraction twice and ethanol (RNase-free absolute alcohol) precipitation and a 70% ethanol wash, half of the recovered RNA is treated with 2.5 U of TAP (tobacco acid pyrophosphatase) in the presence of 20 U of RNasin, 50 mM NaOAc, pH 6, 1% β-mercaptoethanol, 1 mM EDTA, 0.1% Triton X-100 in 20 µl final volume for 60 min at 37 °C. The decapped RNA is phenol-chloroform extracted twice, ethanol-precipitated, and circularized with 20 U of T4 RNA ligase (Pharmacia), 50 mM Tris-HCl, pH 7.5, 10 mM $MgCl_2$, 20 mM DTT, 100 µM ATP, 100 µg/ml acetylated BSA, and 20 U of RNasin in 400 µl volume at 16 °C for 16 h. The circularized RNA is extracted with phenol-chloroform twice and ethanol precipitation as before. Then, cDNA is prepared by reverse transcription with MLV-RT and 10 ng of a gene-specific 18-mer primer. The resulting cDNA reaction mixture should be boiled for 5 min before digestion with a mixture of RNase A and RNase T1. The treated cDNA is recovered by phenol-chloroform extraction once and ethanol precipitation. One percent of the recovered cDNA is used for PCR amplification. The PCR product is gel-purified and cloned into pGEM T Easy vector (Promega). Cloned plasmid DNAs are randomly selected for sequence analysis to determine the 5′- and/or 3′-ends of the mRNAs. We sequenced ∼100 clones to obtain a distribution of the ends of RNA decay intermediates of *gst*D21 mRNA (Akgül and Tu, 2007).

2.5.1. Determination of *gst*D21 mRNA half-lives

D21 mRNA levels are normalized relative to the endogenous RP-49 mRNA levels. All values are subsequently determined relative to the ratio obtained in the control lines (e.g., no HS and/or PB treatment), which is designated as 100% expression. Because PB remains in the gut of the flies and thus continues to induce *gst*D21 expression more than 2 h after the cessation of the PB treatment (Fig. 15.2B), we use the time points between 3 and 10 h for half-life calculations. The log of the relative (% mRNA) level is plotted against time after treatment to obtain a linear regression trend line (Fig. 15.3). The half-life is then calculated from the formula $t_{1/2} = -0.693/k$ (min), where $k = -2.303(m)$. The slope, m, is obtained from the trend line ($y = mx + b$). An example of such a calculation is presented in Fig. 15.3, where log (% mRNA) of HS-induced D21-170-D1 mRNA are plotted against recovery time (in min). The $t_{1/2}$ for the D21-170-D1 chimeric RNA is calculated to be 3.5 h.

To assess whether the native D21 3'-UTR has a *cis*-acting element for stabilizing D21 mRNA in the presence of PB, we prepare two constructs with and without the UTR (e.g., D21-170-D1 vs D21-170-D1 + D21UTR in Fig. 15.1). The $t_{1/2}$ of both transgenic D21 mRNAs is subsequently calculated under HS and HSPB conditions. If the D21 native UTR contains a *cis*-acting stabilizing element, the $t_{1/2}$ is expected to be similar for the reporter transcript without the UTR (D21-170-D1 in Fig. 15.1), whereas the reporter transcript with the UTR (D21-170-D1 + D21UTR

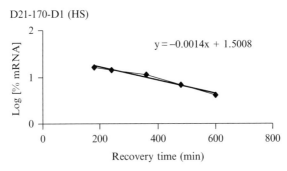

Figure 15.3 Calculations of mRNA half-lives from the time-course data. On the basis of RPA analyses (Fig. 15.2A), we plot log[% mRNA] vs minutes recovery time, defining the value at zero time as 100%. We used linear regression to find a best-fit straight line ($y = mx + b$; m is the slope) for data obtained between 3 and 10 h. The half-life ($t_{1/2}$) is derived from the equation: $t_{1/2} = -0.693/k$, where $k = -2.303m$. This is a representative plot for calculating the $t_{1/2}$ of the HS-induced D21-170-D1 transgenic mRNA (Fig. 15.1). From the best-fit regression line, $t_{1/2} = -0.693/(-2.303)(0.0014) = 213$ (min), where m is equal to 0.0014. We typically use at least three different fly populations for each treatment (e.g., heat shock or combined heat shock and PB treatment) and carry out three RPA analyses from each of two independent RNA isolations (i.e., a total of six RPA analyses).

in Fig. 15.1) would have a greater t (i.e., stabilization) under HSPB conditions compared with HS conditions (Akgül and Tu, 2004; 2007).

Our previous results suggest that PB stabilizes the *gst*D21 mRNA (Tang and Tu, 1995). To identify the *cis*-acting determinant(s) responsible for PB-mediated stabilization, we first deleted the 3′-UTR of the *gst*D21 mRNA (Fig. 15.1, D21L-UTR) and found that this particular chimeric D21 transcript is unresponsive to PB treatment (i.e., no stabilization by PB) (Akgül and Tu, 2004). Reinsertion of the native 3′-UTR of *gst*D21 restored the PB-mediated stabilization (Akgül and Tu, 2004). We also detected a series of decay intermediates, Int-D21L (Fig. 15.1), which lacked the 3′-UTR as mapped by cRT-PCR. These putative decay intermediates were not stabilized by PB, supporting the notion that a PB-responsive element is present in the 3′-UTR of the *gst*D21 mRNA (Akgül and Tu, 2007).

We then used a reporter transcript (Fig. 15.1, D21-170-D1, originally constructed to study translational regulation of *gst*D21 mRNA) to test the effect of the *gst*D21 3′-UTR on the stability of the resulting chimeric D1-D21 mRNA (Fig. 15.1, D21-170-D1 + D21UTR) in the presence of PB. RPA analyses showed that insertion of the *gst*D21 3′-UTR conferred PB-responsiveness to the otherwise PB-nonresponsive D21-170-D1 reporter mRNA (Akgül and Tu, 2007). Although the preceding approach is specifically focused on identifying the PB-responsive cis element in the *gst*D21 mRNA, a similar strategy could be implemented to isolate *cis* elements in any mRNA.

3. Concluding Remarks

Much of our current understanding of the pathways of mRNA degradation in eukaryotic cells came from studies in *S. cerevisiae* and in mammalian cells (Caponigro and Parker, 1996; Isken and Maquat, 2007; Jacobson and Peltz, 1996; Ross, 1995). Commonly used methods for measuring mRNA half-lives include approach to steady-state method, transcriptional pulse-chase, drug inhibition of transcription, inhibition of transcription by regulated promoters, and short-term promoter activation. *In vitro* systems for mRNA turnover have also been developed (Ross, 1993). None of these systems are applicable to investigating mechanisms of drug-induced changes in mRNA stability.

By use of *hsp*70 promoter-based transgenes and transgenic lines, we have established a *Drosophila* system to investigate PB-mediated changes in mRNA decay using the *gst*D21 mRNA. This approach has been quite useful not only in dissecting the *gst*D21 RNA for *cis*-acting element(s) responsible for the changes in mRNA stability by PB treatment but also in explaining alternative polyadenylation of *gst*D21 mRNA (Akgül and

Tu, 2002; 2004; 2007). Our analysis suggests that the *cis*-acting element responsible for the PB-mediated stabilization resides in a 59 nucleotide sequence within the *gst*D21 mRNA 3′-UTR (Akgül and Tu, 2007). Further dissection of this PB-responsive region will allow the identification of *trans*-acting factors essential for the drug's effect on *gst*D21 mRNA at the level of RNA turnover.

ACKNOWLEDGMENT

This project was supported by a grant (ES02678) from the National Institute of Environmental Health Sciences (NIEHS).

REFERENCES

Akgül, B., and Tu, C.-P. D. (2002). Evidence for a stabilizer element in the untranslated regions of *Drosophila* glutathione S-transferase D1 mRNA. *J. Biol. Chem.* **77**, 34700–34707.

Akgül, B., and Tu, C.-P. D. (2004). Pentobarbital-mediated regulation of alternative polyadenylation in *Drosophila* glutathione S-transferase D21 mRNAs. *J. Biol. Chem.* **279**, 4027–4033.

Akgül, B., and Tu, C.-P. D. (2007). Regulation of mRNA stability through a pentobarbital-responsive element. *Arch. Biochem. Biophys.* **459**, 143–150.

Calzone, F. J., Britten, R. S., and Davidson, E. H. (1987). Mapping of gene transcripts by nuclease protection assays and cDNA primer extension. *Methods Enzymol.* **152**, 611–632.

Caponigro, G., Muhlrad, D., and Parker, R. (1993). A small segment of the *MAT* α1 transcript promotes mRNA decay in *Saccharomyces cerevisiae*: A stimulatory role for rare codons. *Mol. Cell Biol.* **13**, 5141–5148.

Cleveland, D. W., and Yen, T. J. (1989). Multiple determinants of eukaryotic mRNA stability. *New Biol.* **1**, 121–126.

Couttet, P., Fromont-Racine, M., Steel, D., Pictet, R., and Grange, T. (1997). Messenger RNA deadenylation precedes decapping in mammalian cells. *Proc. Natl. Acad. Sci. USA* **94**, 5628–5633.

Decker, C. J., and Parker, R. (1993). A turnover pathway for both stable and unstable mRNAs in yeast: Evidence for a requirement for deadenylation. *Genes Dev.* **7**, 1632–1643.

Garneau, N. L., Wilusz, J., and Wilusz, C. J. (2007). The highways and byways of mRNA decay. *Nat. Rev. Mol. Cell Biol.* **8**, 113–126.

Greenberg, J. R. (1972). High stability of messenger RNA in growing cultured cells. *Nature* **240**, 102–104.

Isken, O., and Maquat, L. E. (2007). Quality control of eukaryotic mRNA: Safeguarding cells from abnormal mRNA function. *Genes Dev.* **21**, 1833–1856.

Jacobson, A., and Peltz, S. W. (1996). Interrelationships of the pathways of mRNA decay and translation in eukaryotic cells. *Annu. Rev. Biochem.* **65**, 693–739.

Köhle, C., and Bock, K. W. (2007). Coordinate regulation of Phase I and II xenobiotic metabolisms by the Ah receptor and Nrf2. *Biochem. Pharmacol.* **73**, 1853–1862.

Miller, A. D., Curran, T., and Verma, I. M. (1986). c-fos protein can induce cellular transformation: A novel mechanism of activation of a cellular oncogene. *Cell* **36**, 51–60.

Moreira, R. F., and Noren, C. J. (1995). Minimum duplex requirements for restriction enzyme cleavage near the termini of linear DNA fragments. *Biotechniques* **19**, 58–59.

Nakata, K., Tanaka, Y., Nakano, T., Adachi, T., Tanaka, H., Kaminuma, T., and Ishikawa, T. (2006). Nuclear receptor-mediated transcriptional regulation in Phase I, II, and III xenobiotic metabolizing systems. *Drug Metab. Pharmacokinet.* **21,** 437–457.

Nonet, M., Scafe, C., Sexton, J., and Young, R. (1987). Eucaryotic RNA polymerase conditional mutant that rapidly ceases mRNA synthesis. *Mol. Cell Biol.* **7,** 1602–1611.

O'Connell, P. O., and Rosbash, M. (1984). Sequence, structure and codon preference of the *Drosophila* ribosomal protein 49 gene. *Nucleic Acids Res.* **12,** 5495–5513.

Parsell, D. A., and Lindquist, S. (1994). Heat shock proteins and stress tolerance. *In* "The Biology of Heat Shock Proteins and Molecular Chaperones." (R. I. Morimoto, A. Tissieres, and C. Georgopoulos, eds.), pp. 457–494. Cold Spring Harbor Laboratory Press, Cold Spring Harbor, New York.

Peltz, S. W., Brewer, G., Bernstein, P., Hart, P. A., and Ross, J. (1991). Regulation of mRNA turnover in eukaryotic cell. *Crit. Rev. Eukaryotic Gene Expression* **1,** 99–126.

Rajagopalan, L. E., and Malter, J. S. (1997). Regulation of eukaryotic messenger RNA turnover. *Prog. Nucl. Acid Res.* **56,** 257–286.

Robertson, H. M., Preston, C. R., Phillis, R. W., Johnson-Schlitz, D. M., Benz, W. K., and Engels, W. R. (1988). A stable genomic source of P element transposase in *Drosophila melanogaster*. *Genetics.* **118,** 461–470.

Ross, J. (1995). mRNA stability in mammalian cells. *Microbiol. Rev.* **59,** 423–450.

Ross, J. (1993). mRNA decay in cell-free systems. *In* "Control of mRNA Stability." (J. Belasco and G Brawerman, eds.), pp. 417–448. Academic Press, San Diego, California.

Rubin, G. M. (1988). *Drosophila melanogaster* as an experimental organism. *Science* **240,** 1453–1459.

Rubin, G. M., and Spradling, A. C. (1983). Vectors for P element-mediated gene transfer in *Drosophila*. *Nucleic Acid Res.* **11,** 6341–6351.

Rubin, G. M., and Spradling, A. C. (1983). Genetic transformation of *Drosophila* with transposable element vectors. *Science* **218,** 348–353.

Sambrook, J., and Russell, D. (2001). Molecular cloning: A laboratory manual. Cold Spring Harbor Laboratory Press, Cold Spring Harbor, New York.

Sachs, A. B. (1993). Messenger RNA degradation in eukaryotes. *Cell* **74,** 413–421.

Shaw, G., and Kamen, R. (1986). A conserved AU sequence from the 3′-untranslated region of GM-CSF mRNA mediates selective mRNA degradation. *Cell* **46,** 659–667.

Tang, A. H., and Tu, C.-P. D. (1995). Pentobarbital-induced changes in *Drosophila* glutathione S-transferase D21 mRNA stability. *J. Biol. Chem.* **270,** 13819–13825.

Tang, A. H., and Tu, C.-P. D (1994). Biochemical characterization of *Drosophila* glutathione S-transferases D1 and D21. *J. Biol. Chem.* **269,** 27876–27884.

Tang, H., Liu, Y., Madabusi, L., and Gilmour, D. S. (2000). Promoter-proximal pausing on the hsp70 promoter in *Drosophila melanogaster* depends on the upstream regulator. *Mol. Cell. Biol.* **20,** 2569–2580.

Thummel, C. S., Boulet, A. M., and Lipshitz, H. D. (1988). Vectors for *Drosophila* P-element mediated transformation and tissue culture transfection. *Gene* **74,** 445–456.

Tu, C.-P. D., and Akgül, B. (2005). *Drosophila* glutathione S-transferases. *Methods Enzymol.* **401,** 204–226.

Ullrich, A., Shine, J., Chirgwin, J., Pictet, R., Tischer, E., Rutter, W. J., and Goodman, H. M. (1977). Rat insulin genes; construction of plasmids containing the coding sequences. *Science* **196,** 1313.

CHAPTER SIXTEEN

MEASURING mRNA STABILITY DURING EARLY *DROSOPHILA* EMBRYOGENESIS

Jennifer L. Semotok,[*,†] J. Timothy Westwood,[‡]
Aaron L. Goldman,[*,†] Ramona L. Cooperstock,[*,†] *and*
Howard D. Lipshitz[*,†]

Contents

1. Maternal mRNAs and Early *Drosophila* Development	300
1.1. Dual degradation activities in the early embryo	301
1.2. Studying maternal mRNA decay in unfertilized eggs versus fertilized embryos	302
1.3. Molecular mechanisms regulating maternal mRNA instability	307
1.4. Prerequisites for triggering maternal mRNA destabilization	307
2. Gene-by-Gene Analysis of mRNA Decay	308
2.1. Description and comparison of RNA methods	308
2.2. Analysis of deadenylation	314
2.3. Protocols	315
3. Genome-Wide Analysis of mRNA Decay	318
3.1. RNA isolation	318
3.2. Labeling and hybridization	319
3.3. Microarray platforms	320
3.4. Data normalization	320
3.5. Data analysis, verification, and interpretation	322
3.6. Protocols	325
4. Concluding Remarks	331
Acknowledgments	331
References	332

---

[*] Department of Molecular Genetics, University of Toronto, 1 King's College Circle, Ontario, Canada
[†] Program in Developmental and Stem Cell Biology, Research Institute, Hospital for Sick Children, Toronto, Ontario, Canada
[‡] Department of Cell and Systems Biology and Canadian *Drosophila* Microarray Centre, University of Toronto, Mississauga, Ontario, Canada

Abstract

Maternal mRNAs play a major role in directing early *Drosophila melanogaster* development, and thus, precise posttranscriptional regulation of these messages is imperative for normal embryogenesis. Although initially abundant on egg deposition, a subset of these maternal mRNAs is targeted for destruction during the first 2 to 3 h of embryogenesis. In this chapter, we describe molecular methods to determine the kinetics and mechanisms of maternal mRNA decay in the early *D. melanogaster* embryo. We show how both unfertilized eggs and fertilized embryos can be used to identify maternal mRNAs destined for degradation, to explain changes in decay kinetics over time, and to uncover the molecular mechanisms of targeted maternal mRNA turnover. In the first section, we explore the methods and outcomes of measuring decay on a "gene-by-gene" basis, which involves examination of a small number of transcripts by Northern blotting, RNA dot blotting, and real-time RT-PCR. In the second section, we provide a comprehensive examination of the applications of microarray technology to study global changes in maternal mRNA decay during early development. Genome-wide surveys of maternal mRNA turnover provide a wealth of information regarding the magnitude, temporal regulation, and genetic control of maternal mRNA turnover. Methods that permit the collection and analysis of highly reproducible and statistically robust data in this developmental system are discussed.

1. Maternal mRNAs and Early *Drosophila* Development

Drosophila embryos rely on maternally loaded mRNAs to both activate and orchestrate the first several hours of development. These maternal mRNAs, which represent 55 to 65% of the entire protein-encoding genome (Lecuyer *et al.*, 2007; Tadros *et al.*, 2007a), direct establishment of the embryonic axes, segregation of the future germ line, specification of somatic versus germ-line cell fates, and generation, from a single zygotic nucleus, of the syncytial blastoderm containing 5000 nuclei. During this period, the zygotic genome is virtually silent; with the exception of 30 to 60 early zygotic genes (whose *de novo* synthesis is detected approximately 1.5 h after fertilization), most zygotic gene transcription initiates 2.5 to 3.0 h after fertilization (De Renzis *et al.*, 2007; Edgar and Schubiger, 1986; Lecuyer *et al.*, 2007). Although the first 13 nuclear divisions occur in a syncytium, concomitant with interphase of nuclear cycle 14 and commencement of high-level zygotic transcription, the syncytial blastoderm undergoes cellularization. This is the first developmental process that depends on the newly synthesized zygotic messages and is, thus, termed the midblastula transition (MBT) (Merrill *et al.*, 1988; Wieschaus and Sweeton, 1988).

The maternal coordination of early molecular and cellular events of embryogenesis is accomplished largely through the posttranscriptional regulation of maternal RNAs. Concurrent with fertilization, the *Drosophila* embryo initiates a burst of protein synthesis from a subset of the stored maternal mRNAs. In addition to translational activation, a subset of these maternal mRNAs is subjected to destabilization and elimination. Maternal *Hsp83*, *string*, *polar granule component*, *nanos*, *bicoid*, *smaug*, and *hunchback^{mat}* are examples of transcripts that are abundant in mature oocytes and are targeted for elimination by the MBT (for review see Semotok and Lipshitz, 2007). Indeed, recent genome-wide analyses of early *Drosophila* embryogenesis have estimated that one third of maternally loaded transcripts undergo degradation by the MBT (De Renzis et al., 2007; Tadros et al., 2007a,b). These studies have also determined that 18% of the zygotic genome is transcriptionally activated by the MBT. This transition, during which a subset of maternal mRNAs is removed and specific zygotic mRNAs are synthesized, is termed the maternal-to-zygotic transition (MZT). The MZT marks a shift in the genetic control of development from the posttranscriptionally regulated maternal genome to the transcriptionally directed zygotic genome (Tadros et al., 2007b).

1.1. Dual degradation activities in the early embryo

Two mRNA decay activities have been characterized that target specific maternal mRNAs for degradation in the early embryo (Bashirullah et al., 1999, 2001; Tadros and Lipshitz, 2005). The first to operate is the "maternal degradation activity" (MDA). The MDA is maternally encoded and destabilizes ~21% of maternally expressed transcripts (~1600 on a genome-wide scale) (Bashirullah et al., 1999; Tadros et al., 2007a). The MDA initiates at egg activation, a multifaceted event that is triggered by the passage of the mature oocyte from the ovary to the uterus through the oviduct. Egg activation entails a number of cellular processes such as the cross-linking of the vitelline membrane (an extra-embryonic membrane that surrounds the egg's plasma membrane) to make it impermeable to water, completion of meiosis, condensation of the post-meiotic chromosomes, as well as molecular events that include cytoplasmic polyadenylation and translational activation of a subset of mRNAs, and elimination of specific mRNAs via the MDA (for review see Tadros and Lipshitz, 2005). It is noteworthy that in *Drosophila*, the mature oocyte first passages through the oviduct (where egg activation is triggered), then into the uterus (where fertilization occurs if the female has mated), and is then laid. Since egg activation initiates prior to and independent of fertilization, the MDA also functions independent of fertilization and thus is triggered in both deposited unfertilized and fertilized eggs. The MDA comprises at least two major components, one of which is the highly conserved RNA-binding protein, SMAUG (SMG). SMG targets two-thirds of all maternal transcripts (1,069 of ~1,600 mRNAs) for degradation by the MDA (Tadros et al., 2007a) (also see Figure 16.4 below). The other decay factors, which target the remaining one-third of the unstable transcripts, remain to be identified.

The 'zygotic degradation activity' (ZDA) initiates ~2 hours post-fertilization, just prior to the MBT, and requires zygotic transcription; thus the ZDA can only be studied in fertilized embryos (Bashirullah et al., 1999). The ZDA is less well understood than the MDA; however, a recent genome-scale dissection of maternal and zygotic mRNA expression at the MBT with aneuploid embryos (i.e., embryos that are loaded with the full complement of maternal transcripts but lack the zygotic contribution of a particular chromosome or chromosomal arm) has determined that maternal transcripts are targeted by distinct ZDA components (De Renzis et al., 2007). For example, the zygotic degradation machinery that targets maternal *string* mRNA maps to the X chromosome, whereas a zygotic factor that targets *twine* mRNA maps to the second chromosome.

1.2. Studying maternal mRNA decay in unfertilized eggs versus fertilized embryos

Three major factors influence steady-state mRNA levels in early embryos: the MDA, the ZDA, and *de novo* zygotic transcription. Although the former two activities act to eliminate maternal mRNAs and thus mediate a decrease in mRNA abundance, the latter process counteracts the decay activities by increasing transcript levels. Studying maternal mRNA decay in fertilized embryos is challenging, because these three factors may act individually, or in combination, to influence a maternal mRNA's decay profile (Tadros et al., 2007b). These combinations result in eight possible transcript classes (Table 16.1). Despite this complexity, it is possible to distinguish targets of the MDA alone, of the ZDA alone, of both activities, and of those unstable maternal transcripts that are replaced by zygotic mRNA synthesis during the MZT.

Much of our understanding of the contributions of the maternal and zygotic degradation activities to transcript turnover has come from comparisons between unfertilized and fertilized eggs (Bashirullah et al., 1999; 2001; Semotok et al., 2005; Tadros et al., 2007a). Whereas typical measurements of mRNA decay in eukaryotes (namely, yeast and mammalian cells) require the use of (1) transcriptional inhibitors, (2) conditional RNA polymerase II mutations, and/or (3) inducible promoters fused to a gene of interest to halt *de novo* RNA synthesis, studying mRNA decay in an unfertilized *Drosophila* system has the advantage that it lacks any *de novo* zygotic transcription. Thus, maternal transcript levels can only remain constant or decrease. Also, mRNA decay in unfertilized eggs occurs by means of the MDA alone and is not supplemented by additional decay activities such as the ZDA.

Table 16.1 Deciphering maternal mRNA decay and zygotic transcription. Maternal mRNAs are subject (+) or refractory (−) to factors influencing steady-state mRNA levels, which include the MDA, the ZDA, and/or zygotic transcription

Category of maternal mRNAs	Factors affecting mRNA levels			Outcome	
	MDA	ZDA	Zygotic transcription	UF mRNA profile	F mRNA profile
1^a	+	+	+	Decreasing	NC
2^a	+	−	+	Decreasing	NC
3^b	+	+	−	Decreasing	Decreasing
4^b	+	−	−	Decreasing	Decreasing
5	−	−	+	NC	Increasing
6	−	+	−	NC	Decreasing
7^c	−	+	+	NC	NC
8^c	−	−	−	NC	NC

[a] Category 1 vs 2: Requires the use of embryos aneuploid for the chromosome arm harboring the gene of interest that prevents zygotic transcription that obscures determination of factors influencing mRNA levels (see De Renzis et al., 2007). Compare wild-type UF eggs and aneuploid F embryos at 2 to 3 h AEL. If the UF:$F^{aneuploid}$ mRNA abundance ratio is 1, then the MDA acts alone in F embryos to destabilize the transcript of interest (category 2). If the UF:$F^{aneuploid}$ mRNA abundance ratio is >1, then dual activities are functioning in F embryos (category 1).
[b] Category 3 vs 4: Compare mRNA levels in wild-type UF eggs and wild-type F embryos at 2 to 3 h AEL. If the UF:F mRNA abundance ratio is 1, then the MDA acts alone in F embryos to destabilize the transcript of interest (category 4). If the UF:F mRNA abundance ratio is >1, then dual activities are functioning in F embryos (category 3).
[c] Category 7 vs 8: Requires the use of embryos that are aneuploid for chromosome harboring the gene of interest. Compare aneuploid F embryos and wild-type F embryos at 2 to 3 h AEL. If the $F^{aneuploid}$:F^{WT} mRNA abundance ratio is <1, then the maternal mRNA of interest is targeted by the ZDA (category 7). If the $F^{aneuploid}$:F^{WT} mRNA abundance ratio is 1, then the transcript of interest is not targeted by the ZDA (category 8).
UF, unfertilized eggs; F, fertilized embryos; decreasing, denotes a decrease in mRNA abundance over a 5-h time course after egg-lay (AEL); increasing, denotes an increase in mRNA abundance over a 5-h time course AEL; NC, no change over a 5-h time course AEL.

As outlined in Table 16.1, when maternal mRNA decay rates are examined in unfertilized eggs, all transcript classes that are destabilized by the MDA (e.g., categories 1 to 4) can be readily distinguished from transcript classes that are targeted by the ZDA or by neither degradation activity (e.g., categories 5 to 8). Maternal transcripts such as *Hsp83*, *nanos*, *polar granule component*, and *string* were identified as targets of the MDA because they are degraded in unfertilized eggs (Fig. 16.1) (Bashirullah et al., 1999; 2001).

Although informative, studying maternal mRNA decay exclusively in unfertilized eggs is limiting, because a large number of maternal transcripts is subject to turnover during embryonic development by dual action of the MDA and ZDA or by the ZDA alone; 21% of maternal mRNAs are destabilized in unfertilized eggs, whereas 33% of maternal mRNAs are

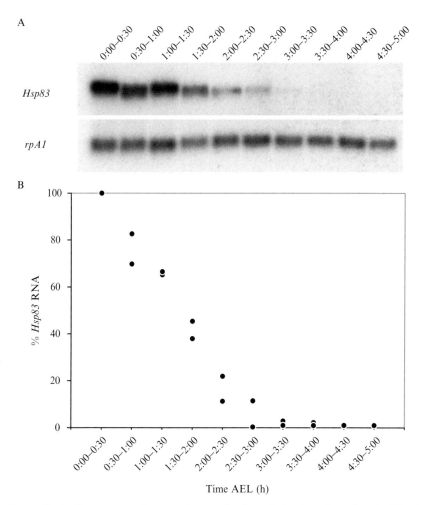

Figure 16.1 Time course of maternal transcript degradation in activated, unfertilized eggs. (A) The same Northern blot probed for *rpA1* (stable) and *Hsp83* (unstable) transcripts. (B) Quantitative analysis of the time course of *Hsp83* transcript degradation. The points represent the percentage of *Hsp83* transcripts remaining relative to the initial (0 to 0.5 h) amount after normalization to stable *rpA1* transcripts. It can be seen that more than 95% of the *Hsp83* transcripts have disappeared by 3.0 to 3.5 h after egg activation. Data from two independent experiments are presented. Half-hour time windows are shown. AEL, after egg lay. Adapted from Bashirullah *et al.* (1999; 2001).

destabilized in fertilized embryos (De Renzis *et al.*, 2007; Tadros *et al.*, 2007a). These additional unstable maternal mRNAs would be missed in analyses with unfertilized eggs alone. For example, in contrast to *Hsp83* mRNA, maternal transcripts such as *bicoid* and *hunchback*^{mat} are stable in unfertilized eggs and thus are not targeted by the MDA (Gamberi *et al.*,

2002; Surdej and Jacobs-Lorena, 1998; Tadros et al., 2007a; Tautz and Pfeifle, 1989). In embryos, these transcripts are destabilized at 2 h AEL and are thus targets of the ZDA alone (categories 6 and 7, Table 16.1; zygotic transcription of *hunchback* contributes little RNA compared with the large amount of maternal mRNA present in eggs and early embryos and thus does not obscure decay of *hunchback^{mat}*). Thus, it is important to compare the decay rate of any particular transcript in both systems.

The use of embryos to tease apart the contributions of the MDA and ZDA to maternal transcript turnover can be challenging. Consider, for example, how to distinguish transcripts that are degraded by the MDA alone (category 4, Table 16.1) from those that are degraded by the MDA + ZDA together (category 3, Table 16.1). Both of these maternal transcript classes are unstable in unfertilized and in fertilized eggs. However, a closer examination of their mRNA decay profiles indicates that a maternal transcript targeted by "MDA + ZDA" will be more abundant at 2 to 3 h in unfertilized eggs than at the same time point in fertilized embryos, where both activities function to accelerate decay. With *Hsp83* mRNA destabilization as an example, *Hsp83* transcripts exhibit a half-life of ∼75 min in 0- to 2-h embryos (when only the MDA is present); this decreases to ∼25 min in 2- to 3-h embryos (when both the MDA and ZDA function) (Bashirullah et al., 1999; 2001). Thus, *Hsp83* transcript levels are lower in 2- to 3-h embryos than in 2- to 3-h unfertilized eggs (category 3, Table 16.1; zygotic transcription contributes little RNA compared with the enormous amount of maternal *Hsp83* mRNA present in eggs and early embryos and thus does not obscure maternal transcript decay kinetics). In contrast, equivalent levels of an mRNA in both fertilized and unfertilized eggs shows that the particular maternal transcript is targeted by the MDA alone (category 4, Table 16.1); *nanos* is an example of such a transcript (Bashirullah et al., 1999).

An additional challenge to investigating maternal mRNA decay in fertilized embryos pertains to the maternal transcripts that are present at constant levels from egg lay through to the MBT (De Renzis et al., 2007; Tadros et al., 2007b). These transcripts may, indeed, be stable maternal messages (category 8, Table 16.1). However, they may also be unstable maternal mRNAs that are targeted by either the MDA, MDA+ZDA, or ZDA alone, but which are replaced by newly synthesized zygotic mRNAs (respectively, categories 2, 1, and 7, Table 16.1). In principle, one possible method to determine whether apparently stable transcripts are, in fact, unstable maternal transcripts that are replaced with zygotic messages, is to inject early embryos with α-amanitin before onset of *de novo* transcription. By preventing the accumulation of new zygotic transcripts, a time-course analysis of maternal mRNA decay through the MBT could be obtained. This experiment, however, cannot uncover maternal transcripts that are targets of the ZDA, because this pathway is, itself, dependent on zygotic transcription (i.e., the level of a ZDA-dependent transcript will remain constant if synthesis of the machinery required for the ZDA is blocked).

An alternate strategy is to examine maternal decay in aneuploid embryos, which provide a system where the normal array of maternal mRNAs is supplied to the embryo, but the zygotic genome lacks entire chromosome arms (De Renzis et al., 2007; Merrill et al., 1988; Wieschaus and Sweeton, 1988). This permits the analysis of maternal mRNA instability in the presence of MDA-mediated maternal mRNA degradation, in the presence of zygotic genome activation and thus ZDA-mediated maternal mRNA degradation, but in the absence of transcription from the absent chromosome or chromosome segment. Aneuploid embryos can be used to decipher two groups of transcripts in early development. The first is represented by maternal transcripts that seem to be stable in unfertilized eggs and through the MBT in fertilized embryos. Maternal transcripts exhibiting such a profile may be either maternal transcripts that are targeted by the ZDA alone and are also zygotically transcribed (category 7, Table 16.1), or maternal transcripts that are refractory to both degradation activities (category 8, Table 16.1). These two categories can be resolved by examining the level of a given transcript at 2 to 3 h AEL in aneuploid versus wild-type embryos: a category 7 mRNA will display a reduction in steady-state level in aneuploid embryos relative to wild-type embryos (because of the action of the ZDA but failure to produce zygotic transcripts), whereas a category 8 mRNA will show no difference in levels.

The second group of transcripts that can be distinguished by use of an aneuploid background is those that decrease in unfertilized eggs but show no change in abundance in fertilized embryos. These are either targets of the MDA alone (category 2, Table 16.1) or targets of the MDA + ZDA (category 1, Table 16.1; which accounts for their instability in unfertilized eggs) but are also zygotically transcribed, therefore masking their destabilization. By comparing the mRNA level for a gene of interest in unfertilized eggs to that in fertilized aneuploid embryos, one avoids obscuring mRNA decay because of zygotic transcription. Those maternal transcripts that are targeted by the MDA alone will show equivalent mRNA levels in both unfertilized eggs and fertilized aneuploid embryos at 2 to 3 h AEL, whereas maternal mRNAs that are targeted by the MDA + ZDA will display reduced levels in fertilized aneuploid embryos relative to unfertilized eggs at the same time point.

Other transcript classes that can be readily distinguished by time courses in both unfertilized eggs and fertilized embryos include those mRNAs that are not targeted by either decay machinery but are zygotically synthesized and, therefore, show an increase in abundance in fertilized embryos relative to unfertilized eggs (category 5, Table 16.1), as well as those maternal messages that are targeted by the ZDA alone but are not zygotically transcribed and, therefore, show a decrease in abundance in fertilized embryos only (category 6, Table 16.1).

1.3. Molecular mechanisms regulating maternal mRNA instability

The half-lives of maternal mRNAs that are destabilized in eggs and embryos vary from several minutes to several hours. This broad range of decay rates is due to multiple *cis*-acting and *trans*-acting factors that act to influence posttranscriptional regulation. Sites affecting mRNA stability in *cis* include the 3'-poly(A) tail (and its length), 5'-cap, and *cis*-acting instability elements within the body of the transcript. The *trans*-acting decay machinery is recruited by these *cis*-acting instability elements. Ultimately, the availability and combination of these factors dictates an mRNA's half-life.

With respect to the molecular pathways that lead to mRNA decay, deadenylation seems to be the first, and rate-limiting, step during early *Drosophila* development (for review see Semotok and Lipshitz, 2007). Maternal *Hsp83* mRNA is one of the best-characterized examples: SMG protein recruits the CCR4-NOT deadenylase complex to *Hsp83* transcripts, thus triggering deadenylation and subsequent degradation of the body of the mRNA (Semotok *et al.*, 2005). Recent examination of the *Drosophila* decapping machinery has also revealed a role for these components in the degradation of maternal transcripts such as *bicoid*, *twine*, and *oskar* (Lin *et al.*, 2006). With respect to postdeadenylation and decapping mechanisms, it is also not know whether decay is accomplished by means of 5' to 3'- and/or 3' to 5'-exonuclease activities or, potentially, by regulated endonucleolytic cleavage.

1.4. Prerequisites for triggering maternal mRNA destabilization

Detailed analyses of the prerequisites for maternal mRNA degradation have assessed the possible relationship between the various processes that initiate on egg activation and triggering of the MDA (Tadros *et al.*, 2003). Egg fragility, caused either by "upstream" oogenesis defects and/or a defective vitelline membrane, results in permeable, flaccid eggs that fail to activate the MDA. A correlation between egg fragility and mRNA decay has also been observed during egg/embryo injection experiments in which overdessication can affect the integrity of the egg membranes and consequently prevent the MDA (J. L. Semotok, A. Karaiskakis, and H. D. Lipshitz, unpublished data). Although fragile eggs fail mRNA decay, cross-linking of the vitelline membrane *per se* is not required to trigger the MDA, because class II *nudel* mutants, which are defective for cross-linking, undergo normal decay

(Tadros *et al.*, 2003). Other aspects of egg activation (i.e., translational activation, chromosome condensation, the S-to-M transition) have also been uncoupled from the trigger for mRNA decay.

2. Gene-by-Gene Analysis of mRNA Decay

2.1. Description and comparison of RNA methods

Three types of molecular assays are commonly used to determine decay kinetics of specific messages in the early embryo: Northern blots, RNA dot (or slot) blots, and real-time RT-PCR. A number of factors determine the assay of choice. These include (1) the amount of total RNA that can be feasibly collected, (2) the abundance and size of the transcript, (3) the number of transcripts to be evaluated, (4) the presence of multiple transcript isoforms, and (5) whether the transcript is not strictly maternal but is also zygotically transcribed. These factors, and additional considerations such as cost and time, are outlined in Table 16.2. Notably, either having some previous knowledge about the transcript of interest (see Berkeley Drosophila Genome Project [BDGP] expression databases, De Renzis *et al.* [2007]; Tadros *et al.* [2007a]) and/or performing some preliminary experiments is essential before these factors can be taken into account.

All three types of molecular assays can be used to determine mRNA abundance on the basis of the same principle. A series of samples is collected over a time course, and total RNA is extracted from each sample. Through the hybridization of radiolabeled or digoxigenin (DIG)-labeled antisense probes to the RNA of interest (i.e., with Northern or dot blots), or through the amplification of cDNA representing the transcript of interest with gene-specific oligonucleotides (i.e., with real-time RT-PCR), a reproducible raw signal can be obtained that is proportional to RNA quantity. By normalizing the amount of the target mRNA to a standard loading control (i.e., dividing the amount of the specific mRNA by the loading control amount, which was derived from the same sample at the same time point), the relative abundance of the specific transcript can be determined. *rpA1* and *rp49* transcripts are excellent loading controls, because they are abundant and stable across a 6-h time course in both unfertilized and fertilized eggs (Bashirullah *et al.*, 1999; Myers *et al.*, 1995; Surdej and Jacobs-Lorena, 1998; Tadros *et al.*, 2007a). As the entire *Drosophila* genome has now been analyzed over this same time course, *rpA1* and *rp49* are representatives of a large number of stable, abundant transcripts that may also be selected to serve as loading controls (De Renzis *et al.*, 2007; Tadros *et al.*, 2007a). Once the data are normalized, they can be graphed and half-lives calculated. This can be done with linear regression analysis in which the slope of the line of best-fit to the semi-log plot of RNA levels represents

Table 16.2 Comparison of RNA methods used to analyze gene-specific mRNA decay rates in *Drosophila* egg/embryos

Considerations	Northern blot	Dot blot	Real-time RT-PCR	FISH	Microarray Affymetrix	Microarray Two-color
Verification of probe specificity	+	−	+	−	+	+/−
Rare transcript	−	−	+	−	−	−
Small amounts of RNA	+/−	+	+	+/−	−	−
Multiple isoforms[a]	+	+	−	−	+/−	+/−
Detecting multiple RNA species simultaneously	+	−	+	+	+++	+++
Detecting multiple RNA species in parallel or sequentially	+	+	+	+	+++	+++
Spatial information	−	−	−	+	−	−
Costs	$	$	$	$	$$$	$$$
Protocol duration (not including sample collection)	2 to 3 days	2 days	1 day	3 to 4 days	2 days	2 days
Data interpretation and analysis	+	+	+	+	+++	+++

[a] If individual isoforms have distinct sequences that can generate isoform-specific probes without cross-hybridization, then it is possible to use methods denoted with (−) in this row.

the decay rate (K_d), and the half-life ($t_{1/2}$) can be calculated from the equation, $t_{1/2} = \ln 2/K_d$ (for a detailed explanation of determining decay rates, see Ross, 1995). Alternately, the empirical half-life can be calculated from the time point at which half of the transcripts have disappeared.

Northern blots resolve RNAs according to length (molecular mass) by denaturing gel electrophoresis. The size-fractionated RNA is transferred to a nylon membrane (either by capillary action or electrotransfer) and is hybridized to a radiolabeled or DIG-labeled antisense probe. After hybridization, the membrane is exposed to either X-ray film, a PhosphorImager screen that is subsequently detected by a laser scanner (radioactive blots), or directly to a chemiluminescence reader (nonradioactive blots). The latter two methods generate a highly sensitive and high-resolution digital signal that can be quantified with software packages such as ImageQuant TL (Amersham Biosciences) or Auto Image Capture (Fluorchem® Imaging System, Alpha Innotech).

A time-course analysis with a Northern blot assay can reveal many details about specific maternal transcripts in addition to decay kinetics. The complete absence of signal in either stage 14 oocytes, 0 to 1 h unfertilized eggs, or 0 to 1 h fertilized embryos would suggest that either the transcript is not present maternally or that its absolute amount is below the level of detection (in the latter case, real-time RT-PCR techniques should be used to resolve these possibilities; see following). If a signal is detectable, one can determine whether there are multiple maternal or zygotic mRNA isoforms. For example, the detection of *string* mRNA in unfertilized eggs reveals the presence of two maternal isoforms that are initially abundant but are subsequently eliminated because of the MDA. Examination of *string* mRNA in early embryos reveals that the shorter isoform is replenished with zygotic messages at the MBT (Bashirullah *et al.*, 1999). Fortunately, the larger, strictly maternal isoform and the smaller maternal + zygotic isoform are distinguishable by size, making the decay kinetics of the strictly maternal *string* RNA amenable to analysis.

An RNA dot blot is similar to a Northern blot except that the RNA sample is not size fractionated before transfer onto a nylon membrane; instead, the entire RNA sample is spotted directly onto the membrane, which is then incubated with a labeled probe. This method is beneficial for small samples such as when (1) few eggs are laid by females (e.g., when females carry a maternal effect mutation that causes a reduction in egg laying ability), (2) few females are available (e.g., because of genetic background or specific zygotic mutations that have an impact on adult survival), or (3) a limiting number of eggs withstand a particular protocol such as injection or permeabilization with drugs, antibodies, nucleic acids, or other compounds. Because the egg deposition rate of virgin females is lower than their mated counterparts, the examination of RNA decay in unfertilized eggs can exacerbate any of the previously listed situations. Fortunately, changes in steady-state RNA levels are readily detectable from single-egg RNA dot blots provided the transcript is not very rare (Fig. 16.2). Because transcripts are not size fractionated in the RNA dot blot protocol, there is a slight modification with respect to the use of loading controls relative to Northern blots. One option is to split each sample onto two blots and then probe these

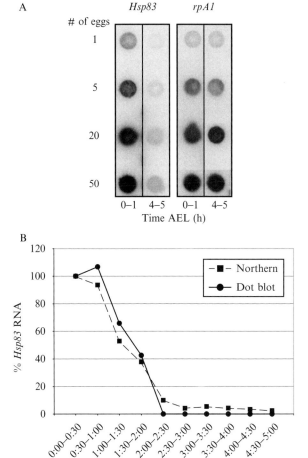

Figure 16.2 Maternal *Hsp83* mRNA degradation detected by dot blot hybridization. (A) Single embryo dot blot. Total RNA was extracted from 1, 5, 20, or 50 embryos collected at the time points indicated (AEL). Each well represents a separately prepared sample that was probed for either *Hsp83* or *rpA1* transcripts. (B) Comparison of *Hsp83* mRNA decay profiles generated by Northern and dot blot hybridizations. Normalized levels of *Hsp83* mRNA were plotted for embryos over a 5-h time course AEL. Data from independent experiments are presented. Half-hour time windows are shown.

separately for the transcript of interest and for the internal control. Accuracy in splitting the sample is essential in this case. The second option is to spot the entire sample and then probe sequentially for the transcript of interest and the loading control. Although this latter option results in an additional hybridization step, it minimizes the variability that can arise from sample splitting.

With respect to protocol limitations, RNA dot blot analysis cannot differentiate between different isoforms when a single gene-specific probe is used. It is, however, possible to bypass this limitation by designing probes that are specific for each isoform; typically, this requires a sequence difference large enough to generate distinct, nonoverlapping probes. With the advent of locked nucleic acid (LNA) probes, unique stretches of ≤ 20 nucleotides can, theoretically, be distinguished. Despite this possible solution, performing a preliminary Northern blot analysis with a single gene-specific probe that is predicted to detect all RNA isoforms is strongly recommended before choosing to use the RNA dot blot. A first-pass Northern blot experiment will verify the correct size of the transcript detected by the gene-specific probe, assess the existence of distinct isoforms, and determine whether any nonspecific bands are detected by the probe. Otherwise, none of these factors are distinguishable when an RNA dot blot is used and will prevent accurate determination of mRNA quantities and, therefore, decay rates.

Transcript abundance is an important factor in deciding which method to select to measure mRNA decay rates. If the mRNA of interest falls below the level of detection on Northern or RNA dot blots, real-time RT-PCR represents a robust and accurate alternative. This technique uses a reverse transcription step with random, gene-specific, or oligo(dT) primers to generate single-stranded cDNA before real-time PCR amplification. Subsequently, the cDNA is amplified in the presence of fluorescent dyes/probes such as SYBR-green reagent, molecular beacons, or Taqman probes (for more real-time RT-PCR information, see Bustin, 2000; 2002). These reagents release a fluorescent signal at each cycle that is proportional to the amount of cDNA present. Plotting these signals over the total number of PCR cycles results in an amplification curve where more abundant mRNAs amplify at earlier cycles than less abundant transcripts. With an independent cDNA reference sample to generate a standard curve, one can extrapolate the abundance of the mRNA of interest. Again, this must be compared with an internal control to determine a normalized value; comparing these normalized amounts over time can be used to generate an mRNA decay profile.

Because most transcripts studied to date in the early egg/embryo are degraded by deadenylation, it is recommended that either gene-specific or random primers be used for the reverse transcription reaction. The traditional use of oligo(dT) primers for first-strand cDNA synthesis may result in an underestimation of transcript abundance, because the mRNA may be present in a fully or partially deadenylated state and thus not be available to serve as a template for the reverse transcription reaction. Furthermore, relative to gene-specific primers, the use of random primers in the reverse transcription reaction has the added benefit of being able to use the same cDNA preparation to examine both the transcript of interest and the internal control. Experimental variation can be introduced when performing multiple independent reverse transcription reactions each using different gene-specific primer sets followed by independent gene-specific PCR amplifications. Thus, it is prudent to perform a single random-primed cDNA synthesis reaction for RNA derived from a single time point and then use this as template for all of the gene-specific real-time PCR analysis.

Without having to switch to microarray-based platforms (see Section 3), assembling transcript decay profiles for a small group of maternal mRNAs is possible with all three of the above-mentioned methods. Northern blots have the advantage that multiple transcripts can be probed for simultaneously (assuming distinguishable size differences between the transcripts). In addition, these membranes can be stripped and reprobed five (or even more) times. It may, however, be more difficult to completely remove probes hybridized to abundant transcripts. Several ways to alleviate this potential problem are: (1) scanning the membrane after stripping the blot to assess stripping efficiency, (2) distinguishing different transcripts according to their varying transcript lengths, and (3) probing for transcripts in the order of "most rare" to "most abundant." RNA dot blots, however, must be stripped and reprobed for each transcript. Examining multiple transcripts through the use of real-time RT-PCR methods is feasible if, as mentioned earlier, a common cDNA pool is used to amplify each transcript of interest. These can be performed as separate gene-specific PCR reactions or by multiplex PCR amplification. The latter has the advantage that the exact amount of cDNA starting material available for each primer set is identical, reducing the experimental variation introduced by pipetting error. Multiplex PCR reactions, however, have the disadvantage that cDNA templates that are more abundant amplify and plateau ahead of the other templates, potentially exhausting the reagents required for the amplification of the other gene-specific amplicons (and thus pushing them into the plateau phase much earlier). Because this may prevent accurate assessment of gene expression levels, it is essential to optimize such parameters as the concentrations of the multiple primer sets, [Mg^{2+}], and deoxynucleotides, as well as the annealing/extension temperatures in these multiplex reactions.

Another consideration with respect to studying maternal mRNA decay in *Drosophila* embryos is that the mRNA may exhibit differential stability in different regions of the shared cytoplasm. *Hsp83*, *hunchback^{mat}*, and *cyclin B* mRNAs are examples of maternal transcripts that are identified as eliminated when total RNA levels are assayed but for which differential spatial stability is revealed by *in situ* hybridization experiments (Ding et al., 1993;

Raff et al., 1990; Tautz and Pfeifle, 1989; Whitfield et al., 1989). Maternal *Hsp83* mRNA is eliminated from the bulk cytoplasm and yet protected in the posterior pole cells, whereas zygotic *Hsp83* mRNA is synthesized in the anterior head region by BICOID-mediated transcription (Ding et al., 1993). The zygotic and posterior-protected maternal *Hsp83* mRNA each represent less than 1% of the initial pool (Bashirullah et al., 1999). Methods such as alkaline phosphatase *in situ* hybridization (Tautz and Pfeifle, 1989) and fluorescent *in situ* hybridization (Lecuyer et al., 2007) can be used to detect spatial differences in mRNA distribution.

2.2. Analysis of deadenylation

Transcript stability is often regulated by changes in the length of the poly(A) tail. Deadenylation tends to be the first, rate-limiting step in eukaryotic mRNA decay, and, thus, shortening of the poly(A) tail correlates with the overall decay of the mRNA body. Because the total length of the poly(A) tail represents a small fraction of the total transcript size, it is difficult to resolve changes in adenylation status by Northern blot analysis alone. To overcome this obstacle, two techniques have been developed that can be used to assess whether a given maternal mRNA is subject to deadenylation in the early embryo. The first technique, the RNase H/oligo (dT)-cleavage assay, works on the principle that by hybridizing an antisense oligomer to a region 200 to 400 nucleotides upstream of the polyadenylation site followed by RNase H addition (which cleaves the RNA fragment of a DNA/RNA hybrid), one can generate a shortened $3'$-fragment of the target mRNA that can be resolved in a denaturing polyacrylamide gel and examined for subtle changes in poly(A) tail length (Decker and Parker, 1993). The inclusion of oligo(dT) along with RNase H and the gene-specific antisense oligomer will generate a cleaved, unadenylated mRNA fragment that can act as a size reference for completely deadenylated transcripts (i.e., A_0). Although the RNase H/oligo(dT)-cleavage method works best for abundant transcripts and/or requires the use of substantial amounts of total RNA, it has the advantage that it quantitatively and directly reads out the distribution of both heterogeneous and homogeneous populations of poly(A) tail lengths. This enables accurate calculation of the weighted mean poly(A) tail length for any particular RNA at any particular stage (Semotok et al., 2005).

The second technique, the poly(A) test (PAT) assay, involves PCR-based amplification of polyadenylated mRNAs with an oligo(dT) primer that can hybridize anywhere along the length of the poly(A) tail and serve as primer for first-strand cDNA synthesis (Salles et al., 1994). Along with a gene-specific primer, amplification of the cDNA can be performed to generate a PCR product representing the range of polyadenylated mRNAs produced by that gene (i.e., an oligo[dT] primer annealed at the very 3′-end of the poly[A] tail will produce a PCR product longer than an oligo[dT] primer annealing at the 5′-end of the poly[A] tail). As a reference for tail-length A_0, cDNA can be primed with a gene-specific primer to the transcript terminus minus the poly(A) tail. Thus, as opposed to the RNase H/oligo(dT) cleavage assay, the PAT assay examines the distribution of amplified DNA fragments and only indirectly the mRNA itself. One version of the PAT assay is RACE-PAT (rapid amplification of cDNA ends), which entails the use of oligo(dT) primers containing GC-rich anchor sequences that act to protect the 3′-end of the poly(A) tail during the PCR reaction. Another modification of PAT is LM-PAT (ligation-mediated-PAT). Here, the poly(A) tail is saturated with phosphorylated oligo(dT) primers that are ligated together with T4 DNA ligase to form a single stretch of dT sequence that coats the entire length of the poly(A) tail. By subsequently annealing with unphosphorylated, but anchored, oligo (dT) primers (which will anneal to any unpaired ends of the poly(A) tail), a second ligation with T4 DNA ligase is performed, and the cDNA synthesis is carried out as for RACE-PAT. Because PAT assays include PCR amplification, it is essential that the reaction is optimized to run within the linear range of the amplification curve to generate the most accurate reflection of poly(A) tail distributions. Both of these PAT assay techniques are relatively quick and sensitive and are beneficial for analyzing rare transcripts or when only small RNA samples are attainable (for detailed protocols, see Salles *et al.*, 1999 or Chapter 24 by Murray and Schoenberg).

2.3. Protocols

2.3.1. Sample collection

To maximize the number of fertilized embryos obtained for a given time point, it is best to select newly enclosed females and place them in egg-laying cages along with fertile males. For the optimal collection of unfertilized eggs, newly enclosed virgin females should be placed in egg-laying cages with sterile males (e.g., $\beta Tub85B^D$ or $Ms(2)M$). Because the flies will take time to adjust to their environment and to produce mature oocytes, egg collections are optimal starting 2 to 3 days after the flies have been placed in the cages. During this time, the cages should be given clean, freshly yeasted egg-laying plates on a daily basis. Once the flies have acclimatized to the cages and begun to produce eggs, collections may begin. For unfertilized eggs, in which decay proceeds only by the MDA, 2-h-window collections are sufficient for preliminary mRNA stability analysis, whereas half-hour or 1-h-window collections are best for fertilized embryos in which both the MDA and ZDA are present. Before the first collection, it is imperative to perform a "prelay." Because females are capable of retaining their eggs during times of stress (e.g., if there is a shortage of food or crowding), eggs or embryos will continue to age *in utero*, preventing accurate developmental staging. Replacing an overnight plate with a freshly yeasted plate once or twice before egg collection (for 30 to 60 min each time) encourages females to shed any retained eggs. To confidently assess the stage of fertilized embryos, one can quickly fix and stain a subset with a DNA stain such as propidium iodide, PicoGreen (Molecular Probes), or DAPI (Sigma).

Small-scale egg collections derived from approximately 50 to 100 females are usually sufficient for preliminary Northern blot analysis of transcript degradation. These eggs/embryos can be collected straight from the egg-laying plate and placed directly into the TRIzol solution for RNA extraction. An important consideration for this approach, however, is determining whether these eggs are fragile or exhibit any other defects in egg activation. Previous analyses have demonstrated that fragile eggs/embryos are also defective in RNA degradation by the MDA (Tadros *et al.*, 2003). Thus, it is important to perform an initial assessment of egg fragility by placing eggs/embryos in 50% bleach for 2 to 3 min; this removes the chorion and exposes the vitelline membrane. If the eggs/embryos survive the bleach treatment (i.e., they do not rupture), then they are not fragile. It is not necessary to bleach the eggs/embryos before RNA extraction from a small-scale egg/embryo collection. For larger-scale collections (>500 females), dechorionation is essential, because the chorion is highly enriched for polysaccharides that form a large, insoluble, glutinous pellet that coprecipitates with the RNA (this pellet is negligible in small-scale preparations). To avoid this, wash the eggs into a sieve, place the sieve in 50% bleach for 2 to 3 min, wash the collected eggs/embryos with 0.7% NaCl, 0.05% Triton X-100, then place them into TRIzol.

2.3.2. Northern blot analysis

The following methods are to be performed with standard molecular biology solutions, reagents, and equipment, as well as standard practices for handling RNA. To minimize contamination of RNA samples with RNases, use RNase-free tips, tubes, and homogenizers; change gloves frequently; rinse equipment with 0.25 N HCl, then 0.5 N NaOH, followed by DEPC-H_2O or use commercially prepared reagents such as RNAzap (Ambion); and prepare RNase-free solutions with DEPC-H_2O or DEPC-treat the solution itself. For detailed information regarding these practices, see Ausubel (1987) and Maniatis (1982).

2.3.2.1. Total RNA extraction with TRIzol For details see Invitrogen: Isolation of RNA with TRIzol reagent.

Tips: Either (1) scrape eggs off of nylon mesh after collecting eggs in sieve or (2) pick eggs off the plate with a small spatula. Place eggs directly into an Eppendorf tube containing 150 µl of TRIzol. Homogenize eggs, leave at room temperature for 5 min, and proceed with the next step of the TRIzol extraction protocol or freeze samples on dry ice/ethanol and store at −80 °C. When ready to extract (thaw samples on ice if needed), add 350 µl of TRIzol (i.e., total volume of TRIzol = 500 µl). For small egg collections (i.e., 50 to 100 eggs), 500 µl total TRIzol volume will be sufficient. For large egg collections (i.e., where the total egg volume ≥50 µl), increase the amount of TRIzol such that the total TRIzol volume is 10 times the total egg volume.

When homogenizing eggs, be sure to homogenize thoroughly. Homogenization ruptures the eggs while leaving intact chorions, which can be mistaken for whole eggs. Examining the homogenate briefly under a dissecting microscope will indicate how well the homogenization was performed.

For the homogenization process, the use of disposable, individually packaged RNase-free pestles and compatible tubes is ideal but can be costly. Alternately, a Teflon pestle that fits standard RNase-free Eppendorf tubes may be used if cleaned and wiped with 100% ethanol in between homogenization of different samples.

For large egg collections, resuspension of the RNA pellet (step 11) may require a larger volume of DEPC-H_2O. Determine RNA concentration of the sample first and then dilute accordingly (aim for a concentration of 1 to 2 µg/µl of total RNA).

2.3.2.2. Formaldehyde gel electrophoresis and Northern blotting
See Ausubel (1987) and Maniatis (1982) for details on preparation of formaldehyde gels, resolution of RNA, and probing by Northern analysis.

Tips: Ribosomal RNAs, which represent ≥95% of total RNA, migrate within the 1.2- to 1.8-kb range. Target mRNAs that also migrate in this range tend to be obscured by the shear abundance of the rRNAs (known as a "white-out" effect). To resolve decay kinetics of transcripts within this range, it may be beneficial to perform an oligo(dT) selection first to generate a poly(A)-enriched, rRNA-free RNA sample before loading on a Northern gel (∼3-5 µg of poly(A)-selected RNA/lane). See, however, the caveats discussed in Section 2.1.

2.3.3. Single-embryo dot blot analysis
See Ausubel (1987) and Maniatis (1982) for details regarding dot blot analysis.

Tips: For the extraction of total RNA from a single embryo, follow the preceding protocol for RNA Extraction with TRIzol (from Invitrogen) but instead of 100% TRIzol, make up a TRIzol + glycogen stock (250 µg glycogen/ml TRIzol). Homogenize single eggs in 150 µl of TRIzol + glycogen, leave at room temperature for 5 min and store at −80 °C. When ready to extract, thaw samples and add 350 μl of TRIzol + glycogen (i.e., total volume of TRIzol = 500 μl) and proceed with extraction protocol as described previously. After the 75% ethanol wash and air-drying steps, dissolve the RNA pellet in 20 μl of DEPC-H$_2$O, pipette up and down, and again incubate at 55 to 60 °C for 10 min, and store at −80 °C.

2.3.4. Poly(A) tail analysis

We prefer the RNase H/oligo(dT) method and use a modified protocol from the Roy Parker laboratory (http://www.mcb.arizona.edu/parker/PROTOCOLS/RnaseHDigestions.htm). Please refer to Chapter 24 by Murray and Schoenberg for a detailed description of the RNase H/oligo (dT) treatment of RNA.

Tips: For good resolution of poly(A) tail lengths with the RNase H method, select an antisense gene-specific primer approximately 200 to 400 nt upstream of the transcript terminus (not including the poly(A) tail itself). This will generate a 3′-cleaved product of the same length (i.e., 200 to 400 nt).

3. Genome-Wide Analysis of mRNA Decay

3.1. RNA isolation

A critical factor in microarray experiments is the preparation of high-quality RNA. Either total RNA or mRNA may be used. However, for RNA degradation studies, there is an added advantage to the use of total RNA. Because most (usually 98% or more) RNA in the sample is not mRNA, when equal amounts of RNA samples by weight are compared, total RNA samples will more accurately reflect the amount of message for a given gene. In contrast, if mRNA samples are used and a substantial fraction of the mRNA pool is undergoing degradation, then comparison of equal amounts of mRNA causes an artifactual increase in the relative fraction of stable messages at the time point at which substantial degradation of a subset of the mRNAs has occurred.

Total RNA may be isolated with a variety of methods, but those most commonly used are purification columns such as Qiagen's RNeasy products or phenol-based reagents such as Invitrogen's TRIzol. Our experience has led us to rely on TRIzol to extract RNA from most samples. On occasion, we have found that RNA from some strains, developmental stage, and/or tissue type requires a secondary clean up (e.g., with MegaClear [Ambion]). In some experiments it may not be possible to obtain enough total RNA to label, and the sample will need to be amplified. We have found Message-AmpII (Ambion) to be able to produce reproducible results so long as only a single round of amplification is used. Whatever type of RNA isolation, purification, and/or amplification method is used it is very important that all

RNA samples are isolated and treated exactly the same or bias will be introduced into the results.

3.1.1. Sample collection protocol. See Section 2.3.1

3.1.2. RNA extraction protocol. See Section 2.3.2

RNA pellets are usually resuspended in 18 MΩ RNase-free water (Sigma) to a minimum concentration of 4 µg/µl. RNA quantity and quality must be confirmed spectrophotometrically. A_{260}/A_{280} ratios should typically be >1.6. Denaturing agarose gel electrophoresis (see Section 2.3.2) can be used to evaluate RNA quality (Sambrook and Russell, 2001).

3.2. Labeling and hybridization

3.2.1. Oligo(dT) versus random priming

In any microarray experiment, the total or mRNA is converted to labeled cDNA with reverse transcriptase. Of importance is the primer used to prime the reverse transcription reaction. In most experiments, some variant of oligo(dT) is used because it will hybridize to the poly(A) tail of mRNA in the production of cDNA. However, when studying RNA processing or degradation, one should consider random primers. The rationale for doing so is that specific transcripts in the process of degradation may have variable length (or no) poly(A) tail, and, therefore, the use of an oligo(dT) primer introduces a bias during the reverse transcription reaction (see Section 2.1 for additional consideration of this point).

Commercially available random primers include random hexamers (Invitrogen), nonamers (New England Biolabs), and decamers (Ambion). One potential drawback to the use of random primers is that, in general, when an equivalent amount of total RNA is used as starting material, they tend to produce less cDNA than if oligo(dT) is used to prime the reverse transcription reaction. However, this problem is largely overcome if longer random primers are used and pentadecamers (15 mers) seem to give the highest yield of cDNA (Stangegaard et al., 2006). One possible reason why shorter random primers (i.e., 6 to 10 mers) do not perform as well as longer ones, is that they do not anneal as well, particularly to AU-rich regions, thus reducing the efficiency of cDNA synthesis on reverse transcription (Stangegaard et al., 2006). When random primers of 18 nucleotides or longer were used, cDNA production was less than for 15 mers (Stangegaard et al., 2006), perhaps because of more secondary structures in the longer oligomers.

3.2.2. "Direct" versus "indirect" labeling methods

For one-sample-per-array systems such as Affymetrix, this is not applicable, because every sample gets labeled in an identical fashion. In standard "two-color" (or two-sample) microarray experiments, cDNA representing both an experimental and reference RNA sample are cohybridized to a single microarray. In the direct labeling approach, different dye-coupled nucleotides are added to the cDNA synthesis reaction and are directly incorporated into the nascent molecules during reverse transcription. This necessity has given rise to the concept of "dye bias": one dye-coupled nucleotide may be more efficiently incorporated than the other by reverse transcriptase, giving rise to a biased pool of labeled cDNA. In many cases, the effects of dye bias may be minimal (Neal and Westwood, 2006; Neal et al., 2003). The potential bias can be largely avoided by the use of the indirect labeling approach. In this method the same modified nucleotide—for example amino-allyl dUTP—is added to the labeling reactions for the respective samples to be compared and a different dye ester is conjugated to each sample in a secondary reaction.

3.3. Microarray platforms

A variety of microarray platforms have successfully been used to study mRNA degradation; these include microarrays based on amplified cDNAs (Tadros et al., 2007a), short oligonucleotides (i.e., Affymetrix) (De Renzis et al., 2007), and long oligonucleotides (Grigull et al., 2004). To date, there does not seem to be a clear advantage of one platform over the others. Potential advantages of the short-oligonucleotide platforms such as Affymetrix is that they generally have multiple evenly spaced probes representing each gene, whereas the probes on cDNA-based arrays are generally 3′-biased and tend to only have one probe per array. Similarly, long-oligonucleotide–based arrays also tend to, but do not always, have only one probe per array. Of course, one disadvantage of Affymetrix and other commercial microarray platforms is the higher cost.

Another point to consider when an array platform is chosen is the reference sample used in the comparisons. Both cDNA and long-oligonucleotide–based array experiments are usually two-color experiments, as described previously. Therefore, in such experiments, one must be sure that there is enough reference sample (which is usually the first time point in the study) available to be labeled and hybridized to each of the experimental samples and their replicates. Extra reference RNA must be available for the repetition of failed arrays and for validation by other techniques (e.g., real-time RT-PCR or Northern blots; see Section 2.1). Also, to reduce the variation in the two-color comparisons, it is advisable to pool all the different isolates of reference-sample RNA so that the exact same reference is used for each array in the study (Neal and Westwood, 2006).

3.4. Data normalization

Unlike most microarray experiments in which relatively few genes change their expression levels between the experimental and reference (control) RNA samples, in studies of maternal mRNA degradation large numbers of genes decrease in expression level. Therefore, proper quantification and normalization of the fluorescent signals on the arrays is crucial.

3.4.1. Scanning and signal quantification

The first step in the data acquisition process is the initial scanning of the microarray. For two-color experiments, it is typical to scan the array with each of the two lasers and adjust either the laser power and/or gain (PMT) such that the total intensity of the signal for the entire array is approximately equal for each of the two channels. Software quantification programs such as GenePix Pro (Molecular Dynamics) have tools to perform these types of comparisons quickly so that additional scans at different settings can be performed to achieve this goal. However, this approach will artificially raise the signals in the experimental channel, because there is less mRNA (per weight of total RNA) in the labeling reaction to begin with. One way to compensate for this during the scanning is to examine the signals for RNA "spike-in" controls (see following) and to use the pair of channel scans that has a ratio of approximately 1.0 for the RNA "spike-ins." Another approach, useful when a given batch of arrays has been labeled and hybridized under identical conditions, is to label the reference sample with each of the two dyes in separate reactions. Then, once laser power and gain settings are found when the total intensity signal for the entire array is approximately equal for each of the two channels, those settings are used to scan the remaining arrays in that hybridization batch. Neither of these approaches is always practical for a given experiment; an alternate is to scan the arrays with a consistent protocol (e.g., routinely scan so the total signal intensity for each channel is equal) and to normalize the signals after scanning. For more information on approaches for scanning, as well as commercial and freely available microarray quantification software, see Neal and Westwood (2006).

3.4.2. Postscanning normalization

To normalize after scanning, a variety of approaches can be used. For more typical microarray experiments, the use of normalization algorithms that use regression or spline-based approaches such as the standard Lowess smoothing method of the experimental data are highly effective, particularly if they are applied on a subarray basis (Yang et al., 2002). However, these types of approaches only work well when there are relatively few changes in the experimental versus reference samples (as discussed previously, in maternal RNA degradation experiments this is not the case). Another approach, which is more suitable for RNA degradation experiments, is to normalize the array signals to genes that are known not to change noticeably over the time course of the experiment. The stability of these transcripts should be confirmed in the experimental and reference samples by some independent method (e.g., Northern blots) before use for normalization. Genes that have transcripts that have been used successfully for normalization in microarray degradation experiments include ribosomal protein subunit genes (e.g., *RpL32* [*rp49*] and *RpLP2* [*rpA1*]) and RNA polymerase subunit genes (e.g., *RpL*) (Tadros *et al.*, 2007a). One of the drawbacks with genes such as these to normalize signals is that, on most array platforms, they do not appear very often on the array; thus, hybridization and other variations that may occur on the array may not be adequately taken into account.

A second approach that can be used to normalize the array signals in degradation experiments is the use of "spike-in" RNAs. "Spike-in" RNAs are usually purchased or generated in an *in vitro* T7- or T3-based transcription reaction. The RNA "spike-ins" can then be added in equal amounts to both the experimental and reference RNA samples before labeling and the signals on the array then normalized to the hybridization signals of the "spike-in" probes. Similar to the use of stable transcript genes for normalization, the use of "spikes-ins" for normalization is most effective when probes for the "spike-ins" occur multiple times on the array, preferably in each subarray.

A third approach that can be used is the rank invariant Lowess extrapolation method (Schadt *et al.*, 2001; Tseng *et al.*, 2001). In this method, each respective channel intensity and the average intensity for each gene within a hybridization group is used to determine a subset of unchanged genes, disregarding relative expression ratios. After iterating through the algorithm, a subset of confidently invariant genes is defined, and the Lowess normalization algorithm is applied to them. The curve generated from the Lowess algorithm is then extrapolated to the rest of the data. The rank invariant Lowess extrapolation algorithm can be used directly within microarray software analysis packages such as GeneTraffic (Stratagene) (Tseng *et al.*, 2001).

3.5. Data analysis, verification, and interpretation

3.5.1. Establishing transcripts present in the reference sample

Before one can identify degraded transcripts and their degradation rates, it is important to carefully identify what transcripts are present in the reference (control) samples. In studies examining the degradation of maternal transcripts during early *Drosophila* embryogenesis, this involves determining which transcripts are maternally loaded and present in the late-stage (14) oocytes (e.g., Tadros *et al.*, 2007a). With the approach taken by Tadros *et al.* (2007a) as an example, one needs first to determine a cutoff value that differentiates between what is maternal expression (i.e., present in the reference sample) and what is background. This can be accomplished by creating a table (spreadsheet) that contains the average of all of the raw signal intensity values for each gene for all the hybridizations in the study (including all of the replicates) of wild-type stage 14 oocyte RNA (i.e., the reference sample used for each hybridization). One can then sort the data on the basis of these averages, from highest to lowest, with the genes at the top of the list having transcripts at the highest levels, whereas the genes at the bottom of the list are expected not to be maternally loaded. In this particular study, of the 10,500 distinct protein-coding genes represented on the microarray, 9362 were analyzed for maternal expression on the basis of a number of filtering criteria. The authors independently verified the expression level of a number of genes found at different positions within the list by scanning the BDGP *in situ* database (http://www.fruitfly.org/cgi-bin/ex/insitu.pl) to determine whether transcripts could be detected for the particular gene in the early embryo, before initiation of high-level mRNA decay (Fig. 16.3). Consistent with the predictions from the microarray data, transcripts from the genes at the top of the table were confirmed to be present by *in situ* hybridization: 90.4% of selected genes with an average raw intensity of >20,000 (out of a possible 64,0000 maximum) were verified by *in situ* data ($n = 73$ genes examined in this intensity range). For genes at the bottom of the list, very few had transcripts that were detectable by *in situ* analysis: only 8.2% of the transcripts with an average raw intensity of <2000 were maternal ($n = 73$). Transcripts with average raw values between 3500 and 5000 were mostly maternal (75%; $n = 76$) and, as the raw values decreased, fewer maternal transcripts could be verified by *in situ* analysis (see Fig. 16.3B and C). With the *in situ* database verification as a guide, the authors chose a cutoff raw intensity value of 3000 to be designated as maternally loaded into wild-type stage 14 oocytes (see Fig. 16.3A). On the basis of that cutoff, it was determined that 54% of the genome was maternally expressed and present in stage 14 oocytes (5097 genes of the 9362 genes analyzed; extrapolated to 7745 genes for the whole genome) (Tadros *et al.*, 2007a).

If *in situ* databases are not available, other means can be used to verify appropriate cutoff limits to establish the transcripts present in a sample. Often other sources of microarray data carried out with the same or similar biological material under similar conditions can be used. For example, the BDGP *in situ* database mentioned earlier also presents transcript levels for individual genes during embryogenesis obtained from Affymetrix Gene-Chip experiments. Alternately, one can select several genes at different raw intensity levels within the table and confirm their presence in the reference sample by use of independent methods such as Northern blot analysis, quantitative real-time PCR, and/or *in situ* hybridization.

3.5.2. Identifying transcripts undergoing degradation

Once each of the transcripts in the reference sample has been identified as "present" or "absent," its stability as a function of time (or treatment) can be determined. Establishing criteria to determine which transcripts show decreased levels is largely up to the experimenter but should be based on

A

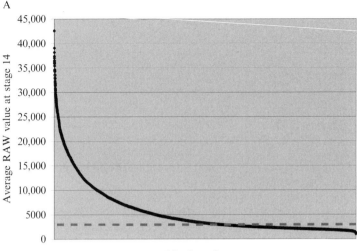

Number of genes

B

Average RAW value	% maternal	n
>20,000	90.4%	73
5000 – 3500	75.0%	76
3500 – 3000	58.7%	172
3000 – 2800	46.9%	160
2800 – 2500	26.2%	80
2500 – 2000	24.7%	81
<2000	8.2%	73

C

Figure 16.3 Identification of maternally deposited transcripts in stage 14 oocytes. (A) The average raw signal intensity of 9362 genes evaluated by microarray analysis in 27 arrays is plotted by descending signal intensity. The dashed line indicates a signal intensity of 3000; genes having signal intensities above this value are considered maternally deposited (see text for details). (B) Groups of genes having signal intensities in the range indicated in the table were examined for their presence or absence in the Berkeley *Drosophila* Genome Project *in situ* database for early development. "n" refers to the number of genes examined in the particular intensity range in the *in situ* database, and "percent maternal" indicates the percentage of those genes that were confirmed to be present in the earliest stage embryos by *in situ* hybridization. (C) Examples of *in situ* hybridizations of early-stage embryos for genes that are maternally deposited (left) and those that are not (right). Data taken from the study conducted by Tadros *et al.* (2007a).

sound statistical criteria. Neal *et al.* (2003) had previously determined for cDNA arrays like those used in the Tadros study that, with triplicate data, the threshold values for the 99% confidence interval were the equivalent of a ±1.3-fold change, far less than the arbitrary twofold threshold used in many earlier microarray studies. With a cutoff of ±1.5-fold, less than 1% of the mRNAs identified as having decreased in amount have a chance of being false positives.

In the Tadros study, consistent with the fact that unfertilized eggs are transcriptionally silent, the amount of up-regulation observed was within the limits of experimental variability: only 88 of 9257 transcripts (less than 1%) showed a greater than 1.5-fold increase by 4 to 6 h after egg activation and only two (0.02%) showed a twofold increase (Fig. 16.4). In contrast, 1069 transcripts (21%) showed a 1.5-fold decrease and 501 (10%) a twofold decrease by 4 to 6 h; 45 transcripts decreased at least fivefold in abundance with *Hsp83*, a well-studied unstable maternal transcript (Bashirullah *et al.*, 1999; 2001; Ding *et al.*, 1993; Semotok *et al.*, 2005; Tadros *et al.*, 2003), being in this group. The most unstable transcript, *draper*, decreased almost tenfold. The observed decreases were biased toward abundant mRNAs, which is consistent with the fact that rare transcripts often reach background levels after undergoing a smaller fold reduction than abundant transcripts and are, therefore, often excluded from analysis by low-threshold filters. It was concluded that more than a fifth of the maternal mRNAs loaded into oocytes during oogenesis (i.e., maternal mRNAs encoded by >1600 distinct genes) are destabilized on egg activation.

3.6. Protocols

3.6.1. Labeling and hybridization

This protocol involves "indirect" labeling of the cDNA and uses amino allyl-dUTP and reactive fluorescent Alexa dyes. Methods that use nucleotides conjugated directly to fluorescent dyes can also be used. See Neal *et al.* (2003) for a detailed version of a "direct" labeling protocol.

3.6.1.1. Solutions required
3.6.1.1.1. Indirect reverse transcription master mix

The Indirect RT master mix may be prepared and stored at −20 °C for ≤2 mo. The volumes below are sufficient for 20 labeling reactions.
160 μl 5× Superscript II buffer (Invitrogen, Cat. no. 18064-014).
30 μl random hexamer primer (3 μg/μl) (Invitrogen, Cat. no. 48190-011).
60 μl dCTP, dATP, dGTP mix (20 mM each in mix).
60 μl 2 mM dTTP.
30 μl 4 mM amino allyl-dUTP (Sigma Cat. no. A 0410).

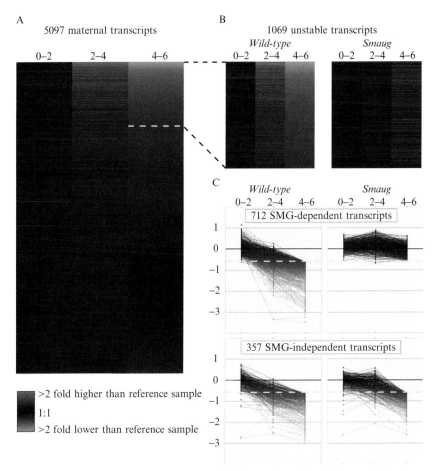

Figure 16.4 Microarray-based gene expression profiling of maternal transcript stability in activated unfertilized eggs from wild-type and *smaug* mutant females. Unfertilized eggs were collected 0 to 2, 2 to 4, and 4 to 6 h after egg laying. (A) Maternal mRNAs (5097) were sorted according to instability at 4 to 6 h in wild type; each is represented by a horizontal bar, with black indicating no change, green a decrease, and red an increase in transcript abundance relative to stage 14 oocytes. The 1069 transcripts above the dashed yellow line (ratio's log base 2 of 0.59 = 1.5-fold decrease) are significantly destabilized in wild type. (B) Effect of *smaug* mutant on the 1069 transcripts that are unstable in wild type. It can be seen that most are stabilized in the mutant (i.e., change from green to black). (C) Kinetics of the effect of the *smaug* mutation on the 1069 transcripts that are unstable in wild type. Two-thirds (712) are stabilized in a *smaug* mutant ("SMG-dependent"), whereas one-third (357) remain unstable in the mutant ("SMG-independent"). Figure from Tadros *et al.* (2007a). (See Color Insert.)

80 µl 0.1 *M* DTT (Invitrogen, Cat. no. 18064-014).
420 µl total volume.
Use 21 µl per labeling reaction.

3.6.1.1.2. Hybridization buffer This solution is prepared fresh. Only make the amount required for each hybridization reaction and the volume prepared will depend on the hybridization chamber or processor being used. With CMT-GAPS hybridization chambers (Corning, 2551), for each array prepare:

75 µl DIG Easy Hyb (Roche, Cat. no.11603558001).
4 µl 10 mg/ml yeast tRNA (Invitrogen, Cat. no. 15401011).
4 µl 10 mg/ml salmon sperm DNA (Sigma, Cat. no. D-9156).
83 µl total volume.
Use 80 µl of hyb buffer per array.

3.6.1.1.3. Procedure

A. Reverse transcription
   1. Add the appropriate volume of RNase-free water to each of the RNA samples to bring the final volume to 19 µl (for 40 to 60 µg total RNA).
   2. Add 21 µl of indirect reverse transcriptase (RT) master mix into each tube.
   3. Incubate the reaction mixture at 65 °C for 5 min.
   4. Incubate the reaction mixture at 42 °C for 5 min.
   5. Add 2.0 µl of Superscript II reverse transcriptase.
   6. Incubate the reaction at 42 °C for 3 h.
B. cDNA clean-up and precipitation
   1. Add 8 µl 1 N NaOH, pipet a few times to mix, give a quick spin.
   2. Incubate at 65 °C for 10 min.
   3. Add 8 µl 1 N HCl, pipet a few times to mix.
   4. Add 4 µl 1 M Tris (pH = 7.5), pipet a few times to mix followed by a quick spin.
   5. Add 38 µl of water to bring total volume to 100 µl.
   6. Purify the amino allyl-cDNA with column purification kit (Qiagen PCR clean up or Invitrogen Purelink purification kit). (*Notes*: Use 80% EtOH for wash buffer; elute 2× 50 µl of water for a total volume of approximately 98 µl.).
   7. Add 10 µl 3 M NaOAc, 1 µl glycogen (20 µg/µl), and 120 µl ice-cold isopropanol, and precipitate at −20 °C for at least 75 min or overnight.
   8. Spin at top speed (>12,000g) for 30 min.
   9. Wash the pellet with 200 µl of 75% EtOH, then spin at top speed for 5 min.
   10. Carefully pipet all the EtOH from the tube, let pellet air-dry for ≤1 min.
   11. Resuspend the probe in 5 µl of water.

C. Dye conjugation
   1. Add 3 µl 0.3 M NaHCO$_3$ to the resuspended amino allyl-cDNA.
   2. Add 2 µl of the reactive dye (Alexa647 or Alexa555; Invitrogen Cat. no. A32755).
   3. Incubate at room temperature in the dark for 1 h.
   4. Add 90 µl of ddH$_2$O to the conjugated cDNA.
   5. Purify as earlier with column purification kit.
   6. Wash with 80% EtOH three times.
   7. Elute 3× 50 µl of water for a total volume of approximately 148 µl.
D. Probe cleanup and precipitation
   1. Combine samples labeled with Alexa647 and Alexa555.
   2. Add 30 µl 3 M NaOAc, 3 µl glycogen (20 µg/µl), and 340 µl ice-cold isopropanol to the labeled probes.
   3. Precipitate at −20 °C for at least 30 min.
   4. Spin at top speed (>12,000g) for 30 min.
   5. Wash the pellet with 200 µl of 75% EtOH, spin at top speed for 5 min.
   6. Carefully pipet all the EtOH from the tube, let pellet air-dry for ≤1 min.
   7. Resuspend the probe in 5 µl of water.
E. Microarray hybridization
   1. Add 80 µl of hybridization buffer to each resuspended probe (total volume 85 µl).
   2. Incubate the mixture at 65 °C for 10 min.
   3. Place the probe on the array and cover with a 24 × 60-mm coverslip. Examples of arrays that have been successfully used for mRNA degradation analysis in *Drosophila* include the 12K cDNA-based array and the 14K long oligo-based array available from the Canadian *Drosophila* Microarray Centre (www.flyarrays.com).
   4. Place in a sealed hybridization chamber (Corning CMT-GAPS hybridization chamber Cat. No. 2551) in a 37 °C waterbath for 16 to 18 h.
   5. Prepare wash buffers as described in the next step.
F. Washing the microarray
   1. Wash the array 3× 15 min in prewarmed 1× SSC, 0.1%SDS.
   2. Wash the array with room temperature 1× SSC for ≤1 min.
   3. Wash the array with room temperature 0.1× SSC for ≤15 sec.
   4. Dry the microarray either by blowing of the residual liquid with clean compressed nitrogen (or air) or by centrifugation.

3.6.2. Scanning and quantification and analysis of microarray data

The scanning and quantification of a given hybridized microarray is very microarray platform specific and experiment specific. The following protocol is essentially what was performed for the microarray study performed by Tadros *et al.* (2007a) in which cDNA-based microarrays were used and each microarray compared with an experimental or reference sample. Most of the following methods can be adapted to similar two-color microarray experiments even if different scanners and software are being used. See also sections 3.4 "Normalization—Scanning and Signal Quantification," 3.5 "Data Analysis, Verification, and Interpretations," and Neal and Westwood (2006) for more explanation.

A. Scanning
   1. Microarrays were scanned with a ScanArray 4000 system (GSI Lumonics/Perkin Elmer) at 10 μm resolution with both blue and red lasers. The scanning software used was ScanArray 2.0 that accompanies the scanner. Laser settings were typically 100% power for both lasers and 70 to 75% gain (PMT) for the blue laser (green channel; Cy3 or Alexa 555 dyes) and 60 to 65% gain for the red laser (red channel; Cy5 or Alexa 647 dyes).
   2. Typically, two or three scans were done with each laser, adjusting the gain between scans. The goal of performing multiple scans is to match as closely as possible the overall intensity of the two channels. This can be assessed in a number of ways and assessments are usually done between scans to determine whether the goal of matching the overall intensities (i.e., to within 10% of each other) has been achieved. One way of assessing the overall intensity is to look at particular spots (genes) that are expected to have equal RNA levels in the experimental and reference samples (e.g., ribosomal protein transcripts) or a "multiple sample pool" that contains a mixture of probes representing several genes. In the case of the Tadros et al. (2007a), the latter was used. Another method that can be used involves software (e.g., GenePix Pro) that has the ability to quickly quantify all of the channel ratios for each of the spots on the array and average them.
B. Quantification
   The two 16-bit TIFF image files generated for each microarray—one for the Cy3 channel and one for the Cy5 channel—were quantified with QuantArray Version 3 (Perkin Elmer). The adaptive quantification algorithm was used in the analysis of all the experiments. The QuantArray program generates an Excel file with the quantified values for each of the image files. After quantification, spots that were not usable because of imperfections in the array or hybridization artifacts were manually flagged in QuantArray. Automatic flagging of spots on the basis of spot morphology and other criteria was also performed in QuantArray.
C. Setting up a microarray data analysis project with specific software (GeneTraffic)
   Numerous commercial and publicly available software packages designed to analyze microarray experiments are available (see Neal and Westwood, 2006 for examples). For this analysis, GeneTraffic 3.2 (now ArrayAssist) (Stratagene) was used for most of the postquantification analysis of the

microarray data. The first step in the process involves the creation of the project in GeneTraffic and the uploading of various files, including an array layout file compatible with the quantification files (in this case an array layout for QuantArray files); the quantification file (i.e., Microsoft Excel file); two TIFF image files associated with each scanned microarray; and an annotation file that contains information for each gene found on the microarray. Array layout and annotation files are often available from the array's manufacturer—in the case of the Tadros study, the Canadian *Drosophila* Microarray Centre (www.flyarrays.com). Once the files have been uploaded into the project, it is advisable to fully annotate the project following Minimal Information About Microarray Experiments (MIAME) conventions.

D. Data flagging and background subtraction

Flagged spots are those that you do not want to consider as being valid. For the (Tadros *et al.*, 2007) project, in addition to the flagging that was performed in QuantArray, the following flagging was performed in GeneTraffic

1. Flag spots with intensity lower than a certain multiple of the local spot background: Labeled Extract—Experimental (or evaluated) (LEX.E) and Labeled Extract—Reference (LEX.R) to background intensity ratio less than 3 for each.
2. Flag spots with intensity lower than a certain multiple of the average background: LEX.E and LEX.R intensity less than three times average background.
3. Flag spots with intensity lower than a given (user defined) value: LEX.E and LEX.R less than 128.
4. Background subtraction: none (i.e., do not subtract local background values from spot intensities).

E. Data normalization

For the Tadros *et al.* project, data were normalized to "Housekeeping Genes" with the ribosomal protein subunit gene *RpL32* (*rp49*) and the RNA polymerase I subunit genes (*RpL*) designated as the "Housekeeping Genes."

F. Identification of differentially expressed genes

To find the transcripts ("genes") that were degraded, differentially expressed genes were identified in the "Gene View" mode of GeneTraffic with the following criteria:

Genes having an average mean $\log^2$ value $\leq$ or ≥ 0.59 (1.5-fold).

Number of valid spots ≥ 4. (i.e., two thirds of spots on triplicate arrays per sample; two spots for each gene on the array).

Coefficient of variance (CoV) <1 (100%).

In addition to the preceding criteria, for a gene to be considered as differentially expressed it also had to be shown to be statistically significant by an independent method. The method used was

Significance Analysis of Microarrays (SAM) (Tusher et al., 2001) version 2.20 (http://www-stat.stanford.edu/~tibs/SAM/). The false discovery rate in the program was set to ≤1.

Further discussion about how maternally expressed genes were identified and which of those underwent degradation can be found in Section 3.5 "Data Analysis, Verification, and Interpretations."

4. Concluding Remarks

This chapter provides the background information and methods required to design and conduct experiments to measure maternal mRNA decay rates and mechanisms during early *Drosophila* embryogenesis. These tools can be used to explain the molecular and genetic components that mediate maternal mRNA instability; by applying the techniques to both wild type and mutants that exhibit defective maternal decay, *trans*-acting factors involved in decay can be identified and analyzed (Bashirullah et al., 1999; Semotok et al., 2005; Tadros et al., 2003; 2007a). Likewise, comparison of the decay kinetics exhibited by engineered transgenic transcripts relative to endogenous ones can be used to map *cis*-elements that are necessary and/or sufficient to mediate transcript stability and instability (Bashirullah et al., 1999; Semotok et al., 2005). In combination with studies in other genetic models such as *C. elegans* (Tenenhaus et al., 2001), zebrafish (Giraldez et al., 2006; Wolke et al., 2002), and mice (Hamatani et al., 2004), these approaches will aid in the explanation of conserved molecular pathways for posttranscriptional control of maternal mRNAs in early metazoan embryos.

ACKNOWLEDGMENTS

We thank Zak Razak for providing information on the normalization method. Trainee support has come from the following sources: a Natural Sciences and Engineering Research Council of Canada (NSERC) Graduate Scholarship (J. L. S.); a Canada Graduate Scholarship from the Canadian Institutes for Health Research (CIHR) (J. L. S.); studentships from the Ontario Student Opportunity Trust-Hospital for Sick Children Foundation Student Scholarship Program (J. L. S. and R. L. C.); a scholarship from the CIHR (R. L. C.); a postdoctoral fellowship from the Research Training Centre of Research Institute at the Hospital for Sick Children (A. L. G.). The Canadian *Drosophila* Microarray Centre is supported by a Major Facilities Access Grant from NSERC to J. T. W., H. D. L., and H. M. Krause. Our research on posttranscriptional regulation of *Drosophila* maternal mRNAs is supported by an operating grant from the CIHR (MOP 14409; to H. D. L.) for gene-by-gene analysis and a CIHR Team Grant in mRNP Systems Biology (to H. D. L., J. T. W., H. M. K., T. R. Hughes, C. A. Smibert, and A. Wilde) for genome-wide analysis. H. D. L. is Canada Research Chair (Tier 1) in Developmental Biology at the University of Toronto.

REFERENCES

Ausubel, F. M. (1987). "Current Protocols in Molecular Biology." Greene Publishing Associates, Brooklyn, N.Y.
Bashirullah, A., Cooperstock, R. L., and Lipshitz, H. D. (2001). Spatial and temporal control of RNA stability. *Proc. Natl. Acad. Sci. USA* **98,** 7025–7028.
Bashirullah, A., Halsell, S. R., Cooperstock, R. L., Kloc, M., Karaiskakis, A., Fisher, W. W., Fu, W., Hamilton, J. K., Etkin, L. D., and Lipshitz, H. D. (1999). Joint action of two RNA degradation pathways controls the timing of maternal transcript elimination at the midblastula transition in *Drosophila melanogaster*. *EMBO J.* **18,** 2610–2620.
Bustin, S. A. (2000). Absolute quantification of mRNA using real-time reverse transcription polymerase chain reaction assays. *J. Mol. Endocrinol.* **25,** 169–193.
Bustin, S. A. (2002). Quantification of mRNA using real-time reverse transcription PCR (RT-PCR): Trends and problems. *J. Mol. Endocrinol.* **29,** 23–39.
De Renzis, S., Elemento, O., Tavazoie, S., and Wieschaus, E. F. (2007). Unmasking activation of the zygotic genome using chromosomal deletions in the *Drosophila* embryo. *PLoS Biol.* **5,** e117.
Decker, C. J., and Parker, R. (1993). A turnover pathway for both stable and unstable mRNAs in yeast: Evidence for a requirement for deadenylation. *Genes Dev.* **7,** 1632–1643.
Ding, D., Parkhurst, S. M., Halsell, S. R., and Lipshitz, H. D. (1993). Dynamic Hsp83 RNA localization during *Drosophila* oogenesis and embryogenesis. *Mol. Cell Biol.* **13,** 3773–3781.
Edgar, B. A., and Schubiger, G. (1986). Parameters controlling transcriptional activation during early *Drosophila* development. *Cell* **44,** 871–877.
Gamberi, C., Peterson, D. S., He, L., and Gottlieb, E. (2002). An anterior function for the *Drosophila* posterior determinant Pumilio. *Development* **129,** 2699–2710.
Giraldez, A. J., Mishima, Y., Rihel, J., Grocock, R. J., Van Dongen, S., Inoue, K., Enright, A. J., and Schier, A. F. (2006). Zebrafish MiR-430 promotes deadenylation and clearance of maternal mRNAs. *Science* **312,** 75–79.
Grigull, J., Mnaimneh, S., Pootoolal, J., Robinson, M. D., and Hughes, T. R. (2004). Genome-wide analysis of mRNA stability using transcription inhibitors and microarrays reveals posttranscriptional control of ribosome biogenesis factors. *Mol. Cell Biol.* **24,** 5534–5547.
Hamatani, T., Carter, M. G., Sharov, A. A., and Ko, M. S. (2004). Dynamics of global gene expression changes during mouse preimplantation development. *Dev. Cell* **6,** 117–131.
Lecuyer, E., Yoshida, H., Parthasarathy, N., Alm, C., Babak, T., Cerovina, T., Hughes, T. R., Tomancak, P., and Krause, H. M. (2007). Global analysis of mRNA localization reveals a prominent role in organizing cellular architecture and function. *Cell* **131,** 174–187.
Lin, M. D., Fan, S. J., Hsu, W. S., and Chou, T. B. (2006). *Drosophila* decapping protein 1, dDcp1, is a component of the oskar mRNP complex and directs its posterior localization in the oocyte. *Dev. Cell* **10,** 601–613.
Maniatis, T., Fritsch, E. F., and Sambrook, J. (1982). "Molecular Cloning: A Laboratory Manual." Cold Spring Harbor Laboratory, Cold Spring Harbor, N.Y.
Merrill, P. T., Sweeton, D., and Wieschaus, E. (1988). Requirements for autosomal gene activity during precellular stages of *Drosophila melanogaster*. *Development* **104,** 495–509.
Myers, F. A., Francis-Lang, H., and Newbury, S. F. (1995). Degradation of maternal string mRNA is controlled by proteins encoded on maternally contributed transcripts. *Mech. Dev.* **51,** 217–226.

Neal, S. J., Gibson, M. L., So, A. K., and Westwood, J. T. (2003). Construction of a cDNA-based microarray for *Drosophila melanogaster*: A comparison of gene transcription profiles from SL2 and Kc167 cells. *Genome* **46**, 879–892.

Neal, S. J., and Westwood, J. T. (2006). Optimizing experiment and analysis parameters for spotted microarrays. *Methods Enzymol.* **410**, 203–221.

Raff, J. W., Whitfield, W. G., and Glover, D. M. (1990). Two distinct mechanisms localise cyclin B transcripts in syncytial *Drosophila* embryos. *Development* **110**, 1249–1261.

Ross, J. (1995). mRNA stability in mammalian cells. *Microbiol. Rev.* **59**, 423–450.

Salles, F. J., Lieberfarb, M. E., Wreden, C., Gergen, J. P., and Strickland, S. (1994). Coordinate initiation of *Drosophila* development by regulated polyadenylation of maternal messenger RNAs. *Science* **266**, 1996–1999.

Salles, F. J., Richards, W. G., and Strickland, S. (1999). Assaying the polyadenylation state of mRNAs. *Methods* **17**, 38–45.

Sambrook, J., and Russell, D. W. (2001). "Molecular Cloning: A Laboratory Manual." Cold Spring Harbor Laboratory Press, Cold Spring Harbor, N.Y.

Schadt, E. E., Li, C., Ellis, B., and Wong, W. H. (2001). Feature extraction and normalization algorithms for high-density oligonucleotide gene expression array data. *J. Cell Biochem. Suppl.* **37**, 120–125.

Semotok, J. L., Cooperstock, R. L., Pinder, B. D., Vari, H. K., Lipshitz, H. D., and Smibert, C. A. (2005). Smaug recruits the CCR4/POP2/NOT deadenylase complex to trigger maternal transcript localization in the early *Drosophila* embryo. *Curr. Biol.* **15**, 284–294.

Semotok, J. L., and Lipshitz, H. D. (2007). Regulation and function of maternal mRNA destabilization during early *Drosophila* development. *Differentiation* **75**, 482–506.

Stangegaard, M., Dufva, I. H., and Dufva, M. (2006). Reverse transcription using random pentadecamer primers increases yield and quality of resulting cDNA. *Biotechniques* **40**, 649–657.

Surdej, P., and Jacobs-Lorena, M. (1998). Developmental regulation of bicoid mRNA stability is mediated by the first 43 nucleotides of the 3′-untranslated region. *Mol. Cell. Biol.* **18**, 2892–2900.

Tadros, W., Goldman, A. L., Babak, T., Menzies, F., Vardy, L., Orr-Weaver, T., Hughes, T. R., Westwood, J. T., Smibert, C. A., and Lipshitz, H. D. (2007a). SMAUG is a major regulator of maternal mRNA destabilization in *Drosophila* and its translation is activated by the PAN GU kinase. *Dev. Cell* **12**, 143–155.

Tadros, W., Houston, S. A., Bashirullah, A., Cooperstock, R. L., Semotok, J. L., Reed, B. H., and Lipshitz, H. D. (2003). Regulation of maternal transcript destabilization during egg activation in *Drosophila*. *Genetics* **164**, 989–1001.

Tadros, W., and Lipshitz, H. D. (2005). Setting the stage for development: mRNA translation and stability during oocyte maturation and egg activation in *Drosophila*. *Dev. Dyn.* **232**, 593–608.

Tadros, W., Westwood, J. T., and Lipshitz, H. D. (2007b). The mother-to-child transition. *Dev. Cell* **12**, 847–849.

Tautz, D., and Pfeifle, C. (1989). A non-radioactive *in situ* hybridization method for the localization of specific RNAs in *Drosophila* embryos reveals translational control of the segmentation gene hunchback. *Chromosoma* **98**, 81–85.

Tenenhaus, C., Subramaniam, K., Dunn, M. A., and Seydoux, G. (2001). PIE-1 is a bifunctional protein that regulates maternal and zygotic gene expression in the embryonic germ line of *Caenorhabditis elegans*. *Genes Dev.* **15**, 1031–1040.

Tseng, G. C., Oh, M. K., Rohlin, L., Liao, J. C., and Wong, W. H. (2001). Issues in cDNA microarray analysis: quality filtering, channel normalization, models of variations and assessment of gene effects. *Nucleic Acids Res.* **29**, 2549–2557.

Tusher, V. G., Tibshirani, R., and Chu, G. (2001). Significance analysis of microarrays applied to the ionizing radiation response. *Proc. Natl. Acad. Sci. USA* **98,** 5116–5121.

Whitfield, W. G., Gonzalez, C., Sanchez-Herrero, E., and Glover, D. M. (1989). Transcripts of one of two *Drosophila* cyclin genes become localized in pole cells during embryogenesis. *Nature* **338,** 337–340.

Wieschaus, E., and Sweeton, D. (1988). Requirements for X-linked zygotic gene activity during cellularization of early *Drosophila* embryos. *Development* **104,** 483–493.

Wolke, U., Weidinger, G., Koprunner, M., and Raz, E. (2002). Multiple levels of posttranscriptional control lead to germ line-specific gene expression in the zebrafish. *Curr. Biol.* **12,** 289–294.

Yang, Y. H., Dudoit, S., Luu, P., Lin, D. M., Peng, V., Ngai, J., and Speed, T. P. (2002). Normalization for cDNA microarray data: a robust composite method addressing single and multiple slide systematic variation. *Nucleic Acids Res.* **30,** e15.

CHAPTER SEVENTEEN

MESSENGER RNA HALF-LIFE MEASUREMENTS IN MAMMALIAN CELLS

Chyi-Ying A. Chen, Nader Ezzeddine, *and* Ann-Bin Shyu

Contents

1. Introduction	336
2. General Considerations of mRNA Half-Life Measurements	337
3. Determining mRNA Decay Constant	338
4. Methods for Measuring mRNA Half-Life	339
4.1. General inhibition of transcription	339
4.2. Use of inducible promoters to specifically promote transient transcription	340
5. Concluding Remarks	354
Acknowledgments	355
References	355

Abstract

The recognition of the importance of mRNA turnover in regulating eukaryotic gene expression has mandated the development of reliable, rigorous, and "user-friendly" methods to accurately measure changes in mRNA stability in mammalian cells. Frequently, mRNA stability is studied indirectly by analyzing the steady-state level of mRNA in the cytoplasm; in this case, changes in mRNA abundance are assumed to reflect only mRNA degradation, an assumption that is not always correct. Although direct measurements of mRNA decay rate can be performed with kinetic labeling techniques and transcriptional inhibitors, these techniques often introduce significant changes in cell physiology. Furthermore, many critical mechanistic issues as to deadenylation kinetics, decay intermediates, and precursor-product relationships cannot be readily addressed by these methods. In light of these concerns, we have previously reported transcriptional pulsing methods based on the *c-fos* serum-inducible promoter and the tetracycline-regulated (Tet-off) promoter systems to better explain mechanisms of mRNA turnover in mammalian cells. In this chapter, we describe and discuss

Department of Biochemistry and Molecular Biology, The University of Texas Medical School at Houston, Houston, Texas, USA in detail different protocols that use these two transcriptional pulsing methods. The information described here also provides guidelines to help develop optimal protocols for studying mammalian mRNA turnover in different cell types under a wide range of physiologic conditions.

1. INTRODUCTION

Regulation of mRNA turnover in the cytoplasm is important for controlling the abundance of cellular transcripts and, in turn, the levels of protein expression (for reviews, see Parker and Song [2004] and Wilusz et al. [2001]). mRNA stability can be regulated at different levels. Under a given physiologic condition, mRNAs display a wide range of stability. For example, c-*fos* proto-oncogene transcript is degraded rapidly in the cytoplasm with a half-life of 10 to 15 min (Shyu et al., 1989; Treisman, 1985), whereas globin mRNA is rather stable and has a half-life of several hours in the same cells (Shyu et al., 1989). Although each individual mRNA has its intrinsic stability under a given condition, stability of an individual mRNA may change in response to a variety of extracellular stimuli. Examples include the autoregulated degradation of tubulin mRNA in response to changes in tubulin concentration (Yen et al., 1988), the iron-dependent destabilization of transferrin receptor mRNA (Casey et al., 1988; Muellner et al., 1989), the DNA synthesis–dependent destabilization of histone mRNA (Pandey and Marzluff, 1987), and the stabilization of lymphokine mRNAs by costimulatory molecules (Lindsten et al., 1989). Thus, modulation of mRNA stability provides a powerful means for controlling gene expression during the cell cycle, cell differentiation, the immune response, as well as many other physiologic transitions.

In mammalian cells, the first major step that triggers mRNA decay is deadenylation (i.e., removal of the 3′-poly(A) tail). All major mRNA decay pathways recognized in mammalian cells, including mRNA decay directed by AU-rich elements (AREs) in the 3′-untranslated region (UTR) (Chen and Shyu, 1995), decay mediated by destabilizing elements in protein-coding regions (Grosset et al., 2000), nonsense-mediated mRNA decay (NMD) (Chen and Shyu, 2003), decay directed by microRNAs (miRNAs) (Wu et al., 2006), and decay of stable mRNAs such as β-globin mRNA (Loflin et al., 1999a; Shyu et al., 1991), are initiated with deadenylation. Mammalian deadenylation exhibits biphasic kinetics. During the first phase, PAN2 poly(A) nuclease, presumably complexed with PAN3, shortens the poly(A) tails to $\sim$110 A nucleotides (nt) (Yamashita et al., 2005). In the second phase, the CCR4-CAF1 poly(A) nuclease complex further shortens the poly(A) tail to oligo(A) (Yamashita et al., 2005). Decapping mediated by the DCP1–DCP2 complex was found to occur after either the first or the second phase of deadenylation (Yamashita *et al.*, 2005). The RNA body can be degraded by the exoribonuclease XRN1 from the 5′-end after decapping (Parker and Song, 2004). Alternately, the mRNA body can also be degraded from the 3′-end after deadenylation by a large protein complex termed the exosome (Parker and Song, 2004).

To unravel the underlying processes of regulated mRNA turnover, a detailed analysis of the major components involved in mRNA turnover is required. The observation that deadenylation is the major trigger for cytoplasmic mRNA degradation in mammalian cells underscores the necessity of explaining the deadenylation step in the process of mRNA turnover. Thus, determination and characterization of the decay mechanisms demand that mRNA decay kinetics and precursor-product relationships be accurately and readily monitored experimentally. The primary emphasis of this chapter is to describe two inducible promoter systems as examples to illustrate how mRNA turnover may be optimally investigated in mammalian tissue culture cells with transient transfection systems. Detailed step-by-step protocols are given so that the half-lives of mRNAs of interest can be determined with experimental systems described here.

2. General Considerations of mRNA Half-Life Measurements

Messenger RNA stability is often studied indirectly by monitoring changes in the steady-state level of mRNA in the cytoplasm. However, changes in mRNA abundance are not necessarily caused by alterations in mRNA stability. For example, mRNA biogenesis in the nucleus (such as transcription, RNA processing, and/or mRNA export) may be fortuitously altered because of changes of the physiologic condition or in response to the environmental stimuli. Thus, alterations in the steady-state level of mRNA may not reflect the changes in mRNA stability. Direct measurements of decay rates of endogenous mRNAs have been performed in a number of ways, including kinetic labeling techniques and the use of transcriptional inhibitors. However, these techniques often introduce significant changes in cell physiology, thereby perturbing the stability of many mRNAs that could potentially lead to secondary consequences on the stability of the tested mRNA (e.g., see Belasco and Brawerman [1993], Harrold *et al.* [1991], and Ross [1995]). As a result, decay rates obtained in experiments that use these methods may not reflect the true stability of the mRNAs.

Aside from the methods described previously for measuring endogenous mRNA stability, one can determine the half-life of an mRNA of interest by transiently transfecting the reporter gene into mammalian tissue culture cells and using a transcriptional pulsing approach to monitor deadenylation and decay kinetics of the reporter transcript without the use of a global transcription inhibitor (Loflin et al., 1999b; Xu et al., 1998). We will detail the procedures and applications of two transcriptional pulsing systems and discuss the advantages of these systems over other methods in terms of investigating mRNA turnover in mammalian cells.

3. Determining mRNA Decay Constant

The turnover rate or stability of mRNA *in vivo* is usually reported as the time required for degrading 50% of the existing mRNA molecules (i.e., the half-life of mRNAs). Before the half-life of a given message can be calculated, the decay rate constant must be determined. Assuming an ideal *in vivo* situation, in which transcription of the mRNA of interest can be turned off completely (or at least to an undetectable level), mRNA decay follows first-order kinetics. The rate of disappearance of mRNA concentration at a given time (dC/dt) is proportional to both the rate constant for decay (k_{decay}) and the cytoplasmic concentration of the mRNA (C). This relation is described by the following equation:

$$dC/dt = -k_{decay} C \quad (17.1)$$

The minus symbol indicates that the mRNA is being degraded rather than synthesized. This relationship leads to the derivation of the equation:

$$\ln(C/C_0) = -k_{decay} t \quad (17.2)$$

where C_0 is the concentration of the mRNA at time 0, before decay starts.

Because we want to determine the half-life ($t_{1/2}$), this means $C/C_0 = 50\%/100\% = 1/2$. Rearrangement of equation (17.2) leads to the following equation:

$$\ln 1/2 = -k_{decay} t_{1/2}$$

from where:

$$t_{1/2} = \ln 2 / k_{decay} \quad (17.3)$$

It is important to note that the half-life of an mRNA ($t_{1/2}$) is inversely proportional to its decay rate constant (k_{decay}).

In a typical time-course experiment, determination of mRNA half-life begins with the analysis of several RNA samples collected at time intervals, and monitoring the loss of a particular message by analyzing equal amounts of these samples with two message-specific probes, one specific for the message of interest and the other specific for an internal control, used to normalize the data and calculating the half-life of the message of interest (Belasco and Brawerman, 1993). With linear regression (least-square) analysis to identify the line that best fits the data, the decay rate constant is obtained from the slope of a semilogarithmic plot of mRNA concentration (C) as a function of time (t). The half-life ($t_{1/2}$) can then be calculated with equation (17.3).

The congruity of mRNA decay to first-order kinetics implies that mRNA molecules of a given type that differ in age are recognized and targeted for degradation in the same way by cellular decay machinery. Although this generally seems to be the case in prokaryotes (Belasco and Higgins, 1988), it is now clear that the deadenylation step is the major step that triggers mRNA decay and determines the ultimate stability of many mRNAs in both yeast and mammalian cells (reviewed in Meyer et al. [2004] and Parker and Song [2004]). Therefore, in eukaryotic cells, many labile or stable mRNAs undergo a period of poly(A) tail shortening during which there is no apparent decay of the transcribed portion of mRNA (e.g., Chen and Shyu [1994] and Muhlrad et al. [1994]). Only after the poly(A) tail is shortened to a certain extent, usually between 10 and 60 nt, does first-order decay of the RNA body ensue. Consequently, eukaryotic mRNAs decay with biphasic kinetics, composed of an initial lag phase during which deadenylation occurs and a second phase during which the body of the mRNA is degraded. As a result, an RNA lifetime cannot meaningfully be described merely in terms of a half-life that obeys first-order kinetics. This caveat concerning decay kinetics further underscores the necessity of monitoring the deadenylation step with a transcriptional pulsing approach.

4. Methods for Measuring mRNA Half-Life

4.1. General inhibition of transcription

A relatively simple way of analyzing mRNA kinetics involves blocking cellular transcription with inhibitors that include actinomycin D (which interferes with transcription by intercalating into DNA) or 5,6-dichloro-1β-1-ribofuranosylbenzimidazole (DRB) (which interacts directly with the RNA polymerase II transcription apparatus) (Harrold et al., 1991). Typically, either actinomycin D (at a concentration of 5 to 10 μg/ml) or DRB (at a final concentration of 20 μg/ml) is added to cells, and the amount of a particular mRNA remaining at various times of treatment is used to calculate the mRNA decay rate. The advantage of this approach is that exogenous genes do not have to be constructed and introduced into cells, and it is an easy way of measuring stability changes of endogenous mRNAs. However, this method suffers the disadvantage that both drugs have a profound impact on cellular physiology and, in fact, have been shown through the years to alter the stability of many mRNAs (Chen et al., 1995; Harrold et al., 1991; Seiser et al., 1995; Shyu et al., 1989; Speth and Oberbaumer, 1993). Of course, this complicates interpretation of the data. In addition, monitoring mRNA decay after a constitutive block in transcription does not allow the determination of deadenylation rates or precursor-product relationships without the use of supplementary methods. Moreover, for experiments involving other factors (such as differentiating agents), transcription inhibitors can result in severe cytotoxicity. These disadvantages limit the application of the seemingly straightforward and convenient use of transcription inhibitors.

4.2. Use of inducible promoters to specifically promote transient transcription

The use of a regulatable promoter subject to transient induction represents a major improvement in the analysis of eukaryotic mRNA decay over other methods (Loflin et al., 1999b). This approach provides a tight and rapid genetic switch to control transcription as required for studying mRNA turnover without the use of transcription inhibitors. This ensures that mRNA degradation is investigated under physiologically undisturbed conditions. The rationale behind this approach, also termed transcriptional pulsing, is rather straightforward: provide a stimulus that activates transcription and leads to a burst of mRNA synthesis, then remove the stimulus to shut off transcription and monitor the decay of mRNA. The success of this method relies on stringent control of the inducible promoter so that induction and silencing of transcription is accomplished within a narrow window of time. In mammalian cells, the c-*fos* promoter has been valuable for this purpose, because it can be induced in response to serum addition quickly and transiently (Greenberg and Ziff, 1984; Treisman, 1985), thereby providing a reliable and simple way of achieving a transient burst in transcription. A more recently developed tetracycline (Tet) regulatory promoter system (Gossen et al., 1993) offers an alternate strategy that further broadens the application of the transcriptional pulsing approach to study mRNA turnover in mammalian cells. The two systems are presented in the following sections.

4.2.1. The c-*fos* serum-inducible promoter system

Our laboratory has extensively used the c-*fos* inducible promoter system to investigate mRNA degradation mediated by AREs in mouse fibroblast NIH3T3 cells (e.g., see Chen et al. [1994], Chen and Shyu [1994], and Grosset et al. [2000]). AREs are found within the 3′ UTRs of many unstable mRNAs and represent the most commonly found RNA destabilizing elements in mammals (Chen and Shyu, 1995; Wilusz et al., 2001). To explain the key functional features of AREs that constitute their destabilizing ability and to decipher the deadenylation and decay kinetics of ARE-containing mRNAs, we first made a chimeric reporter construct. This construct contains a rabbit β-globin gene under the transcriptional control of a 710-base pair (bp) promoter region from the human c-*fos* proto-oncogene (Shyu et al., 1989). The β-globin gene was chosen because it encodes a stable mRNA having a half-life of more than 8 h in NIH3T3 cells (Shyu et al., 1989; 1991). This magnitude of stability facilitates detecting an mRNA destabilizing effect that is mediated by a potential destabilizing element that is experimentally introduced into the β-globin gene. After individually introducing test destabilizing sequences into the β-globin plasmid construct, the construct is delivered to cells by transient transfection. Cells are subsequently serum starved to make them quiescent. After >25 h, β-globin gene transcription driven by the c-*fos* promoter is transiently induced by the addition of serum, which returns to an inactive state after an additional 30 to 40 min (Greenberg and Ziff, 1984; Treisman, 1985). This results in a short burst of chimeric β-globin mRNA synthesis, after which chimeric β-globin mRNA decay can be monitored without the use of transcription inhibitors. To do so, RNA samples are collected at different time points after serum induction, and the decay rate of the chimeric β-globin mRNAs is determined by Northern blot analysis for example (Shyu et al., 1991).

During the time course experiments, in addition to mRNA decay rates, highly synchronized poly(A) shortening has been observed, which makes an unequivocal determination of deadenylation status possible (e.g., see Chen et al. [1994] and Shyu et al. [1991]). Therefore, many critical mechanistic issues, such as deadenylation kinetics, the existence of decay intermediates, and precursor-product relationships can also be readily addressed by this method (Yamashita et al., 2005). In addition, this system is efficient and reproducible, and by allowing the analysis of transcripts that derive from transiently introduced genes, it eliminates the need to establish stably transfected cell clones.

Here, we describe two different protocols for the time-course experiments that our laboratory has been using. The first one is highly cost-effective, which involves calcium phosphate–based transient transfection of NIH3T3 cells and RNA extraction without the use of a commercial RNA purification kit. The second protocol uses Lipofectamine 2000

(Invitrogen) for transfection and RNeasy RNA preparation kit (QIAGENE) for RNA purification.

4.2.1.1. Protocol I: Transient transfection and serum induction

4.2.1.1.1. Materials All solutions should be prepared with extremely pure, glass-distilled water.

1. 2 M $CaCl_2$: Weigh 14.7 g of $CaCl_2$ in a 50-ml conical centrifuge tube and bring H_2O to 50 ml. Dissolve well and use Acrodisc (0.2 μ) to filter-sterilize it. Aliquot the solution to 1.5 ml microcentrifuge tubes and store at $-20\,°C$.

   (*Note*: Different sources and lot numbers of calcium chloride may affect the timing and fineness of the calcium phosphate/DNA precipitate that results, thereby affecting transfection efficiency greatly. More efficient transfection is achieved with fine precipitates.)

2. 2× HBS: NaCl, 274 mM; KCl, 10 mM; Na_2HPO_4, 1.4 mM; glucose, 15 mM; HEPES, 42 mM. Adjust the pH of the solution with NaOH precisely to 7.02. Filter sterilize and aliquot the final solution into 15-ml conical centrifuge tubes and store at 4 °C.

   (*Note*: It is recommended that pH should be checked with two different meters. pH drift to the acidic end will, in subsequent steps, prevent the formation of the fine calcium phosphate/DNA precipitate that gives the solution an opaque look, whereas pH drift to the alkaline end will lead to quick formation of visible calcium phosphate/DNA clumps. Both situations can lower the transfection efficiency by an order of magnitude.)

3. DNA solution: The final concentration of DNA (including the test DNA, control DNA, and carrier DNA) should be 20 μg/0.1 ml in H_2O for a 10-cm cell culture dish. This is made up by empirically determining the optimum amount of transfecting DNA and adding enough carrier DNA to make a total of 20 μg.

   (*Note*: We have been using plasmid vectors as carrier DNA [e.g., pUC18]. We usually use 2 to 3 μg of test DNA [e.g., pBBB + ARE] or its derivatives (Chen and Shyu, 1994; Chen *et al.*, 1994). This amount of DNA seems sufficient to result in the production of an mRNA signal that can be readily detected by RNA blot analysis with an exposure time to X-ray film between 3 and 18 h. We have noticed that when more than 5 μg of test DNA is used in transient transfections, somehow mRNA decay is retarded. For example, instead of degrading with a half-life of 35 min, BBB + ARE mRNA displays a 2- to 3-h half-life. This might represent saturation of the decay machinery.)

4. Bottled Dulbecco's modified Eagle's medium (DMEM), calf serum, L-glutamine (1 mM final concentration), and 1% of penicillin-streptomycin

(prepared with 10,000 units/ml penicillin G sodium and 10 mg/ml streptomycin sulfate in 0.85% saline). Note that all the reagents described here are purchased from GIBCO.
5. 1× Phosphate-buffered saline (PBS).

4.2.1.1.2. Procedures

1. 2×10^6 cells are plated in a 10-cm tissue culture dish 16 to 20 h before transfection and kept in a 5% CO_2 incubator.
2. Prepare calcium phosphate/DNA mixture by combining in the following order (for each plate):
   100 µl   DNA (20 µg) in H_2O
   100 µl   2× HBS
   916 µl   1× HBS
   Then add 84 µl of 2 M $CaCl_2$ solution and mix gently by inverting the tube. Let the mixture sit for 20 min at room temperature.

   (*Note*: If more than one plate is to be transfected, simply multiply each component by the total number of plates and mix them in a 50-ml conical tube. The mixture should become a little opaque right after mixing, and 20 min later it should become fairly opaque, but no precipitation should been seen at this point. Otherwise, the transfection may not work, and we suggest that the pH of HBS solution should be rechecked.)

3. Add 1.2 ml DNA/calcium phosphate mixture directly to the cells that have been incubated in a 5% CO_2 incubator (from step 1) and gently swirl the plate to mix well and then immediately return the plate to the 5% CO_2 incubator.

   (*Note*: Do not move the plates from the 5% CO_2 incubator to the tissue culture hood until ready to add the DNA/calcium phosphate mixture, because the pH of the culture medium will increase because of evaporation of CO_2 when plates are outside the incubator. In a high pH environment, the DNA/calcium phosphate mixture will form coarse precipitates that aggregate into large clumps in the medium. As a result, the transfection efficiency will be significantly reduced. Also, it is important to not swirl the plates again once they are put in the incubator. After a 2-h incubation when examined under the light microscope (20×), one should expect to see fine precipitate of DNA in the culture medium as many tiny black dots. If DNA precipitates can not be seen or form clumps, quit the experiment and check the quality of reagents used as described previously.)

4. Serum starvation

   Twelve to 18 h after transfection, aspirate the medium from the dish and rinse the cells with 5 ml of 1× PBS gently but thoroughly twice. Add 10 ml of DMEM with only 0.5% calf serum and incubate cells in an 8.0% CO_2 incubator for 25 h.

(*Note*: It is normal to see some dead cells floating after 25 h of serum starvation.)

5. Serum induction

After serum starvation for 25 h, remove culture medium from the dish and add 10 ml of fresh culture medium containing 20% calf serum. Harvest cells at desired time points for RNA purification.

(*Note*: We routinely obtained 25 to 35% transfection efficiency with NIH3T3 cells following the preceding protocol.)

4.2.1.2. RNA extraction and Northern blot analysis
4.2.1.2.1. Materials

1. NP-40 lysis buffer:
   10 mM Tris-Cl, 7.4
   10 mM NaCl
   3 mM MgCl$_2$
   0.5 %NP-40 (v/v) (may be sold under the trade name IGEPAL CA-630)
2. 2× PK buffer:
   200 mM Tris-Cl, 7.5
   440 mM NaCl
   25 mM EDTA
   2%   SDS (w/v)

(*Note*: Both solutions should be autoclaved before use.)

4.2.1.2.2. Procedure

A. On ice or at 4 °C
   1. Chill culture dishes (6 to 7 × 10^6 cells per dish) on ice for 2 to 3 min. Aspirate the medium and wash cells twice with 3 ml of ice-cold 1× PBS.
   2. Add 3 ml ice-cold 1× PBS and scrape off cells with a rubber policeman. Transfer cells to a 15-ml disposable conical tube. Salvage the remaining cells with 3 ml 1× PBS.
   3. Pellet cells at 4 °C, 300g for 5 min.
   4. Remove the supernatant, and loosen the cell pellet in the remaining PBS by finger vortexing gently.
   5. Vortex at half-maximal speed as the 200 µl of NP-40 lysis buffer is added. Then vortex at the same speed for another 10 sec.
   6. Incubate the lysed cells on ice for 5 min. Centrifuge at 4 °C, 300g for 5 min to pellet nuclei.
   7. Transfer supernatant to a 1.5-ml microfuge tube containing 10 µl of ribonucleoside-vanadyl complex (200 mM stock green-black solution; New England BioLab). Mix well.

(*Note*: Ribonucleoside-vanadyl complex is a potent inhibitor of various ribonucleases. Removal of the ribonucleoside-vanadyl complex from the RNA can be accomplished by adding 10 equivalents of EDTA before ethanol precipitation.)

8. Centrifuge at 4 °C, 300g for 5 min.

B. At room temperature:
   1. Transfer supernatant to a new 1.5-ml microfuge tube containing 200 µl of 2× PK buffer. Then add 10 µl of protease K (20 mg/ml, Roche), mix well, and incubate at 37 °C for 30 min.
   2. Extract once with phenol-chloroform (P/C) (containing 8-hydroxyquinoline). Precipitate the RNA with 2 vol of absolute ethanol. Store at −70 °C 10 to 15 min or at −20 °C overnight.
      (*Note*: It is not necessary to add extra salt for EtOH precipitation.)
   3. Pellet RNA in a microfuge at 4 °C at maximal speed, remove supernatant as much as possible, and dry the pellet.
   4. Redissolve the RNA pellet in 300 µl of TE (pH 7.4) buffer. Add 3 µl of 1 M MgCl$_2$. Mix well and add 1 µl of RNase-free DNaseI (10 mg/ml; Invitrogen).
   5. Incubate at 37 °C for 20 min.
   6. Add 6 µl of 10% SDS and 12 µl of 0.25 M EDTA. Extract twice with P/C (containing 8-hydroxyquinoline).
   7. Ethanol precipitation, centrifugation, and desiccation of the RNA sample.
   8. Resuspend RNA pellet in 100 µl of RNase-free ddH$_2$O. Take 5 µl to read OD$_{260}$. Yield of RNA should be approximately 30 to 50 µg per plate. RNA samples can be stored at −20 °C for several months to 1 y without any apparent degradation.

(*Note*: It is extremely important to completely digest the transfected plasmid DNA that unavoidably exists in the RNA samples. Otherwise, residual plasmid DNA may be recognized during Northern blot hybridization, resulting in smeary black-looking lanes after autoradiography. Although various RNA purification kits are commercially available, we found often that DNase I-treatment is not very effective with the kits. Therefore, we highly recommend the protocol described previously for DNase I treatment.)

4.2.1.3. Northern blot analysis
4.2.1.3.1. Materials

1. MOPS buffer: 0.2 M MOPS pH7.0, 50 mM Na acetate, 10 mM EDTA.
2. 10× sample loading buffer: 50% glycerol, 1 mM EDTA, 0.4% bromophenol blue (BPB), 0.4% xylene cyanol.

3. Dextran sulfate prehybridization buffer (approximately 20 ml).
   a. 10 ml formamide (deionized and molecular biology grade).
   b. 4 ml 5× P buffer (1% BSA, 1% poly-vinylpyrrolidone, 1% Ficoll, 250 mM Tris pH 7.5, 0.5% sodium pyrophosphate, 5% SDS).
   c. 4 ml 50% Dextran sulfate solution (w/v).
   d. Mix by inversion and then warm in 42 °C waterbath approximately 10 min to help dissolve.
   e. Add 1.16 g NaCl and mix by inversion until completely dissolved. Replace in 42 °C waterbath.
   f. Add 2 ml of 1.5 mg/ml sheared salmon sperm DNA (boiled for 10 min and cooled on ice for 5 min).
   g. Mix by inversion and use immediately for prehybridization. (*Note*: set aside 2 to 3 ml for mixing with the P-32 probe [see steps 12 and 13 of "Procedure" following].)

4.2.1.3.2. Procedure

1. 1.4% agarose/2.2 M formaldehyde gel.
   a. For a 150 ml gel: boil 2.1 g agarose in 123 ml H_2O.
   b. Cool to 60 °C in waterbath.
   c. Add 15 ml 10× MOPS gel buffer and 12 ml formaldehyde.

   (*Note*: Make sure that the pH of formaldehyde is more than 4; pour in hood. Can put EtBr in gel [e.g., 0.66 μg/ml].)

2. Prepare RNA sample (20 μl per sample) by mixing:
   a. RNA (5 to 20 total RNA) in 4.5 μl H_2O.
   10× MOPS   2.0 μl
   Formaldehyde   3.5 μl
   Formamide   10.0 μl
   b. Incubate 55 to 60 °C for 15 min.
   c. Add 2 μl of 10× loading buffer, and load each sample onto gel.
3. Run the gel in 1× MOPS buffer for approximately 3 to 6 h at 5 V/cm. (Approximately 100 to 150 V for 3 to 4 h) or until BPB completely runs into the gel. Be sure to circulate the buffer (e.g., with a peristaltic pump).

   (*Note*: To obtain a resolution sufficient for detecting changes in deadenylation, we recommend that the gel be run overnight by lowering the voltage accordingly.)

4. Remove the gel gently, cut out section to be transferred, and rinse it with water in a large tray two times (15 min each time) to remove formaldehyde.
5. Soak the gel in excess 50 mM NaOH/10 mM NaCl and shake for 30 min.
6. Neutralize the gel by shaking it in 0.1 M Tris (pH 7.5) twice for 15 min each time.

(*Note*: If the RNA to be transferred is relatively small (<2.0 Kb), then omit steps 5 and 6.)

7. Soak the gel in 20× SSC for 1 to 2 h.
8. Blot the gel onto the Gene Screen Membrane (New England Nuclear/DuPont) with one of the commercial vacuum blotting devices (we use the Stratagene Posiblot) for up to 2 h.
9. Wash the membrane after transfer with 2× SSC. Illuminate the membrane and gel with UV to make sure the samples have been transferred. Mark the membrane for the position of rRNA markers or the orientation of the blot.
10. Place the wet membrane on 3MM and irradiate the side containing RNA with the UV Stratalinker (Stratagene) with 1 to 2 cycles to covalently cross-link RNA to the membrane. The wet membrane can directly be subjected to prehybridization.
11. Prehybridize the membrane in 20 ml of dextran sulfate prehybridization buffer with constant agitation for at least 6 h at 42 °C.
12. Prepare probe solution by boiling the probe for 10 min and cooling it immediately on ice for 5 min before mixing it with 3 ml of prehybridization solution that has been set aside when preparing the dextran sulfate prehybridization buffer.
13. For hybridization, add the 3 ml of probe solution from step 12 into the prehybridization buffer, where the membrane has been incubated.

(*Note*: The final concentration of the probe should be less than 10 ng/ml in hybridization solution to avoid high nonspecific background.)

14. Incubate the membrane in hybridization solution with constant agitation for 12 to 16 h at 42 °C.

(*Note*: Various hybridization buffers are commercially available. They may be used to shortened the prehybridization and hybridization steps from 18 to 22 h to 5 to 6 h, in which case the manufacturer's protocol should be followed for both steps.)

15. Remove hybridization solution and wash the membrane in the following solution and condition with constant agitation:
    a. Wash twice with 100 to 200 ml 2× SSC at room temperature for 10 min.
    b. Wash twice with 100 to 200 ml of 2× SSC/1.0% SDS at 65 °C for 15 min.
    c. Wash twice with 100 to 200 ml of 0.1× SSC at room temperature for 15 min.
16. If the membrane is to be rehybridized, do not allow the membrane to dry. Leave the membrane slightly damp and wrap it with plastic wrap before exposure to film for autoradiography. For long-term storage, the membrane should be stripped of probes (see step 17 below) and stored after drying at room temperature.

17. For stripping the membrane, incubate the membrane in approximately 200 ml of wash-off buffer (0.1× Denhardt's soln/5 mM Tris-Cl (pH8.0)/0.2 mM EDTA/0.05% sodium pyrophosphate) at 68 °C for 2 h. Change to à fresh buffer after the first hour of incubation.

4.2.1.4. Protocol II: Transient transfection and serum induction

1. 2.8×10^6 NIH 3T3 cells maintained in DMEM containing 10% calf serum (GIBCO) with 1% of l-glutamine (200 mM stock from GIBCO) are plated in a 10-cm tissue culture dish and kept in an 8.0% CO_2 incubator for 18 h, allowing the cells to reach 90 to 95% confluence.
2. Transfect the cells with Lipofectamine 2000 (Invitrogen)/DNA complexes as follows:
   a. Dilute 60 μl of Lipofectamine 2000 in 1.5 ml of Opti-MEM I medium (GIBCO), vortex for 1 sec, and incubate at room temperature for 5 min.
   b. Dilute 24 μg of DNA in 1.5 ml of Opti-MEM I medium; vortex for 1 sec.
   c. Combine the diluted DNA (from step b) with the diluted Lipofectamine 2000 (from step a), vortex for 10 sec, and incubate for 20 min at room temperature.
3. Add the DNA/Lipofectamine 2000 mixture (total volume 3 ml) gently directly to the cells. Gently rock the plate back and forth so that the DNA/Lipofectamine mixture distributes evenly in the dish, which is then incubated in a 5% CO_2 incubator.
4. 18 h later, aspirate the medium and add 2 ml of 0.25% trypsin-EDTA (GIBCO). Incubate the plate in a 5% CO_2 incubator at 37 °C for 5 min and then add 3 ml of DMEM/10% calf serum to stop trypsinization. Harvest the cells by centrifuging at 300g for 5 min.
5. Split 1.5×10^6 cells to several 6-cm tissue culture dishes (*Note*: the number of dishes depends on the time points required) in 5 ml of DMEM with 0.5% calf serum, 1% of l-glutamine (200 mM stock from GIBCO), 1% of penicillin-streptomycin (prepared with 10,000 U/ml penicillin G sodium (GIBCO), and 10 mg/ml streptomycin sulfate (GIBCO) in 0.85% saline (GIBCO).
6. Serum starvation: incubate cells in an 8.0% CO_2 incubator for 24 h, which will force cells to enter a quiescent (G0) state.
7. Serum induction: after serum starvation for 24 h, remove culture medium from the dish and add 5 ml of fresh culture medium containing 20% calf serum. Harvest cells at desired time points for RNA purification.

(*Note*: After serum induction, transcription from the c-*fos* promoter returns to its original uninduced level in 30 to 40 min [Greenberg and Ziff, 1984].)

8. Refer to Protocol I for cytoplasmic RNA extraction. Alternately, cytoplasmic RNA may be extracted with RNeasy RNA preparation kit (QIAGENE) according to manufacturer's instruction. RNA samples are analyzed by Northern blotting as described previously.

4.2.2. The Tet-off regulatory promoter system

Although the c-*fos* promoter system has been used with success to investigate the decay kinetics and key features of AREs, the system has some limitations that prevent it from being used as a more general approach. Because activation of the c-*fos* promoter requires serum or growth factor induction of quiescent cells, this system has restricted the analysis of mRNA degradation to cells undergoing the G0 to G1 transition. In addition, many transformed cell lines cannot be readily forced to enter a quiescent state by serum starvation. Moreover, the use of serum induction complicates the analysis of regulatory mechanisms or signaling pathways, which may affect decay of mRNA. As an effort to develop a general approach that would be more suitable to study the regulatory aspects of differential and selective mRNA turnover in mammalian cells, we explored the Tet-regulatory promoter system as an alternative. In this system, a chimeric transcription activator termed tTA was generated by fusing the DNA-binding domain of the TN10-derived prokaryotic tetracycline repressor protein (tetR) to the transcription-enhancing domain of VP16 from herpes simplex virus, which binds and strongly activates a minimal promoter containing seven tetracycline operator (tetO) sequences in the absence of tetracycline (Gossen and Bujard, 1992; Gossen *et al.*, 1993). Binding of the tTA to the tetO sequences can be quickly blocked by tetracycline, preventing the activation of the target promoter.

Although its potential application to address the cytoplasmic mRNA turnover seemed obvious, we found that simply blocking constitutive transcription of the Tet-regulatory promoter with tetracycline and then monitoring mRNA decay does not give an accurate measurement of mRNA half-life. This may be due to saturation of cellular decay machineries by the high level of constitutive mRNA expression. In mouse NIH3T3 or human K562 cells, steady-state level of the mRNA driven by the Tet-off promoter is 15- to 30-fold higher than that driven by SV-40 early promoter or CMV promoter (Xu, Loflin, Chen and Shyu, unpublished observations). Moreover, because of the size, heterogeneity of the poly(A) tails resulting from constitutive transcription, deadenylation and decay kinetics, and thus the relationship between deadenylation and decay of the mRNA, cannot be unequivocally determined. In light of these concerns, we have further optimized the systems by modulating the amount of tetracycline and the timing of its addition to or omission from culture medium. We were able to induce a short burst of mRNA synthesis from a reporter gene driven by the Tet-off promoter, which displays kinetics similar to those of the c-*fos* promoter system (Xu *et al.*, 1998).

In this section, we describe a transcriptional pulsing approach developed by the tet-regulatory (Tet-off) system with mouse NIH3T3 cells as an example. With the new strategy, we have demonstrated that AREs can function in mouse NIH 3T3 fibroblasts under growth arrest and density arrest states (Xu *et al.*, 1998). More recently, we have successfully used this new approach to study the destabilizing function of different AREs in various phases of the cell cycle and during blood cell differentiation (Chen *et al.*, 2007). Furthermore, combining this approach with small interference RNA (siRNA)–mediated gene knockdown, we were able to trap different decay intermediates indicative of mechanistic steps of mRNA decay in mammalian cells (Chen *et al.*, 2007; Yamashita *et al.*, 2005). These *in vivo* studies demonstrate the feasibility of this approach to investigate mechanisms underlying differential and selective mRNA turnover in mammalian cells.

4.2.2.1. Establishment of mammalian stable cell lines expressing the tTA
The first critical step toward the successful use of the transcriptional pulsing approach with the Tet-off promoter to monitor mRNA decay kinetics in mammalian cells is to identify or establish a stable line of interest that expresses tTA. In our laboratory, we have been using the stable β-globin mRNA as the reporter message for mRNA turnover study. To adapt this system to Tet regulation, we have constructed a new β-globin reporter plasmid, designated pTet-BBB (Loflin *et al.*, 1999a; Xu *et al.*, 1998). The plasmid pTet-BBB contains the β-globin gene under control of the Tet-off promoter, allowing transcription in the absence of tetracycline. The *Bgl*II site in the 3′ UTR of the β-globin gene remains unique in the pTet-BBB, providing a site into which RNA destabilizing elements, such as different AREs, can be introduced to test their ability to destabilize β-globin mRNA (Loflin *et al.*, 1999a).

Initially, we attempted to test whether tight regulation of transcription from the Tet-off promoter may be obtained by transiently cotransfecting both the pTet-BBB + ARE and the pUHD15-1 plasmid encoding the tTA (Gossen and Bujard, 1992). We observed high basal-level expression of β-globin mRNA bearing the c-*fos* ARE (BBB + ARE) in the presence of tetracycline, indicating inefficient repression when tetracycline was added back to turn off transcription. In addition, the level of BBB + ARE mRNA expression from the Tet-off promoter varies considerably depending on the relative amount of pTet-BBB + ARE and the pUHD15-1 plasmids used in the transient transfections. Therefore, we do not recommend transient cotransfection of the Tet-off promoter–driven reporter plasmid and tTA-encoding plasmid. We refer readers to our previous publications (Chen *et al.*, 2007; Loflin *et al.*, 1999b) for details regarding establishing and characterizing tTA-expressing stable lines that are suitable for the transcriptional pulsing approach.

4.2.2.2. Transcriptional pulse strategy by modulating the amount of tetracycline in culture medium

To obtain a homogeneous population of mRNA, transcription should proceed for a period sufficient to produce a detectable signal by Northern blot analysis, yet brief enough to limit heterogeneity of poly(A) tails. The optimal condition can be determined by a titration experiment. Various concentrations of tetracycline are included in the culture medium to test at which concentration the steady-state level of β-globin mRNA transcribed from the transiently transfected pTet-BBB plasmid will be inhibited >99% compared with that in the absence of tetracycline (Chen *et al.*, 2007; Loflin *et al.*, 1999b; Xu *et al.*, 1998). Depending on the cell lines and sources of tetracycline, the appropriate concentration of tetracycline ranges from 25 ng/ml to 100 ng/ml. After transient transfection of the Tet-off promoter–driven reporter plasmid, the transfected cells should be kept in culture medium containing the identified concentration of tetracycline until transcription induction (transcriptional pulsing). It is striking that all three different mammalian stable lines we established, including mouse NIH3T3 B2A2, human K562 III-2, and human BEAS-2B-19, show the same transcriptional pulsing kinetics in Northern blot analysis (Chen *et al.*, 2007; Loflin *et al.*, 1999b; Xu *et al.*, 1998). A population of mRNA homogenous in size appears in the cytoplasm after removal of tetracycline for 100 to 110 min, making it possible to study deadenylation and decay kinetics and precursor and product relationships during mRNA decay. Typically, the reporter mRNA becomes detectable 90 min after the transfected cells are moved to fresh medium without tetracycline. After another 20 min (i.e., 110 min pulse), a descent signal representing a population of mRNAs with poly(A) tails homogeneous in size is produced. Because this time period is highly reproducible and is also observed in all different stable clones we have tested, this period has been used in our studies that use the Tet-off promoter–driven transcriptional pulsing approach and detailed as follows.

1. After transfection, the cells are grown in 25 to 40 ng/ml tetracycline for a period of 36 to 48 h.
2. Induction of a detectable homogeneous population of β-globin mRNA is accomplished by shifting the cells to fresh culture medium for 110 min.
3. The pulse is quickly terminated by adding 500 ng/ml tetracycline, and RNA samples are isolated at various times afterward.
4. Conduct RNA blot analysis as described previously.

Notes: Several experimental details and caveats are important for the success of transcriptional pulsing strategy. First, the amount of transfected plasmid DNA coding for the reporter mRNA must be empirically optimized with a positive control coding for an unstable message to avoid saturation of the decay machinery with excess mRNA. Second, the optimal amount of transfected DNA may vary when different transient transfection reagents or cells are used and must be optimized empirically. Third, to detect the mRNA synthesized after the removal of tetracycline for 110 min (transcriptional pulsing), it is crucial to select a stable tTA-expressing clone that gives the maximally induced expression level of the reporter gene. Both the NIH3T3 B2A2 and the K562 III-2 clones we selected for our mRNA decay study give more than 10^9 RLU/mg of protein by luciferase assay in transient transfection experiments (Loflin et al., 1999b; Xu et al., 1998). Fourth, to accurately determine deadenylation and decay kinetics, a robust, yet brief, transcription is necessary. Thus, the resumption kinetics of transcription on the removal of tetracycline displayed by the selected stable line is a crucial factor. For example, although we were able to identify a few stable clones that overexpress the transfected reporter genes at an equally high level in the absence of tetracycline, we noticed that not all of them exhibited fast resumption kinetics of transcription after tetracycline removal (data not shown). This difference may be derived from variations in copy number and integration site of the tTA cDNA in stable clones. Because slow resumption kinetics of transcription requires a prolonged pulse, and thus leads to heterogeneity in the size of mRNA molecules, one should pick a clone that is able to give at least 10% of the maximal level of expression after the 110-min pulse period without tetracycline so that a sufficient signal representing an mRNA population homogeneous in size can be generated for kinetic studies. Last, cells should be continually maintained in the presence of tetracycline, except for the short period of induction, because production of tTA for a longer period seems cytotoxic to the cells, and cells that have lost tTA expression may gradually take over the population during prolonged incubation without tetracycline. However, this problem may be avoided if one establishes stable lines with a modified tTA that seems to show little cytotoxicity (Clontech). Also, medium with tetracycline should always be prepared fresh and kept in the dark at 4 °C. A variety of mammalian cell lines harboring the tTA gene have been established, and they are commercially available. They may be tested for appropriateness for mRNA decay kinetic studies as described previously.

4.2.2.3. Transfection procedure

1. 2.6 to 2.8 × 10^6 NIH 3T3 B2A2 cells maintained in DMEM containing 10% calf serum (Gibco), 1% of l-glutamine (200 mM stock from Gibco), and 100 ng/ml tetracycline (CalBiochem) are plated in a 10-cm tissue culture dish and incubated in an 8% CO_2 incubator.

    (*Note*: In previous publications, we used 25 to 30 ng/ml of tetracycline that was purchased from Sigma. We found that higher concentration of tetracycline was needed to achieve the desired transcriptional pulsing when we used the tetracycline purchased from CalBiochem. Thus, the optimal amount of each critical reagent used in the experiment should be tested empirically when it is ordered from different companies or when a new lot is ordered from the same company.)

2. After 18 to 24 h, when cells reach 90 to 95% confluence, prepare the transfect cells and Lipofectamine 2000 (Invitrogen)/DNA complexes and harvest cells as described in Protocol II sections 2 to 4.
3. Split 1.5×10^6 cells to several 6-cm tissue culture dishes (*Note*: the number of dishes depends on the need of each experiment) in 5 ml of DMEM with 10% calf serum, 1% of L-glutamine (200 mM stock from Gibco), 1% of penicillin-streptomycin [prepared with 10,000 U/ml penicillin G sodium (Gibco) and 10 mg/ml streptomycin sulfate (Gibco) in 0.85% saline (Gibco), and 100 ng/ml tetracycline (CalBiochem)]. Incubate cells in an 8.0% CO_2 incubator for 24 h.
4. Remove medium and add 5 ml of complete medium without tetracycline to allow transcription of the reporter gene driven by the Tet-off promoter. Incubate the cells in an 8.0% CO_2 incubator immediately for 110 min.
5. To stop transcription, add complete medium containing tetracycline (500 ng/ml) to each plate.
6. Refer to Protocol I for cytoplasmic RNA extraction. Alternately, cytoplasmic RNA may be extracted with RNeasy RNA preparation kit (Qiagen) according to the manufacture's instruction. RNA samples are analyzed by Northern blotting as described previously.

Notes: The same procedure can also be used to introduce DNA and siRNA simultaneously into NIH 3T3 B2A2 cells. When introducing siRNA together with DNA, up to 320 pmol siRNA should be diluted in the same tube with 6.5 μg of plasmid DNA and 40 μl of Lipofectamine 2000. The ratio between siRNA and DNA is critical, as increasing siRNA amount may reduce the transfection efficiency or expression of the reporter DNA. We also refer readers to one of our previous publications (Chen *et al.*, 2007), in which we describe another protocol that combines a consecutive siRNA knockdown procedure with plasmid DNA transfection with transfectants from Qiagen.

4.2.2.4. Directly measuring mRNA half-life with the Tet-off promoter system without transcriptional pulsing For studies that do not directly deal with mRNA deadenylation and decay kinetics or mechanistic steps (e.g., if one wants to investigate whether a known signaling pathway may alter the stability of the transcript of interest), time-course experiments for measuring mRNA stability can be done without the use of the transcriptional pulsing approach. After transiently transfecting the tTA-expressing cells with the Tet-off promoter–driven reporter plasmid, cells are kept in culture medium without tetracycline to allow constitutive expression of the reporter mRNA. After 48 h, tetracycline is added to the medium to a final concentration of 500 ng/ml to stop transcription from the Tet-off promoter and total RNA is extracted at different time intervals. This approach has been used successfully to study the stability of mRNAs of interest in the presence or absence of a stimulus known to activate a desired signaling pathway (Winzen et al., 1999), as well as to identify stimuli and signaling pathways that modulate stabilities of several cytokine and chemokine mRNAs during various immune responses (reviewed in Stoecklin and Anderson [2006]).

One potential caveat when the aforementioned approach is used to directly measure mRNA half-life is that highly robust and constitutive transcription of the reporter mRNA driven by the Tet-off promoter may saturate the system, thereby impeding proper decay of the transcript. We recommend that a titration experiment should be carried out first to identify the optimal amount of transfecting DNA that does not force the mRNA stabilization but still results in a signal detectable by Northern blot analysis (e.g., by an overnight exposure of an RNA blot to X-ray film).

5. Concluding Remarks

We have described two transcriptional pulsing methods that result in the synthesis of an mRNA population nearly homogeneous in size. These methods have several advantages over other approaches used to measure mRNA half-life. They offer the opportunity to determine deadenylation and decay kinetics, as well as the precursor-product relationship of mRNA turnover (Yamashita et al., 2005). Although the c-fos promoter system is convenient to quickly address the mechanistic steps involved and characterize the cis-acting sequences, the Tet-regulatory system provides the opportunity to study the regulation under physiologically relevant and undisturbed conditions. For example, the role of mRNA turnover in controlling the cell cycle can also be studied with the protocol developed with cells arrested at individual phases of the cell cycle (Chen et al., 2007). In addition, with an increasing collection of tTA-expressing cell lines, the Tet-regulatory promoter system combined with our strategy allows the analysis of mRNA turnover and its regulation under various physiologic conditions including, but not limited to, during cell growth and differentiation, the immune response, and tissue repair. For instance, with the establishment of the Tet-regulatory promoter system in lymphoid cell lines or hematopoietic cell lines representing different hematopoietic stages, explanation of the mechanisms responsible for the regulation of cytokine and chemokine mRNAs are already underway (Chen et al., 2007).

Recently, we have successfully combined the siRNA-mediated gene knockdown and the transcription pulsing systems to decipher mechanistic steps in mammalian mRNA turnover (Yamashita et al., 2005). We showed that deadenylation is the major trigger of mRNA decay in mammalian cells and that decapping does not occur until or after the second phase of deadenylation (Yamashita et al., 2005). By use of the approaches and protocols described here, the applications of transcription pulsing systems can be further expanded to study not only the mechanistic steps of mRNA decay but also their regulation by *trans*-acting factors. For instance, the transcriptional pulsing approach will facilitate our understanding as to how miRNAs tie into the mRNA decay machinery to accomplish gene silencing in mammalian cells, because miRNA-mediated mRNA decay also starts with deadenylation (Wu et al., 2006). The protocols described in this chapter and our approaches to optimize these protocols provide guidelines for more widespread development of new protocols with transcriptional pulsing strategies and their applications to study mRNA turnover under various physiologic conditions.

ACKNOWLEDGMENTS

We thank many past and present members in our laboratory who have contributed in various ways over the years to the development of the approaches and protocols described in this chapter. The work was supported by National Institutes of Health (GM 46454) and in part by the Houston Endowment, Inc., and the Sandler Program for Asthma Research (to A.-B. S.).

REFERENCES

Belasco, J., and Brawerman, G. (1993). Experimental approaches to the study of mRNA decay. In "Control of Messenger RNA Stability." (J. Belasco and G. Brawerman, eds.), pp. 475–491. Academic Press, Inc., San Diego.

Belasco, J. G., and Higgins, C. F. (1988). Mechanisms of mRNA decay in bacteria: A perspective. *Gene* **72,** 15–23.

Casey, J. L., et al. (1988). Iron-responsive elements: Regulatory RNA sequences that control mRNA levels and translation. *Science* **240,** 924–928.

Chen, C.-Y. A., and Shyu, A.-B. (1995). AU-rich elements: Characterization and importance in mRNA degradation. *Trends Biochem. Sci.* **20,** 465–470.

Chen, C.-Y. A., and Shyu, A.-B. (2003). Rapid deadenylation triggered by a nonsense codon precedes decay of the RNA body in a mammalian cytoplasmic nonsense-mediated decay pathway. *Mol. Cell Biol.* **23,** 4805–4813.

Chen, C.-Y. A., et al. (2007). Versatile applications of transcriptional pulsing to study mRNA turnover in mammalian cells. *RNA* **13,** 1775–1786.

Chen, C.-Y. A., et al. (1994). Interplay of two functionally and structurally distinct domains of the c-fos AU-rich element specifies its mRNA-destabilizing function. *Mol. Cell Biol.* **14,** 416–426.

Chen, C.-Y. A., and Shyu, A. B. (1994). Selective degradation of early-response-gene mRNAs: Functional analyses of sequence features of the AU-rich elements. *Mol. Cell Biol.* **14,** 8471–8482.

Chen, C.-Y. A., *et al.* (1995). mRNA decay mediated by two distinct AU-rich elements from *c-fos* and granulocyte-macrophage colony-stimulating factor transcripts: Different deadenylation kinetics and uncoupling from translation. *Mol. Cell Biol.* **15,** 5777–5788.

Gossen, M., *et al.* (1993). Control of gene activity in higher eukaryotic cells by prokaryotic regulatory elements. *TIBS* **18,** 471–475.

Gossen, M., and Bujard, H. (1992). Tight control of gene expression in mammalian cells by tetracycline-responsive promoters. *Proc. Natl. Acad. Sci. USA* **89,** 5547–5551.

Greenberg, M. E., and Ziff, E. (1984). Stimulation of 3T3 cells induces transcription of the *c-fos* proto-oncogene. *Nature* **311,** 433–438.

Grosset, C., *et al.* (2000). A mechanism for translationally coupled mRNA turnover: Interaction between the poly(A) tail and a c-fos RNA coding determinant via a protein complex. *Cell* **103,** 29–40.

Harrold, S., *et al.* (1991). A comparison of apparent mRNA half-life using kinetic labeling techniques vs decay following administration of transcriptional inhibitors. *Anal. Biochem.* **198,** 19–29.

Lindsten, T. C., *et al.* (1989). Regulation of lymphokine messenger RNA stability by a surface-mediated T cell activation pathway. *Science* **244,** 339–343.

Loflin, P. T., *et al.* (1999a). Unraveling a cytoplasmic role for hnRNP D in the *in vivo* mRNA destabilization directed by the AU-rich element. *Genes Dev.* **13,** 1884–1897.

Loflin, P. T., *et al.* (1999b). Transcriptional pulsing approaches for analysis of mRNA turnover in mammalian cells. *Methods. A companion to Methods in Enzymology* **17,** 11–20.

Meyer, S., *et al.* (2004). Messenger RNA turnover in eukaryotes: Pathways and enzymes. *Crit. Rev. Biochem. Mol. Biol.* **39,** 197–216.

Muellner, E. W., *et al.* (1989). A specific mRNA binding factor regulates the iron-dependent stability of cytoplasmic transferrin receptor mRNA. *Cell* **58,** 373–382.

Muhlrad, D., *et al.* (1994). Deadenylation of the unstable mRNA encoded by the yeast MFA2 gene leads to decapping followed by 5′ to 3′-digestion of the transcript. *Genes Dev.* **8,** 855–866.

Pandey, N. B., and Marzluff, W. F. (1987). The stem-loop structure at the 3′-end of histone mRNA is necessary and sufficient for regulation of histone mRNA stability. *Mol. Cell Biol.* **7,** 4557–4559.

Parker, R., and Song, H. (2004). The enzymes and control of eukaryotic mRNA turnover. *Nat. Struct. Mol. Biol.* **11,** 121–127.

Ross, J. (1995). mRNA stability in mammalian cells. *Microbiol. Rev.* **59,** 423–450.

Seiser, C., *et al.* (1995). Effect of transcription Inhibitors on the iron-dependent degradation of transferrin receptor mRNA. *J. Biol. Chem.* **270,** 29400–29406.

Shyu, A.-B., *et al.* (1991). Two distinct destabilizing elements in the c-fos message trigger deadenylation as a first step in rapid mRNA decay. *Genes Dev.* **5,** 221–232.

Shyu, A.-B., *et al.* (1989). The c-fos mRNA is targeted for rapid decay by two distinct mRNA degradation pathways. *Genes Dev.* **3,** 60–72.

Speth, C., and Oberbaumer, I. (1993). Expression of basement membrane proteins: Evidence for complex post-transcriptional control mechanisms. *Exp. Cell Res.* **204,** 302–310.

Stoecklin, G., and Anderson, P. (2006). Posttranscriptional mechanisms regulating the inflammatory response. *Adv. Immunol.* **89,** 1–37.

Treisman, R. (1985). Transient accumulation of c-fos RNA following serum stimulation requires a conserved 5′-element and *c-fos* 3′-sequences. *Cell* **42,** 889–902.

Wilusz, C. W., *et al.* (2001). The cap-to-tail guide to mRNA turnover. *Nat. Rev. Mol. Cell Biol.* **2,** 237–246.

Winzen, R., et al. (1999). The p38 MAP kinase pathway signals for cytokine-induced mRNA stabilization via MAP kinase-activated protein kinase 2 and an AU-rich region-targeted mechanism. *EMBO J.* **18,** 4969–4980.

Wu, L., et al. (2006). MicroRNAs direct rapid deadenylation of mRNA. *Proc. Natl. Acad. Sci. USA* **103,** 4034–4039.

Xu, N., et al. (1998). A broader role for AU-rich element-mediated mRNA turnover revealed by a new transcriptional pulse strategy. *Nucl. Acids Res.* **26,** 558–565.

Yamashita, A., et al. (2005). Concerted action of poly(A) nucleases and decapping enzyme in mammalian mRNA turnover. *Nat. Struct. Mol. Biol.* **12,** 1054–1063.

Yen, T. J., et al. (1988). Autoregulated instability of beta-tubulin mRNAs by recognition of the nascent amino terminus of beta-tubulin. *Nature* **334,** 580–585.

CHAPTER EIGHTEEN

Trypanosomes as a Model to Investigate mRNA Decay Pathways

Stuart Archer, Rafael Queiroz, Mhairi Stewart, *and* Christine Clayton

Contents

1. Introduction	359
2. Genetic Manipulation in Trypanosomes: Down-Regulating Expression of Proteins Involved in mRNA Decay	361
2.1. Summary	361
2.2. Measuring mRNA processing and decay kinetics	363
2.3. Analysis of trypanosome RNA with microarrays	369
Acknowledgments	375
References	375

Abstract

In trypanosomes, individual mRNAs arise by the processing of primary polycistronic transcripts. Consequently, mRNA degradation rates are critical determinants of mRNA abundance. In this chapter, we summarize the various options for genetic manipulation in trypanosomes with the goal of analyzing mRNA stability, including RNA interference. We describe a method for measuring the half-lives of trypanosome mRNAs, including those that are very unstable, and also the isolation of tagged protein–RNA complexes by IgG affinity chromatography. Last, we detail our current methods for RNA analysis with microarrays.

1. Introduction

Current molecular phylogenies divide eukaryotes into six early-branching "supergroups" (Rodriguez-Ezpeleta *et al.*, 2007). Nearly all of the chapters in this volume are devoted to studies of organisms within a single group, the Opisthokonta, which includes both animals and fungi. The exceptions are chapter 21 on plants (Plantae) and this one, which

Zentrum für Molekulare Biologie der Universität Heidelberg (ZMBH), Heidelberg, Germany focuses on trypanosomes. Trypanosomes and their close relatives, the Leishmanias, are unicellular flagellates that are grouped in the Kinetoplastida. The name comes from an unusual assemblage of mitochondrial DNA called the Kinetoplast, located at the base of the flagellum. Kinetoplastids initially attracted attention because of their pathogenicity to both man and his domestic animals. In the past two decades, however, it has become apparent that Kinetoplastids have become unusually dependent on some biochemical pathways, because they lack alternatives that are present in yeast and mammals. This, combined with ease of experimental and genetic manipulation, has enabled them to serve as useful model systems for the study of processes ranging from glycosyl phosphatidylinositol anchor synthesis to RNA editing.

One of the most remarkable features of Kinetoplastids is the transcription of polycistronic pre-mRNAs by RNA polymerase II (Palenchar and Bellofatto, 2006). Most Kinetoplastid genes are also constitutively transcribed (Martinez-Calvillo et al., 2003). Pre-mRNAs are cotranscriptionally processed, with addition of a capped 39-nt spliced leader (SL) to the $5'$-end of each mRNA followed by polyadenylation of the preceding mRNA (Liang et al., 2003). Because a single poly pyrimidine tract often serves to define both the $5'$-end of one mRNA and the $3'$-end of the upstream mRNA (Benz et al., 2005), scope for regulation of mRNA processing is limited. Exceptionally, the genes encoding the most abundant surface proteins of *Trypanosoma brucei* are transcribed by RNA polymerase I and are subject to chromatin-mediated transcriptional control (Borst, 2002). The facility with which functional mRNAs can be generated not only from polymerase I transcripts, but also (in transgenic trypanosomes) from RNAs synthesized by bacteriophage polymerases, implies that mRNA processing is independent of the transcribing polymerase. Genome surveys (Berriman, 2005; Ivens, 2005) correspondingly reveal a remarkable absence of not only regulatory transcription factors but also of the proteins required for coupling of splicing with transcription. As a consequence of this genomic organization, mRNA levels seem to be determined largely by the gene copy number and the mRNA degradation rate; the latter is usually determined by sequences located in the $3'$-untranslated region ($3'$ UTR) (Clayton and Shapira, 2007). Trypanosomes, therefore, exhibit a unique dependence on the enzymes responsible for mRNA decay (Li et al., 2006). The dearth of transcription factors also contrasts with abundant potential RNA-binding proteins containing RNA Recognition Motifs, CCCH zinc fingers, and Pumilio domains.

Most of the studies on mRNA degradation have been carried out with *T. brucei*, and the methods described in the following are for that species. Trypanosomes are commonly grown *in vitro* either as the bloodstream (BS) form (parasites similar to those that grow in mammalian blood and tissue fluids) or as the procyclic (PC) form (similar to the parasites that grow in the tsetse fly midgut). Before giving detailed descriptions of specific methods, we summarize options for trypanosome genetic manipulation.

2. GENETIC MANIPULATION IN TRYPANOSOMES: DOWN-REGULATING EXPRESSION OF PROTEINS INVOLVED IN mRNA DECAY

2.1. Summary

The methods available for genetic manipulation of trypanosomes have been described in detail elsewhere (Clayton et al., 2005) and will only be summarized here. DNA is transfected into the parasites either by conventional electroporation or by nucleófection (Burkard et al., 2007). Nearly all experiments with T. brucei are conducted with permanent cell lines in which the transfected plasmid has integrated into the genome by homologous recombination. Before transfection, plasmids are linearized, and recombination occurs with sequences immediately neighboring the ends thus created. Targeting is possible with homologies as low as 70 nt, but longer fragments seem to increase efficiency. Clones are selected by limiting dilution the day after transfection, at the same time as selection is applied.

2.1.1. Gene knockout

The 5′- and 3′-flanking regions of the gene to be targeted are cloned upstream and downstream, respectively, of a selectable marker gene. The DNA fragment containing the 5′-flanking region-marker-3′-flanking region is excised from the plasmid and transfected into the parasites.

2.1.2. Inducible gene expression

The starting point is trypanosomes expressing the *tet* repressor. Vectors contain a selectable marker under the control of a constitutive promoter, and another, inducible, promoter bearing *tet* operators. Best control is achieved if the inducible promoter used is an RNA polymerase I promoter from trypanosomes. Inducible expression from T7 promoters is also possible; in this case, the parasites must also constitutively express T7 polymerase. The plasmid must integrate into a silent region of the genome. A variety of vectors are available; these differ in the promoters used, the relative orientation of the selectable marker and inducible gene, the nature of the selectable marker, and the site at which the DNA integrates into the genome (see for example Alsford et al. [2005]; Clayton et al. [2005]; DaRocha et al. [2004]; Shi et al. [2000]; and Wickstead et al. [2002]).

2.1.3. *In situ* tagging

In this procedure, a sequence encoding an epitope tag, a fluorescent protein, or an affinity purification tag is integrated into the genome, in frame with the coding region of interest. Many suitable constructs are available for this purpose (e.g., Kelly et al. [2007] and Oberholzer et al. [2006]). Integration at the 5′-end, immediately upstream of the initiation codon, leaves the 3′ UTR intact, making it more likely (though not guaranteed) not to affect the half-life of the mRNA and protein level. However, it should be noted that 5′-integration is not suitable for proteins with N-terminal signal sequences, and tag insertion at any location might affect protein function or half-life.

2.1.4. RNA interference

One reason for choosing *T. brucei* as a model is the presence of an active RNA interference (RNAi) system in this organism (Ngo *et al.*, 1998), whereas the genes for RNAi are lacking in *Trypanosoma cruzi* and *Leishmania major*. RNAi in *T. brucei* can be achieved in three major ways as listed in the following.

2.1.3.1. Synthetic ds RNA   This gives very rapid, but transient, effects (Ngo *et al.*, 1998).

2.1.3.2. Stem loops   Cell lines are made with tetracycline-inducible expression of a stem loop from a trypanosome RNA polymerase I promoter. Cloning the stem loops is time-consuming and can be problematic, but this method tends to give lower background expression of dsRNA and better inducibility than opposing T7 promoters.

2.1.3.3. Opposing T7 promoters   Cell lines are made with tetracycline-inducible expression of dsRNA with opposing T7 promoters. This method is fast and ideal for screening purposes, especially as vectors available for direct cloning of PCR products are available (Alibu *et al.*, 2004). Unfortunately, however, the "background" expression from the T7 promoters can be sufficient to give significant mRNA depletion in the absence of inducer. If the target mRNA is essential, no clones will be obtained. Also anecdotal evidence suggests that these cells are less reliable than stem loop clones after frozen storage.

2.1.3.4. Troubleshooting   The efficiency of RNAi increases with the length of the dsRNA used. Because increasing lengths carry the risk of off-target effects, most experiments are currently done with fragments of approximately 500 bp. For a detailed description see Clayton *et al.* (2005).

Sometimes, RNAi simply does not work for unknown reasons; either no clones with regulation are obtained, or no clones are obtained at all. Trypanosomes seem to be rather flexible so if the RNAi inhibits growth, the effect may be lost after a few days either because of genetic changes (e.g., loss of T7 polymerase) or perhaps epigenetic effects. It is essential to freeze aliquots of clones as soon as possible. Various approaches have been taken to improve the efficiency of RNAi experiments in trypanosomes, with a selection of host-vector combinations (Alibu *et al.*, 2004; Alsford *et al.*,

2005; Wickstead et al., 2002). We usually use vectors with opposing T7 promoters for initial experiments, but construct a stem loop plasmid if we encounter difficulties or wish to do many experiments with the cell lines.

2.1.4. Inducible knockout

If difficulty is experienced in obtaining RNAi lines, the alternative is to undertake a three-step procedure. First, an inducible copy of the gene is inserted into the genome. Then, one of the endogenous genes is knocked out by gene replacement. At this point, the tetracycline inducer is added to turn on the inducible gene. Next the second endogenous copy is knocked out. The inducible gene is then turned off by removal of tetracycline.

Here it is critical that the expression of the protein encoded by the inducible gene in the absence of tetracycline be much lower than the normal level. Essential proteins seem to be present in considerable excess, so reductions of 90% or more may be required to see deleterious effects.

If loss of gene expression inhibits growth, there may again be reversion of the phenotype after a few days.

2.1.5. Induced site-specific recombination

The use of the cre-lox recombinase system in trypanosomes has recently been reported (Scahill et al., 2008). This is an excellent option for creating backgroundless inducible knockout lines. However, applicability of this system has not yet been extensively tested.

2.2. Measuring mRNA processing and decay kinetics

2.2.1. Overview

Each bloodstream-form trypanosome contains approximately 20,000 mRNA molecules; procyclic forms have slightly more than double that number (Haanstra et al., 2008). In bloodstream forms, housekeeping mRNAs are generally stable with half-lives of 45 to 60 min, and the number of mRNA molecules various from 12 mRNA molecules per cell for the single-copy *PGKC* gene to 70 mRNAs per cell for the multicopy tubulin genes. In contrast, highly unstable mRNAs have half-lives of 5 min or less and are present at less than one copy per cell. These numbers influence both the amounts of RNA that have to be prepared to allow detection and the time scale of mRNA degradation experiments. In particular, when testing the half-lives of very unstable RNAs, sample preparation times must be taken into account and minimized.

The tetracycline repressor-operator system would, in theory, allow specific shutoff of the transcription of reporter genes by washing out the inducer. We have attempted this, starting with minimal tetracycline concentrations then washing the cells. Unfortunately, the kinetics of shutoff were far too slow to allow measurement of the mRNA half-life (H. Irmer,

ZMBH, unpublished). The first experiments to analyze mRNA decay in trypanosomes were done by pulse labeling with ^{3}H-adenine, followed by a chase (Ehlers et al., 1987). The half-lives that were obtained were considerably longer than those seen after transcription inhibition. The easiest explanation for the discrepancy is that the turnover of the adenine pool within the cells was slower than the authors thought. This would however be surprising since the pool size had been carefully measured.

The only option is, therefore, to inhibit all transcription. In most studies, actinomycin D has been added, then total RNA isolated after various incubation times and degradation measured by Northern blotting, loading the same amount of total RNA for each time point. This seems to work quite well for highly unstable transcripts. Unfortunately, however, for many longer lived transcripts, the amount of mRNA has been reproducibly found to rise, sometimes up to twofold, in the 15 min after drug treatment (Clayton and Shapira, 2007). This makes kinetic analysis extremely difficult and all interpretation problematic.

Because trypanosome mRNAs are generated by processing of polycistronic transcripts, it is theoretically possible for mRNAs to be made after transcription has been inhibited. Whether this could lead to an increase in mRNA up to 15 min after transcription inhibition is highly dubious, because the very limited measurements done so far suggest that splicing is complete within approximately 2 min (Haanstra et al., 2008), and synthesis of the spliced leader RNA (*SLRNA*) is inhibited by actinomycin D. Nevertheless, the dependence of mRNA synthesis on *trans* splicing did suggest an alternative method to measure half-lives. Addition of sinefungin prevents cap methylation and thereby eliminates splicing-competent *SLRNA* without affecting transcription (McNally and Agabian, 1992; Ullu and Tschudi, 1991). Primer extension experiments have previously shown that 15-min incubation is sufficient to deplete capped *SLRNA* in procyclic forms by approximately 90%, and we have observed similar kinetics in bloodstream forms. Curiously, degradation of mRNAs actually commences much faster, only 5 min after sinefungin addition (Webb et al., 2005). Although sinefungin treatment does not generally result in an increase of mRNAs after drug addition, calculations indicate that 10% residual capping activity could significantly prolong the apparent half-lives of mRNAs. To completely inhibit mRNA synthesis, we, therefore, have adopted a protocol that uses the sequential addition of sinefungin and actinomycin D. This combination gives faster apparent degradation than either actinomycin D or sinefungin alone and is the only protocol we know of that reproducibly yields exponential decay kinetics (Colasante et al., 2007).

2.2.2. Procedure

2.2.2.1. The cells Start with a culture that is in log phase. Between 4 and 8×10^7 cells are required per time point; 4×10^7 is sufficient for identifying mRNA molecules by Northern blotting or RNase protection assay, but double that number should be used for identifying RNA species of low abundance such as precursor RNA. On the day before the experiment is planned, dilute the parasites into warmed medium so that you will have sufficient cells the next day at an appropriate density that should not exceed one quarter of the stationary-phase density. For procyclic form trypanosomes, this is 5×10^6 cells/ml and for highly culture-adapted bloodstream forms, 2×10^6 cells/ml; densities for differentiation-competent "pleomorphic" cells will be 5 to 10 times lower. Keeping the cells in rapid growth is critical, because mRNA levels decrease as the cells approach stationary phase (Häusler and Clayton, 1996). For example, a 500-ml culture of culture-adapted BS form cells seeded at 2×10^5 cells/ml and allowed to grow overnight (18 h) is usually sufficient to provide enough cells at 1 to 2×10^6 for 15 time points of 4 to 8×10^7 cells.

2.2.2.2. Planning the experiment The protocol for the experiment itself depends on the time points to be taken. If time points are very short, treat each tube separately, from addition of inhibitor to resuspension in RNA denaturing solution. For longer time intervals one can either add the inhibitor all at once and centrifuge samples separately or add inhibitor at various intervals and harvest all samples simultaneously.

For developmentally unstable mRNAs that have half-lives in the order of 5 min, time intervals of 2.5 min, 5 min, 7.5 min, and 10 min are appropriate when the indicated times span from the instant of inhibitor addition to cell lysis. The protocol that follows is especially designed for short time intervals and can also be used for analysis of splicing kinetics (Haanstra *et al.*, 2008); the minimum time from actinomycin D addition to cell lysis is approximately 40 sec. For longer lived mRNAs, time points at 10, 20, 30, 45, 60, and perhaps 90 min are useful. In this case, for incubations of 30 min or more, the use of larger volumes with adequate aeration (e.g., normal culture flasks) is preferable.

2.2.2.3. Materials needed Sinefungin (Sigma-Aldrich), 200 µg/ml.
Actinomycin D (Sigma-Aldrich), 1 mg/ml; we try not to store it for more than 2 wk at $-20\,°C$.
TRIzol (Invitrogen) or other favorite RNA preparation solution.
Microfuge, heat block, vortex.
Warmed medium.

2.2.2.4. Time course

1. Sediment the cells at $1000g$ for 10 min, resuspend in warmed medium at 4 to 8×10^7 cells/ml, and aliquot into the appropriate number of Eppendorf tubes (1 ml each, one tube per time point).

2. Incubate for 20 min at 37 °C (bloodstream forms) or 27 °C (procyclic forms) in a heat block to return the cells to a normal growth environment.
3. Save one tube for "no treatment" (time = −5 min).
4. To the remaining tubes, add sinefungin to a final concentration of 2 μg/ml. Incubate 5 min.
5. Take two tubes for the time = 0 samples. (Because this sample is crucial, having two samples is an insurance against accidents later.)
6. To each remaining tube, add actinomycin D to a final concentration of 10 μg/ml. Incubate for desired times.
7. 1 min before the end of the incubation time, pellet the cells at 4000 rpm in microfuge for 30 sec, tip off the medium, then quickly resuspend in RNA preparation solution by vortexing very briefly.
8. Analyze by an appropriate standard method (e.g., Northern blot, RNase protection, real-time PCR). Always check the efficiency of RNA synthesis inhibition with a control of known half-life. (Schwede et al., 2008).

2.2.2.5. Determining the direction of mRNA decay General methods for determining the direction of mRNA decay can be found in chapters by Passos and Parker (Chapter 20) and Murray and Schoenberg (Chapter 24) in this volume. Those that have been used effectively in trypanosomes are as follows:

1. The transcript under study was internally cleaved with specific oligonucleotides and RNase, followed by denaturing polyacrylamide gel electrophoresis and Northern blotting with probes specific to the expected fragments (Irmer and Clayton, 2001).
2. Degradation kinetics was studied after RNAi targeting exoribonucleases (Haile et al., 2003; Li et al., 2006).
3. In trypanosomes, as in *Saccharomyces cerevisiae*, the progress of the exosome and 5′ to 3′-exonuclease is delayed by the presence of a strong secondary structure. A 60-nucleotide stretch of $G_{30}C_{30}$ sequence is sufficient to allow detection of intermediates. This has, however, been successful so far only for one reporter transcript, and it is not clear if it is generally applicable (Haile et al., 2003; Irmer and Clayton, 2001; Li et al., 2006).

2.2.3. Identifying the targets of RNA-binding proteins: Complex purification

2.2.3.1. Overview To determine the targets of RNA-binding proteins, we use a method adapted from Gerber et al. (2004) that involves expression of a TAP-tagged version of the protein of interest purified on an IgG column followed by isolation of the RNA. The RNA is then analyzed either on microarrays or by RT-PCR (Luu et al., 2006). Because overexpression of RNA binding proteins can affect mRNA abundance, the ideal control in both cases is the flow through from the IgG column.

Our attempts to purify complexes between RRM-domain proteins and RNA with other epitope tags (e.g., V5 tag, with elution with the V5 peptide) have so far been less successful (Hartmann et al., 2007).

2.2.3.2. Preparation of the trypanosomes Reserve and precool centrifuge, ultracentrifuge, and rotors. All other tools and PBS should be precooled. For each TAP purification, you need approximately 2×10^9 cells, induced overnight with tetracycline (100 to 1000 ng/ml, depending on the cell line). Do not exceed culture densities of 1.2×10^6 cells/ml (bloodstream forms) or 4×10^6 cells/ml (procyclic forms).

Note: for RT-PCR detection, $\sim 5 \times 10^8$ cells are sufficient. In this case, divide all volumes used by 4 (except for the small aliquots that are taken for Western blot analysis during the procedure) and use smaller columns.

1. Spin cells down 1000g in a normal cell centrifuge at 4 °C for 10 min.
2. Wash cells with 50 ml ice-cold PBS.
3. Optional: cells can be resuspended in 1 ml PBS and UV-cross-linked on ice (400 mJ/cm^2 in a Stratalinker). Place cell suspension in a petri dish on ice and place a $\sim$15 cm^2 ($\sim$4 × 4cm) piece of transparent sheet on top to flatten sample and ensure even exposure.
4. Wash cells in another 50 ml ice-cold PBS. Remove supernatant.
5. Snap-freeze pellets in liquid nitrogen and store at -80 °C or directly proceed with extract preparation.

2.2.4. Preparation of cleared cell lysate
2.2.4.1. Materials

Breakage buffer
10 mM Tris-Cl
10 mM NaCl
0.1% IGEPAL, adjusted to pH 7.8 with HCl
For 50 ml, include one tablet of complete inhibitor (without EDTA, Roche), 200 U RNasin (Promega), and 2 to 5 mM vanadyl ribonucleoside complexes (Sigma)

2.2.4.2. Procedure

1. Break cells in a final volume of 6-ml breakage buffer by passing 15 to 20 times through a 21- to 25-gauge needle. Check on a glass slide to ensure complete breakage.
2. Spin cell lysate at 10,000g for 15 min to remove cell debris.
3. Transfer supernatant to precooled polycarbonate thick wall Beckman centrifuge tubes (13.5 ml capacity, 355630 rec. no.). Spin at 35,000 rpm (109,000g), 4 °C for 45 min.

4. After centrifugation, transfer supernatant to 15-ml plastic centrifuge tube. Measure volume and add NaCl to a final concentration of 150 mM (note that breakage buffer is already 10 mM NaCl, so add 0.035 volumes of 4 M NaCl).
5. Take 25-μl aliquot (this will be a sample of the start material, #1), equivalent to approximately 4×10^6 cells.

2.2.5. TAP purification
2.2.5.1. Materials

IPP150: 10 mM Tris-Cl, pH 7.8, 150 mM NaCl, 0.1% IGEPAL.

TEV cleavage buffer: IPP150 adjusted to 0.5 mM EDTA and 1 mM DTT. Add at least 1/100 volume of 200 mM vanadyl ribonucleoside complexes, freeze.

Note: PMSF and AEBSF (1 mM), TLCK (1 mM), bestatin (1 mg/ml), pepstatin A (1 mM), EDTA (1 mM), and E-64 (3 mg/ml), and "complete" protease inhibitor cocktail (Roche) do not inhibit TEV. However, EDTA may disrupt vanadyl ribonucleoside complexes.

2.2.5.2. Procedure

1. Use a cutoff yellow tip to transfer 200 μl IgG sepharose bead suspension (Fastflow, Amersham Biosciences; equivalent to 100 μl compact beads) into a 0.8×4 cm Poly-Prep column (Bio-Rad). Wash with 10 ml IPP150.
2. Transfer lysate into the column containing the washed beads and rotate for 2 h at 4 °C. Elution is done by gravity flow. Take 25 μl of flow through (IgG flow through, sample #2) and freeze the rest.
3. Wash beads three times in 10 ml of IPP150 and once with 10 ml of TEV cleavage buffer (IPP150 adjusted to 0.5 mM EDTA and 1 mM DTT). Take 42 μl of the 10-ml bead slurry for analysis (beads, sample #3). Be sure to take some beads with the sample, with a cutoff yellow tip.
4. Tap tag cleavage is carried out in the same column by adding 1 ml of TEV cleavage buffer and 100 U of TEV protease (10 μl, Invitrogen). The beads are rotated for 2 h at 16 °C, and the eluate is recovered by gravity flow. (Sample can be frozen at −70 °C at this point.) Take a 4.2 μl sample (IgG eluate, sample #4).
5. Resuspend the beads in 1 ml of IPP150 and take a 4.2 μl sample (#5) of bead slurry to test cleavage/elution efficiency.

Analyze samples 1 to 5 by Western blot with the antiperoxidase antibody (Sigma). *Note:* The C-terminal TAP tag is 20 kDa, of which 15 kDa (including the PAP epitope) is cleaved away by TEV. Boiling the IgG beads yields IgG that may cause extra bands to appear in these samples. For large-scale preparations, the IgG beads may be regenerated with alternate low/high-pH washes (see instructions from Sigma).

2.2.6. RNA isolation

2.2.6.1. Materials required

QIAGEN RNA-easy kit.
QIAGEN RNase-free DNase for on-column digestion.
10 mg/ml proteinase K. Can be preincubated for 10 min at 37 °C to digest RNases.
RNase-free ethanol.

2.2.6.2. Procedure

1. Add ½ sample volume of RLT from the Qiagen RNA-easy kit, mix, and calculate the new volume.
2. Per 1000 μl, add 22 μl of 10 mg/ml proteinase K (PK). Incubate for 10 min at 55 °C.
3. Centrifuge at 12,000g for 2 min at room temperature, take supernatant, add ½ volume of EtOH and mix. Apply to an RNeasy mini spin column; spin several times if necessary to run the entire sample through the column.

The final volume ratios are approximately 4:2:3 for Sample: RLT: EtOH. (e.g., for 500 μl sample, add 250 μl RLT, 11 μl PK (10 mg/ml); incubate, then 375 μl EtOH).

4. Follow kit instructions, including the on-column DNase digestion.

2.3. Analysis of trypanosome RNA with microarrays

2.3.1. Overview

The only types of microarrays that are generally available can be obtained by applying to the pathogen functional genomics resource center at NIAID (current web address is http://www.niaid.nih.gov/dmid/genomes/pfgrc/reqprocess.htm). These contain one 70 mer-oligonucleotide per open reading frame and are very useful for applications that do not involve the analysis of noncoding RNA. (We have, however, found that a few of the oligonucleotides, most notably that for the *PGKB* mRNA, do not give the same results as obtained with Northern blotting.)

We also use custom microarrays based on shotgun genomic clones. The arrays contain amplified PCR products and include noncoding sequences (Brems *et al.*, 2005). They give stronger signals than the oligo arrays. Because the 3′ UTRs of trypanosome mRNAs may consist of several kb, we prefer these arrays when searching for RNAs that are bound to proteins (Luu *et al.*, 2006). We do not, however, have experience in doing such analyses with the oligonucleotide arrays. Because of considerable 3′ UTR lengths, we also use random hexamers for labeling when hybridizing to the oligonucleotide arrays.

2.3.2. Synthesis of the cDNA probe
2.3.2.1. Materials for labeling

Trypanosome RNA: at least 10 μg per labeling reaction. Check the concentration and integrity before starting, for example with a Nanodrop spectrophotometer and running a gel.

Primers: To examine intact mRNA on genomic arrays oligo $(dT)_{12-18}$ (2.5 μg/μl) or a mixture of oligo d(T) and random hexamers can be used. We, however, find that random hexamers (0.5 μg/μl) alone give satisfactory results.

Heating block at 70 °C.
5× 1st Strand buffer (Invitrogen).
0.1 M DTT.
dAGT Mix (dATP, dGTP and dTTP) (10 mM).
dCTP (1 mM).
Cy3-labeled dCTP (25 nmol) (GE Healthcare) (protect from light).
Cy5-labeled dCTP (25 nmol) (GE Healthcare) (protect from light).
RNase Out 40 U/μl (Invitrogen).
Superscript II (or III) reverse transcriptase (Invitrogen).
Oven at 42 °C (available overnight).

2.3.2.2. Labeling procedure

1. For each tube mix:
   a. 10 to 40 μg RNA.
   b. 1 μl primers.
   c. Water to a final volume of 21 μl.
2. Incubate 10 min at 70 °C.
3. Incubate 5 min on ice.
4. Meanwhile make the Master Mix. For each tube you need:

Reagent	μl
5× 1st Strand buffer	8.5
DTT (0.1 mM)	3.5
dAGT Mix (dATP, dGTP and dTTP) (10 mM)	3
dCTP (1 mM)	2
RNase Out	1
Superscript II reverse transcriptase	2
Final volume	20

5. Add to each RNA primer tube:
   a 19 µl of Master mix.
   b 2 µl of labeled dCTP (Cy3/Cy5).
6. Incubate 1 h in the dark at 42 °C (for Superscript II) or 50 °C (for Superscript III); check the manufacturer's instructions.
7. Add 1 µl Superscript II (or III) reverse transcriptase.
8. Incubate 3 to 18 h at 42 °C/50 °C.

2.3.2.3. Materials for probe purification

Heating block at 70 °C.
RNase H (2 U/µl).
DNA purification mini-columns.
PB Buffer from Qiagen PCR purification kit.
PE Buffer from Qiagen PCR purification kit.
RNase-free water or elution buffer.
NanoDrop ND-100.
Speedvac.

2.3.2.4. Procedure for probe purification

1. Remember to protect the probes from light as much as possible.
2. Incubate the reaction for 10 min at 70 °C.
3. Cool to room temperature. Add 1 µl RNase H and incubate 20 min at 37 °C.
4. Add 5 volumes PB buffer and mix (~250 µl) and apply the sample to the mini-columns already with 2-ml collecting tubes.
5. Centrifuge 1 min at 11,000g and discard supernatant.
6. Add 750 µl PE buffer to each column.
7. Centrifuge 1 min at 11,000g.
8. Centrifuge 1 min at 11,000g additionally to dry.
9. Change the collecting tubes to 1.5-ml microfuge tubes.
10. Add 30 µl EB or distilled H_2O, incubate 1 min at room temperature.
11. Centrifuge 1 min at 13,000g.
12. Add 30 µl EB or distilled H_2O, incubate 1 min at room temperature.
13. Centrifuge 1 min at 13,000g.
14. Measure RNA concentration and dye incorporation with NanoDrop® ND-1000 Spectrophotometer.
15. Mix dyes for each hybridization.
16. Speedvac the samples until dry (~40 min).
17. Samples can be stored at 20 °C for 1 mo before being hybridized.

2.3.3. Procedure for hybridizing Quantifoil (self-spotted) slides
2.3.3.1. Material required

Rinsing solution (0.2% SDS).

Distilled H$_2$O (room temperature and at 95 °C).
Prehybridization solution (5× SSC, 0.1% SDS, and 1% BSA).
3× SSC.
EDTA 10 mM.
Ambion hybridization solution (SlideHyb#1) preheated at 68 °C.
Slide holder (to wash the slides).
Three small recipients for washing the slides.
Slide flasks.
Waterbath at 55 °C.
Slide booster hybridization chambers (a specialized hybridization apparatus; Advalytix).
Coupling liquid (Advalytix AS100).
Humidifying solution (AM101 Advahum).
N$_2$ gas.
Heating block at 68 °C.
Heating block at 95 °C.
Cover glasses (e.g., Lifter slips from Implen).

If no specialized hybridization apparatus is available, a waterbath can be used, but we find that the results are not as good.

2.3.3.2. Slide prehybridization

1. Label the slides.
2. 10 sec in rising solution (0.2% SDS) with manual shaking.
3. 10 sec in distilled H$_2$O at room temperature with manual shaking.
4. 3 min in distilled H$_2$O at 95 °C (preheat the water in a microwave oven).
5. Put the samples in slide flasks or hybridization flasks with prehybridization solution.
6. Incubate for 1 h in waterbath at 55 °C. Meanwhile start the sample prehybridization that follows.
7. Rinse 10 sec in distilled H$_2$O at room temperature.
8. Directly dry the slides with N$_2$ (as fast as possible).
9. Place the slides in the hybridization chambers.

2.3.3.3. Sample prehybridization

Start this so that the solution is ready just when the slide preincubation has finished.
Remember to protect from light!

1. Add 10 μl of EDTA 10 mM to each tube containing probe mix.
2. Vortex briefly.
3. Place at 95 °C for 5 min.
4. Place at room temperature for 5 min.
5. Centrifuge briefly to bring any condensate to the bottom of the tube.

6. Add 40 µl of Ambion hybridization solution (SlideHyb#1) (previously heated at 68 °C for 15 min).

2.3.3.4. Hybridization (first day)

1. Prepare hybridization chambers with coupling liquid (Advalytix AS100) and humidifying solution (AM101 Advahum) following the instructions from the manual of Slide Booster (Advalytix).
2. Spread all of the probe mix sample equally on the middle of the slide.
3. Cover the slides carefully with clean cover glasses.
4. Close the chambers well and incubate overnight at 55 °C with a pulse/pause ratio of 3:7.

2.3.3.4.1. Washing materials

Wash A solution at 55 °C (1× SSC, 0.2% SDS).
Wash B solution (0.1× SSC, 0.2% SDS).
Wash C solution (0.1× SSC).
Slide holder (to wash the slides).
Three small recipients for washing the slides.
N_2 gas.
Scanner.

It is important to make sure there is enough of each solution before you begin. You should also make sure the array scanner has been reserved so you can read your results.

2.3.3.4.2. Washing procedure

1. 10 min in Wash A solution (cover it with laminated paper to protect the slides against light).
2. 10 min in Wash B solution (cover it with laminated paper to protect the slides against light).
3. 1 min in Wash C solution (cover it with laminated paper to protect the slides against light).
4. Immediately dry it with N_2.
5. Proceed with scanning process.

2.3.4. Hybridization of TIGR Tb oligo array slides
2.3.4.1. Materials required

Prehybridization buffer (5× SSC, 0.1% SDS, and 1% BSA).
Hybridization buffer (6× SSC, 0.1% SDS, 50% formamide, 0.6 mg/ml herring sperm DNA). (The hybridization buffer should be filtered with a 1-ml syringe and a 0.45-µm filter before adding the salmon sperm DNA.)

Slide holder (to wash the slides).
Three small recipients for washing the slides.
Slide flasks.
Waterbath at 42 °C.
Slide booster hybridization chambers (a specialized hybridization apparatus; Advalytix).
Coupling liquid (Advalytix AS100).
Humidifying solution (AM101 Advahum).
N_2 gas.
Hybridization chambers.
Cover glasses (e.g., Lifter slips from Implen).
Slide prehybridization.

Do not let the slides dry at *any* point throughout this entire process, or there will be unacceptable background.

1. Take note of the slide code bars.
2. Put the samples in slide flasks (or hybridization flasks with blue and green lids) with prehybridization buffer.
3. Incubate for 45 to 60 min in waterbath at 42 °C (meanwhile proceed with the sample prehybridization that follows).
4. Wash three times in distilled H_2O at room temperature (500 ml, 10 sec each).
5. Dip in isopropanol quickly (~1 min).
6. Directly dry the slides with N_2 (working as fast as possible).
7. Place the slides in the hybridization chambers.

2.3.4.2. Sample prehybridization Start so that the probes will just be ready when the slide preincubation has finished. Remember to protect from light!

a. Put 10 µl of distilled H_2O into each tube (the dyes are already mixed).
b. Vortex and spin down for 10 sec.
c. Add 50 µl of hybridization buffer.
d. Place at 95 °C for 5 min.
e. Place on ice for 2 min.
f. Vortex.
g. Again 5 min at 95 °C.
h. Spin down briefly and use immediately. (If the tube is left on ice for too long, the SDS will precipitate. In this case heat briefly.)

2.3.5. Hybridization (first day)

1. Prepare hybridization chambers with coupling liquid (Advalytix AS100) and humidifying solution (AM101 Advahum) following the instructions from the manual of slide booster (Advalytix).

2. Spread all the sample equally on the middle of the slide.
3. Cover it with *clean* cover glasses and try to banish the bigger bubbles.
4. Close the chambers and incubate at 37 °C for 16 to 20 h with a pulse/pause ratio of 3:7.

2.3.5.1. Washing materials

Wash 1 solution (2× SSC; 0.1% SDS; 55 °C).
Wash 2 solution (0.1× SSC; 0.1% SDS).
Wash 3 solution (0.1% SSC).
Three small recipients for washing the slides.
N_2 gas.
Scanner.

2.3.5.2. Washing procedure

a. 10 min in Wash 1 solution (protect against light with laminated paper).
b. 10 min in Wash 2 solution (protect against light with laminated paper).
c. 10 min in Wash 3 solution (protect against light with laminated paper).
d. Immediately dry with N_2.
e. Proceed with scanning process.

ACKNOWLEDGMENTS

Our work was supported by the Deutsche Akademische Austauschdienst, the Deutsche Forschungsgemeinschaft, and the Wellcome Trust. Numerous previous laboratory members, particularly Stefanie Brems, Antonio Estevez, Simon Heile, Chi-Ho Li, and Van-Duc Luu, contributed to developing or adapting the methods described in this chapter.

REFERENCES

Alibu, V. P., Storm, L., Haile, S., Clayton, C., and Horn, D. (2004). A doubly inducible system for RNA interference and rapid RNAi plasmid construction in *Trypanosoma brucei. Mol. Biochem. Parasitol.* **139,** 75–82.

Alsford, S., Kawahara, T., Glover, L., and Horn, D. (2005). Tagging a *T. brucei* RRNA locus improves stable transfection efficiency and circumvents inducible expression position effects. *Mol. Biochem. Parasitol.* **144,** 142–148.

Benz, C., Nilsson, D., Andersson, B., Clayton, C., and Guilbride, D. L. (2005). Messenger RNA processing sites in *Trypanosoma brucei. Mol. Biochem. Parasitol.* **143,** 125–134.

Berriman, M., Ghedin, E., Hertz-Fowler, C., Blandin, G., Renauld, H., Bartholomeu, D. C., Lennard, N. J., Caler, E., Hamlin, N. E., Haas, B., Böhme, U., Hannick, L., *et al.* (2005). The genome of the African trypanosome, *Trypanosoma brucei. Science* **309,** 416–422.

Borst, P. (2002). Antigenic variation and allelic exclusion. *Cell* **109,** 5–8.

Brems, S., Guilbride, D. L., Gundlesdodjir-Planck, D., Busold, C., Luu, V. D., Schanne, M., Hoheisel, J., and Clayton, C. (2005). The transcriptomes of *Trypanosoma brucei* Lister 427 and TREU927 bloodstream and procyclic trypomastigotes. *Mol. Biochem. Parasitol.* **139**, 163–172.

Burkard, G., Fragoso, C., and Roditi, I. (2007). Highly efficient stable transformation of bloodstream forms of *Trypanosoma brucei*. *Mol. Biochem. Parasitol.* **153**, 220–223.

Clayton, C., and Shapira, M. (2007). Post-transcriptional regulation of gene expression in trypanosomes and leishmanias. *Mol. Biochem. Parasitol.* **156**, 93–101.

Clayton, C. E., Estévez, A. M., Hartmann, C., Alibu, V. P., Field, M., and Horn, D. (2005). Down-regulating gene expression by RNA interference in *Trypanosoma brucei*. *In* "RNA Interference." (G. Carmichael, ed.). Humana Press. Totowa, New Jersey.

Colasante, C., Robles, A., Li, C.-H., Schwede, A., Benz, C., Voncken, F., Guilbride, D. L., and Clayton, C. (2007). Regulated expression of glycosomal phosphoglycerate kinase in *Trypanosoma brucei*. *Mol. Biochem. Parasitol.* **151**, 193–204.

DaRocha, W., Otsu, K., Teixeira, S., and Donelson, J. (2004). Tests of cytoplasmic RNA interference (RNAi) and construction of a tetracycline-inducible T7 promoter system in *Trypanosoma cruzi*. *Mol. Biochem. Parasitol.* **133**, 175–186.

Ehlers, B., Czichos, J., and Overath, P. (1987). RNA turnover in *Trypanosoma brucei*. *Mol. Cell. Biol.* **7**, 1242–1249.

Gerber, A., Herschlag, D., and Brown, P. (2004). Extensive association of functionally and cytotopically related mRNAs with Puf family RNA-binding proteins in yeast. *PLoS Biol.* **2**, e79.

Haanstra, J., Stewart, M., Luu, V.-D., van Tuijl, A., Westerhoff, H., Clayton, C., and Bakker, B. (2008). Control and regulation of gene expression: Quantitative analysis of the expression of phosphoglycerate kinase in bloodstream form *Trypanosoma brucei*. *J. Biol. Chem.* **283**, 2495–2507.

Haile, S., Estévez, A. M., and Clayton, C. (2003). A role for the exosome in the initiation of degradation of unstable mRNAs. *RNA* **9**, 1491–1501.

Hartmann, C., Benz, C., Brems, S., Ellis, L., Luu, V.-D., Stewart, M., D'Orso, I., Busold, C., Fellenberg, K., Frasch, A. C. C., Carrington, M., Hoheisel, J., and Clayton, C. E. (2007). The small trypanosome RNA-binding proteins TbUBP1 and TbUBP2 influence expression of F box protein mRNAs in bloodstream trypanosomes. *Eukaryotic Cell* **6**, 1964–1978.

Häusler, T., and Clayton, C. E. (1996). Post-transcriptional control of hsp 70 mRNA in *Trypanosoma brucei*. *Mol. Biochem. Parasitol.* **76**, 57–72.

Irmer, H., and Clayton, C. E. (2001). Degradation of the *EP1* mRNA in *Trypanosoma brucei* is initiated by destruction of the 3′-untranslated region. *Nucleic Acids Res.* **29**, 4707–4715.

Ivens, A. C., Peacock, C. S., Worthey, E. A., Murphy, L., Aggarwal, G., Berriman, M., Sisk, E., Rajandream, M. A., Adlem, E., Aert, R., Anupama, A., Apostolou, Z., *et al.* (2005). The genome of the kinetoplastid parasite, *Leishmania major*. *Science* **309**, 436–442.

Kelly, S., Reed, J., Kramer, S., Ellis, L., Webb, H., Sunter, J., Salje, J., Marinsek, N., Gull, K., Wickstead, B., and Carrington, M. (2007). Functional genomics in *Trypanosoma brucei*: A collection of vectors for the expression of tagged proteins from endogenous and ectopic gene loci. *Mol. Biochem. Parasitol.* **154**, 103–109.

Li, C.-H., Irmer, H., Gudjonsdottir-Planck, D., Freese, S., Salm, H., Haile, S., Estévez, A. M., and Clayton, C. E. (2006). Roles of a *Trypanosoma brucei* 5′ to 3′-exoribonuclease homologue in mRNA degradation. *RNA* **12**, 2171–2186.

Liang, X., Haritan, A., Uliel, S., and Michaeli, S. (2003). Trans and cis splicing in trypanosomatids: Mechanism, factors, and regulation. *Eukaryot. Cell* **2**, 830–840.

Luu, V. D., Brems, S., Hoheisel, J., Burchmore, R., Guilbride, D., and Clayton, C. (2006). Functional analysis of *Trypanosoma brucei* PUF1. *Mol. Biochem. Parasitol.* **150**, 340–349.

Martinez-Calvillo, S., Yan, S., Nguyen, D., Fox, M., Stuart, K., and Myler, P. J. (2003). Transcription of Leishmania major Friedlin chromosome 1 initiates in both directions within a single region. *Mol. Cell* **11,** 1291–1299.

McNally, K. P., and Agabian, N. (1992). Trypanosoma brucei spliced-leader RNA methylations are required for trans splicing *in vivo. Mol. Cell. Biol.* **12,** 4844–4851.

Ngo, H., Tschudi, C., Gull, K., and Ullu, E. (1998). Double-stranded RNA induces mRNA degradation in *Trypanosoma brucei. Proc. Natl. Acad. Sci. USA* **95,** 14687–14692.

Oberholzer, M., Morand, S., Kunz, S., and Seebeck, T. (2006). A vector series for rapid PCR-mediated C-terminal *in situ* tagging of *Trypanosoma brucei* genes. *Mol. Biochem. Parasitol.* **145,** 117–120.

Palenchar, J. B., and Bellofatto, V. (2006). Gene transcription in trypanosomes. *Mol. Biochem. Parasitol.* **146,** 135–141.

Rodrıguez-Ezpeleta, N., Brinkmann, H., Burger, G., Roger, A., Gray, M., Philippe, H., and Lang, B. (2007). Toward resolving the eukaryotic tree: The phylogenetic positions of Jakobids and Cercozoans. *Curr. Biol.* **17,** 1420–1425.

Schwede, A., Ellis, L., Luther, J., Carrington, M., Stoecklin, G., and Clayton, C.E. (2008) A role for Caf1 in mRNA deadenylation and decay in trypanosomes and human cells. *Nucleic Acids Res.* **36,** 3374–3388.

Schwede, A., Ellis, L., Luther, J., Carrington, M., Stoecklin, G., and Clayton, C. E. (2008). A role for Caf1 in mRNA deadenylation and decay in trypanosomes and human cells. *Nucleic Acids Res.* **36,** 3374–3388.

Scahill, M., Pastar, I., and Cross, G. (2008). CRE recombinase-based positive–negative selection systems for genetic manipulation in *Trypanosoma brucei. Mol. Biochem. Parasitol.* **157,** 73–82.

Shi, H., Djikeng, A., Mark, T., Wirtz, E., Tschudi, C., and Ullu, E. (2000). Genetic interference in *Trypanosoma brucei* by heritable and inducible double-stranded RNA. *RNA* **6,** 1069–1076.

Ullu, E., and Tschudi, C. (1991). Trans splicing in trypanosomes requires methylation of the 5′-end of the spliced leader RNA. *Proc. Natl. Acad. Sci. USA* **88,** 10074–10078.

Webb, H., Burns, R., Ellis, L., Kimblin, N., and Carrington, M. (2005). Developmentally regulated instability of the GPI-PLC mRNA is dependent on a short-lived protein factor. *Nucleic Acids Res.* **33,** 1503–1512.

Wickstead, B., Ersfeld, K., and Gull, K. (2002). Targeting of a tetracycline-inducible expression system to the transcriptionally silent minichromosomes of *Trypanosoma brucei. Mol. Biochem. Parasitol.* **125,** 211–216.

CHAPTER NINETEEN

Cell Type–Specific Analysis of mRNA Synthesis and Decay *In Vivo* with Uracil Phosphoribosyltransferase and 4-Thiouracil

Michael D. Cleary

Contents

1. Introduction	380
2. Experimental Design Considerations	382
2.1. Targeted expression of TgUPRT	382
2.2. General RNA tagging with 4-thiouridine	384
2.3. Delivery of 4-thiouracil	385
2.4. Purification and analysis of 4TU-tagged RNA	385
3. Materials	386
3.1. Construction of TgUPRT expression vectors	386
3.2. 4-Thiouracil pulse and uracil chase	386
3.3. RNA preparation	386
3.4. Biotinylation of 4TU-tagged RNA	387
3.5. RNA-blot for detection of 4TU-tagged RNA	387
3.6. Purification of 4TU-tagged RNA	387
4. Methods	388
4.1. Construction of TgUPRT expression vectors	388
4.2. 4TU pulse and uracil chase	388
4.3. RNA preparation	390
4.4. Biotinylation of 4TU-tagged RNA	392
4.5. RNA-blot for detection of 4TU-tagged RNA	393
4.6. Purification of 4TU-tagged RNA	397
4.7. Microarray analysis: Design and normalization considerations	400
4.8. Normalization with RNA blot data	403
4.9. Normalization with "spiked in" RNA transcripts	404
Acknowledgments	405
References	405

University of California, Merced School of Natural Sciences, Merced, California, USA

Methods in Enzymology, Volume 448   © 2008 Elsevier Inc.
ISSN 0076-6879, DOI: 10.1016/S0076-6879(08)02619-0   All rights reserved.

Abstract

Microarray-based analysis of mRNA expression has provided a genome-wide understanding of the genes and pathways involved in many biological processes. However, two limitations are often associated with traditional microarray experiments. First, standard methods of microarray analysis measure mRNA abundance, not mRNA synthesis or mRNA decay, and, therefore, do not provide any information regarding the mechanisms regulating transcript levels. Second, microarrays are often performed with mRNA from a mixed population of cells, and data for a specific cell-type of interest can be difficult to obtain. This chapter describes a method, referred to here as "4TU-tagging," which can be used to overcome these limitations. 4TU-Tagging uses cell type–specific expression of the uracil phosphoribosyltransferase gene of *Toxoplasma gondii* and the uracil analog 4-thiouracil (4TU) to selectively tag and purify RNA. Pulse-labeling of newly synthesized RNA with 4TU followed by a "chase" with unmodified uracil allows *in vivo* measurements of mRNA synthesis and decay in specific cells. Experimental design considerations for applying 4TU-tagging to different systems and protocols for cell type–specific RNA tagging, purification, and microarray analysis are covered in this chapter.

1. INTRODUCTION

In most eukaryotes, steady-state mRNA levels are primarily determined by the transcription rate of each gene and the degradation rate of the resulting mRNA. The balance between these two processes can be altered in response to environmental or developmental signals, allowing for rapid changes in mRNA abundance through changes in synthesis rates, decay rates, or both. Techniques for measuring gene expression, such as microarrays, quantitative reverse transcription-polymerase chain reaction, and Northern blotting, are commonly used to compare steady-state mRNA levels between samples, without determining how differences in mRNA abundance are achieved. To determine the mechanisms responsible for changes in mRNA abundance, it is necessary to measure the relative rates of mRNA synthesis and decay. Many techniques exist for measuring mRNA synthesis (e.g., nuclear run-ons) and mRNA decay (e.g., inactivation of the transcriptional machinery; see chapters by Jeff Coller (chapter 14), Chen, Ezzeddine and Shyu (chapter 17), Passos and Parker (chapter 20)), and some of these techniques have recently been combined with microarray technologies to study mRNA synthesis and decay on a genome-wide level (Fan *et al.*, 2002; Wang *et al.*, 2002).

Microarray analysis of mRNA synthesis and decay provides a significant advance over the traditional "one gene at a time" analysis. However, one technologic hurdle that remains in many cases is the analysis of mRNA synthesis and decay in specific cell types within a multicellular organism or mixed population of cells. To fully understand the regulatory mechanisms that generate unique mRNA profiles in distinct cell types, it is necessary to analyze mRNA synthesis and decay *in vivo* and only in the cells of interest. Several methods for cell type–specific analysis of gene expression have been described, including fluorescence activated cell sorting (Bryant *et al.*, 1999), laser-capture microdissection (Emmert-Buck *et al.*, 1996), and poly-A binding protein expression (Roy *et al.*, 2002). However, these methods either do not allow *in vivo* analysis and/or can be used only to measure mRNA abundance.

This chapter describes a recently developed method that can be used to analyze cell type–specific mRNA synthesis and decay *in vivo*. This method, referred to here as "4TU-tagging" and also known as "*R*NA *a*nalysis by *b*iosynthetic *t*agging" or "RABT" (Zeiner *et al.*, 2007), uses the uracil phosphoribosyltransferase gene of the protozoan parasite *Toxoplasma gondii* (TgUPRT) to convert the modified uracil 4-thiouracil (4TU) into the nucleotide form, 4-thiouridine-monophosphate (4TUMP), for subsequent incorporation into newly synthesized RNA. Multicellular eukaryotes typically lack UPRT activity, and thio-containing nucleotides do not naturally occur in eukaryotic mRNAs, thus targeted expression of TgUPRT allows cell type–specific tagging of mRNAs with 4TU. After 4TU-tagging, RNA from a mixture of cells or a whole organism can be reacted with a thio-specific biotinylation reagent, Biotin-HPDP, and mRNA from the cells of interest can then be purified with streptavidin-coated magnetic beads (Cleary *et al.*, 2005). Short periods of exposure to 4TU can be used to measure mRNA synthesis and the 4TU "pulse" can also be followed by exposure to unmodified uracil (a uracil "chase") to analyze mRNA decay.

4TU-Tagging has little effect on the normal physiology of the cells that incorporate 4TU into RNA (Cleary *et al.*, 2005). Transgenic expression of TgUPRT in human tissue culture cells has no detectable effect on the growth of these cells, and exposure of both *T. gondii* and TgUPRT-transgenic human cells to 4TU has only minor effects on cell growth after long periods of exposure. (Cell growth is slowed only after ≥ 24 h of exposure, which is much longer than the time required for cellular exposure to analyze mRNA synthesis and decay.) Incorporation of 4TU into mRNA was also shown to have no detectable effect on transcription or translation. These results demonstrate that 4TU-tagging does not significantly alter normal cell function and should provide a more accurate measure of gene expression than techniques that rely on physical separation of the cells of interest, drugs to inhibit transcription, or expression of mRNA binding proteins that can interfere with normal mRNA processing.

The possible applications of 4TU-tagging have been described in a study of mRNA synthesis and decay in *T. gondii* (Cleary *et al.*, 2005). First, 4TU-tagging was used to measure mRNA synthesis in two different life cycle stages of the parasite: rapidly dividing tachyzoites and slowly dividing bradyzoites. Comparisons of mRNA synthesis and mRNA abundance (i.e., steady-state mRNA levels) in these two stages identified the mechanisms that regulate differential gene expression between tachyzoites and bradyzoites. For example, genes with equivalent levels of mRNA synthesis in the two stages but increased abundance in bradyzoites are most likely regulated by stage-specific changes in mRNA stability. 4TU-Tagging was also used to directly measure mRNA decay in both stages. These analyses confirmed the predictions made on the basis of the comparisons of mRNA synthesis and abundance. For example, genes that had equal levels of mRNA abundance in both stages but decreased mRNA synthesis after the transition from tachyzoites to bradyzoites were found to have very slow decay rates (i.e., transcripts made in tachyzoites were still detected in bradyzoites, where transcription was shutoff, because of the stability of these mRNAs).

In addition to the analysis of mRNA synthesis and decay in *T. gondii*, several experiments were performed to test the cell type specificity of 4TU-tagging (Cleary *et al.*, 2005). First, because *T. gondii* is an intracellular parasite, the tagging of both the human host cell and parasite RNA was analyzed after growth in the presence of 4TU. With a Northern blot–based method to detect 4TU-tagged RNA, it was shown that only the parasite RNA was tagged, and microarray analysis with purified 4TU-tagged RNA showed that human transcripts were efficiently removed from the RNA pool. The ability to 4TU-tag RNAs in a small population of UPRT-expressing cells within a whole organism was demonstrated by infecting mice with *T. gondii* then injecting 4TU into the mouse. *T. gondii*–infected lymph nodes were then harvested, and RNA was extracted from these tissues. Analysis of this RNA showed that only the RNA from UPRT-expressing cells (in this case the parasites) was 4TU-tagged. Finally, the potential to 4TU-tag RNAs in cells that normally lack UPRT activity was demonstrated by generating TgUPRT-transgenic human and mouse cell lines and showing that these transgenic cells, and not wild-type cells, were capable of incorporating 4TU into newly synthesized RNA.

These previous studies suggest that 4TU-tagging should be applicable to any experimental system where transgenic expression of TgUPRT can be engineered. This chapter outlines experimental design considerations for developing 4TU-tagging in other systems, as well as the protocols required for performing the types of experiments described previously. Figure 19.1 shows the major steps involved in 4TU-tagging and some of the potential applications of this technique.

2. Experimental Design Considerations

2.1. Targeted expression of TgUPRT

The application of 4TU-tagging to organisms other than *T. gondii* requires two things:

Cell-Specific RNA Tagging with 4-Thiouracil

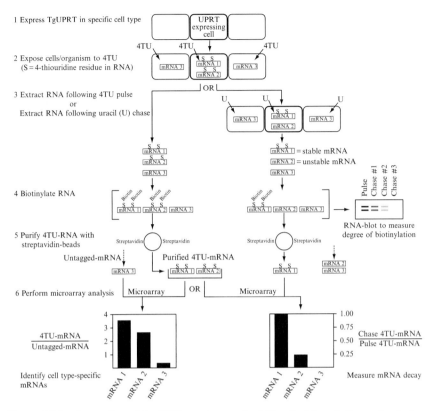

Figure 19.1 An outline of the 4TU-tagging method. Detailed descriptions of each step are provided in the main text. Three types of analysis are shown: detection of 4TU-tagged RNA with RNA blotting (RNA blot), microarray identification of cell type–specific mRNAs, and cell type–specific microarray analysis of mRNA decay.

There must be a method for expressing the TgUPRT cDNA in cells of interest. In most cases, this will involve placing the TgUPRT cDNA under the control of a cell type–specific promoter. The type of vector used to introduce the TgUPRT transgene will depend on the system; however, two general considerations should be kept in mind. First, expression of TgUPRT should be tightly restricted to the cells of interest. Avoid "leaky" promoters and screen multiple lines containing random TgUPRT transgene insertions to identify those with little or no nonspecific TgUPRT expression. TgUPRT expression does not need to be inducible/repressible because the timing of 4TU-tagging is controlled by the addition 4TU. Second, the strongest possible cell type–specific promoter should be used to ensure high levels of TgUPRT expression. Weak promoters may not result in sufficient 4TU-tagging.

For cell type–specific 4TU-tagging to work, an organism must not express its own UPRT enzyme or other enzymes that efficiently convert uracil to uridine-monophosphate (UMP). UPRT enzymatic activity is typically restricted to unicellular eukaryotes and bacteria, and various assays have shown that human and mouse cells are unable to convert uracil to UMP (Cleary et al., 2005; Pfefferkorn and Pfefferkorn, 1977). However, BLAST comparisons of the TgUPRT amino acid sequence to proteins in other organisms will identify homologs with varying degrees of sequence similarity, and these proteins are often annotated as UPRTs. For example, genes annotated as UPRT orthologs exist in humans, mice, and *Drosophila*. A recent analysis of these sequences found that all of the "UPRT" proteins in multicellular animals lack several amino acid residues that are highly conserved in the protozoan, yeast, and bacterial UPRT proteins, including two residues in the uracil-binding domain (Li et al., 2007). On the basis of this sequence divergence and the biochemical data, it is likely that any UPRT orthologs that lack these residues will have minimal or no UPRT activity and, therefore, the presence of such genes should not preclude the use of 4TU-tagging via introduction of TgUPRT.

Although the absence of a functional UPRT is a good indicator that cell type–specific 4TU-tagging will be possible, there are other biochemical pathways that can carry out the conversion of uracil to UMP. These pathways are typically much less efficient than the UPRT reaction (Carter et al., 1997). Conversion of uracil to UMP in mammals can be performed by orotate phosphoribosyltransferase (OPRT), the penultimate enzyme in the *de novo* pyrimidine synthesis pathway, although this activity is rare and occurs only at alkaline pH. Uridine phosphorylase and uridine kinase can also sequentially convert uracil to UMP, again at a very low rate compared with UPRT. These enzymes are responsible for 5-fluorouracil (5-FU) toxicity in mammalian tumors. These alternative pathways have not proven to be a problem in previous applications of 4TU-tagging, but the possibility that other mechanisms of 4TU to 4TUMP conversion may exist underscores the importance of performing some of the negative controls described in this chapter.

2.2. General RNA tagging with 4-thiouridine

For researchers who are interested in measuring mRNA synthesis and decay in a homogeneous population of cells, it is not necessary to use transgenic expression of TgUPRT. A similar approach is possible with 4-thiouridine (the nucleoside form of uracil). RNA tagging with 4-thiouridine relies on the activity of uridine kinase to convert the labeled nucleoside to 4-thiouridine-monophosphate. Uridine kinase is ubiquitously expressed in most organisms, and 4TU-tagging will, therefore, occur in all cell types of an organism or mixed population of cells exposed to 4-thiouridine. RNA tagging with 4-thiouridine for the analysis of mRNA synthesis in cell lines and animals has been previously described (Johnson *et al.*, 1991; Kenzelmann *et al.*, 2007).

Non-cell–specific RNA tagging with 4-thiouridine should be used as a control when developing cell type–specific mRNA tagging by means of TgUPRT and 4TU. The tagging of RNA in all cells with 4-thiouridine provides a good positive control for the mRNA biotinylation and purification steps described in the following, because large quantities of 4-thiouridine–tagged RNA can be prepared and processed in parallel to tagged RNA prepared with 4TU. (See the tips and controls segment at the end of each methods section for guidelines on the use of 4-thiouridine and other controls.)

2.3. Delivery of 4-thiouracil

The method of delivering 4TU depends on the type of experiment and the system being used. In tissue culture experiments, 4TU can be added to the medium and will readily cross cell membranes. In whole organisms, it should be possible to either feed 4TU or inject it into the animal (as previously described for *T. gondii* infected mice [Cleary *et al.*, 2005]). As a guideline, any previously described protocols used for bromodeoxyuridine (BrdU) labeling of DNA in cells or animals should be able to be modified for 4TU-tagging by replacing BrdU with 4TU. One important consideration, particularly for feeding 4TU, is that the media must be free of yeast or bacteria that could convert the 4TU to 4TUMP by the microorganism's UPRT enzyme. If these microorganisms are then ingested by the animal being studied, it is possible that the 4TUMP from the yeast or bacteria could be transferred to the animal and incorporated into the RNA of all cell types by means of the activity of uridine kinase.

Other parameters that will be unique for each system are the concentration of 4TU to use and the length of the 4TU exposure required to achieve sufficient RNA-tagging. These parameters must be determined empirically, and various concentrations and exposure times can be compared with the RNA-blots described in section 4.5.

2.4. Purification and analysis of 4TU-tagged RNA

Whenever TgUPRT is expressed in a small percentage of cells within an organism or tissue, it is likely that the yields of 4TU-tagged mRNA will be small. This is due to both the low number of cells and the fact that after a short 4TU pulse (≤ 1 h), only approximately 10% of the transcripts (only the newly synthesized population) in the UPRT-expressing cells will be 4TU-tagged. (The 10% value is based on experiments in *T. gondii* [Cleary *et al.*, 2005]).

In these cases, it may be necessary to perform some type of physical enrichment of the cells of interest immediately after the 4TU exposure and before extracting RNA. 4TU Tagging and uracil chase steps can still be performed *in vivo*; the physical enrichment step is added only at the end of the experiment to help overcome the limitation of low 4TU-tagged RNA yields. Physical enrichment could involve surgical techniques or methods such as fluorescence-activated cell sorting.

The three primary applications of 4TU-tagging are identification of mRNAs that are enriched in a specific cell type, analysis of mRNA synthesis in specific cell types, and analysis of mRNA decay in specific cell types. Depending on the type of questions one wants to address, one or all of these applications may be relevant. Each type of analysis is described in more detail at the end of this chapter, with a focus on the use of 4TU-tagging for microarray analysis. This technique can be used, however, to analyze mRNA synthesis and decay with other methods such as quantitative RT-PCR and Northern blots.

3. Materials

Materials required for each step are listed in the following. Some materials are used in more than one step but only listed under the procedure where they will first be used.

3.1. Construction of TgUPRT expression vectors

Primers for amplifying the TgUPRT cDNA are listed in the following. Add the desired restriction enzyme sequences for cloning into an expression vector.

5′-Sense: ATGGCGCAGGTCCCAGCG
3′-Antisense: CTACATGGTTCCAAAGTACC

3.2. 4-Thiouracil pulse and uracil chase

4-Thiouracil (Sigma-Aldrich): store at room temperature in the dark.
Uracil (Sigma-Aldrich).
4-Thiouridine (Sigma-Aldrich).
DMSO (Sigma-Aldrich).

3.3. RNA preparation

TRIzol (Invitrogen).
Chloroform.

Isopropanol.
75% Ethanol.
RNase-free water.
mRNA purification kit (e.g., Oligotex from Qiagen).
RNeasy RNA purification kit (Qiagen): optional.

3.4. Biotinylation of 4TU-tagged RNA

EZ-Link Biotin-HPDP (Pierce).
Dimethylformamide (Pierce).
1 M Tris-HCL, pH 7.4.
0.5 M EDTA, pH 8.0.
5 M NaCl.

3.5. RNA-blot for detection of 4TU-tagged RNA

UltraPure Agarose (Invitrogen).
10× TAE buffer (10 mM Tris-Acetate, pH 8.5, 1 mM EDTA).
10 mg/ml Ethidium bromide (Invitrogen).
6× Loading dye (0.25% [weight/volume] bromophenol blue, 30% glycerol, 70% RNase-free water).
SYBR green dye (Invitrogen).
10× SSC buffer (1.5 M sodium chloride, 150 mM sodium citrate).
Hybond-N$^+$ membrane (GE Healthcare Life Sciences).
Blot blocking solution (125 mM NaCl, 17 mM Na$_2$HPO$_4$, 7.3 mM NaH$_2$PO$_4$, 1% SDS [w/v]). Blocking solution may need to be heated to 37 °C to solubilize SDS.
Blot wash #1 (1:10 [v/v] dilution of blocking solution in water).
Blot wash #2 (10× stock is 100 mM Tris, 100 mM NaCl, 21 mM MgCl$_2$, pH to 9.5 with HCl).
Streptavidin-HRP (Perkin Elmer).
Streptavidin-AP (Perkin Elmer).
ECL reagent (GE Healthcare Life Sciences).
ECF reagent (GE Healthcare Life Sciences).
Kodak Biomax MR film (Kodak).
A gel and blot imaging system that is capable of measuring fluorescence intensity (for example, a Molecular Dynamics STORM system).

3.6. Purification of 4TU-tagged RNA

MPG-streptavidin beads (PureBiotech).
Magnetic particle separator (PureBiotech).
Gene elute linear polyacrylamide (Sigma).

MPG buffer (1 M NaCl, 10 mM EDTA, 100 mM Tris-Cl pH 7.4, in RNase-free H_2O).
2-mercaptoethanol.

4. METHODS

4.1. Construction of TgUPRT expression vectors

The exact method for introducing transgenic expression of TgUPRT depends on the type of organism or types of cells being studied. The cDNA encoding TgUPRT has been cloned into several different expression vectors, including the retroviral vector LNCX for expression in mammalian tissue culture cells (Cleary et al., 2005) and pPUAST for the generation of transgenic *Drosophila* (M. Cleary, unpublished data). As described in section 2, vectors that allow cell type–specific expression of TgUPRT at the highest possible levels should be used. A plasmid for subcloning TgUPRT into different expression vectors is not currently available; instead, PCR amplification of TgUPRT with primers containing the desired restriction sites can be used. A plasmid containing the TgUPRT cDNA for use as a template in the PCR reaction is available from the author. The primers listed in the previous materials section will amplify the 735-bp TgUPRT cDNA (RH strain, GenBank Accession number U10246).

4.1.1. Tips and controls for the construction of TgUPRT expression vectors

Epitope tags may be attached to the TgUPRT protein to allow analysis of protein expression by either immunofluorescence or Western blotting. Both an N-terminal Myc tag and a C-terminal HA tag have been added to TgUPRT without affecting the activity of the enzyme.

4.2. 4TU pulse and uracil chase

4.2.1. 4TU pulse

Prepare a stock solution of either 200 mM or 400 mM 4TU (MW = 128.15) in DMSO. The 400 mM solution may need to be heated at 37 °C to make sure all of the 4TU goes into the solution. This higher concentration may be desirable when injecting animals with 4TU, because smaller volumes can be used. Make small aliquots and store them at −20 °C. Thaw each aliquot only once before use (repeated freeze– thaw cycles may degrade the 4TU). 4TU is minimally light sensitive, and stock solutions, media containing 4TU, and even RNA samples containing 4TU-tagged RNAs should be protected from light as much as possible.

The amount of 4TU to use for 4TU-tagging must be determined empirically. Previously reported experiments provide concentration ranges that can serve as a guide:

1. To tag parasite mRNA in cultures of *T. gondii*–infected human foreskin fibroblast cells, 4TU was added to culture media at a final concentration of 20 μM (Cleary et al., 2005).
2. To tag parasite RNA in infected mice, 100 μl of 200 mM 4TU was injected into the peritoneum of an adult mouse followed by another 100-μl injection of 400 mM 4TU 4 h later. RNA was then extracted 4 h after the second 4TU injection (Cleary et al., 2005).

Too high a concentration of 4TU can result in nonspecific labeling of RNA in other (UPRT-negative) cell types and may also have adverse effects on normal cell physiology or health of animals treated with 4TU.

Another parameter that needs to be determined empirically is the length of the 4TU pulse necessary to achieve sufficient RNA labeling. For measurements of mRNA synthesis, the shortest possible labeling time is desirable, because only the 4TU-tagged mRNAs made during a short (≤ 1 h) pulse can reasonably be considered part of the "newly synthesized" population of transcripts. For measurements of mRNA abundance (or "steady-state" levels) and for the 4TU pulse before a uracil chase, the labeling time can be longer to ensure that all mRNAs are 4TU-tagged.

4.2.2. Uracil chase

The concentration of uracil required to chase away the 4TU also needs to be determined empirically. As an example, in the analysis of *T. gondii* mRNA decay, a stock solution of 400 mM uracil was prepared and diluted to a concentration of 12 mM in the culture media during chase experiments (Cleary et al., 2005). This final uracil concentration was approximately 1000 times higher than the concentration of 4TU used in the pulse. In some systems, it may be more efficient to use uridine instead of uracil for chase experiments, because this nucleoside may be taken up by cells and converted to the nucleotide form more quickly than uracil.

4.2.3. Tips and controls for the 4TU pulse and uracil chase

As a *positive control*, pulse–chase analysis with 4-thiouridine can be performed. As described in section 2.2, 4-thiouridine will be incorporated into the RNA of all cells. A stock solution of 4-thiouridine can be prepared at 200 mM in DMSO. The same concentrations of 4-thiouridine should give higher levels of RNA tagging relative to 4TU, because all cells will incorporate 4-thiouridine into RNA.

As a *negative control*, parallel 4TU treatments should be performed in cells or animals that are not expressing TgUPRT. The "background" levels of RNA tagging in the absence of any TgUPRT can then be analyzed with the techniques described in the following. If 4TU-tagging of RNA is detected in these negative control samples, it will be necessary to test a range of 4TU concentrations to identify the conditions that give the highest levels of UPRT-specific RNA tagging and the lowest levels of nonspecific tagging.

4.3. RNA preparation

4.3.1. Total RNA extraction

Total RNA extraction should be performed with TRIzol reagent. As described in section 2.4, it may be necessary to perform a physical enrichment of the cells of interest after the 4TU pulse or uracil chase before RNA extraction. Tissues should be thoroughly homogenized in TRIzol reagent with mechanical tissue grinders or similar methods. Sufficient amounts of TRIzol for the number of cells or mass amount of tissue should be used (according to the manufacturer's instructions). Tissue should be homogenized in TRIzol as quickly as possible; plan experiments to minimize the number of steps between ending a 4TU pulse or uracil chase and the addition of TRIzol. Samples homogenized in TRIzol can be stored at $-80\,^{\circ}$C until ready to perform the RNA extraction. For the purification of 4TU-tagged RNA from small populations of cells when the yield of tagged RNA is expected to be low, samples can be stored in TRIzol at $-80\,^{\circ}$C until a sufficient amount of material is collected. The frozen samples can then be thawed and combined for a single large-scale RNA extraction at a later time.

RNA extraction from samples in TRIzol is performed according to the manufacturer's instructions with the following recommendations. First, after homogenizing tissues in TRIzol, spin the sample at $12,000g$ for 10 min at $4\,^{\circ}$C. This will pellet out any insoluble material, and the TRIzol solution can then be transferred to a new tube. Second, after transferring the "prespun" sample to a new tube, let the sample sit at room temperature for 10 min to ensure that all proteins are dissociated from the RNA. Finally, to resuspend the pelleted RNA, add the necessary amount of RNase-free water and incubate the sample at $55\,^{\circ}$C for 5 min. Pipette the sample repeatedly to ensure that all the RNA is completely solubilized. The final concentration of RNA should be $\geq 0.4\ \mu g/\mu l$.

TRIzol extractions need to be of high purity, with A_{260}/A_{280} ratios of ≥ 2.0. Protein contamination can give false-positive signals in the Northern blot assay described in the following because of biotinylation of proteins bound to the RNA. If the A_{260}/A_{280} ratio is <2.0, use RNeasy columns to remove proteins. The use of these columns may result in a final RNA concentration that is $<0.4\ \mu g/\mu l$. In this case, the RNA should be isopropanol precipitated and resuspended in a smaller volume. The precipitation is performed as follows:

1. Add 2 μl of linear polyacrylamide.
2. Add 1/10 the sample volume of 5 M NaCl.
3. Add an equal volume of isopropanol.
4. Mix well and let the sample sit at room temperature for 5 min.
5. Spin at maximum speed in a refrigerated microcentrifuge for 20 min.
6. Decant the supernatant and add 75% ethanol.
7. Spin at maximum speed in a refrigerated microcentrifuge for 10 min.
8. Decant the ethanol and resuspend the RNA pellet in the appropriate volume of RNase-free water. Heat the sample at 55 °C for 5 min and pipette repeatedly to ensure the pellet is completely solubilized.

4.3.2. mRNA preparation

When preparing samples for purification and microarray analysis of 4TU-tagged RNAs, a small aliquot of total RNA should be saved (2 to 20 μg to be used in the RNA-blot assay described in section 4.5), and the remaining RNA should be used for the preparation of polyA-selected mRNA. PolyA+ mRNA can be purified with any desired method; the only important consideration is that the final mRNA concentration should be ≥ 0.4 μg/μl. The isopropanol precipitation protocol outlined earlier can be used to concentrate the RNA sample if necessary. This polyA-selection step is only recommended as a way to remove the excess ribosomal RNA. The benefit of removing the ribosomal RNA is that smaller amounts of the streptavidin-magnetic beads can be used in the purification step described in section 4.6.

4.3.3. Tips and controls for RNA preparation

As for any RNA work, it is important to ensure that all reagents are RNase free. A 5% SDS solution may be used to inactivate any RNases on surfaces such as bench tops.

Although the yields of 4TU-tagged RNA obtained after the purification step described in the following will have to be determined empirically, it is worth estimating the yields to decide how much starting RNA will be required. A general guideline for RNA yields comes from experiments with *T. gondii*. In experiments in which approximately 50% of the cells in a culture are *T. gondii* and 50% are host cells, a 1-h pulse with 4TU resulted in 4TU-tagged mRNA yields of 10 to 16% of the total mRNA extracted from the mixture of cells. Longer 4TU exposure times will result in higher yields of tagged mRNAs, but yields from tissues in which the UPRT-expressing cells are very rare may be below the limit of detection or not above "background" levels. As mentioned previously, this is the reason that physical enrichment of the cells of interest after a 4TU-pulse or uracil-chase may be necessary in some cases.

4.4. Biotinylation of 4TU-tagged RNA

Biotinylation of 4TU-tagged RNA is used to measure the amount of RNA tagging in a Northern blot–based assay (section 4.5) and for purification of the 4TU-tagged RNA (section 4.6).

4.4.1. Biotinylation reaction

1. Prepare a 1 mg/ml solution of EZ-Link Biotin-HPDP in dimethylformamide (DMF). After addition of the EZ-Link Biotin HPDP powder, heat the solution at 55 °C for 5 to 10 min, vortex well, and check to make sure all of the powder is solubilized. Make small aliquots of the Biotin HPDP solution (typically 20 μl) and store at -20 °C. Thaw and use only once.
2. Prepare a 10× T.E. buffer for the biotinylation reaction: 10× T.E. = 100 mM Tris-HCL (pH 7.4), 10 mM EDTA, in RNase-free water.
3. Heat the RNA to be biotinylated at 70 °C for 3 min, then place on ice for 1 min.
4. For the biotinylation of both total RNA and polyA-selected mRNA, use 2 μl of the 1 mg/ml biotin-HPDP solution per microgram of RNA. Do not add biotin-HPDP at this time. First, determine the amount of RNase-free water to add to the RNA so that the amount of biotin-HPDP in the reaction will be equal to 30% of the final volume. Add the RNase-free water, T.E. (1:10 dilution), and biotin-HPDP in that order (i.e., always add biotin-HPDP last).
5. For example, to biotinylate 20 μg of RNA, the reaction mixture would be:
RNA (0.4 μg/μl): 50.0 μl.
RNase-free water: 10.0 μl.
10× T.E: 13.3 μl.
Biotin-HPDP solution: 40.0 μl.
6. Incubate at 25 °C for 3 h. Protect the reaction from light.

4.4.2. Precipitation of biotinylated RNA

1. Add 1/10 the reaction volume of 5 M NaCl.
2. Add an equal volume of isopropanol and mix well by pipetting.
3. Incubate at room temperature for 5 min.
4. Spin at maximum speed in a refrigerated microcentrifuge for 20 min.
5. Decant supernatant and add 75% ethanol. Vortex briefly, then spin at maximum speed in a refrigerated microcentrifuge for 5 min.

6. Resuspend RNA in RNase-free water at a concentration of 0.5 µg/µl. Incubate at room temperature for 5 min and pipette repeatedly to ensure RNA is completely solubilized.
7. Store RNA at −80 °C until ready to use.

4.4.3. Tips and controls for the biotinylation of 4TU-tagged RNA

The following positive and negative controls should be biotinylated for subsequent analysis on the RNA-blots described in section 4.5:

1. RNA from cells or organisms exposed to 4-thiouridine. This serves as a positive control for the biotinylation of thio-groups.
2. RNA from cells or organisms that were not exposed to 4TU or 4-thiouridine. This serves as a negative control for nonspecific biotinylation of RNAs. Biotinylation of these samples would indicate that the RNA is contaminated with proteins or that the RNA molecules contain endogenous thio- or thiol-groups.
3. RNA from cells or organisms that do not express TgUPRT but are exposed to 4TU. This serves as a negative control for nonspecific 4TU-tagging of RNA. Biotinylation of these samples would indicate that alternative pathways exist for the conversion of 4TU to 4TUMP or that the organism being studied has endogenous UPRT activity.

The amount of biotin-HPDP in the biotinylation reaction can be reduced to as low as 20% if the concentration of RNA in the sample is low (<0.4 µg/µl). In these cases, the biotinylation reaction should be performed for 4 to 6 h instead of 3 h.

4.5. RNA-blot for detection of 4TU-tagged RNA

The procedure described here is a variation of the traditional Northern blot method. In this case, membrane-bound RNA is probed with streptavidin-HRP for the detection of biotinylated 4TU residues (instead of a nucleic acid probe). Another difference is that these blots are performed with standard agarose gels (not formaldehyde gels), because the exact size of the RNAs is typically not important. However, this technique does work with formaldehyde gels and MOPS buffer if one wants to determine the exact size of tagged RNAs. The RNA-blots described here are essential for optimizing the 4TU-tagging conditions, because multiple parameters can be compared on the same blot. RNA-blots are also used to quantify the degree of RNA decay during pulse–chase experiments, and these data can be used for normalizing microarray measurements (described in section 4.7). The instructions that follow include variations for either qualitative or quantitative measurements of 4TU-tagging.

4.5.1. Run biotinylated RNA on agarose gel

1. Pour a 1.0% agarose gel in TAE with either ethidium bromide (for qualitative analysis) or with no nucleic acid dye (for later staining with SYBR green and quantitative measurement of RNA amounts). Use a large gel box (not a mini-gel box) and fill completely with 1× TAE running buffer. The large volume of buffer will help keep the temperature low during electrophoresis.
2. The amount of RNA that needs to be loaded to detect a signal varies depending on the type of sample. For a 4TU pulse in which ≥5% of the cells is expressing UPRT, 1 μg is typically enough for detection of the biotinylated ribosomal RNA. If the percentage of UPRT-expressing cells is lower, more RNA should be loaded. When analyzing RNA collected during a pulse–chase experiment, 10 μg from each time point (the initial 4TU pulse and the subsequent uracil chases) should be loaded so that a "smear" of mRNAs can be detected. Loading a large amount of RNA from pulse–chase samples also ensures that there is sufficient biotinylated RNA present in the chase samples to provide a detectable signal. (This should at least be true for the early time points before most of the 4TU-tagged RNA decays.)
3. Before loading the gel, spin the tubes containing the RNA at maximum speed in a refrigerated microcentrifuge for 5 min. This will pellet any free biotin-HPDP.
4. Load RNA with the 6× RNase-free loading buffer.
5. Run the gel at 200 volts for 20 to 30 min. Monitor heat in the gel apparatus by watching for steam condensing on the cover. If the gel apparatus is getting hot, place ice packs around the apparatus to keep the buffer cool. Running the gel at a high voltage for a short time allows sufficient separation of the RNA molecules before any degradation of the RNA occurs.
6. If the gel is for a nonquantitative blot, photograph the ethidium-stained RNA according to standard procedures.

If the blot will be used for quantification of the amount of RNA loaded, stain the gel in SYBR green dye according to the manufacturer's instructions. After staining, wrap the gel in plastic wrap and smooth out any wrinkles or bubbles. Place the gel in an imaging system that is capable of measuring fluorescence intensity (for example a Molecular Dynamics STORM system). To measure the amount of ribosomal RNA in each lane, draw equal size boxes around the large ribosomal RNA band and an area of the gel where there is no RNA present (typically somewhere above the RNA "smear" and below the sample well). After subtracting the background fluorescence, these measurements can be used to determine the relative amounts of RNA loaded in each sample. These data are used to correct for variability in the amount of RNA present when quantifying the degree of biotinylation of each sample (as described in the following).

4.5.2. Transfer RNA to nylon membrane
Set up the transfer of the RNA from the gel to a nylon membrane with traditional Northern-blotting techniques. The only important parameters for this method are:

1. The transfer buffer should be 10× SSC.
2. Use Hybond-N+ nylon membrane.

4.5.3. Detect biotinylated RNA

1. After transfer to the nylon membrane, remove the membrane and mark the position and orientation of the lanes with a pencil.
2. Cross-link the RNA to the membrane by exposing to 1200 joules of ultraviolet light.
3. Incubate the membrane in "blot blocking solution" on an orbital shaker for 30 min at low speed. Ensure that the membrane is not exposed to cold air (e.g., from an air conditioning vent) during this step or the SDS in the blocking solution may precipitate.
4. Remove the blocking solution and incubate with streptavidin–HRP (for qualitative analysis) or streptavidin–AP (for quantitative analysis). Dilute the streptavidin conjugates 1:5000 to 1:10,000 in blot-blocking solution. Incubate for 5 min on an orbital shaker at low speed.
  Wash in "blot wash #1" for 20 min on orbital shaker.
  Repeat wash in "blot wash #1" for 20 min on orbital shaker.
  Wash in "blot wash #2" for 5 min on orbital shaker.
  Repeat wash in "blot wash #2" for 5 min on orbital shaker.
  Transfer the membrane to a new container and add either ECL substrate (for streptavidin–HRP probed blots) or ECF substrate (for streptavidin–AP probed blots). Incubate for 1 min, then remove the substrate solution.
  Expose streptavidin–HRP probed blots to radiographic film. Perform exposures of varying length to provide a qualitative measure of the relative amounts of biotinylation in each sample.
  Quantify the fluorescent signal from streptavidin–AP probed blots with an image capture system that can measure fluorescence.

4.5.4. Quantification of amounts of 4TU-tagged RNA
The detection of biotinylated 4TU-tagged RNA with streptavidin-AP and ECF allows quantitative measurements of the relative amount of biotinylated RNA in each sample. In a pulse–chase experiment, this allows the decay of 4TU-tagged ribosomal RNAs and messenger RNAs (visible on the blot as a "smear" of varying length transcripts) to be measured. The method for calculating these turnover values is described in the following,

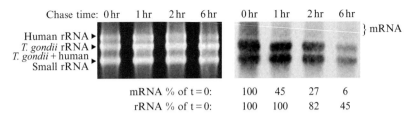

Figure 19.2 Analysis of RNA decay with RNA blotting; 10 μg of biotinylated RNA from *T. gondii*–infected human cells was analyzed from the following samples: a 1-h 4TU-pulse (chase time = 0) and three uracil-chase time points. The amount of biotinylated 4TU-tagged ribosomal RNA and mRNA in each sample was quantified as described in the main text. The SYBR green-stained agarose gel is shown on the left, and the streptavidin-AP probed blot incubated with ECF substrate is shown on the right. The percent decay of both the mRNA pool and the large rRNA was calculated for each chase sample. This figure also demonstrates the specific 4TU-tagging of the *T. gondii* rRNA and the absence of any tagging of the human rRNA. The faint biotinylated band above the *T. gondii* large rRNA is *T. gondii* precursor rRNA. (First published as supplementary data in Cleary *et al.* [2005]. *Nat. Biotechnol.* **23**, 232–237.)

and an example of a blot used for these types of calculations is shown in Fig. 19.2.

Use the image analysis software provided with your gel imaging system to draw equal size measurement boxes around:

1. The large ribosomal RNA band.
2. A region of the messenger RNA "smear" above the large rRNA band.
3. An area of the gel where there is no RNA present (typically somewhere above the mRNA "smear" and below the sample well).

After subtracting the background signal (measurement 3) and normalizing for differences in the amount of RNA loaded (as determined with the SYBR green quantification described earlier), these measurements can be used to determine the relative amounts of biotinylated RNA present in each sample. The mRNA decay calculations can then be used to normalize microarray data, as described in section 4.7.

4.5.5. Tips and controls for RNA blots

The positive and negative controls described at the end of section 4.4 should be analyzed by RNA blotting. An additional negative control is to load RNA that has not been incubated with biotin-HPDP. This will serve as a control for any nonspecific binding of the streptavidin probe to nonbiotinylated RNA.

In cases in which there is a mixture of cells from different species and the rRNA molecules from each species can be distinguished from each other, different sized rRNAs provide an ideal negative control, because only the rRNA from the UPRT-expressing cells should be biotinylated. This type of UPRT-specific rRNA tagging is observed during *T. gondii* infection of mammalian host cells, as shown in Fig. 19.2.

Another way to analyze the efficiency and specificity of 4TU-tagging is to compare samples that were exposed to 4TU for increasing amounts of time. There should be a correlation between the length of the 4TU exposure and the degree of RNA biotinylation. Comparing samples in which an increasing percentage of cells express TgUPRT provides a similar type of control: samples with the highest percentage of UPRT-expressing cells should have the strongest signal on the blot.

4.6. Purification of 4TU-tagged RNA

This protocol works for both total RNA and polyA-selected mRNA. Purification and microarray analysis are typically performed with polyA-selected mRNA, but for simplification the generic term "RNA" will be used here.

When purifying 4TU-tagged RNA for pulse–chase analysis, be sure to use the same mass amount of RNA (i.e., micrograms of "input" RNA) for each sample.

4.6.1. Purify biotinylated 4TU-tagged RNA with streptavidin-magnetic beads

1. Use 2 μl of MPG streptavidin beads per μg of RNA. The biotinylated RNA should be at a concentration of 0.5 μg/μl as described in section 4.4.
2. Perform the bead purification in a tube that is appropriately sized for the sample volume but large enough to hold at least five times the sample volume. Volumes are typically small (10 to 100 μl) and the purification should be performed in 0.2-ml to 0.5-ml tubes.
3. Add 1 μg of yeast transfer RNA (tRNA) per 5 μl of MPG slurry. Incubate for 20 min at room temperature with rotation. This tRNA incubation serves as a blocking step for nonspecific RNA binding to the MPG beads.
4. Place the tube in a magnetic stand and collect the beads for 1 min. Decant the supernatant.
5. Wash the beads in three changes of MPG buffer. For each wash, fill the tube with MPG buffer and invert the tube five times. Collect the beads in the magnetic stand for 1 min before decanting the wash solution.
6. Centrifuge the biotinylated RNA sample that will be added to the beads for 2 min at maximum speed in a refrigerated microcentrifuge. This will pellet out any free biotin-HPDP. Transfer the RNA to a new tube.
7. Heat the RNA at 70 °C for 3 min then place on ice for 1 min.

8. Add the RNA to the washed MPG beads.
9. Add a volume of MPG buffer equal to half the volume of the RNA sample. This will make the final solution one third MPG and two thirds RNA sample (in water).
10. Incubate the tube at room temperature, with rotation, for 20 min.
11. Although the bead binding proceeds, place a tube of MPG buffer at 65 °C to warm for a subsequent wash step. Warm a volume of MPG buffer equal to 10 times the binding reaction volume from step 1.9.
12. Place the tube in the magnetic stand and collect the beads for ≥1 min. Carefully remove the liquid by pipetting without disturbing the beads. *Save this sample*; it contains the nonbiotinylated/non-4TU–tagged RNA. This RNA includes genes expressed in the UPRT-negative cells, as well as some nontagged mRNAs from the UPRT-positive cells, and can be used in microarray experiments to compare gene expression in the "whole sample" to gene expression in the UPRT-positive cells.
13. Wash the beads according to the five steps that follow, adding the wash solution and incubating for the indicated time. After each wash, collect the beads for ≥1 min before removing the liquid and adding the next wash solution:

    Wash 1: Wash in MPG buffer equal to five times the original sample volume (from step 9). Incubate at room temperature for 5 min with rotation.

    Wash 2: Repeat wash 1 but only incubate for 1 min.

    Wash 3: Wash in 65 °C MPG buffer equal to five times the original sample volume. Mix by pipetting, let sit 1 min, then collect the beads.

    Wash 4: Repeat wash 3.

    Wash 5: Repeat wash 2. *Save this last wash.* This wash sample can be used to check for the presence of any RNA (as described later). This serves as a measure of the efficiency of the washing because there should not be any RNA present in this last wash.
14. Elute the bound RNA with freshly prepared 5% 2-mercaptoethanol. (Dilute 100% 2-mercaptoethanol in RNase-free water to make the 5% solution.)
15. Incubate the beads for 10 min at room temperature with rotation.
16. Collect the beads for ≥1 min and *save this sample*. This sample contains the 4TU-tagged mRNAs.

4.6.2. Precipitate the RNA

1. Precipitate RNA from the following samples:
    a. Unbound/non-4TU–tagged RNA sample (step 12).
    b. Final wash sample (step 13).
    c. Eluted RNA/4TU-tagged RNA sample (step 14).

2. To the unbound and wash samples, add the following:
   a. Equal volume of isopropanol.
   b. 1 ml linear acrylamide.
3. To the eluted RNA sample, add the following:
   a. Equal volume of isopropanol.
   b. 1 ml linear acrylamide.
   c. 1/10 volume 5 M NaCl.
4. Incubate all samples at room temperature for 5 min then spin at maximum speed in a refrigerated microcentrifuge for 30 min.
5. Decant the solution and wash the precipitated RNA in 75% ethanol. Spin at maximum speed in a refrigerated microcentrifuge for 10 min.
6. Remove the 75% ethanol by pipetting. The pelleted RNA in the 4TU-tagged sample will typically be very small. Keep track of the orientation of the tube during the centrifugation step to know where the pellet is located even if it is not visible. Remove all liquid with a pipette; do not air dry.
7. Resuspend the pelleted RNA in RNase-free water. The 4TU-tagged RNA is typically resuspended in a small volume (5 to 10 μl). For pulse–chase experiments, be sure to resuspend each sample in the same volume. Heat the tube at 55 °C for 2 min and pipette repeatedly to ensure the RNA is completely solubilized.
8. Place the tube back in the magnetic stand and let sit for ≥ 1 min to collect any beads that are still present in the sample. Carefully remove the RNA sample and place in a new tube.
9. Determine RNA yields by UV spectrophotometry. Use a Nanodrop or similar device that allows measurements with very small volumes. The 4TU-tagged RNA sample can typically be diluted 1:3 to 1:5 in 1.0 μl of water. A_{260}/A_{280} ratios may be low when analyzing the 4TU-tagged RNA. If this happens, spin the sample for 2 min in a refrigerated microcentrifuge and repeat the reading with RNA taken from the upper portion of the sample. Sometimes the spectrophotometry readings give high A_{280} values because of beads or bead material that is still in the sample.

4.6.3. Tips and controls

Some of the same positive and negative controls described for the RNA blot should be tested in the RNA purification procedure as well. Specifically, samples prepared with 4-thiouridine should be used as a positive control, and samples prepared from UPRT-negative cells or organisms should be used as a negative control. The negative control samples should yield either no detectable RNA or "background" levels that are significantly lower than the 4TU experimental samples.

Another type of control that can be performed is to "spike in" heterologous RNAs that are not from the cells or organism being studied. This could be accomplished by *in vitro* transcribing RNAs encoding bacterial genes, such as the set of *Bacillus subtilis* genes that are often used as normalization controls in microarray experiments (Hill *et al.*, 2001). Some heterologous genes could be *in vitro* transcribed in the presence of 4-thiouridine, and some could be *in vitro* transcribed in the absence of 4-thiouridine. Addition of these RNAs to samples before the biotinylation procedure described in section 4.5 could provide positive and negative controls for both the biotinylation and purification steps. These types of "spike in" controls could also be used to normalize microarray experiments, as discussed in section 4.7.

One way to assay the efficiency of capturing biotinylated RNAs is to run both the noncaptured RNA fraction (step 12) and each wash fraction (all washes from step 13) on an RNA blot to see how much biotinylated RNA is present. Equivalent volumes (after isopropanol precipitation and resuspension in water, as described earlier) should be loaded for each sample. Also load an "input RNA" sample containing the same mass amount of biotinylated 4TU-tagged RNA used in the purification. There is typically some biotinylated RNA present in the noncaptured RNA fraction and the first wash fraction, but the relative amount should not exceed 10% of the input. If more than 10% of the input is not captured, it may be necessary to use more beads or to increase the incubation time.

It is also possible to determine how much RNA remains stuck to the streptavidin-magnetic beads after the 2-mercaptoethanol incubation. To do this, suspend the beads in water after the final wash step and heat the bead solution to 95 °C for 5 min. Collect the beads in the magnetic stand, then remove the liquid and perform the isopropanol precipitation described previously. Measure the amount of RNA in the sample by spectrophotometry and also check the RNA on an agarose gel. (The spectrophotometer readings are sometimes inaccurate because of the presence of impurities from the beads.) Typically, some RNA can be recovered from the beads with this treatment. The amount of RNA is typically only 10 to 20% of what is eluted with 2-mercaptoethanol. If the amount that remains bound to the beads is much greater than 10 to 20% of the eluted fraction, there may be a problem with the 2-mercaptoethanol or some other aspect of the purification procedure.

4.7. Microarray analysis: Design and normalization considerations

This section describes guidelines for the use of purified 4TU-tagged mRNA for microarray analysis. Specific instructions for performing microarray experiments are not provided, because standard microarray techniques are suitable and are well described elsewhere. One important consideration, however, is that yields of 4TU-tagged mRNA are typically very low, and amplification of the purified mRNA will likely be necessary. Eberwine-based linear RNA amplification is recommended and can be performed with either protocols available on the internet (for example: http://www.flychip.org.uk/protocols/gene_expression/amplification2.php) or commercially available kits.

4.7.1. Microarray analysis to identify cell type–specific mRNAs

Before performing any analysis of mRNA synthesis or decay, it is necessary to confirm that the 4TU-tagging technique is working properly. This can be tested with microarray analysis to confirm the enrichment of known cell type–specific mRNAs in the 4TU-tagged mRNA sample. The two samples that should be compared on a microarray are:

1. The 4TU-tagged mRNA purified in step 14 of section 4.6.
2. The nontagged mRNA collected in step 12 of section 4.6.

This design allows mRNA from the same sample to be used for both the 4TU tagged pool (i.e., mRNA from the UPRT-expressing cells) and the nontagged pool (i.e., mRNA from all other cells in the sample). Equivalent mass amounts of mRNA from both pools should be used to prepare the probes for microarray hybridization.

4.7.2. Expected results

Any genes that are known to be expressed predominately in the UPRT-expressing cells should be reproducibly and significantly enriched in the 4TU-tagged mRNA sample. Statistically significant enrichment, without the use of arbitrary fold-change values, can be identified with microarray analysis software such as Significance Analysis of Microarrays, or SAM (available at: http://www-stat.stanford.edu/~tibs/SAM/). Depending on the length of the 4TU pulse, the enrichment of some genes may be relatively minor if the gene is transcribed at a slow rate (i.e., few 4TU-tagged mRNAs will be made during a short 4TU pulse). Genes that are transcribed in both the UPRT-expressing cells and other cells in the sample (e.g., "housekeeping" genes) should not have any increased representation in the purified mRNA pool. Depletion of mRNAs that are known to be absent from the cells of interest is another potential measure of the specificity of the purification. The depletion of any "spiked in" heterologous RNAs that are not biotinylated can also be used to determine the quality of the purification procedure.

4.7.3. Experimental design for the analysis of mRNA synthesis versus abundance

Analysis of mRNA synthesis versus abundance is based on the principle that during a short 4TU pulse only newly synthesized mRNAs will be tagged, whereas after a long 4TU pulse, the entire "steady-state" pool of mRNAs will be tagged. These types of experiments are particularly useful for identifying the regulatory mechanisms that control steady-state mRNA levels in two different types of cells. The basic design for these types of experiments is as follows.

First, 4TU-tagged mRNAs purified after a short 4TU pulse are compared on a microarray to identify differences in mRNA synthesis. Second, 4TU-tagged mRNAs purified after a long 4TU exposure are compared on a microarray to identify differences in the "steady-state" abundance of all mRNAs. Comparing the results of the "synthesis microarrays" and the "abundance microarrays" can identify the potential regulatory mechanisms controlling expression of individual genes. For example, genes with increased synthesis in only one cell type and a corresponding increase in mRNA abundance in that cell type are likely regulated by differences in transcription rates. Another possibility (seen in studies of *T. gondii*) is the identification of genes with similar levels of mRNA synthesis in both cell types but increased mRNA abundance in only one cell type. These genes are likely regulated by cell type–specific differences in mRNA stability.

4.7.4. Experimental design for the analysis of mRNA decay

The analysis of mRNA decay with 4TU-tagging is performed with a standard pulse–chase approach. Special considerations for analyzing 4TU-tagged mRNA samples from pulse–chase experiments on microarrays are described in the following.

4.7.4.1. Direct comparison of samples versus hybridization against a common reference

For the analysis of mRNA decay, 4TU-tagged mRNA levels at each chase time point are compared with the levels of 4TU-tagged mRNA after the 4TU pulse. These comparisons can be made in one of two ways: as either a "type I" experiment, where samples are directly compared with each other on the microarray, or a "type II" experiment, where each sample is hybridized to the microarray in combination with a common reference sample and direct comparisons are then made by dividing the chase/reference signal by the pulse/reference signal. The type I approach requires a large amount of 4TU-tagged mRNA from the pulse sample to ensure there is enough material for direct comparisons to each chase sample. The type II approach is advantageous because it does not require an excess of the 4TU pulse sample.

There are several ways to prepare a reference sample for microarray experiments. One commonly used method is to perform large-scale RNA extractions from all relevant experimental conditions then pool these RNA samples. The pooled RNA then serves as the template for preparation of a Cy-dye–labeled reference sample. Another method for preparing a reference sample is to Cy-label a fragment of DNA or RNA that will hybridize to a sequence that is common to all spots on the microarray. For example, for microarrays spotted with PCR-amplified cDNAs, it is possible to use PCR or *in vitro* transcription to generate Cy-dye–labeled probes that recognize the primer and vector sequences common to all PCR products spotted on the microarray (Cleary *et al.*, 2002).

4.7.4.2. Standardization of the amount of mRNA used for microarray analysis Because yields of 4TU-tagged mRNA decrease in each chase sample, it is important to use equivalent volumes (not equivalent masses) of purified sample from the pulse and chase time points. For example, if 5 μl of 4TU-tagged RNA from the pulse sample is used to prepare Cy-dye labeled probe, use 5 μl of each chase sample as well.

4.7.4.3. Normalization of pulse–chase microarrays Normalization of the fluorescence intensities measured in pulse–chase microarrays cannot rely on the common "mean-normalization" approach (where the signal intensities are adjusted to make the mean Cy5/Cy3 ratio equal to 1.0), because there is typically an increasingly small number of spots with detectable signals in each chase sample. It is not safe to assume that the mean Cy5/Cy3 ratio in the chase microarrays should be 1.0, because there is a bias toward a unique set of stable mRNAs. However, normalization is still necessary to correct for variability in both the purification of the 4TU-tagged RNAs and the preparation of Cy-labeled probes. Two ways of normalizing pulse–chase microarrays are described in the following.

4.8. Normalization with RNA blot data

The quantitative RNA-blot analysis described in section 4.5 can be used to measure the percentage of the entire mRNA pool that is degraded at each chase time point. Microarrays also provide a measure of the percentage of the entire mRNA pool that is degraded at each time point. The average mRNA decay values from the RNA blot and the microarrays can, therefore, be compared, and the values from the RNA-blot can be used to normalize the microarray signal intensities. For example, in the analysis of mRNA decay in *T. gondii*, RNA blot measurements determined that the percent mRNA decay at three chase time points was 45%, 27%, and 6% (Fig. 19.2). The same samples were analyzed on duplicate microarrays, and the percent mRNA decay at each time point (measured as the average for all genes) ranged from 32 to 54%, 25 to 30%, and 10 to 12%. The signal intensities for the microarrays were then normalized so that the average percent mRNA decay values for all spots on the microarrays matched the percent mRNA decay values determined by the RNA blot.

4.9. Normalization with "spiked in" RNA transcripts

The "spiked in" heterologous RNA transcripts described in section 4.6 could provide another means of normalizing the pulse–chase microarrays. This type of normalization is commonly used in microarray experiments but has not previously been used in 4TU-tagging experiments. The signal intensities of the control spots on each microarray could be used to control for both the purification steps (because the RNAs would be "spiked in" before the biotinylation and purification of 4TU-tagged RNAs) and the microarray probe preparation step. Normalization values that make the signal intensities of the control spots constant across all microarrays could be applied to each sample.

4.9.1. Analysis of pulse–chase microarray data

The analysis of mRNA decay with 4TU-tagging allows quantitative measurements of mRNA stability in specific cells, but the technique is not well suited for making precise calculations of mRNA half-lives ($t_{1/2}$). The main limitation in calculating mRNA half-lives is that the kinetics of nucleotide movement through different "pools" will significantly affect decay measurements (Ross, 1995). This limitation applies to all pulse–chase techniques that use precursors to measure product stability. Nucleotide pools compartmentalize nucleotides in different molecular forms, as well as in distinct locations within the cell. Uridine, for example, is present as UDP, UMP, and UTP, depending on the activity of multiple enzymes involved in uridine synthesis and metabolism. The different forms of uridine are physically separated within compartments of the cell, such as nuclear versus cytoplasmic pools, and UMP is separated into different macromolecules, including tRNA, mRNA, and rRNA. In the experiments described here, pool kinetics will influence the relative amounts of 4-thiouridine and uridine incorporated into mRNA. This makes precise measurements of mRNA half-lives difficult because the efficiency of the "chase" is influenced by the uridine pool kinetics.

For example, pulse–chase analysis of mRNA decay in *T. gondii* showed that the entire pool of mRNAs in bradyzoites decayed at half the rate of the mRNA pool in tachyzoites (Cleary *et al.*, 2005). Therefore, it appeared that the half-life of all mRNAs in bradyzoites was approximately twice as long as the half-life in tachyzoites. However, this result was likely due to differences in nucleotide pool kinetics. For this reason, the data were used to compare relative mRNA stabilities between tachyzoites and bradyzoites as opposed to attempting to calculate mRNA half-lives. The comparison of relative mRNA stabilities can be performed by dividing the percent decay values of individual mRNAs by the mean percent decay value for all mRNAs at that chase time point. The use of such "mean-normalized decay" values allows genes to be categorized on the basis of the relative stability of their transcripts (e.g., "high stability," "average stability," "low stability") in different cell types regardless of variations in uridine pool kinetics.

ACKNOWLEDGMENTS

I thank the numerous people who have assisted in the development of the 4TU-tagging methods described here, particularly Christopher Meiering, Eric Jan, Kristin Robinson, Gus Zeiner, John Boothroyd, Chris Doe, and members of the Boothroyd and Doe laboratories.

REFERENCES

Bryant, Z., Subrahmanyan, L., Tworoger, M., LaTray, L., Liu, C. R., Li, M. J., van den Engh, G., and Ruohola-Baker, H. (1999). Characterization of differentially expressed genes in purified Drosophila follicle cells: Toward a general strategy for cell type-specific developmental analysis. *Proc. Natl. Acad. Sci. USA* **96,** 5559–5564.

Carter, D., Donald, R. G., Roos, D., and Ullman, B. (1997). Expression, purification, and characterization of uracil phosphoribosyltransferase from Toxoplasma gondii. *Mol. Biochem. Parasitol.* **87,** 137–144.

Cleary, M. D., Singh, U., Blader, I. J., Brewer, J. L., and Boothroyd, J. C. (2002). Toxoplasma gondii asexual development: Identification of developmentally regulated genes and distinct patterns of gene expression. *Eukaryot. Cell.* **1,** 329–340.

Cleary, M. D., Meiering, C. D., Jan, E., Guymon, R., and Boothroyd, J. C. (2005). Biosynthetic labeling of RNA with uracil phosphoribosyltransferase allows cell-specific microarray analysis of mRNA synthesis and decay. *Nat. Biotechnol.* **23,** 232–237.

Emmert-Buck, M. R., Bonner, R. F., Smith, P. D., Chuaqui, R. F., Zhuang, Z., Goldstein, S. R., Weiss, R. A., and Liotta, L. A. (1996). Laser capture microdissection. *Science* **274,** 998–1001.

Fan, J., Yang, X., Wang, W., Wood, W. H., 3rd, Becker, K. G., and Gorospe, M. (2002). Global analysis of stress-regulated mRNA turnover by using cDNA arrays. *Proc. Natl. Acad. Sci. USA* **99,** 10611–10616.

Hill, A. A., Brown, E. L., Whitley, M. Z., Tucker-Kellogg, G., Hunter, C. P., and Slonim, D. K. (2001). Evaluation of normalization procedures for oligonucleotide array data based on spiked cRNA controls. *Genome Biol* **2,** RESEARCH0055 (online).

Johnson, T. R., Rudin, S. D., Blossey, B. K., Ilan, J., and Ilan, J. (1991). Newly synthesized RNA: Simultaneous measurement in intact cells of transcription rates and RNA stability of insulin-like growth factor I, actin, and albumin in growth hormone-stimulated hepatocytes. *Proc. Natl. Acad. Sci. USA* **88,** 5287–5291.

Kenzelmann, M., Maertens, S., Hergenhahn, M., Kueffer, S., Hotz-Wagenblatt, A., Li, L., Wang, S., Ittrich, C., Lemberger, T., Arribas, R., Jonnakuty, S., Hollstein, M. C., *et al.* (2007). Microarray analysis of newly synthesized RNA in cells and animals. *Proc. Natl. Acad. Sci. USA* **104,** 6164–6169.

Li, J., Huang, S., Chen, J., Yang, Z., Fei, X., Zheng, M., Ji, C., Xie, Y., and Mao, Y. (2007). Identification and characterization of human uracil phosphoribosyltransferase (UPRTase). *J. Hum. Genet.* **52,** 415–422.

Ross, J. (1995). mRNA stability in mammalian cells. *Microbiol. Rev.* **59,** 423–450.

Roy, P. J., Stuart, J. M., Lund, J., and Kim, S. K. (2002). Chromosomal clustering of muscle-expressed genes in Caenorhabditis elegans. *Nature* **418,** 975–979.

Wang, Y., Liu, C. L., Storey, J. D., Tibshirani, R. J., Herschlag, D., and Brown, P. O. (2002). Precision and functional specificity in mRNA decay. *Proc. Natl. Acad. Sci. USA* **99,** 5860–5865.

Zeiner, G. M., Cleary, M. D., Fouts, A. E., Meiering, C. D., Mocarski, E. S., and Boothroyd, J. C. (2008). RNA Analysis by biosynthetic tagging (RABT) using 4-thiouracil and uracil phosphoribosyltransferase. *Methods Mol. Biol.* **419,** 135-146.

SECTION FIVE

DEFINING DEGRADATIVE ACTIVITIES

CHAPTER TWENTY

Analysis of Cytoplasmic mRNA Decay in *Saccharomyces cerevisiae*

Dario O. Passos *and* Roy Parker

Contents

1. Introduction	410
2. Measuring mRNA Half-Life	410
2.1. *In vivo* labeling	413
2.2. General inhibition of transcription with drugs or mutations	413
2.3. Regulatable promoter systems	413
3. Determination of mRNA Decay Pathways	415
3.1. Trapping mRNA decay intermediates	416
3.2. Determining precursor–product relationships in a transcriptional pulse–chase	419
3.3. Analysis of decay pathways through mutations in *trans*-acting factors	422
4. Combination of the Preceding Approaches	425
References	425

Abstract

The yeast, *Saccharomyces cerevisiae,* is a model system for the study of eukaryotic mRNA degradation. In this organism, a variety of methods have been developed to measure mRNA decay rates, trap intermediates in the mRNA degradation process, and establish precursor–product relationships. In addition, the use of mutant strains lacking specific enzymes involved in mRNA destruction, or key regulatory proteins, allows one to determine the mechanisms by which individual mRNAs are degraded. In this chapter, we discuss methods for analyzing mRNA degradation in *S. cerevisiae*.

1. INTRODUCTION

An important step in the control of eukaryotic gene expression is the process of mRNA turnover in the cytoplasm. The study of mRNA turnover often requires knowledge of the rates of mRNA degradation and the pathway by which a particular mRNA is being degraded. Therefore, in this chapter we describe experimental procedures that can be used to determine the rates of mRNA turnover in yeast, as well as the specific pathway of degradation for any given transcript. Although the methods described here are routinely used to examine cytoplasmic mRNA decay events, they can also be used in similar manners to monitor nuclear mRNA turnover.

An understanding of the pathways of mRNA degradation in yeast, and other eukaryotes, has facilitated these approaches (Fig. 20.1) (reviewed in Parker and Song, 2004). In yeast, mRNA decay is generally initiated by shortening of the 3′-poly(A) tail in a process referred to as deadenylation. Deadenylation is primarily followed by decapping and rapid 5′ to 3′-exonucleolytic decay. Alternately, cytoplasmic transcripts can be subject to 3′ to 5′-exonucleolytic decay after deadenylation.

Additional specialized pathways of decay exist that target subsets of mRNAs, including those with translational abnormalities (Fig. 20.1B). First, mRNAs with premature termination codons or extended 3′ UTRs are degraded by decapping in a deadenylation-independent manner by rapid 5′ to 3′-decay (Muhlrad and Parker, 1994). In addition, some normal mRNAs can also be degraded independently of deadenylation (Badis et al., 2004). Second, mRNAs lacking translation termination codons recruit the exosome complex for very rapid 3′ to 5′-degradation without a distinct and slow deadenylation process (Van Hoof et al., 2002). Finally, mRNAs with strong stalls to translation elongation can be subject to endonucleolytic cleavage (Doma and Parker, 2006). An unresolved issue is how many normal yeast mRNAs are subject to these deadenylation-independent pathways of mRNA degradation. Thus, investigating the mechanism of mRNA degradation can include determining whether a given transcript is subject to any of these specialized mechanisms of mRNA degradation.

2. MEASURING mRNA HALF-LIFE

An mRNA half-life is routinely measured for an mRNA of interest such that the overall stability of that mRNA can be analyzed and quantitated. Several different experimental procedures can be used to measure the decay rates of individual mRNAs in the yeast *Saccharomyces cerevisiae*. In the

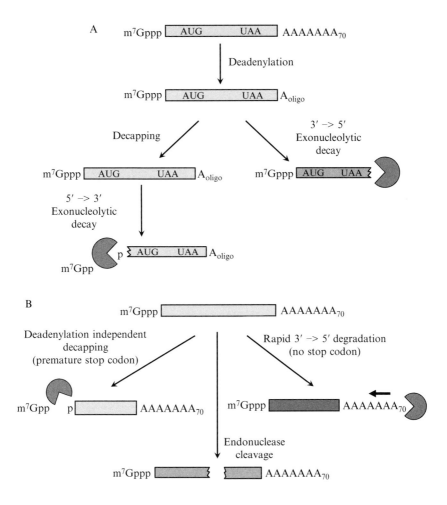

Figure 20.1 Pathways of mRNA decay in yeast. (A) Model showing the two major routes for general mRNA decay in yeast. Normally, mRNAs are initially subjected to deadenylation followed either by decapping and 5' to 3'-decay by Xrn1p or by 3' to 5'-degradation by the exosome. (B) Three pathways known in yeast for bypassing the deadenylation requirement for mRNA turnover include nonsense-mediated mRNA decay (left), which causes deadenylation independent decapping, no-go decay (middle), which leads to endonucleolytic cleavage, and nonstop decay (right), which leads to rapid and continuous 3' to 5'-degradation.

following sections, we discuss the main procedures developed to determine the mRNA half-lives in yeast and their advantages and disadvantages (summarized in Table 20.1).

Table 20.1 Methods for measuring mRNA half-lives in yeast

Method		Advantage	Disadvantage
In vivo labeling	Approach to steady-state pulse–chase	Minimal cell perturbation No need for special strain or construct Can monitor the half-life of many mRNAs simultaneously	Requires large amounts of radioactive material Poor signal-to-noise ratio for mRNA of low mRNA abundance
Transcriptional inhibition	With drugs (thiolutin, 1,10-phenanthroline)	No need for special strain or construct Can monitor the half-life of many mRNAs simultaneously	May cause a loss of labile factors May alter the decay of specific mRNAs May alter other cellular pathways (i.e., transcription or translation)
Transcriptional inhibition	Temperature-sensitive RNA polymerase II mutant (rpb1-1 mutation)	No need for special construct Can monitor the half-life of many mRNAs simultaneously	May cause a loss of labile factors Not useful with other conditional mutants Possible secondary complications caused by heat shock Requires a special strain
Transcriptional control	Regulatable GAL promoter	Minimal cell perturbation Applicable to transcriptional pulse–chase experiments	Requires a special construct Allows only mRNAs under control of GAL promoter to be analyzed Changing the carbon source may alter mRNA stability
Transcriptional control	Regulatable "Tet-off"	Minimal cell perturbation Applicable to transcriptional pulse–chase experiments Allows an accurate control of inhibition or induction	Requires a special construct Allows only mRNAs under control of Tet-promoter to be analyzed The antibiotics used may alter cellular metabolism, influencing the gene under control

2.1. In vivo labeling

Transcripts can be radiolabeled *in vivo* and the rate of mRNA decay determined from either the disappearance of specific mRNAs during a chase (pulse chase) or from the kinetics of the initial labeling (approach to steady state). The main advantage of the *in vivo* labeling methods is the limited cell perturbation. Nevertheless, several disadvantages include a requirement for large quantities of radioactively labeled nucleic acid precursors, poor signal-to-noise ratios for low abundance mRNAs, and a failure to provide information on mRNA integrity during the course of an experiment (Herrick *et al.*, 1990). Given such disadvantages, we will emphasize protocols that measure mRNA decay rates in yeast with methods to control transcription.

2.2. General inhibition of transcription with drugs or mutations

A simple way to measure mRNA decay rates is to use methods to inhibit transcription and then follow the loss of the mRNA of interest over time. In the following section, we discuss procedures by which mRNA synthesis can be inhibited, either in general or for specific genes, and, at various times after such inhibition, the abundance of particular mRNAs is monitored by simple techniques (e.g., Northern blotting or RNase protection). Inhibition of total mRNA synthesis is accomplished by the use of either drugs that globally inhibit transcription in yeast (thiolutin; 1,10-phenanthroline) or with a strain with a temperature-sensitive allele of RNA polymerase II and shifting the culture to the restrictive temperature (Herrick *et al.*, 1990; see also chapter 14 by Coller). These approaches are generally applicable, straightforward, and also provide information on mRNA integrity. The potential disadvantages of these protocols include the nonspecific side effects of drugs used to inhibit transcription and the potential loss of labile turnover factors in the absence of ongoing transcription. It should be noted that actinomycin D, a drug commonly used to inhibit transcription in mammalian cells, does not effectively inhibit transcription in yeast at reasonable doses, presumably because it does not efficiently cross the cell wall. The ability to globally repress transcription in *Saccharomyces cerevisiae* can also been used in conjunction with microarrays to analyze the rates of decay of the entire mRNA transcriptome (Duttagupta *et al.*, 2005; Wang *et al.*, 2002).

2.3. Regulatable promoter systems

2.3.1. Use of the galactose promoter

An alternate approach to the use of global transcriptional repression is to control the transcription of the mRNA of interest, which can be done by placing the corresponding gene under the control of a regulatable promoter.

The GAL promoter is often used because it is one of the strongest inducible promoters in *S. cerevisiae*. In the presence of galactose, the GAL promoter is induced, and the mRNA under its control is highly expressed. Conversely, in the presence of glucose, transcription is promptly inhibited, allowing a quick switch between induction and inhibition of the mRNA to be studied. In addition, genes regulated by the GAL promoter can be used in the transcriptional pulse–chase protocol as described later. One disadvantage of the use of the GAL promoter to control transcription is that this method subjects the cells to a change in carbon source, which might alter cell physiology and thereby affect mRNA decay processes.

A procedure for measuring the half-life of an mRNA with the GAL promoter, which can also be used for the use of the temperature-sensitive allele of RNA polymerase II (rpb1-1), is described as follows:

1. Grow 200 ml of cells in the appropriate medium (for the GAL promoter, the medium should contain 2% galactose) until the cells reach midlog phase (OD_{600} 0.3 to 0.4).
2. Harvest the cells by centrifugation and resuspend in appropriate medium. For rpb1-1 shutoffs, resuspend in 20 ml of medium preheated to 37 °C. For shutoffs of the GAL promoter, use medium containing 2% glucose. Immediately remove a 2-ml aliquot of cells as a zero time point. This and all subsequent aliquots are centrifuged for 10 sec in a microfuge, the medium supernatant is removed by suction, and the resulting cell pellet is quickly frozen in dry ice.
3. Continue growing the cells taking subsequent time points as described by the individual protocol. Initially, time points at 5, 10, 20, 30, 40, 50, and 60 min are commonly used.
4. RNA is isolated from these cells (Caponigro *et al.*, 1993). Northern blotting, dot blotting, or RNase protection procedures are useful for quantitating the amount of a particular mRNA at each time point. A semilog plot of the percent of mRNA remaining versus time allows for the assessment of mRNA half-life (Kim and Warner, 1983; Losson and Lacroute, 1979; Parker *et al.*, 1991).

2.3.2. Use of the tetracycline-regulatable promoter system

An alternate method for repressing transcription of a specific yeast mRNA and thereby measure of its decay rate is to use the "Tet-Off" system, which allows the addition of the antibiotic tetracycline (Tc) or its derivative doxycycline (Dox) to selectively repress transcription of an mRNA placed under the control of the "Tet-off" promoter (Hilleren and Parker, 2003). The principle of this technique is based on the regulatory operon system present in *Escherichia coli,* which has been adapted to the yeast system (Gari *et al.*, 1997). The main advantage of the "Tet-off" system over the use of the GAL promoter is that addition of Dox to cells does not elicit a nonspecific cellular response, and, therefore, the transcriptional repression is specific to only the genes expressed from the "Tet-off" promoter. However, it should be noted that a few yeast strains are sensitive to Tc, and this method should not be used in those genetic backgrounds (Blackburn and Avery, 2003). An experimental procedure with the Tet-off system to analyze mRNA decay rates is described as follows:

1. Prepare a construct with the gene of interest under the control of the Tet-off promoter. Note that if is the 5′ UTR portion of the mRNA is to be analyzed, care must be taken to ensure that the normal site of transcription is used.
2. Grow strains expressing the mRNA under the control of the Tet-off promoter in 200 ml of medium containing 2% glucose to midlog (OD_{600} of 0.3 to 0.4).
3. Pellet the cells and resuspend in 20 ml of the same medium used to grow the cells. This allows the cells to be more concentrated and facilitates the rapid harvesting of aliquots.
4. Add Dox to a final concentration of 2 μg/ml to block transcription.
5. Immediately remove a 2-ml aliquot of cells as a zero time point. This and all subsequent aliquots should be centrifuged for 10 sec in a microfuge, the medium supernatant removed by suction, and the resulting cell pellet quickly frozen in dry ice.
6. Continue growing the cells taking subsequent time points as described by the individual protocol. Initially, time points at 5, 10, 20, 30, 40, 50, and 60 min are commonly used. If the mRNA of interest is highly unstable, then time points of shorter duration should be used.
7. RNA is isolated from these cells (Caponigro et al., 1993). Northern blotting, dot blotting, or RNase protection procedures are useful for quantitating the amount of a particular mRNA at each time point. A semilog plot of the percent of mRNA remaining versus time allows for the assessment of mRNA half-life (Kim and Warner, 1983; Losson and Lacroute, 1979; Parker et al., 1991).

3. Determination of mRNA Decay Pathways

The experimental procedures that were used to define the pathways of mRNA turnover in yeast can now be applied to any transcript to determine the specific pathway(s) of degradation an mRNA undergoes. Three main experimental approaches allow for determining the mRNA decay mechanism. First, inserting a strong secondary structure within the mRNA allows for trapping mRNA decay intermediates, which are useful for determining the directionality of decay. Second, putting the gene of interest under a regulatable promoter allows for pulse–chase experiments, thereby allowing the determination of precursor-product relationships during the mRNA decay process. Third, examining mRNA decay in various strains defective in individual steps in the mRNA turnover pathways can help define the specific turnover pathway(s) that target the mRNA of interest. Although each experimental approach has its limitations, when used in combination, clear information about the mechanism through which a transcript is degraded can often be obtained.

3.1. Trapping mRNA decay intermediates

Understanding the mechanism responsible for degrading a specific mRNA can be facilitated through trapping and structurally analyzing mRNA decay intermediates. Normally, mRNA transcripts are rapidly degraded after an initiating event (e.g., decapping), and decay intermediates cannot be detected. However, introduction of a strong secondary structure into a transcript that inhibits exonucleases traps intermediates in mRNA turnover (Fig. 20.2; Beelman and Parker, 1994; Decker and Parker, 1993; Muhlrad et al., 1994; 1995 Vreken and Raue, 1992).

Introducing a poly(G) tract of at least 18 G nucleotides into a yeast mRNA is commonly used as a means for introducing secondary structure (Beelman and Parker, 1994; Decker and Parker, 1993; Muhlrad et al., 1994; 1995). The strong secondary structure conferred by the addition of a series of G nucleotides is likely a result of the formation of a G-quartet structure arising from hydrogen-bonding of four G nucleotides in a planar array (Williamson et al., 1989). The poly(G) tract serves as a partial block to

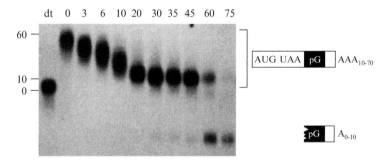

Figure 20.2 Example of a transcriptional pulse–chase experiment. The figure shows Northern blot analysis of a transcriptional pulse–chase experiment with PGK1 mRNA containing a poly(G) tract inserted into its 3′ UTR. After transcriptional repression, samples were taken at different times shown on the top. The first lane (dT) specifies sample treated with oligo d(T) and RNase H to remove the poly(A) tail. At later time points, an intermediate of mRNA decay trapped by the presence of the poly(G) tract in the 3′ UTR can be seen to accumulate.

both 5′ to 3′ and 3′ to 5′-exonucleases, thus causing the accumulation of distinct mRNA decay intermediates (see Fig. 20.2, last two lanes, 60 and 75 min).

3.1.1. Determining the directionality of a decay pathway

The exact nature of the accumulated mRNA decay intermediate depends on the directionality of mRNA degradation. Any appropriate combination of Northern blot analyses, RNase protection assays, and primer extension assays can be used to determine the structure of a trapped intermediate and infer the nucleolytic steps that produced the intermediate. For example, an mRNA undergoing 5′ to 3′-decay will accumulate an mRNA fragment beginning with the poly(G) tract and extending to the 3′-end of that RNA (Fig. 20.2) (Decker and Parker, 1993; Muhlrad et al., 1994; 1995). Alternately, an RNA undergoing 3′ to 5′-decay will accumulate an mRNA fragment from the 5′-end to the 3′-side of the poly(G) tract (Anderson and Parker, 1998).

The placement of a poly(G) tract within an mRNA requires careful consideration. Placing the poly(G) tract in the 5′ UTR blocks ribosome scanning and alters translation of the mRNA. Because translation and mRNA degradation are interrelated (Coller and Parker, 2004), an inhibition of translation is likely to alter the decay of that mRNA and is, therefore, inappropriate for the analysis of mRNA turnover. A poly(G) tract placed within the open reading frame of an mRNA may actually stall translation elongation and trigger no-go decay, on which the mRNA is then subject to endonucleolytic cleavage (Doma and Parker, 2006). Given these complications with the 5′ UTR and coding regions, the poly(G) tract is often placed in 3′ UTR, because this positioning does not generally interfere with mRNA translation or turnover. However, additional analyses are necessary to ensure that insertion of the poly(G) tract does not alter the rate or mechanism of mRNA decay by possibly disrupting an important regulatory element found in the 3′ UTR.

A limitation of inferring decay mechanisms from the structures of trapped intermediates is that the initial nucleolytic event cannot be identified. For example, a poly(G)→3′-end fragment can be generated by decapping and then 5′ to 3′-decay (Muhlrad et al., 1994) or by an initial endonucleolytic cleavage followed by 5′ to 3′-degradation (Doma and Parker, 2006). To distinguish between these two possibilities, one needs to capture early decay intermediates to examine the initial nucleolytic events. One way to capture early decay intermediates is to insert multiple poly(G) tracts (Muhlrad and Parker, 1994). When a transcript is degraded first by decapping and followed by 5′ to 3′-degradation, the transcript will first be degraded to the 5′-poly(G) tract. In contrast, an initial endonucleolytic cleavage between the poly(G) tracts will produce two different fragments. Therefore, these two events can be distinguished by examining the structures of the intermediates derived from a transcript with two poly(G) insertions. Importantly, it should be noted that different mechanisms leading to the same decay intermediates could also be distinguished with specific *trans*-acting mutations (see section 3.3).

3.1.2. The use of decay intermediates to determine the role of deadenylation in decay

Trapped decay intermediates can also be examined directly to determine the role of deadenylation in mRNA decay. The predominant decay pathway accumulates poly(G)→3′-end fragment with an oligo (short) A tail. This is because turnover proceeds through first deadenylation, then decapping, and finally rapid 5′ to 3′-decay. However, decapping can occur without poly(A)-tail shortening, as has been seen with aberrant mRNAs containing premature translation termination codons (Cao and Parker, 2003; Muhlrad and Parker, 1994). In this case, because decapping occurs before deadenylation, the trapped intermediate initially contains a longer poly(A)-tail, as determined by measuring the poly(A)-tail length. Poly(A)-tail length is measured by hybridizing a portion of isolated RNA with oligo d(T), followed by RNase H cleavage; this removes the poly(A)-tail that is not hybridized to oligo d(T) (Fig. 20.2, lane dT). Separation, through a 6% polyacrylamide/8 M urea gel, of isolated RNAs both treated and untreated with oligo d(T)/RNase H allows for direct determination of the poly(A)-tail state of the RNA (Decker and Parker, 1993).

It should be noted that mRNA fragments with long poly(A) tails can be hard to detect in a steady-state population for the following reason. After their production, the poly(G)→3′-end fragments with a long poly(A) tail undergo both deadenylation and 3′ to 5′-decay (Muhlrad and Parker, 1994). Because the 3′ to 5′-decay rate is typically slower than deadenylation rate, most fragments originally with a long poly(A) tail will deadenylate rapidly and accumulate at steady state with short or no poly(A) tails. Given this limitation, the best experiment to examine whether an mRNA decay fragment is initially produced with a long poly(A) tail is to use a transcriptional induction or pulse-chase (see later) to generate a pool of newly synthesized poly(G)→3′-end fragments, whose poly(A) tails lengths can then be examined.

3.1.3. Trapped decay intermediates as a simple assay for 3′ to 5′-mRNA decay

Examining 3′ to 5′-mRNA decay can be specifically examined with the poly(G)→3′-end fragment. Investigating 3′ to 5′-decay on the full-length mRNA can be difficult, because 3′ to 5′-decay occurs slower than 5′ to 3′-decay, and very little 5′-end→poly(G) fragment is formed (Anderson and Parker, 1998; Beelman and Parker, 1994; Decker and Parker, 1993; Muhlrad *et al.*, 1994; 1995; Williamson *et al.*, 1989). Thus, the poly(G)→3′-end fragment can be used as a starting mRNA species for monitoring 3' to 5'-mRNA decay; decay of this RNA will result in a fragment containing only the poly(G) tract (Anderson and Parker, 1998). However, determining the rate of 3' to 5'-decay on the poly(G)→3'-end fragment requires that the production of this RNA species be inhibited (as discussed in section I for determining mRNA half-life). Production of the poly(G)→3'-end fragment can be stopped by inhibiting decapping through the addition of 0.1 mg/ml cycloheximide to cultures. Cycloheximide added to cultures at a concentration of 0.1 mg/ml also inhibits translation elongation; the inhibition of mRNA decapping is likely caused by an indirect effect of cycloheximide addition (Beelman and Parker, 1994). Cycloheximide can be added at the time mRNA transcription is inhibited, and cells can be collected as indicated previously (Beelman and Parker, 1994). Alternately, the dcp1-2 temperature-sensitive allele of the decapping enzyme can be used to rapidly block decapping at the nonpermissive temperature (Tharun and Parker, 1999).

3.1.4. Limitations in the interpretations of trapped decay intermediates

Trapping and analyzing mRNA decay intermediates can ultimately lead to an understanding of the mechanism responsible for mRNA degradation, but this method does have limitations. The trapped intermediates may not reflect the major mRNA decay pathway for a particular mRNA. Specifically, mRNAs that undergo decay through more than one pathway may accumulate intermediates from a minor pathway. Two main aspects should be considered in such a case. First, the relative amounts of an accumulated intermediate should be proportional to the initial amount of full-length mRNA, although this can be misleading if the intermediate itself is very unstable. Second, analyzing the decay rate of an mRNA in cells where one decay mechanism is defective (see section 3.3) will aid in understanding the predominant decay mechanism. A mutation within the main pathway responsible for degrading an mRNA results in a more stable mRNA than mutations that alter a minor or secondary decay pathway. Another limitation of the methods described previously is that the intermediates trapped in the steady state provide limited information about how an individual transcript is degraded. The experiments described in the following overcome this limitation by revealing precursor-product–relationships during mRNA decay.

3.2. Determining precursor–product relationships in a transcriptional pulse–chase

A powerful method to analyze the pathway of mRNA degradation is a transcriptional pulse–chase experiment. In this experiment a regulatable promoter is used to produce a homogeneous population of mRNA transcripts, which are then followed during the subsequent steps in mRNA degradation. This reveals the order in which decay events occur and precursor-product relationships.

In *S. cerevisiae*, the GAL promoter has been used to perform transcriptional pulse–chase experiments (Decker and Parker, 1993; Muhlrad and Parker, 1994; Muhlrad et al., 1994; 1995). A gene of interest is put under GAL control so that a short period of transcription can be accomplished by changing carbon source. This is performed by first growing cells in medium containing a neutral carbon source, usually raffinose, which does not suppress or induce the GAL promoter. Subsequently, the promoter is induced by addition of galactose followed by suppression by glucose after a short time. Raffinose can breakdown to yield galactose at low pH, so when pregrowing cultures in raffinose, care must be taken to keep the pH above 6.0. This is mostly an issue when yeast cells are grown in synthetic media, which can be addressed by adjusting the pH of the medium (see later). For galactose induction, yeast cells can also be pregrown in medium with sucrose as a carbon source, which generally allows better growth and limits preinduction because of raffinose breakdown. However, sucrose partially represses the GAL promoter so the level of induction after galactose addition is not optimal.

After a brief period of transcriptional induction, the GAL promoter is inhibited by the addition of glucose. This creates a "pulse" of relatively homogeneous transcripts whose pathway of degradation can then be analyzed as a sequence of events. For instance, poly(A) shortening and disappearance of the full-length mRNA can be monitored to calculate the deadenylation rate and whether mRNA degradation begins before or after deadenylation. When transcripts with a poly(G) tract are used as mRNA decay precursors, the fragments trapped by the poly(G) tract will provide even more information. The level of full length mRNA and poly(G)→3′-end fragments can be monitored as can the appearance and disappearance of poly(G)→3′-end fragments and the poly(A) tail length of mRNA whose transcribed region is intact. These data provide information about the precursor-product relationships and thus the sequence of events. A detailed experimental procedure is outlined in the following.

1. Pregrow cells in 5 ml of medium containing 2% raffinose or 2% sucrose. Use this culture to inoculate 200 ml of medium containing 2% raffinose or 2% sucrose until cells reach early-log phase (OD_{600}: 0.3 to 0.4). Because low pH results in acid hydrolysis of raffinose into galactose and sucrose, the pH of the medium should be adjusted to pH 6.5 with NH_4OH. In some cases, overgrowth of the culture grown in raffinose can result in a lowering of the pH and, therefore, premature induction of transcription.

2. Pellet cells in four 50-ml Falcon tubes by spinning for 2 min at top speed in a tabletop centrifuge, resuspend cells in 10 ml of medium containing 2% raffinose (or 2% raffinose + 2% sucrose), return to incubator and shake for 10 min.
3. Transcriptional induction is accomplished by adding 0.5 ml of 40% galactose (a final concentration of 2%).
4. Immediately after adding galactose, transfer an aliquot of cells (usually 1 to 2 ml) to a 2-ml Eppendorf tube, briefly centrifuge 10 sec at top speed in a microcentrifuge, and remove the medium/supernatant by aspiration. Rapidly freeze cell pellets in crushed dry ice. This is the preinduction sample.
5. After a short time of induction (typically 8 to 10 min), an equal amount of medium containing 4% glucose is added. Immediately remove and quickly harvest an aliquot of cells as described above. This is the t_0 sample. (Transcriptional repression can also be carried out by temperature shift from 24 °C to 36 °C with a strain harboring a Rpb1-1 mutation [a temperature-sensitive allele of RNA polymerase II, as described in section 1].) With Rpb1-1 mutants, addition of glucose to repress the GAL promoter and temperature shift to repress global transcription can be applied simultaneously to achieve tighter transcriptional repression (Decker and Parker, 1993).
6. Additional cell aliquots are harvested at different time points (as described previously).
7. mRNA is isolated from the cells. Typically 10 to 40 μg of total-cell RNA is separated through a 6% polyacrylamide/8 M urea gel (this gel is 20-cm long and 1-mm thick), which is typically subject to 300 volts for 7.5 h at room temperature. These gels allow direct determination of the poly(A) tail length by comparing the size of the RNA before and after hybridizing to oligo d(T) and treating with RNase H. RNAs too long for polyacrylamide gel analysis can be cleaved with oligonucleotides that specifically anneal to the 3′-end and subsequently treated with RNase H before loading.

Note: RNase H digestion of RNAs requires 10 μg of RNA plus 300 ng of oligo dried in a speed vac. Resuspend the pellet in 10 μl Hyb mix (25 mM Tris [pH 7.5], 1 mM EDTA, 50 mM NaCl), heat for 10 min at 68 °C, cool slowly to 30 °C and spin down. Next, add 10 μl of 2× RNase H buffer (40 mM Tris [pH 7.5], 20 mM MgCl$_2$, 100 mM NaCl, 2 mM DTT, 60 μg/ml BSA, and 1 unit of RNase H), and incubate 30 °C for 1 h. Stop the reaction by adding 130 μl of stop mix (0.04 mg/ml tRNA, 20 mM EDTA and 300 mM NaOAC). Prepare the RNA for electrophoresis through a polyacrylamide gel by extracting with phenol/chloroform and, subsequently, chloroform, precipitating with ethanol, washing the pellet with 70% ethanol and resuspending in 10 μl of formamide gel loading dye (samples are heated to 100 °C for 3 min before loading).

Northern blot analysis with a [^{32}P]-labeled oligonucleotide probe requires transfer of the RNA from the gel to a nitrocellulose membrane. After transfer, the membrane is washed in 0.1× SSC/0.1% SDS for 1 h at 65 °C. The blot is incubated for at least 1 h with prehybridization buffer (10× Denhardt's, 6× SSC, 0.1% SDS) at a temperature 15 °C below the Tm of the oligonucleotide probe. Then the blot is hybridized to the labeled probe at the same temperature for at least 6 h. The blot is washed three times with 6× SSC, 0.1% SDS for 5 min at room temperature, and once for 20 min at 10 °C below the Tm of the oligonucleotide probe, dried, and exposed to X-ray film or a PhosphorImager.

3.3. Analysis of decay pathways through mutations in *trans*-acting factors

With *S. cerevisiae* as a model organism, the stability of a particular mRNA can be examined in a strain deficient in a specific mechanism of decay. An alteration of an mRNAs half-life caused by a specific mRNA decay defect directly implicates the defective mRNA turnover pathway in the decay of the mRNA of interest. Analysis of mRNA stability in a strain deficient for a particular mRNA decay pathway should be coupled with direct analysis of the decaying mRNA. The combined data will limit any possible indirect defects on mRNA decay.

3.3.1. Strains defective in deadenylation

Specific yeast mutants can be used to determine whether an mRNA requires deadenylation for degradation. This is based on the demonstration that the major cytoplasmic mRNA deadenylase requires the products of the CCR4 and CAF1 genes. In either ccr4Δ or caf1Δ strains, mRNAs that require deadenylation show slower rates of mRNA turnover, although it is most pronounced in ccr4Δ strains (Tucker *et al.*, 2001). In contrast, mRNAs that undergo deadenylation-independent decapping show no alteration in mRNA turnover (Badis *et al.*, 2004). In addition, it is possible that some mRNAs will be primarily deadenylated by a second deadenylase known as Pan2 (Brown *et al.*, 1996). Thus, examining whether deadenylation is required for the degradation of a specific mRNA should involve examining the decay rate of the transcript in both ccr4Δ and pan2Δ mutant strains.

3.3.2. Strains defective in 5' to 3'-decay

Mutations in several genes has been identified that affect mRNA decapping or 5' to 3'-exonucleolytic decay (Table 20.2). These genes encode Dcp2p, which is the catalytic subunit of the decapping enzyme (Dunckley and Parker, 1999; Steiger *et al.*, 2003), and Dcp1p, which directly activates

Table 20.2 Genes that affect mRNA decay

Pathway	Gene	Function of protein	Accumulating intermediate in loss of function (mutant)
Deadenylation	CCR4, CAF1, Pan2	Major deadenylases Minor deadenylase	Adenylated capped mRNA
5' to 3'-Decay	Dcp2	Decapping enzyme	
	Dcp1 Dhh1 Lsm1–7 Pat1	Activator of decapping enzyme	Deadenylated capped mRNA
	Xrn1	5' to 3'-exonuclease	Deadenylated decapped mRNA
3' to 5'-Decay	Ski6 (RRP41), Ski4 (CSL4) RRP4 Ski2, Ski3, Ski7, Ski8	Components of exosome	Poly(G)-3'-end fragment and a series of 3' to 5'-decay intermediates
		Modulate 3' to 5'-exonuclease activity	
Specialized mRNA decay	Upf1 Upf2 Upf3	Required for nonsense-mediated decay	
	Hbs1 Dom34	Required for no-go decay	
	Ski7	Specifically blocks nonstop decay	

Dcp2p (She et al., 2006). Furthermore, several proteins, including Dhh1p, Pat1p, Edc3p, and the Lsm1-7 complex, have been identified to enhance the rate of decapping of some, if not most, mRNAs (Coller and Parker, 2004). In addition, Xrn1p is the nuclease responsible for 5′ to 3′-degradation of mRNAs after decapping. Mutations in any of the corresponding genes cause a stabilization of transcripts that are degraded by decapping followed by 5′ to 3′-exonuclease digestion. Because Dcp2p and Xrn1p have clear biochemical roles, an ideal first experiment to see if an mRNA is subject to 5′ to 3′-degradation is to determine its rate of degradation in dcp2Δ and xrn1Δ strains compared with a wild-type strain. Such an experiment can also be followed up by determining whether degradation of the mRNA is affected in dhh1Δ, pat1Δ, lsm1Δ, and edc3Δ strains, which will provide information as to the decapping activators that are most effective on that particular mRNA.

In principle, xrn1Δ and dcp2Δ mutants could be used to distinguish endonucleolytic cleavage followed by 5′ to 3′-decay from decapping followed by 5′ to 3′-decay. In a case of endonucleolytic cleavage, a dcp2Δ mutation would be predicted to have no effect, and an xrn1Δ mutant would accumulate the products of the initial cleavage. In contrast, in a case of decapping, a dcp2Δ mutation would stabilize the transcript of interest, and an xrn1Δ mutant would accumulate full-length decapped RNA.

3.3.3. Strains defective in 3′ to 5′-decay

It is also possible to use *trans*-acting mutations to determine whether a transcript is primarily degraded in a 3′ to 5′-direction. Several proteins have been identified that are required for efficient cytoplasmic 3′ to 5′-decay of mRNA, including the Ski2p, Ski3p, Ski4p, Ski6p/Rrp41p, Ski7p, Ski8p, and Rrp4p (reviewed in Parker and Song, 2004). Rrp4p, Skip6/Rrp41p, and Ski4p/Csl4p are components of a multiprotein complex termed the exosome (Mitchell et al., 1997) and are likely to be part of the actual nucleolytic complex that can degrade the mRNA body 3′ to 5′- (Anderson and Parker, 1998). Ski2p, Ski3p, Ski7p, and Ski8p do not seem to be nucleases and are likely to modulate the exosome activity on mRNA substrates. Although all transcripts examined to date are not stabilized significantly in mutants solely defective in 3′ to 5′-decay, it is possible that there will be specific mRNAs, or specific conditions, wherein the 3′ to 5′-decay pathway is predominant. Thus, to determine whether an mRNA is primarily degraded 3′ to 5′, its decay rate should be examined in some combination of the ski2, ski3, ski4, ski7, ski8, rrp4, and ski6 mutants.

3.3.4. Strains specifically affecting specialized mRNA decay pathways

Yeast strains exist that allow the determination of whether a specific mRNA is targeted to one of the specialized pathways of mRNA decay (Fig. 20.1B). For example, in yeast, the Upf1p, Upf2p, and Upf3p have been shown to be specifically required for the nonsense-mediated mRNA decay (NMD) pathway (He and Jacobson, 1995; Lee and Culbertson, 1995; Leeds et al., 1991; 1992). Thus, to determine whether an mRNA is subject to NMD, its rate of degradation can be examined in Upf mutant strains compared with wild-type strains. Whether an mRNA is subject to nonstop decay (NSD) can also be determined on the basis of the finding that NSD requires a GTPase domain in the Ski7p that is not required for 3′ to 5′-degradation of normal mRNAs (Van Hoof et al., 2002). Thus, mRNAs that are stabilized in a ski7Δc mutant are, in theory, candidate NSD substrates. Finally, mRNAs that are subject to no-go decay (NGD) require Hbs1p and Dom34p for efficient endonucleolytic cleavage (Doma and Parker, 2006). Given this, mRNAs that are stabilized in these mutant strains are NGD candidates. Taken together, these mutant strains provide a mechanism for determining whether an mRNA is subject to any of these specialized pathways of destruction.

4. Combination of the Preceding Approaches

A combination of the preceding approaches provides a powerful analysis of the mechanism responsible for the decay of a specific mRNA. A useful first step in combining the experimental approaches described is to place the gene of interest under a regulatable promoter. This allows for both easy measurements of the mRNA decay rate and the ability to perform transcriptional pulse–chase experiments if the GAL promoter is used. Next, a poly(G) tract can be inserted into the 3′ UTR. Insertion of the poly(G) tract allows for the detection of mRNA decay intermediates, which enables analysis of the role of deadenylation and the directionality of decay. Finally, examining the rate and mechanism of mRNA decay in yeast strains defective for specific mechanisms of mRNA decay can link the turnover to a specialized decay pathway. However, it should be noted that all the approaches described here are specific for defining the decay of an mRNA of interest within the context of the known mRNA decay pathways. The discovery of new mRNA decay pathways will rely on different experimental approaches and thereby will increase the number of variables to consider.

REFERENCES

Anderson, J. S. J., and Parker, R. (1998). The 3′ to 5′-degradation of yeast mRNAs is a general mechanism for mRNA turnover that requires the SKI2 DEVH box protein and 3′ to 5′-exonucleases of the exosome complex. *EMBO J.* **17,** 1497–1506.

Badis, G., Saveanu, C., Fromont-Racine, M., and Jacquier, A. (2004). Targeted mRNA degradation by deadenylation-independent decapping. *Mol. Cell* **15,** 5–15.

Blackburn, A. S., and Avery, S. V. (2003). Genome-wide screening of *Saccharomyces cerevisiae* to identify genes required for antibiotic insusceptibility of eukaryotes. *Antimicrob. Agents Chemother.* **47,** 676.

Beelman, C. A., and Parker, R. (1994). Differential effects of translational inhibition in cis and in trans on the decay of the unstable yeast MFA2 mRNA. *J. Biol. Chem.* **269,** 9687.

Brown, C. E., Tarun, S. Z., Boeck, R., and Sachs, A. B. (1996). PAN3 encodes a subunit of the Pab1p-dependent poly(A) nuclease in Saccharomyces cerevisiae. *Mol. Cell. Biol.* **16,** 5744–5753.

Cao, D., and Parker, R. (2003). Computational modeling and experimental analysis of nonsense-mediated decay in yeast. *Cell* **133,** 533–545.

Caponigro, G., Muhlrad, D., and Parker, R. (1993). A small segment of the MAT alpha 1 transcript promotes mRNA decay in *Saccharomyces cerevisiae*: A stimulatory role for rare codons. *Mol. Cell. Biol.* **13,** 5141–5148.

Coller, J., and Parker, R. (2004). General translational repression by activators of mRNA decapping. *Cell* **122,** 875–886.

Coller, J., and Parker, R. (2005). General translational repression by activators of mRNA decapping. *Cell* **122,** 875–886.

Decker, C. J., and Parker, R. (1993). A turnover pathway for both stable and unstable mRNAs in yeast: Evidence for a requirement for deadenylation. *Genes Dev.* **7,** 1632–1643.

Doma, M. K., and Parker, R. (2006). Endonucleolytic cleavage of eukaryotic mRNAs with stalls in translation elongation. *Nature* **440,** 561–564.

Dunckley, T., and Parker, R. (1999). The DCP2 protein is required for mRNA decapping in *Saccharomyces cerevisiae* and contains a functional MutT motif. *EMBO J.* **18,** 5411–5422.

Duttagupta, R., Tian, B., Wilutsz, C. J., Khounh, D. T., Soteropoulos, P., Ouyang, M., Dougherty, J. P., and Peltz, S. W. (2005). Global analysis of Pub1p targets reveals a coordinate control of gene expression through modulation of binding and stability. *Mol. Cell. Biol.* **25,** 5499–5513.

Gari, E., Piedrafita, L., Aldeia, M., and Herrero, E. (1997). A set of vectors with a tetracycline-regulatable promoter system for modulated gene expression in *Saccharomyces cerevisiae. Yeast* **13,** 837–848.

He, F., and Jacobson, A. (1995). Identification of a novel component of the nonsense-mediated mRNA decay pathway by use of an interacting protein screen. *Genes Dev.* **9,** 437–454.

Herrick, D., Parker, R., and Jacobson, A. (1990). Identification and comparison of stable and unstable mRNAs in *Saccharomyces cerevisiae. Mol. Cell. Biol.* **10,** 2269–2284.

Hilleren, P. J., and Parker, R. (2003). Cytoplasmic degradation of splice-defective pre-mRNAs and intermediates. *Mol. Cell* **12,** 1453–1465.

Kim, C. H., and Warner, J. R. (1983). Messenger RNA for ribosomal proteins in yeast. *J. Mol. Biol.* **165,** 79–89.

Lee, B. S., and Culbertson, M. R. (1995). Identification of an additional gene required for eukaryotic nonsense mRNA turnover. *Proc. Natl. Acad. Sci. USA* **92,** 10354–10358.

Leeds, P., Peltz, S. W., Jacobson, A., and Culbertson, M. R. (1991). The product of the yeast UPF1 gene is required for rapid turnover of mRNAs containing a premature translational termination codon. *Genes Dev.* **5,** 2303–2314.

Leeds, P., Wood, J. M., Lee, B. S., and Culbertson, M. R. (1991). The product of the yeast UPF1 gene is required for rapid turnover of mRNAs containing a premature translational termination codon. *Mol. Cell. Biol.* **12,** 2165–2177.

Losson, R., and Lacroute, F. (1979). Interference of nonsense mutations with eukaryotic messenger RNA stability. *Proc. Natl. Acad. Sci. USA* **76**, 5134–5137.

Mitchell, P., Petfalski, E., Shevchenko, A., Mann, M., and Tollervey, D. (1997). The exosome: A conserved eukaryotic RNA processing complex containing multiple 3′ to 5′-exoribonucleases. *Cell* **91**, 457–466.

Muhlrad, D., and Parker, R. (1994). Premature translational termination triggers mRNA decapping. *Nature* **370**, 578–581.

Muhlrad, D., Decker, C. J., and Parker, R. (1994). Deadenylation of the unstable mRNA encoded by the yeast MFA2 gene leads to decapping followed by 5′ to 3′-digestion of the transcript. *Genes Dev.* **8**, 855–866.

Muhlrad, D., Decker, C. J., and Parker, R. (1995). Turnover mechanisms of the stable yeast PGK1 mRNA. *Mol. Cell. Biol.* **15**, 2145–2156.

Parker, R., and Song, H. (2004). The enzymes and control of eukaryotic mRNA turnover. *Nat. Struct. Mol. Biol.* **11**, 121–127.

Parker, R., Herrick, D., and Peltz, S.W (1991). Measurement of mRNA decay rates in *Saccharomyces cerevisiae*. *Methods Enzymol.* **194**, 415–423.

She, M., Decker, C. J., Chen, N., Tumati, S., Parker, R., and Song, H. (2006). Crystal structure and functional analysis of Dcp2p from *Schizosaccharomyces pombe*. *Nat. Struct. Mol. Biol.* **13**, 163–70.

Steiger, M., Carr-Schmid, A., Schwartz, D. C., Kiledjian, M., and Parker, R. (2003). Analysis of recombinant yeast decapping enzyme. *RNA* **9**, 231–238.

Tharun, S., and Parker, R. (1999). Analysis of mutations in the yeast mRNA decapping enzyme. *Genetics* **151**, 1273–1285.

Tucker, M., Valencia-Sanchez, M. A., Staples, R. R., Chen, J., Denis, C. L., and Parker, R. (2001). The transcription factor associated Ccr4 and Caf1 proteins are components of the major cytoplasmic mRNA deadenylase in *Saccharomyces cerevisiae*. *Cell* **104**, 377–386.

Van Hoof, A., Frischmeyer, P. A., Dietz, H. C., and Parker, R. (2002). Exosome-mediated recognition and degradation of mRNAs lacking a termination codon. *Science* **295**, 2262–2264.

Vreken, P., and Raue, H. A. (1992). The rate-limiting step in yeast PGK1 mRNA degradation is an endonucleolytic cleavage in the 3′-terminal part of the coding region. *Mol. Cell. Biol.* **12**, 2986–2996.

Wang, Z., Jiao, X., Carr-Schmid, A., and Kiledjian, M. (2002). The hDcp2 protein is a mammalian mRNA decapping enzyme. *Proc. Natl. Acad. Sci. USA* **99**, 12663–12668.

Williamson, J. R., Raghuraman, M. K., and Cech, T. R. (1989). Monovalent cation-induced structure of telomeric DNA: The G-quartet model. *Cell* **59**, 871–880.

CHAPTER TWENTY-ONE

Transcriptome Targets of the Exosome Complex in Plants

Dmitry Belostotsky

Contents

1. Exosome: At the Nexus of the Cellular RNA Transactions	429
2. Unique Features of the Plant Exosome	432
3. Resources for the Mutational Analyses of the Plant Exosome	434
4. Transcriptome-wide Mapping of Targets of the Plant Exosome Complex	436
Acknowledgments	440
References	440

Abstract

The exosome complex is endowed with the capabilities to conduct 3′-end RNA processing, 3′-end degradation, and surveillance of various RNA substrates. Although the exosome is present in both eukaryotes and archaea, where it plays a central role in cellular RNA metabolism, a substantial degree of variability exists in its modus operandi in the different domains of life. This chapter discusses features of the exosome in plants that distinguish it from the exosome in archaea, fungi, and animals, as well as reviews the resources and tools that are needed to identify and catalog plant exosome targets on a transcriptome-wide scale.

1. Exosome: At the Nexus of the Cellular RNA Transactions

The exosome complex is an evolutionarily conserved macromolecular assembly with 3′ to 5′-exoribonucleolytic activity that is present in both nuclear and cytoplasmic cellular compartments. What the necessary brevity of this definition conceals is an overwhelming complexity and variability

Division of Molecular Biology and Biochemistry, School of Biological Sciences, University of Missouri, Kansas City, Missouri, USA of exosome architecture and function. For example, the exosome can (1) process 3′-extended precursor RNAs to their mature products, (2) completely degrade other types of RNA substrates, and (3) selectively eliminate aberrant (e.g., misfolded) RNAs without affecting their normal counterparts. How substrate fate is determined in individual cases remains largely unknown. Moreover, which of the individual subunits of the exosome complex is responsible for its catalytic activity seems to vary across the phylogenetic spectrum, and depletion of the individual subunits from the exosome core engenders distinct sets of molecular phenotypes, depending on the species, tissue, and cell type. Hence, it is necessary to comprehensively define the RNA substrates that are processed or degraded by the exosome in various circumstances to fully grasp the impact of this macromolecular complex on gene expression, as well as to understand the reasons why its function is essential for viability.

The eukaryotic exosome core is an assembly of 9 or 10 polypeptides containing two main classes of subunits. One such class, represented by six polypeptides, has pronounced sequence similarity to *Escherichia coli* RNase PH (although these RNase PH-like polypeptides in fungi and animals lack detectable exoribonucleolytic activity). The second class contains proteins that possess S1- and KH-type RNA binding domains. In some, but not all, eukaryotic species, the exosome core also contains Dis3/Rrp44, a hydrolytic exoribonuclease. The crystal structures of the exosome from archaea revealed that the RNase PH type subunits are organized into a trimer of dimers, in a manner analogous to the prokaryotic phosphorolytic RNases (Lorentzen *et al.*, 2005). The archaeal exosome has three phosphorolytic exoribonuclease active sites within its core, although the width of its central channel seems too narrow to accommodate three RNA molecules simultaneously. Instead, the multiplicity of active sites may play a significant role in the processivity of the RNA degradation (Lorentzen *et al.*, 2007). Early ideas about the structural organization of the eukaryotic exosome likewise envisioned a heterohexameric ring of RNase PH-like subunits organized into a RNA processing chamber, with an RNA binding adaptor module containing the S1 and KH domain proteins, hereafter referred to as "cap", added on top of it (Aloy *et al.*, 2002). The modularity implicit in this model offered a potentially appealing solution to the problem of correctly choosing the substrate's fate whereby the events in the processing chamber would depend on the mode, of interaction of the RNA substrate and the adaptor module. Namely, depending on the stability of secondary structures in the RNA substrate and/or presence of RNA helicases in the accessory complexes (see below), the substrate would be either fully degraded or undergo a limited processing, followed by dissociation and release of the reaction product. However, as opposed to the apparent modularity of the exosome complex in archaea, experimentally determined crystal structure of the reconstituted human exosome (Liu et al., 2006 and chapter 10 by Greimann and Lima in this volume) shows that structural symmetry has considerably degenerated during evolution, such that every subunit in the human complex interacts with the other subunits through a unique interface. One corollary of this finding is that the integrity (and consequently function) of the eukaryotic exosome should strictly require the simultaneous presence of all subunits. Although this view is, indeed, consistent with results of genetic depletion experiments in yeast and *Trypanosama brucei*, it is not the case in plants, as discussed later.

A number of auxiliary factors interact with the exosome to facilitate its functions. The nuclear form of the exosome is distinguished by the presence of an additional subunit, RRP6, associated with the exosome core. The nuclear exosome is remarkably versatile and able to carry out either $3'$-end-processing of RNA, such as in the case of the 5.8S rRNA precursor (Allmang et al., 1999), or complete RNA degradation, as in the case of external transcribed rRNA spacers (Allmang et al., 2000), aberrant pre-rRNAs, pre-mRNAs, pre-tRNAs (Bousquet-Antonelli et al., 2000; Kadaba et al., 2004; 2006; Libri et al., 2002; Torchet et al., 2002), or normal mRNAs trapped in the nucleus when mRNA export is blocked (Das et al., 2003). Although the RNase D-like protein RRP6 is required for all activities of the nuclear exosome, a subset of these activities requires distinct auxiliary factors. For example, the putative RNA binding protein LRP1 participates in the processing of stable RNAs (Mitchell et al., 2003; Peng et al., 2003). On the other hand, the degradative activity of the nuclear exosome is linked to the *TRF4/5-AIR1/2-MTR4* polyadenylation (TRAMP) complex, which helps recruit the exosome to aberrant or misfolded structural RNAs, cryptic unstable transcripts (CUTs), as well as to mRNAs that fail to complete splicing, $3'$-end processing, or transport steps of mRNA biogenesis (Houseley and Tollervey, 2006; Houseley et al., 2007; Kadaba et al., 2004; 2006; LaCava et al., 2005; Vanacova et al., 2005; Wyers et al., 2005).

In the cytoplasm, most of the activities of the exosome involve mRNA degradation (e.g., homeostatic mRNA turnover, rapid decay of unstable mRNAs, nonsense-mediated mRNA decay, degradation of mRNAs lacking stop codons and substrates of no-go mRNA decay [reviewed in Houseley et al., 2006], and removal of $5'$-products of RISC-mediated endonucleolytic mRNA cleavage) (Orban and Izaurralde, 2005). These reactions are mediated by the SKI2/SKI3/SKI8 complex and the SKI7 protein (Wang et al., 2005). In addition, the exosome down regulates the levels of uncapped, unadenylated transcripts of LA virus, which is a cytoplasmic dsRNA virus present in many laboratory strains of yeast (Brown and Johnson, 2001). Hence, one major function of the cytoplasmic exosome is to destroy aberrant mRNAs and viral transcripts.

The 3' to 5'-decay pathway mediated by the cytoplasmic exosome is a major mRNA decay pathway in mammals (Mukherjee *et al.*, 2002; Wang and Kiledjian, 2001). Specific RNA-binding proteins often play an important role in this process by acting as mediators of the exosome recruitment (e.g., as in the case of ARE [AU-rich element]) binding proteins (Chen *et al.*, 2001; Gherzi *et al.*, 2004). Interestingly, mammalian ZAP protein recruits the exosome to degrade viral mRNAs, thus suggesting that the exosome in mammalian cells also has an antiviral function (Guo *et al.*, 2007).

One important unresolved aspect of the coordination of mRNA degradation events in the cytoplasm concerns the spatial distribution of mRNA degradation reactions carried out by the exosome relative to those that are initiated by decapping. Although the decapping enzymes are concentrated in P bodies—distinct compartments for mRNA sequestration, storage, or decay (Sheth and Parker, 2003)—the exosome is not known to localize to P bodies (Brengues *et al.*, 2005). On the other hand, in the cultured Drosophila cells, some (but not all) of the exosome subunits are enriched in the cytoplasmic foci and, moreover, different subunits exhibit distinct localization patterns (Graham *et al.*, 2006). Perhaps this indicates a certain degree of plasticity in exosome composition and function. For example, one might envision that in higher cells there exist multiple kinds of specialized mRNA degradation subcompartments as opposed to just a single type in yeast. Therefore, different classes of mRNA may be routed to distinct sets of degradation sites, possibly depending on the cell type and/or physiological state. A more precise definition of spatially distinct exosome subcomplexes and a comprehensive global identification of their RNA substrates are required to address these issues.

2. Unique Features of the Plant Exosome

Although the "parts list" of the exosome is largely shared between plants and animals (and even archaea), the structure–function relationships in this macromolecular complex have been evolving independently for a long time. Indeed, the last common ancestor of plants and animals seems to have existed approximately 1.6 billion years ago (i.e., long before a clear fossil record of multicellular eukaryotes). Therefore, many informative lessons may result from comparative examination of the exosome composition, structure, localization, and RNA targets across the phylogenetic spectrum.

One major feature that distinguishes the plant exosome from its animal and fungal cousins is the presence of an active site in the RNase PH heterohexameric ring. Contrary to earlier proposals that were largely based on the extrapolations from structure–function studies of the archaeal exosome complex (Aloy et al., 2002), the RNase PH–like subunits of the yeast and mammalian complexes are catalytically inactive, and the hydrolytic subunit Dis3/Rrp44 is solely responsible for its activity (Dziembowski et al., 2007). Concomitant with the addition of Dis3/Rrp44 to the complex in the course of evolution, exosomal RNase PH–type subunits have lost their catalytic competence through a divergence of amino acid residues responsible for the catalysis and/or binding to RNA and/or binding inorganic phosphate (which is essential for the catalytic mechanism of phosphorolytic enzymes). It has been hypothesized that RNase Dis3/Rrp44 acquisition may have decreased the selective pressure to maintain the phosphorolytic sites (Wahle, 2007). Yet, in contrast to the animal and yeast exosomes, all plants (and even the unicellular green alga *Chlamydomonas reinhardtii*), maintain the single phosphorolytic active site in the PH ring that resides in the RRP41 subunit (Chekanova et al., 2000). In addition, no Dis3/Rrp44-like subunit copurified with the affinity-tagged *Arabidopsis* exosome (Chekanova et al., 2007), which is also consistent with the above view.

Although the X-ray crystal structure of the human exosome (Liu et al., 2006 and chapter 10 by Greimann and Lima in this volume) suggests that the structural integrity of the complex requires the simultaneous presence of every core subunit, a null allele of CSL4 (which is one of the RNA binding "cap" subunits in *Arabidopsis* has no apparent phenotype. Furthermore, although genetic depletion of any exosomal core subunit in *Saccharomyces cerevisiae* and *Trypanosoma brucei* results in virtually identical effects, at least on the RNA substrates examined so far, depletions of RRP4 and RRP41 in *Arabidopsis* produced overlapping yet distinct molecular signatures (and their null alleles led to different developmental phenotypes).

Taken together, these findings point to an unexpectedly high degree of functional plasticity in the plant exosome core, as well as call for broader examination of the structure–function relationships and targets of plant exosome complex. For example, it is important to explain the significance of the plant-specific conservation of the phosphorolytic active site in the RRP41. Moreover, the distinct effects resulting from the loss of the few tested subunits of the heterohexameric "core" and the RNA binding "cap" may reflect functional subfunctionalization of the complex into two submodules. This hypothesis can be tested by examining the phenotypes of additional representative core and cap subunits, as well as by studying the transcriptome-wide effects of their depletion. Finally, it remains to be established to what extent the unexpected features of the plant exosome can be generalized, particularly to monocot plants, which diverged ~230 million years ago from the dicot lineage, and of which *Arabidopsis* is a member. The tools and approaches necessary for such studies are discussed in the subsequent sections.

3. Resources for the Mutational Analyses of the Plant Exosome

The key prerequisite for inquiry into the function of the plant exosome is the availability of loss-of-function alleles. At present, the most commonly used access point for such alleles in *Arabidopsis* is the web site of the Salk Institute Genomic Analysis Laboratory (SIGnAL; signal.salk.edu/cgi-bin/tdnaexpress). The SIGnAL site represents an extensive, multifunctional web environment that not only provides convenient access to the SIGnAL's own sequence-indexed T-DNA insertional mutant library but also links to the multiple collections of *Arabidopsis* insertional mutants worldwide. The centerpiece of the SIGnAL's interface is a genome browser that contains links to a wide variety of relevant data for each of the annotated genome entries (e.g., transcriptome, methylome, small RNA), as well as a color-coded graphic display of the insertional events. Each insertion site symbol is a clickable link to information about which insertional mutagen was used, the flanking genomic sequence, and an interface for ordering seed stocks. A useful set of links to the individual papers and projects that more fully describe each set of tagged lines is provided at the bottom of the browser screen. As of late 2007, there were close to 400,000 tagged insertion lines publicly available. For a 119-Mb genome, this translates into a >96% chance of having a "hit" in every 1-Kb genomic segment. The single largest collection of the insertional mutants is the SIGnAL's own population of ~150,000 lines, produced with T-DNA insertions in the Columbia genetic background. In addition, SIGnAL has been generating a sub library of homozygous insertion lines for all nonessential genes, termed a "phenome-ready" genome set (i.e., ready for phenotyping). Although all core exosome subunit genes are essential in most species, CSL4 is dispensable in *Arabidopsis*, and this may hold true for some other core and noncore subunits. As of 2008, 17,637 homozygous insertion lines that represent 12,872 individual genes are available from ABRC.

Other sequence-indexed collections of *Arabidopsis* insertional mutants are available. The major ones include Syngenta (so-called SAIL lines) and GABI-Kat (Bielefeld University, Germany) T-DNA collections, the University of Wisconsin and Cold Spring Harbor Laboratory populations of stabilized *Ds* transposon insertions (all of the above, like the SIGnAL lines, are in the Columbia genetic background), FLAGdb T-DNA lines from INRA, France (in the WS background), as well as RIKEN (Japan) and the European EXOTIC consortium Ds element-based collections (in Nossen and Landsberg backgrounds, respectively). Information about the insertion sites for all of these resources can be accessed through SIGnAL. Although the distribution policies and procedures vary, most of mutant line seed stocks are distributed for a nominal fee by one of two major public repositories" - the *Arabidopsis* Biological Resource Center (ABRC) at the Ohio State University or the Nottingham Arabidopsis Stock Centre (NASC) at the University of Nottingham.

It is worth noting that SIGnAL's own collection has been created with a vector that contains the complete 35S promoter from Cauliflower mosaic virus, including the enhancer. Therefore, insertion events may occasionally result in "activation tagging" instead of (or in addition to) the disruption of the normal expression of the nearby gene. Instances of unexpected and/or abnormal expression of flanking genomic regions (e.g., of antisense transcripts) have been reported (Ren et al., 2004; and our unpublished observations). In addition, in the case of the SIGnAL lines, the kanamycin resistance gene that was used for the initial selection of transformants often gets silenced in subsequent generations and, hence, should not be used for selection or cosegregation analyses. This is less of a concern with hygromycin or BASTA resistance markers used in some of the other insertional collections. One final issue is that multiple T-DNA insertions per genome, as well as totally unrelated mutations (not linked to the T-DNA), are not uncommon and should be accounted for in interpreting the phenotypes.

Great strides have been made toward developing sequence-indexed collections of insertional mutants for other model plants. Among monocots, rice is the most advanced, where some 100,000 T-DNA insertion lines have been generated by Gynheung An's group in POSTECH (www.postech.ac.kr/life/pfg/risd; Jeong et al. [2002]). Additional T-DNA collections are available through Genoplante (France) and RIFGP (China), and Tos17 transposon insertion lines are available from NIAS (Japan). All of these are linked to the SIGnAL web site through the RiceGE (Rice Functional Genomic Mapping Tool) interface. The ~100,000 POSTECH lines are readily accessible by signing the MTA and provide an ~29.5% probability of a "hit" in any given rice gene on the basis of a genome size of 430 Mb, an average gene size of 3 kb, and 1.4 insertions per line (based on the relationship $P = 1 - [1 - f]^n$, where P is the probability of an insertion in any given gene, f is an inverse of the number of targets, and n is the total number of available T-DNA insertion events) (Clarke and Carbon, 1976).

A complementary resource is based on TILLING (*targeted induced local lesions in genomes*) technology. TILLING is a method of identifying EMS-induced point mutations in a gene of interest that relies on PCR amplification followed by the detection of heteroduplexes formed upon annealing of the mutant and wild-type PCR products using cleavage with a highly mismatch-specific endonuclease, CEL I (McCallum et al., 2000). An Arabidopsis TILLING resource at Fred Hutchinson Cancer Center (tilling.fhcrc.org:9366) operates a high-throughput screening service on a fee-for-service basis. As of this year, the TILLING facility processed upward of 600 orders and reported nearly 8000 point mutations. Parallel TILLING resources have been established for rice (Till et al., 2007), maize

(Till *et al.*, 2004), wheat (Slade *et al.*, 2005), and tomato (see tilling.ucdavis.edu/index.php/Main_Page). However, at present, TILLING is available as a public service only for maize (genome.purdue.edu/maizetilling) and rice (presently at the beta testing stage at the UC Davis Genome Center). Drawbacks of TILLING that should be kept in mind are its relatively high cost and the need for multiple outcrossing (because every genome experiences multiple "hits" during the typical mutagenesis procedure). Another approach to create mutants applicable to many plant species uses fast neutron mutagenesis, which tends to generate small deletions in the genome, coupled to a PCR-based method to identify abnormally small amplicons (Li *et al.*, 2001). It remains to be seen whether populations with a sufficient density of deletions can be generated to make this process efficient.

Another alternative approach involves the targeted engineering of a gene knockdown, which may be the only solution if an insertion or other type of mutation in the gene of interest is unavailable or not useful. For example, one could engineer an artificial microRNA (amiRNA) according to established thermodynamic constraints. This strategy has been successful in *Arabidopsis*, as well as rice (Ossowski *et al.*, 2008; Schwab *et al.*, 2006; Warthmann *et al.*, 2008), and it is particularly useful when one attempts to specifically target just one gene in a multigene family that consists of several closely related members. However, when discriminating among the members of a gene family is not of particular concern, higher efficiencies of gene knockdown can be achieved by creating transgenic plants that express relatively large (several hundred bp-long) double-stranded RNA hairpins, with the two arms of the hairpin separated by an intron or a nonspliced spacer (Smith *et al.*, 2000).

4. TRANSCRIPTOME-WIDE MAPPING OF TARGETS OF THE PLANT EXOSOME COMPLEX

A caution is warranted regarding use of constitutive loss-of-function mutations to study the transcriptome-wide effects of exosomal subunits. Such mutations (or any constitutive mutations, for that matter) typically exhibit a complex mixture of primary and secondary effects (i.e. those that are caused by the chain of events whereby primary change in the expression of gene X in turn causes altered expression of gene Y etc) that are quite challenging to distinguish from one another, particularly on a transcriptome-wide scale. Usually, special considerations during the design stage and/or extensive follow-up efforts—both bioinformatic and experimental—are required to distinguish the primary effects from the secondary ones. This problem becomes particularly acute when deep sequencing or tiling arrays are used for the analysis, because of the sheer volume of data generated by such approaches.

The most universal and radical approach that allows to bypass this issue is the engineering of a conditional allele, because in this case one can impose restrictive conditions at will and then monitor the real-time progression of the ensuing changes in the transcriptome, specifically focusing on the earliest time points of the response to the inactivation of the exosome. The expectation here is that the primary changes (e.g., those that directly result from the inactivation of the exosome complex) would develop early, whereas the secondary effects will become evident only after the significant alterations of the immediate substrates of the exosome accumulate (first in the mRNA, and subsequently in the protein levels). Another prediction would be that inactivation of a ribonucleolytic activity associated with the exosome would primarily result in an increase of immediate targets that are normally degraded by this complex, rather than their downregulation. On the other hand, those RNA species that normally are processed (rather than degraded) by the exosome would exhibit extensions beyond the point that normally correspond to the mature $3'$-end of the molecule.

Conditional plant mutants in the core exosome subunit or its auxiliary factor can be constructed by means of an inducible expression of an artificial miRNA (Ossowski *et al.*, 2008; Schwab *et al.*, 2006), or by engineering an inducible RNAi (iRNAi) (Chekanova *et al.*, 2007; Chen *et al.*, 2003; Ohashi *et al.*, 2003; Smith *et al.*, 2000) to trigger a conditional knockdown of the desired target. One such iRNAi system successfully implemented in *Arabidopsis* is based on the estradiol-regulated chimeric transactivator XVE (Zuo *et al.*, 2000). To this end, transgenic *Arabidopsis* plants are engineered to express long (several hundred base pairs, if feasible) segments of the target gene as a pair of inverted repeats separated by an intron. Such intron-spliced hairpin RNAs tend to give higher silencing efficiency than those containing a nonspliced spacer between the arms of the hairpin (Stoutjesdijk *et al.*, 2002). Growing such exosome iRNAi plants on estradiol-containing media induces the RNAi-mediated knockdown of the respective targets and leads to growth arrest and subsequent death of seedlings. Importantly, a highly characteristic molecular phenotype characteristic of exosome malfunction (accumulation of the $3'$-underprocessed 5.8S rRNA intermediates) was observed before any obvious effect on growth in both $RRP4^{iRNAi}$ and $RRP41^{iRNAi}$ lines (Chekanova *et al.*, 2007; Mitchell *et al.*, 1997). Such observations of characteristic molecular events that precede any overt effects on growth can be used to validate the transgenic iRNAi system. Hence, the corresponding time point of the depletion time course was used to harvest the plant tissues for extracting the RNA for the microarray experiments. In contrast, a similar analysis of the constitutive null mutant of CSL4 revealed massive amounts of secondary effects. This can seriously confound interpretations and limit the usefulness of the resulting data set (Chekanova *et al.*, 2007).

It should be noted that not all inducible RNAi systems are created equal, and hence a good deal of consideration must be given to their relative advantages, limitations, and potential associated artifacts. For example, iRNAi under the control of the popular glucocorticoid receptor/VP16/GAL4 DNA binding domain system (GVG) was found to trigger the expression of defense response genes on its own; similar observations have been reported by others in several plant species (Amirsadeghi *et al.*, 2007; Andersen *et al.*, 2003; Kang *et al.*, 1999). An excellent comprehensive review that evaluates numerous conditional promoter systems in multiple transgenic plants species has been recently published and should be consulted (Moore *et al.*, 2006).

The estradiol-regulated conditional RNAi knockdown lines were successfully used in conjunction with the whole genome tiling microarrays to comprehensively define the exosome targets in *Arabidopsis* (Chekanova *et al.*, 2007) and can be considered as a framework for future experiments (e.g., in other plants). The empty-vector control line and the iRNAi lines targeting the exosome components of choice are treated by estradiol for several days to deplete the respective exosome subunit. In our experience, germination for 5 days on half-strength MS media containing 8 μM of 17 β-estradiol (Sigma) results in approximately fivefold depletion in the level of the respective exosome subunit (measured by Q-PCR). Verification of the efficiency of the depletion is followed by the second RNA quality control to verify the $3'$-under processing of the 5.8S rRNA by standard Northern blotting. RNA samples passing these quality control checks are then used to synthesize targets for the microarray hybridization. In addition to the estradiol-treated iRNAi lines and iRNAi lines exposed to just the solvent (DMSO) as a control (subsequently referred to as $\pm$ estradiol for brevity), the empty vector-transformed line should be processed similarly (hereafter referred to as WT for brevity). At present, most of the reagents and kits that are available off the shelf for *Arabidopsis* and other plants are designed for oligo(dT)-primed synthesis of the target. However, the exosome is believed to enter the pathway of the $3'$ to $5'$-mRNA decay only after the poly(A) tail is removed by one of several specialized deadenylating enzymes. Thus, many, if not most, of the mRNA decay intermediates accumulating as a result of the exosome depletion would be expected to lack the poly(A) sequence and, hence, would be missed in experiments with an oligo(dT)-primed target. This problem could be addressed by randomly priming the target instead of the use of the oligo(dT) priming, but doing so requires prior subtraction of the highly abundant ribosomal RNA. Although the reagents to do so are available for nonplant systems (e.g., Ribominus kit from Invitrogen), these are yet to be developed for subtracting plant ribosomal RNA.

The choice of an array platform for the studies of the plant exosome targets depends on a number of considerations (such as the availability of commercial arrays for a given genome, equipment, core facilities, financial, and logistical constraints), but the whole genome tiling arrays are preferable because of a large number of noncoding RNAs regulated by the exosome (Chekanova et al., 2007) that are largely missing from the annotation-based arrays. The Affymetrix whole-genome chip for *Arabidopsis* tiles both strands of its ~119 Mb genome end-to-end, with the average distance of 35 bases between the center points of the neighboring 25 base-long probes. Like most Affymetrix gene chips, each probe set consists of perfect match and mismatch probes (PM and MM), which presents the investigator with a choice of the use of MM signal as a measure of background probe binding. In practice, however, it has been observed that MM probes tend to still detect real signal in addition to nonspecific binding, hence subtracting the MM value from the PM signal value often adds to the noise instead of providing a gain in specificity (discussed in Irizarry et al. [2003]).

The transcriptome signals obtained by interrogating the tiling arrays with the exosome depleted line-derived targets can be analyzed with TileMap (Ji and Wong, 2005). The distinctive feature of the Tilemap algorithm is that it does not combine probes into groups to compute a summary metric for a given gene or a given region, but rather treats every probe as a separate entity. Hence, this software produces an unbiased identification of any genomic regions showing significantly altered expression level. The raw data are processed as follows.

The feature intensities for all replicates for the given knockdown line and the control line are quantile normalized and subjected to a multi-comparative TileMap analysis. For example, in the case of $rrp4^{RNAi}$, the data for a total of six tiling arrays (e.g., two replicates for WT treated with estradiol, two replicates for $rrp4^{RNAi}$ treated with estradiol, and two replicates for $rrp4^{RNAi}$ not treated with estradiol) from one strand were all quantile normalized at the same time. Differentially expressed regions are then identified by comparing the WT sample treated with estradiol and $rrp4^{RNAi}$ both with and without estradiol treatment, for both Watson and Crick strands. This generates two lists: (1) $rrp4^{RNAi}$ treated with estradiol > WT treated with estradiol as well as rrp4RNAi treated with estradiol > $rrp4^{RNAi}$ not treated with estradiol, and (2) $rrp4^{RNAi}$ treated with estradiol < WT treated with estradiol as well as $rrp4^{RNAi}$ treated with estradiol < $rrp4^{RNAi}$ not treated with estradiol. The posterior probability setting of 0.7, within a maximal region of 100 bases, can be used as a reasonable starting point in these analyses. The [$rrp4^{RNAi}$ treated with estradiol > $rrp4^{RNAi}$ not treated with estradiol] comparison identifies regions with increased expression because of the knockdown of Rrp4 protein levels, and the [rrp4^{RNAi} treated with estradiol > WT treated with estradiol] comparison removes regions with increased expression only because of estradiol alone.

To estimate the false-positive rate (i.e., transcripts that are induced or repressed by the estradiol treatment per se), replicates for WT sample with and without estradiol treatment are quantile normalized and subjected to the TileMap analysis. Differentially expressed regions are identified by generating lists (WT with estradiol > WT no estradiol and WT treated with estradiol < WT with no estradiol) at posterior probability 0.7 within a maximal region of 100 bases (i.e., the same parameters as used for the comparisons of the knockdown lines). In our experience, this analysis yielded no differentially expressed regions. Moreover, even at a lower posterior probability value of 0.5, only six genomic regions could be identified (one upregulated and five downregulated). This is taken as evidence that any effects of estradiol alone can be considered negligible.

Application of the preceding tools in *Arabidopsis* has allowed the construction of the first high-resolution genome-wide atlas of the targets of the exosome complex in a multicellular eukaryote (Chekanova *et al.*, 2007). This analysis produced evidence for widespread polyadenylation-mediated and exosome-mediated quality control of plant RNA, as well as revealed the major categories of the plant exosome targets, including multiple classes of stable structural RNAs, a select subset of mRNAs, primary microRNA (pri-miRNA) processing intermediates, tandem repeat-associated siRNA precursor species, and numerous noncoding RNAs, many of which can only be revealed through repressing the exosome activity, because they remain below the level of detection in WT situation. This discovery of hundreds of noncoding RNAs that have not been previously described and hence belong to a "deeply hidden" layer of the transcriptome suggests that modulating the exosome activity, combined with further improvements in the sensitivity of detection beyond the one afforded by tiling microarrays, may help peel another layer off the transcriptome. This can be accomplished with deep sequencing technologies, such as Illumina 1G, ABI SOLiD, and/or 454 Life Sciences platforms.

ACKNOWLEDGMENTS

Research support from the grants from NIH, USDA, NSF, and BARD is gratefully acknowledged.

REFERENCES

Allmang, C., Kufel, J., Chanfreau, G., Mitchell, P., Petfalski, E., and Tollervey, D. (1999). Functions of the exosome in rRNA, snoRNA and snRNA synthesis. *EMBO J.* **18,** 5399–5410.

Allmang, C., Mitchell, P., Petfalski, E., and Tollervey, D. (2000). Degradation of ribosomal RNA precursors by the exosome. *Nucleic Acids Res.* **28,** 1684–1691.

Aloy, P., Ciccarelli, F. D., Leutwein, C., Gavin, A. C., Superti-Furga, G., Bork, P., Bottcher, B., and Russell, R. B. (2002). A complex prediction: Three-dimensional model of the yeast exosome. *EMBO Rep.* **3,** 628–635.

Amirsadeghi, S., McDonald, A. E., and Vanlerberghe, G. C. (2007). A glucocorticoid-inducible gene expression system can cause growth defects in tobacco. *Planta* **226,** 453–463.

Andersen, S. U., Cvitanich, C., Hougaard, B. K., Roussis, A., Gronlund, M., Jensen, D. B., Frokjaer, L. A., and Jensen, E. O. (2003). The glucocorticoid-inducible GVG system causes severe growth defects in both root and shoot of the model legume *Lotus japonicus*. *Mol. Plant Microbe Interact.* **16,** 1069–1076.

Bousquet-Antonelli, C., Presutti, C., and Tollervey, D. (2000). Identification of a regulated pathway for nuclear pre-mRNA turnover. *Cell* **102,** 765–775.

Brengues, M., Teixeira, D., and Parker, R. (2005). Movement of eukaryotic mRNAs between polysomes and cytoplasmic processing bodies. *Science* **310,** 486–489.

Brown, J. T., and Johnson, A. W. (2001). A cis-acting element known to block 3′-mRNA degradation enhances expression of polyA-minus mRNA in wild-type yeast cells and phenocopies a ski mutant. *RNA* **7,** 1566–1577.

Chekanova, J. A., Gregory, B. D., Reverdatto, S. V., Chen, H., Kumar, R., Hooker, T., Yazaki, J., Li, P., Skiba, N., Peng, Q., Alonso, J., Brukhin, V., *et al.* (2007). Genome-wide high-resolution mapping of exosome substrates reveals hidden features in the *Arabidopsis* transcriptome. *Cell* **131,** 1340–1353.

Chekanova, J. A., Shaw, R. J., Wills, M. A., and Belostotsky, D. A. (2000). Poly(A) tail-dependent exonuclease AtRrp41p from Arabidopsis thaliana rescues 5.8 S rRNA processing and mRNA decay defects of the yeast ski6 mutant and is found in an exosome-sized complex in plant and yeast cells. *J. Biol. Chem.* **275,** 33158–33166.

Chen, C. Y., Gherzi, R., Ong, S. E., Chan, E. L., Raijmakers, R., Pruijn, G. J., Stoecklin, G., Moroni, C., Mann, M., and Karin, M. (2001). AU binding proteins recruit the exosome to degrade ARE-containing mRNAs. *Cell* **107,** 451–464.

Chen, S., Hofius, D., Sonnewald, U., and Bornke, F. (2003). Temporal and spatial control of gene silencing in transgenic plants by inducible expression of double-stranded RNA. *Plant J.* **36,** 731–740.

Clarke, L., and Carbon, J. (1976). A colony bank containing synthetic Col El hybrid plasmids representative of the entire *E. coli* genome. *Cell* **9,** 91–99.

Das, B., Butler, J. S., and Sherman, F. (2003). Degradation of normal mRNA in the nucleus of Saccharomyces cerevisiae. *Mol. Cell. Biol.* **23,** 5502–5515.

Dziembowski, A., Lorentzen, E., Conti, E., and Seraphin, B. (2007). A single subunit, Dis3, is essentially responsible for yeast exosome core activity. *Nat. Struct. Mol. Biol.* **14,** 15–22.

Gherzi, R., Lee, K. Y., Briata, P., Wegmuller, D., Moroni, C., Karin, M., and Chen, C. Y. (2004). A KH domain RNA binding protein, KSRP, promotes ARE-directed mRNA turnover by recruiting the degradation machinery. *Mol. Cell* **14,** 571–583.

Graham, A. C., Kiss, D. L., and Andrulis, E. D. (2006). Differential distribution of exosome subunits at the nuclear lamina and in cytoplasmic foci. *Mol. Biol. Cell* **17,** 1399–1409.

Guo, X., Ma, J., Sun, J., and Gao, G. (2007). The zinc-finger antiviral protein recruits the RNA processing exosome to degrade the target mRNA. *Proc. Natl. Acad. Sci. USA* **104,** 151–156.

Houseley, J., Kotovic, K., El Hage, A., and Tollervey, D. (2007). Trf4 targets ncRNAs from telomeric and rDNA spacer regions and functions in rDNA copy number control. *EMBO J.* **26,** 4996–5006.

Houseley, J., LaCava, J., and Tollervey, D. (2006). RNA-quality control by the exosome. *Nat. Rev. Mol. Cell. Biol.* **7,** 529–539.

Houseley, J., and Tollervey, D. (2006). Yeast Trf5p is a nuclear poly(A) polymerase. *EMBO Rep.* **7,** 205–211.

Irizarry, R. A., Hobbs, B., Collin, F., Beazer-Barclay, Y. D., Antonellis, K. J., Scherf, U., and Speed, T. P. (2003). Exploration, normalization, and summaries of high density oligonucleotide array probe level data. *Biostatistics* **4**, 249–264.

Jeong, D. H., An, S., Kang, H. G., Moon, S., Han, J. J., Park, S., Lee, H. S., An, K., and An, G. (2002). T-DNA insertional mutagenesis for activation tagging in rice. *Plant Physiol.* **130**, 1636–1644.

Ji, H., and Wong, W. H. (2005). TileMap: Create chromosomal map of tiling array hybridizations. *Bioinformatics* **21**, 3629–3636.

Kadaba, S., Krueger, A., Trice, T., Krecic, A. M., Hinnebusch, A. G., and Anderson, J. (2004). Nuclear surveillance and degradation of hypomodified initiator tRNAMet in *S. cerevisiae*. *Genes Dev.* **18**, 1227–1240.

Kadaba, S., Wang, X., and Anderson, J. T. (2006). Nuclear RNA surveillance in Saccharomyces cerevisiae: Trf4p-dependent polyadenylation of nascent hypomethylated tRNA and an aberrant form of 5S rRNA. *RNA* **12**, 508–521.

Kang, H. G., Fang, Y., and Singh, K. B. (1999). A glucocorticoid-inducible transcription system causes severe growth defects in *Arabidopsis* and induces defense-related genes. *Plant J.* **20**, 127–133.

LaCava, J., Houseley, J., Saveanu, C., Petfalski, E., Thompson, E., Jacquier, A., and Tollervey, D. (2005). RNA degradation by the exosome is promoted by a nuclear polyadenylation complex. *Cell* **121**, 713–724.

Li, X., Song, Y., Century, K., Straight, S., Ronald, P., Dong, X., Lassner, M., and Zhang, Y. (2001). A fast neutron deletion mutagenesis-based reverse genetics system for plants. *Plant J.* **27**, 235–242.

Libri, D., Dower, K., Boulay, J., Thomsen, R., Rosbash, M., and Jensen, T. H. (2002). Interactions between mRNA export commitment, 3′-end quality control, and nuclear degradation. *Mol. Cell. Biol.* **22**, 8254–8266.

Liu, Q., Greimann, J. C., and Lima, C. D. (2006). Reconstitution, activities, and structure of the eukaryotic RNA exosome. *Cell* **127**, 1223–1237.

Lorentzen, E., Dziembowski, A., Lindner, D., Seraphin, B., and Conti, E. (2007). channelling by the archaeal exosome. *EMBO Rep* **8**, 470–476.

Lorentzen, E., Walter, P., Fribourg, S., Evguenieva-Hackenberg, E., Klug, G., and Conti, E. (2005). The archaeal exosome core is a hexameric ring structure with three catalytic subunits. *Nat. Struct. Mol. Biol.* **12**, 575–581.

McCallum, C. M., Comai, L., Greene, E. A., and Henikoff, S. (2000). Targeting induced local lesions IN genomes (TILLING) for plant functional genomics. *Plant Physiol.* **123**, 439–442.

Mitchell, P., Petfalski, E., Houalla, R., Podtelejnikov, A., Mann, M., and Tollervey, D. (2003). Rrp47p is an exosome-associated protein required for the 3′-processing of stable RNAs. *Mol. Cell. Biol.* **23**, 6982–6992.

Mitchell, P., Petfalski, E., Shevchenko, A., Mann, M., and Tollervey, D. (1997). The exosome: A conserved eukaryotic RNA processing complex containing multiple 3′ to 5′-exoribonucleases. *Cell* **91**, 457–466.

Moore, I., Samalova, M., and Kurup, S. (2006). Transactivated and chemically inducible gene expression in plants. *Plant J.* **45**, 651–683.

Mukherjee, D., Gao, M., O'Connor, J. P., Raijmakers, R., Pruijn, G., Lutz, C. S., and Wilusz, J. (2002). The mammalian exosome mediates the efficient degradation of mRNAs that contain AU-rich elements. *EMBO J.* **21**, 165–174.

Ohashi, Y., Oka, A., Rodrigues-Pousada, R., Possenti, M., Ruberti, I., Morelli, G., and Aoyama, T. (2003). Modulation of phospholipid signaling by GLABRA2 in root-hair pattern formation. *Science* **300**, 1427–1430.

Orban, T. I., and Izaurralde, E. (2005). Decay of mRNAs targeted by RISC requires XRN1, the Ski complex, and the exosome. *RNA* **11**, 459–469.

Ossowski, S., Schwab, R., and Weigel, D. (2008). Gene silencing in plants using artificial microRNAs and other small RNAs. *Plant J.* **53,** 674–690.

Peng, W. T., Robinson, M. D., Mnaimneh, S., Krogan, N. J., Cagney, G., Morris, Q., Davierwala, A. P., Grigull, J., Yang, X., Zhang, W., et al. (2003). A panoramic view of yeast noncoding RNA processing. *Cell* **113,** 919–933.

Ren, S., Johnston, J. S., Shippen, D. E., and McKnight, T. D. (2004). Telomerase activator1 induces telomerase activity and potentiates responses to auxin in *Arabidopsis*. *Plant Cell* **16,** 2910–2922.

Schwab, R., Ossowski, S., Riester, M., Warthmann, N., and Weigel, D. (2006). Highly specific gene silencing by artificial microRNAs in *Arabidopsis*. *Plant Cell* **18,** 1121–1133.

Sheth, U., and Parker, R. (2003). Decapping and decay of messenger RNA occur in cytoplasmic processing bodies. *Science* **300,** 805–808.

Slade, A. J., Fuerstenberg, S. I., Loeffler, D., Steine, M. N., and Facciotti, D. (2005). A reverse genetic, nontransgenic approach to wheat crop improvement by TILLING. *Nat. Biotechnol.* **23,** 75–81.

Smith, N. A., Singh, S. P., Wang, M. B., Stoutjesdijk, P. A., Green, A. G., and Waterhouse, P. M. (2000). Total silencing by intron-spliced hairpin RNAs. *Nature* **407,** 319–320.

Stoutjesdijk, P. A., Singh, S. P., Liu, Q., Hurlstone, C. J., Waterhouse, P. A., and Green, A. G. (2002). hpRNA-mediated targeting of the *Arabidopsis* FAD2 gene gives highly efficient and stable silencing. *Plant Physiol.* **129,** 1723–1731.

Till, B. J., Cooper, J., Tai, T. H., Colowit, P., Greene, E. A., Henikoff, S., and Comai, L. (2007). Discovery of chemically induced mutations in rice by TILLING. *BMC Plant Biol.* **7,** 19.

Till, B. J., Reynolds, S. H., Weil, C., Springer, N., Burtner, C., Young, K., Bowers, E., Codomo, C. A., Enns, L. C., Odden, A. R., et al. (2004). Discovery of induced point mutations in maize genes by TILLING. *BMC Plant Biol.* **4,** 12.

Torchet, C., Bousquet-Antonelli, C., Milligan, L., Thompson, E., Kufel, J., and Tollervey, D. (2002). Processing of 3′-extended read-through transcripts by the exosome can generate functional mRNAs. *Mol. Cell* **9,** 1285–1296.

Vanacova, S., Wolf, J., Martin, G., Blank, D., Dettwiler, S., Friedlein, A., Langen, H., Keith, G., and Keller, W. (2005). A new yeast poly(A) polymerase complex involved in RNA quality control. *PLoS Biol.* **3,** e189.

Wahle, E. (2007). Wrong pH for RNA degradation. *Nat. Struct. Mol. Biol.* **14,** 5–7.

Wang, L., Lewis, M. S., and Johnson, A. W. (2005). Domain interactions within the Ski2/3/8 complex and between the Ski complex and Ski7p. *RNA* **11,** 1291–1302.

Wang, Z., and Kiledjian, M. (2001). Functional link between the mammalian exosome and mRNA decapping. *Cell* **107,** 751–762.

Warthmann, N., Chen, H., Ossowski, S., Weigel, D., and Herve, P. (2008). Highly specific gene silencing by artificial miRNAs in rice. *PLoS ONE* **3,** e1829.

Wyers, F., Rougemaille, M., Badis, G., Rousselle, J. C., Dufour, M. E., Boulay, J, Regnault, B., Devaux, F., Namane, A., Seraphin, B., et al. (2005). Cryptic pol II transcripts are degraded by a nuclear quality control pathway involving a new poly(A) polymerase. *Cell* **121,** 725–737.

Zuo, J., Niu, Q. W., and Chua, N. H. (2000). Technical advance: An estrogen receptor-based transactivator XVE mediates highly inducible gene expression in transgenic plants. *Plant J.* **24,** 265–273.

CHAPTER TWENTY-TWO

Sensitive Detection of mRNA Decay Products by Use of Reverse-Ligation–Mediated PCR (RL-PCR)

Thierry Grange

Contents

1. Introduction	446
2. Footprinting of RNA–Protein Interaction	450
2.1. Procedure	451
3. RL-PCR with Ligation of an RNA Linker	452
3.1. Synthesis of the RNA linker	454
3.2. Preliminary treatments of the 5′-ends of the total cellular RNA	455
3.3. RL-PCR	456
4. Circularization RL-PCR to Analyze mRNA Decay Involving Modification of the 5′- and 3′-Ends	458
4.1. Transcriptional pulsing with a tetracycline-regulated promoter and RNA preparation	460
4.2. Circularization RL-PCR	462
5. Concluding Remarks	465
Acknowledgments	465
References	465

Abstract

Ligation-mediated PCR allows the detection and mapping of cleavage products of specific nucleic acid molecules out of complex nucleic acid mixtures. It can be applied to the detection of either degradation products of exogenously added nucleases in footprinting applications or natural decay products or reaction intermediates. We have developed various ligation-mediated PCR approaches to analyze mRNAs, all relying on RNA ligation, followed by reverse-transcription and PCR amplification. We have termed these approaches reverse-ligation–mediated PCR (RL-PCR). The ligation event involves either an RNA linker added to the 5′-end of cleaved RNA or RNA circularization, allowing, respectively, the mapping and quantification of the cleavage points or the

Institut Jacques Monod du CNRS, Université Paris 7, Paris, France

Methods in Enzymology, Volume 448   © 2008 Elsevier Inc.
ISSN 0076-6879, DOI: 10.1016/S0076-6879(08)02622-0   All rights reserved.

simultaneous analysis of the presence or absence of the 5′-cap structure and the length of the poly(A) tail. These methods enabled us to develop a very efficient 5′-RACE procedure to map mRNA 5′-ends, to footprint in permeabilized cells the interaction of regulatory proteins with RNA, to detect the products of cellular ribozyme action and to analyze cellular decay pathways that involve deadenylation and/or decapping. I review herein the methodologic aspects and protocols of the various RL-PCR procedures we have developed.

1. INTRODUCTION

Ligation-mediated PCR (LM-PCR) was first developed to perform *in vivo* footprinting experiments to detect the interaction of proteins with DNA in cells (Mueller and Wold, 1989). The exponential amplification of molecules with an unknown and variable end is achieved through the ligation of a known linker of discrete length to these ends, allowing the amplification of a sequence ladder with high selectivity, specificity, and fidelity (Grange *et al.*, 1997). We have used LM-PCR for many years to analyze in mammalian cells aspects as diverse as the dynamics of transcription factor interactions with regulatory sequences (Espinás *et al.*, 1995; Rigaud *et al.*, 1991), the regulation of DNA methylation (Thomassin *et al.*, 2001), the precision of nucleosome positioning (Flavin *et al.*, 2004), as well as to detect intermediates in the DNA demethylation pathway (Kress *et al.*, 2006). The power of the approach prompted us to adapt it to the analysis of events involving RNA. We have developed reverse-ligation–mediated PCR (RL-PCR) to map and quantify unknown heterogeneous 5′-ends of a specific RNA sequence with sufficient selectivity and sensitivity to analyze low-abundance RNAs in mammalian cells (Bertrand *et al.*, 1993). As schematized in Fig. 22.1, left panel, an RNA linker of discrete length is ligated to the unknown 5′-ends, the RNA sequence of interest is selectively reverse-transcribed with an RNA-specific primer (primer 1), and the cDNAs are PCR-amplified with a second nested RNA-specific primer (primer 2) and a linker primer whose sequence corresponds to that of the RNA linker. Visualization of products is performed with a 5′-^{32}P–labeled primer that, depending on the abundance of the RNA sequence and the performance of the previous primers, can be either primer 2 or a third RNA-specific primer (primer 3 in Fig. 22.1).

We used RL-PCR to map various types of unknown 5′-ends. We first developed RL-PCR to map cellular RNase cleavage points in *in vivo* footprinting applications (Bertrand *et al.*, 1993). We applied the procedure to assess the RNase T1 footprint of iron-regulatory proteins (Rouault, [2006] for a review) with two target sequences in the 3′ UTR of the human transferrin receptor mRNA (Fig. 22.2A). We also applied the

Reverse-Ligation–Mediated PCR

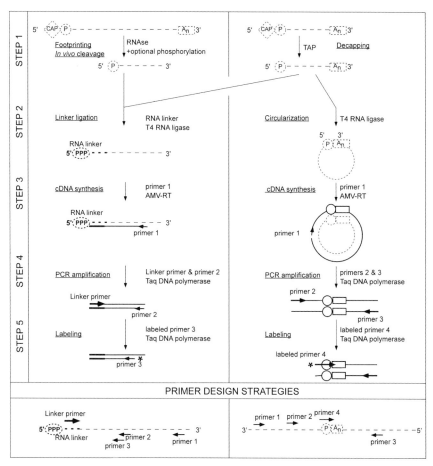

Figure 22.1 Comparative analysis of the flow scheme of the various RL-PCR procedures involving ligation of the 5′-P ends either to an RNA linker (left panel) or to the 3′-OH ends in an intramolecular manner (circularization RL-PCR, right panel). RNA molecules are represented as dotted lines; DNA molecules by continuous lines. The arrows indicate the 3′-ends of the primers. The cap and poly(A) tail (An) are indicated by a sideways square and a rectangle, respectively. The circle indicates the 5′-phosphate group of the most 5′-base involved in the ligation event. For a number of the events we have analyzed, it was possible to perform the RL-PCR procedure with three instead of four RNA-specific primers. In those cases, for the linker-dependent procedure, the amplification and labeling steps were performed simultaneously with 5′-^{32}P–labeled primer 2, whereas for the circularization procedure, the reverse-transcription primer 1 was also used instead of primer 2 during PCR amplification. Lower panel, Arrangement of the primers used. The junction point of the circularized RNA is shown in the right panel. To preserve the relative arrangement of the features with the upper right panel, the RNA molecule represented on the right panel is in the 3′ to 5′-orientation. To ensure that the analysis is detecting decay product and not uncapped mRNA precursors (i.e., pre-mRNA), primer 1 can overlap an exon junction.

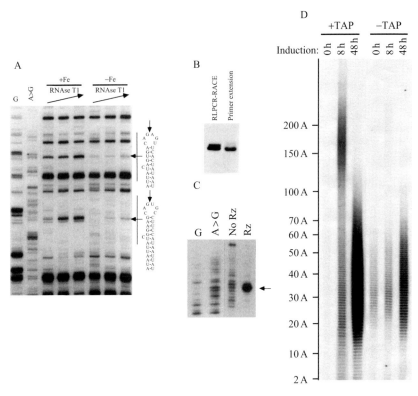

Figure 22.2 Examples of results obtained with the various RL-PCR strategies we have developed. (A) RNase T1 footprinting of the iron-dependent interaction of iron regulatory proteins with two neighboring target sites in the 3′ UTR of the human transferrin receptor mRNA (Bertrand et al., 1993). Human hepatoma cells were treated with either hemin (lanes +Fe) or desferrioxamine (lanes −Fe) before treatment of permeabilized cells with varying concentrations of RNase T1. Lanes G and A > G, Total cytoplasmic RNA was chemically cleaved at base-specific locations (Peattie, 1979) before RL-PCR analysis. Bars to the right indicate the positions of the IRE stem-loop structures (represented as nucleotide sequences). Arrows specify bands corresponding to the phosphodiester bonds whose susceptibility to cleavage by RNase T1 is altered on changing the iron level. (B) Comparative analysis of RL-PCR RACE and primer extension to analyze the 5′-end of transiently expressed CAT mRNA (Fromont-Racine et al., 1993). The quantity of material loaded for RL-PCR RACE corresponded to 3 ng of total cytoplasmic RNA and was visualized with autoradiography (6-h exposure). The primer extension analysis was performed directly with labeled primer 1. The quantity of material loaded corresponded to 5 μg of total cytoplasmic RNA and was visualized with autoradiography (5-day exposure). (C) Detection of hammerhead ribozyme-cleavage products after transient expression of either the substrate alone (No Rz) or the substrate and the ribozyme (Rz) (Bertrand et al., 1994). The arrow indicates the cleaved bond. (D) Variation in the poly(A) tail length of serum amyloid A mRNA at various times after induction of the acute phase response in mouse liver (Couttet et al., 1997). Capped mRNA species were analyzed by RL-PCR after TAP treatment (+TAP), whereas uncapped mRNA species were analyzed directly (−TAP). Samples were PCR amplified for 20 (+TAP) or 25 cycles (−TAP). The length of the poly(A) tail is indicated.

procedure to detect ribozyme cleavage products (Bertrand et al., 1994; 1997), Fig. 22.2C). Simultaneously, we adapted the procedure to map the natural 5′-ends of the mRNA by first decapping the mRNAs with tobacco acid pyrophosphatase (TAP) (Fromont-Racine et al., 1993) (Fig. 22.2B). This was the first description of a procedure that later turned out to be the most powerful 5′-RACE (rapid amplification of cDNA ends) method (Scotto-Lavino et al., 2006). While setting up this adaptation, we detected trace amounts of capless mRNAs with 5′-ends that were identical to those of their capped counterparts. Since it was known at the time that the major mRNA decay pathway in yeast involves a deadenylation-dependent decapping pathway (Parker and Song, 2004 for a review), we adapted the RL-PCR procedure to simultaneously analyze the 5′- and 3′-ends of the same mRNA molecule (schematized in the right panel of Fig. 22.1). In this adaptation, the RNA is circularized by ligation of the decapped 5′-end to the 3′-end, reverse transcribed across the ligated termini and the poly(A) tail, and the surrounding sequences are PCR amplified (Couttet et al., 1997). This allows measurement of the poly(A) tail length of the decapped mRNA, and thus comparison of the length of this tail in naturally decapped species and in capped species that were decapped *in vitro* with TAP (Fig. 22.2D). Because the naturally decapped species are by far less abundant than the capped ones, additional PCR cycles are necessary to detect them. Analysis of the highly inducible serum-amyloid A mRNA in mouse liver showed that decay involved first deadenylation, followed by the production of a decapped intermediate with a relatively short poly(A) tail (20 to 60 As, Fig. 22.2D and Couttet et al., 1997). We also observed for several other mRNAs that the poly(A) tail was shorter for the decapped than for the capped species and that the distribution of the poly(A) tail length varied for each species. Therefore, we could conclude that decapping is preceded by a shortening of the poly(A) tail in mammalian cells, as it is in yeast, and that for each mRNA there are differences in the relative rates of deadenylation and the subsequent degradation steps (Couttet et al., 1997).

We then applied this circularization RL-PCR procedure to the analysis of nonsense-mediated mRNA decay (NMD) (Isken and Maquat, 2007 for a review). We used a transcriptional pulse strategy relying on a tetracycline-regulated promoter to study the decay of premature termination codon (PTC)-containing β-globin mRNA in human cells (Couttet and Grange, 2004). This allowed us to show that a PTC enhances the rate of mRNA decapping. We also observed that the PTC in the β-globin mRNA increases the deadenylation rate and that decapping does not occur before deadenylation has proceeded to a certain extent. Thus, for this mRNA, NMD involves an increase in both the rate of deadenylation and decapping, while keeping a certain dependence of decapping on deadenylation (Couttet and Grange, 2004). Because the level of PTC-containing mRNA was always lower than that of the wild-type counterpart, even at the earliest induction time point, NMD must act very early during mRNA metabolism. Indeed, numerous studies indicate that protein-encoding RNAs are scrutinized at every stage of their biogenesis and function (reviewed in Isken and Maquat, 2007).

Here, I describe the RL-PCR methods we have used to analyze mRNA–protein interactions and mRNA decay products.

2. Footprinting of RNA–Protein Interaction

The success of footprinting with intact and permeabilized cells depends on the ability to use an RNA modifying agent that reacts differently with the RNA as a function of their interaction with regulatory proteins without disrupting the RNA–protein complex during the course of the analysis. Furthermore, because the RL-PCR procedure allows the mapping of $5'$-P ends, it is also necessary to use footprinting agents that either directly produce such ends or modify the RNA in such a way that $5'$-P ends can be generated at the positions that have reacted. Optimal sensitivity is obtained when the reactivity of the RNA toward the footprinting agent is extensively altered upon protein binding. Because many RNA footprinting reagents are nucleases (Ehresmann et al., 1987), it is necessary to permeabilize cells so the nuclease has access to the RNP complex. Permeabilization could destabilize the complex or release endogenous nucleases. Furthermore, it is necessary to use nuclease quantities that do not cause extensive degradation of the RNA molecule, because RNA cleavage is expected to modify RNP stability; ideally, single-hit conditions should be used. This requires careful titration of nuclease quantities. The exogenous nuclease should, however, be responsible for the largest proportion of cleavage to allow data interpretation, which may be difficult to achieve if endogenous nucleases are active and/or activated during the treatment. As a rule of thumb, it is advisable to incubate the permeabilized cells in the cold for a short time with a relatively high quantity of the footprinting nuclease, because this increases the probability that cleavage will be largely due to the added nuclease.

When analyzing the interaction of iron-regulatory proteins with the $3'$ UTR of transferrin receptor mRNA, we have obtained interpretable footprints essentially only with RNase T1. RNase T1 cleaves GpX phosphodiester bonds in single-stranded RNA (Donis-Keller et al., 1977). We used this enzyme to analyze both secondary structures and protein interactions in a region with two neighboring target sequences characterized by a stem-loop structure (Fig. 22.2A). In the presence of iron, the *in vivo* RNase T1 footprinting pattern does not reveal all of the expected guanosines. Strikingly, those that are involved in double-stranded regions in the proposed secondary structure are not cleaved, suggesting that such a secondary structure is adopted *in vivo*. When the iron level is decreased under conditions favoring the binding of iron-regulatory proteins, a bond located in the middle of the loop of each iron-responsive element is protected, revealing protein interaction. Careful analysis of the images obtained indicates that the intensity of the corresponding bands varies as a function of the quantity of nuclease at the two iron levels. This suggests that if the extent of RNase T1 cleavage is high, the interaction of the proteins with RNA could be destabilized, or that if the complex is in rapid equilibrium, the dissociated state could contribute to the pattern. In this experiment, the ability to observe the mRNA in cells under two different conditions that affect the RNA–protein interaction was instrumental to the ability to interpret the footprints. In general, when *in vivo* footprinting is used, the quality of the interpretation relies on the combined use of various reagents and an extensive preliminary *in vitro* analysis (Grange *et al.*, 1997). This necessarily restricts this approach to the analysis of well-characterized systems. However, given the mechanistic insights that have been gained from the use of *in vivo* footprinting to transcription regulation (Grange *et al.*, 2001), this approach can be rewarding.

2.1. Procedure

Cells from a 100-mm plate (10^7 cells) are rinsed with 10 ml of PBS and overlaid with 1 ml of a trypsin/EDTA solution. The excess liquid is removed before cells begin to detach, and the plate is incubated for a few minutes at room temperature until cells detach. Collect cells in 10 ml of serum-containing medium (the serum serves to stop trypsin action), centrifuge gently, rinse first with 10 ml of PBS then with 1 ml of physiologic buffer (11 mM KH$_2$PO$_4$/K$_2$HPO$_4$, pH 7.4, 108 mM KCl, 22 mM NaCl, 1 mM MgCl$_2$, 1 mM DTT, 1 mM ATP).

Resuspend cells in 600 μl of physiologic buffer. Dilute 100 μl of the cell suspension with 100 μl of ice-cold physiologic buffer supplemented with 0.2% NP40 and varying amounts of RNase T1 (a range going from 25 to 250 units is a useful start). Incubate 3 min on ice and pellet the nuclei by a 15-sec centrifugation at 5500g.

Transfer the supernatant to a tube containing 200 μl of stop buffer (10 mM Tris-HCl, pH 7.5, 150 mM NaCl, 5 mM EDTA, 1% SDS) and 400 μl of water-saturated phenol/isoamyl alcohol (1/30). Immediately mix the contents of the tube thoroughly. Extract the aqueous phase two more times with phenol/isoamyl alcohol and then with chloroform. Precipitate the RNA by adding 20 μl of 3 M sodium acetate, pH 5.5, and 1 ml of 100% ethanol. Incubate on ice for 30 min, pellet the RNA by centrifugation, rinse with 70% ethanol and air-dry briefly.

Resuspend the RNA in 100 μl of water. Measure the concentration of the solution with a microspectrophotometer (Nanodrop). Precipitate the

RNA again with ethanol and resuspend it in water at an approximate concentration of 1 mg/ml.

Comments: The method is described for RNase T1 treatment of a human hepatoma cell line: Hep G2. To use other RNases and other cell lines, it is necessary to adjust empirically the RNase amounts. It is useful to verify the extent of cleavage of rRNA on an agarose gel or with a Bioanalyzer (Agilent). If some nuclei were lysed, DNA contamination can occur that will adversely affect the procedure by causing some background during RL-PCR analysis. In that case, the RNAs can be treated with DNase I as follows: To the RNA in water, add the adequate amount of $10\times$ DNase buffer (500 mM Tris-HCl, pH 7.5, 100 mM MgCl$_2$), 1 unit of RNase-free DNase I, and 20 U of human placental ribonuclease inhibitor. Incubate 30 min at 37 °C. Extract with an equal volume of water-saturated phenol/isoamyl alcohol and then with chloroform. Precipitate the RNA with ethanol.

3. RL-PCR WITH LIGATION OF AN RNA LINKER

RL-PCR is a multistep enzymatic procedure involving various primers, the design of which is critical to ensure results of good quality (see primer arrangement in the lower left panel of Fig. 22.1). Successful RL-PCR relies on the specificity and fidelity of the various enzymatic reactions, because trustworthy results require the faithful reproduction of the original distribution of the 5′-ends of the various molecules corresponding to a given RNA species diluted in a complex RNA mixture. The overall specificity of the procedures depends on the specificity of each step. Because in RL-PCR, the linker is ligated unspecifically to all 5′-P ends available in the mixture, specificity is achieved by the use of a series of nested primers corresponding to the 3′-side of the RNA of interest. One key parameter for specific amplification of the desired products is that primer 1 anneal with site specificity during the reverse transcription step ensuring that only the linker primer attached to the desired RNA ends is reverse transcribed (Fig. 22.1). Unspecifically, reverse-transcribed molecules will likely be amplified linearly during the subsequent PCR, causing background PCR products and reducing the amount of linker primer available for exponential priming. Indeed, primer 1 concentration and hybridization temperature during reverse transcription are critical to minimize background. Provided this reverse transcription step is specific, the specificity of the PCR amplification step is not as difficult to achieve as is the case with LM-PCR. Indeed, because only DNA molecules participate in the PCR, there is in RL-PCR a much lower amount and complexity of material that could be responsible for background at that step, because the only DNA molecules present are the reverse-transcribed molecules. In contrast, with

LM-PCR, the whole genome can interfere with the PCR reaction. Thus, in most cases, it is not even necessary to use three nested primers, as are required for LM-PCR (Grange et al., 1997), and the amplification and labeling steps can be performed simultaneously with 5′-^{32}P labeled primer 2. For primer design, we use the assistance of a computer program (Oligo^{TM} or Primer 3) and follow a few rules to maximize the chances of success: (1) Primers are nested so that they cover the longest sequence possible to maximize the specificity. i.e., primer 1 hybridizes upstream of primer 2 so there is no overlap and when a 3-primer design is used, primer 2 and primer 3 overlap by 15 bases, in order to minimize interference between primers (Mueller and Wold, 1989). (2) The Tm of the primers should increase in the order in which they are used (i.e., $1 < 2 < 3$). We use the nearest neighbor method to estimate the Tms. These Tms are chosen to be compatible with the optimal temperature of the enzyme used for primer extension (we usually perform the hybridization at the Tm minus 5 °C). (3) The primers are chosen so that their 3′-ends do not tend to form a stable intermolecular structure so as to prevent the formation of hairpins or primer dimers, among themselves, with the linker primer, or with nonspecific RNA sequences (i.e., the ΔG of the 3′-end is as low as possible).

The final amplified material must reflect accurately the initial distribution of the various 5′-ends of the sequence of interest. This requires that the efficiency of the reaction at each step of the procedure is equivalent for all ends. In principle, the efficiency of the ligation of the linker may vary, depending on the nucleotides at the end, in particular the linker ligation efficiency could be affected by the secondary structure that the end can adopt (Romaniuk and Uhlenbeck, 1983). However, we have observed neither gaps in the sequencing ladders that drastic variations of efficiency should produce nor modifications of the appearance of the ladders on addition of components that are expected to minimize the inhibitory effect of secondary structure, like DMSO or protein 32 of T4 phage. To test for both specificity and fidelity, we recommend setting up the experimental conditions with each primer set with the total cellular RNA population chemically cleaved at base-specific locations with a purine and a pyrimidine specific reaction (Peattie, 1979). The presence and nature of all bases of the region of interest should be observed.

The PCR provides very high sensitivity to the RL-PCR procedure, and the lowest amount of material that can be used is set only by statistical considerations. If this amount is too low, the number of starting molecules that are responsible for a band on the final ladder might not be sufficient to allow reproducible intensity of the band because of statistical fluctuations. When analyzing the transferrin receptor mRNA with RL-PCR, the minimal amount of material that we have amplified was 350 ng of total cytoplasmic RNA. This corresponds approximately to 3×10^5 molecules of a low abundance mRNA (30 copies per cell). Such a quantity, however, cannot be used as an absolute rule, because the critical minimal quantity that can be used depends on the relative abundance of the analyzed RNA and the extent of cleavage by the footprinting agent. Differences in the quantity of starting material might not result in differences in the intensity of the ladders, because these differences may be hidden by the so-called plateau effect that is encountered at the end of the PCR. Under the conditions that we currently use for optimal and reproducible visualization of the footprints, the reactions are at, or near, the plateau of saturation. As a consequence, bands of similar intensities can be obtained with samples that contain different initial quantities of the molecules of interest. To obtain results that reflect more accurately the initial relative amount of molecules in various samples, one should use PCR conditions that allow incorporation of only a small proportion (probably not more than 5%) of the labeled primer. One should be aware that a decrease in the number of cycles would be at the expense of sensitivity and that conditions suitable for both footprint detection and quantitative analysis require manipulation of high quantities of radioactivity.

3.1. Synthesis of the RNA linker

The RNA linker can be either chemically synthesized by a company as 5′-unphosphorylated RNA or synthesized with *in vitro* transcription, generating a 5′-triphosphate RNA. Both types of 5′-ends are suitable for the RNA linker used in RL-PCR, because neither end allows the concatenation of several linker molecules during the ligation step. I describe here the *in vitro* transcription synthesis that uses two synthetic oligonucleotides that, after annealing, form a hemiduplex allowing synthesis of the RNA linker by *in vitro* transcription with T7 RNA polymerase (Milligan et al., 1987). As long as RNA molecules are involved, all reagents are prepared and all reactions are performed with distilled water that has been treated with DEPC and autoclaved. When preparing the RNA linker for the first time, it is best to test the synthesis conditions on a small scale before the use of the large-scale reaction conditions described here.

3.1.1. Procedure

Mix in a reaction tube: 4 μg of the bottom oligo (TTTCAGCGAGGGT-CAGCCTATGCCCTATAGTGAGTCGTATTA), 1.8 μg of the top oligo (TAATACGACTCACTATAG), 4 μl of 10× annealing buffer (100 mM Tris-HCl, pH 8.3, 50 mM MgCl$_2$), and H$_2$O to a volume of 40 μl. Heat the tube at 95 °C for 1 min, and then incubate at 42 °C for 15 min. Place the tube at room temperature, and add in this order: 160 μl of 10 mM rNTPs, 20 μl of 0.1 M DTT, 90 μl of dH$_2$O, 40 μl of 10× transcription buffer (400 mM Tris-HCl, pH 8.3, 10 mM spermidine, 0.1% Triton, 80 mM MgCl$_2$), 10 μl of RNase inhibitor (200 U), and 40 μl of T7 RNA polymerase (1000 U). Let the reaction proceed for 3 to 4 h at 42 °C.

While the reaction proceeds, prepare a 12% polyacrylamide 8 M urea gel (with DEPC-treated H$_2$O). The gel is 3-mm thick and 20-cm long with a 5-cm wide well.

Precipitate the nucleic acids by adding 4 µl of 3 M Na acetate, pH 5.5, and 100 µl of 100% ethanol. Spin for 10 min in a microfuge and wash the pellet extensively with 70% ethanol. Dry briefly, and resuspend the pellet in 100 µl of 50% formamide in 1× TBE.

Heat the sample at 95 °C for 1 min, put immediately on ice, and load the sample onto the preparative gel. In a neighboring well, load a molecular weight marker (for example, a DNA oligonucleotide of 20 to 30 bases) diluted in sequencing gel-loading buffer (with xylene cyanol and bromophenol blue). Run the gel at 50 °C (250 V for a 15 × 20-cm gel) until the bromophenol blue reaches the bottom of the gel.

Visualize the bands by UV shadowing with shortwave UV light (254 nm) and a fluor-coated TLC plate; 2 to 4 consecutive bands should be seen with a size close to 25 bases. (Do not mistake the xylene cyanol and bromophenol blue as RNA.) This heterogeneity is due to the RNA polymerase, which sometimes adds 1 to 3 nucleotides at the end of the expected transcript (Milligan et al., 1987). The band corresponding to the shortest product of these consecutive bands is cut, and eluted overnight at 37 °C in 2 ml of elution buffer (10 mM Tris-HCl, pH 8, 100 mM NaCl, 1 mM EDTA, 0.1% SDS).

Purify the RNA linker by reverse-phase chromatography on Sep-Pack C18 column (Waters) as follows: Before use, wash the column with 4 ml of 100% methanol, and then with 4 ml of H_2O. Apply the eluted sample to the column. Wash the column twice with 3 ml of H_2O. Elute the RNA linker with 1.5 ml of 60% methanol.

Evaporate the solvent in a speedvac, and resuspend the RNA linker in 100 µl of H_2O. Measure the quantity of linker on a microspectrophotometer. In these conditions, it should be in the order of 10 to 30 µg. Adjust the linker concentration to 100 ng/µl, and store at −80 °C.

Comments: The synthesized RNA can precipitate sometimes in the course of RNA synthesis, because of its high concentration and the presence of spermidine. In this case, it is possible to recover the precipitated RNA as follows: spin down the precipitate in a microfuge, resuspend the pellet in 200 µl of 3 M ammonium sulfate, and dilute it 10-fold with H_2O before ethanol precipitation. A lower concentration of nucleotides and magnesium (2 mM and 6 mM, respectively) can also be used to circumvent the problem, but the yield will be lower.

3.2. Preliminary treatments of the 5′-ends of the total cellular RNA

In RL-PCR, the RNA linker can be ligated only to RNA molecules with a 5′-P end. Depending on the application, different preliminary treatments might be necessary to obtain such ends. In the 5′-RACE application, the cap at the 5′-end is removed by a TAP treatment, which removes the cap, leaving a 5′-P end. This treatment can be preceded by an alkaline phosphatase treatment to remove traces of degradation products with 5′-P ends that can be present in the RNA preparation (Fromont-Racine et al., 1993). In footprinting applications in which the footprinting reagent is an enzyme like RNase T1 that generates 5′-OH ends, these ends must be phosphorylated to permit linker ligation.

3.2.1. Procedure for 5′-RACE

3.2.1.1. Optional prior dephosphorylation of the uncapped ends Mix in a reaction tube: 10 µg RNA, 1 µl of 10× SAP buffer (0.2 M Tris-HCl, pH 8, 0.1 M MgCl$_2$), 20 U of RNase inhibitor, 1 U of shrimp alkaline phosphatase (SAP), and H$_2$O to a volume of 10 µl. Incubate for 1 h at 37 °C, and for 15 min at 65 °C (SAP inactivation step). Spin briefly.

3.2.1.2. Decapping of the capped RNAs Mix in a reaction tube: 10 µg RNA (or 10 µl of the previous dephosphorylation reaction), 5 µl of 10× TAP buffer (500 mM sodium acetate, pH 6, 10 mM EDTA, 10% β-mercaptoethanol, 1% Triton X-100), 20 U of RNase inhibitor, 2.5 units of tobacco acid pyrophosphatase (TAP, Epicentre), and H$_2$O to a volume of 50 µl. Incubate for 1 h at 37 °C. Extract the reaction with water-saturated phenol-chloroform (3/1), extract again with chloroform, and precipitate the RNA with 5 µl of 3 M sodium acetate, pH 5.5, and 125 µl of 100% ethanol. Spin for 10 min in a microfuge and wash the pellet with 70% ethanol. Dry briefly, and resuspend the pellet in 10 µl H$_2$O.

3.2.2. Procedure for footprinting

Phosphorylation of the 5′-ends. Mix in a reaction tube: 10 µg RNA, 1 µl of 10× kinase buffer (500 mM Tris-HCl, pH 7.6, 100 mM MgCl$_2$, 50 mM DTT, 1 mM spermine, 1 mM EDTA), 1 µl of 10 mM ATP, 20 U of RNase inhibitor, 10 U of T4 polynucleotide kinase, and H$_2$O to a volume of 10 µl. Incubate for 15 min at 37 °C. Stop the reaction by freezing.

3.3. RL-PCR

Most of the time, it is sufficient to use only two nested primers to obtain good results with RL-PCR, one for the reverse transcription step (primer 1), and one, labeled at the 5′-end with ^{32}P for the PCR amplification step (primer 2). The standard protocol is thus given with a single PCR step. We have used primer 1 having a length ranging generally from 16 to 18 nucleotides and a Tm (calculated with the nearest neighbor method as available in Oligo 4.0) ranging from 46 °C to 67 °C. Most of the time, we used a hybridization temperature of 42 °C, but for primers that had a Tm higher than 54 °C, we raised the temperature up to 56 °C, generally with a hybridization temperature 10 to 12 °C below that of the estimated Tm.

The quantity of primer 1 used is rather low to minimize background, thus the hybridization time is rather long. We have used primers 2 having a length ranging generally from 21 to 25 nucleotides and a Tm (calculated with the nearest neighbor method as available in oligo 4.0) ranging from 64 °C to 66 °C with a hybridization temperature of 59 °C. The sequence of the linker primer corresponds to that of the RNA linker (GGGCATAGGCTGACCCTCGCTGAAA).

3.3.1. Procedure

3.3.1.1. Ligation of the RNA linker In a reaction tube, mix 1 µg of 5′-P RNAs with 1 µl of RNA linker (0.1 µg), 1 µl of 10× ligase buffer (500 mM Tris-HCl, pH 7.5, 100 mM MgCl$_2$), 1 µl of 200 mM DTT, 1 µl of 1 mM ATP, 1 µl of RNase-free BSA (1 µg), 1 µl of RNase inhibitor (20 U), 1 µl of RNA ligase (20 U), and H$_2$O to a volume of 10 µl. Incubate the tube overnight at 17 °C. Add 40 µl of H$_2$O, extract the reaction with water-saturated phenol-chloroform (3/1), extract again with chloroform, and precipitate the RNA with 5 µl of 3 M sodium acetate, pH 5.5, and 125 µl of 100% ethanol. Spin for 10 min in a microfuge and wash the pellet with 70% ethanol. Dry briefly, and resuspend the pellet in 12 µl H$_2$O.

3.3.1.2. cDNA synthesis Place 6 µl of the ligated RNA from the previous step in a reaction tube that fits in a PCR apparatus. Incubate 1 min at 95 °C to denature the RNA. Put at 42 °C and add 1 µl of primer 1 (10 ng), 1 µl of 10× RT/Taq buffer (650 mM Tris-HCl, pH 8.8, 100 mM β-mercaptoethanol, 165 mM (NH$_4$)$_2$SO$_4$), and 1 µl of 60 mM MgCl$_2$. Mix and let the primer hybridize by incubating 30 min at 42 °C (in case of background, try to raise the temperature of this reaction gradually up to 50 °C; if unsuccessful, try another primer 1). Without taking the tube out of the PCR apparatus, add 1 µl of a mixture containing 5 mM of each dNTPs and 3 U of AMV reverse transcriptase. Mix and incubate 30 min at 42 °C. Stop the reaction by incubating 5 min at 95 °C. Place the tube on ice.

3.3.1.3. Primer labeling for the analysis of 10 samples (scale up if necessary) While the reverse transcription is proceeding, label primer 2 as follows: In a reaction tube, mix 1 µl of primer 2 (100 ng), 2.5 µl of [γ-^{32}P] ATP (50 µCi, 7 pmol), 2 µl of 10× kinase buffer (500 mM Tris-HCl, pH 7.6, 100 mM MgCl$_2$, 50 mM DTT, 1 mM spermine, 1 mM EDTA), 1 µl of T4 polynucleotide kinase (10 U), and H$_2$O to a volume of 20 µl. Incubate 30 min at 37 °C. Stop the reaction by incubating 5 min at 95 °C. The primer is used without further purification.

To the reverse transcription reaction, add 1 µl of DNA linker primer (1 µg), 2 µl of the labeling reaction (corresponding to 10 ng of primer 2), 1 µl of 10× RT/Taq buffer, 2 µl of DNase-free BSA (4 µg), and 3 µl of H$_2$O. Mix and overlay the sample with paraffin oil. Incubate 3 min at 95 °C.

While maintaining the tube at 95 °C, add 1 μl of a mixture containing 5 mM of each dNTPs and 1 U of Taq DNA polymerase. Mix and amplify for 25 cycles: 30 sec at 95 °C/3 min at Tm primer 2 to 5 °C/3 min at 74 °C. Finish by an extension step of 10 min at 74 °C.

Add 80 μl of 0.3 M sodium acetate, pH 5.5, containing tRNA at 0.1 μg/μl. Extract with 100 μl of Tris-buffered phenol/chloroform (3/1) and precipitate with 300 μl of 100% ethanol. Incubate on ice for 30 min. Spin down for 10 min, rinse the pellet with 70% ethanol, and dry. Resuspend the pellet in 8 μl of formamide-loading buffer (95% formamide, 10 mM EDTA, 0.05% bromophenol blue, 0.05% xylene cyanol), incubate 3 min at 95 °C, and load 2 μl on a 6% sequencing gel.

Comments: The PCR amplification step should be performed with a hot start protocol. If the PCR machine is with a heated lid, or if one wants to avoid the paraffin oil procedure described here, a hot start Taq DNA polymerase can be used. In that case, the dNTP and enzyme are present before the denaturation step. When there is undesired background, we first recommend trying to perform the reverse transcription under more stringent conditions by raising gradually the hybridization and reaction temperature (up to 50 °C). If this does not fix the problem, we recommend that another primer 1 be designed, or use a third nested primer (primer 3 in Fig. 22.1, left panel) during the PCR, (i.e., to perform a first PCR with unlabeled primers and a second PCR for the labeling step). In this case, 100 ng of unlabeled primer 2 is used during the amplification step. The labeling of primer 3 is performed as described for primer 2. The PCR products are labeled as follows: mix 5 μl of the amplification reaction with 5 μl of a mix that contains per reaction: 0.5 μl of the primer labeling reaction, 0.5 μl of 10× RT/Taq buffer, 0.5 μl of 30 mM MgCl$_2$, 0.5 μl of 5 mM dNTP, 0.5 μl of DNase-free BSA (1 μg), 0.5 to 1 U of Taq DNA polymerase, completed to 5 μl with H$_2$O. Amplify as follows: 2 min at 94 °C, then 40 sec at 94 °C/3 min at Td$_{primer}$ 3 to 5 °C/5 min at 76 °C for 5 cycles. Background can also be caused by DNA contaminating the initial RNA solution, and a DNase I treatment of the sample, as described in the comment of Section 2, can solve the problem in that case.

4. Circularization RL-PCR to Analyze mRNA Decay Involving Modification of the 5'- and 3'-Ends

Because some mRNA decay pathways involve cross-talk between the ends of RNA molecules, ligation of the ends of a molecule leading to its circularization allows the simultaneous analysis of the status of the two ends by reverse transcription and PCR amplification across the ligated termini.

Because of the sensitivity that can be achieved with PCR amplification, it is possible to detect minor quantities of molecules, and thus to detect decay intermediates, provided RNAs are prepared and handled in a way that does not produce higher quantities of artifactually degraded molecules. To infer that the detected products are decay intermediates, it is important to be able to use a pulse-chase strategy. In mammalian cells, it is best to analyze RNAs that are produced from inducible genes, because transcription inhibitors cause unwanted side effects altering the mRNA decay pathways (Loflin et al., 1999). We have analyzed decay products with PCR-based approaches with either naturally inducible endogenous genes (Couttet et al., 1997; Kress et al., 2006) or tetracycline-regulated transfected genes (Couttet and Grange, 2004). The use of the latter allows testing of the effect of mutations on the decay pathway. With a circularization RL-PCR strategy, the detection of specific decay products relies on the combination of enzymatic treatments and the primer design strategy (Fig. 22.1, right panel). The method requires that the cap structure at the 5'-end be absent or removed by TAP (step 1). By comparing results obtained with or without this TAP treatment, the status of the 3'-ends of the capped and uncapped RNA species can be compared. In the second step, the 5'- and 3'-ends of the RNA are intramolecularly ligated with T4 RNA ligase at a low RNA concentration. The circularized RNA of interest is specifically reverse transcribed with primer 1 (step 3, Fig. 22.1, right panel). This allows for the synthesis of a copy of the poly(A) tail that is ligated to the 5'-end. This poly(A) tail and surrounding sequences are amplified with Taq DNA polymerase and 2 primers that hybridize on both sides of the junction region (step 4). The pool of amplified PCR products is visualized on a denaturing polyacrylamide gel after extension of a 5'-end labeled nested primer (step 5). The size distribution of the poly(A) tail is deduced by comparison to a size standard and subtraction of the length of the sequence surrounding the poly(A) tail. The contribution of the poly(A) tail to the length of the PCR product can be verified with a combined oligo(dT)-RNase H treatment that reduces the length of the tail between 0 to 11 A's under the conditions described here.

The primer design ensures that only the mRNAs with the structure under investigation are analyzed (Fig. 22.1, lower right panel). To analyze selectively spliced mature mRNAs, reverse transcription can be performed with a primer 1 overlapping the exon 1–exon 2 boundary (Couttet and Grange, 2004). Selective PCR amplification of the cDNA originating from the circularized mRNA can be performed with primers 2 and 3 hybridizing close to the 3'- and 5'-ends, respectively, of the mRNA. Finally, only products issued from the full-length decapped species can be visualized during the labeling step with a primer 4 that overlaps the junction point (i.e., that is complementary to the most 5'-bases of the mRNA and contains in addition at its 3'-end several thymidine residues that are complementary to the end of the poly(A) tail) (we used two to three Ts for that purpose [Couttet and Grange, 2004; Couttet et al., 1997]).

This strategy allowed us to show that the decay of several mammalian mRNAs involved first deadenylation, followed by the production of a decapped intermediate with a relatively short poly(A) tail (Couttet et al., 1997) (Fig. 22.2D) and that the NMD pathway involves enhanced rates of deadenylation and decapping in human cells (Couttet and Grange, 2004). I described here protocols for the transcriptional pulsing analysis of mRNA decay with transient transfection of a tetracycline-regulated promoter and the analysis of decay products with cRL-PCR.

4.1. Transcriptional pulsing with a tetracycline-regulated promoter and RNA preparation

The original tetracycline-responsive regulator adapted to mammalian cells, a fusion of the prokaryotic tet-regulator to the activation domain of the VP16 protein of the herpes simplex virus, binds to the tet-operator and activates transcription in the absence of tetracycline in the culture medium (Gossen and Bujard, 1992). In our hands, the original system, called tet-off, yielded more consistent induction to analyze mRNA decay with transient transfection of a tet-regulated promoter, than the subsequently developed tet-on regulator that activates transcription in the presence of tetracycline (Gossen et al., 1995), presumably because the DNA-binding properties of the original protein are more optimal for tight regulation. The original HeLa cell clone H-tTa-1 expressing the tet-off-VP16 chimera (Gossen and Bujard, 1992) gave excellent induction levels provided the amount of transfected plasmid carrying the tet-regulated promoter was carefully titrated. In transient transfection of a well-transfectable cell line like HeLa cells, a relatively high number of plasmid molecules can be found in the nuclei of transfected cells. This excess of plasmid can give rise to significant basal expression even if the promoter is not activated by the VP16 domain. Furthermore, the promoter is not fully activated under these conditions, presumably because the amounts of regulator expressed in the cells are not high enough to activate all promoter molecules. Dilution of the tet-regulated plasmid with a neutral plasmid allows reduction of the amount of plasmid molecules present in each nucleus to a level optimal for strong induction. We used derivatives of the pUHD10-3 vector that harbor seven copies of the tet operator upstream the minimal CMV promoter (http://www.zmbh.uni-heidelberg.de/bujard/reporter/pUHD10-3.html).

4.1.1. Procedure

On day 1, seed 100-mm diameter plates with HeLa cell clone H-tTa-1 with, per plate, 10^6 cells in 10 ml of 8% FCS-DMEM medium containing 1 ng/ml doxycycline (Dox; Sigma).

In the morning of day 2, refresh the medium (1 ng/ml Dox).

Approximately 3 h later, mix in a tube at room temperature with 1 ml of 250 mM $CaCl_2$ 2 μg of the Tet-regulated reporter plasmid, 1 μg of a luciferase expression vector as an internal control of transfection efficiency, and 37 μg of a carrier plasmid devoid of eukaryotic promoters. In a 10-ml tube, add 1 ml of room temperature–equilibrated HBS2× solution (HEPES 50 mM; NaCl 280 mM; $NaHPO_4$ 1.5 mM, pH adjusted precisely to 6.95). While vortexing the HBS-containing tube gently and continuously, add drop wise the DNA-$CaCl_2$ solution. This is sufficient for two 100-mm plates. If more plates are to be used, it is better to perform several independent precipitates and to mix them afterwards because the relative volume of the solutions and of the tube is a critical parameter.

Let the DNA-$CaPO_4$ precipitate for 5 min at room temperature and add 1 ml of the homogenized precipitate suspension onto the cells in the 100-mm plate, distributing it drop wise throughout the plate.

Incubate the cells for approximately 16 h in 5% CO_2 and on day 4 refresh the medium (1 ng/ml Dox).

On day 5, rinse cells three times with PBS (37 °C) and incubate cells with Dox-free medium for 3 h to induce the expression of the reporter gene.

Stop transcription by adding Dox to the medium to a final concentration of 50 ng/ml.

At various time points after transcription arrest of the reporter gene, collect cells by trypsinization and centrifugation for RNA preparation.

Homogenize the cell pellet in 750 μl of GTC solution (4 M guanidinium thiocyanate, 0.1 M sodium acetate [pH 5.0], 5 mM EDTA [pH 8.0], 2% sarkosyl, 0.14 M 2-mercaptoethanol) with a 19-gauge needle to break DNA. This is a slight modification of a classical protocol (Chomczynski and Sacchi, 2006) that allows the use of 1.5-ml Eppendorf tube that is easier to handle. Add 750 μl of water-saturated phenol, shake vigorously, incubate 15 min on ice, shake vigorously again, and centrifuge for 10 min at 4 °C at max speed of a microfuge. Extract with chloroform and add 750 μl of isopropanol. Incubate at −20 °C at least 1 h. Centrifuge for 20 min at 4 °C at max speed of a microfuge, rinse the pellet with 70% ethanol, dry it gently, and resuspend in 100 μl of H_2O. Determine the amount and verify the quality with a Nanodrop and a Bioanalyzer if available.

Remove traces of DNA as follows: In a reaction tube, mix 50 μg of RNA with 10 μl of 10× DNase buffer (500 mM Tris-HCl, pH 7.5, 100 mM $MgCl_2$), 20 U of RNase-free DNase I, 20 U of ribonuclease inhibitor, and H_2O to a volume of 100 μl. Incubate 20 min at 37 °C. Extract with an equal volume of water-saturated phenol and then with chloroform. Precipitate the RNA by adding 7 μl of 3 M sodium acetate, pH 5.5, and 250 μl of 100% ethanol. Incubate on ice for 30 min, pellet the RNA by centrifugation, rinse with 70% ethanol, dry briefly, and resuspend in 50 μl of H_2O.

Comments: The concentration of Dox used before induction, here 1 ng/ml, is the minimal concentration that ensures sustained repression. It is important to avoid the use of too much Dox at this stage, because traces of the remaining Dox can prevent maximal induction by the tet-off regulator. This concentration might require some adjustment if working with another cell line. The quantity of plasmid in the precipitate might also require adjustment when another cell line is used. For RNA preparation, the recent detailed description of the classical procedure contains a number of additional worthwhile recommendations (Chomczynski and Sacchi, 2006).

4.2. Circularization RL-PCR

The circularization RL-PCR procedure is similar to the linker ligation RL-PCR procedure, but it differs in two salient aspects. The ligation is performed without linker in a 10-fold larger volume, compared with the linker ligation conditions, to favor monomolecular ligation events. Because the PCR amplification step involves two RNA-specific primers, instead of only one in ligation-mediated PCR, the specificity of this step is as easy to achieve as for any PCR. Thus, the use of several different nested primers is mostly dictated by the need to detect specifically certain types of decay products. For a number of the RNAs we analyzed in one study (Couttet *et al.*, 1997), we used during the PCR the same primer as the one used for the reverse transcription (primer 1 and 2 were identical on the scheme presented on the right panel of Fig. 22.1). In those cases, the Tm of the primer used during the reverse transcription step was higher than described previously so that it could withstand higher annealing temperature during the PCR: typically approximately 60 °C and above, so that it could be used at 56 °C and above during PCR. We would not recommend the use of primers with such elevated Tm in the linker RL-PCR procedure, because the specificity of the reverse transcription is critical for the specificity of the subsequent PCR. In some instances, in particular when we prepared total rather than cytosolic RNAs, we wanted to ensure that we were detecting only decay products and that intermediates in the biosynthesis of the mRNA were not participating. Thus, we used for reverse transcription a primer overlapping exon 1–2 junction to detect only spliced RNAs (Couttet and Grange, 2004). This was performed to avoid the participation of decapped species that could have corresponded to precursors of the mature RNA. Indeed, mRNA capping precedes and stimulates splicing (Proudfoot *et al.*, 2002). In this case, we used during PCR another primer (primer 2) hybridizing to the 5′-portion of the mRNA, close to the mRNA 5′-end, in order to reduce the size of the PCR products, to minimize potential biases that favor the amplification of shorter products that could have affected measurements of the poly(A) tail length. Because in our applications of circularization RL-PCR to the analysis of RNA decay, we were interested in detecting the decapped RNAs that had not yet undergone degradation by the 5' to 3'-exonucleases of the Xrn1p family (Parker and Song, 2004), we used for the labeling step a primer (primer 4) that overlapped the junction between the full length 5'-end and the poly(A) tail. To permit detection of mRNAs with very short poly(A) tail, we used only an extension with 2 As. The Tm of primer 4 should be well above the Tm of the previous primers to avoid that the unlabeled primers compete with the labeled one during the labeling step. We recommend Tm above 68 °C. The PCR primer (primer 3) hybridizing to the 3'-portion of the mRNA was also chosen close to the poly(A) tail to minimize the length biases previously mentioned and to avoid interference with potential mRNA decay products that were degraded by 3' to 5'-exonucleases (Parker and Song, 2004). Other primer combinations could be used to detect other decay products.

4.2.1. Procedure

4.2.1.1. Intramolecular ligation In a reaction tube, mix 4 μg of total cellular RNAs with 40 μl of 10× ligase buffer (500 mM Tris-HCl, pH 7.5, 100 mM $MgCl_2$), 40 μl of 200 mM DTT, 40 μl of 1 mM ATP, 40 μl of RNase-free BSA (40 μg), 1 μl of RNase inhibitor (20 U), 3 μl of T4 RNA ligase (60 U), and H_2O to a volume of 400 μl. Incubate the tube overnight at 17 °C. Extract the reaction with water-saturated phenol-chloroform (3/1), extract again with chloroform, and precipitate the RNA with 40 μl of 3 M sodium acetate, pH 5.5, and 1 ml of 100% ethanol. Spin for 20 min in a microfuge and wash the pellet with 70% ethanol. Dry briefly, and resuspend the pellet in 12 μl H_2O.

4.2.1.2. cDNA synthesis Place 6 μl of the ligated RNA from the previous step in a reaction tube that fits in a PCR apparatus. Incubate 1 min at 95 °C to denature the RNA. Put at 42 °C and add 1 μl of primer 1 (10 ng), 1 μl of 10× RT/Taq buffer (650 mM Tris-HCl, pH 8.8, 100 mM β-mercaptoethanol, 165 mM $(NH_4)_2SO_4$), and 1 μl of 60 mM $MgCl_2$. Mix and let the primer hybridize by incubating 30 min at 42 °C (in case of background, in particular when a primer with a higher Tm is used, the temperature of this reaction can be raised up to 50 °C). Without taking the tube out of the PCR apparatus, add 1 μl of a mixture containing 5 mM of each dNTPs and 3 U of AMV reverse transcriptase. Mix and incubate 30 min at 42 °C (or higher if required). Stop the reaction by incubating 5 min at 95 °C. Place the tube on ice and spin briefly.

4.2.1.3. PCR amplification To the reverse transcription reaction, add 2 μl (100 ng) of primer 2, 2 μl (100 ng) of primer 3, 1 μl of 10× RT/Taq buffer, 2 μl of DNase-free BSA (4 μg), 1 μl of a mixture of the 4 dNTP (5 mM each), 1 U of Hot start Taq DNA polymerase (AmpliTaq Gold, Applied Biosystems), and 2 μl of H_2O. After 10 min at 95 °C, cycle 25 times (30 sec at 94 °C, 3 min at TmPrimers-5 °C, 3 min at 74 °C). Finish by an extension step of 10 min at 74 °C.

4.2.1.4. Primer labeling for the analysis of 20 samples While the PCR is proceeding, label primer 4 as follows: In a reaction tube, mix 1 μl of primer 4 (50 ng), 2.5 μl of [γ-^{32}P] ATP (50 μCi, 7 pmol), 2 μl of 10× kinase buffer (500 mM Tris-HCl, pH 7.6, 100 mM $MgCl_2$, 50 mM DTT, 1 mM spermine, 1 mM EDTA), 1 μl of T4 polynucleotide kinase (10 U), and H_2O to a volume of 20 μl. Incubate 30 min at 37 °C. Stop the reaction by incubating 5 min at 95 °C. The primer is used without further purification.

PCR products labeling: 8 μl of the PCR reaction were mixed with 5 μl of a solution containing 1 μl of the primer 4 labeling reaction, 0.5 μl of 10× Taq buffer, 0.5 μl of a mixture of the 4 dNTP (5 mM each), 1 μl of DNase-free BSA (2 mg/ml), 0.5 μl of 30 mM $MgCl_2$, 2.5 μl of H_2O, and 0.5 U of Taq DNA polymerase. After a 5-min denaturation step at 95 °C, the reaction was cycled 5 times (30 sec at 94 °C, 3 min at 68 °C, 3 min at 74 °C). The hybridization temperature should be adapted to the Tm of primer 4, but it is best to use primers compatible with the temperature indicated here.

Add 80 μl of 0.3 M sodium acetate, pH 5.5, containing tRNA at 0.1 μg/μl. Extract with 100 μl of Tris-buffered phenol/chloroform (3/1) and precipitate with 300 μl of 100% ethanol. Incubate on ice for 30 min. Spin down for 10 min, rinse the pellet with 70% ethanol, and dry. Resuspend the pellet in 6 μl of formamide-loading buffer (95% formamide, 10 mM EDTA, 0.05% bromophenol blue, 0.05% xylene cyanol), incubate 3 min at 95 °C and load 3 μl on a 6% sequencing gel.

Comments: When the circularization RL-PCR is used to measure the poly(A) tail length of capped mRNAs, it is necessary to perform prior decapping with TAP exactly as described herein in Section 3.2.1.2 for the 5′-RACE RL-PCR procedure. A control to verify that the length of the products are, indeed, due to the poly(A) tail can be performed by digesting this poly(A) tail by means of an oligo(dT)/RNase H treatment before RL-PCR analysis. For that purpose, mix 10 μg of RNA with 500 ng of oligo(dT)$_{12-18}$ with 2 μl of 2 M KCl, 10 mM EDTA, pH 8, and H_2O to 20 μl, incubate 2 min at 90 °C and 10 min at 25 °C. Then add 4 μl of 10× RNase H buffer (200 mM Tris-HCl, pH 8, 280 mM $MgCl_2$, 1 M KCl, 5 mM EDTA), 20 U of RNase A inhibitor, 3 units of RNase H. Incubate 30 min at 37 °C. Extract with an equal volume of water-saturated phenol/

isoamyl alcohol, and then with chloroform. Precipitate the RNA with ethanol.

5. Concluding Remarks

The ligation-mediated PCR approaches proved very powerful to observe a number of regulatory events taking place in mammalian cells. They allowed the analysis of protein–nucleic interactions and could be used to detect minor amounts of intermediary reaction products in either mRNA decay pathway as described here, or in the DNA demethylation pathway triggered by transcriptional activators (Kress *et al.*, 2006). Because the strategy relies on ligation to 5′-P ends of various molecules and on the PCR detection of the products under investigation, other combinations of previous enzymatic treatments and primer design strategies could be used to detect other reaction intermediates, provided the proper biologic systems are used to demonstrate the presence of these intermediates. The use of transcriptional pulse strategies combined with RL-PCR should provide other insights in the mRNA decay pathways active in mammalian cells.

ACKNOWLEDGMENTS

This work was supported in part by the CNRS and grants from the Association de Recherche sur le Cancer and the Ligue Nationale contre le Cancer. I thank Eva-Maria Geigl for the critical reading of the manuscript.

REFERENCES

Bertrand, E., Fromont-Racine, M., Pictet, R., and Grange, T. (1993). Visualization of the interaction of a regulatory protein with RNA *in vivo*. *Proc. Natl. Acad. Sci.USA* **90**, 3496–3500.

Bertrand, E., Fromont-Racine, M., Pictet, R., and Grange, T. (1997). Detection of ribozyme cleavage products using reverse ligation mediated PCR (RL-PCR). In "Ribozyme Protocols," (P. C. Turner, ed.), Vol. 74, pp. 311–323. Humana Press Inc., Totowa, NJ.

Bertrand, E., Pictet, R., and Grange, T. (1994). Can hammerhead ribozymes be efficient tools for inactivating gene function. *Nucleic Acids Res.* **22**, 293–300.

Chomczynski, P., and Sacchi, N. (2006). The single-step method of RNA isolation by acid guanidinium thiocyanate-phenol-chloroform extraction: Twenty-something years on. *Nat. Protoc.* **1**, 581–585.

Couttet, P., Fromont-Racine, M., Steel, D., Pictet, R., and Grange, T. (1997). mRNA deadenylylation precedes decapping in mammalian cells. *Proc. Natl. Acad. Sci. USA* **94**, 5628–5633.

Couttet, P., and Grange, T. (2004). Premature termination codons enhance mRNA decapping in human cells. *Nucleic Acids Res.* **32**, 488–494.

Donis-Keller, H., Maxam, A. M., and Gilbert, W. (1977). Mapping adenines, guanines, and pyrimidines in RNA. *Nucl. Acids Res.* **4,** 2527–2538.

Ehresmann, C., Baudin, F., Mougel, M., Romby, P., Ebel, J. P., and Ehresmann, B. (1987). Probing the structure of RNA in solution. *Nucl. Acids Res.* **15,** 9109–9128.

Espinás, M. L., Roux, J., Pictet, R., and Grange, T. (1995). Glucocorticoids and protein kinase A coordinately modulate transcription factor recruitment at a glucocorticoid-responsive unit. *Mol. Cell. Biol.* **15,** 5346–5354.

Flavin, M., Cappabianca, L., Kress, C., Thomassin, H., and Grange, T. (2004). Nature of the accessible chromatin at a glucocorticoid-responsive enhancer. *Mol. Cell. Biol.* **24,** 7891–7901.

Fromont-Racine, M., Bertrand, E., Pictet, R., and Grange, T. (1993). A highly sensitive method for mapping the 5′-termini of mRNAs. *Nucl. Acids Res.* **21,** 1683–1684.

Gossen, M., and Bujard, H. (1992). Tight control of gene expression in mammalian cells by tetracycline-responsive promoters. *Proc. Natl. Acad. Sci. USA* **89,** 5547–5551.

Gossen, M., Freundlieb, S., Bender, G., Muller, G., Hillen, W., and Bujard, H. (1995). Transcriptional activation by tetracyclines in mammalian cells. *Science* **268,** 1766–1769.

Grange, T., Bertrand, E., Espinas, M. L., Fromont-Racine, M., Rigaud, G., Roux, J., and Pictet, R. (1997). *In vivo* footprinting of the interaction of proteins with DNA and RNA. *Methods a companion to methods in Enzymology* **11,** 151–163.

Grange, T., Cappabianca, L., Flavin, M., Sassi, H., and Thomassin, H. (2001). *In vivo* analysis of the model tyrosine aminotransferase gene reveals multiple sequential steps in glucocorticoid receptor action. *Oncogene* **20,** 3028–3038.

Isken, O., and Maquat, L. E. (2007). Quality control of eukaryotic mRNA: Safeguarding cells from abnormal mRNA function. *Genes Dev.* **21,** 1833–1856.

Kress, C., Thomassin, H., and Grange, T. (2006). Active cytosine demethylation triggered by a nuclear receptor involves DNA strand breaks. *Proc. Natl. Acad. Sci. USA* **103,** 11112–11117.

Loflin, P. T., Chen, C. Y., Xu, N., and Shyu, A. B. (1999). Transcriptional pulsing approaches for analysis of mRNA turnover in mammalian cells. *Methods a companion to methods in Enzymology* **17,** 11–20.

Milligan, J. F., Groebe, D. R., Witherell, G. W., and Uhlenbeck, O. C. (1987). Oligoribonucleotide synthesis using T7 RNA polymerase and synthetic DNA template. *Nucl. Acids Res.* **15,** 8783–8798.

Mueller, P. R., and Wold, B. (1989). *In vivo* footprinting of a muscle specific enhancer by ligation-mediated PCR. *Science* **246,** 780–786.

Parker, R., and Song, H. (2004). The enzymes and control of eukaryotic mRNA turnover. *Nat. Struct. Mol. Biol.* **11,** 121–127.

Peattie, D. A. (1979). Direct chemical method for sequencing RNA. *Proc. Natl. Acad. Sci. USA* **76,** 1760–1764.

Proudfoot, N. J., Furger, A., and Dye, M. J. (2002). Integrating mRNA processing with transcription. *Cell* **108,** 501–512.

Rigaud, G., Roux, J., Pictet, R., and Grange, T. (1991). *In vivo* footprinting of the rat TAT gene: Dynamic interplay between the glucocorticoid receptor and a liver-specific factor. *Cell* **67,** 977–986.

Romaniuk, P. J., and Uhlenbeck, O. C. (1983). Joining of RNA molecules with RNA ligase. *Methods Enzymol.* **100,** 52–59.

Rouault, T. A. (2006). The role of iron regulatory proteins in mammalian iron homeostasis and disease. *Nat. Chem. Biol.* **2,** 406–414.

Scotto-Lavino, E., Du, G., and Frohman, M. A. (2006). Amplification of 5′-end cDNA with 'new RACE'. *Nat. Protoc.* **1,** 3056–3061.

Thomassin, H., Flavin, M., Espinas, M. L., and Grange, T. (2001). Glucocorticoid-induced DNA demethylation and gene memory during development. *EMBO J.* **20,** 1974–1983.

CHAPTER TWENTY-THREE

Tethering Assays to Investigate Nonsense-Mediated mRNA Decay Activating Proteins

Niels H. Gehring,[*,†] Matthias W. Hentze,[†,‡] *and* Andreas E. Kulozik[*,†]

Contents

1. Introduction	468
2. Plasmid Cloning	470
2.1. Effector plasmid	470
2.2. Reporter plasmid	472
2.3. Insertion of boxB or MS2 sites into the reporter mRNA	473
2.4. Transfection control plasmid	474
2.5. Detection of the effector protein	479
2.6. Control experiments	479
References	480

Abstract

Nonsense-mediated mRNA decay (NMD) is activated by exon–junction complexes (EJCs) that are located downstream of the termination codon of the substrate mRNAs. This situation can be imitated by tethering components of the EJC to the 3′ untranslated region (3′ UTR) of a reporter mRNA. Here we describe the detailed use of two analogous tethering systems that are based on the coat protein of bacteriophage MS2 or on the 22 amino acid RNA-binding domain of the bacteriophage λ-antiterminator protein N (λN-peptide). These polypeptides are fused as tags to proteins of interest. Their respective RNA binding sites are inserted into reporter mRNAs. This enables recruitment of the NMD activity of the fusion protein to an NMD-activating position, bypassing the requirement for splicing. In this chapter we explicate the cloning of appropriate reporter plasmids and the setup of a tethering experiment with the necessary control experiments. Advantages of the different systems and tags are discussed.

---

[*] Department of Pediatric Oncology, Hematology and Immunology, University of Heidelberg, Heidelberg, Germany
[†] Molecular Medicine Partnership Unit, EMBL/University of Heidelberg, Heidelberg, Germany
[‡] European Molecular Biology Laboratory, Heidelberg, Germany

1. Introduction

Mammalian nonsense-mediated mRNA decay (NMD) critically depends on pre-mRNA splicing, which imprints the mRNA with exon junction complexes (EJCs). EJCs are deposited on the mRNA in the nucleus and remain bound after mRNA export to the cytoplasm until they are removed by translating ribosomes. They consist of multiple proteins, some of which have been shown to enable recognition of premature termination codons (PTCs) and activation of NMD (eIF4A3, BTZ, Y14, MAGOH, and RNPS1) (Chan et al., 2004; Degot et al., 2004; Ferraiuolo et al., 2004; Gehring et al., 2003; Kataoka et al., 2000, 2001; Kim et al., 2001; Le Hir et al., 2001a,b; Lykke-Andersen et al., 2000, 2001; Palacios et al., 2004; Shibuya et al., 2004).

Termination codons more than 55 nucleotides (nt) upstream of the 3′-most exon–exon junction elicit NMD (Hentze and Kulozik, 1999; Maquat, 2004; Thermann et al., 1998; Zhang et al., 1998a,b). The NMD factor UPF1 is thought to physically link the EJC to the terminating ribosome by means of its interactions with EJC-bound UPF2 and the release factors (eRF1 and eRF3) (Kashima et al., 2006). Hence, an EJC downstream of a termination codon activates NMD in an EJC-dependent and UPF1-dependent manner.

It has been shown previously that introducing a spliceable intron into the 3′ UTR of the β-globin mRNA activates NMD of this otherwise normal mRNA (Thermann et al., 1998). This is due to the recruitment of an EJC downstream of the normal termination codon of the β-globin mRNA (Fig. 23.1, top panel). The same situation that occurs during recognition of a nonsense codon can be mimicked without the need for splicing by tethering components of the EJC to an NMD activating position of a suitable reporter mRNA (Gehring et al., 2003, 2005; Lykke-Andersen et al., 2000, 2001). Tethering is accomplished by fusing NMD-activating factors (e.g., components of the exon junction complex) to an RNA-binding polypeptide (referred to as a tethering tag). Binding sites for the tethering tag are inserted into the 3′ UTR of a reporter mRNA and serve to directly recruit the essential NMD factors to the reporter mRNA (Fig. 23.1, middle and bottom panel). Activation of NMD by a tethered factor leads to a reduced abundance of the reporter mRNA that can be monitored, for example, by Northern blotting.

The tethering approach is not restricted to the analysis of NMD activating proteins. It enables analysis of the posttranscriptional activities of a number of different RNA-binding proteins. Tethered function analyses have been successfully used, for example, to decipher the molecular mechanisms of translational silencing and mRNA degradation by the miRNA-machinery

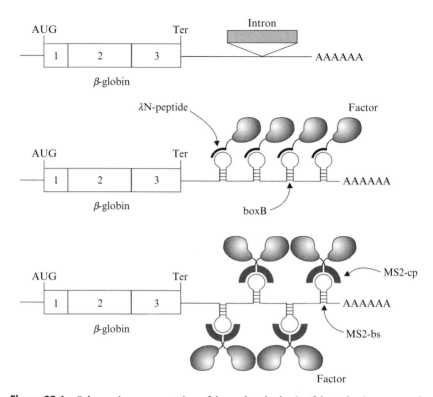

Figure 23.1 Schematic representation of the molecular basis of the tethering approach. Experimentally inserting an intron into the 3' UTR of the β-globin gene, can activate the NMD of β-globin mRNA when translation terminates normally (at Ter), because an exon–junction complex (EJC) forms upstream of the resulting exon–exon junction (top panel). This situation is mimicked by tethering components of the EJC (factor) as either λN or MS2 coat protein fusion proteins to the 3' UTR of a β-globin reporter mRNA by means of the interaction of the λN-peptide (middle panel) or the MS2 coat protein (MS2-cp; bottom panel) with its cognate RNA binding sequence boxB or MS2-binding sequence (MS2-bs), respectively.

(Behm-Ansmant et al., 2006; Kiriakidou et al., 2007; Pillai et al., 2004, 2005), to recapitulate the stimulation of translation by the poly(A) binding protein (Gray et al., 2000), by components of the exon–junction complex (Nott et al., 2004; Wiegand et al., 2003), or by SR-proteins (Sanford et al., 2004) to analyze the mRNA degrading activities of ARE-binding proteins (Lykke-Andersen and Wagner, 2005; Yamasaki et al., 2007); or to explain the mechanism of CRM1-Rev mediated export of HIV-1 mRNAs (Yi et al., 2002). Furthermore, tethering of fluorescent proteins like GFP or its variants has been used to visualize the localization of mRNAs in living cells (Bertrand et al., 1998; Fusco et al., 2003; Sheth and Parker, 2003).

Two different tethering tags have been commonly used for the analysis of NMD by tethering approaches: the coat protein of bacteriophage MS2 and the 22 amino acid RNA-binding peptide of the bacteriophage λ-antiterminator protein N (λN-peptide). Here we provide detailed protocols for tethering assays with these two different systems.

2. Plasmid Cloning

We and others have previously reported experimental systems to analyze NMD by tethering (Gehring *et al.*, 2003, 2005; Kim *et al.*, 2005; Kunz *et al.*, 2006; Nott *et al.*, 2004; Lu and Cullen, 2004; Lykke-Andersen *et al.*, 2000, 2001). All these systems consist of three components:

1. An effector plasmid that expresses the protein to be analyzed fused to an RNA-binding domain (MS2 coat protein or N-peptide).
2. A reporter plasmid that expresses a reporter mRNA with several binding sites for the tethering tag within the 3' UTR.
3. An expression control plasmid that expresses an mRNA used to control for comparable transfection efficiencies and to normalize the expression levels of the reporter mRNAs.

Here we describe the generation of systems that use the MS2 coat protein or the λN-peptide, respectively, to tether NMD proteins to a β-globin–based reporter mRNA.

2.1. Effector plasmid

Because of its short open reading frame, the sequence encoding the λN-peptide can be (1) amplified by polymerase chain reaction (PCR) from parental vectors, (2) cloned with appropriate annealed oligonucleotides, or (3) directly integrated into the PCR primers for the amplification of the NMD protein to be subcloned. Mutations of the λN-peptide that enhance its binding affinity to the boxB sequence have been described (Austin *et al.*, 2002). Such mutated λN-peptides (e.g., D2N, Q4R) can improve the readout of tethering experiments (Fig. 23.2). To enhance the detection of the fusion protein, it is recommended to fuse the λN-peptide to an epitope tag. Successful combinations described so far are λN-HA and λN-V5 (Gehring *et al.*, 2005; Pillai *et al.*, 2004). The high amount of functionally important positively charged amino acids within the λN-peptide should be considered when potential fusion partners are selected. We have generated cDNA for a λN-V5 fusion tag (λN: MNARTRRRERRAEKQAQWKAAN; V5:GKPIPNPLLGLDST) with suitable oligonucleotides and inserted it into the *NheI* and *XhoI* sites of the pCI-neo mammalian expression vector (Promega).

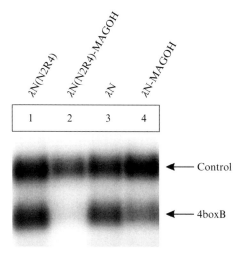

Figure 23.2 A mutant λN-peptide with higher affinity for its RNA binding site boxB improves the efficiency of the tethering assay. Northern blot analysis of RNA isolated from HeLa cells transfected with three plasmids. The first two either serve as the transfection efficiency control (control) or encode β-globin 4boxB reporter mRNA (4boxB). The third encodes the effector protein. Shown is unfused λN (either mutated D2N, Q4R, lane 1, or wild type, lane 3), or λN-MAGOH (either mutated, lane 2, or wild type, lane 4).

The longer open reading frame of the MS2 coat protein has to be subcloned from a parental plasmid, preferentially by PCR. MS2 coat protein mutants have been described that decrease the tendency of the MS2 coat protein to oligomerize but retain the original affinity for RNA (LeCuyer et al., 1995). In addition, mutants that exhibit a higher RNA affinity exist (Lim and Peabody, 1994). So, a combination of mutations is likely to improve the performance of the MS2 coat protein in the tethering assay. Notably, we have not observed major differences between the MS2 wild-type and two oligomerization-incompetent mutants we have tested in our preparatory experiments (Fig. 23.3). MS2 coat proteins bind their cognate RNA binding sequence as homodimers (Nagai, 1996; Valegard et al., 1994). This doubles the amount of recruited proteins per binding site in the reporter mRNA. If homodimerization is disadvantageous, a tandem, head-to-tail dimer of the MS2 coat protein can be used (Hook et al., 2005; Peabody and Lim, 1996). In our MS2 tethering system, the cDNA for the MS2 coat protein was inserted between the *NheI* and *XhoI* sites of pCI-neo.

cDNAs encoding different NMD factors were PCR amplified from HeLa cDNA and inserted in frame with the tethering tag into the pCI-λNV5 or pCI-MS2 expression vectors.

Why do we use two different tags for our tethering assays? We observed discrepancies for certain NMD factors when we used the two different tags

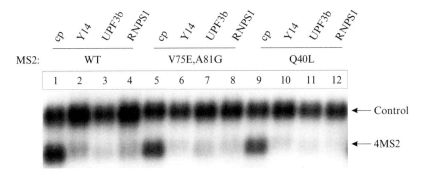

Figure 23.3 Comparison of mutated MS2 coat proteins. Northern blot analysis of RNA isolated from HeLa cells transfected with plasmids for the transfection efficiency control (control), the β-globin 4MS2 reporter, and the indicated proteins (Y14 lanes 2, 6, 10; UPF3b lanes 3, 7, 11; RNPS1 lanes 4, 8, 12; unfused MS2 coat protein lanes 1, 5, 9) fused to different mutants of the MS2 coat protein (wild type, lanes 1 to 4, V75E, A81G, lanes 5 to 8, Q40L lanes 9 to 12).

(Gehring et al., 2005). This may be caused by the different sizes of the tags (22 amino acids for the λN-peptide, and 120 amino acids for the MS2 coat protein). Consistent with this notion, we found that the λN-peptide gives stronger readouts than the MS2 coat protein with the small proteins MAGOH and Y14 (Gehring et al., 2003, 2005). By contrast, a small tag size may represent a disadvantage when fused to larger proteins, because it may be less accessible in that situation. Furthermore, other characteristics of the tethering tags could influence the performance in the assay. It has been reported that an MS2-UPF3b fusion protein is localized predominantly in the cytoplasm, whereas the wild-type UPF3b is predominantly nuclear in the steady state (Lu and Cullen, 2004). Thus, the tag can influence the subcellular distribution of the fusion protein and may target it to a cellular compartment that does not support NMD. To avoid false-negative readouts as a result of inappropriate localization or inaccessibility of the tag a previously characterized NMD protein should be tested as a positive control in both systems (N-peptide and MS2 coat protein).

2.2. Reporter plasmid

For the analysis of NMD, a reporter mRNA that is well characterized as an NMD substrate should be used. In addition, the reporter mRNA should display a long half-life in the absence of the tethered factor. This improves the readout of the tethering system. To avoid interference (e.g., cross-hybridization) of endogenous transcripts with the reporter mRNA, the gene encoding the reporter RNA should not be actively expressed in the cells used for the assays. In consideration of these specifications we decided to use β-globin mRNA as reporter mRNA. It is not expressed in most human cells used for transfection experiments (e.g., HeLa, HEK, 293), has a long half-life and a high expression level, and possesses an appropriate length for Northern blot analysis. In addition, β-globin mRNA represents an extensively characterized model system for the study of NMD and is also one of the best studied examples for the medical significance of NMD (Danckwardt et al., 2002; Holbrook et al., 2004).

2.3. Insertion of boxB or MS2 sites into the reporter mRNA

It has been shown that increasing the number of binding sites can increase the sensitivity of the tethering assay (Gehring et al., 2003; Lykke-Andersen et al., 2000). Thus, the number of binding sites required for an appropriate readout has to be determined experimentally. We have generated reporter constructs containing 2, 4, 6, and 8 boxB or MS binding sites. Depending on the tethered factor, we found that the reporter mRNAs with 4 or 6 tethering sites are robustly reduced in abundance to levels comparable to PTC-containing NMD substrates (i.e., 5 to 20% of the negative control) (Fig. 23.4).

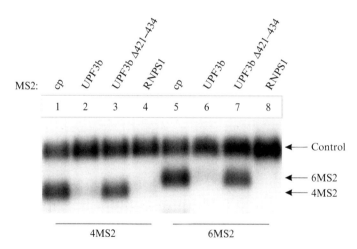

Figure 23.4 Determining the optimal number of binding sites in a reporter mRNA. Analysis of tethering reporters with 4 or 6 MS2 binding sites. Northern blot of RNA isolated from HeLa cells transfected with plasmids for the transfection efficiency control (control), the β-globin 4MS2 (lanes 1 to 4) or 6MS2 reporter (lanes 5 to 8), and the unfused MS2 (cp, lanes 1 and 5), MS2-UPF3b (lanes 2 and 6), MS2-UPF3b Δ421 to 434 (lanes 3 and 7), or MS2-RNPS1 (lanes 4 and 8). The UPF3b mutant Δ421 to 434 has been shown previously to be unable to activate NMD in the tethering assay and serves as an additional specificity control.

BoxB or MS2 sites were inserted with the same strategy into the 3′ UTR of a β-globin expression vector. We inserted an *XhoI* restriction site at a position 50 bp downstream of the termination codon by site-directed mutagenesis. An intron at this position was shown to activate NMD by functionally redefining the normal termination codon as a PTC (Thermann *et al.*, 1998). This supported the notion that tethering EJC constituents at the selected position within the β-globin 3′ UTR would support NMD. Binding sites were inserted in tandem with DNA oligonucleotides encoding the respective binding sites and 5′-*SalI* and 3′-*XhoI* sites. The oligos were annealed and amplified by PCR, and the PCR products were digested with *SalI* and *XhoI*. The ligation of this insert into the *XhoI* site of the β-globin expression plasmid enables directional cloning of multiple tandem sites, because this strategy generates a construct with two binding sites and a single 3′-*XhoI* site for further cloning steps if the product is inserted in the correct orientation (5′-*SalI*, 3′-*XhoI*). Constructs with more binding sites were generated by repeated insertions of tandem binding sites. A schematic representation of the resulting 3′ UTR sequences is shown in Fig. 23.5.

2.4. Transfection control plasmid

The activation of NMD by the tethered protein is expected to lead to a reduced abundance of the reporter mRNA. This effect needs to be controlled for with an mRNA that is not affected by the tethered factor but that can ideally be detected with the same probe by Northern blotting. For this purpose, we generated an expression plasmid for an elongated β-globin

```
A  TAAGCTCGCTTTCTTGCTGTCCAATTTCTATTAAAGGTTCCTTTGTTCCCTAAGCT
   CGACCAAAGGTTCCTTTGTGGCCCTGAAAAAGGGCCAAATTGGTGGCTGGTGT
   GGCTAATGCCCTATGGCCCTGAAAAAGGGCCACTGGAGGATATTCATGCACTC
   GACCAAAGGTTCCTTTGTGGCCCTGAAAAAGGGCCAAATTGATGGCTGGTGTG
   GCTAATGCCCTATGGCCCTGAAAAAGGGCCACTGGAGGATATTCATGCACTCGA
   GTACTAAACTGGGGGATATTATGAAGGGCCTTGAGCATCTGGATTCTGCCTAAT
   AAAA

B  TAAGCTCGCTTTCTTGCTGTCCAATTTCTATTAAAGGTTCCTTTGTTCCCTAAGCT
   CGACCAAAGGTTCCTTTGTCACATGAGGATCACCCATGTGAAGTGGTGGCTGGT
   GTGGCTAATGCCCTGACATGAGGATCACCCATGTCTGGAGGATATTCATGCACT
   CGACCAAAGGTTCCTTTGTCACATGAGGATCACCCATGTGAAGTGGTGGCTGGT
   GTGGCTAATGCCCTGACATGAGGATCACCCATGTCTGGAGGATATTCATGCACT
   CGAGTACTAAACTGGGGGATATTATGAAGGGCCTTGAGCATCTGGATTCTGCCT
   AATAAAA
```

Figure 23.5 Sequence of the 3′ UTRs of the β-globin 4boxB and 4MS2 reporter plasmids. The plasmid β-globin 4boxB (Gehring *et al.*, 2005) contains 4 consecutive λN-binding sites (boxB). The sequence shown starts with the termination codon (TAA) of β-globin and ends with the poly(A) signal (AATAAA). A single boxB site has the sequence TGGCCCTGAAAAAGGGCCA. (B) The plasmid β-globin 4MS2 (Gehring *et al.*, 2005) contains 4 consecutive MS2-binding sites. (Sequence coverage as in panel A). A single MS2 binding site has the sequence ACATGAGGATCACCCATGT.

mRNA that contains an insertion of β-globin sequences in the second and third exon. The mRNA expressed from this control plasmid differs in size from the 4 and 6 boxB or MS2 reporter mRNAs and thus allows for the clear separation of the two RNAs in the same gel, which can subsequently be subjected to Northern blotting.

2.4.1. Buffers and solutions

2× BBS: 50 mM BES (N,N-bis(hydroxyethyl)-2-aminoethansulfonat), 1.5 mM Na$_2$HPO$_4$, 280 mM NaCl; adjust the pH exactly to 6.96. Calibrate the pH meter before use. Filter-sterilize through a 0.22-μm filter and store in aliquots at $-20\,°C$.
RNA lysis buffer: 50 mM Tris, pH 7.2, 0.5% NP-40 (or Triton X100), 150 mM NaCl, 5 mM ribonucleoside-vanadyl complex (VRC), 0.1% deoxycholate, 1 mM DTT; add protease inhibitors (e.g., PMSF) if required.
10× MOPs buffer: 0.2 M 3-(N-morpholino)propanesulfonic acid (MOPS), pH 7.0, 80 mM sodium acetate, 10 mM EDTA.
Northern RNA buffer: mix 500 μl formamide, 200 μl formaldehyde (pH > 4), 100 μl 10× MOPs buffer, and 1 μl ethidium bromide (10 mg/ml).
RNA loading buffer: 50% glycerol, 1 mM EDTA, 0.25% bromophenol blue.
20× SSC: 3 M NaCl, 0.3 M trisodium-citrate, pH 7.0.
Church buffer: 0.5 M Na$_2$HPO$_4$, pH 7.2, 1 mM EDTA, 7% SDS.
Wash buffer 1: 2× SSC, 0.1% SDS.
Wash buffer 2: 0.2× SSC, 0.1% SDS.
Transfection of HeLa cells.

Titration experiments should be performed to determine the optimal concentrations for transfecting cells with the control, reporter, and effector plasmids. In our preparatory experiments, we found reproducible readouts for our plasmids with concentrations of 2 μg (reporter), 0.8 to 1.0 μg (effector), and 0.5 μg (control) per 6-well plate (i.e. 2.0-2.4 × 10^5 cells). Practically, the best concentrations will depend on the particular experimental setup, including transfection method used, the resulting transfection efficiencies, and the particular plasmid expression levels in the selected cell line.

24 h before transfection, seed HeLa cells in an appropriate density (2.0 to 2.4 × 10^5 cells per well) in 6-well plates. On the next day cells must not be more than 70% confluent.
1 to 4 h before transfection change the cell-culture medium (2ml).
Mix 2 μg of the reporter plasmid DNA, 0.5 μg of the transfection efficiency control plasmid, 0.8 to 1 μg of the effector plasmid and 0.2 μg of a GFP expression plasmid (optional). Add autoclaved water to a final volume of 90 μl and mix. Add 10 μl of a 2.5 M CaCl$_2$ stock solution (final

concentration is 250 mM) and mix. From a sterile stock of 2× BBS (stored in aliquots at −20 °C) add 100 μl to the DNA-CaCl$_2$ solution and mix thoroughly. Incubate for 15 min at room temperature.

Add the transfection mix to the cell-culture dishes and mix by gently shaking the dish.

Incubate the cells for 18 to 24 h in an incubator at 37 °C and 3% CO$_2$.

Wash the cells twice with prewarmed TBS and add fresh medium. Incubate cells at 37 °C and 5% CO$_2$ for 24 h for optimal expression levels.

2.4.2. Cell lysis

Wash cells twice with PBS, and then add 300 μl of lysis buffer to each well. Harvest the cells with a scraper and transfer them into a 1.5-ml microcentrifuge tube.

Incubate the lysates for 5 min on ice.

Pellet the nuclei and membranes of the lysed cells in a microcentrifuge (8000 rpm, 4 °C, 10 min). Transfer the supernatant to a fresh microcentrifuge tube, take a small aliquot to analyze the expression of the MS2- or λN-fusion protein (effector).

2.4.3. Preparation of total-cytoplasmic or total-cell RNA

For total-cell RNA preparations (including nuclear RNAs), add TRIzol* directly to the cell culture plates (1 ml of TRIzol per well), transfer the homogenate to microcentrifuge tubes, and proceed with step 2 of the protocol.

For preparation of total-cytoplasmic RNA, add three volumes (750 μl) of TRIzol LS reagent to the supernatants (approximately 250 μl) of the cytoplasmic lysis and mix by pipetting up and down.

Incubate for 5 min at room temperature.

Add 200 μl of chloroform per 750 μl TRIzol LS reagent (or 1 ml TRIzol). Shake the tubes vigorously by vortexing. Incubate for 2 to 5 min at room temperature.

Centrifuge the samples (12,000g, 4 °C, 15 min) to separate the phases.

Transfer the upper aqueous phase to a fresh microcentrifuge tube. Be careful not to disturb the interphase. Precipitate the RNA by mixing with isopropanol. Use 500 μl isopropanol per 750 μl TRIzol LS reagent (or 1 ml TRIzol) used for the initial homogenization. Mix by vortexing and incubate for 5 min at room temperature.

Centrifuge the samples (12,000g, 4 °C, 15 min) to pellet the RNA.

Wash the pellet with 0.8 ml ice-cold 75% ethanol and spin again (12,000g, 4 °C, 5 min).

---

* RNA extraction can be performed with comparable reagents. e.g. TRI reagent, RNAzol etc.

Aspirate the supernatant completely, but do not overdry the RNA pellet at this stage.

Resuspend the RNA in 15 to 20 µl H$_2$O by incubating the tubes for 5 min at 50 °C.

Measure the A$_{260}$ of a 1:100 dilution of the RNA. Calculate the RNA content with the A$_{260}$ value for RNA (40 µg/ml).

2.4.4. RNA gel electrophoresis

Melt 2.1 g agarose (for a 180 ml gel) in 120 ml H$_2$O.

Add 18 ml of 10× MOPs buffer and 42 ml formaldehyde (pH >4.0) under a fume hood.

Cast the gel, and allow the gel to set for at least 30 min at room temperature.

Mix 2 to 5 µg RNA with 15 µl of Northern RNA buffer. Incubate the samples for 15 min at 65 °C in a heating block, and then cool on ice. Add 2 µl RNA loading buffer.

Load the gel and let run in 1× MOPs buffer for 16 h at 50 volts.

2.4.5. Capillary transfer

After the gel run has completed, take a picture of the gel on a UV-transilluminator, and cut off the parts that are not required.

Put the gel into a clean plastic box and shake gently in 200 ml of autoclaved H$_2$O for 8 min.

Repeat the wash step.

Remove the H$_2$O and add 200 ml of freshly prepared 50 mM NaOH to the gel. Shake gently for 5 min.

Discard the NaOH.

Add 200 ml of 20× SSC to the gel. Equilibrate the gel for at least 40 min in the 20× SSC.

To set up the capillary transfer, put the gel-tray upside down into a baking dish.

Fill dish with 250 ml of 20× SSC.

Cut a piece of Whatman 3MM paper as wide as the gel tray and long enough to cover the tray and extend into the 20× SSC to serve as wicks.

Wet the paper with 20× SSC and place it on the gel tray.

Place the gel upside down on the Whatman paper. Remove bubbles between gel and paper by rolling with a pipette.

Surround the gel with plastic wrap, leaving no space between the gel and the surrounding paper (to prevent contact between the paper towels and the bottom Whatman paper or the 20× SSC).

Cut a nylon membrane (Nytran N, 0.45 μm, Schleicher and Schuell) slightly larger than the gel. Prewet the membrane by floating it in RNase-free water. Then place the membrane on top of gel. Remove bubbles by rolling with a sterile plastic pipette. Be careful to handle the filter solely by the edges.

Add two pieces of Whatman 3MM cut to the size of gel. Remove bubbles.

Add approximately 10 cm of paper towels.

Add a glass plate (or gel tray etc.) to distribute weight.

Add weight to make the stack tight (e.g., a full 500-ml bottle or a catalog).

Let transfer proceed for 4 h.

Disassemble the transfer setup and immobilize the RNA on the nylon filter by crosslinking the RNA side (or better both sides) in a UV cross linker (1200 mJ in a Stratalinker).

2.4.6. Hybridization

Prehybridize the nylon membrane for 2 h in 20 to 35 ml (depending on the size) Church buffer at 65 °C.

Please note that the plasmid that serves as a template for *in vitro* transcription has to be linearized by restriction digestion at the desired site of transcription termination before transcription.

Prepare the transcription reaction by mixing the following: 2 μl transcription buffer (10× conc.), 4 μl H_2O, 0.5 μl 0.1 M DTT, 1.2 μl 10 mM ATP, 1.2 μl 10 mM CTP, 1.2 μl 10 mM UTP, 0.7 μl RNasin© (40 U/μl), 1.2 μl SP6 RNA-polymerase, 3 μl linearized plasmid DNA (0.1 to 0.4 μg/μl), 5 μl α-^{32}P-GTP

Incubate the transcription reaction for 40 to 60 min at 40 °C. Then add 1.5 μl DNase I (RNase-free, 1 U/μl), mix and incubate for 20 min at 37 °C.

Add 29 μl H_2O to the reaction (total volume 50 μl) and purify the labeled transcripts on Quick-spin columns (Roche; see manufacturer's recommendations).

Remove the prehybridization solution and add 8 to 10 ml of prewarmed Church buffer to the membrane. Add the flow-through to the hybridization buffer and incubate over night at 65 °C.

Wash the membrane at 65 °C: 2× 15 min with wash buffer 1; 2× 15 min with wash buffer 2.

Dry the membrane briefly and wrap in plastic wrap. Expose to Phosphor-Imager screen or X-ray film.

2.4.7. Alternative detection methods

Northern blotting represents the optimal detection method for tethering experiments, because it allows analyzing both, expression levels and correctly processing of the reporter mRNA at the same time. Generally, the detection limits of Northern blots are sufficient for many transiently expressed mRNAs. Alternative detection methods such as RNase protection assays or quantitative PCR methods could be used, but their technical limitations should be considered.

2.5. Detection of the effector protein

The fusion proteins (either the λNV5- or MS2-tagged effector) are detected with the same extracts that were also used for the isolation of total-cell or cytoplasmic RNA. This procedure enables a direct correlation of the effects of the tethered factor with its expression level. This is particularly important when different effector mutations are analyzed, because an equal expression level of all proteins is required for the interpretation of results.

Determine the protein concentration of the cell lysates.
Resolve equal amounts of protein lysates (10 to 20 μg of total protein) on an SDS-containing polyacrylamide gel.
Blot the gel onto polyvinylidene fluoride (PVDF) membrane with standard procedures.
Block the membrane with 5% nonfat skimmed milk in TBS-Tween (0.1%) for 1 h at room temperature.
Incubate with primary antibody: anti-V5 (λNV5 tagged proteins, 1:5000, Sigma) or anti-MS2 (MS2 tagged proteins, 1:2000) at room temperature for 1 h or at 4 °C overnight.
Wash four times with TBS-Tween.
Incubate with secondary antibody: anti-rabbit IgG HRP (e.g., Sigma, 1:10,000) for 1 h at room temperature.
Wash four times with TBS-Tween.
Detect with chemiluminescent detection method (e.g., ECL, Amersham).

2.6. Control experiments

The tethering assay requires a number of control experiments for validation. The aim of these control experiments is to confirm that the observed effect is not only due to binding of the effector to the reporter mRNA but also recapitulates the characteristics of NMD. We suggest testing the reporter (with the integrated binding sites) together with effectors that are not fused to an RNA-binding tag (but may be fused to another epitope tag such as FLAG or V5). A clear difference should be observed between the effector that binds and the effector that does not bind to the reporter mRNA. To test for NMD criteria, several approaches are conceivable. The degradation of NMD substrates depends on the translation of the mRNA. Thus, blocking the translation of the reporter mRNA should be tested, because this is expected to abolish the destabilizing effect of the tethered effector (Gehring *et al.*, 2003; Lykke-Andersen *et al.*, 2000). A dominant-negative version of the central NMD factor UPF1 has been characterized previously (Sun et al., 1998). This inhibits not only the NMD of PTC-containing mRNAs but also the NMD that is activated by tethering the UPF2 or UPF3b NMD factor (Lykke-Andersen et al., 2000). Similarly, siRNA-mediated depletion of NMD factors, in particular of the central NMD factor UPF1, stabilizes PTC-containing mRNAs and mRNA with tethered NMD-proteins (Gehring et al., 2003, 2005; Kim et al., 2005).

REFERENCES

Austin, R. J., Xia, T., Ren, J., Takahashi, T. T., and Roberts, R. W. (2002). Designed arginine-rich RNA-binding peptides with picomolar affinity. *J. Am. Chem. Soc.* **124,** 10966–10967.

Behm-Ansmant, I., Rehwinkel, J., Doerks, T., Stark, A., Bork, P., and Izaurralde, E. (2006). mRNA degradation by miRNAs and GW182 requires both CCR4:NOT deadenylase and DCP1:DCP2 decapping complexes. *Genes Dev.* **20,** 1885–1898.

Bertrand, E., Chartrand, P., Schaefer, M., Shenoy, S. M., Singer, R. H., and Long, R. M. (1998). Localization of ASH1 mRNA particles in living yeast. *Mol. Cell* **2,** 437–445.

Chan, C. C., Dostie, J., Diem, M. D., Feng, W., Mann, M., Rappsilber, J., and Dreyfuss, G. (2004). eIF4A3 is a novel component of the exon junction complex. *RNA* **10,** 200–209.

Danckwardt, S., Neu-Yilik, G., Thermann, R., Frede, U., Hentze, M. W., and Kulozik, A. E. (2002). Abnormally spliced beta-globin mRNAs: A single point mutation generates transcripts sensitive and insensitive to nonsense-mediated mRNA decay. *Blood* **99,** 1811–1816.

Degot, S., Le Hir, H., Alpy, F., Kedinger, V., Stoll, I., Wendling, C., Seraphin, B., Rio, M. C., and Tomasetto, C. (2004). Association of the breast cancer protein MLN51 with the exon junction complex via its speckle localizer and RNA binding module. *J. Biol. Chem.* **279,** 33702–33715.

Ferraiuolo, M. A., Lee, C. S., Ler, L. W., Hsu, J. L., Costa-Mattioli, M., Luo, M. J., Reed, R., and Sonenberg, N. (2004). A nuclear translation-like factor eIF4AIII is recruited to the mRNA during splicing and functions in nonsense-mediated decay. *Proc. Natl. Acad. Sci. USA* **101,** 4118–4123.

Fusco, D., Accornero, N., Lavoie, B., Shenoy, S. M., Blanchard, J. M., Singer, R. H., and Bertrand, E. (2003). Single mRNA molecules demonstrate probabilistic movement in living mammalian cells. *Curr. Biol.* **13,** 161–167.

Gehring, N. H., Neu-Yilik, G., Schell, T., Hentze, M. W., and Kulozik, A. E. (2003). Y14 and hUpf3b form an NMD-activating complex. *Mol. Cell* **11,** 939–949.

Gehring, N. H., Kunz, J. B., Neu-Yilik, G., Breit, S., Viegas, M. H., Hentze, M. W., and Kulozik, A. E. (2005). Exon-junction complex components specify distinct routes of nonsense-mediated mRNA decay with differential cofactor requirements. *Mol. Cell* **20,** 65–75.

Gray, N. K., Coller, J. M., Dickson, K. S., and Wickens, M. (2000). Multiple portions of poly(A)-binding protein stimulate translation *in vivo*. *EMBO J.* **19,** 4723–4733.

Hentze, M. W., and Kulozik, A. E. (1999). A perfect message: RNA surveillance and nonsense-mediated decay. *Cell* **96,** 307–310.

Holbrook, J. A., Neu-Yilik, G., Hentze, M. W., and Kulozik, A. E. (2004). Nonsense-mediated decay approaches the clinic.. *Nat. Genet.* **36,** 801–808.

Hook, B., Bernstein, D., Zhang, B., and Wickens, M. (2005). RNA-protein interactions in the yeast three-hybrid system: affinity, sensitivity, and enhanced library screening. *RNA* **11**, 227–233.

Kashima, I., Yamashita, A., Izumi, N., Kataoka, N., Morishita, R., Hoshino, S., Ohno, M., Dreyfuss, G., and Ohno, S. (2006). Binding of a novel SMG-1-Upf1-eRF1-eRF3 complex (SURF) to the exon junction complex triggers Upf1 phosphorylation and nonsense-mediated mRNA decay. *Genes Dev.* **20**, 355–367.

Kataoka, N., Yong, J., Kim, V. N., Velazquez, F., Perkinson, R. A., Wang, F., and Dreyfuss, G. (2000). Pre-mRNA splicing imprints mRNA in the nucleus with a novel RNA-binding protein that persists in the cytoplasm. *Mol. Cell* **6**, 673–682.

Kataoka, N., Diem, M. D., Kim, V. N., Yong, J., and Dreyfuss, G. (2001). Magoh, a human homolog of *Drosophila mago nashi* protein, is a component of the splicing-dependent exon-exon junction complex. *EMBO J.* **20**, 6424–6433.

Kim, V. N., Kataoka, N., and Dreyfuss, G. (2001). Role of the nonsense-mediated decay factor hUpf3 in the splicing-dependent exon-exon junction complex. *Science* **293**, 1832–1836.

Kim, Y. K., Furic, L., Desgroseillers, L., and Maquat, L. E. (2005). Mammalian Staufen1 recruits Upf1 to specific mRNA 3′ UTRs so as to elicit mRNA decay. *Cell* **120**, 195–208.

Kiriakidou, M., Tan, G. S., Lamprinaki, S., De Planell-Saguer, M., Nelson, P. T., and Mourelatos, Z. (2007). An mRNA m7G cap binding-like motif within human Ago2 represses translation. *Cell* **129**, 1141–1151.

Kunz, J. B., Neu-Yilik, G., Hentze, M. W., Kulozik, A. E., and Gehring, N. H. (2006). Functions of hUpf3a and hUpf3b in nonsense-mediated mRNA decay and translation. *RNA* **12**, 1015–1022.

Le Hir, H., Gatfield, D., Izaurralde, E., and Moore, M. J. (2001a). The exon-exon junction complex provides a binding platform for factors involved in mRNA export and nonsense-mediated mRNA decay. *EMBO J.* **20**, 4987–4997.

Le Hir, H., Gatfield, D., Braun, I. C., Forler, D., and Izaurralde, E. (2001b). The protein Mago provides a link between splicing and mRNA localization. *EMBO Rep.* **2**, 1119–1124.

LeCuyer, K. A., Behlen, L. S., and Uhlenbeck, O. C. (1995). Mutants of the bacteriophage MS2 coat protein that alter its cooperative binding to RNA. *Biochemistry* **34**, 10600–10606.

Lim, F., and Peabody, D. S. (1994). Mutations that increase the affinity of a translational repressor for RNA. *Nucleic Acids Res.* **22**, 3748–3752.

Lu, S., and Cullen, B. R. (2004). Nonsense mediated decay induced by tethered human UPF3B is restricted to the cytoplasm. *RNA Biol.* **1**, 42–47.

Lykke-Andersen, J., Shu, M. D., and Steitz, J. A. (2000). Human Upf proteins target an mRNA for nonsense-mediated decay when bound downstream of a termination codon. *Cell* **103**, 1121–1131.

Lykke-Andersen, J., Shu, M. D., and Steitz, J. A. (2001). Communication of the position of exon-exon junctions to the mRNA surveillance machinery by the protein RNPS1. *Science* **293**, 1836–1839.

Lykke-Andersen, J., and Wagner, E. (2005). Recruitment and activation of mRNA decay enzymes by two ARE-mediated decay activation domains in the proteins TTP and BRF-1. *Genes Dev.* **19**, 351–361.

Maquat, L. E. (2004). Nonsense-mediated mRNA decay: Splicing, translation and mRNP dynamics. *Nat. Rev. Mol. Cell. Biol.* **5**, 89–99.

Nagai, K. (1996). RNA-protein complexes. *Curr. Opin. Struct. Biol.* **6**, 53–61.

Nott, A., Le Hir, H., and Moore, M. J. (2004). Splicing enhances translation in mammalian cells: An additional function of the exon junction complex. *Genes Dev.* **18**, 210–222.

Palacios, I. M., Gatfield, D., St. Johnston, D., and Izaurralde, E. (2004). An eIF4AIII-containing complex required for mRNA localization and nonsense-mediated mRNA decay. *Nature* **427,** 753–757.

Peabody, D. S., and Lim, F. (1996). Complementation of RNA binding site mutations in MS2 coat protein heterodimers. *Nucleic Acids Res.* **24,** 2352–2359.

Pillai, R. S., Artus, C. G., and Filipowicz, W. (2004). Tethering of human Ago proteins to mRNA mimics the miRNA-mediated repression of protein synthesis. *RNA* **10,** 1518–1525.

Pillai, R. S., Bhattacharyya, S. N., Artus, C. G., Zoller, T., Cougot, N., Basyuk, E., Bertrand, E., and Filipowicz, W. (2005). Inhibition of translational initiation by Let-7 MicroRNA in human cells. *Science* **309,** 1573–1576.

Sanford, J. R., Gray, N. K., Beckmann, K., and Caceres, J. F. (2004). A novel role for shuttling SR proteins in mRNA translation. *Genes Dev.* **18,** 755–768.

Sheth, U., and Parker, R. (2003). Decapping and decay of messenger RNA occur in cytoplasmic processing bodies. *Science* **300,** 805–808.

Shibuya, T., Tange, T. O., Sonenberg, N., and Moore, M. J. (2004). eIF4AIII binds spliced mRNA in the exon junction complex and is essential for nonsense-mediated decay. *Nat. Struct. Mol. Biol.* **11,** 346–351.

Sun, X., Perlick, H. A., Dietz, H. C., and Maquat, L. E. (1998). A mutated human homologue to yeast Upf1 protein has a dominant-negative effect on the decay of nonsense-containing mRNAs in mammalian cells. *Proc. Natl. Acad. Sci. USA* **95,** 10009–10014.

Thermann, R., Neu-Yilik, G., Deters, A., Frede, U., Wehr, K., Hagemeier, C., Hentze, M. W., and Kulozik, A. E. (1998). Binary specification of nonsense codons by splicing and cytoplasmic translation. *EMBO J.* **17,** 3484–3494.

Valegard, K., Murray, J. B., Stockley, P. G., Stonehouse, N. J., and Liljas, L. (1994). Crystal structure of an RNA bacteriophage coat protein-operator complex. *Nature* **371,** 623–626.

Wiegand, H. L., Lu, S., and Cullen, B. R. (2003). Exon junction complexes mediate the enhancing effect of splicing on mRNA expression. *Proc. Natl. Acad. Sci. USA* **100,** 11327–11332.

Yamasaki, S., Stoecklin, G., Kedersha, N., Simarro, M., and Anderson, P. (2007). T-cell intracellular antigen-1 (TIA-1)-induced translational silencing promotes the decay of selected mRNAs. *J. Biol. Chem.* **282,** 30070–30077.

Yi, R., Bogerd, H. P., and Cullen, B. R. (2002). Recruitment of the Crm1 nuclear export factor is sufficient to induce cytoplasmic expression of incompletely spliced human immunodeficiency virus mRNAs. *J. Virol.* **76,** 2036–2042.

Zhang, J., Sun, X., Qian, Y., LaDuca, J. P., and Maquat, L. E. (1998a). At least one intron is required for the nonsense-mediated decay of triosephosphate isomerase mRNA: A possible link between nuclear splicing and cytoplasmic translation. *Mol. Cell. Biol.* **18,** 5272–5283.

Zhang, J., Sun, X., Qian, Y., and Maquat, L. E. (1998b). Intron function in the nonsense-mediated decay of beta-globin mRNA: Indications that pre-mRNA splicing in the nucleus can influence mRNA translation in the cytoplasm. *RNA* **4,** 801–815.

CHAPTER TWENTY-FOUR

Assays for Determining Poly(A) Tail Length and the Polarity of mRNA Decay in Mammalian Cells

Elizabeth L. Murray *and* Daniel R. Schoenberg

Contents

1. Introduction: Poly(A) Tail Length Assays	484
1.1. Poly(A) length assay	484
1.2. Ligation-mediated poly(A) test (LM-PAT)	488
1.3. RNase H assay	490
2. Introduction: Invader RNA Assay	492
2.1. Materials: Invader RNA assay	494
2.2. Methods: Invader RNA assay	499
Acknowledgments	504
References	504

Abstract

This chapter describes several methods for measuring the length of the mRNA poly(A) tail and a novel method for measuring mRNA decay. Three methods for measuring the length of a poly(A) tail are presented: the poly(A) length assay, the ligation-mediated poly(A) test (LM-PAT), and the RNase H assay. The first two methods are PCR-based assays involving cDNA synthesis from an oligo(dT) primer. The third method involves removing the poly(A) tail from the mRNA of interest. A major obstacle to studying the enzymatic step of mammalian mRNA decay has been the inability to capture mRNA decay intermediates with structural impediments such as the poly(G) tract used in yeast. To overcome this, we combined a standard kinetic analysis of mRNA decay with a tetracycline repressor–controlled reporter with an Invader® RNA assay. The Invader RNA assay is a simple, elegant assay for the quantification of mRNA. It is based on signal amplification, not target amplification, so it is less prone to artifacts than

Department of Molecular and Cellular Biochemistry and the RNA Group, The Ohio State University, Columbus, Ohio, USA

Methods in Enzymology, Volume 448
ISSN 0076-6879, DOI: 10.1016/S0076-6879(08)02624-4

© 2008 Elsevier Inc.
All rights reserved.

other methods for nucleic acid quantification. It is also very sensitive, able to detect attomolar levels of target mRNA. Finally, it requires only a short sequence for target recognition and quantitation. Therefore, it can be applied to determining the decay polarity of a mRNA by measuring the decay rates of different portions of that mRNA.

1. Introduction: Poly(A) Tail Length Assays

1.1. Poly(A) length assay

This is a straightforward and fast assay that can be completed in a day. The first step involves the synthesis of cDNA from the RNA sample with an oligo(dT) primer. The next step is to perform PCR on the cDNA with a ^{32}P end–labeled target mRNA–specific primer. After PCR, samples are resolved in a polyacrylamide gel. mRNAs with short tails yield a compact band, whereas mRNAs with long tails yield PCR products of a variety of lengths that appear as a smear on the gel. Enough cDNA is produced for the PCR step to be repeated a number of times.

1.1.1. Materials for the poly(A) length assay

Two oligonucleotides: An oligo(dT) primer/adapter (5′-GGGGATCCGC GGTTTTTTTTT) and an mRNA-specific primer. The mRNA-specific primer should be located upstream of the polyadenylation site at a location that will give a convenient size for resolution on a 6% polyacrylamide gel, approximately 100 to 300 bases.
M-MLV reverse transcriptase (Invitrogen).
T4 polynucleotide kinase (Roche).
Standard molecular biology reagents and equipment (heat block, thermocycler, PCR reagents, equipment for polyacrylamide gel electrophoresis).

1.1.2. Methods for the poly(A) length assay

1.1.2.1. RNA preparation RNA may be purified from cells with any number of techniques. However, it is important that contaminating DNA is removed from the sample before cDNA synthesis and that the final RNA sample is sufficiently concentrated. In our laboratory, we typically isolate RNA with TRIzol (Invitrogen), followed by treatment with DNase, purification by organic extraction, and precipitation with isopropanol.

A 60-mm dish of cells should give enough RNA to repeat the poly(A) assay several times. Before harvesting the cells, wash the plate two or three times with PBS. Then add 1 ml TRIzol, and incubate the plate on a shaker or rocker for 5 min at room temperature to lyse the cells directly on the plate. Transfer the TRIzol to a microcentrifuge tube and continue to purify the

RNA according to the manufacturer's directions. Glycogen may be added during the isopropanol precipitation step to improve the yield of the RNA precipitation. Dissolve the RNA pellet in RNase-free water and quantitate it by measuring the absorbance at 260 nm. Next, RNA preparations should be treated with DNase to remove any contaminating DNA. To do this, bring the volume of the RNA preparation to 300 μl and add 30 μl 10× DNase reaction buffer and 10 μl RQ DNase (RNase-free, Promega) then incubate at 37 °C for 15 min. After incubation, purify the RNA by adding 30 μl 5 M ammonium acetate and 350 μl phenol/chloroform/isoamyl alcohol, followed by vortexing for 10 sec, and centrifugation at maximum speed for 5 min to separate the phases. Remove the top layer containing the RNA and place it in a clean microcentrifuge tube. Add an equal volume of chloroform/isoamyl alcohol and repeat the extraction. After centrifugation, transfer the top layer to a clean tube and precipitate the RNA by adding an equal volume of isopropanol to the RNA. Place the tube at −20 °C for at least 15 min and then pellet the RNA by centrifuging it at maximum speed for 20 min. After centrifugation, remove the supernatant and discard it. Wash the pellet by adding 1 ml of 70% ethanol to the tube, briefly vortexing, and then centrifuging again at maximum speed for 5 min. After centrifugation, remove the supernatant from the pellet and discard. Centrifuge the tube briefly to collect any remaining ethanol from the sides of the tube and then remove it by pipetting. Briefly air-dry the pellet and then dissolve it in RNase-free water. The typical RNA recovery from this treatment is 60%, so dissolve the pellet in a volume of water to yield a final concentration greater than 1 μg RNA per μl and then quantitate the RNA sample.

1.1.2.2. cDNA synthesis In this step, the oligo(dT) primer/adapter is annealed to polyadenylated mRNAs and then extended in a cDNA synthesis reaction. Those RNAs with short poly(A) tails that have a limited number of sites for oligo(dT) binding will give cDNA products of a uniform size. Those RNAs with long poly(A) tails will have multiple sites for binding the oligo(dT) primer/adapter and will give cDNA products of heterogeneous size. For this assay, we have found that the M-MLV reverse transcriptase gives satisfactory results. However, engineered reverse transcriptases (i.e., Superscript RT) may give better results in some circumstances; therefore, this is a factor to consider when optimizing the assay.

To synthesize the cDNA, first denature the RNA and anneal the oligo(dT) primer/adapter to the sample. Combine 10 μg (or more) RNA and 300 ng oligo(dT) primer/adapter in RNase-free water to a volume of 16.5 μl and heat to 85 °C for 5 min and then cool quickly on ice. For the reverse transcription reaction, add 6 μl 5× first-strand buffer, 3 μl 0.1 M DTT, 2 μl 10 mM dNTPs, 0.5 μl RNase Inhibitor (Invitrogen), and 2 μl M-MLV reverse transcriptase (Invitrogen) and incubate at 42 °C for 1.5 h. After the incubation, inactivate the reverse transcriptase by heating the reaction to 85 °C for 10 min. The cDNA may be used immediately as the template for PCR or stored at −20 °C.

1.1.2.3. Second primer end labeling   In this step, the upstream, target-specific primer for PCR is 5′-end labeled with γ-^{32}P–ATP by T4 polynucleotide kinase (PNK). The primer should be located approximately 100 nt upstream of the polyadenylation site so that the PCR products will be between 100 nt and 300 nt, a size easily resolved in a 6% polyacrylamide gel. Primer labeling can be performed during the cDNA synthesis; it is convenient to label the size marker at this time also.

To perform the end-labeling reaction, combine 250 ng primer, 2.5 μl 10× polynucleotide kinase buffer (supplied with the enzyme), 3 μl γ-^{32}P–ATP, 1 μl T4 PNK (Roche), and RNase-free water to 25 μl and incubate the reaction at 37 °C for 15 to 60 min. After incubation, heat-inactivate the PNK by incubating the reaction at 65 °C for 5 min. The labeled oligo may be purified with a G25 column or a silica spin-column (i.e., Qiagen Nucleotide Removal Kit), but purification is not required, because any unincorporated γ-^{32}P–ATP will run far ahead of the PCR products in a gel and not interfere with their visualization. The labeled oligo may be used immediately for PCR or stored at −80 °C.

1.1.2.4. PCR   The cDNA is used as a template for amplification by a PCR reaction with the 5′-end labeled target-specific primer and any unincorporated oligo(dT) primer/adapter. As noted previously, cDNAs from mRNAs with short poly(A) tails will give PCR products of a more uniform size because of the limited sites for oligo(dT) priming, whereas cDNAs from mRNAs with long heterogeneous tails will yield PCR products with heterogeneous sizes because of the availability of multiple sites for oligo(dT) priming. An example of this is shown in Fig. 24.1A, which uses this assay to show the impact of a poly(A)-limiting element (PLE, Das Gupta *et al.*, 1998) on the length of the poly(A) tail on luciferase mRNA in transfected murine fibroblasts. This PCR is performed in hot-start tubes and is the step in the poly(A) assay that requires the most optimization. Some important variables to consider include the location of the target mRNA–specific oligonucleotide, the amount of cDNA template in the reaction, the concentration of the MgCl$_2$ in the reaction, the kind of Taq DNA polymerase in the reaction, the annealing temperature of the cycling reaction, the length of the extension time in the cycling reaction, and the number of cycles of the PCR reaction. When optimizing the PCR reaction, it is helpful to have *in vitro* transcripts of the target mRNA with A20 and A100 tails to serve as positive controls for short and long tails, respectively.

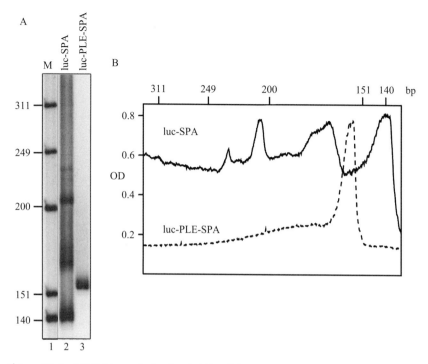

Figure 24.1 RT-PCR analysis of poly(A) tail length. RNA was recovered from cells expressing luciferase mRNA with a synthetic polyadenylation element (luc-SPA) or luciferase mRNA with a poly(A) limiting element (luc-PLE-SPA) that restricts the length of the poly(A) tail to <20 nucleotides. The length of the poly(A) tail on each mRNA was assayed by RT-PCR with an upstream radiolabeled primer that anneals to luciferase mRNA and the oligo(dT) primer adapter. The resulting PhosphorImager analysis is shown in (A). (B) A graphic view of the data in (A) obtained with the graphing function within ImageQuant. Data reveal that the poly(A) tail of luc-SPA mRNA analysis of the tail shows is heterogeneous in length with peaks corresponding to the 25-nucleotide spacing for poly(A)-binding protein. In contrast, the poly(A) tail of luc-PLE-SPA mRNA manifests a discrete length.

The following protocol serves as a good starting point. In a hot-start tube, combine 3 μl cDNA, 2.5 μl 10× PCR buffer, 3 μl 50 mM MgCl$_2$, and 3 μl 1.25 mM dNTPs. Heat the tube at 85 to 95 °C for a few seconds to melt the wax, then put the tube on ice to allow the wax to solidify. After the wax solidifies, add 2 μl 5′-labeled primer, 11 μl RNase-free water, and 0.5 μl Taq DNA Polymerase (Invitrogen). Incubate the reaction at 94 °C for 2 min, then perform 25 cycles of 94 °C for 1 min, 60 °C for 30 sec, and 72 °C for 2 min. Finish the PCR reaction with a final extension at 72 °C for 2 min. PCR products may be immediately run in a polyacrylamide gel or stored at −20 °C.

1.1.2.5. Resolution of PCR products by electrophoresis The final step is to combine 3 µl of each PCR reaction with 3 µl formamide loading buffer (95% formamide, 0.025% xylene cyanol and bromophenol blue, 18 mM EDTA and 0.025% SDS), heat to 95 °C for 5 min, and resolve in a 6% polyacrylamide gel. For proper resolution of the PCR products, the gel itself should be the length and thickness of a sequencing gel. Prerun the gel in 1× TBE at 60 W for 30 min before adding the samples, and run the gel at 60 W for approximately 70 min after the samples are loaded, or until the lower dye front is at the bottom of the gel and the upper dye front is midway down the gel. Remove one glass plate and transfer the gel to a piece of Whatman paper; dry the gel under heat (80 °C) and vacuum for approximately 70 min. When the gel is dry, expose it to film or a phosphor screen overnight. If one uses a storage phosphor screen and an instrument such as a PhosphorImager, the graphing function of ImageQuant can be used to obtain a graphic distribution of poly(A) tail lengths that is useful for determining the length distribution of the poly(A) tail (see Fig. 24.1B). mRNAs with a short tail will give a discrete peak, whereas an mRNA with a long tail will generate PCR products in a range of sizes that can appear as a smear of bands either with or without several peaks at 25 nucleotide intervals, which correspond in size to the distribution of poly(A)-binding protein.

1.2. Ligation-mediated poly(A) test (LM-PAT)

This assay differs from the poly(A) length assay in several ways. First, the molar ratio of RNA to oligo(dT) is much lower so that the entire length of each poly(A) tail (with the exception of some nucleotides at the 5′-most and 3′-most ends) is saturated with oligo(dT) primers. Second, the oligo(dT) primer has a 5′-phosphate to allow the ligation of adjacent primers. Third, an additional oligo(dT) primer-adapter at the 5′-end is ligated to the 3′-most oligo(dT) primer. Although there are a few additional steps compared with the poly(A) length assay, this is still a relatively fast assay, and, like the poly (A) length assay, enough cDNA is synthesized to repeat the PCR analysis several times. This assay should be less biased to short tails than the poly(A) length assay.

1.2.1. Materials: LM-PAT assay

Three oligonucleotides; oligo(dT)$_{12-18}$, the oligo(dT) primer/adapter (see 1.1.1), and a target mRNA-specific primer.
Superscript II reverse transcriptase (Invitrogen),
T4 DNA ligase (Invitrogen).
Standard equipment and reagents for molecular biology (heat blocks, thermocycler, reagents for PCR, equipment for polyacrylamide gel electrophoresis).

1.2.2. Methods: LM-PAT assay

1.2.2.1. RNA preparation The preparation of RNA suitable for LM-PAT analysis is the same as the preparation of RNA for the poly(A) length assay (see 1.1.2.1).

1.2.2.2. Primer annealing and primer ligation A broad range of RNA concentrations can be used with this assay (typically 20 ng to 1 µg), but the optimum concentration of RNA for a given experiment may be empirically determined by performing several LM-PAT assays with different concentrations of the same RNA sample. The addition of too much sample RNA will interfere with the saturation of the poly(A) tail with oligo(dT) binding and ligation of adjacent oligo(dT) primers, which can cause a "laddering" effect on electrophoresis, or poor PCR amplification (Salles and Strickland, 1999).

To anneal oligo(dT)$_{12-18}$ to polyadenylated RNA, dilute the RNA sample to a concentration of 180 to 200 ng/µl and combine 1 µl diluted RNA, 4.5 µl RNase-free water, and 2 µl oligo(dT)$_{12-18}$ (at 10 ng/µl, with 5′-phosphorylated end) in a microcentrifuge tube. Heat-denature the reaction to 65 °C for 10 min and then immediately transfer it to 42 °C. While the tube is at 42 °C, add 4 µl 5× first-strand buffer, 1 µl RNase Inhibitor (Invitrogen), 2 µl 0.1 M DTT, 1 µl 10 mM dNTP, 1 µl 10 mM ATP, 3 µl RNase-free water, and 1.5 µl T4 DNA ligase (Invitrogen) to ligate the adjacent oligo(dT)$_{12-18}$ that are annealed to the poly(A) tails. These reagents should be mixed together and prewarmed before adding to the RNA-oligo(dT) sample. A large amount of DNA ligase is required in this step to offset the enzyme's half-life at this temperature (Salles and Strickland, 1999). After the addition of the mix, combine the reagents thoroughly by pipetting and continue to incubate the reaction for an additional 30 min. While the reaction is still at 42 °C, ligate the oligo(dT) primer/adapter to the 5′-phosphorylated oligo(dT)$_{12-18}$ primer by adding 1 µl of the oligo(dT) primer/adapter (200 ng/µl) to the reaction and immediately transferring it to 12 °C. After a few minutes at 12 °C, centrifuge the tube briefly to collect the entire reaction at the bottom of the tube and incubate it at 12 °C for 2 h. The 10-fold excess of the oligo(dT) primer/adapter and the low temperature favor the oligo(dT) primer/adapter over the oligo(dT)$_{12-18}$ primer for annealing and ligation to the 3′-end of the poly(A) tail, such that most cDNA products will have the GC-rich adapter sequence at the 5′-end.

1.2.2.3. Reverse transcription The ligation reaction is placed at 42 °C for 2 min, followed by addition of 1 µl Superscript II reverse transcriptase (Invitrogen) and incubation at 42 °C for 1 h. The brief incubation at 42 °C before addition of the reverse transcriptase serves to decrease, but not eliminate, annealing of the oligo(dT) primer/adapter to the 5′-end of the poly(A) tail. Both ligase and the reverse transcriptase are inactivated by incubating the reaction at 65 °C for 20 min. The cDNA synthesis is now complete, and the cDNA is ready for use as a template in PCR amplification or it may be stored at −20 °C.

1.2.2.4. Target-specific primer end labeling, PCR amplification, and resolution of PCR products by polyacrylamide gel electrophoresis These steps are similar to those described for the poly(A) length assay (see 1.1.2).

1.3. RNase H assay

The previously described methods for poly(A) length measurement rely on cDNA synthesis from an oligo(dT) primer and PCR amplification of the cDNA products, and are thus indirect measurements of the poly(A) tail length. In contrast, the RNase H assay involves the removal of the poly(A) tail from the mRNA and the direct comparison of the deadenylated mRNA with its polyadenylated control by Northern blotting. Although this assay is more time-consuming than the poly(A) length assay or the LM-PAT assay, it can be useful, in particular when the PCR step in the previous two methods is proving problematic.

RNase H is an endoribonuclease that specifically hydrolyzes the phosphodiester bonds of RNA in RNA/DNA duplexes. In this application, the poly(A) tail is removed by the action of RNase H on the oligo(dT)/poly(A) duplex. RNase H does not degrade single-stranded or double-stranded RNA or DNA. The size of a poly(A) tail on a given mRNA can be determined by running reactions with and without RNase H side by side on a gel and comparing the mobilities of the two samples. If the size of the mRNA of interest is so large that removing the poly(A) tail does not give a measurable shift in electrophoretic mobility, a second deoxyoligonucleotide can be annealed upstream of the poly(A) tail, and the products generated by RNase H cleavage at this site are compared side by side in identical reactions ±oligo(dT).

1.3.1. Materials: RNase H assay

Two oligonucleotides: oligo(dT)$_{12-18}$ and a target mRNA-specific primer (optional). The target mRNA-specific oligo should be located no more than 400 nt upstream of the polyadenylation site such that cleavage of the RNA/DNA hybrid between the primer and the mRNA leaves an mRNA product that is large enough to be visualized by Northern blot analysis but small enough that removal of the poly(A) tail makes an observable difference in its electrophoretic mobility.
RNase H (United States Biochemicals, Inc.).

Standard molecular biology equipment and reagents (heat blocks, equipment for agarose gel electrophoresis, RNA transfer, and Northern blotting).

1.3.2. Methods: RNase H assay

1.3.2.1. RNA purification RNA is purified for RNase H assay as for the poly(A) length assay or LM-PAT.

1.3.2.2. Primer annealing and RNase H reaction One or two oligonucleotides may be annealed in this step, depending on the length of the mRNA body (e.g., whether or not it is necessary to shorten the length of the mRNA body to see the effect of removing the poly(A) tail on its electrophoretic mobility). All reactions should be prepared in duplicate; one with the addition of oligo(dT), the other without.

The RNA is first heat denatured, followed by a slow annealing of the oligonucleotides, and then RNase H is added to degrade the RNA/DNA hybrids. Combine 10 μg RNA, 5 μl oligo(dT) primer (at 600 ng/μl), 5 μl internal primer (at 600 ng/μl, optional) and 4 μl 5× RNase H buffer in a volume of 18.8 μl. Heat the reaction to 85 °C for 5 min to denature the RNA, then incubate it at 42 °C for 10 min followed by slow cooling to 32 °C. This is readily accomplished by performing the 42 °C incubation in a water-filled 250-ml beaker in a waterbath and then transferring the beaker from the waterbath to the benchtop where it should cool at a rate of 1 °C per min. When the reaction reaches 32 °C, add 0.5 μl RNase Inhibitor (Invitrogen) and 0.7 μl RNase H (USB) and incubate it at 37 °C for 1 h. The choice of RNase inhibitor is important here, because some inhibit a wide variety of RNases and may, therefore, interfere with the degradation of the specific RNA/DNA duplexes.

1.3.2.3. Purify and concentrate the RNA by organic extraction and ethanol precipitation To purify the RNA from the RNase H reaction, first bring the volume of the RNase H reaction to 150 μl with RNase-free water and add an equal volume of phenol/chloroform/isoamyl alcohol. Extract the RNA by vortexing the reactions for 10 sec and then centrifuge them for 5 min at maximum speed to separate the layers. After centrifugation, transfer the top layer containing the RNA to a clean microcentrifuge tube and add 1 μl glycogen (20 μg/μl), 10% volume 5 M ammonium acetate, and 2.5× volume 95% ethanol. Precipitate at −20 °C for at least 20 min and then centrifuge the samples for 20 min to pellet the RNA. After centrifugation, remove the supernatant and dissolve the RNA pellet in a volume of buffer appropriate for electrophoresis.

1.3.2.4. Visualization of RNA products RNase H degradation products are visualized by Northern blotting, and if one uses an upstream primer in the cleavage reaction to generate a smaller 3′-product, it is best to use a hybridization probe that is specific for the 3′-end of the mRNA. (Chapter 13, Section 2.1.1.2.2) describes an asymmetric PCR protocol that is particularly useful for generating probes that are specific to a given portion of the mRNA. The products of RNase H digestion are separated on either agarose or polyacrylamide gels, transferred to a nylon membrane, and hybridized with a radioactively labeled probe. As with the PCR-based assays, the difference in the sizes of the poly(A) tails can be most easily determined by use of a PhosphorImager and analyzing differences in mobility of the fragments with and without the poly(A) tail with ImageQuant. Note that appropriate molecular weight standards should be run on the same gel. RNA from which the poly(A) tail has been removed with RNase H plus oligo(dT) will appear as a tight band corresponding to the distance between the upstream primer binding site and the 3′-end of the mRNA. Excluding oligo(dT) from the reaction will result in an RNase H cleavage product corresponding to the same 3′ sequence with poly(A). The difference between the mobility of this fragment and the fragment lacking poly(A) results from the presence or absence of the poly(A) tail and from this one deduces its size.

2. Introduction: Invader RNA Assay

The methods available to quantify mRNA for studying mRNA decay include Northern blotting, ribonuclease protection assay (RPA), RT-PCR, and Invader RNA chemistry. All these methods have different advantages and limitations. The advantage of Northern blotting is the ability to monitor changes in the full-length mRNA (e.g., deadenylation) and, in some cases, to detect intermediates in the decay process. RT-PCR and RPA are more sensitive than Northern blotting and are, therefore, useful for assaying less abundant transcripts, or when only a small amount of sample is available. When used with multiple probe sets to different portions of a target mRNA, an Invader assay approach (Third Wave Technologies, Inc., www.twt.com) combines high sensitivity with the ability to offer insights into decay mechanics by quantifying changes in specific portions of mRNA.

The Invader assay (diagrammed in Fig. 24.2) consists of two reactions that are run sequentially in the same well at the same temperature (Eis et al., 2001). In the primary reaction, three oligonucleotides (stacking oligonucleotide, primary probe oligonucleotide, and invasive oligonucleotide) specifically bind a region of the target mRNA. The primary probe oligonucleotide has two parts, an assay-specific region (ASR) that is complementary to a portion of the target mRNA and a "flap" that is not complementary to the target mRNA. The invasive oligonucleotide binds the target mRNA immediately downstream of the bound ASR of the primary probe oligonucleotide. The 3′-most nucleotide of the invasive

Characterization of mRNA Decay

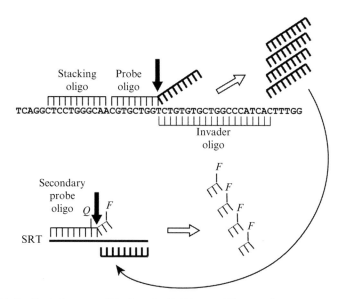

Figure 24.2 Basic features of the Invader RNA assay. The Invader RNA assay uses two isothermal reactions. In the first reaction, the target mRNA is bound by a stacking oligonucleotide that serves to stabilize the binding of a probe oligonucleotide and an invasive oligonucleotide whose Tm is greater than that of the primary probe bound immediately downstream. This has a one-nucleotide overlap that distorts the duplex between the primary probe and the target, creating a site for cleavage by the Cleavase enzyme (filled arrow). Cleavage generates a generic flap sequence (shown in bold), which serves as the invasive oligonucleotide in a second reaction in which a FRET-labeled probe oligonucleotide is bound to a secondary reaction template (SRT). Cleavage at the overlap in this complex generates a fluorescent signal. Shown are the β-globin exon 3 probe sets from Murray and Schoenberg (2007).

oligonucleotide does not bind the mRNA target or the primary probe, but instead pushes on the 5′-most bond between the primary probe and the target mRNA to create a bulged secondary structure. The thermostable Cleavase enzyme recognizes the specific one-base overlap structure and cleaves the primary probe oligonucleotide to release the 5′-flap. The stacking oligonucleotide binds upstream of the complementary portion of the primary probe oligonucleotide and serves to modulate the Tm of the primary probe to 60 °C. At the reaction temperature of 60 °C, the primary probe oligonucleotide freely cycles on and off the target mRNA, while the invasive oligonucleotide, which has a higher Tm, remains bound. Thus, multiple primary probes are cleaved per target, and the number of free flaps produced in the primary reaction is a product of the length of incubation and target abundance.

In the secondary reaction, three additional oligonucleotides are added: an arrestor oligonucleotide (not shown), a FRET-labeled probe oligonucleotide (containing both a fluorophore and a quencher molecule), and a secondary reaction template (SRT) oligonucleotide (a synthetic sequence). The cleaved flap of the primary probe from the primary reaction acts as an invasive oligonucleotide in conjunction with the SRT and the FRET-labeled probe oligonucleotide in another cleavage reaction in which the FRET-labeled probe oligonucleotide is cleaved to separate the fluorophore on the flap from the quencher molecule. As in the primary reaction, the cleaved 5′-probe flap from the primary probe (acting as an invasive oligonucleotide) remains bound to the SRT, while the FRET-labeled probe oligonucleotide cycles on and off, resulting in linear fluorescence signal generation. The arrestor oligonucleotide binds uncleaved primary probe oligonucleotide. This assay can be run in biplex format to allow the simultaneous detection of two targets from any one sample, which is helpful for the use of internal controls.

2.1. Materials: Invader RNA assay

2.1.1. *In vitro* synthesized transcript of the gene of interest

The *in vitro* synthesized transcript of the gene of interest serves as a control for the Invader RNA assay. It is used to optimize the assay and also to perform the standard curves that must always be performed when running the assay. The transcript ideally spans the entire length of the mature mRNA and includes a poly(A) tail. The length of the tail may be short (e.g., A20) so that the concentration of the transcript may be accurately determined. The plasmid serving as the template for *in vitro* transcription should contain the cDNA of the gene of interest downstream of a promoter that will transcribe the cDNA in the sense direction and upstream of a restriction site for plasmid linearization.

Before *in vitro* transcription, the plasmid template must be linearized by restriction enzyme digestion. It is essential to check for complete linearization by running a portion of the product in an agarose gel and staining with ethidium bromide or other nucleic acid staining reagent before UV visualization. The digestion product must be gel purified if it is not a single band. After restriction enzyme digestion, the linearized plasmid should be purified. This can be accomplished with a spin column kit such as the PCR Clean up kit (Qiagen).

We use the Megascript kit (Ambion) for *in vitro* transcription. For each transcription reaction, combine 1 μg template DNA, 2 μl 10× RNA polymerase buffer, 2 μl each ATP, CTP, GTP, or UTP solution, 2 μl RNA polymerase, and RNase-free water to 20 μl and incubate overnight at 37 °C. The plasmid template can be removed from the *in vitro* synthesized transcript by adding 1 μl DNase I to the transcription reaction and incubating it for 15 min at 37 °C. After DNase treatment, the transcript should be purified by organic extraction and concentrated by precipitation. To do this, first bring the volume of the transcription reaction up to 100 μl with

RNase-free water, then add 100 μl phenol/chloroform/isoamyl alcohol, vortex for 10 sec and centrifuge for 5 min at maximum speed. Remove the top (RNA-containing) layer to a clean microcentrifuge tube and precipitate with isopropanol. After centrifugation to pellet the RNA, remove the supernatant and dissolve the pellet in 50 μl RNase-free water. Check the integrity of the *in vitro* transcript by running an aliquot of it in an agarose gel and quantify the *in vitro* transcript by measuring the absorbance at 260 nm. Dilute the transcript to a concentration of 100 amol/μl and store aliquots at −80 °C.

2.1.1.1. Primary reaction oligonucleotide design, synthesis, and preparation The primary probe oligonucleotide is composed of two regions, an assay-specific region (ASR) and a flap region. The ASR is ≥ 10 bases in length to ensure specificity, and it anneals to the target mRNA. The flap region does not hybridize to the target mRNA and is a synthetic sequence designed only for use in the assay. The stacking oligonucleotide binds the target mRNA at the 3′-end of the probe ASR and stacks coaxially with it. The function of the stacking oligonucleotide is to improve assay performance by increasing the Tm of the primary probe, and thus the assay reaction temperature. Together, the primary probe and stacking oligonucleotide should have a Tm of 60 °C. The Tm may be slightly higher, but a Tm of ≥65 °C is detrimental to the assay performance. Because the Tm of the primary probe oligonucleotide is the temperature at which the assay is run, the primary probe oligonucleotide cycles on and off the target freely during the reaction. As noted previously, the Tm of the invasive oligonucleotide is ∼80 °C to ensure that it is stably bound to the mRNA target throughout the assay. Because the difference in Tms between the stacking and primary probe oligonucleotides and the invasive oligonucleotide is key to the success of the assay, the difference between them should always be ∼20 °C.

The first step in designing the three oligonucleotides used in the primary reaction is to choose the region on the target mRNA one wishes to quantitate. To study the polarity of mRNA decay one needs at least three Invader probe sets; one toward the 5′-end of the target mRNA, one in the middle, and one in the 3′-end. The polarity of the decay process is determined by comparing the rate of decay of each of these against an internal standard (we use GAPDH). The data in Fig. 24.3 (Murray and Schoenberg, 2007) show this approach used to study the decay of human β-globin mRNA without or with the *c-fos* AU-rich instability element added to the 3′ UTR. The sites within each of these regions that are chosen for each probe set should avoid known splice variants or SNPs unless one wishes to include these in the study. If the surrounding ∼50-nt region has strong secondary or tertiary structure, this will interfere with oligonucleotide binding and assay performance, and these sites should be avoided.

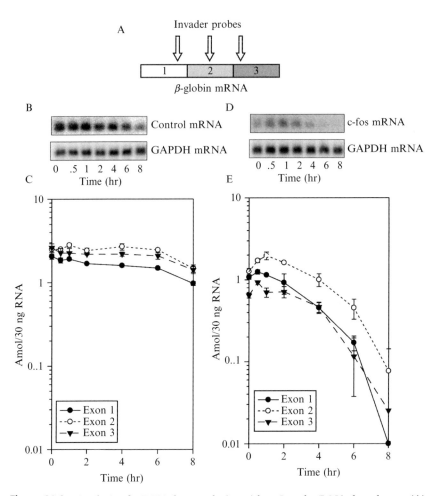

Figure 24.3 Analysis of mRNA decay polarity with an Invader RNA–based assay. (A) Invader probe sets were prepared to each of the 3 exons of human β-globin mRNA. Tetracycline was added at time = 0 to a murine fibroblast cell line that stably expresses the human β-globin gene (B and C) or the human β-globin gene with the c-fos ARE in the 3′ UTR (D and E) under tetracycline repressor control. RNA from triplicate cultures was isolated at the indicated times, and pooled samples were assayed by Northern blotting (B and D), and individual samples were analyzed by Invader RNA assay. The results show that stable mRNA decays with no evidence for polarity, whereas mRNA with the c-fos ARE decays faster from both ends than the middle. These data are abstracted from Murray and Schoenberg (2007).

Methods to determine secondary and tertiary structure include computer modeling and RT-ROL (reverse transcription of random oligonucleotide libraries), an experimental method (Allawi et al., 2001). In practice, one usually needs to design and test probe sets for several possible sites. The length of sequence of the target mRNA each oligonucleotide needs to cover to get the appropriate Tm of 60 °C (for stacking and primary probe oligonucleotides) or 80 °C (for invasive oligonucleotide) determines the sequence of each oligonucleotides, because these are strictly a function of its Tm. The final step in probe set design is to check that each is free of the following pitfalls: the length is too short to be sequence-specific (this can happen in GC-rich regions), primer/dimer formation, secondary structure in the oligonucleotide (e.g., GGGG), 2 base invasion (the nucleotide upstream of the cleavage nucleotide in the probe [located in the flap region] must not bind to the target molecule), and a G at the cleavage site in the probe, because this is less efficiently cleaved. The preferred nucleotide at this location is a C (would be a G in the mRNA sequence). Primary probe oligonucleotides require a flap sequence at the $5'$-end and published probe flap sequences are available (Eis et al., 2001, Wagner et al., 2003).

These oligonucleotides should be synthesized at 1 μm scale and purified by HPLC, because the purity of the oligonucleotide preps is essential for good results. The primary probe is a DNA oligonucleotide and has a $3'$-amino modifier C7. The invasive oligonucleotide is also a DNA oligonucleotide; the stacking oligonucleotide is an RNA oligonucleotide and is $2'$ O-methylated throughout. We have had success with oligonucleotides from IDT, Inc. Because EDTA inhibits the Invader RNA assay, these are dissolved in 300 μl of buffer containing only 0.1 mM EDTA (Te buffer 10 mM Tris-HCl, pH 8.0, 0.1 mM EDTA). The solution is diluted 1:200 and the concentration of each oligonucleotide is determined by its absorbance at 260 nm. The primary oligonucleotide mix consists of 40 μM primary probe oligonucleotide, 20 μM invasive oligonucleotide, and 12 μM stacking oligonucleotide diluted in Te.

2.1.1.2. Secondary reaction oligonucleotide design, synthesis, and preparation The secondary reaction is performed by adding three more oligonucleotides to the primary reaction. These oligonucleotides are the arrestor oligonucleotide, the FRET-labeled probe oligonucleotide, and the synthetic reaction template (SRT). In the secondary reaction, the arrestor oligonucleotide binds up any uncleaved primary probe oligonucleotide remaining from the primary reaction. Then, the primary probe flap released in the first reaction becomes the invasive oligonucleotide for cleavage of the FRET-labeled probe simultaneously bound to the SRT (Fig. 24.2). Cleavase separates the fluorescent tag from the quencher (which is located in the ASR for this reaction) to generate a fluorescent signal that can be quantitated with a fluorescence plate reader.

The arrestor oligonucleotide should be designed to be complementary to the uncleaved primary probe oligonucleotide. This is an RNA oligonucleotide and is $2'$-O methylated throughout, and purified by desalting. The FRET-labeled probe oligonucleotide is a DNA oligonucleotide containing a fluorescent tag on the 3′-end (we use 6-FAM) and a quencher molecule at the third position (we use Eclipse Dark Quencher). Published sequences for the design of the FRET-labeled probe oligonucleotide are available (Eis et al., 2001; Wagner et al., 2003). This oligonucleotide should be HPLC purified. The SRT should be designed to hybridize with both the FRET-labeled probe oligonucleotide and the primary probe flap oligonucleotide to create the secondary structure recognized by Cleavase as a cleavage substrate. This is a DNA oligonucleotide and the 3′-most 3 to 5 nucleotides should be 2′-O methylated. There are two oligonucleotide mixes to prepare for the secondary reaction of the Invader RNA assay; one contains the arrestor oligonucleotide diluted to 53.4 μM in Te buffer and the other contains 2.0 μM SRT, and 13.4 μM FRET-labeled probe oligonucleotide in Te buffer.

2.1.1.3. Experimental design considerations The essence of studying mRNA decay polarity is to terminate transcription and use the Invader RNA assay to quantify the decay of the 5′-, middle, and 3′-portions of the mRNA (Murray and Schoenberg, 2007). This is best done with a tetracycline repressor–based system, where one can rapidly terminate transcription by adding antibiotic to the medium. One variable is whether to run a transcription pulse assay or a transcription turn-off assay. The transcription pulse assay has the advantage of generating a homogeneous population of transcripts that should decay in a homogeneous fashion, provided that the pulse length is much shorter than the half-life of the mRNA (Loflin et al., 1999). Alternately, one can examine decay from steady state by adding tetracycline to the medium of cells stably expressing the reporter gene. The major determining factor is the overall degree of reporter gene expression. The background signal increases when the amount of reporter mRNA reaches the lower limit of detection, a problem we encountered with transcriptional pulsing. This is obviated by the higher starting levels of reporter mRNA present in experiments after decay from steady state. Because of its ubiquitous use as a reporter for studying mRNA decay, we used human β-globin mRNA as the target for developing Invader probe sets and quantifying the impact of several instability elements on mRNA decay (Murray and Schoenberg, 2007).

Another experimental consideration when reporter genes are used is whether to use transiently transfected cells or stably transfected cells. For experiments with the tetracycline repressor, the starting cell line must stably express the tTA protein, but whether or not to construct stable cells that also express the reporter gene is another question. We have used transiently transfected cells, and these typically give higher expression of the gene of interest; however, stably transfected cells show a more consistent level of expression and provide greater reproducibility. Also the use of stable lines makes it easier to analyze the consequence of overexpressing a regulatory protein or RNAi knockdown to decay.

Finally, there are some experimental considerations for setting up the assay itself. Every time the Invader RNA assay is run, there must be a standard curve for each RNA target with an *in vitro* synthesized transcript. This serves to establish the linear range of the assay, which can vary between experiments even for the same oligonucleotides and target. To facilitate statistical analysis of the data, an Invader RNA assay should be run in triplicate or quadruplicate. Statistical analysis becomes a greater issue when transiently transfected cells are used, because each experiment must also quantify changes in a cotransfected control and one should also examine an endogenous loading control. This is significantly easier with stable cell lines, because the target and endogenous loading control (e.g., β-actin, GAPDH) can be assayed at the same time by doing the Invader RNA assay in biplex format. Until more probes are available for commonly used cotransfected controls (e.g., luciferase, β-galactosidase, GFP), these must be quantified by an alternate method, such as Northern blotting or ribonuclease protection assay.

2.2. Methods: Invader RNA assay

At all times, procedures should be performed in a manner that is consistent with preventing RNase contamination and protein degradation. If applicable, treat solutions with DEPC; wipe surfaces with RNaseAway (Invitrogen), wear gloves, use plastic ware that is certified RNase-free, and perform all manipulations on ice. After RNA isolation, the Invader RNA assay itself can be run in one day.

2.2.1. RNA isolation

Only a small number of cells are needed for each point because the Invader RNA assay requires a small amount of input RNA. A 60-mm dish of cells yields enough RNA to run the Invader assay multiple times with enough RNA remaining to analyze by RPA or Northern blot analysis to confirm the Invader results or run a cotransfection control. When preparing RNA for use in the Invader RNA assay, it may be important to separate cells into nuclear and cytoplasmic fractions and use the cytoplasmic fraction, because the assay will not differentiate between spliced and unspliced RNAs because of the small size of the target sequence. However, the kit includes a buffer for lysing cells directly on the plate for analysis of RNA without RNA isolation. To harvest the cells and prepare nuclear and cytoplasmic fractions, first remove the growth medium from the cells and wash the plate twice with PBS. Add 1 ml PBS to the plate and remove the cells by scraping. Gently mix the cells by pipetting and then transfer the cell suspension from the plate to a microcentrifuge tube. Pellet the cells by centrifuging at $1000g$,

4 °C for 1 min. Remove the supernatant and resuspend the cell pellet in 200 ml cytoplasmic extraction buffer. (Cytoplasmic isolation buffer: 0.14 M NaCl, 1.5 mM MgCl$_2$, 0.01 M Tris-HCl, pH 8.0, 0.0025% NP20, 0.01 M DTT [dithiothreitol], RNase OUT [Invitrogen], in RNase-free water to desired volume. DTT and RNase OUT should be added immediately before use. Store the buffer at 4 °C.) Incubate the mixture on ice for 10 min then centrifuge at 12,000g at 4 °C for 5.5 min to pellet the nuclei. Remove the supernatant containing the cytoplasmic contents and transfer it to a clean microcentrifuge tube. The cytoplasmic contents and cell nuclei may be stored frozen at −80 °C until further use at this point. If it is important to have a clean nuclear fraction, wash the nuclear pellet with sodium deoxycholate instead of freezing. To do this, after the removal of the supernatant containing the cytoplasmic fraction, resuspend the nuclear pellet in hypotonic buffer (10 mM NaCl, 10 mM Tris-HCl, pH 7.4, 1.5 mM MgCl$_2$, protease inhibitor, and RNaseOUT, with the last two items added immediately before use) and then add sodium deoxycholate to a concentration of 0.5%. Centrifuge the nuclei for 5.5 min at 12,000g at 4 °C, remove the supernatant, and resuspend the pellet in TE. The nuclei may be frozen at −80 °C for storage or processed immediately. RNA may be purified from the cytoplasmic or nuclear fractions with an RNA purification protocol of choice. For experiments involving quantitation of small mRNA decay products, the RNA purification method should recover small RNAs. Finally, quantify RNA yield by its absorbance at 260 nm with a spectrophotometer.

2.2.2. Invader RNA assay

2.2.2.1. Preparation of the standard curve Samples for the standard curve are diluted and then aliquoted to an 8-tube strip for convenient loading into a 96-well dish later. A typical standard curve is composed of *in vitro* transcript at concentrations of 0, 0.01, 0.02, 0.04, 0.16, 0.64, 2.5, 5.0, 10.0. and 20.0 amol/well. Note for precise quantitation, all dilutions and standard curves should be based on molar amounts of RNA, not micrograms. Because the volume of each diluted *in vitro* transcript added to each well is always 5 μl, the *in vitro* transcripts must be diluted to one fifth of what will be in the reaction well. For example, to add 5.0 amol of an *in vitro* synthesized transcript, the *in vitro* synthesized transcript must be diluted to 1.0 amol/μl; 5 μl of this is 5.0 amol in the well. *In vitro* synthesized transcripts must be diluted in 20 ng/μl tRNA because of the very low concentration of the transcripts. Be sure to thoroughly vortex and centrifuge down each sample between dilutions. Each standard should be measured in triplicate or quadruplicate, so at least 15 to 20 μl of each standard concentration is required, plus extra volume for pipetting error, typically 10% volume. Transfer each standard to one tube of an 8-tube strip, and store on ice.

2.2.2.2. Preparation of samples Sample RNAs are also diluted and then aliquoted to an 8-tube strip for convenient loading into a 96-well plate. Prepare samples by diluting RNA in water or in 20 ng/ml tRNA. RNA samples should be diluted in water if the expression of the target gene is expected to be low such that a large amount of RNA will be added to each well (50 to 100 ng). RNA samples should be diluted in tRNA if the expression of the gene of interest is expected to be high such that a small amount of RNA will be added to each well ($\leq$50 ng). It is not advisable to use more than 100 ng cellular RNA per well. The volume of sample RNA that will be added to each well is always 5 μl. Therefore, cellular RNA must be diluted to a concentration one fifth of the amount that will be added to each well. For example, to add 30 ng of cellular RNA to a well, the cellular RNA will be diluted to a concentration of 6 ng/μl, such that the addition of 5 μl to a well gives 30 ng of sample RNA in the well. Remember that each sample will be read in triplicate or quadruplicate, so prepare 15 to 20 μl of each sample, plus 10% extra volume for pipetting error. Thoroughly vortex each sample, briefly centrifuge contents down, and transfer each sample to a well of an 8-well tube strip, and store on ice.

2.2.2.3. Preparation of reaction mixes It is convenient to prepare both primary and secondary reaction mixes before assembling the assay reactions and store them both on ice until needed. The reagents required for the preparation of the reaction mixes are provided in the Invader RNA assay generic kit (Third Wave Technologies, Inc.). Invader RNA assay kits for popular housekeeping genes such as GAPDH and β-actin are also available and include primary and secondary oligonucleotide mixes and *in vitro* transcripts for controls. For each reaction, prepare the primary reaction mix by combining 4 μl RNA primary buffer, 0.25 μl primary oligonucleotide mix for the mRNA of interest, 0.25 μl primary oligonucleotide mix for the housekeeping mRNA (optional as loading control, otherwise add Te), and 0.5 μl Cleavase per well. Cleavase is added last. The reaction mix should be vortexed, briefly spun down, and stored on ice. The volume of primary buffer mix to prepare depends on the number of wells being assayed plus extra for pipetting error. If preparing a small number of wells ($\leq$20), use 20% extra volume. If preparing a larger number of wells, it is sufficient to prepare enough extra volume for 4 wells. It is not usually necessary to prepare as large a volume as suggested in the assay manual.

To prepare the secondary reaction mix for one reaction, combine 2 μl RNA secondary buffer, 0.75 μl secondary oligonucleotide mix (i.e., arrestor) for the mRNA of interest, 0.75 μl FRET-labeled probe oligonucleotide mix (i.e., the FRET-labeled probe plus SRT) and for reactions run in biplex mode, 1.5 μl of a secondary oligonucleotide mix for the housekeeping gene. The latter is a packaged mixture from Third Wave Technologies and contains all three oligonucleotides. If one is not assaying a housekeeping gene in the same reaction, this should be substituted with 1.5 μl of Te per well so that the total volume of the reaction is 5 μl. The secondary reaction mix should be vortexed, spun down, aliquoted to an 8-tube strip, and stored on ice. The secondary reaction mix can be prepared before the assay or during the first incubation.

2.2.2.4. Perform assay The reactions should be assembled on ice in a 96-well microtiter plate with round-bottomed wells. First, 5 μl primary reaction mix is added to each well with an electronic repeat pipetter for speed and accuracy. Then, 5 μl *in vitro* transcript standard or 5 μl sample RNA is added to the plate with an 8-channel micropipettor. Finally, each reaction is overlaid with 10 μl Chill-Out wax with an electronic repeat pipetter. The plate should be covered with adhesive film and incubated at 60 °C for 90 min either in a thermocycler with a 96-well plate format-heating element or in a hybridization oven (particularly useful for multiple plates). After 90 min, the plate is removed (if a hybridization oven is used), the adhesive film is removed and 5 μl secondary reaction mix is added to each sample with an 8-channel micropipettor. The adhesive film is replaced, and the plate is further incubated at 60 °C for 60 min to 90 min, depending on the desired signal strength, because longer incubation times give a stronger signal. Note that this is a particular advantage of the Invader assay, because the signal amplification afforded by increased reaction time is not affected by issues of target amplification common to PCR. The plate is then removed from the thermocycler or hybridization oven, the adhesive film removed, and stop buffer (10 mM Tris-HCl, pH 8.0, 10 mM EDTA) can be added. The addition of stop buffer is optional. The plate should be read immediately in a fluorescence plate reader. We use a Tecan Genios instrument with Magellan II software. Some preliminary settings to try include setting "gain" to 50, "# flashes" to 10, "lag time" to 0 μsec, "integration time" to 50 μsec, and "plate definition file" as NUN96ft.pdf. The gain should be adjusted so that the "0 amol/well" sample of the standard curve has a value of approximately 100.

2.2.3. Data analysis, standard curves

The first step in analyzing the data from the Invader RNA assay is to draw the standard curves. This will serve as a control for the sensitivity of the assay and demonstrate the linear range of the assay, which is important for data analysis. First, the data should be exported to Microsoft Excel™. For each set of triplicate or quadruplicate points on the standard curve, the mean, standard deviation and percentage coefficient of variation ([(standard deviation)/(average raw signal)] × 100) should be calculated. The value of the percent coefficient of variation, which is a measure of the variation between replicates, should be <10%, and preferably <5%. The presence of an outlier can cause this value to be very high (≥10%). To generate net values for each of the standard samples >0 amol *in vitro* transcript (IVT)/well, subtract the mean value for the 0 amol IVT/well from each of the other standard curve mean values. For all standard samples >0 amol IVT/well, the FOLD > 0 is calculated by dividing the average gross signal for any particular sample by the average gross signal for the 0 amol IVT/well sample. The SD > 0 for all standard samples >0 amol IVT/well is calculated by dividing the net signal for any particular sample by the standard deviation of the 0 amol IVT/well sample. For all standard samples >0 amol IVT/well, a Student's t test is calculated comparing that sample to zero. This gives the confidence that the value is different than zero. The Student t test function in Excel can be used, including all 4 zero signal values and all 4 sample signal values, and selecting 2 tails, type 1. For all standard samples, 95% confidence values should be calculated with the confidence function in Excel. When the confidence function in Excel is used, "alpha" should be chosen as 0.05, "size" should be chosen as 4 (quadruplicates) or 3 (triplicates), and also choose the worksheet cell containing the standard deviation of the sample which is being analyzed. For all samples >0.01 amol a Student's t test should be calculated to compare it with the next lowest standard sample. The t test function in Excel can be used as previously, but the 4 signal values for the sample being analyzed should be selected, as well as the 4 signal values for the sample below it. Other selections may remain the same.

After the statistical analysis of the data for the standard curve is finished, plot the standard curve as net counts vs amol IVT/well with a linear scale (not logarithmic). Low points that have either t test scores >0.05 or FOLD > 0 scores less than 1.15 should be discarded. The lowest value that can be plotted is the limit of detection (LOD) of the assay. High points that have plateaued should also be discarded. The linear range of the assay is the range of values between the lowest and highest plotted points. The data points should be fitted with a linear trendline and the R^2 value should be calculated.

2.2.4. Data analysis, mRNA decay

The standard curve should serve as a guideline for which RNA samples to include for analysis. That is, only signal values that fall within the linear range of the standard curves run simultaneously to the experiment should be used. To calculate net signal values for all experimental values, subtract the signal value from the 0 amol IVT/well sample on the standard curve. For biplexed assays, the signal values for the mRNA of interest can be normalized for loading by dividing them by the control signal values. For each set of quadruplicates or triplicates, the mean and standard deviation should be calculated. For mRNA decay experiments where transcription is inhibited at time = 0 (by the addition of a transcriptional inhibitor), calculate the percent remaining by use of time = 0 h as 100% (Percent remaining = [(signal of sample at time = x)/(signal of sample at time = 0)] × 100).

Finally, plot the data as percent remaining (y axis, logarithmic) vs time (x axis, linear), fit the line, and calculate the half-life for the mRNA. For experiments studying the polarity of mRNA decay (Fig. 24.3), the decay curves for each probe set should be plotted on the same graph. By quantifying changes in the absolute amount of each portion of the target mRNA as a function of time, this graph will directly indicate the polarity of the decay process, because that portion that decays most rapidly will have a steeper slope than more slowly decaying parts of the mRNA.

ACKNOWLEDGMENTS

This work was supported by PHS grant R01 GM38277 to D. R. S. We thank the members of the Schoenberg laboratory for helpful discussions. Invader® Cleavase and Third Wave® are registered trademarks of Third Wave Technologies, Inc.

REFERENCES

Allawi, H. T., Dong, F., Ip, H. S., Neri, B. P., and Lyamichev, V. I. (2001). Mapping of RNA accessible sites by extension of random oligonucleotide libraries with reverse transcriptase. *RNA* **7,** 314–327.

Das Gupta, J., Gu, H., Chernokalskaya, E., Gao, X., and Schoenberg, D. R. (1998). Identification of two cis-acting elements that independently regulate the length of poly (A) on Xenopus albumin pre-mRNA. *RNA* **4,** 766–776.

Eis, P. S., Olson, M. C., Takova, T., Curtis, M. L., Olson, S. M., Vener, T. I., Ip, H. S., Vedvik, K. L., Bartholomay, C. T., Allawi, H. T., Ma, W.-P., Hall, J. G., *et al.* (2001). An invasive cleavage assay for direct quantitation of specific RNAs. *Nature Biotechnol.* **19,** 673–676.

Harrold, S., Genovese, C., Kobrin, B., Morrison, S. L., and Milcarek, C. (1991). A comparison of apparent mRNA half-life using kinetic labeling techniques vs decay following administration of transcriptional inhibitors. *Anal. Biochem.* **198,** 19–29.

Loflin, P. T., Chen, Y. A, C., Xu, N., and Shyu, A.-B. (1999). Transcriptional pulsing approaches for analysis of mRNA turnover in mammalian cells. *Methods: A Companion to methods in Enzymology* **17,** 11–20.

Murray, E. L., and Schoenberg, D. R. (2007). A+U-rich instability elements differentially activate 5' to 3'-and 3' to 5'-mRNA decay. *Mol. Cell Biol.* **27,** 2791–2799.

Salles, F. J., and Strickland, S. (1999). Analysis of poly(A) tail lengths by PCR: The PAT assay. *In* "Methods in Molecular Biology, vol. 118, RNA-Protein Interaction Protocols," (S.R HaynesIn, ed.), Humana Press. Totowa, New Jersey.

Wagner, E. J., Curtis, M. L., Robson, N. D., Baraniak, A. P., Eis, P. S., and Garcia-Blanco, M. A. (2003). Quantification of alternatively spliced FGFR2 RNAs using the RNA invasive cleavage assay. *RNA* **9,** 1552–1561.

SECTION SIX

CELL BIOLOGY OF RNA DECAY (I.E., TRANSLATIONAL REPRESSION/ P BODIES)

CHAPTER TWENTY-FIVE

ANALYZING P-BODIES IN *SACCHAROMYCES CEREVISIAE*

Tracy Nissan *and* Roy Parker

Contents

1. Introduction	508
2. Determining Whether a Specific Protein Can Accumulate in P-Bodies	508
2.1. Markers of P-bodies	508
2.2. Preparation of samples	512
3. Monitoring Messenger RNA in P-Bodies	514
4. Determining Whether a Mutation or Perturbation Affects P-Body Size and Number	515
4.1. Conditions to observe increases or decreases in P-bodies	515
4.2. Interpreting alterations in P-body size and number	517
5. Quantification of P-Body Size and Number	518
Acknowledgments	519
References	519

Abstract

Cytoplasmic processing bodies, or P-bodies, are RNA-protein granules found in eukaryotic cells. P-bodies contain non-translating mRNAs and proteins involved in mRNA degradation and translational repression. P-bodies, and the mRNPs within them, have been implicated in mRNA storage, mRNA degradation, and translational repression. The analysis of mRNA turnover often involves the analysis of P-bodies. In this chapter, we describe methods to analyze P-bodies in the budding yeast, *Saccharomyces cerevisiae*, including procedures to determine whether a protein or mRNA can accumulate in P-bodies, whether an environmental perturbation or mutation affects P-body size and number, and methods to quantify P-bodies.

The University of Arizona, Department of Molecular and Cellular Biology and Howard Hughes Medical Institute, Tucson, Arizona, USA

1. Introduction

The control of translation and mRNA degradation can involve conserved cytoplasmic RNA granules. One class of such RNA granules is the cytoplasmic processing body, or P-body. P-bodies are dynamic aggregates of untranslating mRNAs in conjunction with translational repressors and proteins involved in deadenylation, decapping, and $5'$ to $3'$-exonucleolytic decay (Parker and Sheth, 2007). P-bodies and the mRNPs assembled within them are of interest because they have been implicated in translational repression (Coller and Parker, 2005; Holmes *et al.*, 2004), normal mRNA decay (Cougot *et al.*, 2004; Sheth and Parker, 2003), nonsense-mediated mRNA decay (Sheth and Parker, 2006; Unterholzner and Izaurralde, 2004), miRNA-mediated repression (Liu *et al.*, 2005; Pillai *et al.*, 2005), and mRNA storage (Bhattacharyya *et al.*, 2006; Brengues *et al.*, 2005). At a minimum, P-bodies serve as markers proportional to the concentration of mRNPs complexed with the mRNA decay machinery and may have additional biochemical properties that affect the control of mRNA translation and/or degradation.

Given the concentration of mRNAs and mRNA decay factors in P-bodies, the analysis of mRNA turnover can involve examining aspects of P-body composition and function. In this chapter, we describe methods to analyze P-bodies in the budding yeast, *Saccharomyces cerevisiae*, to determine: (1) Does a given protein or mRNA accumulate in P-bodies? (2) Does a specific perturbation (e.g., mutation, overexpression, or environmental cue) change P-body size or number? and (3) Is there a quantifiable change in the number and size of P-bodies in a population of cells?

2. Determining Whether a Specific Protein Can Accumulate in P-Bodies

2.1. Markers of P-bodies

A common experimental goal is determining whether a given protein accumulates in P-bodies. Previous work has identified many proteins enriched in yeast P-bodies (summarized in Table 25.1). These include a conserved core of proteins found in P-bodies from yeast to mammals that consists of the mRNA decapping machinery. These core P-body components include the decapping enzyme, Dcp1p/Dcp2p; proteins associated with decapping: Dhh1p, Pat1p, Scd6p, Edc3p, and the Lsm1p-7p complex; and the $5'$ to $3'$-exonuclease, Xrn1p. Some proteins observed in yeast P-bodies, such as proteins involved in nonsense-mediated decay (Sheth and Parker, 2006), are only observed in P-bodies under certain mutant,

Table 25.1 Protein components of P-bodies

Core components	Function
Dcp1p	Decapping enzyme subunit[a]
Dcp2p	Catalytic subunit of decapping enzyme[a]
Dhh1p	DEAD box helicase required for translational repression; decapping activator[a]
Edc3p	Decapping activator[b]
Lsm1p-7p	Sm-like proteins involved in decapping[a]
Pat1p	Decapping activator and translational repressor[a]
Scd6p	Protein containing Sm-like and FDF domain; Involved in translation repression[c]
Xrn1p	5' to 3'-exonuclease[a]
Ccr4p/Pop2p/ Not1-5p	Major cytoplasmic deadenylase[a,d]
Proteins involved in NMD	**Function**
Upf1p	ATP-dependent helicase required for NMD, accumulates in yeast P bodies in dcp1Δ, xrn1Δ, dcp2Δ, upf2Δ, and upf3Δ strains[e]
Upf2p	Component required for NMD, accumulates in P-bodies in dcp1Δ, dcp2Δ, and xrn1Δ mutants[e]
Upf3p	Component required for NMD, accumulates in P bodies in dcp1Δ, dcp2Δ, and xrn1Δ mutants[e]
Ebs1p	Putative ortholog of human Smg7, accumulates in glucose deprivation[f]
Translation and Translational Repression	**Function**
Cdc33p	eIF4E: mRNA m7G-cap binding protein[g,h]
Pab1p	Predominant poly(A) binding protein[g,h]
Rbp1p	RNA-binding protein, localizes to P bodies during stress[i]
Sbp1p	Facilitates mRNA decapping[j]
Tif4631p	eIF4G1: Component of eIF4F initiation factor[g,h]
Tif4632p	eIF4G2: Component of eIF4F, induced at stationary phase[g,h]
Additional Components	**Function**
Dsc2p	Nutrient stress–dependent regulator of the scavenger enzyme Dcs1[k]
Pby1p	Putative tubulin tyrosine ligase[l]
Rpb4p	Subunit of RNA polymerase II[m]
Rpm2p	Protein component of the mitochondrial RNaseP[n]

cell type, stress, or overexpression conditions (Table 25.1). These proteins may normally rapidly transit through P-bodies, but under some conditions accumulate to detectable levels.

To determine whether a given protein can accumulate in P-bodies one simply needs to examine the subcellular distribution of the protein relative to other P-body components. In the absence of good antisera, the simplest way to do this is to use the relevant protein tagged with GFP, or some other fluorescent protein, and then examine the potential colocalization of the candidate with one or more of the core P-body components. For this type of experiment, many of the core P-body components are available as fusions to fluorescent proteins on yeast plasmids (Table 25.2). Thus, to examine whether a protein accumulates in yeast P-bodies one should take the following steps:

1. Obtain an expression vector for a fluorescent-tagged version of the protein of interest. Note that most of the yeast (ORFs) fused to GFP are available from a genomic collection (Huh *et al.*, 2003) and can be purchased from Invitrogen. Use of the native promoter in fusion strains integrated into the genome should avoid mislocalization due to overexpression of the tagged protein.

---

[a] Sheth and Parker, 2003.
[b] Kshirsagar, M., and Parker R. (2007). Identification of Edc3p as an enhancer of mRNA decapping in Saccharomyces cerevisiae.*Genetics* **166**, 729–739.
[c] Barbee, S. *et al.* (2006). Staufen- and FMRP-containing neuronal RNPs are structurally and functionally related to somatic P bodies. *Neuron* **52**, 997–1009.
[d] Mulhrad, D., and Parker R. (2005). The yeast EDC1 mRNA undergoes deadenylation-independent decapping stimulated by Not2p, Not4p, and Not5p. *EMBO J.* **24**, 1033–1045.
[e] Sheth, U., and Parker R. (2006). Targeting of aberrant mRNAs to cytoplasmic processing bodies. *Cell* **125**, 1095–1109.
[f] Luke, B., Azzalin, C. M., Hug, N., Deplazes, A., Peter, M., and Lingner, J. (2007). *Saccharomyces cerevisiae* Ebs1p is a putative ortholog of human Smg7 and promotes nonsense-mediated mRNA decay. *Nucleic Acids Res. Adv. Access* published on November 4, 2007, DOI 10.1093/nar/gkm912.
[g] Brengues and Parker, 2007.
[h] Hoyle, N. P., Castelli, L. M., Campbell, S. G., Holmes, L. E., and Ashe, M. P. (2007). Stress-dependent relocalization of translationally primed mRNPs to cytoplasmic granules that are kinetically and spatially distinct from P-bodies. *J Cell Biol.* **179**, 65–74.
[i] Jang, L. T., Buu, L. M., and Lee, F. J. (2006). Determinants of Rbp1p localization in specific cytoplasmic mRNA-processing foci, P-bodies. *J. Biol. Chem.* **281**, 29379–293790.
[j] Segal, S. P., Dunckley, T., and Parker, R. (2006). Sbp1p affects translational repression and decapping in *Saccharomyces cerevisiae*. *Mol. Cell. Biol.* **26**, 5120–5130.
[k] Malys, N., and McCarthy, J. E. (2006). Dcs2, a novel stress-induced modulator of m7GpppX pyrophosphatase activity that locates to P bodies. *J. Mol. Biol.* **363**, 370–382.
[l] Sweet, T. J., Boyer, B., Hu, W., Baker, K. E., and Coller, J. (2007). Microtubule disruption stimulates P-body formation. *RNA* **13**, 493–502.
[m] Lotan, R., Bar-On, V. G., Harel-Sharvit, L., Duek, L., Melamed, D., and Choder M. (2005). The RNA polymerase II subunit Rpb4p mediates decay of a specific class of mRNAs. *Genes Dev.* **19**, 3004–3016.
[n] Stribinskis, V., and Ramos, K. S. (2007). Rpm2p, a protein subunit of mitochondrial RNase P, physically and genetically interacts with cytoplasmic processing bodies. *Nucleic Acids Res.* **35**, 1301–1311.

Table 25.2 Core P-body components available on plasmids as fluorescent protein fusions

Protein	Fluorescent tag	Plasmid marker	Lab plasmid number
Dcp2p full length	GFP	LEU2	pRP1175[a]
	GFP	TRP1	pRP1316[b]
	GFP	URA3	pRP1315[c]
Dcp2p full length	RFP	LEU2	pRP1155[d]
	RFP	TRP1	pRP1156[e]
	RFP	URA3	pRP1186[f]
Dcp2p truncated (1-300)	RFP	LEU2	pRP1167[g]
	RFP	TRP1	pRP1165[h]
	RFP	URA3	pRP1152[h]
Dhh1p	GFP	LEU2	pRP1151[g]
Edc3p	mCherry	URA3	pRP1574[i]
	mCherry	TRP1	pRP1575[i]
Lsm1p	GFP	LEU2	pRP1313[a]
	GFP	URA3	pRP1314[j]
Lsm1p	mCherry	LEU2	pRP1400[k]
Lsm1p	RFP	URA3	pRP1084[l]
	RFP	LEU2	pRP1085[l]
Pat1p	GFP	URA3	pRP1502[m]

[a] Coller and Parker, 2005.
[b] Segal, S. P., Dunckley, T., and Parker R. (2006). Sbp1p affects translational repression and decapping in *Saccharomyces cerevisiae*. *Mol. Cell. Biol.* **26**, 5120–5130.
[c] Unpublished, Parker Lab.
[d] Teixeira *et al.*, 2005.
[e] Sheth and Parker, 2006.
[f] Teixeira *et al.*, 2005.
[g] Sweet, T. J., Boyer, B., Hu, W., Baker, K. E., and Coller J. (2007). Microtubule disruption stimulates P-body formation. *RNA* **13**, 493–502.
[h] Muhlrad, D., and Parker, R. (2005). The yeast EDC1 mRNA undergoes deadenylation-independent decapping stimulated by Not2p, Not4p, and Not5p. *EMBO J.* **24**, 1033–1045.
[i] Unpublished, Buchan, R, and Parker, R.
[j] Tharun, S., Muhlrad, D., Chowdhury, A., and Parker, R. (2005). Mutations in the Saccharomyces cerevisiae LSM1 gene that affect mRNA decapping and 3′-end protection. *Genetics* **170**, 33–46.
[k] Beckham, C. J., Light, H. R., Nissan, T. A., Ahlquist, P., and Parker, R., Noueiry A. (2007). Interactions between brome mosaic virus RNAs and cytoplasmic processing bodies. *J. Virol.* **81**, 9759–9768.
[l] Sheth and Parker, 2003.
[m] Pilkington G. R., and Parker R. (2008). Pat1 contains distinct functional domains that promote P-body assembly and activation of decapping. *Mol. Cell. Biol.* **4**, 1298-1312.

2. Determine whether the fusion protein is functional by some criteria.
3. Compare the localization of the tagged protein of interest with a core P-body marker tagged with a different fluorescent protein as described

in the following. In our experience, the most reliable and brightest components of yeast P-bodies are Dcp2p and Edc3p.

2.2. Preparation of samples

To determine whether a protein can accumulate in P-bodies, we recommend examining its subcellular distribution in mid log phase and also under a condition where P-bodies are enhanced such as glucose deprivation, hyperosmotic stress, high optical density (OD), or when decapping is inhibited (Sheth and Parker, 2003; Teixeira et al., 2005; Fig. 25.1). Note that it is important to be careful when preparing cells for examining P-bodies microscopically, because P-bodies are dynamic and can change rapidly in response to a variety of stresses (Teixeira et al., 2005). Therefore, proper aeration and, if required, a carbon source are essential. In all cases, care should be taken to reduce centrifugation and handling times, because variations can alter P-bodies. A final caveat is that growth conditions can alter P-body composition; for example, different results can be obtained from growth in rich versus minimal media for non-core proteins. Protocols for examining P-bodies in mid log cultures or with glucose deprivation or osmotic stress are described in the following.

2.2.1. Examination of P-bodies in midlog growth

1. Grow a 5- to 10-ml yeast culture in rich or minimal medium as appropriate to mid log phase, with an absorbance between 0.3 and 0.6 at 600 nm.
2. Pellet 5 ml of the cells in a clinical centrifuge at room temperature for 1 min.

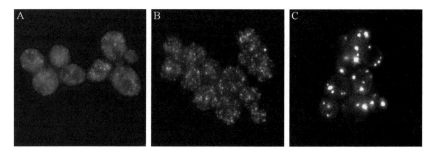

Figure 25.1 P-bodies under different growth and induction conditions. The P-body marker used is the genomically integrated DCP2-GFP (yRP2162). The cells were grown in YEPD to (A) mid log growth phase, then treated with (B) hyperosmotic (1 M KCl) stress for 15 minutes in complete minimal media with glucose as described in the text or (C) glucose deprivation for 10 minutes in complete minimal media (without glucose).

3. Resuspend in 1 ml minimal medium supplemented with amino acids and the same carbon source that the cells were originally grown in, then transfer the sample to a microfuge tube.
4. Pellet 1 ml of cells in a tabletop microfuge for 20 sec, aspirate off the supernatant, and resuspend the cells in 50 to 100 μl of minimal medium supplemented as in the preceding step. Keep in a microfuge tube with constant flicking by hand to maintain aeration.
5. Add the suspension to the slide at the microscope, then place a coverslip on the sample and examine immediately. This is important, because cells under a coverslip lack aeration, which can eventually induce a stress response and artifactually increase the presence of P-bodies.
6. Take the pictures rapidly to allow accurate assessment of the P-body state of the cells. Because two channels need to be taken to colocalize the P-body markers with the protein of interest, it is optimal to use a microscope that splits the beam to record the two channels simultaneously.
7. If cells are moving, which can make colocalization difficult, then it may be necessary to immobilize the cells by coating the coverslips with the lectin concanavalin A, which binds to the yeast cell wall. The protocol we use to coat coverslips is to wash coverslips overnight in sterile filtered 1 M NaOH. After washing well with distilled water, add concanavalin A solution (0.5 mg/ml Sigma #L7647, 10 mM phosphate buffer, pH 6, 1 mM CaCl$_2$, 0.02% azide) for 20 min with gentle shaking. After removal of the solution, rinse once in distilled water, pour off the liquid, and let dry overnight. The coverslips can be stored at room temperature after coating.
8. If longer exposures or time lapse data are required, we use an inverted Deltavision microscope with concanavalin A–coated glass-bottom microwell dishes (MatTek Corporation #P35G-1.5-14-C) and the coverslip immersed in enough minimal medium (supplemented as appropriate for the experiment) to fully cover it. This avoids both the problem of cell movement and the lack of aeration.

2.2.2. Examination of P-bodies under glucose deprivation or osmotic stress

1. Grow a 5- to 10-ml yeast culture in rich or minimal medium as appropriate to mid log phase, with an absorbance between 0.3 and 0.6 at 600 nm.
2. Pellet 5 ml of the cells in a clinical centrifuge at room temperature for 1 min.
3. After pelleting the cells, aspirate off the supernatant and wash with 1 ml minimal medium to be used for the induction. For example, in a glucose-deprivation experiment, the medium should be identical to that used for the initial growth but lacking glucose. For hyperosmotic

stress, the medium should be the starting medium but with KCl added to a final concentration of 1 M.

4. Pellet the cells and resuspend them in 1 ml of induction medium and transfer into 4 ml of prewarmed induction media in a small conical flask with shaking to maintain aeration. Grow for the appropriate induction time with shaking at the temperature that cells were grown before induction. The induction time should be determined experimentally; for example, during glucose depletion or osmotic stress, P-bodies are substantially increased after 10 min of stress (Teixeira et al., 2005).
5. Pellet 1.5 ml of cells in a tabletop microfuge for 20 sec, aspirate off the supernatant and resuspend them in 50 to 100 μl of minimal media supplemented as previously described. Keep in a microfuge tube with constant flicking by hand to maintain aeration.
6. Add the suspension to the slide at the microscope, then place a coverslip on the sample and examine immediately. This is important because yeast cells under a coverslip lack aeration, which can influence the induction or presence of P-bodies.
7. Take the pictures rapidly to allow accurate assessment of the P-body state of the cells. Because two channels need to be taken to colocalize the P-body markers with the protein of interest, it is optimal to use a microscope that splits the beam to record the two channels simultaneously, although not necessary. If need be, cells can be immobilized and time lapse images can be taken as described previously.

If a protein shows substantial overlap in its subcellular distribution with core P-body proteins, it can be inferred to accumulate in, or near, P-bodies. We recommend confirming and extending such possible interactions by use of standard approaches to perform an immunoprecipitation of the P-body component and probing for the experimental protein.

3. Monitoring Messenger RNA in P-Bodies

In some cases, it is useful or important to determine whether bulk mRNA or a specific transcript is accumulating in P-bodies. There are two general approaches to determine whether mRNAs are accumulating in P-bodies. First, one can use *in situ* hybridization techniques to monitor the presence of bulk mRNA with an oligo(dT) probe, or specific mRNAs with sequence specific probes. Although such approaches have worked well in mammalian cells (Franks and Lykke-Andersen, 2007; Pillai et al., 2005), robust techniques for *in situ* hybridization of yeast P-bodies have not yet been described.

In an alternate method, one can use "GFP-tagged" RNAs to follow the localization of specific transcripts in yeast and determine whether they accumulate in P-bodies. To visualize specific mRNAs, binding sites for a RNA binding protein fused to a fluorescent protein are inserted into the 3' UTR of the mRNA of interest, allowing its subcellular distribution to be examined by following the location of the RNA binding protein fused to the fluorescent protein. Most commonly, the well-characterized U1A or the MS2 binding sites are inserted into the 3' UTR of the mRNA of interest (Bertrand et al., 1998; Brodsky and Silver, 2000). These mRNA constructs are then coexpressed with either the U1A-GFP (Brodsky and Silver, 2000) or the MS2 coat protein fused to GFP (Bertrand et al., 1998). Both of these have pico to nano molar affinity for their respective binding sites, allowing detection of the mRNA.

Several of these types of engineered mRNAs have been used to demonstrate the accumulation of specific yeast mRNAs in P-bodies. Available plasmids expressing "tagged" versions of the stable PGK1 and the unstable MFA2 mRNA are described in Table 25.3. In addition, variants of the tagged PGK1 mRNAs are available with premature nonsense codons in specific positions, which can be used for examining the accumulation of mRNA in P-bodies as a result of the action of NMD (Sheth and Parker, 2006). A variety of plasmids expressing the MS2 or U1A proteins fused to GFP are also available (Table 25.3).

4. Determining Whether a Mutation or Perturbation Affects P-Body Size and Number

4.1. Conditions to observe increases or decreases in P-bodies

A common experimental issue is determining whether a mutation, protein overexpression, or an environmental cue affects the size and number of P-bodies. To examine whether P-bodies are altered under a certain condition, we make three suggestions. First, one should use multiple markers of P-bodies to ensure that any differences seen are not unique to a single protein. Second, because a specific mutation may affect P-bodies only under certain conditions, we recommend that P-bodies be examined under multiple conditions (e.g., mid log growth, glucose deprivation, osmotic stress, high cell density). Finally, we recommend quantification of the amount and number of P-bodies by computational methods to allow unbiased calculation of the size and number of P-bodies present in a given situation.

In practice, there are different methods for examining whether a perturbation reduces or increases P-body size or number. To determine whether a perturbation reduces P-bodies, it is most convincing to examine conditions where P-bodies are large and easily detectable. For example, P-bodies are large under stress conditions such as glucose deprivation, high cell density, and when decapping or 5' to 3'-degradation is inhibited (Teixera et al., 2005). Therefore,

Table 25.3 Plasmids for localizing mRNA in yeast cells: GFP fusion proteins that bind to specific binding sites in mRNA engineered in their 3′ UTR

Protein + Tag	Plasmid marker	Promoter			Lab plasmid number
MS2 CP-GFP	HIS3	Met25			pRP1094[a]
U1A-GFP	TRP1	GPD			pRP1187[b]
U1A-GFP	LEU2	GPD			pRP1194[c]

RNA +	Binding Seq.	Plasmid marker	Promoter	Description[h]	Lab plasmid number
MFA2	pG MS2	URA3	GPD	two MS2 sites 3′ to poly(G) tract in 3′ UTR	pRP1081[d]
MFA2	MS2	URA3	GPD	two MS2 sites in 3′ UTR	pRP1083[e]
MFA2	U1A	URA3	GPD	PGK1 3′ UTR with sixteen U1A binding sites	pRP1193[c]
MFA2	U1A	URA3	Tet-Off	PGK1 3′ UTR with sixteen U1A binding sites	pRP1291[c]
PGK1	pG MS2	URA3	PGK1	two MS2 sites 3′ to poly(G) tract in 3′ UTR	pRP1086[c]
PGK1	U1A	URA3	PGK1	sixteen U1A sites in 3′ UTR	pPP1354[b]
PGK1early PTC	U1A	URA3	PGK1	PGK1 U1A with nonsense mutation at position 22	pRP1295[f]
PGK1late PTC	U1A	URA3	PGK1	PGK1 U1A with nonsense mutation at position 225	pRP1296[f]
PGK1	U1A	URA3	GAL	sixteen U1A sites in 3′ UTR	pRP1303[g]

[a] Beach, D. L., and Bloom, K. (2001). ASH1 mRNA localization in three acts. *Mol. Biol. Cell.* **12**, 2567–2577.
[b] Brodsky and Silver, 2000.
[c] Brengues et al., 2005.
[d] M. Valencia-Sanchez and R. Parker, unpublished.
[e] Sheth and Parker, 2003; has short polyG tract in 3′ UTR that does not inhibit exonucleolytic decay.
[f] Sheth and Parker, 2005.
[g] U. Sheth and R. Parker, unpublished.
[h] All mRNA constructs have their native 5′- and 3′ UTR except where noted.

these are ideal conditions to examine whether P-bodies are reduced by a mutation or condition (for examples see Decker *et al.*, 2007, and Teixera and Parker, 2007). Conversely, P-bodies are small when cells are undergoing mid log growth (Teixeira *et al.*, 2005), which makes this condition an ideal situation to see whether P-bodies increase (Teixeira and Parker, 2007).

4.2. Interpreting alterations in P-body size and number

Any alteration in P-body size and number observed as a result of a specific mutation or alteration in growth can be due to a variety of effects and should be interpreted with care. This is because a variety of changes in cell physiology will affect P-body size and number because of the flux of mRNP in and out of these structures (Table 25.4). For example, P-body size and number can be increased by defects in mRNA decapping or 5' to 3'-degradation, which increase the pool of mRNPs in P-bodies by decreasing the destruction of mRNAs in this compartment, or by defects in translation initiation, which increase the pool of untranslating mRNPs associated with P-bodies. Alternately, P-bodies can be reduced in size and number by inhibiting translational repression, by removing interactions that promote aggregation of the individual mRNPs into larger structures, or by reductions in the level of the P-body marker being examined. Note that to be confident of the underlying mechanism affecting P-body size and number, additional experiments should be performed to identify the true cause of the defect (Table 25.4).

Table 25.4 Dissection of effects on P-body size and number

Observation	Possible cause	Follow-up experiments
Increase in P-body size and number	1) Inhibition of decapping or 5' to 3'-degradation	1) Examine mRNA degradation rate and whether mRNAs are degraded accurately
	2) Defects in translation initiation	2) Measure the rate of protein synthesis by polysomes or ^{35}S incorporation.
Decrease in P-body size and number	1) Decreases in translation elongation	1) Polysome analysis to determine if their size is increased
	2) Enhanced translation	2) Measure the rates of protein synthesis by polysomes or ^{35}S incorporation.
	3) Reduced expression of marker proteins	3) Western blot to ensure P-body expression levels

5. Quantification of P-Body Size and Number

A common and important issue is determining whether there are differences in P-body size and number between different conditions or mutants. When comparing two conditions or mutants to determine whether there is a difference in P-body size and number it is important to use unbiased approaches to determine the effects. One method is to score P-bodies under the two conditions by comparing a large number of cells (e.g., 100) by visual inspection, recording the number and approximate size distribution in each cell. However, given that such an approach can introduce experimental bias, we recommend that if P-bodies are scored by visual means, it be done in a blind manner with one experimenter preparing the images and a second scoring them without knowing the identity of any image. In addition, we recommend that manual scoring only be used for dramatic differences in P-body size and number.

An unbiased and more quantitative approach is to use semiautomated software programs for image analysis on the basis of computational algorithms. This can be accomplished with ImageJ, a free downloadable software from the NIH (Abramoff et al., 2004). On the ImageJ web site, there are instructions and downloadable plug-ins, which allow direct loading of raw images from the microscope. The quantification of P-bodies is accomplished by setting a threshold mask, which allows regions of the image above a certain intensity to be scored as "on" and the rest of the image "off." The P-bodies are counted by the number of "on" regions, and their area and number can be computed. To further reduce bias, the masking is performed automatically with the Otsu Thresholding Filter (Otsu, 1979). Once the images, ImageJ, and the Otsu plug-in are obtained, the image can be quantified. A protocol for quantifying such images is detailed as follows:

1. Open the image in ImageJ to quantify. Normally, an entire Z-stack is used that captures the entire thickness of the cell. Once the stack is collapsed, the P-bodies can be analyzed.
2. In the ImageJ program, go to Process, and select ("Smooth"). Then go to Process, select Math, and then select Subtract to remove the background. This is the most important step in the process and will probably need to be modified for different microscopes and marker proteins or RNA. It should be noted that because of threshold subtractions, such quantifications are not necessarily absolute numbers or areas of P-bodies but provide a systematic and unbiased measure of relative P-body number and area within experiments.
3. Go to Plug-ins, select Filters, and then select "Otsu Thresholding." This option will only be available if you have downloaded the plug-in and placed it in the correct directory on your computer.

4. Go to Image, select Lookup Tables, and then select "Invert LUT." This will reverse the image, allowing the P-bodies to be considered "on."
5. Go to analyze and select "Analyze Particles" to set the pixel size range to be counted. In our experience, it should be between 7 and 500, but it will vary depending on samples. Optimization is required so random speckles do not count as P-bodies and large fluorescent regions unrelated to P-bodies are not counted (for example, those outside of the cells).
6. Select show Masks, which will graphically demonstrate the calculated P-bodies compared with the P-bodies' thresholded image. This procedure generates a table that lists the number of foci, average area, and total area.
7. The number of cells in each image must be calculated manually; however, all steps except cell counting can be automated by writing a macro to perform them to obtain more high-throughput analyses.

ACKNOWLEDGMENTS

We thank the members of the Parker laboratory for helpful discussions, especially Carolyn Decker for discussions and images and Guy Pilkington for critical review of the manuscript. NIH grant (R37 GM45443) and funds from the Howard Hughes Medical Institute supported this work. T. N. was supported in part by T32 CA09213.

REFERENCES

Abramoff, M. D., Magelhaes, P. J., and Ram, S. J. (2004). Image Processing with ImageJ. *Biophotonics Int.* **11,** 36–42.

Bertrand, E., Chartrand, P., Schaefer, M., Shenoy, S. M., Singer, R. H., and Long, R. M. (1998). Localization of ASH1 mRNA particles in living yeast. *Mol. Cell* **2,** 437–245.

Bhattacharyya, S. N., Habermacher, R., Martine, U., Closs, E. I., and Filipowicz, W. (2006). Relief of microRNA-mediated translational repression in human cells subjected to stress. *Cell* **125,** 1111–1124.

Brengues, M., Teixeira, D., and Parker, R. (2005). Movement of eukaryotic mRNAs between polysomes and cytoplasmic processing bodies. *Science* **310,** 486–489.

Brodsky, A. S., and Silver, P. A. (2000). Pre-mRNA processing factors are required for nuclear export. *RNA* **6,** 1737–1749.

Coller, J., and Parker, R. (2005). General translational repression by activators of mRNA decapping. *Cell* **122,** 875–886.

Cougot, N., Babajko, S., and Seraphin, B. (2004). Cytoplasmic foci are sites of mRNA decay in human cells. *J. Cell. Biol.* **165,** 31–40.

Decker, C. J., Teixeira, T., and Parker, R. (2007). Edc3p and a glutamine/asparagine-rich domain of Lsm4p function in processing body assembly in *Saccharomyces cerevisiae*. *J. Cell. Biol.* **179,** 437–449.

Franks, T. M., and Lykke-Andersen, J. (2007). TTP and BRF proteins nucleate processing body formation to silence mRNAs with AU-rich elements. *Genes Dev.* **21,** 719–735.

Holmes, L. E., Campbell, S. G., De Long, S. K., Sachs, A. B., and Ashe, M. P. (2004). Loss of translational control in yeast compromised for the major mRNA decay pathway. *Mol. Cell. Biol.* **24,** 2998–3010.

Huh, W. K., Falvo, J. V., Gerke, L. C., Carroll, A. S., Howson, R. W., Weissman, J. S., and O'Shea, E. K. (2003). Global analysis of protein localization in budding yeast. *Nature* **425,** 686–691.

Liu, J., Rivas, F. V., Wohlschlegel, J., Yates, J. R., 3rd, Parker, R., and Hannon, G. J. (2005). A role for the P-body component GW182 in microRNA function. *Nat. Cell. Biol.* **7,** 1261–1266.

Otsu, N. (1979). Threshold selection method from gray-level histograms. *IEEE Trans. Syst. Man Cybern.* **9,** 62–66.

Parker, R., and Sheth, U. (2007). P bodies and the control of mRNA translation and degradation. *Mol. Cell* **25,** 635–646.

Pillai, R. S., Bhattacharyya, S. N., Artus, C. G., Zoller, T., Cougot, N., Basyuk, E., Bertrand, E., and Filipowicz, W. (2005). Inhibition of translational initiation by Let-7 MicroRNA in human cells. *Science* **309,** 1573–1576.

Sheth, U., and Parker, R. (2003). Decapping and decay of messenger RNA occur in cytoplasmic processing bodies. *Science* **300,** 805–808.

Sheth, U., and Parker, R. (2006). Targeting of aberrant mRNAs to cytoplasmic processing bodies. *Cell* **125,** 1095–1109.

Teixeira, D., Sheth, U., Valencia-Sanchez, M. A., Brengues, M., and Parker, R. (2005). Processing bodies require RNA for assembly and contain nontranslating mRNAs. *RNA* **11,** 371–382.

Teixeira, T., and Parker, R. (2007). Analysis of P-body assembly in *Saccharomyces cerevisiae*. *Mol. Biol. Cell* **18,** 2274–2287.

Unterholzner, L., and Izaurralde, E. (2004). SMG7 acts as a molecular link between mRNA surveillance and mRNA decay. *Mol. Cell* **16,** 587–96.

CHAPTER TWENTY-SIX

Real-Time and Quantitative Imaging of Mammalian Stress Granules and Processing Bodies

Nancy Kedersha, Sarah Tisdale, Tyler Hickman, *and* Paul Anderson

Contents

1. Introduction	522
2. Experimental Rationale	524
2.1. Choice of SG and PB marker proteins	524
3. Experimental Considerations	524
3.1. Choice of fluorescent tag	524
4. Selection Criteria	527
4.1. Protocol to obtain stable cell lines	528
5. Transfection	528
5.1. Drug selection	529
5.2. Testing for expression	530
5.3. Subcloning procedure	531
6. Properties of Representative Stable Lines	533
6.1. Imaging live cells	541
7. Environmental Control	541
7.1. Room temperature	541
7.2. Heated chamber	542
7.3. Microscope incubator	542
7.4. Additional notes	543
8. Microscope Hardware: Widefield vs Confocal	543
8.1. Widefield	545
8.2. Laser scanning confocal	545
8.3. Spinning disk confocal	546
8.4. General tips for improving PB images	547
9. Useful Microscopy Internet Resources	547
9.1. Collection of protocols and guides	547
9.2. Fluorescence spectra viewers	547

Division of Rheumatology, Immunology and Allergy, Brigham and Women's Hospital, Boston, Massachusetts, USA

9.3. General microscopy information	547
9.4. Nature microscopy submission guideline	547
10. Sample Protocols	547
10.1. Live cell imaging for tracking processing bodies	547
10.2. Live cell imaging for single-image quantification	549
10.3. Relative fluorescence protein localization in PBs	550
10.4. Acquiring FRAP/FLIP images	550
10.5. Analyzing FRAP/FLIP images	551
11. Conclusions	551
References	551

Abstract

Nuclear mRNA domains such as nucleoli, speckles, Cajal bodies, and gems demonstrate that RNA function and morphology are inextricably linked; granular mRNA structures are self-generated in tandem with metabolic activity. Similarly, cytoplasmic compartmentalization of mRNA into mRNP structures such as stress granules (SGs) and processing bodies (PBs) reiterate the link between function and structure; the assembly of SGs and PBs requires mRNA released from disassembling polysomes on translational arrest. SGs contain mRNA still associated with some of the translational machinery, specifically 40S subunits and a subset of translation initiation factors including eIF3, eIF4F, eIF4B, and PABP. PBs also contain mRNA and eIF4E but lack other preinitiation factors and contain instead a number of proteins associated with mRNA decay such as DCP1a, DCP2, hedls/GE-1, p54/RCK. Many other proteins (e.g., argonaute, FAST, RAP-55, TTP) and microRNAs are present in both SGs and PBs, sometimes shepherding specific mRNA transcripts between the translation and decay machineries. Recently, we described markers and methods to visualize SGs and PBs in fixed cells (Kedersha and Anderson, 2007), but understanding the dynamic nature of SGs and PBs requires live cell imaging. This presents unique challenges, because it requires the overexpression of fluorescently tagged SG/PB marker proteins, which can shift the mRNA equilibrium toward SGs or PBs, thus obscuring the result. We describe stably expressed, fluorescently tagged SG and PB markers that exhibit similar behavior to their endogenous counterparts, thus allowing real-time imaging of SGs and PBs.

1. INTRODUCTION

Cells reprogram their translation state in response to developmental cues, cell cycle, hormonal or other physiologic signals, or environmental changes. In contrast to genetically programmed developmental changes, environmental changes are unpredictable. Survival requires that these changes trigger a rapid response, in which ongoing translation is arrested, polysomes are disassembled, and translation machinery is reprioritized. Although some level of mRNA turnover occurs continuously, severe stresses activate the eIF2α kinases (e.g., PKR, PERK, HRI, and GCN2), which reduce levels of eIF2-GTP-tRNA$_i^{Met}$ ternary complex, thereby preventing assembly of the 48S preinitiation complexes (Anderson and Kedersha, 2002, 2006). Stalled initiation drives polysome disassembly as ribosomes terminate and run off, leaving mRNA transcripts bound to abortive 48S complexes. The sudden influx of disassembled polysomes overloads the cell's ability to process and remodel these transcripts, so they are temporarily packaged into SGs and PBs. It is important to realize that proteins and mRNA transcripts are not brought "to" SGs or PBs. Instead, mRNA transcripts and proteins participate in the assembly of SGs and PBs, much as water flows into and defines a river. Both SGs and PBs are dynamic structures whose contents are in continuous flux. Although P-bodies appear continually present in some cell lines such as COS or HeLa, we observe that quiescent cultures (e.g., cells stably arrested in G0 because of serum starvation) exhibit few or no PBs and no SGs. However, when stressed, even quiescent cells respond by forming SGs and PBs, indicating that the basal levels of translation in arrested cells are sufficient to require stress-induced reprogramming.

SGs are transient structures that are assembled rather hastily as a consequence of interrupted RNA translation. SGs contain several classes of components, any of which may legitimately be used to follow SG dynamics, but these exhibit different kinetics and some may also be present in PBs. SG components include mRNA, SG-associated mRNA binding proteins (such as argonaute, ataxin-2, BRF-1, CPEB, FMRP, FXR1, HuR, PABP1, smaug, staufen, TIA-1, TIAR, TTP, pumillio, ZPG1), components of the translation initiation machinery itself (such as eIF3, eIF4F, eIF4B, and small ribosomal subunits), and other proteins associated with SGs in less obvious ways (FAST, ORF1p, SMN, SRC3). Live studies that used tagged TIA-1, TIAR, PABP, G3BP, TTP, FAST (Kedersha et al., 2000, 2005), GFP-Ago2 (Leung et al., 2006), and hnRNPA1 (Guil et al., 2006) have revealed that different SG and PB components exhibit different residence time within SGs and PBs.

Transient expression of many SG-associated proteins results in the spontaneous appearance of seeming SGs in the absence of stress. In some cases, as in the case of enforced expression of phospho-eIF2α mimetic S51A (Kedersha et al., 1999) or of eIF2α kinases such as PKR or HRI (our unpublished results), the "spontaneous" SGs are logically the result of eIF2α phosphorylation and thus stalled initiation. In other cases, overexpression of translational silencers such as FMRP, FXR1, TIA-1, TIAR, CPEB, p54/RCK, or argonaute (Khandjian et al., 2004; 2007; Wilczynska et al., 2005) may recruit specific subsets of mRNAs to assemble SGs from their high-affinity mRNA targets. In other cases, the mechanism whereby overexpression of SG-associated proteins produces SGs is not clear (notably SMN and FAST [Hua and Zhou, 2004; Kedersha et al., 2005]). Regardless of the cause, the use of a tagged protein that induces SGs by its overexpression does not allow one to study the SG assembly process itself. Stable cell lines expressing tagged proteins can be obtained that no longer display SGs in the absence of stress, as was recently shown by Leung and Sharp (2006), who stably expressed GFP-Ago2 in HeLa cells. However, in other cases, overexpression of tagged PB marker proteins (e.g., hedls/GE-1, DCP1a) can result in giant PBs and increases the expression of other PB components (Fenger-Gron et al., 2005). In the case of DCP1a, stable cells expressing low levels of FLAG-tagged DCP1a were obtained, which were used for biochemical studies of PB subcomplexes (Fenger-Gron et al., 2005). We, therefore, undertook the development of cell lines stably expressing fluorescently tagged SG or PB markers to allow the study of SGs and PBs.

2. Experimental Rationale

2.1. Choice of SG and PB marker proteins

Table 26.1 indicates some of the proteins we have successfully tagged and stably expressed in U2OS osteosarcoma cells. Two different classes of proteins are possible for use as SG markers: RNA-binding proteins associated with mRNA and components of the translation machinery itself. In our system, we are able to stably express tagged versions of the RNA-binding proteins TIA-1, TIAR, PABP1, eIF4E, and G3BP1. The resulting cell lines display normal SG assembly and disassembly, and the tagged proteins exhibit behavior similar to their endogenous counterparts. Tagging components of the translation machinery is more complicated, because it involves tagging a subunit of a multiprotein complex (eIF3) or of the ribosome; however, we have successfully obtained stable cells expressing different tagged subunits of eIF3 and are in the process of analyzing their behavior in more detail. Regarding suitable markers of PBs, stable lines overexpressing the N-terminally tagged (nonfluorescent) versions of the decapping enzyme DCP1a have previously been described by the Lykke-Andersen laboratory (Fenger-Gron et al., 2005). Similarly, stable lines expressing GFP-Ago2 have been obtained in HeLa cells, and these studies established that Ago2 could move from SGs to PBs in a micro-RNA-dependent manner (Leung et al., 2006). We, therefore, undertook to generate stable U2OS cell lines expressing mRFP-tagged DCP1 and GFP-Ago2. In addition, we have also developed lines stably expressing GFP-hedls/GE-1, GFP-GW182, FAST, and mRFP-RCK. The osteosarcoma cell line U2OS was chosen for its relatively stable nature (unlike many immortal lines, it has wild-type p53), ability to undergo cell cycle arrest on serum withdrawal, and relatively flat morphology suitable for imaging studies.

3. Experimental Considerations

3.1. Choice of fluorescent tag

Fluorescent tags are now available in a wide spectrum of colors, stability, and brightness (see Shaner et al. [2007] for a helpful review). Earlier versions of GFP or dsRFP are not optimal for SG or PB studies, because these

Table 26.1 SG/PB markers stably tolerated in U2OS cells

Protein	Position of tag	Expression levels*	Localization/behavior	Function/nature
Ago1	N	High	PBs, SGs, near-normal	MicroRNA
Ago2	N	High	PBs, SGs, near-normal	MicroRNA
Ago3	N	High	PBs, SGs, near-normal	MicroRNA
Ago4	N	High	PBs, SGs, near-normal	MicroRNA
DCP1a	N, not C	High	PBs, near-normal	mRNA decapping
eIF3b/p116	N	High	cyt, SGs normal	eIF3 core subunit
eIF3e/INT6	N	Low	SGs, SGs	eIF3 noncore subunit
eIF3g/p44	N	Medium	SGs, SGs	eIF3 core subunit
eIF4E	N or C	Low	PBs, SGs, nuclear	mRNA cap-binding
eIF5a	N or C	High	SGs	Translation, other
FAST	C	Low	SGs, PBs, nuclear	Multifunctional
G3BP-1	N	High	SGs, cytoplasm	Multifunctional
GW128	N, not C	Low	Small PBs	MicroRNA
hedls/GE-1	N	Grossly high	Giant static PBs	mRNA decapping
ORF1p	N	High	SGs and PBs	Transposon
PABP	N	Low	SGs, nuclear	Translational enhancer
RAP-55	N	High	PBs and SGs	mRNA silencing
RCK	N	High	Normal SGs, PBs	RNA helicase
SMN	N	High	Gems, cyt, SGs	mRNP assembly
TIA-1	N	Low	Nuclear, SGs	Multifunctional
TIAR	N	Low	Nuclear, SGs	Multifunctional

*Low, 1-2X above endogenous levels; Medium 2-5X above endogenous levels; High, 5-10X above endogenous levels; Grossly high, >20X endogenous levels.

proteins tend to form dimers and tetramers—obviously not desired properties for tagging protein markers of structures whose assembly is regulated by aggregation. The choice of fluorescent tag should be compatible with the filter sets available on the particular fluorescent microscope used for image analysis—this should be determined in advance, because not all microscopes are similarly equipped. The data shown here were obtained with eGFP (Clontech), eYFP (Clontech), or mRFP obtained from the Tsien laboratory (Campbell *et al.*, 2002). More recent tags are available that are brighter and more photostable (Shaner *et al.*, 2007).

3.1.1. Reporter construct design

The first step in generating a stable cell line expressing the target protein is constructing a fluorescently-tagged fusion protein. Some essential considerations are: (1) choice of fluorescent tag (as mentioned previously), (2) choice of antibiotic resistance gene to use for drug selection, and (3) the position of the tag (e.g., N-terminal or C-terminal). The presence of an easily selectable antibiotic resistance gene within the same vector as the tagged protein is essential. We have used neomycin/G418 or kanamycin/G418 and puromycin. Note that cells derived from knockout animals may already have an antibiotic resistance gene, because many mice are generated with neomycin resistance, and unless this resistance was genetically removed, their cells are similarly resistant. If so, a different drug needs to be used, such as hygromycin, bleomycin, or blasticidin. A key step for effective drug selection is determining the threshold of the parent cell line to the particular antibiotic before beginning selection, and investing the time making constructs with a drug resistance gene that may not be appropriate. Different cell lines react differently to antibiotics, so it is necessary to determine the drug threshold for each different cell type used. A selection concentration sufficient to kill 99.99% of the cells is optimal.

Determining which end of the protein to tag is not completely arbitrary. Inserting a tag at the end of a protein where a natural binding or localization domain resides can alter the protein's normal function and localization but is generally less disruptive than inserting the tag internally into a protein. Although tagging at either end may not alter the localization of some proteins, in others it can completely disrupt the protein's subcellular localization and/or stability. If unsure of which end to tag, it is best to make two separate constructs tagged at each end and compare localizations. Because SGs and PBs are dynamic and transient entities, the presence of normal functional domains is essential. Successfully tagged proteins should exhibit behavior similar to endogenous protein. In the case in which the localization of the endogenous protein is not known, "correct" tagging is harder to assess. Minimally, any SGs or PBs induced by expression of a tagged protein should exhibit dynamic behavior (e.g., if spontaneous aggregation occurs on transient overexpression, it should disperse on treatment with emetine). Standard cloning strategies should be used to generate a fluorescently tagged fusion protein. Most importantly, the fusion protein and the protein of choice must be in the same reading frame. In addition, if placing the tag at the amino terminal end of the protein, a stop codon should be retained. Some investigators choose to remove the first methionine when making N-terminally tagged proteins, but this is not mandatory. If the tag is placed at the carboxyl terminal end of the protein, the protein's natural stop codon must be removed. Leaving the stop codon intact will result in an untagged protein.

The preinitiation factor eIF3 is a complex composed of at least 11 protein subunits (Hinnebusch, 2006). To complicate the picture, some of these subunits may be present in distinct eIF3 subcomplexes (Zhou et al., 2005). Tagging eIF3 required choosing one subunit. To avoid biasing our detection toward one subcomplex, we chose to tag eIF3b, one of the core scaffolding subunits that is present in both subcomplexes. We have used antibodies against endogenous eIF3b as a marker for SGs and found that it is a more robust marker than other components of the translational machinery, such as protein components of the small ribosomal subunit. A structural model of the proteins within eIF3 was recently published (Siridechadilok et al., 2005) that suggested the N-terminus of eIF3b was a better end to tag than the C-terminus, owing to the proximity of other eIF3 subunits that bind eIF3b at or near its C-terminal end. A cloning strategy was devised to generate N-terminally tagged eIF3b. A cDNA clone of full-length eIF3b was obtained (MGC: 38325) and used as a template for PCR amplification. The full-length coding region was cloned into the pE-YFP-C1 fluorescent reporter vector in three separate pieces, which were ultimately ligated into a single molecule. This strategy allowed us to clone a molecule too large for a single PCR reaction, and also generated truncations that have dominant-negative effects on SG assembly that may prove useful in mechanistic studies.

4. Selection Criteria

Cells expressing tagged SG/PB markers should exhibit normal behavior, defined as follows: (1) cells should exhibit no SGs and few PBs under normal conditions, (2) cells should display rapid SG/PB assembly upon treatment with sodium arsenite or other stresses, and (3) cells should completely disassemble PBs and SGs when treated with emetine or cycloheximide, agents that stabilize polysomes (see Table II of Kedersha and Anderson [2007] for a list of different stresses that induce SGs). To confirm that the tagged marker protein is behaving normally, we compare its localization in normal and stressed cells with that of the endogenous protein. We also counterstain the stable cells for independent markers (see Table I, Kedersha and Anderson [2007] for suitable markers).

4.1. Protocol to obtain stable cell lines

4.1.1. An overview of the procedures involved in obtaining stable cell lines

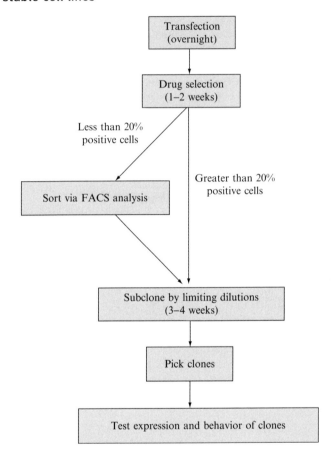

5. TRANSFECTION

We use SuperFect® (Qiagen) as our transfection agent; any reliable agent may be used.

Plate cells in a 6-well tissue culture plate to be nearly 80% confluent. For U2OS cells, a concentration of 5.0×10^5 cells per well is sufficient.

Once cells have spread and appear to be growing and dividing, approximately 6 h later, they are ready to be transfected.

In a 4-ml polystyrene tube, aliquot 100 μl of serum-free DMEM. The use of a serum-free medium here is imperative, because the presence of serum may inhibit the DNA complexes from forming.

Aliquot 2 μg of plasmid DNA per tube.

Add 6 μl of SuperFect® to each tube and mix. Let tubes sit at room temperature for 15 min to allow the DNA complexes to form.

Meanwhile, remove medium from the cells in the 6-well plate, and replace with 1 ml of serum-containing medium.

Once the DNA complexes have formed, add 1 ml of serum-containing medium to each tube, mix, and add directly to the cells. Each well will have a final volume of 2.1 ml.

Incubate cells at 37 °C overnight, or for approximately 18 h.

After 12 to 18 h, the transfected cells need to be replated. Trypsinize each transfected well with 1 ml trypsin and replate into a 10-cm tissue culture dish.

Transfection efficiency can vary widely, depending on transfection agent, cell line, and protein to be expressed. To generate a stable cell line, a cell population positive for the transfected protein must be isolated. A selectable antibiotic resistance gene encoded in the expression vector facilitates this process. Cells are grown in the presence of the appropriate antibiotic, allowing only positively transfected cells to survive. We describe in detail our standard procedure for generating a stable cell line. It should be noted that these conditions are optimized for stable lines made with U2OS cells.

5.1. Drug selection

Approximately 24 h after the transfected cells have been replated onto a 10-cm dish, they should be roughly 50% confluent. Remove medium and replace with medium containing the appropriate antibiotic. Note that different antibiotics work at varying rates. We have found that neomycin/G418 (0.5 mg/ml) works slowly, taking several days until significant cell death is observed. Puromycin (2.0 μg/ml) works much more quickly and can show effects within 24 h. Do not be alarmed by the percentage of cells that die while on drug selection—it is normal to see a sea of devastation! Typically, anywhere from 90 to 99.9% of the cells will die. The presence of a mock-transfected control becomes essential in this step to gauge antibiotic susceptibility. Similarly, a positive control is useful to ensure that you are not using a drug concentration that is so high it will overcome the drug resistance gene.

Antibiotic-containing medium should be changed frequently to replenish the drug and to remove dead or dying cells. After approximately 1 to 2 wk, drug-resistant colonies will begin emerging amid the dying cells. The colonies should be allowed to grow until the dish is at least 30% confluent, which can take up to 2 wk. During this time, some of the colonies may become dense and necrotic in the centers. If this becomes noticeable, trypsinize and replate cells back into the same dish, allowing the cells to spread out.

5.2. Testing for expression

The drug-resistant population of cells should now be tested for expression of the desired protein. Trypsinize, spin down, and resuspend the cells in 5 ml of medium. Replate most of the cells into a small T-25 flask, while plating the rest into a few wells of a 24-well plate with serum-conditioned coverslips. The general protocol is to plate 1.0×10^5 cells per well, to be confluent the next day. For the purpose of simply testing expression levels, the concentration does not have to be as accurate. Simply add 0.5 to 1.0 ml of resuspended cells to each well, which should suffice. At least two coverslips should be plated for testing: one as a control and one treated with 500 μM sodium arsenite to test for SG and PB formation. If greater than 20% of the cells are positive, the drug-selected population can be subcloned directly. If less than 20% of the cell are positive, consider sorting the drug-selected population by FACS analysis before subcloning to reduce the number of clones that must be screened.

5.2.1. Subcloning

By making limiting dilutions to obtain single colonies, the drug-selected population can now be subcloned. Single colonies represent unique clones, which are the basis for homogenously expressing stable cell lines. Single colonies should be allowed to grow until they occupy roughly one third of the well. This typically takes between 2 and 3 wk. See the following for detailed protocol.

5.2.2. Picking clones

Single colonies representing unique clones must be trypsinized, expanded, and tested for expression. Special 96-well flat bottom plates are available from Corning (product #3614) that have an ultra thin plastic bottom to allow direct prescreening of colonies with an inverted fluorescent microscope. This can be a very timesaving tool, especially when dealing with a drug-selected population in which most cells do not express the target protein. Whether a prescreening step is used or not, the single colonies within the wells need to be individually trypsinized and replated as described in the following into two wells, one containing a coverslip for screening and the sister well for possible expansion.

5.2.3. Testing expression and behavior

Each clone picked should be tested for expression and behavior in response to various SG- and PB-inducing drugs. Treatment with 500 μM sodium arsenite for 30 min strongly induces both SGs and PBs and is a good starting point in observing behavior. We typically treat all new clones first with sodium arsenite to determine whether the tagged protein localizes to, or blocks, the formation of SGs or PBs. Uniformity of expression of each clone should also be noted. Some clones can express the fusion protein quite uniformly from cell to cell. Others show a heterogeneous mix of expression levels, ranging from strong to barely visible levels. Although sorting by FACS analysis can help in obtaining a more positive population, in some cases the heterogeneity is due to cell-cycle variations in protein localization. For example, PBs exhibit size variation throughout the cell cycle (Eystathioy et al., 2003), and neither PBs nor SGs are present in mitotic cells (Sivan et al., 2007).

5.2.4. Antibiotic selection

For U2OS cells, the working concentrations for drug selection are 0.5 mg/ml for G418 and 2.0 μg/ml for puromycin. We find it convenient to make aliquots of antibiotic stock solutions and keep them in the freezer until needed. To make antibiotic stocks: Weigh out antibiotic powder and dissolve it in sufficient cell culture medium to reach the final stock concentration. For G418 our stock concentration is 100 mg/ml, and for puromycin we use a 10 mg/ml stock. Note that many drugs are not 100% pure as supplied by the manufacturer; often G418 is only $\sim$70% pure. In this case, compensate by using proportionally more drug; for example, if 70% pure, calculate the amount needed (10 mg/ml divided by 0.70; 14.28 mg/ml).

The addition of the antibiotic will change the pH of the medium. Correct the pH of the solution so that it is approximately pH 7.0. If phenol red is included in your medium, you can avoid the use of the pH meter by adding drops of either NaOH or HCl until the normal color returns.

Take the dissolved antibiotic solution into a sterile hood. Sterile-filter the solution through a 0.20-μm syringe filter into prelabeled 15-ml tubes. For a 550-ml bottle of medium, 2.75 ml of a 100 mg/ml G418 stock is required. Therefore, aliquot 5.5 ml of stock solution into each 15-ml tube, which will make two bottles of antibiotic-containing medium.

Store the aliquots of stock solutions at $-20\,^\circ$C until needed.

5.3. Subcloning procedure

Set up five 15-ml tubes and label them dilutions A through E.
Trypsinize, resuspend, and count cells with a hemocytometer and trypan blue, scoring only live cells.

Add the appropriate volume of resuspended cells to dilution A to give 5×10^3 total cells in 5 ml of medium.
For dilution B, mix 1 ml of dilution A with 9 ml of medium.
For dilution C, mix 2.5 ml of dilution B with 2.5 ml of medium.
For dilution D, mix 2.5 ml of dilution C with 2.5 ml of medium.
For dilution E, mix 1.5 ml of dilution B with 13.5 ml of medium.
Set up two 96-well flat bottom plates. With a multichannel pipette, aliquot 100 μl of medium into each well.
Add 100 μl per well of each dilution using the following outline:
For plate 1:
Column 1: Dilution A (100 cells/well)
Columns 2 and 3: Dilution B (10 cells/well)
Columns 4 and 5: Dilution C (5 cells/well)
Columns 6 and 7: Dilution D (2 to 3 cells/well)
Columns 8, 9, 10, 11, and 12: Dilution E (1 cell/well)
For plate 2:
All columns: Dilution E (1 cell/well)

Place the plates in the incubator, to be checked occasionally for the next 2 to 3 wk to monitor colony growth. During this incubation time, some of the wells close to the edges can dry out. To avoid this, place the plates in sealable plastic bags within the incubator. Once the plates have equilibrated with the atmosphere of the incubator, partially seal the bags to prevent moisture loss.

5.3.1. Clone expansion procedure

When individual colonies within the wells grow to cover roughly 30% of the well, they are ready to be harvested. Prescreening the plates for positive cells with an inverted fluorescence microscope (if available) will reduce subsequent work.
Set up a 12-well tissue culture plate containing 2 ml of medium per well, and a 24-well tissue culture plate with conditioned circular coverslips in 1 ml of medium.
Aspirate the medium from each colony-containing well to be harvested. Wash the cells with a few drops of HBSS (without Ca^{++} and Mg^{++}), aspirate, and add 50 μl of trypsin. Incubate the plate at 37 °C for a minute or two.
When the cells have started to round up, neutralize the trypsin by adding approximately 150 μl of medium. Pipette up and down several times within the well to fully detach the cells.
With the same pipette tip, add approximately two thirds of the suspended cell solution to a coverslip in the 24-well plate, and add the rest to the correspondingly labeled well in the 12-well plate. The cells plated on the coverslip serve to test for expression and behavior of the clone, whereas the cells in the 12-well plate will be expanded into stocks if positive.

Once the cells on coverslips are roughly 50 to 80% confluent, they need to be tested for expression and behavior. Treat each coverslip with 500 μM sodium arsenite for 30 to 45 min. Fix cells by incubation in 4% paraformaldehyde in PBS for 15 min, followed by a methanol permeabilization step for 5 min. The addition of a DNA-staining dye such as Hoechst or DAPI is helpful to identify cells. Ideally, clones to be used further should be uniform, 100% expressing, and exhibit the correct behavior (e.g., assembly into SGs or PBs or both, as appropriate).

Once positively expressing clones have been identified, the corresponding 12-well culture serves as the initial clone stock to be expanded, frozen, and used for experiments.

Western blot analysis should be performed to verify that the protein present in each clone is the correct size. In rare cases, rearrangement of the cDNA occurs on its insertion into the DNA of the cells, resulting in a truncated protein that may express the tag but not the full-length fusion protein. Blotting with an antibody directed against the endogenous protein also allows a determination of the ratio of the overexpressed protein (e.g., containing the fluorescent tag, and thus migrating more slowly than the endogenous protein) versus the endogenous protein.

6. Properties of Representative Stable Lines

Stable overexpression of the classical SG markers TIA-1 and TIAR is difficult because of the translational silencing activities of these proteins that inhibit cell growth. After drug selection, a very low percentage of cells (<5%) express any fluorescence, and those that do express have a dim signal unsuitable for viewing over time (e.g., as required for time-lapse imaging). After subcloning, only half of the YFP-positive clones exhibited normal recruitment to SGs because of rearrangement of the cDNA and resulting truncation of the expressed protein. Our attempts to make stable lines overexpressing tagged TTP or 4E-T were even less successful; after drug selection, no cells expressed even the GFP tag. Because the SG-associated protein G3BP1 is naturally overexpressed in a number of tumor cell lines (Irvine et al., 2004), we reasoned that it should be possible to develop stable cell lines expressing high levels of GFP-G3BP1. Endogenous G3BP1 (green), DCP1a (red), and eIF3b (blue) are shown in untreated (Fig. 26.1A), arsenite-treated (Fig. 26.1B), or emetine-treated (Fig. 26.1C) parental U2OS cells, which were fixed and stained by the previously described methods (Kedersha and Anderson, 2007). Both G3BP-1 (green) and eIF3 (blue) are diffuse and cytoplasmic in unstressed cells (Fig. 26.1A), relocalize into SGs on arsenite stress (Fig. 26.1B, white arrows), and are diffusely cytoplasmic in emetine-treated cells (Fig. 26.1C). The PB marker

DCP1a is largely diffuse in unstressed cells (Fig. 26.1A, red), assembles into PBs on arsenite treatment (Fig. 26.1B, red, yellow arrowheads in enlarged views), and appears diffuse in emetine-treated cells (Fig. 26.1C).

Transient overexpression of GFP-G3BP1 has been reported to induce SGs without additional stress; this signature property of G3BP1 has been exploited in a number of studies (Cande et al., 2004; Kedersha et al., 2005; Tourriere et al., 2003) and used as a criterion to determine whether other proteins are recruited to SGs. As shown in Fig 26.1D, transient overexpression of GFP-tagged G3BP-1 (green) induces abnormally large "spontaneous" SGs in the absence of stress (white arrow), which poorly colocalize with endogenous eIF3 (blue, shown white in enlarged view). Transient overexpression of GFP-G3BP1 does not promote PB assembly, as shown by the diffuse distribution of the endogenous PB marker DCP1a (Fig. 26.1D, red). Arsenite treatment of GFP-G3BP1 transient transfectants (Fig. 26.1E) does not alter the GFP-G3BP1 pattern but increases the recruitment of eIF3 to SGs (white arrows), and facilitates PB assembly (DCP1a, red, yellow arrowheads). Despite their large size, SGs induced by overexpression of G3BP1are disassembled when cells are treated with inhibitors of protein translation such as cycloheximide or emetine (Fig. 26.1F; as reported previously in Tourriere et al. [2003]), thus establishing that the SGs induced by GFP-G3BP1 overexpression are dynamic and not nonspecific inert aggregates induced by overexpression. This interpretation is consistent with photobleaching data revealing that that transiently overexpressed GFP-G3BP1 exhibits a very rapid shuttling rate into SGs (Kedersha et al., 2005) similar to GFP-TIA-1 and faster than GFP-tagged PABP (Kedersha et al., 2002).

Once the transiently transfected cells shown in Fig. 26.1D to F were subjected to drug selection, approximately 50% of the cells expressed relatively high levels of GFP. Unlike transient transfectants, the drug-selected GFP-positive cells did not exhibit spontaneous SGs but instead displayed diffusely cytoplasmic GFP signal that only formed SGs on stress. On subcloning, 50% of the clones screened expressed full-length GFP-G3BP1 as confirmed by Western blotting, with only very rare clones expressing truncated forms of GFP-G3BP1. We then introduced mRFP-DCP1a (in a puromycin-resistant vector) with transfection into the GFP-G3BP1 cell line, selected with puromycin, and cloned the cells by limiting dilution. As reported previously (Fenger-Gron et al., 2005), cells tolerate stable DCP1a overexpression, and more than 50% of the drug-selected population expressed mRFP that localized into PBs. After subcloning, we obtained stable cell lines expressing both GFP-G3BP-1 and mRFP-DCP1a, in which the tagged proteins exhibited very similar behavior as their endogenous counterparts in the parental U2OS cells as shown in Fig. 26.1G to H. Without stress, GFP-G3BP1 is cytoplasmic and dispersed (Fig. 26.1G, green), similar to endogenous eIF3b (blue). Overexpressed RFP-DCP1a forms small

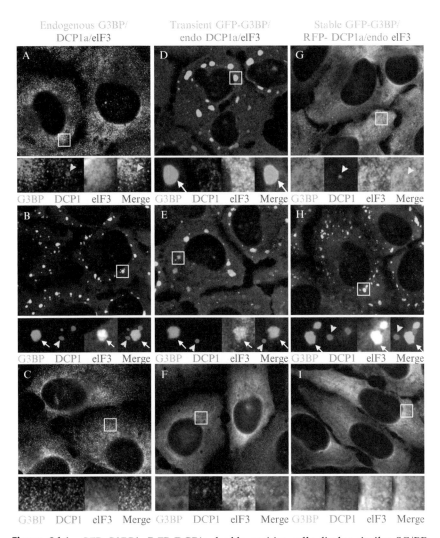

Figure 26.1 GFP-G3BP/mRFP-DCP1a double-positive cells display similar SG/PB assembly as untransfected cells. (A to C) Parental U2OS stained for endogenous G3BP1 (green), DCP1a (red), and eIF3b (blue); (D to F), U2OS cells transiently transfected with GFP-tagged G3BP1 (green), counterstained for endogenous DCP1a (red) and eIF3b (blue); (G to I), U2OS cells stably expressing GFP-G3BP1 (green) and mRFP-DCP1a (red), counterstained for endogenous eIF3b (blue). Cells were untreated (top row, A, D, and G), exposed to 500 μM sodium arsenite for 1 h (middle row, B, E, and H), or treated with 40 μg/ml emetine for 1 h (bottom row, C, F, I). Enlarged views of boxed regions are displayed underneath the corresponding panels, separately showing the views of the same field. Note that endogenous G3BP1 and stably expressed GFP-G3BP1 appear dispersed and cytoplasmic in unstressed (A, G) and emetine-treated (C, I) cells but is entirely recruited into SGs on arsenite treatment (B, H). Similarly, endogenous DCP1a (A to C) and stably expressed mRFP-DCP1a (G to I) display few small PBs in the absence of stress (A, G), numerous PBs on arsenite treatment (B, H), and no PBs in

PBs (red, yellow arrowheads) visible in approximately 40% of unstressed cells. Arsenite (Fig. 26.1H) and other stresses (not shown) result in the assembly of both SGs (green and blue, white arrowheads) and PBs (mRFP-DCP1a, yellow arrowheads), whereas emetine treatment (Fig. 26.1I) results in the complete disassembly of both SGs and PBs. The double-positive cells thus exhibit behavior similar to that of the parental cell line despite the fact that both GFP-G3BP1 and RFP-DCP1a are expressed at about 10-fold higher levels relative to the endogenous proteins. However, the presence of overexpressed mRFP-DCP1a increases the percentage of cells containing PBs under normal growing conditions from approximately 10% to approximately 40%.

Despite the utility of GFP-G3BP1 as a SG marker, our lack of understanding regarding its function at SGs makes the use of it as a "SG-only" marker somewhat worrisome. For example, we had earlier determined that FXR1 localizes exclusively to SGs and not PBs (Kedersha et al., 2005), but a recent report describes other conditions (serum starvation) that result in the localization of FXR1 to PBs (Vasudevan et al., 2007). In case the same should prove true with G3BP1, we sought to create a tagged large subunit of eIF3, part of the core translation initiation machinery itself, which has not been found to associate with PBs from yeast to mammals, and would, therefore, be truly SG-specific. As shown in Fig. 26.2A to C, endogenous eIF3b (green) localizes only to SGs (white arrows), not PBs (yellow arrowheads). In contrast to results obtained with GFP-G3BP, transient overexpression of YFP-tagged eIF3b did not induce spontaneous SGs, and only a small fraction was recruited to SGs when cells were treated with arsenite (Fig. 26.2, panel E, green, white arrows). However, once we obtained stable cell lines coexpressing YFP-eIF3b and mRFP-DCP1a (Fig. 26.2, panels G-I), the YFP-tagged eIF3b exhibited similar SG-targeting behavior similar to the endogenous protein (compare panel H with endogenous protein shown in panel B). Studies are ongoing to further analyze the properties of this line in detail.

Our success with DCP1a as a marker for PBs notwithstanding, we then tried to use the large scaffolding protein hedls/GE-1 as a PB marker. We previously described a commercial antibody that detects cytoplasmic hedls (Kedersha and Anderson, 2007) in addition to staining nuclear speckles. As shown in Fig. 26.3A to C (cytoplasmic signal, green), hedls is a robust marker of PBs and exhibits less diffuse cytoplasmic signal than endogenous DCP1a (compare Fig. 26.3A, green, with 26.1A, red), suggesting it might make a cleaner marker for PBs than DCP1a. As with endogenous DCP1a (Fig. 26.1A to C), endogenous hedls (Fig. 26.3A to C, cytoplasmic green signal) is present

---

emetine-treated cells (C, I). In contrast, transiently overexpressed GFP-G3BP1G3BP1 nucleates abnormally large SGs without stress (D) and in response to arsenite (E) but is still disassembled in response to emetine (F). (See Color Insert.)

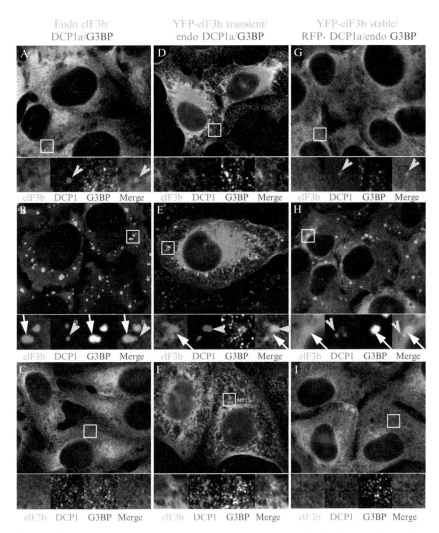

Figure 26.2 YFP-eIF3b/mRFP-DCP1a double-positive cells display similar SG/PB assembly as untransfected cells. (A to C) Parental U2OS stained for endogenous eIF3b (green), DCP1a (red), and G3BP1 (blue); (D to F), U2OS cells transiently transfected with YFP-tagged eIF3b (green), counterstained for endogenous DCP1a (red) and G3BP1 (blue); (G to I), U2OS cells stably expressing YFP-eIF3b (green) and mRFP-DCP1a (red), counterstained for endogenous G3BP1 (blue). Cells were untreated (top row, A, D, and G), exposed to 500 μM sodium arsenite for 1 h (middle row, B, E, and H), or treated with 40 $\mu g/ml$ emetine for 1 h (bottom row, C, F, I). Boxed regions appear enlarged underneath each panel. Note that endogenous eIF3b and stably expressed eIF3b appear dispersed and cytoplasmic in unstressed (A, G) and emetine-treated cells (C, I) but are entirely recruited into SGs on arsenite treatment (B, H). Similarly, endogenous DCP1a (A to C) and stably expressed mRFP-DCP1a (G to I) display few small PBs in the absence of stress (A, G), numerous PBs on arsenite treatment (B, H), and no PBs in emetine-treated cells (C, I). In contrast, transiently overexpressed YFP-eIF3b appears less uniform in untreated (D) and in emetine-treated cells (F), and only weakly recruited to SGs in response to arsenite (E). (See Color Insert.)

in few PBs in unstressed U2OS cells, many PBs in arsenite-treated cells, and few very small PBs in emetine-treated cells (Fig. 26.3C, green). Transiently overexpressed YFP-tagged hedls (Fig. 26.3, D to F, green) assembles into huge, irregular PBs, the largest of which appear to exclude DCP1a in unstressed cells (Fig. 26.3D, inset, compare green to red). Arsenite treatment (Fig. 26.3E) does not alter the size of these giant PBs but does promote recruitment of endogenous DCP1a. Emetine treatment, however, fails to promote their disassembly; in some cases, DCP1a is retained in these abnormal PBs (Fig. 26.3F, inset, white). Stable clones were obtained, but all of them greatly overexpressed YFP-hedls, which assembled into giant PBs in the absence of stress (Fig. 26.3G, green, yellow arrowheads). Arsenite treatment did not increase the number or size of the YFP-hedls PBs (Fig. 26.3H), nor did emetine treatment enforce their disassembly. Interestingly, however, the hedls-containing giant PBs completely disassembled in mitotic cells (not shown). With YB-1 as a marker of mRNA content (Fig. 26.3G to I, blue) indicates that the giant hedls-PBs contain mRNA in unstressed cells (Fig. 26.3G, compare green to white in inset, yellow arrowheads), but not in arsenite-treated cells (Fig. 26.3H, yellow arrowheads) nor in emetine-treated cells (Fig. 26.3I, yellow arrowheads). We therefore conclude that YFP-hedls nucleates giant pseudo-PBs by means of protein–protein interactions, unlike bona fide PBs, whose assembly requires mRNPs. This line is not suitable for the study of normal PB dynamics, but its detailed characterization may allow us to investigate some aspects of PB disassembly that occur during mitosis.

GFP-Ago2 stably expressed in HeLa cells was reported to move from PBs into SGs on stress (Leung et al., 2006). At the time of these studies, no antibody suitable for the detection of endogenous Ago2 was available. With a new antibody against Ago2 from Wako Chemicals (product #016-20861), we show that endogenous Ago2 (Fig. 26.4A to C, green) in U2OS cells colocalizes with YB-1 (Fig. 26.4A to C, blue) in few small PBs in unstressed cells (Fig. 26.4A), but largely relocalizes into eIF3-positive SGs (red, white arrows) on arsenite treatment (Fig. 26.4B). Emetine treatment (Fig. 26.4C) disassembles all PB-sized foci, leaving small foci that contain Ago2 but lack YB-1 (Fig. 26.4C). Transiently overexpressed GFP-Ago2 (Fig. 26.4D to F) assembles into PBs, some of which are very large (Fig. 26.4D), but arsenite treatment partially relocalizes GFP-Ago2 into SGs, as previously reported in HeLa cells (Leung et al., 2006). Emetine treatment does not completely disassemble foci containing transiently overexpressed GFP-Ago2 (Fig. 26.4F), which at higher magnification (Fig. 26.4F, inset) do not perfectly colocalize with endogenous DCP1a (Fig. 26.4F, white). When stably expressed, the GFP-Ago2 line exhibits more basal PBs than parental U2OS, which are more resistant to emetine disassembly (Fig. 26.4I, green) but which colocalize with DCP1a under all conditions. Arsenite treatment causes marked movement of GFP-Ago2 from PBs into SGs, although the relocalization is not as complete

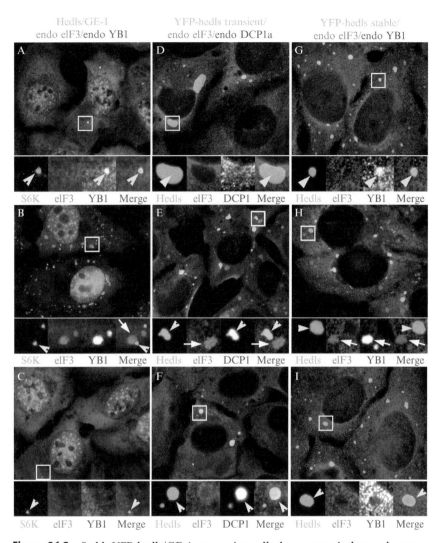

Figure 26.3 Stable YFP-hedls/GE-1–expressing cells do not recapitulate endogenous hedls/GE-1 behavior. Endogenous hedls/GE-1 (A to C, cytoplasmic signal, green), transiently overexpressed YFP-hedls/GE-1 (D to F), and stably expressed YFP-hedls/GE-1 (G to I), counterstained for endogenous YB1 (A to C; G to I) or DCP1a (D to F). Top row (A, D, G), no drug treatment; middle row (B, E, H), cells treated with sodium arsenite for 1 h; and bottom row (C, F, I), cells treated with 40 μg/ml emetine for 1 h. Although endogenous hedls exhibits PB assembly on arsenite treatment (B, green) and disassembly when treated with emetine, both transiently expressed (D to F) and stably expressed (G to I) YFP-hedls/GE-1 form giant PBs that are not affected by drug treatments. (See Color Insert.)

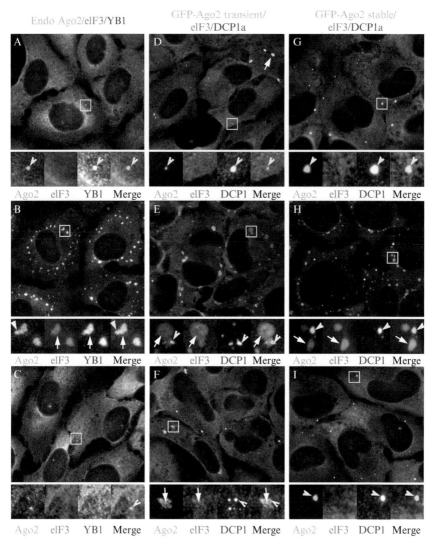

Figure 26.4 GFP-Ago2–expressing stable cells recapitulate behavior of endogenous Ago2. Endogenous Ago2 (A to C, green) colocalizes with endogenous YB1 (A to C, blue) in PBs (yellow arrowheads) in unstressed cells but is present in both PBs (yellow arrowheads) and SGs (white arrows, endogenous eIF3b, red) when cells are treated with arsenite (4B). Transient overexpression of GFP-Ago2 (D to F) induces SGs in some unstressed cells (D, white arrows) but is present in PBs in others (D, boxed region, yellow arrowheads). Arsenite treatment (E) induces movement of some GFP-Ago2 into SGs (enlarged view, white arrows), but a significant amount is retained in PBs (yellow arrowheads), costaining with DCP1a (red). Emetine treatment (F) is insufficient to disassemble SGs and PBs in the transient transfectants. Stably expressed GFP-Ago2 (G–I) exhibits behavior identical to that of endogenous Ago2 (A to C). Enlarged views of boxed regions are displayed underneath the corresponding panels, separately showing the views of the same field. (See Color Insert.)

movement as shown by endogenous Ago2. Taken together, these data confirm the findings reported by Leung and Sharp (2006). The more complete movement of endogenous Ago2 into SGs than overexpressed GFP-AGo2 supports their contention that this movement requires microRNA, which is likely limiting when GFP-Ago2 is overexpressed.

6.1. Imaging live cells

Successful imaging of live cells requires addressing several issues that are not relevant to imaging fixed cells. We consider the obstacles and challenges presented by various environmental conditions, bleaching, depth of field, image capture time, and the pros and cons of widefield versus confocal microscopy in detail in the following. As different microscopes, confocal systems, and software are available, a detailed description of all systems is beyond the scope of this chapter. The following is intended as an overview, and the protocols may need modification-depending on the reader's specific equipment. A list of useful web sites appears at the end of this chapter. We highlight the parameters most important to SG/PB behavior.

7. Environmental Control

Environmental control is critical to imaging live cells, especially because heat shock can induce SGs. Furthermore, although photobleaching measurements taken on preformed SGs at room temperature are similar to those obtained at 37 °C, SG assembly is dramatically slowed at room temperature, and movies of SG assembly may not recapitulate normal progression. Experimental temperature is often difficult to control. Although it is ideal to run a live cell experiment at 37 °C, adding an external heat source can complicate image acquisition. Even small fluctuations in temperature causes the coverglass to contract, expand, or bend. This causes the specimen to move up or down, out of the focal plane (which is referred to as focal drift) and is especially problematic when viewing cells over time. To minimize focal drift, the coverglass must remain a consistent temperature. Samples can be viewed at room temperature or using some sort of heated chamber.

7.1. Room temperature

7.1.1. Advantages

- Temperature is very stable.
- Kinetics are slowed down to track fast-moving PBs or record quickly recovering proteins in fluorescence recovery after photobleaching (FRAP).

- Medium evaporates less quickly.
- No additional cost.

7.1.2. Disadvantages

- Temperature is not physiologic.
- Aggresomes can form at lower temperatures.
- Kinetics are slower than usual, SG formation/disassembly is severely retarded/may not occur.
- CO_2 concentration is not controlled, which may adversely affect pH.
- Prolonged incubation at room temperature can promote cell detachment from substrate.

7.2. Heated chamber

7.2.1. Advantages

- Cells can be kept at 37 °C.
- Temperature can be raised to induce heat shock.
- Kinetics of PB formation and slow recovery after FRAP is faster than at room temperature.
- Cells are usually viable with HEPES-containing medium for several hours.
- Relatively inexpensive.

7.2.2. Disadvantages

- Temperature is constantly fluctuating to remain above ambient temperatures, greatly increasing focal drift.
- Oil objectives pull heat away from the viewed region of the specimen.
- Objective heaters significantly add to the cost.
- Kinetics of quickly moving proteins may be too fast for FRAP using some confocal systems.
- Medium evaporates faster, causing osmotic shock. This can be rectified by using a thin Teflon film to allow gas exchange but not water vapor loss.
- CO_2 not controlled.

7.3. Microscope incubator

7.3.1. Advantages

- Cells maintained at 37 °C.
- Temperature is very stable.
- CO_2 control possible as an option.

- Medium evaporation can be minimized by humidifying the chamber or by using Teflon film to allow gas exchange but prevent water vapor loss.
- Cells can be viewed for over 24 h.

7.3.2. Disadvantages

- CO_2 addition requires additional stage adaptors.
- Hot-moving air evaporates medium quickly if humidity is not controlled.
- Can be bulky and can interfere with microscope controls.
- Expensive.

7.4. Additional notes

- Raising the temperature of the room will help to reduce the negative effects of having a lower ambient temperature (such as focal drift with a heated stage).
- Humidifying the entire room will help to reduce medium evaporation, and is especially recommended during dry months with a microscope incubator.
- Although it is often suggested, we *do not* recommend adding mineral oil to the top of medium to prevent evaporation/gas exchange. It has been shown in our studies to induce a stress response.
- Adding HEPES to medium is recommended if your cells will tolerate it (U2OS cells are unaffected by 25 mM HEPES) even in a CO_2-controlled incubator.

8. Microscope Hardware: Widefield vs Confocal

An ideal live cell video will have high x, y, and z resolution, rapid frame rate for time or temporal resolution, and high signal-to-noise ratio (signal:noise), with minimal light exposure. Unfortunately, some aspects must be optimized at the expense of others. High-resolution images, for example, take longer to acquire, thereby decreasing the frame rate, which in turn increases bleaching of the fluorophores and thus decreases signal:noise and damages the cells with intense light. In some cases, however, such as when acquiring a single image for quantitative analysis, resolution and signal:noise become more important. Image acquisition methods must, therefore, be carefully thought out and possibly adjusted according to the samples used and information desired. One of the first considerations is whether to use widefield or confocal microscopy. Although each method captures fluorescent images, differences between widefield and confocal microscopy are significant. Figure 26.5 compares these techniques.

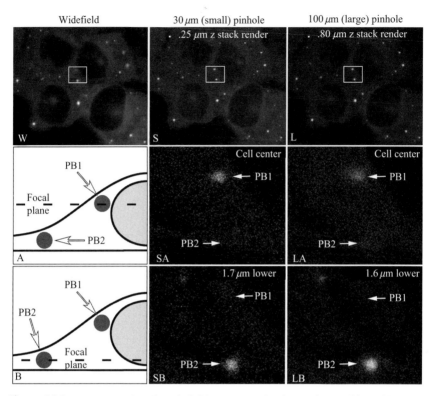

Figure 26.5 PBs viewed with widefield versus confocal z stack. Double-stable GFP-G3BP/mRFP-DCP1a U2OS cells (shown in Fig. 26.1) were treated with arsenite (0.5 mM, 45 min) and then fixed with PFA (4%, 15 min) without permeabilization. Only the mRFP-DCP1a fluorescence is shown. Top, Three images of the same field were collected with widefield (W, 100× objective lens) or confocal with either a small (S) or large (L) pinhole setting to acquire a confocal z stack. The widefield image (W) shows that only a fraction of the total PBs throughout the cell is clearly resolved. The small (S) and large (L) pinhole confocal z stacks were rendered to display the pixels collected in the entire z stack for each x and y location. Both z stacks allow PBs throughout the cell to be clearly resolved. Although the large pinhole z stack allows a larger step size, it was collected faster with less photobleaching, making this setting preferable for live imaging. Middle, When a confocal image is acquired, only photons from a thin section in z are collected. A cartoon (A, left) diagrams how the photons from PB1 are collected from a focal plane in the center of a cell, whereas the photons from PB2, below the focal plane, are omitted. A single z slice acquired with a small pinhole (SA) and large pinhole (LA) clearly shows photons emitted from a PB in the center of the cell (PB1) through which the focal plane passes. Photons from a second PB (PB2), which lies below the focal plane, do not pass through the pinhole and are not detected. Bottom, When the focal plan is shifted lower in the cell, only photons from that focal plane are collected. The cartoon (B) shows how a lower PB (PB2) now resides within the focal plane and can, therefore, be detected. The single z slice collected with a small (SB) and large (LB) pinhole now shows a lower PB (PB2), whereas the PB clearly seen in a higher focal plane (PB1) is no longer detected. A small pinhole (SA, SB) resolves the punctate pattern of DCP1 within the PB more clearly than a larger pinhole (LA, LB).

8.1. Widefield

A widefield microscope uses a mercury lamp and filters to excite fluorophores and a charge-coupled device (CCD or camera) to detect photons emitted by the specimen.

8.1.1. Advantages

- With bright fluorescence, has very fast frame rates and thus good temporal resolution.
- Very efficient at detecting emitted photons.
- When used with binning, can increase signal:noise and frame rate, especially important in viewing dimmer specimens (see Waters et al. [2007] for more information).
- Can be used with deconvolution to acquire z resolution.
- Photo bleaches less than confocal.
- Less expensive than confocal hardware.

8.1.2. Disadvantages

- Out-of-focus fluorescence is captured, removal of which requires stack aquisition hardware and expensive deconvolution software.
- PBs can move in and out of focus during real-time imaging.

8.2. Laser scanning confocal

Laser scanning confocal microscopy uses lasers to excite fluorophores on a specific focal plane. Emitted fluorescence is filtered through an adjustable pinhole to detect thick or thin optical slices. Photons allowed through the pinhole are detected by a photomultiplier tube (PMT), which amplifies the signal and converts it into a gray value, creating a pixel. These pixels are collected point to point and are combined to create an image.

8.2.1. Advantages

- Can obtain very high-resolution images, capable of resolving areas within PBs.
- Pinhole can be adjusted for ideal resolution with different wavelengths or to brighten dimmer fluorescence.
- Out-of-focus fluorescence is omitted.
- Z slice can be very thin.
- Z stacks can capture all PBs within a cell.
- 3D rendering possible.

- Laser power can be adjusted for FRAP or FLIP (fluorescence loss in photobleaching) experiments.
- Images can be averaged to increase signal:noise.

8.2.2. Disadvantages

- Laser scanning light bleaches more than widefield or spinning disk confocal.
- Frame rate is usually slower than widefield or spinning disk confocal.
- High-resolution images take longer to acquire.
- Z stacks take longer to acquire.
- PMT can be noisy, especially at higher gain.

8.3. Spinning disk confocal

Spinning disk confocal microscopy allows for much faster confocal image acquisition, with less photobleaching of the sample than laser scanning confocal microscopy. Instead of having a single laser point and pinhole scanning across a specimen, a spinning disk confocal moves a series of pinholes and lenses while the laser remains stationary. The photons are collected by a CCD as with widefield, not point-by-point with a PMT as with laser scanning confocal.

8.3.1. Advantages

- Much faster image acquisition time than laser scanning confocal.
- Out-of-focus fluorescence is omitted.
- Bleaches less than laser scanning confocal.
- Ability to z stack.
- Z stacks acquired much faster than laser scanning confocal, allowing tracking of fast moving PBs, or rapid recovery in FRAP.
- 3D rendering is possible.
- Laser power can be adjusted for FRAP or FLIP experiments.
- Images can be averaged to increase signal:noise.

8.3.2. Disadvantages

- Requires an additional laser for FRAP or FLIP; not all spinning disc systems are equipped for this.
- Less efficient at detecting emitted photons than widefield.
- Pinhole size is not adjustable; less-optimal x, y, z resolution than laser scanning confocal (Murray et al., 2007).
- Resolution is lower than laser scanning confocal.
- Each disk has many pinholes, allowing some out of focus fluorescence to pass through neighboring pinholes.

8.4. General tips for improving PB images

- Decreasing the ambient light (by turning off the room lights or turning a monitor away from the microscope) will reduce background fluorescence.
- Focusing the mercury lamp will increase signal:noise and evenly light the field.
- When using widefield, reducing the field diaphragm to include only the area around the cell will make PBs clearer.
- Oil objectives produce clearer images.
- When tracking fast-moving PBs, frame rate is more important than resolution.
- When capturing a single image for quantification, the importance of signal:noise > resolution > image acquisition time.

9. Useful Microscopy Internet Resources

9.1. Collection of protocols and guides

http://confocal.bwh.harvard.edu

9.2. Fluorescence spectra viewers

http://www.invitrogen.com/site/us/en/home/support/Research-Tools/Fluorescence-SpectraViewer.reg.us.html
http://fluorescence.nexus-solutions.net/frames6.htm

9.3. General microscopy information

http://www.microscopyu.com/
http://www.olympusconfocal.com/resources/index.html

9.4. Nature microscopy submission guideline

http://www.nature.com/nature/authors/submissions/images/index.html

10. Sample Protocols

10.1. Live cell imaging for tracking processing bodies

The goal is to create a high-frame-rate movie that is high enough resolution to track PBs, without excessively photobleaching the specimen.

Day 1:

1. Select a suitable glass-bottom cell chamber (We use cat: FD35PDL-100 from World Precision Instruments, http://www.wpiinc.com/products/cell-tissue/FluoroDish/). Add enough medium to completely cover the glass in each well without evaporating (1 to 3 ml).

Day 2:

1. Replace medium with HBSS in chambers that will not be used immediately. Add cells to 90% confluence by adding several different concentrations (e.g., 3×10^5, 4×10^5, and 5×10^5) in three different chambers. This helps ensure that at least one sample will have ~90% confluence and will leave extra samples on hand in case one is contaminated or ruined for other unforeseen reasons.

   Allow cells to spread overnight.

Day 3:

1. Remove phenol red–containing medium and replace with phenol red–free medium. If floating dead cells are present, rinse cells with HBSS before adding fresh medium.
2. Raise the temperature of the room and turn on incubator at least 5 h prior to experiment.
3. Place cells in incubator or heated chamber. If possible, choose cells in a nearly confluent area, without dead cells or other distractions. Choose cells of medium brightness, because highly-overexpressing cells may yield irreproducible results. Expose cells to as little light as possible when focusing and selecting specimens under widefield.
4. Switch to confocal, change pinhole to large size, and adjust the gain to approximately 90% saturation.
5. Set up z stack with 0.8-μm steps and 1 to 2 extra steps above and below the specimen to allow for some focal drift. More steps may prove necessary.
6. Zoom on 4 to 8 cells with resolution between 256×256 and 512×512 (less for dimmer samples/longer experiments) to obtain between 100 x 100 and 200 x 200 micron field size.
7. Acquire a test z stack (render maximum). Gain may need to be reduced, because granules will be brighter than diffuse labeling, or increased to account for photobleaching.
8. Adjust the frame rate to allow at least 10 sec in which to adjust focus between frames in case focal drift is a problem. One frame per minute is sufficient.
9. Carefully add pre-warmed medium containing drug, trying not to move the specimen. It is helpful to add a large volume of diluted drug (e.g. add equal volume of 2X drug) so that no additional mixing is necessary. Begin acquiring images as soon as possible after drug addition

in case focus adjustments must be made before the cells respond to the drug.
10. Monitor focus, making sure the cell remains within the z stack. Monitor brightness with real-time rendering (render maximum value, not additive).
11. Save raw data and archive.
12. Render maximum value over time, and save files as avi, mov, or a series of tiffs (the higher the bit depth, the better).
13. If necessary, use software (we use Photoshop CS3) to make gradual contrast adjustments or to reduce the appearance of photobleaching. Adjust for nonquantitative uses only.

10.2. Live cell imaging for single-image quantification

The goal is to create high-resolution images with linear fluorescence detection, without taking superfluous steps. Here we use room temperature, as it inhibits P-body movement, and thus allows slower image acquisition and better resolution.

Days 1 and 2

Plate cells as described in the previous protocol "Live Cell Imaging for Tracking Processing Bodies."

Day 3

1. Remove medium containing phenol red and replace with phenol red–free medium. Remove any floating dead cells by rinsing with HBSS before adding fresh medium.
2. Allow cells to equilibrate for several hours.
3. Allow cells to cool to room temperature.
4. Choose a field containing cells with medium brightness by use of widefield; use minimal light to minimize photobleaching.
5. Switch to confocal and adjust the pinhole size to small.
6. Adjust the gain in each channel to just below saturation. If the gain is too high/grainy for a channel, increase the pinhole size for that channel.
7. Select a smaller field containing 10 to 20 cells, and acquire a test image.
8. If the test image is acceptable, adjust the resolution to ≈150 nm/pixel, set the average to 3 to 5× and acquire the image.★
9. Save the image in raw format. Additional saves may be made in 12-bit/channel tiff to keep bit depth linear.

★ If a 3D render is desired, increase pixel dwell threefold and set the z step size to 0.25 μm.

10.3. Relative fluorescence protein localization in PBs

1. Acquire images as previously described.
2. If there are many z slices, choose a z slice that includes many PBs.
3. Use software to mask PBs, or manually select individual PBs.
4. Measure and record the average fluorescence intensity in each PB.
5. Measure and record the fluorescence intensity in the area around each PB.
6. Subtract the PB intensity from the background to get the net intensity.
7. Repeat for different z slices, or for a rendered z stack. Results should show differences in the labeled protein's localization in PBs with different stimuli.

Different fluorescence channels can be compared simultaneously.

10.4. Acquiring FRAP/FLIP images

FRAP and FLIP images must be scanned rapidly to measure the fastest recovery, yet allow sufficient resolution to measure PB intensity over several pixels.

Days 1 and 2

Pretreat glass and spread cells as described in "Live Cell Imaging for Tracking Processing Bodies."

Day 3

1. Remove phenol red–containing medium and replace with 4 ml phenol red–free medium. If floating dead cells are abundant, rinse cells with HBSS before adding fresh medium.
2. Raise the temperature of the room/incubator for at least 5 h prior to experiment.
3. Remove 2 ml conditioned medium from a well and reserve it (if extra conditioned medium is a, this is not necessary).
4. Treat the well with a drug such as SA (0.5 mM) to induce PB formation.
5. Remove SA, and replace with 2 ml conditioned medium.
6. Immediately place the cells in the heated chamber or microscope incubator.★
7. Choose 2 to 3 similar neighboring cells (1 experimental and 1 to 2 controls) and a target PB.
8. Use confocal with a large pinhole to see most of the PB. If recovery is slow, PBs will move in and out of focus. A z stack 0.8-μm step size with a few steps around the PB should be used to measure initial intensity and recovery versus loss because of photobleaching.

★ If kinetics are too fast to measure, or focal drift is a problem, all FRAP/FLIP experiments can be carried out at room temperature and compared.

9. Bleach the PB (for FRAP), or bleach the entire cytosol except the PB (for FLIP). If FRAP is used, bleach the PB to 15 to 30% initial intensity. Multiple passes may be needed to thoroughly bleach for FRAP and FLIP. If FLIP is used, do not bleach the very edge of the cell, because it may also overlap and bleach the neighboring control cells.
10. Record recovery/loss.

10.5. Analyzing FRAP/FLIP images

1. Create a region of interest (ROI) around the bleached PB.
2. Choose control PBs in neighboring cells, and create ROIs for each.
3. Measure the control PBs' fluorescence, and average for each time point.
4. Measure the bleached PBs' fluorescence at every time point.
5. Control PBs are used to normalize for photobleaching. Divide the average control PB fluorescence at each time point by the initial average control PB fluorescence to get a normalizing factor.
6. Multiply the bleached PBs' fluorescence by the normalizing factor at each time point to account for photobleaching.

11. CONCLUSIONS

SGs and PBs exhibit several types of dynamic behavior: (1) assembly and disassembly, (2) motility within cells and docking with each other, and (3) turnover of various contents, especially RNA and RNA-binding proteins. Each of these dynamics can be altered by overexpression of individual SG/PB markers, yet only by overexpression of fluorescently tagged proteins can we view these dynamic processes in real time. Stably expressing cell lines are superior to transient lines for imaging studies and also are amenable to biochemical studies with antibodies against the GFP/mRFP tags. With these increasingly refined tools, better microscopic images will yield integrated global visions, as we decipher the many dance steps between translation and decay.

REFERENCES

Anderson, P., and Kedersha, N. (2002). Stressful initiations. *J. Cell Sci.* **115**, 3227–3234.
Anderson, P., and Kedersha, N. (2006). RNA granules. *J. Cell Biol* **172**, 803–808.
Campbell, R. E., *et al.* (2002). A monomeric red fluorescent protein. *Proc. Natl. Acad. Sci. USA* **99**, 7877–7882.
Cande, C., *et al.* (2004). Regulation of cytoplasmic stress granules by apoptosis-inducing factor. *J. Cell Sci.* **117**, 4461–4468.
Eystathioy, T., *et al.* (2003). The GW182 protein colocalizes with mRNA degradation associated proteins hDcp1 and hLSm4 in cytoplasmic GW bodies. *RNA* **9**, 1171–1173.

Fenger-Gron, M., et al. (2005). Multiple processing body factors and the ARE binding protein TTP activate mRNA decapping. *Mol. Cell* **20,** 905–915.

Guil, S., et al. (2006). hnRNP A1 relocalization to the stress granules reflects a role in the stress response. *Mol. Cell Biol.* **26,** 5744–5758.

Hinnebusch, A. G. (2006). eIF3: A versatile scaffold for translation initiation complexes. *Trends Biochem. Sci.* **31,** 553–562.

Hua, Y., and Zhou, J. (2004). Survival motor neuron protein facilitates assembly of stress granules. *FEBS Lett.* **572,** 69–74.

Irvine, K., et al. (2004). Rasputin, more promiscuous than ever: a review of G3BP. *Int. J. Dev. Biol.* **48,** 1065–1077.

Kedersha, N., and Anderson, P. (2007). Mammalian stress granules and processing bodies. *Methods Enzymol.* **431,** 61–81.

Kedersha, N., et al. (2000). Dynamic shuttling of TIA-1 accompanies the recruitment of mRNA to mammalian stress granules. *J. Cell Biol.* **151,** 1257–1268.

Kedersha, N., et al. (2005). Stress granules and processing bodies are dynamically liked sites of mRNP remodeling. *J. Cell Biol.* **169,** 871–884.

Kedersha, N. L., et al. (1999). RNA-binding proteins TIA-1 and TIAR link the phosphorylation of eIF-2a to the assembly of mammalian stress granules. *J. Cell Biol.* **147,** 1431–1441.

Khandjian, E. W., et al. (2004). Biochemical evidence for the association of fragile X mental retardation protein with brain polyribosomal ribonucleoparticles. *Proc. Natl. Acad. Sci. USA* **101,** 13357–13362.

Leung, A. K., et al. (2006). Quantitative analysis of argonaute protein reveals microRNA-dependent localization to stress granules. *Proc. Natl. Acad. Sci. USA* **103,** 18125–18130.

Shaner, N. C., et al. (2007). Advances in fluorescent protein technology. *J. Cell Sci.* **120,** 4247–4260.

Siridechadilok, B., et al. (2005). Structural roles for human translation factor eIF3 in initiation of protein synthesis. *Science* **310,** 1513–1515.

Sivan, G., et al. (2007). Ribosomal slowdown mediates translational arrest during cellular division. *Mol. Cell Biol.* **19,** 6639–6646.

Tourriere, H., et al. (2003). The RasGAP-associated endoribonuclease G3BP assembles stress granules. *J. Cell Biol.* **160,** 823–831.

Vasudevan, S., et al. (2007). Switching from repression to activation: MicroRNAs can up-regulate translation. *Science* **318,** 1931–1934.

Wilczynska, A., et al. (2005). The translational regulator CPEB1 provides a link between dcp1 bodies and stress granules. *J. Cell Sci.* **118,** 981–92.

CHAPTER TWENTY-SEVEN

CELL BIOLOGY OF mRNA DECAY

David Grünwald,* Robert H. Singer,*,† and Kevin Czaplinski†,1

Contents

1. Introduction 554
2. FISH Probe Design 556
3. Hybridization 557
4. Image Acquisition 557
5. FISH Protocol 558
   - 5.1. Probe mixture 559
   - 5.2. Preparation of cells 559
   - 5.3. Hybridization 559
   - 5.4. Wash 559
   - 5.5. Preparation of humidified chamber 560
6. Colabeling Protein with IF and RNA with FISH 560
7. IF-FISH Protocol 561
   - 7.1. Primary antibody incubation 561
   - 7.2. FISH hybridization 561
   - 7.3. FISH wash 562
   - 7.4. Secondary antibody 562
8. Following mRNA in Living Cells 562
9. Live Single-Molecule Detection 563
10. Single mRNA Data Analysis; What You Can Observe 564
11. How Do You Know That You See Single Molecules? 566
12. The Secret to Getting Good Data: More Photons, Less Noise 568
13. Setting Up a Microscope for Single Molecule Detection 569
14. Conclusion 571
Acknowledgments 571
Appendix 571
References 575

---

* Albert Einstein College of Medicine, Anatomy and Structural Biology and Gruss-Lipper Biophotonics Center, Bronx, Albert Einstein College of Medicine, New York, USA
† Department of Biochemistry and Cell Biology, Stony Brook University, Center for Molecular Medicine 542, Stony Brook, New York, USA
1 Corresponding author

Abstract

Studying single mRNA molecules has added new dimensions to our understanding of gene expression and the life cycle of mRNA in cells. Advances in microscopes and detection technology have opened access to single molecule research to most researchers interested in molecular biology. Here we provide an overview technique for single molecule studies of RNA in either fixed samples or in living cells. As part of a volume on mRNA turnover, it is increasingly relevant, because many of the recent advances in studies of mRNA turnover have suggested that there is non-homogeneous distribution of turnover factors in the cell. For this reason, understanding of spatial relationships between mRNA and mRNA turnover factors should enrich our understanding of this process.

1. Introduction

Studies of mRNA localization within the cytoplasm have driven the development of microscopy techniques to evaluate the spatial distribution of mRNA within the cell. Consequently, these discoveries have also made the techniques more relevant to studies of mRNA turnover. The discoveries of processing (P)-bodies in yeast and similar P-bodies or GW bodies in mammalian cells, as well as stress granules in mammalian cells, have increased the importance of collecting spatial information for understanding the processing and degradation of mRNA within the cytoplasm of eukaryotic cells (Anderson and Kedersha, 2008; Garneau et al., 2007; Kedersha and Anderson, 2002; Parker and Sheth, 2007; Sheth and Parker, 2003). P-bodies and stress granules are observed in the eukaryotic cytoplasm as local enrichments of factors that are involved in the processing of RNA, many of these being factors involved in mRNA degradation (Eulalio et al., 2007; Fenger-Gron et al., 2005; see chapters by Nissan and Parker; Kedersha and Anderson). Similar observations have been made in nuclear processing and turnover pathways (Houseley et al., 2006; LaCava et al., 2005; Wagner et al., 2007). In each case, mRNA is likely to interact with the proteins found within these structures, but how the mRNA interaction with these structures contributes to regulation of mRNA in the cytoplasm is poorly understood. Therefore, studies of individual mRNAs should contribute significantly to studies of these structures and other processes regulating mRNA.

When an entire culture of cells or a piece of tissue is homogenized and extracted to isolate RNA, DNA or protein, subpopulations are averaged and only observations occurring within a significant proportion of these cells in the whole population can be detected. The most widespread approaches for studying individual macromolecules rely on examination of the pooled samples of these molecules from whole cultures or tissue. These approaches have limitations when only subpopulations are being sought, such as cells in some particular phase of the cell cycle, or some specific cell type within a tissue (Levsky and Singer, 2003). Sorting individual cells within these cultures enhances the ability to look at subpopulations of cells; however, the requirements for large numbers of cells limit this approach when cells of interest are sparse. Cultures can be manipulated to enrich for subpopulations, but these manipulations may have cryptic effects on the results obtained. For these reasons, microscopic analysis of individual cells has become commonplace to examine cellular phenomena. The ability to examine particular subpopulations of cells in a population with markers allows one even to evaluate cells that are rare within a population.

These same advantages of microscopic analysis use can be applied at the molecular level with single molecule imaging approaches to evaluate populations of molecules individually rather than in bulk. One major advantage to this approach is that spatial information is retained that would normally be lost on sample homogenization. As more sensitive quantitative microscopy equipment becomes accessible to most researchers, it is feasible for most laboratories to perform single molecule analysis, and RNA is one of the most amenable subjects to apply single molecule approaches. With *in situ* hybridization using fluorescently labeled oligonucleotide probes (fluorescence *in situ* hybridization, FISH), one single protocol is flexible enough to examine almost any RNA species microscopically. Any unique RNA sequence can be examined by synthesis of a complementary deoxy oligonucleotide probe. The availability of many spectrally distinguishable fluorophores even allows examination of many different mRNAs within the same cell (Capodieci *et al.*, 2005; Levsky *et al.*, 2002).

Our laboratory has previously established methods for detecting RNA molecules in fixed mammalian cells and tissues with FISH and has shown that the technique can successfully detect individual mRNAs with commonly available microscopy equipment and software (Capodieci *et al.*, 2005; Femino *et al.*, 1998; 2003; Shav-Tal *et al.*, 2004). We refer readers interested in detailed description of FISH to these articles to become more familiar with the approach. In the first part of this article, we will provide a brief presentation of the important aspects of single molecule mRNA detection with FISH for mammalian cells and present a basic protocol. More details about the protocols for FISH can be found in the previous references and at the following url: www.Singerlab.org. Recent years have seen the development of techniques for visualizing single mRNAs in real time in living cells, and we will devote the second half of this chapter to the analysis of live cell single mRNA detection.

2. FISH Probe Design

The first step is to design probes for hybridization. Oligo DNA probes are our preferred probe; therefore, the protocol presented here is designed for such probes. Parameters for probe design have been previously elaborated, and our standard FISH protocol has been worked out with 40 to 50 mer oligonucleotide probes that have a 50% G–C base pair content. Higher or lower GC content can be accommodated if need be but will necessitate changes in hybridization temperature or formamide concentration that should be empirically determined. When hybridizing multiple probes, it is simplest to have all probes with very similar GC content so that all the probes can be hybridized in one reaction. Oligo probe sequences should be selected such that there is no extensive complementarity to other RNA sequences in the transcriptome, and advances in genomic databases have made the selection of very specific probes more attainable. Sequences against exons will hybridize to mRNA everywhere; however, sequences against introns can be used to very specifically localize sites of transcription, because this sequence should only be present in the nascent transcript before splicing (Zhang et al., 1994).

The probes are detectable by the covalent modification of the synthesized DNA oligo with fluorescent dyes. We have used two methods for labeling probes. The first is to synthesize the oligo probe with modified amino-allyl T residues spaced approximately 10 to 15 nt apart. This spacing is important to prevent quenching of fluorophores that are too close but to accommodate enough dyes to detect a single probe clearly. In this way, a sequence can be modified with three to five fluorophores. Labeling with three dyes has been shown to be sufficient, depending on the dye, but results from a single probe are more robust with five dye labels. A single modified T residue that is not complementary can be added at either the 5′ or 3′-end without affecting the hybridization of the whole probe, increasing flexibility of choosing the probe sequence. After synthesis and probe purification, the amino-allyl T residues are available to any dye that can couple through the free amine group, and most commercially popular dyes are available ready for amino coupling. Many scientific reagent suppliers now offer oligos prelabeled with fluorophores; therefore, oligo probes can be synthesized with the fluorophore, eliminating a separate labeling step. Although there are limitations to the numbers of fluorophores that commercial providers can add. When designing probes to be used against multiple, different RNA targets simultaneously, ensure that the two different probes are each labeled with fluorophores that have nonoverlapping spectra. Efficient probe labeling is a critical parameter for robust FISH detection. Significant populations of poorly labeled or unlabeled probes will compete directly with well-labeled probes, reducing the amount of mRNAs that can be detected by the technique. For negative controls, a no probe FISH can control for autofluorescence, and a 50-mer randomized probe sequence with evenly spaced amino-ally T residues can control for nonspecific probe binding.

3. Hybridization

The procedure for FISH is quite similar to that for standard immunofluorescence (IF) and, therefore, should not require any specialized equipment that cannot be found in any typical molecular/cellular biology laboratory. However, specific nucleic acid hybridization necessitates quite different buffer conditions than IF. For the 50% GC content probes, we present a basic protocol that can be used for the *in situ* hybridization. This protocol can be optimized for alterations in GC content by changes in formamide concentration or temperature. Higher GC contents may require either higher formamide concentrations or temperature of hybridization, and, conversely, lower GC contents may require lower formamide or temperature of hybridization. In cases in which GC content for multiple probes to be hybridized to the same sample cannot be similar, it should be possible to perform two sequential hybridizations, with the higher stringency conditions being performed first, followed by the lower stringency conditions.

For studies of mRNA localization, the distribution of individual mRNAs can be described from these images, and the influence of experimental variables on this distribution can be analyzed in parallel samples. Similarly, by use of IF-FISH as described in the following, the spatial distribution of mRNA can be analyzed quantitatively in relationship to cellular landmarks that are colocalized with the IF channel. The spatial relationship of mRNAs relative to protein markers for particular cellular structures can also be measured under different experimental conditions to examine the effect of these variables on interaction of mRNA with these structures.

4. Image Acquisition

The signal of three to five fluorophores on a single oligo DNA probe should be sensitive enough to clearly observe FISH signal, but the intensity of these signals may, nonetheless, not be very high above the autofluorescent background of the cell. This may result in a skewed appearance of the raw FISH image, because photons that come from autofluorescence and from out of focus probes can add to the photons in voxels containing the

FISH probe. In this way, regions of the cell with higher autofluorescence (which is not typically uniform throughout the cell) or concentrations of FISH probe may appear to have more intense mRNA particles than regions with lower autofluorescence or more sparse distribution of mRNA. However, the photons derived from the FISH probes in these regions are equivalent, but may not appear so, making it a challenge to find an appropriate scale to display images. For this reason, deconvolution is used to analyze FISH images, because the difference between the autofluorescence and FISH probe point signals become more apparent. Deconvolution is a mathematical algorithm designed to correct for the unavoidable distortion of the optics (Wang, 2007). The algorithm uses the point-spread function of the optical system to remove out-of-focus light, which reduces noise. Constrained iterative deconvolution algorithms will reassign the out-of-focus light to its source position, thereby improving signal. Several commercial deconvolution software packages are available. One should acquire serial optical sections throughout the Z-axis of the cell to perform deconvolution.

To acquire images for quantitative analysis, exposure times and excitation intensities for each channel must be carefully set independently, such that the images for the FISH and the negative controls are comparable. The exposure time and intensity required to obtain significant signal from the hybridized samples can be determined first, and the no-probe containing samples can be acquired under the same excitation and exposure conditions. Both control and FISH samples can be deconvolved, and the resulting images analyzed; 60× and 100× magnification high numerical aperture (>1.3 N.A.) objectives combined with standard scientific grade cameras that have pixel size between 6 and 8 μm provide sufficient spatial resolution for image analysis, 100× providing slightly higher spatial over sampling.

5. FISH Protocol

This protocol has been established for cultured cells grown on glass coverslips. Fixation with 4% paraformaldehyde (PFA) is compatible with FISH, but most routine methods of fixation are also suitable. We routinely use alcohol permeabilization after fixation (80% methanol or ethanol), because samples can be stored for days at -20 °C after this treatment, but detergent permeabilization works too if cells will be used immediately. Extended storage of fixed cells may result in loss of hybridization sites. The FISH step will prevent phalloidin staining, so this reagent cannot be used to label actin filaments. DAPI remains effective after FISH. When performing FISH on samples expressing fluorescent proteins, both alcohol permeabilization and the FISH buffer will denature fluorescent proteins, eliminating the fluorescence produced by the protein, but the fluorescent protein can still be detectable by IF with primary antibodies against the fluorescent protein that are compatible with FISH. (See IF-FISH protocol.)

5.1. Probe mixture

50 μl of probe mixture is enough for one 18-mm coverslip
24 μl formamide
24 μl 20% dextran sulfate in 4× SSC
0.5 μl 20 mg/ml BSA fraction V (or acetylated BSA)
0.5 μl 10 mg/ml sheared salmon sperm DNA
0.5 μl 10 mg/ml *Escherichia coli* RNase-free tRNA
0.5 μl 10 ng/μl labeled oligonucleotide probe (for multiple probes one mixture of all probes at 10 ng/μl for each probe)

5.2. Preparation of cells

Fix: 20 min in 4% PFA/PBS
Quench: 20 min in PBS w/0.1 M glycine (PBSG)
Permeabilize: 80% methanol 10 min
Rehydrate: 6 serial twofold dilutions of the methanol in PBSG
Final rinse 5 min in PBSG
Equilibrate in 50% formamide/2× SSC (2× 5-min incubations)

5.3. Hybridization

Denature probe by heating at 65 °C for 5 min.
Hybridize: Place coverslips face down onto the probe solution spotted onto a strip of parafilm in a *humidified chamber* to prevent evaporation.
Incubate 37 °C for 2 h.

5.4. Wash

Transfer coverslips face up into the wash vessel with 2× SSC/50% formamide at 37 °C.
10 min at 37 °C in 2× SSC/50% formamide.
Change buffer and incubate another 10 min at 37 °C.
Change buffer to 1× SSC/50% formamide prewarmed to 37 °C.
Wash 15 min 37°.
Change buffer and wash an additional 15 min.
Change buffer to 1× SSC (no formamide, room temperature).
15 min room temp.
Change buffer to 0.5× SSC.
15 min room temperature.
Add 0.5× SSC with and DAPI.

5 min room temperature.

Rinse 3×, 1 min each with 2× SSC.

Mount coverslips in mounting medium and follow the appropriate instructions for the mounting medium used, then the cells are ready for imaging.

5.5. Preparation of humidified chamber

To prepare a simple humidified chamber, spread a piece of parafilm in the bottom of a plastic culture dish.

Place the probe solution in one drop per coverslip on the parafilm; leave enough distance between so that coverslips will not contact each other during incubation.

Place coverslips face down onto drops of probe.

Add a falcon tube capful of PBS in the corner to keep humidity in the chamber during incubation.

Place the lid on the culture dish and seal the vessel by wrapping with parafilm around all the edges.

6. COLABELING PROTEIN WITH IF AND RNA WITH FISH

FISH is also compatible with colabeling by immunofluorescence; however, the potential for RNase contamination in antiserum or antibody preparations necessitates very careful control of the FISH steps, because degradation of the RNA during any step will result in the loss of hybridization sites in the fixed cells. Ribonucleoside vanadyl complex (RVC; New England Biolabs Inc.) is an effective inhibitor of RNase activity residing in serum or antibody preparations, but these must be used according to the manufacturer's instructions, paying attention to the effective concentrations, methods for resuspending them, and buffer conditions (these are inactivated by EDTA). If RVCs cause some low-level background, adding EDTA to the final washes may eliminate this. When performing IF and FISH simultaneously, the high concentration of formamide is denaturing to some antigens. The order of steps is flexible but will be dictated by the antigen–antibody combination. The potential to lose antibody or target antigen during the harsh FISH step may not be compatible for some antibody–antigen combinations. In this case, a brief PFA cross-linking step after primary antibody incubation can help retain IF signal during FISH. However, for each particular antigen-antibody that works in IF, we suggest that the IF steps without the FISH hybridization steps be run on a parallel sample to control for the loss of antigen during the FISH step when performing IF-FISH. We present a basic protocol for IF-FISH that performs the primary antibody incubation, followed by a brief cross-link and then FISH, followed by the secondary antibody before mounting and imaging cells. The protocol allows for the primary antibody incubation to occur as normal before the use of the harsh FISH conditions to minimize damage to sensitive antigens and facilitate their detection. Good antigen–primary antibody combinations can be probed after the FISH is performed first; however, contaminating RNase activity can release hybridized signals and must still be controlled. The secondary antibody incubation is performed subsequent to FISH to prevent denaturation of the secondary antibody during FISH, including RNase inhibitors in case secondary antibody is also contaminated with RNase activity.

7. IF-FISH Protocol

This protocol uses fixed cells prepared as earlier for FISH and the same probe solution. RNase inhibitor (RVC) is added to all IF incubation solutions to control for potential RNase contaminations that will eliminate the FISH signal. CAS block (Zymed/Invitrogen) is a commercially available blocking agent that is compatible with FISH. Because this is not routinely assayed for RNase activity, we add RVCs to this during blocking, and any typical blocking solution should work with RVCs added.

7.1. Primary antibody incubation

(This protocol starts after rehydration of the methanol permeabilized cells earlier.)

Block: CAS block with 10 mM RVC for 30 min at room temperature.
Primary: Incubate coverslips with an appropriate dilution of primary antibody in CAS block with 10 mM RVC (timing and temperature conditions appropriate for particular antibody-antigen combination).
Wash: 10 min PBST (PBS with 0.1% Tween 20).
Fix primary: 10 min PBS-4% PFA.
Quench: 20 min PBSGT.
Equilibrate in 50% formamide/2× SSC (2× 5-min incubations).

7.2. FISH hybridization

(Probe preparation and FISH is as described in section 5, FISH protocol.)

Denature probe by heating at 65 °C for 5 min.
Hybridize: Place coverslips face down onto the probe solution spotted onto a strip of parafilm in a humidified chamber to prevent evaporation.
Incubate at 37 °C for 2 h.

7.3. FISH wash

Transfer coverslips face up into the wash vessel with 2× SSC/50% formamide at 37 °C.
10 min at 37 °C in 2× SSC/50% formamide.
Change buffer and incubate another 10 min at 37 °C.
Change buffer to 1× SSC/50% formamide prewarmed to 37 °C.
Wash 15 min 37 °C.
Change buffer and wash 15 min more.
Change buffer to 1× SSC (no formamide, room temperature).
15 min at room temperature.
Change buffer to 0.5× SSC.
15 min room temperature.

7.4. Secondary antibody

Secondary: Incubate coverslips with an appropriate dilution of secondary antibody in CAS block with 10 mM RVC (timing and temperature conditions appropriate for particular secondary antibody).
Wash: 4× 15-min each with PBST (add 1 mM EDTA if RVC background is observed).
DAPI: 5 min in PBST with DAPI.
Wash 3× 1-min with PBS.
Mount coverslips in mounting medium and follow the appropriate instructions for the mounting medium being used, then the cells are ready for imaging.

8. Following mRNA in Living Cells

The use of fluorescent proteins has allowed researchers to observe the behavior of proteins in real time in living cells. Exploiting very specific RNA–protein interactions has enabled application of fluorescent protein technology to studies of RNA by introducing tandem RNA tag sequences that can be specifically recognized by corresponding fluorescent protein chimeras (Fusco *et al.*, 2003; Janicki *et al.*, 2004; Shav-Tal *et al.*, 2004; Bertrand *et al.*, 2008). The concentration of these chimeras on coexpressed tagged RNAs enables microscopic visualization of the mRNA in real time in the living cells. Similar to the FISH methods, we have recently published detailed descriptions of the use of the bacteriophage MS2 coat protein (MCP) as a fluorescent probe for tagging mRNAs in living cells and suitable microscopic requirements for these experiments (Chao *et al.*, 2008; Wells *et al.*, 2007). The MCP fused to a fluorescent protein (MCP-FP) can bind an

RNA sequence from the MS2 genome (the MCP-binding site, MBS). When the MBS is multimerized, many copies of the MCP-FP are targeted to an individual mRNA, and the label is significantly stronger than the background of individual MCP-FP monomers. Because we recently published detailed descriptions of this system, we suggest readers refer to those articles for detailed descriptions and protocols.

The primary data that are acquired from these types of experiments are time-lapse images (movies) of the mRNA signal in the cell. Critical in being able to generate movies of mRNA behavior is to obtain sufficiently strong signal from the multiply tagged mRNA target to detect the mRNAs as distinct particles above the background of untethered fluorescent protein probes consistently over multiple frames. The next critical parameter is the acquisition time. mRNA particles are likely to be moving within the cell, either by diffusion or directed, and to successfully generate movies that can capture these movements, exposure times need to be short and frame rates fast enough to capture these particles over multiple frames. Longer exposures and slower frame rates have the effect of allowing the faster moving particle to blur over many pixels, thus only slower moving particles can be reliably analyzed, and the fastest moving population of particles is missed. The ultimate goal of these efforts is to achieve movies from which temporal measurements can be obtained. We have only summarized the protocols for FISH in the first part, because there are several detailed articles regarding this technique in press. However, because the technical parameters to use this approach for live imaging of mRNA are not as well described, we provide a more detailed discussion of this than in the first part. We will discuss parameters that can be extracted from these movies that apply to the MCP-MBS system, but these approaches apply to analysis of any moving particles.

9. LIVE SINGLE-MOLECULE DETECTION

The observation of single molecules has become a beneficial tool for biologists during the last decade, and it has been demonstrated that single particle tracking can resolve diffusion times of proteins similar to fluorescence correlation spectroscopy (Grünwald et al., 2006a; Zlatanova and van Holde, 2006). Many assorted techniques are used in single molecule studies, and some are more suited for work in living cells than others (Serdyuk et al., 2007). Of special interest in cell biology are fluorescence-based approaches because of easy application to fixed and live cells in culture allowing observation in the native biologic environment of the molecules under study (Grünwald et al., 2006b). In both scenarios (fixed and live cells), observation of single molecules allows one to extract quantitative data about the number of observed molecules and their position in three-dimensional space. Counting the number of individual molecules is done by summing up the total of all observed signals, and intensity analysis allows one to determine how many molecules are present within each observed signal. Signal intensity is directly proportional to the number of single molecules present within a discrete signal only when the probe signal intensity is homogenous and interaction between probe and target is uniform and specific. The positions of each discrete signal within the 3D space can be extracted by fitting the intensity distribution, and methods for this have been described in the literature (Schmidt *et al.*, 1996; Thompson *et al.*, 2002). As long as discrete signals are sufficiently separated in space (i.e., no overlap of signals within the limits of optical resolution), these can be reordered in a spatially separated manner, and the fitting process will report their position with sub-wavelength accuracy. For a detailed discussion see Thompson *et al.* (2002) and Yildiz and Selvin (2005). The use of multiple colors (multiplexing) in single molecule experiments promises new quantitative accuracy for colocalization data.

The analysis of live cells makes it possible to study the biogenesis and decay of single molecules. Imaging fixed cells provides a clear precision in determining localization within 3D space at an instant in time, because subjects are not moving, but imaging of single molecules within living cells offers the added benefit of observing behavior (transport, diffusion, confined mobility) of the molecules. Also, single molecule analysis in real time circumvents the problems inherent with analyzing ordered processes in heterogeneous cell populations, where a typically ordered series of events under study are initiating and proceeding asynchronously throughout the population. These individual events can be visualized under more physiologically relevant conditions than if artificial conditions need to be enforced to synchronize all of these events within the population. Because each observation is made on its own time scale, it is possible to pool data from many single molecule events adding significant statistical relevance.

10. Single mRNA Data Analysis; What You Can Observe

The ultimate goal of most single molecule live experiments involves the tracking of an individual molecule for a given time to describe its behavior. For instance, the dynamic behavior of an RNA molecule can be tracked as it diffuses away from the transcription site in the nucleus, exports through the nuclear pore complex (NPC) into the cytoplasm, transports to discrete cytoplasmic locations, and finally degrades. Technologies for single molecule imaging in live cells are still developing, and even more challenging is extending these technologies to studies of RNA. The major limitation in single molecule imaging is the detection of signal. The available signal from a single molecule depends on the intensity of the fluorophore it carries. Efficiency of target labeling, photobleaching during fluorescent excitation, and, in live cell experiments, proper exposure time to allow imaging of the moving particles, all present limitations to establishing single molecule sensitivity. Adding multiple labels can decrease the detection time but must be weighed against the possible influence of the label on the observed behavior and function of the labeled molecule. This is in contrast to fixed cells where no dynamic processes are observed and the exposure time is limited by photobleaching. For dynamic processes, such as can be observed in living cells, the integration time is necessarily dictated by the kinetics of the process under observation. For slow-moving processes, longer integration times are possible, but for a process (e.g., free diffusion) that involves rather fast motion, very short integration times (<100 msec) may be necessary.

Several parameters can be obtained from even simple time-lapse observations of single molecules. In analyzing movements from single focal planes, extended single path traces are often difficult to obtain in many cell types because of movement of particles out of the focal plane, but jump distance (the distance a particle travels between consecutive frames) is simple to derive, requiring only a few frames of continuous observation, and collecting multiple events delivers the necessary statistics. Dwell time is the amount of time a molecule spends in one place, for example, the amount of time a molecule remains associated with the NPC while it traverses the nuclear envelope during export (Kubitscheck *et al.*, 2005). Analyzing multiple individually observed molecules allows one to build a histogram of the timing to describe the average timing of the event. The particle brightness as measured by fluorescent intensity is another parameter that can easily be quantified for multiple objects and plotted to collect many individual particles. Several criteria that we will discuss next are required to extract valid data from these experiments.

For single traces, a detailed analysis showed that the variance of the mean value is a function of the square root of the number of measured positions (Qian *et al.*, 1991). Therefore, traces must be observed through enough frames to become statistically significant for mobility patterns to be evaluated. To obtain an accuracy of 10%, Qian *et al.* suggest a "trace length of 100 observations." This limit can be reached for processes that are bound to a two-dimensional structure or in very flattened out cellular cytoplasm. But because rapid observation in cells is often limited to a single focal plane, it is much harder to fulfill this requirement, because molecules can move in all three dimensions, including out of the focal plane. The signal of a particle fades into noise very rapidly when it moves out of the focal plane. However, very bright particles can cast light into multiple focal planes. Hence, in single molecule imaging, the effective thickness of the focal plane depends not only on the objective but also on the signal intensity of the single molecule (Kues and Kubitscheck, 2002).

Imaging a single molecule for a sufficiently long time is not only necessary to evaluate its mobility, but even more important for determining many biological functions. For many cell biological questions, the ultimate application of single molecule tracking with live cell microscopy is to image a complex, multistep process, such as the lifetime of an RNA from its birth at the transcription site in the nucleus to its degradation in the cytoplasm. One major obstacle in imaging such a process includes photobleaching the fluorescent label before the molecule reaches the final stage of its journey. Another challenge is to optimize the probability of tracking the molecule in 3D space. In fact, the higher the mobility of the molecule under observation, the faster it will move out of the focal plane (Kues and Kubitscheck, 2002). A molecule with a mobility of 1 $\mu m^2/sec$ has an average jump distance of 550 nm in any direction within 50 msec. Detecting enough signal within a single visual plane to visualize this molecule over this time is already difficult, but taking serial optical z-sections on the same time scale to capture the z movement of the molecule can become limiting for the experiment. When interactions between a single molecule of interest and a structure (e.g., P-bodies speckles or the NPC) are observed, the less mobile entity will limit detectable movement. This can help determine the dwell time of the more mobile elements by restricting observation to the volume containing the less mobile structure of interest.

11. How Do You Know That You See Single Molecules?

Commonly, three criteria can be used to support the observation of single molecules. These are digital *bleaching*, *blinking*, and *intensity* calibrations. If a signal of a particular intensity disappears in one step (digital, meaning either on or off), the source of the signal must have been a single emitter. Digital bleaching of single fluorophores occurs in one step. Blinking has been used to argue for the observation of single quantum dots, where fluorescent emissions cease temporarily before being observed again. Working with single emitters in fixed cells, blinking and/or bleaching will provide valuable evidence supporting the observation of single molecules. Enhancing the signal by multiple labeling is a powerful strategy to improve signal intensities for imaging. Here, bleaching can still provide good support if multiple irreversible stepwise decreases of fluorescence intensity are observed. Observing a subsequent stepwise bleaching of immobilized fluorophores known quantities of multiple labels is a strong indicator for observing single molecules. When multiply labeling, autoquenching of dyes is possible, and, if so, the total steps observed for a given molecule might not represent the total labeling ratio.

These criteria are challenging to meet during live observation, because a particle that leaves the focal plane will leave the same signature as a bleached emitter. A moving emitter that blinks will not be seen while in the dark state and cannot be analyzed. To calibrate mobile particles, intensity calibrations can be used to identify single molecules. For calibration of fixed samples, such as described in the FISH section, individual probes can be immobilized to measure the average intensity, and then the average intensity of this signal is used to generate a fluorescence intensity standard curve against which to compare mobile molecules. A slightly lower intensity for these molecules during live imaging is expected, because their movement will blur their signals during image acquisition. With one or few fluorophores, this intensity calibration approach works well as in FISH quantification. However, the use of many fluorophores introduces much more variation to the intensity averages of single particles and, therefore, presents challenges for calibrating the genetically encoded MCP-MBS particles. In this case, an intensity distribution–based argument can be made. Digital photobleaching is accepted as support for observing single dyes. If the intensity distribution of all the observed objects in a field shows discrete steps separated by uniform integral intensity differences, this argues for the presence of a fundamental base unit, and the steps then most likely represent multiples of this base unit.

Can this be used to argue for the observation of single molecules with a uniform multiplexed label? If sensitivity does not allow detection of single molecules, the observed base intensity (the lowest intensity value of all observed objects) could be any multiple of the true single molecule. However, the least complex possibility is that the lowest intensity particles are, in fact, single molecules when sensitivity is sufficient to detect them. If this is true, objects observed in the same frame or movie with an intensity that is an integral multiple of the lowest intensity found must be complexes of the exact stoichiometry that agrees with this integral difference. In plain English if the lowest intensity objects are singles, then objects twofold this intensity are dimers, threefold this intensity are trimers, etc. With multimeric single molecules (e.g., proteins that exist constitutively as multimeric) then this property will obviously be inherently present in the value of the base unit. Nonetheless, all such objects will still be equivalent in intensity. When the nature of the molecule under study is unknown, then this possibility merely leads to some uncertainty about the nature of the single molecule, not that it is a single molecule. mRNA does not multimerize nonspecifically, although complementary sequence can lead to dimerization in some cases. FISH studies have confirmed that an individual mRNA species does not generally multimerize (Femino *et al.*, 1998). Therefore, there is precedent that the base unit for MCP-MBS labeled mRNA is, in fact, single species.

A potential weakness of the argument with integral multiples comes from the imaging method. Brightly labeled single molecules by multiplexing its labels increases the focal depth that the particle is observable in a far-field microscope. Bright complexes that are out of the focal plane could have the same or any intermediate intensity. As a consequence, z-sections and photon reassignment could be used to minimize artifacts as in deconvolution for FISH (Femino *et al.*, 1998). Alternately, 3D tracking (Levin and Gratton, 2007) or optical sectioning would help. The use of confocal or TIRF microscopes could provide such sectioning (different microscopes for single molecule detection will be described later). By use of an epi-illuminated microscope, blurring of out-of-focus signal may be used to define a threshold for a diffraction-limited signal that has a higher probability of originating from the optical plane of the objective. Theoretically, it is even possible to analyze diffraction patterns to determine the exact 3D position (Speidel *et al.*, 2003).

The imaging, and especially the dynamic imaging of single molecules, requires a critical reevaluation of the way most biological imaging is performed. How many molecules can be tracked at the same time? How fast do they exchange? How precisely can their positions be determined? How much signal must be integrated? How long can a single molecule be imaged before it bleaches? What time resolution is needed to track them? What signal sampling is required? There are no uniform answers to these questions, because each experiment will require optimization of these parameters. To keep the tone of this chapter consistent, we summarize a descriptive framework as an appendix to this chapter that should help researchers in finding reasonable start parameters on the basis of established principles. The appendix also contains a separate section on suitable controls for microscope function and data acquisition, which are important, because microscope stability can have a large impact on the results of single molecule experiments.

12. THE SECRET TO GETTING GOOD DATA: MORE PHOTONS, LESS NOISE

The best images for extracting the types of data discussed require low background and as high a signal as possible at sufficient sampling rates in both time and space. These experiments face limited observation area and low signal intensities as major challenges. Reducing background signal and increasing photon output of the fluorescent labels may improve the signal available from single molecules. Under these conditions it is also important to recognize that too many molecules within a volume field hinder efforts to analyze individual molecule behavior. With the extremely sensitive detection required for detecting single molecules, signal produced from out-of-focus particles can significantly add to the background of any image; therefore, it is important that the proper number of molecules is present to be able to image them. Use of laser light is very beneficial, as will be discussed in the technical section, but adds to the cost of the setup.

13. SETTING UP A MICROSCOPE FOR SINGLE MOLECULE DETECTION

Point scanning confocal, spinning disc confocal, TIRF, and epi-illuminated microscopes are the most common equipment available in imaging facilities or laboratories. Depending on the experimental conditions, each of them has specific advantages. Detailed discussions of these advantages can be found in the literature (Pawley and Masters, 2008). Confocal setups provide contrast enhancement and hence SNR improvements, because they reduce out-of-focus background in the image with a pinhole or slit. Although point scanning leads to either long integration times or very short excitation of each pixel, spinning disc or slit-based confocal microscopes are faster while offering longer integration times per pixel. New resonant scanners can be used for very fast point scanning but do not change the dwell time of the laser on each pixel. Spinning disc confocal microscopes, TIRF microscopes, and epi-illuminated microscopes use CCD cameras for detection. The quantum efficiency of these detectors can be significantly higher than 90% and hence better than the efficiency of photomultiplier detectors used in point scanning microscopes. Although spinning disc imaging allows optical sectioning, TIRF and epi-illuminated microscopes will be more sensitive for signal detection. Descanning optics (these are the optical elements necessary to guide the detected light to the detector with the scanning element) in confocal and spinning disc microscopes limit the total transmission efficiency, whereas TIRF and epi-illuminated microscopes can be setup with a minimal number of lenses consisting of the objective and a matched tube lens. Although TIRF microscopes provide outstanding signals close to the cover glass surface, the penetration depth of the evanescent field is limited to approximately 100 to 200 nm above the glass surface. Low angle oblique or highly inclined thin illumination (Sako and Yanagida, 2003; Tokunaga et al., 2008) can be used to increase the z-axis penetration depth of the excitation light but even epi-illuminated alignment is possible.

Standard epi-illuminated microscopes often already exist in a laboratory and can be upgraded for single molecule work. The technological step from normal fluorescence imaging to imaging of single molecules can be rather small and will depend mainly on the brightness of the single molecule signal and the required acquisition speed. Most laboratory front-illuminated CCD cameras with low noise over long integration times can be used for imaging of fixed cells and multiplexed labeling of single molecules. The use of sputtered fluorescent filter sets can enhance the signal-to-noise ratio, because these filter sets have nearly perfect transmission characteristics. The next step of improving such a setup is to add a highly sensitive back illuminated CCD. Recently, electron multiplying (EM) has been introduced as a standard feature for fast and sensitive CCDs by all major companies. These EMCCDs provide a major step in detection sensitivity, but because most are based on a 16-μm pixel size, a standard 512-pixel chip is already approximately 8 mm in size. For precisely localizing single molecules with 60× or 100× objectives, these CCDs do not provide Nyquist sampling. This can be addressed by use of a magnification lens in front of the CCD, by adjusting the focal length of the tube lens, or by use of additional magnification provided by the microscope stand (e.g., the 1.6× magnifying lens on the microscope stand).

If these changes do not result in strong enough signals (e.g., because the required acquisition time is short), changing the excitation source can be beneficial. Although fluorescence lamps deliver a total of 50 to 300 W over the entire spectrum, their effective power at the sample is limited by the excitation filter that allows only the needed spectral region. Although the total power is given for the whole spectrum of the bulb (approximately 250 nm to 1100 nm), the power resulting from any 40-nm wide band-pass filter will be only a couple of mW. Although many single emitter experiments are done with excitation powers in the kW/cm^2 range, the power of a lamp-based fluorescent microscope will be in the W/cm^2 range. For very sensitive imaging, background reduction is often more efficient than boosting the signal. Any excitation filter will have a certain not negligible bandwidth introducing a low additional background. In this sense, even the background of the excitation filter (which is due to very good but imperfect optical densities of the filter for the blocking range) can be a problem. The use of laser excitation is advantageous because the band-width of the excitation light is in the sub-nm range, eliminating stray light from non-exciting wavelengths. Moreover the power delivered to the microscope is almost identical to the output power of the laser, and by the use of high-magnification objectives, the applied power at the sample can be adjusted to the kW/cm^2 range while excitation background is reduced to a bare minimum. Laser merge modules are commercially available but costly. For integration of a laser into an imaging system, it is necessary to have fast enough shuttering ability. The use of acoustic-optical devices such as acoustic-optical tunable filters (AOTF) allows for precise control of the excitation light. It is important to understand that, whereas a fluorescent lamp will be collimated to provide a nearly homogenous intensity distribution for the field of view, a laser will come with a narrow Gaussian beam shape. Flat top optics or expanding the beam with an adjustable iris to limit the illuminated area can be used to provide a homogenous excitation profile. Although the flat top optics are costly, they provide the benefit that the total loss of energy is small and, hence, high-power densities can be achieved with low-power lasers, producing less heat and are cheaper than high-power versions of the same laser. Because flat top optics are very carefully calculated, one should keep in mind that the use of multiple laser lines is a feature that needs very careful planning and integration.

14. Conclusion

Analysis of single RNAs has proven its potential to help our understanding of regulatory processes in cells in greater detail than possible by ensemble measurements. Technology development in recent years has opened avenues into this research field by making the necessary equipment available to many laboratories. Here we provide tools, references, and background discussion that are needed before starting these kinds of experiments, both at the level of fixed and living cells. It becomes clear that light microscopy today has become an interdisciplinary research field that allows addressing questions from synthesis to decay of mRNA inside of cells with a temporal and spatial resolution unavailable before. Numbers of RNA molecules and distinct positions can be analyzed and experiments designed that allow looking at changes in these numbers and positions as a result of biological changes. In living cells, it becomes possible to investigate interaction times and mobility features of RNA. However, as with any new technology, this approach has its own features, which sometimes do not come naturally to the biologically oriented researchers. It is the goal of this chapter to address the more hidden aspects of single molecule microscopy while presenting the benefits of RNA research to the reader.

ACKNOWLEDGMENTS

This work was supported by NIH grants to R. H. S. and a DFG postdoctoral fellowship to D. G. (GR 3388/1-1). We thank Amber Wells and Shailesh Shenoy for critical reading of the manuscript.

Appendix

The following presents a starting point for how to define the parameters necessary to generate meaningful data from single molecule approaches. Two major effects limit the number of single molecules that can be observed at the same time: Resolution and mobility. Imaging is diffraction limited, which means the point spread function of the emission signal will be larger than the actual molecule by a factor given by the resolution of the microscope. (What you see in the image is much larger than the actual molecule you are observing because of distortions introduced by the imaging process.) To separate single molecules, the molecules have to be isolated well enough so that the individual diffraction limited signals can be evaluated. If individual molecules come too close to each other, they will not be resolvable. The lateral resolution of a light microscope is described in Eq. (27.1).

$$r_{x,y} = 0.61\lambda/N.A. \tag{27.1}$$

Here λ is the wavelength of the emitted light in nm and $N.A.$ is the numerical aperture of the objective. As discussed earlier, fitting the signal precisely requires a sufficient spread of the signal over a couple of pixels (Thompson et al., 2002). The fit becomes more stable if more than the absolute minimum of pixels is used. Practically, the resolution limit is too small to be used as minimum criteria for interparticle distances in this approach, because the localization precision is better the more isolated the signals are. The use of photoswitchable labels with techniques like PALM and STORM, temporal distances between signals allow construction of high-resolution images of labels that are in closer spatial proximity than the resolution limit (Betzig et al., 2006; Hess et al., 2006; Rust et al., 2006). Although some groups use a small number of large pixels (Schmidt et al., 1995), other groups use higher spatial oversampling (Thompson et al., 2002; Yildiz and Selvin, 2005). To test the experimental procedure, test samples can be prepared with dyes or beads at different concentrations and then imaged in an immobilized state. Signals should show a full-width at half-maximum (FWHM) corresponding to the resolution of the optical system. The FWHM is a very descriptive measure. It is the width of the signal distribution at 50% of the intensity level. A Gaussian function is used to describe the signal distribution. Each axis can be fitted independently, that is, along the x and y coordinates of the image. The width of the fit is described by the standard deviation (σ):

$$y = y_0 + A \, \exp(-0.5((x - x_c)/\sigma)^2) \tag{27.2}$$

The intensity (y) is a function of the position (x), the parameter A, which scales for the Area under the fit curve, x_c indicating the center position of the distribution and y_0, which is related to background offsets in the image if no flat fielding is used in prior image processing. For experimental data, it is noteworthy that the FWHM can be calculated from the fit data with the standard deviation (σ) of the fit accordingly.

$$FWHM = 2(2\ln(2))^{1/2}\sigma \tag{27.3}$$

This is also valuable data to measure the localization precision and its dependence on the signal-to-noise ratio.

Once the molecules start moving, their mobility will be limiting for the number of observable molecules per image. The limit for being able to track particles is related to the question of how to make sure a particle observed in one frame is coming from the same molecule in the next frame. If molecules have a high chance to exchange for each other (when there are many mobile molecules close together), tracking will become arbitrary. Axial mobility in the z-axis will also show this problem. The relation between the diffusion coefficient (D) and the jump distance (r) of a molecule is given by:

$$D = \langle r^2_{x,y}\rangle/(2f\Delta t_{lag}) \qquad (27.4)$$

The parameter f scales Eq. (27.4) for the dimensions in which movement is observed. The time between two frames is indicated by Δt_{lag}. Imaging is normally two-dimensional, because the detector is planar, and although the particle moves in 3D, only the 2D projection of the movement is seen on the detector. The assumption behind this is that the molecule experiences an isotropic environment. Here a movement in any direction can be written as a vector with x, y, and z components. The assumed probability distribution for each component (x, y, and z) is equal, and, hence, any of these components represents the mobility sufficiently. Measuring the jump from one frame to the next will give a uniform time t_{lag} that equals the frame rate of detection. Leaving one frame out will double the lag time, leaving two frames out will triple it. Linearity over such time scales indicates diffusion, and divergence from linearity can be related to corralled movement or active transport (Saxton and Jacobson, 1997).

The diffusion constant of a sphere can be estimated as a function of the temperature (T in [K]), the Boltzmann constant (k_B), the radius of the sphere (r) and the viscosity of the environment (η) by:

$$D = k_B T/(2f\pi\eta r) \qquad (27.5)$$

The dimensionality of movement that normally occurs in all three dimensions is represented by the f factor. Here f is three because mobility is in 3D. In Eq. (27.4) it will still be two, because the observation is two dimensional. For a detailed discussion on how to deal with other geometric shapes, see Berg (1993).

By use of the molecular mass (m) of a protein and a mass density (ρ) of 1.2 [g cm^{-3}] (Andersson and Hovmoller, 1998), the radius can be estimated by:

$$r = ((3m/N_A)/(4\pi\rho))^{1/3} \qquad (27.6)$$

With this small theoretical framework, it becomes possible to estimate the influence of mobility on the imaging conditions. The diffusion coefficient is a parameter that shows scattering. The mean square $\langle x^2 \rangle$ of the jump distance in Eq. (27.4) clearly indicates this. It becomes even more obvious when the probability function for a two-dimensional diffusion process is used. This function $(P(r,t))$ describes the jump distance (r) at time (t):

$$P(r,t) = 1/(4Dt) * \exp(r^2/(4Dt)) * r \qquad (27.7)$$

It is used to evaluate jump distance histograms and can be written to account for multiple mobility classes:

$$P(r,t) = \sum_i N_i/(4D_it) * \exp(r^2/(4D_it)) * r \qquad (27.8)$$

More descriptively, this means that the mean jump distance is the center value of a distribution of jump lengths, and longer jumps will happen. Determining how many mobile molecules can be observed at the same time is related to this distribution and by the probability that molecules can exchange places with each other between two frames and, hence, cannot be identified individually. Because exchange can happen within the x, y plane but also along the optical axis (z), this is a profound limitation that reduces the number of observable molecules dramatically as the mobility increases. For tracking, one can define a criteria of minimum distances between two molecules (e.g., 3σ of the width of the jump distance distribution) and terminate tracking if molecules are identified below this distance to each other. (Imagine a circle with the radius of this value around each molecule; if circles from different molecules overlap in one frame, tracking is terminated for the next frame.) For identification of the molecule in the next frame, this circle can be used to search for the molecule. If more than one molecule is found, one needs to either discard the track or use other criteria to discriminate the positions into individual tracks. Nearest neighbor criteria are widely used (Saxton, 1997; Thompson et al., 2002), but lately other criteria have emerged in the field, for example, with signal correlation or flow conservation (Gennerich and Schild, 2005; Vallotton et al., 2003; 2005). On the basis of the assumptions made, each criterion will present a bias to the data. For example, the use of the nearest neighbor will select for short jump distances and underestimate mobility.

Experimental Controls

Immobilized dyes or beads are standard tools for calibrating imaging systems. Imaging of immobilized beads allows users to define the localization precision and drifts of the microscope. Imaging beads or dyes for a few hundred frames is performed for the following two controls, calculating *average position* and *standard deviation* of the signal. In these control images, imaging parameters should be identical to those used in the experimental conditions and must be performed at signal-to-noise ratios (SNR) similar to the experimental levels. To calibrate localization precision for different SNRs, only the intensity of the excitation light needs to be regulated; frame rate and other parameters should remain constant. The standard deviation of the position of the beads is the localization precision at this SNR. To check for physical drift of the microscope stage, the bead should be imaged for a period at least the duration of a normal experiment, longer if possible. For this, one can take fewer images over longer times at high SNRs, resulting in a time stack with low bleaching and more precisely localized single bead signals over time. The position of these beads can be plotted as a function of time, and this directly reflects the positional stability of the microscope. Because most microscope stages show some drift, it will either be necessary to carefully choose observation times for which drifts are smaller than the localization precision or for which the total precision is still acceptable. Alternately, internal controls, like beads or quantum dots (Qdots), could be imaged simultaneously during the experiment to allow image correction after acquisition. Qdots can present large Stoke shifts. This is an advantage that can be used to excite them with the same wavelength used for imaging and detect them concurrently with a different channel than the fluorophore being studied. The use of an immobilized signal on a glass surface to calibrate intensities to be observed in living cells will represent an upper limit for precision. Changes in the refractive index within the cell and light scattering on membranes affects the signals obtained and affects the precision. Calibrating the test slide signals at different intensities and, hence, different SNRs, compensates partially for these effects.

REFERENCES

Anderson, P., and Kedersha, N. (2008). Stress granules: The Tao of RNA triage. *Trends Biochem. Sci.* **33,** 141–150.

Andersson, K. M., and Hovmoller, S. (1998). The average atomic volume and density of proteins. *Zeitschrift Fur Kristallographie.* **213,** 369–373.

Berg, H. C. (1993). "Random Walks in Biology." Princeton, New Jersey: Princeton University Press, Princeton, New Jersey.

Bertrand, E., *et al.* (1998). "Localization of ASH1 mRNA Particles in Living Yeast." *Mol Cell* **2**(4):437–445.

Betzig, E., *et al.* (2006). Imaging intracellular fluorescent proteins at nanometer resolution. *Science* **313,** 1642–1645.

Capodieci, P., *et al.* (2005). Gene expression profiling in single cells within tissue. *Nat. Methods* **2,** 663–665.

Chao, J., Czaplinski, K., and Singer, R. H. (2008). Using the bacteriophage MS2 coat protein-RNA binding interaction to visualise RNA in living cells. *In* "Probes and Tags to Study Biomolecular Function." (P. R.Selvin and T.Ha, eds.), Wiley-VCH, Weinheim.

Eulalio, A., *et al.* (2007). P bodies: At the crossroads of post-transcriptional pathways. *Nat. Rev. Mol. Cell. Biol.* **8**, 9–22.

Femino, A. M., *et al.* (1998). Visualization of single RNA transcripts *in situ*. *Science* **280**, 585–590.

Femino, A. M., *et al.* (2003). Visualization of single molecules of mRNA *in situ*. *Methods Enzymol.* **361**, 245–304.

Fenger-Gron, M., *et al.* (2005). Multiple processing body factors and the ARE binding protein TTP activate mRNA decapping. *Mol. Cell.* **20**, 905–915.

Fusco, D., *et al.* (2003). Single mRNA molecules demonstrate probabilistic movement in living mammalian cells. *Curr. Biol.* **13**, 161–167.

Garneau, N. L., *et al.* (2007). The highways and byways of mRNA decay. *Nat. Rev. Mol. Cell. Biol.* **8**, 113–126.

Gennerich, A., and Schild, D. (2005). Sizing-up finite fluorescent particles with nanometer-scale precision by convolution and correlation image analysis. *Eur. Biophys. J. Biophys. Lett.* **34**, 181–199.

Grünwald, D., *et al.* (2006a). Direct observation of single protein molecules in aqueous solution. *Chemphyschem.* **7**, 812–815.

Grünwald, D., *et al.* (2006b). Intranuclear binding kinetics and mobility of single native U1 snRNP particles in living cells. *Mol. Biol. Cell* **17**, 5017–5027.

Hess, S. T., *et al.* (2006). Ultra-high resolution imaging by fluorescence photoactivation localization microscopy. *Biophys. J.* **91**, 4258–4272.

Houseley, J., *et al.* (2006). RNA-quality control by the exosome. *Nat. Rev. Mol. Cell. Biol.* **7**, 529–539.

Janicki, S. M., *et al.* (2004). From silencing to gene expression: Real-time analysis in single cells. *Cell* **116**, 683–698.

Kedersha, N., and Anderson, P. (2002). Stress granules: Sites of mRNA triage that regulate mRNA stability and translatability. *Biochem. Soc. Trans.* **30**, 963–969.

Kubitscheck, U., *et al.* (2005). Nuclear transport of single molecules: Dwell times at the nuclear pore complex. *J. Cell Biol.* **168**, 233–243.

Kues, T., and Kubitscheck, U. (2002). Single molecule motion perpendicular to the focal plane of a microscope: Application to splicing factor dynamics within the cell nucleus. *Single Molecules* **3**, 218–224.

LaCava, J., *et al.* (2005). RNA degradation by the exosome is promoted by a nuclear polyadenylation complex. *Cell* **121**, 713–724.

Levi, V., and Gratton, E. (2007). Exploring dynamics in living cells by tracking single particles. *Cell Biochem. Biophys.* **48**, 1–15.

Levsky, J. M., *et al.* (2002). Single-cell gene expression profiling. *Science* **297**, 836–840.

Levsky, J. M., and Singer, R. H. (2003). Fluorescence *in situ* hybridization: Past, present and future. *J. Cell Sci.* **116**, 2833–2838.

Parker, R., and Sheth, U. (2007). P bodies and the control of mRNA translation and degradation. *Mol. Cell* **25**, 635–646.

Pawley, J. B., and Masters, B. R. (2008). Handbook of biological confocal microscopy, third edition. *J. Biomed. Opt.* **13**, 1–16.

Qian, H., *et al.* (1991). Single particle tracking. Analysis of diffusion and flow in two-dimensional systems. *Biophys. J.* **60**, 910–921.

Rust, M. J., *et al.* (2006). Sub-diffraction-limit imaging by stochastic optical reconstruction microscopy (STORM). *Nat. Methods* **3**, 793–795.

Sako, Y., and Yanagida, T. (2003). Single-molecule visualization in cell biology. *Nat. Rev. Mol. Cell. Biol.* **Suppl**, SS1–5.

Saxton, M. J. (1997). Single-particle tracking: The distribution of diffusion coefficients. *Biophys. J.* **72**, 1744–1753.

Saxton, M. J., and Jacobson, K. (1997). Single-particle tracking: Applications to membrane dynamics. *Annu. Rev. Biophys. Biomol. Struct.* **26**, 373–399.

Schmidt, T., et al. (1996). Imaging of single molecule diffusion. *Proc. Natl. Acad. Sci. USA* **93**, 2926–2929.

Schmidt, T., et al. (1995). Characterization of photophysics and mobility of single molecules in a fluid lipid membrane. *J. Phys. Chem.* **99**, 17662–17668.

Serdyuk, I. N., et al. (2007). "Methods in Molecular Biophysics: Structure, Dynamics, Function." Cambridge, University Press, Cambridge, England and NY, New York.

Shav-Tal, Y., et al. (2004). Dynamics of single mRNPs in nuclei of living cells. *Science* **304**, 1797–1800.

Sheth, U., and Parker, R. (2003). Decapping and decay of messenger RNA occur in cytoplasmic processing bodies. *Science* **300**, 805–808.

Speidel, M., et al. (2003). Three-dimensional tracking of fluorescent nanoparticles with subnanometer precision by use of off-focus imaging. *Optics Lett.* **28**, 69–71.

Thompson, R. E., et al. (2002). Precise nanometer localization analysis for individual fluorescent probes. *Biophys. J.* **82**, 2775–2783.

Tokunaga, M., et al. (2008). Highly inclined thin illumination enables clear single-molecule imaging in cells. *Nat. Methods* **5**, 159–161.

Vallotton, P., et al. (2005). Tracking retrograde flow in keratocytes: News from the front. *Mol. Biol. Cell* **16**, 1223–1231.

Vallotton, P., et al. (2003). Recovery, visualization, and analysis of actin and tubulin polymer flow in live cells: A fluorescent speckle microscopy study. *Biophys. J.* **85**, 1289–1306.

Wagner, E., et al. (2007). An unconventional human Ccr4-Caf1 deadenylase complex in nuclear Cajal bodies. *Mol. Cell. Biol.* **27**, 1686–1695.

Wang, Y. L. (2007). Computational restoration of fluorescence images: Noise reduction, deconvolution, and pattern recognition. *Methods Cell Biol.* **81**, 435–445.

Wells, A., Condeelis, J. S., Singer, R. H., and Zenklusen, D. (2007). Imaging real-time gene expression in living systems. In "Single-Molecule Techniques: A Laboratory Manual." (S. P.a.H. T., ed.), pp. 209–238. Cold Spring Harbor Laboratory Press, Cold Spring Harbor, New York.

Yildiz, A., and Selvin, P. R. (2005). Fluorescence imaging with one nanometer accuracy: Application to molecular motors. *Acc. Chem. Res.* **38**, 574–582.

Zhang, G., et al. (1994). Localization of pre-mRNA splicing in mammalian nuclei. *Nature* **372**, 809–812.

Zlatanova, J., and van Holde, K. (2006). Single-molecule biology: What is it and how does it work? *Mol. Cell* **24**, 317–329.

Author Index

A

Abramoff, M. D., 518
Accornero, N., 469
Achsel, T., 50, 60, 62, 67, 70, 169
Agabian, N., 364
Ahlquist, P., 511
Akrigg, A., 216
Albrecht, M., 58
Aldea, M., 275
Aldeia, M., 414
Alibu, V. P., 361, 362, 363
Allawi, H. T., 492, 496, 497, 498
Allmang, C., 187, 212, 220, 221, 229, 431
Alm, C., 300, 313
Aloy, P., 430, 433
Alpy, F., 468
Alsford, S., 361, 363
Altman, S., 209, 221
Amberg, D. C., 168
Amirsadeghi, S., 438
Amrani, N., 168
An, G., 435
An, K., 435
An, S., 435
Anantharaman, V., 43, 58
Andersen, S. U., 438
Anderson, J., 229, 431
Anderson, J. R., 70, 82, 140, 152
Anderson, J. S., 269
Anderson, J. S. J., 5, 417, 418, 419, 424
Anderson, J. T., 187, 212, 431
Anderson, K., 270
Anderson, P., 212, 213, 221, 354, 469, 521, 522, 528, 534, 538, 554
Andersson, B., 360
Andersson, K. M., 573
Andre, B., 270
Andresson, T., 121
Andrulis, E. D., 432
Anstrom, J., 79, 80, 81, 83, 84
Antonellis, K. J., 439
Antoniou, M., 243, 248
Aoyama, T., 437
Araki, Y., 228
Aramaki, T., 15
Aranda, A., 141
Aravind, L., 43, 58, 187
Archer, S., 359

Arndt-Jovin, D. J., 60, 62, 169
Arora, C., 231
Arribas, R., 385
Artus, C. G., 469, 470, 508, 514
Ashe, M. P., 508, 510
Astrom, A., 79, 80, 81, 83, 84
Astrom, J., 81, 83, 84, 87
Astromoff, A., 270
Austin, R. J., 470
Ausubel, F. M., 316, 317
Avery, S. V., 415
Avis, J. M., 43, 58, 61, 63
Azzalin, C. M., 510

B

Babajko, S., 5, 508
Babak, T., 109, 300, 313
Bachi, A., 50, 60, 67, 70
Bachinger, H. P., 58
Badis, G., 212, 410, 422, 431
Baggs, J. E., 82, 83, 87
Bahler, J., 78, 80
Bai, X., 237
Baker, K. E., 268, 510, 511
Baker, R. E., 228, 231
Bakker, B., 363, 364, 365
Balatsos, N. A., 81, 87
Baldwin, J., 80
Ballantyne, S., 121, 128, 130
Ballmer-Hofer, K., 61, 62, 64, 67, 70
Bangham, R., 270
Baraniak, A. P., 497, 498
Barbee, S., 510
Barbot, W., 79, 80
Barlat, I., 263
Barnard, D. C., 121, 128, 130
Bar-On, V. G., 510
Barreau, C., 108
Bartholomay, C. T., 492, 497, 498
Bashirullah, A., 109, 301, 303, 304, 305, 309, 310, 313, 325, 331
Bashkirov, V. I., 168
Basquin, J., 43
Basyuk, E., 469, 508, 514
Bates, E. J., 141
Batra, S. K., 221
Beach, D. L., 516
Beazer-Barclay, Y. D., 439

Bechhofer, D. H., 174
Becker, K. G., 42, 141, 380
Beckham, C. J., 511
Beckmann, K., 469
Beelman, C. A., 24, 416, 418, 419
Beggs, J. D., 41, 42, 43, 44, 45, 50, 58, 60
Behlen, L. S., 471
Behm-Ansmant, I., 469
Beilharz, T. H., 78, 80
Belagaje, R. M., 63
Belasco, J. G., 108, 337, 339
Belin, D., 126, 128
Bell, L., 50
Belle, A., 78
Belles, J. M., 181
Bellofatto, V., 82, 109, 115, 360
Belostotsky, D. A., 81, 429, 433, 437, 438, 439, 440
Benard, L., 167, 181
Benito, R., 270
Benoit, P., 80, 87, 88
Benz, C., 360, 364, 367
Berg, H. C., 573
Berger, I., 72
Bergman, N., 70, 141
Bergsten, S. E., 109
Bernard, L., 168, 175, 180
Bernstein, D. S., 82, 471
Berriman, M., 360
Berthelot, K., 5
Bertrand, E., 446, 448, 449, 469, 508, 514, 515
Besharse, J. C., 152
Bessman, M. J., 5, 25
Betzig, E., 572
Bevan, E. A., 237
Beyer, A., 78
Bhattacharyya, S. N., 469, 508, 514
Bianchin, C., 79, 81, 82, 87, 88
Bieganowski, P., 6
Bilger, A., 120
Bindereif, A., 59
Birren, B., 80
Black, D. L., 50
Blackburn, A. S., 415
Blader, I. J., 403
Blanchard, J. M., 469
Blank, D., 212, 229, 431
Blazquez, M., 121
Bloch, D. B., 5
Bloch, K. D., 5
Bloom, K., 516
Blossey, B. K., 385
Boeck, R., 42, 81, 87, 422
Boeke, J. D., 270
Bogerd, H. P., 469
Bonisch, C., 115
Bonner, R. F., 381

Boon, K. L., 58, 60
Boothroyd, J. C., 381, 382, 384, 385, 388, 389, 396, 403, 404
Bordonne, R., 60
Bork, P., 430, 433, 469
Bornke, F., 437
Borst, P., 360
Botstein, D., 274
Bottcher, B., 430, 433
Boulay, J., 212, 431
Bousquet-Antonelli, C., 212, 431
Bouveret, E., 42, 43, 44, 45, 58, 87, 259
Bower, K., 78
Bowers, E., 436
Boyer, B., 510, 511
Boylan, A. M., 42
Brachat, A., 270
Bragado-Nilsson, E., 43, 44, 58, 62, 87, 259
Brahms, H., 43, 50, 58, 59, 60, 61, 67, 70
Bramham, C. R., 78
Bratkowski, M. A., 208
Braun, I. C., 468
Brawerman, G., 337, 339
Brechemmier-Baey, D., 180
Breit, S., 468, 470, 472, 474, 480
Bremer, K., 243, 248, 260
Brems, S., 366, 367, 369, 370
Brengues, M., 432, 508, 510, 511, 512, 514, 515, 516, 517
Brennan, R. G., 43, 44, 58, 72
Brenner, C., 6
Brewer, G., 109, 141, 157
Brewer, J. L., 403
Briata, P., 141, 432
Briggs, M. W., 228
Brinkmann, H., 359
Brodsky, A. S., 515, 516
Brookes, A. J., 245
Brouwer, R., 220, 221
Brown, B. A., 140
Brown, C. E., 42, 81, 85, 86, 87, 88, 422
Brown, E. L., 400
Brown, J. T., 228, 237, 431
Brown, P., 366
Brown, P. O., 380
Brunet-Simon, A., 43
Bryant, Z., 381
Buerstedde, J. M., 168
Bujard, H., 349, 350, 460
Bullions, L. C., 5
Burchmore, R., 366, 370
Burger, G., 359
Burkard, K. T., 207, 228
Burke, D., 233, 271
Burkhard, G., 361
Burns, R., 364
Burtner, C., 436
Bushell, M., 80

Busold, C., 367
Busseau, I., 109
Bussey, H., 270
Bustin, S. A., 312
Butler, J. S., 207, 212, 228, 230, 431
Butler, S., 208
Buttner, K., 187, 189
Buu, L. M., 510
Buzek, S. W., 260

C

Caceres, J. F., 469
Cagney, G., 431
Callahan, K., 208
Callebaut, I., 180
Camier, S., 5
Campbell, R. E., 526
Campbell, S. G., 508, 510
Cande, C., 534
Cano, A., 140
Cantor, C. R., 50
Cao, D., 269, 418
Capodieci, P., 555
Caponigro, G., 5, 24, 168, 236, 414, 415
Capony, J. P., 263
Carbon, J., 435
Carlson, M., 274
Carpousis, A. J., 187
Carrington, M., 361, 364
Carroll, A. S., 78, 510
Carr-Schmid, A., 5, 6, 10, 43, 141, 156, 413, 422
Carter, D., 384
Carter, M. G., 331
Cascio, D., 43
Casey, J. L., 336
Caspary, F., 43, 44, 87, 259
Castelli, L. M., 510
Cech, T. R., 26, 416, 418
Century, K., 436
Cerovina, T., 300, 313
Chami, M., 61, 62, 64, 67, 70
Chan, C. C., 468
Chan, E. L., 141, 154, 212, 432
Chan, R., 50
Chanfreau, G., 168, 187, 212, 229, 431
Chang, T. C., 87
Chao, J., 562
Chartrand, P., 469, 515
Cheadle, C., 42, 141
Chebli, K., 263
Chekanova, J. A., 81, 433, 437, 438, 439, 440
Chemokalskaya, E., 486
Chen, A.C.-Y., 336, 341
Chen, C. Y., 87, 141, 154, 212, 335, 339, 340, 341, 342, 432
Chen, C.-Y. A., 336, 350, 351, 353, 354

Chen, H., 433, 436, 437, 438, 439, 440
Chen, J., 79, 81, 82, 83, 84, 87, 88, 108, 169, 384, 422
Chen, K., 80, 81
Chen, N., 5, 6, 25, 181, 424
Chen, S., 230, 437
Chen, Y. A. C., 498
Cheng, C., 190
Chernokalskaya, E., 242, 261
Chernukhin, I. V., 169
Chiang, Y. C., 79, 81, 82, 83, 84, 87, 88
Chiba, Y., 81
Chicoine, J., 80, 87, 88
Cho-Chung, Y. S., 42, 141
Choder, M., 510
Chomczynski, P., 461, 462
Chou, T. B., 307
Chowdhury, A., 42, 44, 45, 46, 49, 50, 51, 52, 53, 141, 511
Chu, G., 330
Chua, G., 230
Chua, N. H., 437
Chuaqui, R. F., 381
Ciccarelli, F. D., 430, 433
Clark, A., 227
Clarke, L., 435
Claverys, J. P., 5
Clayton, C., 224, 359, 360, 362, 363, 364, 365, 366, 367, 369, 370
Clayton, C. E., 340, 361, 362, 365, 366
Cleary, M. D., 379, 381, 382, 384, 385, 388, 389, 396, 403, 404
Clement, S. L., 79, 82, 83, 84, 88
Closs, E. I., 508
Codomo, C. A., 436
Cohen, L. S., 141, 156
Colasante, C., 364
Colbert, J. T., 269
Cole, D. J., 42
Cole, S., 42
Colegrove-Otero, L. J., 78
Coller, J., 4, 5, 24, 42, 168, 230, 234, 235, 267, 268, 269, 270, 278, 413, 417, 424, 469, 508, 510, 511
Colley, A., 43
Collin, F., 439
Collins, B. M., 43
Colowit, P., 435
Comai, L., 435
Condeelis, J. S., 562
Condon, C., 167, 168, 175, 180, 181
Conti, E., 168, 187, 189, 191, 206, 208, 228, 230, 430, 433
Cooper, J., 435
Cooperstock, R. L., 109, 116, 299, 301, 303, 304, 305, 307, 309, 310, 313, 314, 325, 331
Copeland, P. R., 81, 84, 128

Corbo, L., 79, 80, 81, 82, 87, 88
Costa-Mattioli, M., 468
Cougot, N., 5, 24, 469, 508, 514
Couttet, P., 448, 449, 459, 460, 462
Crater, D., 141
Crittenden, S. L., 78
Cross, G., 363
Cubeddu, L., 43
Culbertson, M. R., 425
Cullen, B. R., 469, 470, 472
Cunningham, K. S., 242
Curmi, P. M., 43
Curtis, M. L., 492, 497, 498
Cusack, S., 81, 87
Cvitanich, C., 438
Czaplinski, K., 553, 562
Czichos, J., 364

D

Dahanukar, A., 109
Danckwardt, S., 473
Daniels, S., 169
Daou, R., 168, 175, 180
Darnell, J. E., Jr., 78
DaRocha, W., 361
Darrow, A. L., 126, 128
Das, B., 212, 431
Das Gupta, J., 486
Daugeron, M. C., 82, 83, 84, 85, 86, 88
Davierwala, A. P., 230, 431
Davis, K., 169, 170
Davis, R. E., 141, 156
Dawes, I. W., 43
Dawson, D., 233, 271
Day, N., 14
Deardorff, J. A., 80, 81, 84, 86, 87
de Bradandere, V., 59, 61
Decker, C. J., 4, 5, 6, 155, 228, 234, 269, 270, 314, 416, 417, 418, 420, 421, 424, 517
Decourty, L., 43
Degot, S., 468
DeHaven, K., 141
Dehlin, E., 42, 81, 86, 87, 88, 128
de la Cruz, J., 228, 230
de la Fortelle, E., 43, 58, 61, 63
De Long, S. K., 508
DeMaria, C. T., 141
Demarini, D. J., 270
de Moor, C. H., 80
Denis, C. L., 79, 81, 82, 83, 84, 87, 88, 422
Dephoure, N., 78
De Planell-Saguer, M., 469
Deplazes, A., 510
De Renzis, S., 300, 301, 302, 303, 305, 307, 309, 319, 320
Desai, A., 134

Desgroseillers, L., 470, 480
Deshmukh, M. V., 24, 25, 30
Deters, A., 468, 474
Dettwiler, S., 212, 229, 431
Deutscher, M. P., 79
Devaux, F., 212, 431
de Villartay, J. P., 180
Devon, K., 80
Dewar, K., 80
Dey, S., 169
Di, P. F., 108
Dichtl, B., 181
Dickson, K. S., 469
Diem, M. D., 468
Dietz, H. C., 228, 232, 235, 410, 425, 479
Ding, D., 313, 325
Ding, F., 208
Dix, I., 43
Dixon, M., 27
Djikeng, A., 361
Dlakic, M., 79
Do, L., 42, 141
Doerks, T., 469
Doma, M. K., 228, 410, 417, 425
Dominski, Z., 181
Dompenciel, R. E., 242, 261
Donald, R. G., 384
Donelson, J., 361
Dong, F., 496
Dong, S., 168
Dong, X., 436
Donis-Keller, H., 450
D'Orso, I., 367
Dostie, J., 59, 468
Dougherty, J. P., 413
Dower, K., 235, 431
Doyle, M., 80
Dreyfuss, G., 59, 468
Dubell, A. N., 242
Dudoit, S., 321
Duek, L., 510
Dufour, M. E., 212, 431
Dunckley, T., 5, 24, 422, 510
Dunn, M. A., 331
Dupressoir, A., 79, 80
Dutko, J. A., 81
Duttagupta, R., 413
Dziembowski, A., 189, 206, 208, 228, 230, 430, 433

E

Eckmann, C. R., 121, 128, 130
Edgar, B. A., 300
Egberts, W. V., 212, 220, 221
Eggert, C., 59, 72
Ehlers, B., 364
Ehrenberg, M., 81, 86, 88

Author Index

Ehrenfeld, E., 140
Ehrenman, K., 141
Ehresmann, C., 450
Eis, P. S., 492, 497, 498
Eisenberg, D. S., 43, 61
Elbashir, S. M., 213
Elemento, O., 300, 301, 302, 303, 305, 307, 309, 319, 320
El Hage, A., 431
Ellis, B., 322
Ellis, L., 361, 364, 367
Emmert-Buck, M. R., 381
Engel, A., 61, 62, 64, 67, 70
Enns, L. C., 436
Enright, A. J., 108, 331
Erdjument-Bromage, H., 221
Eritja, R., 121
Ersfeld, K., 361, 363
Espinás, M. L., 446
Estevez, A. M., 224, 340, 361, 362, 366
Etkin, L. D., 109, 301, 303, 304, 305, 309, 310, 313, 325, 331
Eulalio, A., 25, 554
Evans, D. H., 169
Even, S., 180
Evguenieve-Hackenberg, E., 187, 189, 191, 206, 430
Eystathioy, T., 531
Ezzeddine, N., 87, 335

F

Fabrega, C., 6, 141, 158
Fabrizio, P., 43, 58
Facciotti, D., 436
Falvo, J. V., 78, 510
Fan, J., 42, 108, 141, 380
Fan, S. J., 307
Fang, Y., 438
Fei, X., 384
Fellenberg, K., 367
Femino, A. M., 555, 567, 568
Feng, W., 468
Fenger-Gron, M., 5, 24, 524, 536, 554
Ferby, I., 121
Ferraiuolo, M. A., 468
Fersht, A. R., 26
Field, M., 361, 362
Filipowicz, W., 469, 470, 508, 514
Fillman, C., 5
Fink, G. R., 88
Fire, A., 140
Fischer, U., 58, 59, 61, 62, 72
Fisher, W. W., 109, 301, 303, 304, 305, 309, 310, 313, 325, 331
Fitzgerald, D. J., 72
FitzHugh, W., 80
Flavin, M., 446

Ford, L. P., 81, 82, 84, 85, 86, 109, 115, 140, 141, 152, 155
Forler, D., 468
Fortner, D. M., 24
Foulaki, K., 43, 58
Fouts, A. E., 381
Fox, C. A., 120
Fox, M., 360
Fragoso, C., 361
Fraig, M. M., 42
Franch, T., 58
Francis-Lang, H., 309
Franks, T. M., 514
Frasch, A. C. C., 367
Fraser, M. M., 42
Frede, U., 468, 473, 474
Freese, S., 340, 366
Freudenreich, D., 82, 109, 110, 112, 113, 114, 115
Fribourg, S., 63, 187, 189, 191, 206, 430
Frick, D. N., 5
Fried, M. G., 96, 97
Friedlein, A., 212, 229, 431
Friedrich, H., 113
Friesen, W. J., 59
Frischmeyer, P. A., 228, 232, 235, 410, 425
Fritsch, E. F., 280, 316, 317
Fritz, D. T., 81, 82, 86
Frokjaer, L. A., 438
Fromont-Racine, M., 43, 62, 410, 422, 448, 449, 456
Fu, W., 109, 301, 303, 304, 305, 309, 310, 313, 325, 331
Fuerstenberg, S. I., 436
Funke, R., 80
Furic, L., 470, 480
Furuichi, Y., 15, 168
Fusco, D., 469, 562

G

Gage, D., 80
Gallie, D. R., 174
Gallouzi, I. E., 263
Gamberi, C., 80, 87, 88, 303
Ganesan, R., 168
Gangloff, Y. G., 63
Gant, T. W., 80
Gao, G., 432
Gao, M., 5, 81, 86, 115, 154, 155, 221, 269, 432
Gao, X., 486
Garbarino-Pico, E., 152
Garcia-Blanco, M. A., 497, 498
Garcia-Martinez, J., 141
Garfield, D., 468
Gari, E., 275, 414
Garneau, N. L., 4, 78, 79, 140, 152, 154, 242, 554
Gatfield, D., 168, 244

Gavin, A. C., 430, 433
Gavis, E. R., 109
Gebauer, F., 108, 110
Gefter, M. L., 140
Gehring, N. H., 467, 468, 470, 472, 473, 474, 479, 480
Gennerich, A., 574
Genovese, C., 504
Gerber, A., 366
Gergen, J. P., 109, 314
Gerke, L. C., 78, 510
Ghaemmaghami, S., 78
Gherzi, R., 141, 154, 212, 432
Ghosh, S., 168
Ghosh, T., 24, 25
Gibson, M. L., 319, 324, 326
Gilbert, W., 169
Giles, K. M., 42
Gilman, M., 86
Gingery, M., 61
Giraldez, A. J., 108, 331
Glover, D. M., 313
Glover, L., 361, 363
Goldman, A. L., 109, 299
Goldstein, S. R., 381
Goldstrohm, A. C., 77, 78, 79, 80, 81, 82, 84, 85, 87, 88, 91, 93, 95, 96, 99, 100
Gonzales, F. A., 230
Gonzalez, C., 313
Gottlieb, E., 303
Gorgoni, B., 78, 108
Gorospe, M., 42, 141, 380
Gossen, M., 340, 349, 350, 460
Graber, J. H., 50
Graham, A. C., 187, 432
Gram, H., 108
Grange, T., 445, 446, 449, 451, 453, 459, 460, 462
Granneman, S., 213
Gratton, E., 568
Gray, M., 359
Gray, N. K., 78, 108, 469
Green, A. G., 436, 437
Green, C. B., 82, 83, 87, 152
Green, P. J., 81, 168, 169
Greenberg, M. E., 340, 341, 348
Greene, E. A., 435
Gregory, B. D., 433, 437, 438, 439, 440
Greimann, J. C., 185, 189, 190, 191, 201, 206, 207, 208, 213, 228, 431, 433
Gretz, N., 385
Grigull, J., 80, 320, 431
Grimmler, M., 58, 59
Grocock, R. J., 108, 331
Groebe, D. R., 28
Gröne, H. J., 385
Gronlund, M., 438
Gross, J. D., 23

Grosset, C., 336, 341
Grünwald, D., 553, 563
Gu, H., 486
Gu, M., 6, 25, 141, 158
Gubler, P., 126, 128
Gudjonsdottir-Planck, D., 340, 366
Guil, S., 523
Guilbride, D. L., 360, 364, 366, 370
Gulick, T., 5
Gull, K., 361, 362, 363
Guo, X., 432
Guranowski, A., 6
Guth, S., 108
Guthrie, C., 88
Guymon, R., 381, 382, 384, 385, 388, 389, 396, 404

H

Haanstra, J., 363, 364, 365
Habara, Y., 80, 87, 88
Habermacher, R., 508
Hagan, K. W., 168
Hagemeier, C., 468, 474
Haile, S., 340, 362, 363, 366
Hake, L. E., 120, 121
Hall, J. G., 492, 497, 498
Halsell, S. R., 109, 301, 303, 304, 305, 309, 310, 313, 325, 331
Hamatani, T., 331
Hamilton, D. A., 81
Hamilton, J. K., 109, 301, 303, 304, 305, 309, 310, 313, 325, 331
Han, G. W., 25
Han, J., 108
Han, J. J., 435
Han, W., 65
Handa, H., 140
Hannon, G. J., 508
Hanson, M. N., 242, 249, 251
Harborth, J., 213
Harel-Sharvit, L., 510
Haritan, A., 360
Harmann, C., 367
Harris, K., 80
Harrold, S., 337, 339, 340, 504
Harrop, S. J., 43
Hartmann, C., 361, 362
Hartmuth, K., 50
Hatfield, L., 24
Hattan, S., 169
Häusler, T., 365
Haynes, J., 230
He, F., 168, 270, 425
He, L., 303
He, W., 41, 42, 44, 45, 50, 58
Heidmann, T., 79, 80
Hellman, L. M., 96, 97

Hellman, U., 81, 83, 84, 87
Hendrickson, W. A., 6
Henikoff, S., 435
Henry, Y., 168
Hentze, M. W., 108, 110, 467, 468, 470, 472, 473, 474, 479, 480
Henzel, W. J., 109
Heo, I., 242
Hergenhahn, M., 385
Hermann, H., 43, 58
Hernandez, N., 140
Herrero, E., 275, 414
Herrick, D., 277, 280, 413, 414, 415
Herschlag, D., 26, 366, 380
Herve, P., 436
Hess, S. T., 572
Heyer, W. D., 168, 169, 170
Hickman, T., 521
Higgins, C. F., 339
Higgs, D. C., 269
Hilgers, V., 279
Hill, A. A., 400
Hill, K., 80
Hilleren, P. J., 275, 414
Hinnebusch, A. G., 229, 431, 527
Hirose, S., 15
Hitomi, M., 230
Hobbs, B., 439
Hochleitner, E., 187
Hoepfner, R., 243, 248
Hofius, D., 437
Hoheisel, J., 366, 367, 369, 370
Hojrup, P., 58
Holbrook, J. A., 473
Hollams, E. M., 42
Hollien, J., 242, 243
Hollstein, M. C., 385
Hollunder, J., 78
Holmes, L. E., 508, 510
Hook, B. A., 77, 80, 81, 82, 84, 85, 87, 88, 91, 93, 95, 96, 99, 100, 471
Hooker, T., 433, 437, 438, 439, 440
Hopfner, K.-P., 187, 189
Hopper, J. E., 274
Horn, D., 361, 362, 363
Hornig, H., 43, 58
Hoshino, S., 79, 84, 228, 468
Hoshino, T., 15
Hotz-Wagenblatt, A., 385
Houalla, R., 228, 431
Hougaard, B. K., 438
Houseley, J., 141, 190, 212, 228, 229, 431, 554
Hovmoller, S., 573
Howard, J. T., 245
Howson, R. W., 78, 510
Hoyle, N. P., 510
Hsu, C. L., 168, 269
Hsu, J. L., 468

Hsu, W. S., 307
Hu, W., 510, 511
Hua, Y., 523
Huang, S., 108, 384
Huang, Y. N., 169
Huarte, J., 126, 128
Huebner, K., 6
Hug, N., 510
Hughes, T. R., 80, 109, 300, 313, 320
Huh, W. K., 78, 510
Hunter, C. P., 400
Hurlstone, C. J., 436, 437
Hus, H. A., 208
Hyman, L., 78

I

Igarashi, K., 15
Ikeda, H., 168
Ilan, J., 385
Ingelfinger, D., 60, 62, 169
Inoue, K., 108, 331
Ip, H. S., 492, 496, 497, 498
Irizarry, R. A., 439
Irmer, H., 340, 366
Irvine, K., 524
Isham, K. R., 168
Isken, O., 449, 450
Ittrich, C., 385
Ivens, A. C., 360
Izaurralde, E., 141, 168, 169, 244, 431, 468, 469, 508
Izumi, N., 468

J

Jacobs, E., 230
Jacobs Anderson, J. S., 212, 228, 230, 231, 234
Jacobson, A., 81, 168, 270, 277, 280, 413, 425
Jacobson, K., 573
Jacquier, A., 212, 229, 410, 422, 431
Jan, E., 381, 382, 384, 385, 388, 389, 396, 404
Jang, L. T., 510
Janicki, S. M., 562
Jelicic, B., 270
Jensen, D. B., 438
Jensen, E. O., 438
Jensen, T. H., 431
Jeong, D. H., 435
Jeong, J., 65
Jeske, M., 82, 107, 109, 110, 112, 113, 114, 115
Ji, C., 384
Ji, H., 439
Jiang, T., 209, 221
Jiao, X., 3, 5, 6, 10, 11, 16, 28, 33, 42, 86, 141, 156, 157, 181, 413
Jing, Q., 108
Johnson, A. D., 85, 86
Johnson, A. W., 5, 24, 169, 170, 228, 237, 431

Johnson, K. A., 25
Johnson, M. A., 81
Johnson, T. R., 385
Johnston, J. S., 435
Johnston, S. A., 274
Johnstone, O., 108
Joly, N., 43
Jonard, L., 87
Jones, B. N., 23
Jones, C. L., 5, 115
Jones, G. H., 187, 206, 212, 228
Jones, K., 213
Jones, P., 80
Jonnakuty, S., 385
Jushaz, P., 169

K

Kadaba, S., 212, 229, 431
Kadowaki, T., 230
Kadyk, L. C., 121, 128, 130
Kadyrova, L. Y., 80, 87, 88
Kahrs, A., 61, 62
Kajiho, H., 228
Kambach, C., 43, 57, 58, 59, 61, 62, 63, 64, 67, 70
Kandels-Lewis, S., 43, 58, 62
Kang, H. G., 435, 438
Karaiskakis, A., 109, 301, 303, 304, 305, 309, 310, 313, 325, 331
Karin, M., 141, 154, 212, 432
Kashima, I., 468
Kasprzak, W., 42
Kastenmayer, J. P., 168, 169
Kastner, B., 50, 60, 67, 70
Katada, T., 79, 84, 228
Kataoka, N., 468
Kaufman, R. J., 140
Kawahara, T., 361, 363
Ke, A., 208
Kearsey, S., 168
Kedersha, N., 243, 469, 521, 522, 523, 528, 534, 536, 538, 554
Kedersha, N. L., 523
Kedinger, V., 468
Kee, K., 231
Keene, D. R., 58
Keene, J. D., 80, 82, 85, 86, 109, 115, 141
Keeney, S., 231
Keith, G., 212, 229, 431
Kellems, R. E., 140
Keller, W., 85, 86, 140, 212, 229, 431
Kelley, J. R., 42
Kelly, S., 361
Kenzelmann, M., 385
Kerr, K., 109
Kervestin, S., 168
Kerwitz, Y., 113

Khabar, K. S., 78
Khainovski, N., 169
Khandijian, E. W., 523
Khanna, R., 5, 6, 42, 141, 156
Khounh, D. T., 413
Khusial, P., 58
Kiledjian, M., 3, 4, 5, 6, 10, 11, 14, 15, 16, 18, 24, 28, 29, 33, 42, 43, 86, 141, 156, 157, 158, 181, 207, 269, 413, 422, 432
Kim, C. H., 414, 415
Kim, H. E., 65
Kim, H. K., 242
Kim, H. S., 242
Kim, J. H., 80, 88, 119, 121, 126, 128, 137
Kim, K. H., 242
Kim, K. J., 65
Kim, O., 242
Kim, S. D., 62
Kim, S. J., 242
Kim, S. K., 242, 381
Kim, V. N., 242, 468
Kim, Y. K., 470, 480
Kim, Y. S., 242
Kimble, J., 78, 82, 108, 121, 128, 130
Kimblin, N., 364
Kim Do, J., 242
Kiriakidou, M., 469
Kirschner, M. W., 134
Kirsebom, L. A., 81, 88
Kiss, D. L., 432
Klein, M. G., 6
Kloc, M., 109, 301, 303, 304, 305, 309, 310, 313, 325, 331
Klug, G., 187, 189, 191, 206, 430
Knoth, A., 113
Ko, M. S., 331
Kobayashi, T., 228
Kobrin, B., 504
Koch, G., 15
Kolodner, R. D., 168, 169, 170
Kon, R., 15
Koonin, E. V., 5, 187
Koprunner, M., 331
Korner, C. G., 42, 81, 83, 84, 86, 87, 88, 128
Kornfeld, G. D., 43
Kornstadt, U., 61, 62
Kostka, S., 61
Kotovic, K., 431
Kozhukhovsky, A., 61
Kramer, S., 361
Krause, H. M., 300, 313
Krecic, A. M., 229, 431
Kreig, P. A., 85, 86
Kress, C., 159, 446, 465
Kressler, D., 228, 230
Krogan, N. J., 25, 431
Krueger, A., 229, 431
Kshirsagar, M., 510

Kubitscheck, U., 565, 566
Kueffer, S., 385
Kues, T., 566
Kufel, J., 187, 212, 229, 431
Kuhn, U., 78, 113, 152
Kulozik, A. E., 467, 468, 470, 472, 473, 474, 479, 480
Kumagai, C., 230
Kumagai, H., 15
Kumar, R., 433, 437, 438, 439, 440
Kunkel, G., 141, 156
Kunz, J. B., 468, 470, 472, 474, 480
Kunz, S., 361
Kuperwasser, N., 235
Kuroda, S., 168
Kurup, S., 438
Kwak, J. E., 121, 128, 130

L

Labas, V., 180
Labourier, E., 263
LaCava, J., 141, 190, 212, 228, 229, 431, 554
Lackner, D., 78, 80
Lacroute, F., 414, 415
LaDuca, J. P., 468
LaGrandeur, T. E., 5, 24
Lamprinaki, S., 469
Lander, E. S., 80
Lane, W. S., 59
Lang, B., 359
Langen, H., 212, 229, 431
Lapeyre, B., 42
Larimer, F. W., 168
Lasko, P., 80, 87, 88, 108
Lassner, M., 436
LaTray, L., 381
Lavoie, B., 469
Lecuyer, E., 300, 313
LeCuyer, K. A., 471
Lee, B. G., 242
Lee, B. S., 425
Lee, C. S., 468
Lee, F. J., 510
Lee, H. H., 242
Lee, H. S., 435
Lee, K. H., 65
Lee, K. Y., 141, 432
Lee, S. J., 242
Lee, T. H., 80, 87, 88
Leedman, P. J., 42
Leeds, P., 425
Legagneux, V., 109
Legrain, P., 43, 44, 62
Le Hir, H., 12, 18, 468, 469, 470
Lehmann, R., 109
Lehner, B., 212, 224
Leibowitz, M. J., 236, 237

Lejeune, F., 168, 213, 221
Lemberger, T., 385
Lendeckel, W., 213
Lengauer, T., 58
Lennertz, P., 41, 42, 44, 45, 50, 58, 212, 229
Ler, L. W., 468
Le Romancer, M., 87
Leung, A. K., 523, 524, 538
Leutwein, C., 430, 433
Levin, V., 568
Levsky, J. M., 555
Lewer, C. E., 137
Lewis, M. S., 431
Li, C., 168, 322
Li, C.-H., 340, 364, 366
Li, J., 43, 58, 61, 63, 384
Li, L., 385
Li, M. J., 381
Li, P., 433, 437, 438, 439, 440
Li, X., 168, 213, 221, 436
Li, Y., 15
Liang, H., 270
Liang, S., 230
Liang, X., 360
Liao, J. C., 322
Libri, D., 212, 431
Lidder, P., 81
Lie, Y. S., 109
Lieberfarb, M. E., 109, 314
Light, H. R., 511
Lilie, H., 113
Liljas, L., 471
Lim, F., 471
Lima, C. D., 6, 25, 141, 158, 185, 189, 190, 191, 200, 201, 206, 207, 208, 213, 228, 431, 433
Lin, M. D., 307
Lin, S. C., 108
Linder, P., 228, 230
Lindner, D., 189, 206, 208, 430
Lindsten, T. C., 336
Lingner, J., 510
Linton, L. M., 80
Liotta, L. A., 381
Lipshitz, H. D., 109, 116, 299, 301, 302, 303, 304, 305, 306, 307, 309, 310, 313, 314, 325, 331
Littlepage, L. E., 121
Liu, C. L., 380
Liu, C. R., 381
Liu, H., 4, 5, 6, 16, 24, 141, 156, 158, 181
Liu, J., 508
Liu, Q., 189, 190, 191, 201, 206, 207, 208, 213, 228, 431, 433, 436, 437
Liu, S. W., 3, 6, 25, 28, 33, 86, 141, 156, 157, 158
Liu, X., 243
Liu, Y., 6, 181

Liu, Z., 169
Loeffler, D., 436
Loflin, P. T., 336, 338, 340, 350, 351, 352, 459, 498
Lohr, D., 270
Loireau, M. P., 79, 80
Long, L., 141
Long, R. M., 469, 515
Longtine, M. S., 270
Lopez, A., 230
Lorentzen, E., 187, 189, 191, 206, 208, 228, 230, 430, 433
Losson, R., 414, 415
Lotan, R., 510
Lottspeich, F., 187
Lowell, J. E., 79, 80, 81, 83, 84, 85, 86, 87
Lu, S., 469, 470, 472
Lucke, S., 59
Lührmann, R., 43, 50, 58, 59, 60, 61, 62, 63, 67, 70, 169, 213
Luisi, B. F., 187, 206, 212, 228
Luke, B., 510
Lukowiak, A. A., 213
Lund, J., 381
Luo, M. J., 468
Lutz, C. S., 154, 155, 221, 269, 432
Luu, D. M., 321
Luu, V.-D., 363, 364, 365, 366, 367, 370
Lyamichev, V. I., 496
Lykke-Andersen, J., 5, 79, 80, 82, 83, 84, 88, 468, 469, 470, 479, 514

M

Ma, J., 432
Ma, W.-P., 492, 497, 498
Ma, Y., 59
Mabbutt, B. C., 43
Macdonald, P. M., 109
Maertens, S., 385
Magelhaes, P. J., 518
Magnasco, M., 78
Maleki, S., 231
Malys, N., 510
Mandal, S. S., 221
Mangus, D. A., 168
Maniatis, T., 280, 316, 317
Manley, J. L., 121, 128, 130, 140
Mann, M., 59, 87, 141, 154, 187, 212, 213, 221, 222, 228, 229, 424, 431, 432, 437, 468
Mao, Y., 384
Maquat, L. E., 24, 168, 213, 221, 243, 248, 449, 450, 468, 470, 479, 480
Marchese, J. N., 169
Marguerat, S., 78, 80
Marinsek, N., 361
Mark, T., 361

Martin, G., 85, 86, 212, 229, 431
Martine, U., 508
Martinez, J., 81, 83, 84, 86, 87, 88
Martinez-Calvillo, S., 360
Marzluff, W. F., 181, 336
Massenet, S., 59
Masters, B. R., 569
Mata, J., 78, 80
Mathy, N., 167, 168, 175, 180
Mattox, S. A., 213
Maupin, M. K., 24
Maurier, F., 263
Mauxion, F., 79, 81, 82, 83, 84, 85, 86, 88
Mayer, C., 43
Mayes, A. E., 41, 42, 43, 44, 45, 50, 58
Mayo, T., 212, 213, 221
Mazza, C., 81, 87
McCallum, C. M., 435
McCarthy, J. E., 5, 510
McConnell, T. S., 26
McDonald, A. E., 438
McKenzie, A. III, 270
McKnight, T. D., 435
McNally, K. P., 364
Meaux, S., 227, 228, 232, 233, 235, 237
Meheus, L., 59, 61
Meiering, C. D., 381, 382, 384, 385, 388, 389, 396, 404
Meijer, H. A., 80
Meister, G., 58, 59, 72
Mejean, V., 5
Melamed, D., 510
Mendez, R., 120, 121
Menzies, F., 109
Merrikh, H., 235
Merrill, P. T., 300, 305
Meyer, S., 5, 42, 82, 84, 85, 86, 108, 109, 110, 112, 113, 114, 115, 339
Michaeli, S., 360
Mifflin, R. C., 140
Mikhli, C., 141, 156
Milcarek, C., 504
Mildvan, A. S., 25
Milligan, J. F., 28, 454, 455
Milligan, L., 431
Milone, J., 82, 109, 115
Min, H., 50
Minshall, N., 78
Mishima, Y., 108, 331
Mitchell, P., 187, 212, 213, 221, 222, 228, 229, 230, 424, 431, 437
Mitchison, T. J., 134
Miyasaka, T., 82, 85
Mnaimneh, S., 80, 230, 320, 431
Mocarski, E. S., 381
Moffat, J., 230
Mohr, S. C., 50
Molin, L., 81

Molineux, C., 169, 170
Moller, T., 43, 44, 58, 72
Moon, S., 435
Moore, C., 78, 140
Moore, I., 438
Moore, M. J., 24, 468, 469, 470
Moraes, K. C., 70, 80, 81
Morand, S., 361
Moras, D., 63
Morel, A. P., 79, 80, 87
Morelli, G., 437
Morishita, R., 468
Morita, M., 82, 85
Moritz, B., 115
Moritz, M., 110
Mornon, J. P., 180
Moroni, C., 141, 154, 212, 432
Morris, Q., 431
Morrison, P. T., 169
Morrison, S. L., 504
Morrissey, J. P., 168
Morse, D., 230
Moshous, D., 180
Moskaitis, J. E., 242, 260
Moss, B., 24, 25
Mossessova, E., 190, 191, 200
Motoyama, A., 108
Mourelatos, Z., 469
Muckenthaler, M., 81, 87, 88, 128
Mueller, P. R., 446, 453
Muellner, E. W., 336
Muhlrad, D., 4, 24, 42, 44, 81, 82, 87, 88, 168, 175, 228, 234, 236, 269, 270, 339, 410, 414, 415, 416, 417, 418, 420, 511
Mukherjee, D., 154, 155, 221, 269, 432
Mukhopadhyay, J., 42, 44, 45, 46, 49, 50, 51, 52, 53, 141
Mulhrad, D., 510
Muller-Auer, S., 81, 87
Mura, C., 43, 61
Murguia, J. R., 181
Murray, A. W., 134
Murray, E. L., 141, 243, 366, 483, 493, 495, 496, 498, 547
Murray, J. B., 471
Myers, F. A., 309
Myler, P. J., 360

N

Nagai, K., 43, 58, 59, 61, 63, 471
Naidoo, N., 43
Nakamura, T., 82, 85
Namane, A., 212, 431
Nasheuer, H. P., 78
Neal, S. J., 319, 320, 321, 324, 326, 328, 329
Nebreda, A. R., 121
Nelson, P. S., 42

Nelson, P. T., 469
Neri, B. P., 496
Neu-Yilik, G., 468, 470, 472, 473, 474, 479, 480
Newbury, S. F., 168, 169, 309
Ngai, J., 321
Ngo, H., 362
Nguyen, D., 360
Nilsson, D., 360
Nilsson, P., 81, 86, 87, 88
Nishikawa, M., 15
Nissan, T., 169, 507, 554
Nissan, T. A., 511
Niu, Q. W., 437
Niu, S., 152
Noe, D. A., 62
Nogales, E., 208
Nonet, M., 279
Norrild, B., 5
Nott, A., 469, 470
Noueiry, A., 511
Nusbaum, C., 80
Nuss, D. L., 15

O

Oberbaumer, I., 340
Oberholzer, M., 361
O'Connell, M. L., 126, 128
O'Connor, J. P., 154, 155, 221, 269, 432
Odden, A. R., 436
Oh, B. H., 65
Oh, M. K., 322
O'Handley, S. F., 5
Ohara, O., 108, 109
Ohashi, Y., 437
Ohn, T., 81, 82, 88
Ohno, M., 468
Ohno, S., 468
Oka, A., 437
Oliveira, C. C., 230
Olson, M. C., 492, 497, 498
Olson, S. M., 492, 497, 498
Omilli, F., 109
Ong, S. E., 141, 154, 212, 432
Ongkasuwan, J., 109, 121
Opyrchal, M., 82, 140, 141, 152
Orban, T. I., 141, 169, 431
Orr-Weaver, T., 109
Osborne, H. B., 108, 109
O'Shea, E. K., 78, 510
Ossowski, S., 436, 437
Otsu, K., 361
Otsu, N., 518
Otsuka, Y., 241, 243
Oussenko, I. A., 174
Ouyang, M., 413
Overath, P., 364

P

Pace, H. C., 6
Padmanabhan, K., 134
Page, A. M., 169, 170
Paillard, L., 108
Pain, V. M., 137
Palacios, I. M., 468
Palenchar, J. B., 360
Palfi, Z., 59
Paliouras, M., 80, 87, 88
Palmer, A., 121
Pandey, B. B., 336
Pannone, B. K., 62
Pardi, A., 208
Paris, J., 120
Park, C. M., 242
Park, S., 435
Parker, F., 263
Parker, K., 169
Parker, R., 4, 5, 6, 24, 25, 41, 42, 43, 44, 45, 50, 53, 58, 60, 78, 79, 80, 81, 82, 83, 84, 86, 87, 88, 108, 141, 155, 168, 169, 181, 212, 228, 229, 230, 231, 232, 234, 235, 236, 268, 269, 270, 275, 277, 278, 279, 280, 314, 336, 337, 339, 366, 409, 410, 413, 414, 415, 416, 417, 418, 419, 420, 421, 422, 424, 425, 432, 449, 463, 469, 507, 508, 509, 510, 511, 512, 514, 515, 516, 517, 554
Parker, R. P., 5, 212, 268, 269
Parkhurst, S. M., 313, 325
Parrish, S., 24, 25
Parthasarathy, N., 300, 313
Pascolini, D., 168, 169, 170
Passos, D. O., 230, 234, 235, 269, 336, 366, 409
Pastar, I., 363
Pastori, R. L., 242, 260
Patrick, T. D., 137
Paushkin, S., 59
Pawley, J. B., 569
Peabody, D. S., 471
Pearson, R. F., 43, 44, 72
Peattie, D. A., 448, 453
Peculis, B. A., 24, 25
Pelechano, V., 277
Pellegrini, O., 167, 168, 175, 180
Peltz, S. W., 5, 81, 115, 168, 413, 414, 415, 425
Peng, Q., 433, 437, 438, 439, 440
Peng, V., 321
Peng, W. T., 230, 431
Peng, Y., 243, 254, 256, 257
Perez-Ortin, J. E., 141, 277
Perkinson, R. A., 468
Perlick, H. A., 479
Pesiridis, G. S., 59
Peter, M., 510
Peterson, D. S., 303

Petfalski, E., 168, 187, 212, 213, 220, 221, 222, 228, 229, 230, 424, 431, 437
Pfeifle, C., 303, 313
Pham, A. D., 221
Philippsen, P., 270
Phillippe, H., 359
Phillips, M., 61
Phillips, S., 230
Piccirillo, C., 5, 6, 24, 141, 156
Piedrafita, L., 275, 414
Pilkington, G. R., 511
Pillai, R. S., 58, 59, 469, 470, 508, 514
Pillai, S., 169
Pinder, B. D., 109, 116, 303, 307, 314, 325
Piwnica-Worms, H., 120
Plaag, R., 58
Plessel, G., 59, 61, 67
Pluk, H., 213
Podtelejnikov, A., 221, 228, 229, 431
Poole, T. L., 168
Pootoolal, J., 80, 230, 320
Possenti, M., 437
Poterszman, A., 63
Prabhu, V. P., 169
Preiss, T., 78, 80, 108
Presutti, C., 212, 431
Preusser, C., 59
Pringle, J. R., 270
Proudfoot, N. J., 462
Pruijn, G. J., 141, 154, 155, 187, 211, 212, 213, 220, 221, 223, 269, 432
Puig, O., 43, 44, 87, 259
Puisieux, A., 81
Purkayastha, S., 169
Putzer, H., 180

Q

Qian, H., 565
Qian, Y., 468
Quan, Y., 168
Quang-Dang, D.-U., 23
Queiroz, R., 359

R

Raats, J. M. H., 212, 213, 221
Raff, J. W., 313
Ragheb, J. A., 42
Raghuraman, M. K., 416, 418
Raijmakers, R., 141, 154, 155, 187, 212, 213, 220, 221, 269, 432
Rain, J. C., 43, 62
Rajewsky, N., 78, 80, 81
Raker, V. A., 43, 50, 58, 59, 61, 63
Ram, S. J., 518
Ramirez, C. V., 5
Ramos, K. S., 510

Rappsilber, J., 59, 468
Raue, H. A., 416
Ray, J., 42, 141
Raz, E., 331
Reddy, R., 24
Reed, J., 361
Reed, R., 468
Regnault, B., 212, 431
Rehwinkel, J., 24, 469
Reijns, M. A., 58, 60
Reinberg, D., 221
Reinhart, U., 169
Remigy, H., 61, 62, 64, 67, 70
Ren, J., 470
Ren, S., 435
Ren, Y. G., 81, 83, 84, 86, 87, 88
Reverdatto, S. V., 81, 433, 437, 438, 439, 440
Reynolds, S. H., 436
Ricard, F., 43
Richards, W. G., 126, 315
Richmond, T. J., 72
Richter, J. D., 80, 88, 108, 119, 120, 121, 126, 128, 130, 134, 137
Rieger, M., 81, 87
Riester, M., 436, 437
Rigaud, G., 446
Rigaut, G., 42, 43, 44, 45, 58, 87, 259
Rihel, J., 108, 331
Rimokh, R., 87
Rio, M. C., 468
Rivas, F. V., 508
Roberts, R. W., 470
Robinson, M. D., 80, 320, 431
Robles, A., 364
Robson, N. D., 497, 498
Rockmill, B., 168
Rodgers, N. D., 5, 6, 16, 181, 269
Roditi, I., 361
Rodrigues-Pousada, R., 437
Rodriguez-Ezpeleta, N., 359
Roeder, G. S., 168
Roger, A., 359
Rohlin, L., 322
Rollag, M. D., 152
Romaniuk, P. J., 453
Romier, C., 63
Ronald, P., 436
Roos, D., 384
Rosbash, M., 235, 431
Ross, J., 108, 109, 309, 337, 404
Ross, P. L., 169
Rostan, M. C., 87
Rouault, T. A., 446
Rougemaille, M., 212, 431
Rousselle, J. C., 212, 431
Roussis, A., 438
Roux, K. H., 245
Roy, P. J., 381

Ruberti, I., 437
Ruderman, J. V., 121
Rudin, S. D., 385
Rudner, D. Z., 79, 80, 81, 83, 84, 85, 86, 87
Ruegsegger, U., 107
Ruiz-Echevarria, M. J., 168
Ruohola-Baker, H., 381
Ruppert, T., 224
Russell, D. W., 114, 318
Russell, R. B., 430, 433
Rust, M. J., 572
Rutz, B., 43, 44, 87, 259
Ryan, K., 121, 128, 130

S

Sacchi, N., 461, 462
Sachs, A., 81
Sachs, A. B., 42, 79, 80, 81, 83, 84, 85, 86, 87, 88, 422, 508
Sako, Y., 569
Salgado-Garrido, J., 43, 58, 62
Salje, J., 361
Salles, C., 5
Salles, F. J., 109, 126, 314, 315, 489
Salm, H., 340, 366
Samalova, M., 438
Sambrook, J., 114, 280, 316, 317, 318
Sanchez, R., 174
Sanchez-Herrero, E., 313
Sanderson, C. M., 212, 224
Sanford, J. R., 469
Sato, Y., 168
Saveanu, C., 212, 229, 410, 422, 431
Sawaya, M. R., 43
Saxton, M. J., 573, 574
Scafe, C., 279
Scahill, M., 363
Scarsdale, J. N., 24, 25
Schadt, E. E., 322
Schaefer, A. W., 141
Schaefer, M., 469, 515
Schaeffer, D., 227
Schell, T., 468, 470, 472, 473, 479, 480
Scherf, U., 439
Scherthan, H., 168
Scheuermann, T., 113
Schier, A. F., 108, 331
Schild, D., 574
Schilders, G., 187, 211, 212, 213, 220, 221, 223
Schmid, W., 385
Schmidt, T., 564
Schneider, 81, 87, 88
Schneider, C., 187
Schneider, R. J., 70
Schneider, S., 128
Schneiter, R., 230

Schoenberg, D. R., 141, 241, 242, 243, 248, 249, 251, 254, 255, 256, 257, 260, 261, 366, 483, 486, 493, 495, 496, 498
Schoner, B. E., 63
Schoner, R. G., 63
Schroeder, M., 78
Schubert, F., 78, 80
Schubiger, G., 300
Schumacher, M. A., 43, 44, 72
Schumperli, D., 58, 59
Schütz, G., 385
Schwab, R., 436, 437
Schwartz, D. C., 5, 43, 81, 422
Schwarz, E., 113
Schwede, A., 364
Scotto-Lavino, E., 449
Seago, J. E., 169
Seay, D. J., 80, 81, 82, 84, 85, 87, 88, 91, 93, 95, 96, 99, 100
Seebeck, T., 361
Segal, S. P., 510
Seiser, C., 340
Seko, Y., 42
Seli, E., 78
Selvin, P. R., 564
Semotok, J. L., 109, 116, 299, 301, 303, 306, 307, 314, 325
Sentis, S., 79, 81, 82, 87, 88
Séraphin, B., 5, 6, 12, 18, 24, 25, 42, 43, 44, 45, 58, 62, 79, 81, 82, 83, 84, 85, 86, 87, 88, 189, 206, 208, 212, 228, 230, 259, 430, 431, 433, 468, 508
Serdyuk, I. N., 563
Serrano, R., 181
Sexton, J., 279
Seydoux, G., 331
Shah, N. G., 270
Shaner, N. C., 524, 526
Shapira, M., 360, 364
Shapiro, B. A., 42
Sharov, A. A., 331
Sharp, P. A., 140
Shatkin, A. J., 15, 24
Shav-Tal, Y., 555, 562
Shaw, R. J., 433, 437, 438, 439, 440
She, M., 5, 6, 24, 424
Shen, V., 6
Shenk, T., 157
Shenoy, S. M., 469, 515
Sherman, F., 212, 431
Sheth, U., 24, 25, 60, 81, 141, 169, 432, 469, 508, 509, 510, 511, 512, 514, 515, 516, 517, 554
Shevchenki, A., 187
Shevchenko, A., 42, 43, 44, 45, 58, 87, 213, 222, 229, 424, 437
Shi, H., 361
Shibuya, T., 468
Shillinglaw, W., 109

Shimamoto, A., 168
Shippen, D. E., 435
Shobuike, T., 168
Shoemaker, D. D., 270
Shors, T., 24
Shu, M. D., 468, 469, 470, 479
Shuman, S., 8, 190
Shyu, A.-B., 87, 335, 336, 339, 340, 341, 342, 498
Silver, P. A., 515, 516
Simarro, M., 469
Simon, E., 5, 88
Simonelig, M., 80, 82, 84, 85, 86, 87, 88, 109
Simons, A. M., 169
Singer, R. H., 469, 515, 553, 555, 562
Singh, K. B., 438
Singh, S. P., 436, 437
Singh, U., 403
Singleton, D., 230
Siridechadilok, B., 527
Sission, J. C., 110
Sivan, G., 531
Skiba, N., 433, 437, 438, 439, 440
Slade, A. J., 436
Slonim, D. K., 400
Smibert, C. A., 109, 116, 303, 307, 314, 325
Smith, G. R., 168
Smith, L. D., 126
Smith, N. A., 436, 437
Smith, P. D., 381
Smith, T. F., 50
So, A. K., 319, 324, 326
Sokoloski, K. J., 14, 82, 139, 140, 152
Solinger, J. A., 168, 169, 170
Solomon, E., 245
Somers, J. M., 237
Sonenberg, N., 136, 468
Song, H., 5, 6, 24, 78, 79, 80, 86, 108, 168, 181, 242, 336, 337, 339, 410, 424, 449, 463
Song, M., 14, 15
Song, Y., 436
Sonnewald, U., 437
Sopta, M., 270
Soteropoulos, P., 413
Souret, F. F., 168, 169
Spatrick, P., 168
Speckmann, W. A., 213
Speed, T. P., 321, 439
Speidel, M., 568
Speth, C., 340
Spiller, M. P., 58, 60
Springer, N., 436
St. Johnston, D., 468
Standart, N., 78, 109, 121
Staples, R. R., 42, 79, 81, 82, 83, 84, 87, 88, 228, 231, 422
Stark, A., 469
Stark, H., 50

Steams, T., 233
Stearns, T., 271
Steiger, M., 5, 24, 43, 422
Steine, M. N., 436
Steinman, R. A., 42
Steitz, J. A., 24, 78, 109, 121, 468, 469, 470, 479
Stevens, A., 24, 168, 170, 171, 173, 175, 181, 243, 248, 260, 269
Stewart, M., 359, 363, 364, 365, 367
Stockley, P. G., 471
Stoecklin, G., 141, 154, 212, 213, 221, 354, 432, 469
Stoll, I., 468
Stonehouse, N. J., 471
Storey, J. D., 380
Storm, L., 362, 363
Stoutjesdijk, P. A., 436, 437
Straight, S., 436
Strayer, C. A., 152
Stribinskis, V., 510
Strickland, S., 109, 126, 128, 314, 315, 489
Stuart, J. M., 381
Stuart, K., 360
Stutz, A., 126, 128
Subrahmanyan, L., 381
Subramaniam, K., 331
Suck, D., 43, 79
Sugimoto, M., 168
Suh, S. W., 242
Sumpter, V., 61, 62
Sun, J., 432
Sun, X., 468, 479
Sunter, J., 361
Superti-Furga, G., 430, 433
Suzuki, T., 82, 85, 230
Svitkin, Y. V., 136
Sweet, T. J., 510, 511
Sweeton, D., 300, 305
Swenson, K., 120
Symmons, M. F., 187, 206, 212, 228
Szankasi, P., 168

T

Tadros, W., 109, 301, 302, 305
Tai, T. H., 435
Takahashi, S., 228
Takahashi, T. T., 470
Takimoto, K., 108, 109
Takova, T., 492, 497, 498
Tan, G. S., 469
Tange, T. O., 468
Tartakoff, A. M., 230
Tarun, S. Z., 422
Tarun, S. Z., Jr., 81, 85, 86, 87, 88
Tautz, D., 303, 313
Tavazoie, S., 300, 301, 302, 303, 305, 307, 309, 319, 320

Tazi, J., 263
Teixeira, D., 279, 432, 508, 511, 512, 514, 515, 516, 517
Teixeira, S., 361
Teixeira, T., 517
Temme, C., 42, 82, 84, 85, 86, 108, 109, 110, 112, 113, 114, 115
Tempst, P., 221
Tenenhaus, C., 331
Terns, M. P., 213
Terns, R. M., 213
Tharun, S., 41, 42, 44, 45, 46, 49, 50, 51, 52, 53, 58, 60, 141, 419, 511
Thermann, R., 468, 473, 474
Thomassin, H., 446
Thompson, B., 108
Thompson, E., 212, 229, 431
Thompson, R. E., 564
Thomsen, R., 431
Thomson, A. M., 42
Thore, S., 43, 79
Thuresson, A. C., 81, 83, 84, 87
Tian, B., 413
Tibshirani, R., 330
Tibshirani, R. J., 380
Till, B. J., 435, 436
Tisdale, S., 521
Tishkoff, D. X., 168
Tocque, B., 263
Todeschini, A. L., 181
Toh, E. A., 228, 236
Tokunaga, M., 569
Tollervey, D., 141, 168, 181, 187, 190, 212, 213, 220, 221, 222, 228, 229, 230, 424, 431, 437
Tomancak, P., 300, 313
Tomasetto, C., 468
Tomasevic, N., 24, 25
Torchet, C., 431
Toto, I., 43
Tourriere, H., 534
Traven, A., 270
Treisman, R., 336, 340, 341
Trice, T., 229, 431
Trifillis, P., 14
Trochesset, M., 230
Tschudi, C., 361, 362, 364
Tseng, G. C., 322
Tucker, M., 42, 79, 81, 82, 83, 84, 87, 88, 422
Tucker-Kellogg, G., 400
Tumati, S., 5, 6, 424
Tuschl, T., 111, 213
Tusher, V. G., 330
Tworoger, M., 381

U

Uchida, N., 79, 84
Uhlenbeck, O. C., 28, 453, 471

Uliel, S., 360
Ullman, B., 384
Ullu, E., 361, 362, 364
Unterholzner, L., 508
Urlaub, H., 61

V

Valegard, K., 471
Valencia-Sanchez, M. A., 42, 79, 81, 82, 83, 84, 87, 88, 422, 511, 512, 514, 515, 517
Valentin-Hansen, P., 43, 44, 58, 72
Vallotton, P., 574
van Aarssen, Y., 220, 221
Vanacova, S., 212, 229, 431
van den Engh, G., 381
van Dijk, E., 5, 12, 18, 24, 25, 212, 213, 220, 221, 223
Van Dongen, S., 108, 331
Van Duyne, G. D., 59
van Erp, H., 81
van Holde, K., 563
van Hoof, A., 212, 227, 228, 229, 231, 232, 233, 235, 237, 269, 410, 425
Vanlerberghe, G. C., 438
van Nimwegen, E., 78
van Tuijl, A., 363, 364, 365
van Venrooij, W. J., 212, 213, 220, 221
Vardy, L., 109
Vari, H. K., 109, 116, 303, 307, 314, 325
Varnold, R. L., 126
Vassalli, J. D., 126, 128
Vasudevan, S., 78, 536
Vedvik, K. L., 492, 497, 498
Velazquez, F., 468
Vener, T. I., 492, 497, 498
Venkov, P., 270
Venrooij, W. J., 213
Verdone, L., 44
Viegas, M. H., 468, 470, 472, 474, 480
Vihn, J., 180
Vilela, C., 5
Virtanen, A., 79, 80, 81, 83, 84, 86, 87, 88
Viswanathan, P., 81, 82, 83, 84, 87, 88
Voeltz, G. K., 109, 121
Vogel, J. T., 81
Vogelzangs, J., 213
Vogt, V. M., 245
Voncken, F., 364
Vreken, P., 416

W

Wach, A., 270
Wagner, B. J., 141
Wagner, E., 79, 80, 82, 83, 84, 88, 469, 554
Wagner, E. J., 497, 498

Wahle, E., 5, 42, 78, 81, 82, 83, 84, 85, 86, 87, 88, 107, 108, 109, 110, 112, 113, 114, 115, 120, 128, 152, 433
Wakiyama, M., 108, 109
Walczak, C. E., 134
Walke, S., 43, 58, 59, 61, 63
Walker, J. A., 109
Walsh, M. A., 6, 181
Walter, P., 187, 189, 191, 206, 430
Wan, L., 59
Wang, C., 109
Wang, F., 468
Wang, H.-W., 208
Wang, J., 50, 208
Wang, L., 121, 128, 130, 431
Wang, M. B., 436, 437
Wang, P., 59
Wang, S., 385
Wang, W., 380
Wang, X., 212, 431
Wang, Y., 243, 248, 380
Wang, Y. L., 558
Wang, Z., 5, 6, 10, 11, 14, 18, 24, 29, 33, 42, 141, 156, 207, 269, 413, 432
Ward, J., 245
Warner, J. R., 414, 415
Warthmann, N., 436, 437
Waterhouse, P. M., 436, 437
Watkins, N. J., 213
Watson, D. K., 42
Watson, J., 82, 85, 86, 109, 115, 141
Watson, J. N., 245
Watson, P. M., 42
Watt, S., 78, 80
Webb, H., 361, 364
Weber, K., 213
Wegmuller, D., 141, 432
Wehr, K., 468, 474
Weidinger, G., 331
Weigel, D., 436, 437
Weil, C., 436
Weischenfeldt, J., 24
Weiss, R. A., 381
Weissman, J. S., 78, 242, 243, 510
Welch, S., 3, 28, 33, 86, 156, 157
Wells, A., 562
Wells, D. G., 78
Wen, T., 168, 175, 180
Wendling, C., 468
Wenig, K., 187, 189
Werner, T., 42, 141
Werten, S., 63
Westerhoff, H., 363, 364, 365
Westwood, J. T., 109, 299, 301, 302, 305, 319, 320, 321, 324, 326, 328, 329
Wharton, R. P., 80, 87, 88, 109
Whitfield, W. G., 313

Whitley, M. Z., 400
Wickens, M., 77, 78, 79, 80, 81, 82, 84, 85, 87, 88, 91, 93, 95, 96, 99, 100, 108, 120, 121, 128, 130, 469, 471
Wickner, R. B., 228, 236, 237
Wickstead, B., 361, 363
Wiegand, H., 469
Wieschaus, E. F., 300, 301, 302, 303, 305, 307, 309, 319, 320
Wilczynska, A., 523
Wilhelm, T., 78
Wilkinson, G. W., 216
Will, C. L., 58, 59
Williamson, B., 169
Williamson, J. R., 416, 418
Willis, A. E., 80
Wills, M. A., 433, 437, 438, 439, 440
Wilm, M., 42, 43, 44, 45, 50, 58, 60, 67, 70, 87, 259
Wilson, G. M., 157
Wilson, J. E., 109
Wilson, M. A., 233, 235
Wilusz, C. J., 4, 5, 14, 43, 70, 78, 79, 80, 81, 82, 115, 139, 140, 152, 154, 157, 242, 336, 341, 413
Wilusz, J., 4, 14, 43, 70, 78, 79, 80, 81, 82, 84, 85, 86, 109, 115, 139, 140, 141, 152, 154, 155, 221, 242, 269, 432
Winkler, F. K., 61, 62, 64, 67, 70
Winzeler, E. A., 270
Winzen, R., 354
Wirtz, E., 361
Wisniewska, J., 230
Witherell, G. W., 28
Wohlschlegel, J., 508
Wold, B., 446, 453
Wolf, J., 212, 229, 431
Wolin, S. L., 62
Wolke, U., 331
Wollard, A., 169
Wong, W. H., 322, 439
Wood, H., 168
Wood, J. M., 425
Wood, W. H. III, 380
Woolf, Y. I., 187
Wormington, M., 42, 81, 84, 86, 87, 88, 128
Wreden, C., 109, 314
Wu, L., 108, 336, 355
Wyce, A., 59
Wyers, F., 212, 431

X

Xia, T., 470
Xie, Y., 384
Xu, N., 338, 350, 351, 352, 498
Xu, W. L., 126

Y

Yalcin, A., 213
Yamamoto, T., 82, 85
Yamasaki, S., 469
Yamashita, A., 87, 336, 337, 350, 354, 355, 468
Yamashita, Y., 87
Yan, S., 360
Yanagida, T., 569
Yang, E., 78
Yang, F., 243, 254, 255, 256, 257
Yang, W. H., 5
Yang, X., 230, 380, 431
Yang, X. C., 181
Yang, Y. H., 321
Yang, Z., 384
Yates, J. R. III, 508
Yazaki, J., 433, 437, 438, 439, 440
Yen, T. J., 336
Yi, M. J., 65
Yi, R., 469
Yildiz, A., 564
Yokoyama, K., 82, 85
Yokoyama, S., 108, 109
Yong, J., 59, 468
Yoon, H. J., 242
Yoon, J. Y., 242
Yoshida, H., 300, 313
Young, H. A., 78
Young, K., 436
Young, R., 43, 58, 61, 63, 279
Yu, J. H., 5

Z

Zaessinger, S., 82, 84, 85, 86, 109
Zanchin, N. I., 230
Zaric, B. L., 57, 61, 62, 64, 67, 70
Zarubin, T., 108
Zavolan, M., 78
Zeiner, G. M., 381
Zenklusen, D., 562
Zer, C., 168
Zhang, B., 471
Zhang, G., 556
Zhang, J., 243, 248, 468
Zhang, W., 141, 230, 431
Zhang, Y., 436
Zhang, Z., 169
Zhao, J., 78
Zheng, D., 87
Zheng, M., 384
Zheng, Y., 221
Zhong, Z., 87
Zhou, J., 523
Zhu, B., 221
Zhu, W., 87
Zhuang, Z., 381

Zieve, G. W., 58
Ziff, E., 340, 341, 348
Zig, L., 180
Zlatanova, J., 270, 563

Zody, M. C., 80
Zoller, T., 469, 508, 514
Zuo, J., 437
Zuo, Y., 79

Subject Index

A

Adenylation, *see* Deadenylation, messenger RNA; Polyadenylation, messenger RNA; Poly(A) tail, messenger RNA

B

Biotinylation, *see* 4-Thiouracil tagging

C

5′-Cap, *see* Decapping, messenger RNA
c-fos serum-inducible promoter system, messenger RNA decay analysis
  Northern blot analysis, 345–349
  overview, 340–342
  RNA extraction, 344–345
  transient transfection and serum induction, 342–344, 348
Confocal microscopy
  processing body and stress granule studies
    laser scanning confocal microscopy, 544–546
    spinning disk confocal microscopy, 546–547
    single-molecule detection of messenger RNA in live cells, 569–571
CPEB, *see* Deadenylation, messenger RNA
Csl4, *see* RNA exosome
Cytoplasmic processing body, *see* P-body, yeast; Processing body

D

Dcp2
  assay, 12–15
  human protein preparation
    endogenous enzyme, 11–12
    recombinant bacteria, 10
    recombinant human cells, 10–11
  Nudix motif, 5–6
  regulation, 25
  single-turnover analysis
    assay
      buffers, 37–38
      curve fitting, 33–34
      enzyme quantification and elution, 32
      metal activation, 34–36
      overview, 30–31
      substrate mix, 32
      thin-layer chromatography, 32–33
    Dcp1/2 complex coexpression and purification, 30
    kinetic equations, 25–28
    substrate preparation and characterization, 28–30
  substrate preparation
    cap-labeling, 8
    gel purification, 8–9
    transcription *in vitro*, 7–8
  substrate specificity, 5, 25
DcpS
  assay, 16–17
  recombinant protein preparation, 16
  regulation, 25
  structure, 6
  substrate preparation, 16
  substrate specificity, 25
Deadenylation, messenger RNA
  deadenylase features, 79–80
  Drosophila embryo extract system
    deadenylation assay, 113–114
    extract preparation, 110–112
    overview, 107–109
    RNA substrate preparation, 112–113
    sequence-dependent deadenylation analysis, 115–116
  Drosophila embryogenesis messenger RNA degradation analysis, 313–315, 317
  HeLa cell extracts for messenger RNA decay analysis
    advantages, 140–141
    deadenylation rate assay, 152–153
    decapping activity assay, 156–157
    exonuclease activity assay, 153–156
    extract preparation
      overview, 142–143
      S100 extract preparation, 143–145
      standardization of activity, 145–148
    immunodepletion studies, 161
    protein–RNA interaction analysis
      immunoprecipitation, 160
      ultraviolet crosslinking, 157–160
    RNA substrate preparation
      cap labeling, 151
      transcription, 148–151
  length and polarity assays, *see* Poly(A) tail, messenger RNA

Deadenylation, messenger RNA (cont.)
  overview, 4, 78–79
  pathways, 268–269
  Xenopus systems for
    polyadenylation–deadenylation analysis,
    see Polyadenylation, messenger RNA
  yeast *in vitro* system
    advantages, 80–81
    assay-regulated deadenylation
      interpretation, 99–100
      overview, 97–98
      reaction assembly, 99
    deadenylation buffers, 93, 96
    gel electrophoresis and interpretation,
      93–94
    historical perspective, 81–82
    optimization, 82–84
    Pop2p complex purification
      cell culture and harvesting, 88
      characterization, 91–93
      immunoaffinity purification, 90–91
      lysate preparation, 90
      overview, 87–88
      yeast strain, 88
    protein–protein interaction analysis, 97
    Puf4p regulator protein
      assay, 94–96
      dissociation constant determination,
        96–97
      purification, 94
    substrates
      capping, 86
      design, 84–85
      3′-labeling, 86
      5′ labeling, 85–87
      purification, 86
    troubleshooting
      buffer conditions, 101
      enzyme concentration, 100
      regulator concentration, 100
      ribonuclease contamination, 101
Decapping, messenger RNA
  cap analog thin-layer chromatography
    inorganic phosphate, 19
    m^7GDP preparation, 18
    m^7GMP preparation, 19
    m^7GpppG preparation, 18
    m^7GTP preparation, 18
    overview, 17–18
  Dcp2
    assay, 12–15
    human protein preparation
      endogenous enzyme, 11–12
      recombinant bacteria, 10
      recombinant human cells, 10–11
    Nudix motif, 5–6
    regulation, 25
    structure, 24–25

  substrate preparation
    cap-labeling, 8
    gel purification, 8–9
    transcription *in vitro*, 7–8
  substrate specificity, 5, 25
  DcpS
    assay, 16–17
    recombinant protein preparation, 16
    regulation, 25
    structure, 6
    substrate preparation, 16
    substrate specificity, 25
  HeLa cell extracts for messenger RNA
    decay analysis
    advantages, 140–141
    deadenylation rate assay, 152–153
    decapping activity assay, 156–157
    exonuclease activity assay, 153–156
    extract preparation
      overview, 142–143
      S100 extract preparation, 143–145
      standardization of activity, 145–148
    immunodepletion studies, 161
    protein–RNA interaction analysis
      immunoprecipitation, 160
      ultraviolet crosslinking, 157–160
    RNA substrate preparation
      cap labeling, 151
      transcription, 148–151
  human activator complex, *see* Sm/LSm
    complex
  overview, 4
  single-turnover analysis
    assay
      buffers, 37–38
      curve fitting, 33–34
      enzyme quantification and elution, 32
      metal activation, 34–36
      overview, 30–31
      substrate mix, 32
      thin-layer chromatography, 32–33
    Dcp1/2 complex coexpression and
      purification, 30
    kinetic equations, 25–28
    substrate preparation and characterization,
      28–30
  yeast activator complex, *see* Lsm1p-7p-Pat1p
    complex
  yeast decapping complex, 5
Diffusion coefficient
  jump distance relationship, 573
  sphere, 573
DNA microarray
  Drosophila embryogenesis messenger RNA
    degradation analysis
    gene-by-gene analysis, 308, 312–313
    genome-wide analysis
      data normalization, 320–322, 330

Subject Index

degraded transcript identification, 324–326, 330
labeling and hybridization, 318–319, 326–328
platforms, 319–320
quantification, 329
reference sample transcript establishment, 322–324, 329–330
RNA isolation and extraction, 318
scanning, 328–329
limitations in RNA synthesis/decay studies, 380–381
plant exosome analysis, 439–440
4-thiouracil tagging analysis
cell type-specific RNA identification, 401
experimental design
messenger RNA decay, 402–403
messenger RNA synthesis versus abundance, 401–402
RNA blots for normalization, 403
spiked transcripts for normalization, 404–405
overview, 400–401
trypanosome messenger RNA decay analysis
complementary DNA probe synthesis
labeling, 370–371
materials, 370–371
purification, 371
overview, 369–370
Quantifoil slides
hybridization and washing, 373
materials, 371–372
prehybridization, 372–373
TIGR Tb oligo array slides
hybridization and washing, 374–375
materials, 373–374
prehybridization, 374

E

Exosome, see RNA exosome

F

FISH, see Fluorescence in situ hybridization
Fluorescence in situ hybridization, messenger RNA
cell preparation, 559
colabeling of proteins with immunofluorescence
hybridization and washing, 561–562
overview, 560–561
primary antibody incubation, 561
secondary antibody incubation, 562
controls, 574–575
equations, 571–574
hybridization, 557, 559
imaging, 557–558

overview, 554–555
probe
design, 556–557
mixture, 559
washing, 559–560
Fluorescence recovery after photobleaching, processing body and stress granule studies, 550–551
Fluorescence resonance energy transfer, Invader RNA assay, 493–494
FRAP, see Fluorescence recovery after photobleaching
FRET, see Fluorescence resonance energy transfer
Full-width at half-maximum, calculation in light microscopy, 572
FWHM, see Full-width at half-maximum

G

Glycerol gradient sedimentation
PMR1 protein–protein interaction analysis, 256–257
RNA exosome
antibodies, 220–221
digitonin extraction, 219
gradient preparation, 219–220
immunoprecipitation, 222
overview, 218–219
RNA analysis, 220, 222
Western blot, 220

H

Half-life assays, see Messenger RNA half-life

I

Immunoprecipitation
PMR1 complexes, 259–260
protein–RNA interaction analysis in HeLa cell extracts, 160
RNA exosome glycerol gradients, 222
Invader RNA assay, messenger RNA decay polarity analysis
data analysis, 502–504
experimental design, 498–499
primary probe oligonucleotide preparation, 495–497
principles, 492–494
reaction mixes, 501–502
RNA isolation, 499–500
sample preparation, 501
secondary probe oligonucleotide preparation, 497–498
standard curve preparation, 500
transcript synthesis, 494–495

J

Jump distance
  diffusion coefficient relationship, 573
  equation, 574

L

Ligation-mediated poly(A) test, *see* Poly(A) tail, messenger RNA
Ligation-mediated reverse transcription–polymerase chain reaction
  circularization technique for RNA end analysis
    amplification reaction, 464
    complementary DNA synthesis, 463
    intramolecular ligation, 463
    overview, 458–460, 462–463
    primer labeling, 464–465
    transcriptional pulsing with tetracycline-regulated promoter and RNA preparation, 460–462
  footprinting of RNA–protein interactions, 450–452
  ligation of RNA linker
    complementary DNA synthesis, 457
    ligation of linker, 457
    linker synthesis, 454–455
    overview, 452–454
    primer labeling, 457–458
    treatment of 5' ends
      footprinting, 456
      RACE, 456
  PMR1 messenger RNA 5'-cleavage product identification with ligation-mediated reverse transcription–polymerase chain reaction
    overview, 249
    primer
      labeling, 250
      ligation, 249–250
    reverse transcription and amplification, 250–251
  principles and applications, 446–450
Live cell imaging
  processing bodies and stress granules
    confocal microscopy
      laser scanning, 544–546
      spinning disk, 546–547
    fluorescence recovery after photobleaching, 550–551
    heated chamber, 542–543
    Internet resources, 547–548
    microscope incubator, 543
    overview, 541
    processing body
      relative fluorescence protein localization, 550
      tracking, 548–549
    room temperature, 542
    single image quantification, 549–550
    widefield fluorescence microscopy, 543–544
  single-molecule detection of messenger RNA in live cells
    applications, 563–564
    controls, 574–575
    critical parameters, 563
    equations, 571–574
    fluorescent probe, 562–563
    microscopy, 569–571
    nuclear pore complex observations, 564–566
    optimization, 568–569
    verification, 566–568
LMPCR, *see* Ligation-mediated reverse transcription–polymerase chain reaction
Lsm1p-7p-Pat1p complex
  conservation, 44
  function, 42
  protein–protein interactions, 45
  purification
    affinity chromatography, 47–49
    cell lysate preparation, 46–47
    solutions, 45–46
    tags, 44–45
    tandem affinity purification, 43–44
  RNA-binding analysis with gel shift assay, 50–53
LSm/Sm complex, see Sm/LSm complex

M

Mammalian two-hybrid system, protein–protein interaction analysis in RNA exosome
  calculations, 216
  false-negative results, 216, 218
  false-positive results, 216
  overview, 213–215
  transfection, 214
Maternal degradation activity, *Drosophila*, 301–306
MDA, *see* Maternal degradation activity
Messenger RNA
  deadenylation, *see* Deadenylation, messenger RNA
  decapping, *see* Decapping, messenger RNA
  half-life and decay analysis, *see* Messenger RNA half-life
  nonsense-mediated decay, *see* Nonsense-mediated messenger RNA decay
  nucleases, *see* PMR1; RNA exosome; Xrn1
  poly(A) tail, *see* Deadenylation, messenger RNA; Poly(A) tail, messenger RNA
  polyadenylation, *see* Polyadenylation, messenger RNA
Messenger RNA half-life

Subject Index

decay product analysis, see Ligation-mediated reverse transcription–polymerase chain reaction
Drosophila embryogenesis analysis
  deadenylation analysis, 313–315, 317
  DNA microarray, 308, 312–313
  genome-wide analysis with DNA microarray
    data normalization, 320–322, 330
    degraded transcript identification, 324–326, 330
    labeling and hybridization, 318–319, 326–328
    platforms, 319–320
    quantification, 329
    reference sample transcript establishment, 322–324, 329–330
    RNA isolation and extraction, 318
    scanning, 328–329
  maternal degradation activity, 301–306
  maternal messenger RNA
    decay in unfertilized eggs versus fertilized embryos, 302–306
    destabilization prerequisites, 307
    dual degradation activities in early embryo, 301
    functional overview, 300–301
    molecular mechanisms regulating instability, 306–307
  Northern blot
    formaldehyde gel electrophoresis and blotting, 317
    overview, 307–310
    RNA extraction, 316–317
  reverse transcription–polymerase chain reaction, 307–310, 312
  RNA dot blot, 307–311, 317
  sample collection, 315–316
  zygotic degradation activity, 301–306
Drosophila glutathione S-transferase D21 messenger RNA stability assay *in vivo*
  applications, 295–296
  calculations, 294–295
  decay intermediate and end analysis, 293
  heat shock treatment, 290
  materials, 287
  overview, 286
  pentobarbital treatment, 290
  ribonuclease protection assay, 291–293
  RNA isolation, 291
  transgenic constructs
    microinjection and establishment of transgenic lines, 289
    nomenclature, 287–289
mammalian cell assays
  c-fos serum-inducible promoter system
    Northern blot analysis, 345–349
    overview, 340–342
    RNA extraction, 344–345

    transient transfection and serum induction, 342–344, 348
  decay constant determination, 338–339
  general considerations, 337–338
  mechanisms of decay, 336–337
  tetracycline-off regulatory promoter system
    direct measurement without transcriptional pulsing, 353–354
    overview, 349–350
    stable transfection, 350–351
    transcriptional pulse strategy, 351–353
  transcription inhibitors, 339–340
4-thiouracil tagging, see 4-Thiouracil tagging
yeast assays
  calculations, 280
  combining of assays in analysis, 425
  comparison of assays, 410–412
  defective strains in decay
    deadenylation defects, 422
    3' to 5' defects, 424
    5' to 3' defects, 422, 424
    genes in messenger RNA decay, 423
    specialized pathway defects, 424–425
  galactose promoter assays, 413–414
  inducible promoters
    GAL1 upstream activating sequence, 269–270, 275
    galactose metabolism overview, 270
  labeling *in vivo* assay, 413
  Northern blot, 280
  pathway determination
    deadenylation analysis of decay intermediates, 418
    decay intermediate trapping, 416–419
    directionality determination, 417–418
    mutations in *trans*-acting factors, 422–425
    overview, 410, 415–416
    transcriptional pulse–chase analysis, 419–422
  RNA extractions
    overview, 277–278
    RNase H cleavage of 3'-untranslated regions, 278–279
    solutions, 278–279
  RNA polymerase II thermally labile alleles, 276
  tetracycline-off system, 275, 414–415
  thiolutin assay, 277
  transcription inhibition with drugs or mutations, 413
  transcriptional pulse–chase
    medium preparation, 274
    overview, 272–275
  transcriptional shut-off
    incubation and analysis, 272–274
    medium preparation, 271–272
    overview, 270–271, 275
Mtr3, see RNA exosome

N

NMD, *see* Nonsense-mediated messenger RNA decay
Nonsense-mediated messenger RNA decay
  exon junction complexes, 468
  P-body components, 509
  tethering assays for activating protein identification
    control experiments, 479–480
    effector protein detection with Western blot, 479
    overview, 468–470
    plasmid cloning
      boxB and MS2 site incorporation into reporter RNA, 473–474
      effector plasmid, 470–472
      reporter plasmid, 472–473
    transfection control plasmid analysis
      cell lysis, 476
      materials, 475–476
      Northern blotting of reporter messenger RNA, 477–478
      overview, 474–475
      RNA preparation, 476–477
Northern blot
  Drosophila embryogenesis messenger RNA degradation analysis
    formaldehyde gel electrophoresis and blotting, 317
    overview, 307–310
    RNA extraction, 316–317
  messenger RNA half-life determination
    mammalian cells, 345–349
    yeast, 280
  poly(A) length assay with RNase H, 491–492
  tethering assay for activating protein identification in nonsense-mediated messenger RNA decay, 477–478
  4-thiouracil tagging analysis
    agarose gel electrophoresis, 394–395
    biotinylated RNA detection, 395
    blotting, 395
    controls, 396
    materials, 387
    quantitative analysis, 395–396
NPC, *see* Nuclear pore complex
Nuclear pore complex, single-molecule detection of messenger RNA in live cells, 564–566

P

P-body, yeast
  functions, 508
  messenger RNA analysis, 514–516
  protein accumulation assay
    fluorescent fusion proteins, 511
    glucose deprivation studies, 513–514
    markers, 508–512
    midlog growth sample examination, 512–513
    osmotic stress studies, 513–514
    sample preparation, 512–514
  protein components, 509
  size and number analysis
    interpretation, 517
    microscopy, 518–519
    overview, 515, 517
  assembly, 523
  fluorescent tagging of marker proteins
    reporter construct design, 526–527
    tag selection, 524, 526
  function, 522–524
  live cell imaging
    confocal microscopy
      laser scanning, 544–546
      spinning disk, 546–547
    fluorescence recovery after photobleaching, 550–551
    heated chamber, 542–543
    Internet resources, 547–548
    microscope incubator, 543
    overview, 541
    processing body
      relative fluorescence protein localization, 550
      tracking, 548–549
    room temperature, 542
    single image quantification, 549–550
    widefield fluorescence microscopy, 543–544
  marker proteins, 524–525
  properties of stable cell lines expressing markers, 533–541
  transfection
    drug selection, 529–530
    expression testing
      antibiotic selection, 531–532
      clone expression testing and behavior, 531
      clone picking, 530–531
      subcloning, 530
    materials, 529
    stable transfectant selection, 527–528
    subcloning, 532–533
  yeast, *see* P-body, yeast
PCR, *see* Polymerase chain reaction
PMR1
  assays
    in vitro, 261–262
    in vivo, 260–261
  functional overview, 242–243
  messenger RNA 5'-cleavage product identification with ligation-mediated reverse transcription–polymerase chain reaction
    overview, 249
    primer

Subject Index

  labeling, 250
  ligation, 249–250
  reverse transcription and amplification, 250–251
messenger RNA endonuclease cleavage site identification
  primer extension assay, 247–249
  S1 nuclease protection assay
    DNA primer end-labeling, 245–246
    incubation conditions, 246–247
    overview, 244–245
  phosphorylation, 243
  protein–protein interactions
    analysis
      glycerol gradient sedimentation, 256–257
      postmitochondrial extract preparation, 252–253
      sucrose density gradient centrifugation, 253–256
      transfection, 252
    immunoglobulin G–Sepharose selection
      equilibration and blocking of beads, 258
      sample application and elution, 258–259
    immunoprecipitation, 259–260
    overview, 243, 251
Polyadenylation, messenger RNA
  length and polarity assays, see Poly(A) tail, messenger RNA
  Xenopus systems for polyadenylation–deadenylation analysis
    components and functions, 120–121
    cyclin B1 messenger RNA deadenylation–readenylation assay
      incubation conditions and gel electrophoresis, 129–130
      materials, 129
      principles, 128–129
    cyclin B1 messenger RNA polyadenylation assay in oocytes
      gel electrophoresis, 125
      nuclear and cytoplasmic polyadenylation assay, 126–128
      oocyte preparation, injection, and probe retrieval, 122–125
      principles, 121
      probe preparation, 122–124
    cyclin B1 messenger RNA polyadenylation–deadenylation assay in egg extracts
      extract preparation, 133–135
      polyadenylation–translation assay, 134, 137
      principles, 133
      protein depletion with antibody and polyadenylation assay, 133, 135–136
      protein depletion with interacting protein and polyadenylation assay, 134, 136–137

  messenger RNA overexpression and polyadenylation–deadenylation analysis, 132
  poly(A) polymerase activity detection, 130–131
Poly(A) tail, messenger RNA
  deadenylation, see Deadenylation, messenger RNA
  Invader RNA assay for decay polarity analysis
    data analysis, 502–504
    experimental design, 498–499
    primary probe oligonucleotide preparation, 495–497
    principles, 492–494
    reaction mixes, 501–502
    RNA isolation, 499–500
    sample preparation, 501
    secondary probe oligonucleotide preparation, 497–498
    standard curve preparation, 500
    transcript synthesis, 494–495
  length assays
    ligation-mediated poly(A) test
      materials, 488
      overview, 488
      polymerase chain reaction, 490
      primer annealing and ligation, 489
      reverse transcription, 489–490
      RNA preparation, 489
    poly(A) length assay
      complementary DNA synthesis, 485–486
      gel electrophoresis of amplification products, 488
      materials, 484
      overview, 484
      polymerase chain reaction, 486–487
      RNA preparation, 484–485
      second primer end-labeling, 486
    RNase H assay
      materials, 490–491
      overview, 490
      primer annealing, 491
      RNA purification and Northern blotting, 491–492
      RNase H reaction, 491
  synthesis, see Polyadenylation, messenger RNA
Polymerase chain reaction, see also Ligation-mediated reverse transcription–polymerase chain reaction
  Drosophila embryogenesis messenger RNA degradation analysis with reverse transcription–polymerase chain reaction, 307–310, 312
  poly(A) tail length assays, 486–487, 490
Pop2p complex, see Deadenylation, messenger RNA
Puf4p, see Deadenylation, messenger RNA

Q

Quantifoil slide, *see* DNA microarray
Quantum dot, live cell imaging controls, 575

R

Resolution, light microscope, 572
Reverse ligation-mediated polymerase chain reaction, *see* Ligation-mediated reverse transcription–polymerase chain reaction
Reverse transcription–polymerase chain reaction, *see* Ligation-mediated reverse transcription–polymerase chain reaction; Polymerase chain reaction
Ribonuclease protection assay
 Drosophila glutathione S-transferase D21 messenger RNA stability assay *in vivo*, 291–293
 PMR1 endonuclease cleavage site identification
  primer extension assay, 247–249
  S1 nuclease protection assay
   DNA primer end-labeling, 245–246
   incubation conditions, 246–247
   overview, 244–245
RL-PCR, *see* Ligation-mediated reverse transcription–polymerase chain reaction
RNA analysis by biosynthetic tagging, *see* 4-Thiouracil tagging
RNA dot blot, *Drosophila* embryogenesis messenger RNA degradation analysis, 307–311, 317
RNA exosome
 Archaea, 187–188
 functional overview, 430
 glycerol gradient sedimentation
  antibodies, 220–221
  digitonin extraction, 219
  gradient preparation, 219–220
  immunoprecipitation, 222
  overview, 218–219
  RNA analysis, 220, 222
  Western blot, 220
 human exosome
  architecture, 189–190
  comparative assays for eukaryotic exosomes, 207–208
  exoribonuclease assay, 206–207
  recombinant protein expression and purification
   cloning strategies, 190–191
   Csl4, 94, 201
   induction, 198
   Mtr3, 194, 200
   nickel affinity chromatography, 198–199
   polymerase chain reaction and subcloning, 191, 194
   Rrp4, 194, 201
   Rrp6, 194
   Rrp40, 194, 201
   Rrp41, 194, 199
   Rrp42, 194, 200
   Rrp43, 194, 200
   Rrp45, 194, 199
   Rrp46, 194, 200–201
  reconstitution and purification
   nine-protein exosome, 205
   overview, 201
 mammalian two-hybrid system for protein–protein interaction analysis
  calculations, 216
  false-negative results, 216, 218
  false-positive results, 216
  overview, 213–215
  transfection, 214
 plant exosome
  mutational analysis, 434–436
  overview of features, 432–433
  transcriptome-wide mapping of targets, 436–440
 protein components in humans and yeast, 186–189, 212–213, 228, 430–432
 RNA interference studies, 222–224
 yeast exosome
  comparative assays for eukaryotic exosomes, 207–208
  exoribonuclease assay, 206–207
  function, 228–229
  growth analysis of aberrant transcript degradation, 232–233
  killer assay of cytoplasmic exosome activity, 236–237
  mutant analysis *in vivo*
   core exosome mutants, 229–231
   cytoplasmic cofactor mutants, 231
   nuclear cofactor mutants, 231
  recombinant protein expression and purification
   cloning strategies, 190–191
   Csl4, 193, 197
   induction, 195
   Mtr3, 193, 196
   nickel affinity chromatography, 195–196
   polymerase chain reaction and subcloning, 191–194
   Rrp4, 193, 197
   Rrp6, 193–194, 198
   Rrp40, 193, 197
   Rrp41, 191, 193, 196
   Rrp42, 193, 196
   Rrp43, 193, 197
   Rrp44, 193, 197
   Rrp45, 191, 193, 196
   Rrp46, 193, 197
  reconstitution and purification
   eleven-protein complex, 203, 205

Subject Index

nine-subunit exosome, 202
overview, 201
ten-protein complex, 202
RNA stability assays *in vivo*
aberrant transcript degradation, 235–236
normal transcript degradation, 236
synthetic lethality studies, 234
RNA interference
plant exosome analysis, 437–438
RNA exosome studies, 222–224
RNA polymerase II, thermally-labile alleles and messenger RNA half-life determination, 276
RNase H
cleavage of 3′-untranslated regions for messenger RNA half-life analysis, 278–279
poly(A) tail length assay
materials, 490–491
overview, 490
primer annealing, 491
RNA purification, 491
RNA purification and Northern blotting, 491–492
RNase H reaction, 491
RPA, *see* Ribonuclease protection assay
Rrp proteins
purification, *see* RNA exosome
ribosome biosynthesis role, 186–187

S

SG, *see* Stress granule
Single-molecule detection, messenger RNA in live cells
applications, 563–564
controls, 574–575
critical parameters, 563
equations, 571–574
fluorescent probe, 562–563
microscopy, 569–571
nuclear pore complex observations, 564–566
optimization, 568–569
verification, 566–568
Sm/LSm complex
characterization, 60, 70–71
function, 58
LSm protein types, 58–59
prospects for study, 72
reconstitution of human proteins *in vitro*
cloning, 61–65
protein expression and purification, 65–68
reconstitution, 68–70
ribonucleoprotein complex assembly, 59–60
structure, 58
Stress granule
assembly, 523
fluorescent tagging of marker proteins

reporter construct design, 526–527
tag selection, 524, 526
function, 522–524
live cell imaging
confocal microscopy
laser scanning, 544–546
spinning disk, 546–547
fluorescence recovery after photobleaching, 550–551
heated chamber, 542–543
Internet resources, 547–548
microscope incubator, 543
overview, 541
room temperature, 542
single image quantification, 549–550
widefield fluorescence microscopy, 543–544
marker proteins, 524–525
transfection
drug selection, 529–530
expression testing
antibiotic selection, 531–532
clone expression testing and behavior, 531
clone picking, 530–531
subcloning, 530
materials, 529
stable transfectant selection, 527–528
subcloning, 532–533
Sucrose density gradient centrifugation, PMR1 protein–protein interaction analysis
linear gradients, 253–254
step gradients, 255–256

T

Tandem affinity purification
Lsm1p-7p-Pat1p complex, 43–44
trypanosome RNA–protein complex purification, 368–369
TAP, *see* Tandem affinity purification
Targeted induced local lesions in genomes, plant exosome analysis, 435–436
Tetracycline-off regulatory promoter system, messenger RNA half-life determination
mammalian cells
direct measurement without transcriptional pulsing, 353–354
overview, 349–350
stable transfection, 350–351
transcriptional pulse strategy, 351–353
yeast, 275, 414–415
Thin-layer chromatography
Dcp2 single-turnover analysis, 32–33
messenger RNA cap analogs
inorganic phosphate, 19
m^{7}GDP preparation, 18
m^{7}GMP preparation, 19
m^{7}GpppG preparation, 18

Thin-layer chromatography (cont.)
  m^{7}GTP preparation, 18
  overview, 17–18
Thiolutin, messenger RNA half-life determination in yeast, 277
4-Thiouracil tagging
  biotinylation of tagged RNA
    controls, 393
    incubation conditions, 392
    materials, 387
    precipitation, 392–393
  DNA microarray analysis
    cell type-specific RNA identification, 401
    experimental design
      messenger RNA decay, 402–403
      messenger RNA synthesis versus abundance, 401–402
      RNA blots for normalization, 403
      spiked transcripts for normalization, 404–405
    overview, 400–401
  general RNA tagging with 4-thiouridine, 384–385
  principles, 381–382
  purification and analysis of tagged RNA
    controls, 399–400
    materials, 387–388
    overview, 385–386
    RNA precipitation, 398–399
    streptavidin-magnetic bead purification, 397–398
  RNA blot for detection of tagged RNA
    agarose gel electrophoresis, 394–395
    biotinylated RNA detection, 395
    blotting, 395
    controls, 396–397
    materials, 387
    quantitative analysis, 395–396
  RNA preparation
    controls, 391–392
    materials, 386–387
    messenger RNA preparation, 391
    total RNA extraction, 390–391
  4-thiouracil
    delivery, 385
    pulse and uracil chase
      chase, 389
      controls, 389–390
      materials, 386
      pulse, 388–389
  uracil phosphoribosyltransferase from *Toxoplasma gondii*
    expression vector construction
      cloning, 388
      materials, 386
    targeted expression in cells, 382–384
TILLING, see Targeted induced local lesions in genomes

TLC, see Thin-layer chromatography
Trypanosome messenger RNA decay
  cell culture, 364–365
  directionality assay, 366
  DNA microarray analysis
    complementary DNA probe synthesis
      labeling, 370–371
      materials, 370–371
      purification, 371
    overview, 369–370
    Quantifoil slides
      hybridization and washing, 373
      materials, 371–372
      prehybridization, 372–373
    TIGR Tb oligo array slides
      hybridization and washing, 374–375
      materials, 373–374
      prehybridization, 374
  genetic manipulation
    gene knockout, 361
    inducible gene expression, 361
    inducible knockout, 363
    RNA interference, 362–363
    site-specific recombination, 363
    tagging *in situ*, 361–362
  materials, 366
  planning experiments, 365
  RNA–protein complex purification
    lysate preparation, 367–368
    RNA isolation, 369
    tandem affinity purification, 368–369
    trypanosome preparation, 367
  transcription overview, 360, 363–364
Two-hybrid system, see Mammalian two-hybrid system

U

Uracil phosphoribosyltransferase, see 4-Thiouracil tagging

W

Western blot, tethering assay for activating protein identification in nonsense-mediated messenger RNA decay, 479

X

Xenopus oocyte, see Polyadenylation, messenger RNA
Xrn1
  conservation, 168–169
  exoribonuclease assays
    directionality assays
      doubly-labeled RNA degradation, 176–179
      single end-labeled RNA, 179
    fluorescence-based assays, 176

gel-based assays, 174–175
trichloroacetic acid-based assays, 173–174
function, 168
mutation studies, 168–169
prospects for study, 180–181
purification from yeast, 170–171

RNA substrate synthesis, 171–173

Z

ZDA, *see* Zygotic degradation activity
Zygotic degradation activity, *Drosophila*, 301–306

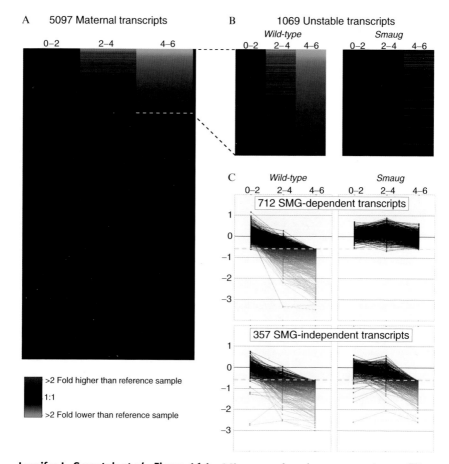

Jennifer L. Semotok et al., Figure 16.4 Microarray-based gene expression profiling of maternal transcript stability in activated unfertilized eggs from wild-type and *smaug* mutant females. Unfertilized eggs were collected 0 to 2, 2 to 4, and 4 to 6 h after egg laying. (A) Maternal mRNAs (5097) were sorted according to instability at 4 to 6 h in wild type; each is represented by a horizontal bar, with black indicating no change, green a decrease, and red an increase in transcript abundance relative to stage 14 oocytes. The 1069 transcripts above the dashed yellow line (ratio's log base 2 of $0.59 = 1.5$-fold decrease) are significantly destabilized in wild type. (B) Effect of *smaug* mutant on the 1069 transcripts that are unstable in wild type. It can be seen that most are stabilized in the mutant (i.e., change from green to black). (C) Kinetics of the effect of the *smaug* mutation on the 1069 transcripts that are unstable in wild type. Two-thirds (712) are stabilized in a *smaug* mutant ("SMG-dependent"), whereas one-third (357) remain unstable in the mutant ("SMG-independent"). Figure from Tadros *et al.* (2007a).

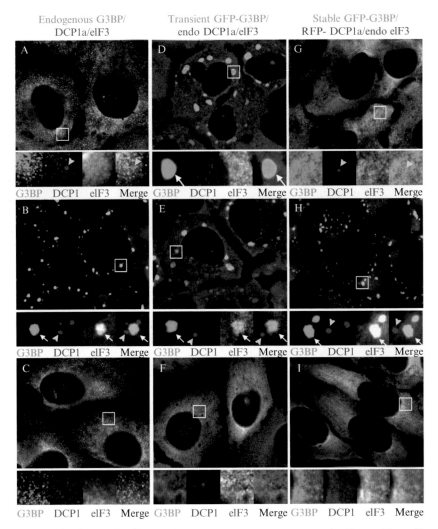

Nancy Kedersha *et al.*, Figure 26.1 GFP-G3BP/mRFP-DCP1a double-positive cells display similar SG/PB assembly as untransfected cells. (A to C) Parental U2OS stained for endogenous G3BP1 (green), DCP1a (red), and eIF3b (blue); (D to F), U2OS cells transiently transfected with GFP-tagged G3BP1 (green), counterstained for endogenous DCP1a (red) and eIF3b (blue); (G to I), U2OS cells stably expressing GFP-G3BP1 (green) and mRFP-DCP1a (red), counterstained for endogenous eIF3b (blue). Cells were untreated (top row, A, D, and G), exposed to 500 μM sodium arsenite for 1 h (middle row, B, E, and H), or treated with 40 μg/ml emetine for 1 h (bottom row, C, F, I). Enlarged views of boxed regions are displayed underneath the corresponding panels, separately showing the views of the same field. Note that endogenous G3BP1 and stably expressed GFP-G3BP1 appear dispersed and cytoplasmic in unstressed (A, G) and emetine-treated (C, I) cells but is entirely recruited into SGs on arsenite treatment (B, H). Similarly, endogenous DCP1a (A to C) and stably expressed mRFP-DCP1a (G to I) display few small PBs in the absence of stress (A, G), numerous PBs on arsenite treatment (B, H), and no PBs in emetine-treated cells (C, I). In contrast, transiently overexpressed GFP-G3BP1G3BP1 nucleates abnormally large SGs without stress (D) and in response to arsenite (E) but is still disassembled in response to emetine (F).

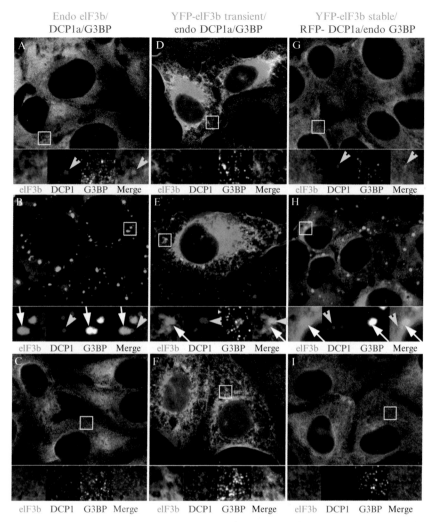

Nancy Kedersha et al., Figure 26.2 YFP-eIF3b/mRFP-DCP1a double-positive cells display similar SG/PB assembly as untransfected cells. (A to C) Parental U2OS stained for endogenous eIF3b (green), DCP1a (red), and G3BP1 (blue); (D to F), U2OS cells transiently transfected with YFP-tagged eIF3b (green), counterstained for endogenous DCP1a (red) and G3BP1 (blue); (G to I), U2OS cells stably expressing YFP-eIF3b (green) and mRFP-DCP1a (red), counterstained for endogenous G3BP1 (blue). Cells were untreated (top row, A, D, and G), exposed to 500 μM sodium arsenite for 1 h (middle row, B, E, and H), or treated with 40 $\mu g/ml$ emetine for 1 h (bottom row, C, F, I). Boxed regions appear enlarged underneath each panel. Note that endogenous eIF3b and stably expressed eIF3b appear dispersed and cytoplasmic in unstressed (A, G) and emetine-treated cells (C, I) but are entirely recruited into SGs on arsenite treatment (B, H). Similarly, endogenous DCP1a (A to C) and stably expressed mRFP-DCP1a (G to I) display few small PBs in the absence of stress (A, G), numerous PBs on arsenite treatment (B, H), and no PBs in emetine-treated cells (C, I). In contrast, transiently overexpressed YFP-eIF3b appears less uniform in untreated (D) and in emetine-treated cells (F), and only weakly recruited to SGs in response to arsenite (E).

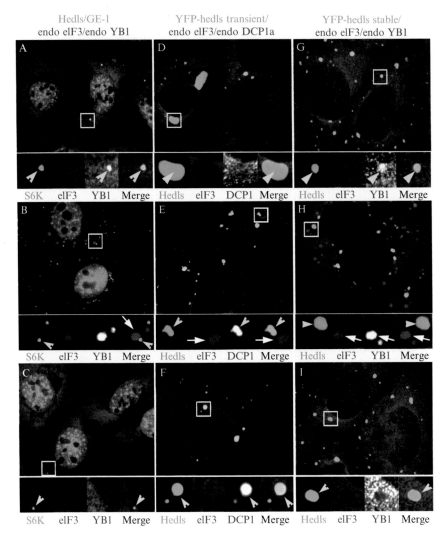

Nancy Kedersha et al., Figure 26.3 Stable YFP-hedls/GE-1–expressing cells do not recapitulate endogenous hedls/GE-1 behavior. Endogenous hedls/GE-1 (A to C, cytoplasmic signal, green), transiently overexpressed YFP-hedls/GE-1 (D to F), and stably expressed YFP-hedls/GE-1 (G to I), counterstained for endogenous YB1 (A to C; G to I) or DCP1a (D to F). Top row (A, D, G), no drug treatment; middle row (B, E, H), cells treated with sodium arsenite for 1 h; and bottom row (C, F, I), cells treated with 40 μg/ml emetine for 1 h. Although endogenous hedls exhibits PB assembly on arsenite treatment (B, green) and disassembly when treated with emetine, both transiently expressed (D to F) and stably expressed (G to I) YFP-hedls/GE-1 form giant PBs that are not affected by drug treatments.

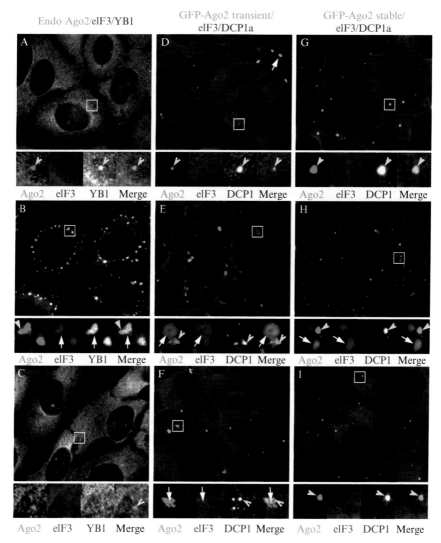

Nancy Kedersha et al., Figure 26.4 GFP-Ago2–expressing stable cells recapitulate behavior of endogenous Ago2. Endogenous Ago2 (A to C, green) colocalizes with endogenous YB1 (A to C, blue) in PBs (yellow arrowheads) in unstressed cells but is present in both PBs (yellow arrowheads) and SGs (white arrows, endogenous eIF3b, red) when cells are treated with arsenite (4B). Transient overexpression of GFP-Ago2 (D to F) induces SGs in some unstressed cells (D, white arrows) but is present in PBs in others (D, boxed region, yellow arrowheads). Arsenite treatment (E) induces movement of some GFP-Ago2 into SGs (enlarged view, white arrows), but a significant amount is retained in PBs (yellow arrowheads), costaining with DCP1a (red). Emetine treatment (F) is insufficient to disassemble SGs and PBs in the transient transfectants. Stably expressed GFP-Ago2 (G-I) exhibits behavior identical to that of endogenous Ago2 (A to C). Enlarged views of boxed regions are displayed underneath the corresponding panels, separately showing the views of the same field.